Biotechnology

Second Edition

Volume 4

Measuring, Modelling, and Control

VCH

Biotechnology

Second Edition

Fundamentals

Volume 1
Biological and Biochemical Fundamentals

Volume 2
Genetic Fundamentals and
Genetic Engineering

Volume 3
Bioprocessing

Volume 4
Measuring, Modelling, and Control

Products

Volume 5
Genetically Engineered Proteins and
Monoclonal Antibodies

Volume 6
Products of Primary Metabolism

Volume 7
Products of Secondary Metabolism

Volume 8
Biotransformations

Special Topics

Volume 9
Enzymes, Biomass, Food and Feed

Volume 10
Special Processes

Volume 11
Environmental Processes

Volume 12
Patents, Legislation, Information Sources,
General Index

© VCH Verlagsgesellschaft mbH, D-6940 Weinheim (Federal Republic of Germany), 1991

Distribution:

VCH, P.O. Box 101161, D-6940 Weinheim (Federal Republic of Germany)

Switzerland: VCH, P.O. Box, CH-4020 Basel (Switzerland)

United Kingdom and Ireland: VCH (UK) Ltd., 8 Wellington Court, Cambridge CB1 1HZ (England)

USA and Canada: VCH, Suite 909, 220 East 23rd Street, New York, NY 10010–4606 (USA)

ISBN 3-527-28314-5 (VCH, Weinheim)　　　　　　　　ISBN 1-56081-154-4 (VCH, New York)

A Multi-Volume Comprehensive Treatise

Biotechnology
Second, Completely Revised Edition

Edited by
H.-J. Rehm and G. Reed
in cooperation with
A. Pühler and P. Stadler

Volume 4

Measuring, Modelling, and Control

Edited by
K. Schügerl

VCH

Weinheim · New York · Basel · Cambridge

Series Editors:
Prof. Dr. H.-J. Rehm
Institut für Mikrobiologie
Universität Münster
Corrensstraße 3
D-4400 Münster

Prof. Dr. A. Pühler
Biologie VI (Genetik)
Universität Bielefeld
P.O. Box 8640
D-4800 Bielefeld 1

Dr. G. Reed
1016 Monmouth Ave.
Durham, NC 27701
USA

Dr. P. J. W. Stadler
Bayer AG
Verfahrensentwicklung Biochemie
Leitung
Friedrich-Ebert-Straße 217
D-5600 Wuppertal 1

Volume Editor:
Prof. Dr. K. Schügerl
Institut für Technische Chemie
Universität Hannover
Callinstraße 3
D-3000 Hannover 1

Published jointly by
VCH Verlagsgesellschaft mbH, Weinheim (Federal Republic of Germany)
VCH Publishers Inc., New York, NY (USA)

Editorial Director: Dr. Hans-Joachim Kraus
Editorial Manager: Christa Maria Schultz
Production Director: Maximilian Montkowski
Production Manager: Peter J. Biel

Library of Congress Card No.: 91-16462

British Library Cataloguing-in-Publication Data:
Biotechnology Second Edition
  Biotechnology: Vol 4. Measuring, modelling and
  control.
  Vol. Ed. Schügerl, K.
  620.8
  ISBN 3-527-28314-5

Die Deutsche Bibliothek - CIP-Einheitsaufnahme
Biotechnology : a multi volume comprehensive treatise / ed. by
H.-J. Rehm and G. Reed. In cooperation with A. Pühler and P.
Stadler. – 2., completely rev. ed. – Weinheim; New York;
Basel; Cambridge: VCH.
NE: Rehm, Hans J. [Hrsg.]

2., completely rev. ed.
Vol. 4. Measuring, modelling, and control / ed. by K. Schügerl.
  – 1991
  ISBN 3-527-28314-5 (Weinheim)
  ISBN 1-56081-154-4 (New York)
NE: Schügerl, Karl [Hrsg.]

Composition and Printing: Zechnersche Buchdruckerei, D-6720 Speyer, Bookbinding: Klambt-Druck GmbH, D-6720 Speyer
Printed in the Federal Republic of Germany

Preface

In recognition of the enormous advances in biotechnology in recent years, we are pleased to present this Second Edition of "Biotechnology" relatively soon after the introduction of the First Edition of this multi-volume comprehensive treatise. Since this series was extremely well accepted by the scientific community, we have maintained the overall goal of creating a number of volumes, each devoted to a certain topic, which provide scientists in academia, industry, and public institutions with a well-balanced and comprehensive overview of this growing field. We have fully revised the Second Edition and expanded it from ten to twelve volumes in order to take all recent developments into account.

These twelve volumes are organized into three sections. The first four volumes consider the fundamentals of biotechnology from biological, biochemical, molecular biological, and chemical engineering perspectives. The next four volumes are devoted to products of industrial relevance. Special attention is given here to products derived from genetically engineered microorganisms and mammalian cells. The last four volumes are dedicated to the description of special topics.

The new "Biotechnology" is a reference work, a comprehensive description of the state-of-the-art, and a guide to the original literature. It is specifically directed to microbiologists, biochemists, molecular biologists, bioengineers, chemical engineers, and food and pharmaceutical chemists working in industry, at universities or at public institutions.

A carefully selected and distinguished Scientific Advisory Board stands behind the series. Its members come from key institutions representing scientific input from about twenty countries.

The present volume, fourth in the series, reflects the enormous impact of computer technology on biotechnology, especially in the areas of measurement and control. It describes monitoring of the biotechnological process with sophisticated analytical techniques, use of the resulting data by means of mathematical models, and computer-aided closed loop control for improvement of the productivity of biotechnological processes. While Volume 4 can be used independently, Volume 3 "Bioprocessing" is recommended as a companion volume.

The volume editors and the authors of the individual chapters have been chosen for their recognized expertise and their contributions to the various fields of biotechnology. Their willingness to impart this knowledge to their colleagues forms the basis of "Biotechnology" and is gratefully acknowledged. Moreover, this work could not have been brought to fruition without the foresight and the constant and diligent support of the publisher. We are grateful to VCH for publishing "Biotechnology" with their customary excellence. Special thanks are due Dr. Hans-Joachim Kraus and Christa Schultz, without whose constant efforts the series could not be published. Finally, the editors wish to thank the members of the Scientific Advisory Board for their encouragement, their helpful suggestions, and their constructive criticism.

May 1991

H.-J. Rehm
G. Reed
A. Pühler
P. Stadler

Scientific Advisory Board

Contents

Contributors

Dr. Graham F. Andrews
EG & G
Idaho National Engineering Laboratory
Idaho Falls, ID 83415, USA
Chapter 13

Dr. Rakesh Bajpai
Department of Chemical Engineering
University of Missouri
Columbia, MO 65203, USA
Chapter 10

Prof. Dr. Karl-Heinz Bellgardt
Institut für Technische Chemie
Universität Hannover
Callinstraße 3
D-3000 Hannover 1, FRG
Chapters 9 and 12

Dr. Irving J. Dunn
Biological Reaction Engineering Group
Chemical Engineering Department
Eidgenössische Technische Hochschule (ETH)
Universitätsstraße 6
CH-8092 Zürich, Switzerland
Chapter 2

Prof. Dr. Aarne Halme
Automation Technology Laboratory
Helsinki University of Technology
Electrical Engineering Building
Otakaari 5A
SF-02150 Espoo, Finland
Chapter 19

Dr. Elmar Heinzle
Biological Reaction Engineering Group
Chemical Engineering Department
Eidgenössische Technische Hochschule (ETH)
Universitätsstraße 6
CH-8092 Zürich, Switzerland
Chapter 2

Prof. Dr. Nazmul Karim
Department of Agricultural and
Chemical Engineering
Colorado State University
Fort Collins, CO 80523, USA
Chapter 19

Dr. Jeong-Yoon Kim
Department of Chemical Engineering
University of California
Davis, CA 95616, USA
Chapter 15

Dr. Kyu-Sung Lee
Biochemical Engineering Program
University of California
Irvine, CA 92717, USA
Chapter 16

Dr. Sun Bok Lee
Pohang Institute of Technology
Pohang, Korea
Chapter 15

Prof. Dr. Henry C. Lim
Biochemical Engineering Program
University of California
Irvine, CA 92717, USA
Chapter 16

Priv.-Doz. Dr. Andreas Lübbert
Institut für Technische Chemie
Universität Hannover
Callinstraße 3
D-3000 Hannover 1, FRG
Chapters 4 and 17

Prof. Dr. Bo Mattiasson
Department of Biotechnology
Chemical Center
University of Lund
P.O. Box 124
S-22100 Lund, Sweden
Chapter 3

Prof. Dr. José C. Merchuk
Department of Chemical Engineering
Program of Biotechnology
Ben-Gurion University of the Negev
Beer Sheva, Israel
Chapter 11

Prof. Dr. Axel Munack
Institut für Biosystemtechnik
Bundesforschungsanstalt für Landwirtschaft
Bundesallee 50
D-3300 Braunschweig-Völkenrode, FRG
Chapter 8

Dr. Seujeung Park
Department of Chemical Engineering
Massachusetts Institute of Technology
Cambridge, MA 02139, USA
Chapter 7

Dr. Kenneth F. Reardon
Colorado State University
Fort Collins, CO 80523, USA
Chapter 6

Prof. Dr. Matthias Reuss
Institut für Bioverfahrenstechnik
Universität Stuttgart
Böblinger Straße 72
D-7000 Stuttgart 1, FRG
Chapter 10

Prof. Dr. Dewey D. Y. Ryu
Department of Chemical Engineering
University of California
Davis, CA 95616, USA
Chapter 15

Priv.-Doz. Dr. Thomas H. Scheper
Institut für Technische Chemie
Universität Hannover
Callinstraße 3
D-3000 Hannover 1, FRG
Chapter 6

Prof. Dr. Karl Schügerl
Institut für Technische Chemie
Universität Hannover
Callinstraße 3
D-3000 Hannover 1, FRG
Chapters 1 and 5

Dr. Dirk Schürbüscher
Am braunen Berg 9
D-6477 Limeshain-Himbach, FRG
Chapter 14

Prof. Dr. Suteaki Shioya
Department of Fermentation Technology
Faculty of Engineering
Osaka University, Suita
Osaka 565, Japan
Chapter 18

Prof. Dr. Gregory Stephanopoulos
Department of Chemical Engineering
Massachusetts Institute of Technology
Cambridge, MA 02139, USA
Chapter 7

Prof. Dr. Ken-ichi Suga
Department of Fermentation Technology
Faculty of Engineering
Osaka University, Suita
Osaka 565, Japan
Chapter 18

Prof. Dr. Christian Wandrey
Institut für Biotechnologie 2
Forschungszentrum Jülich
Postfach 1913
D-5170 Jülich, FRG
Chapter 14

Dr. Jingqi Yuan
Institute for Automatic Control
East China University
of Chemical Technology
130 Meilong Lu
Shanghai, People's Republic of China
Chapter 12

Introduction

KARL SCHÜGERL

Hannover, Federal Republic of Germany

The fourth volume of the second edition of "Biotechnology" presents a survey on an increasingly important field of biotechnology: monitoring of the biotechnological process with sophisticated analysis techniques, use of the resulting data by means of mathematical models, and (computer-aided) closed loop control for improvement of the productivity of biotechnological processes.

The bottleneck in biotechnological process control is the on-line measurement of controlled process variables. Except for temperature, impeller speed (for stirred tank reactors), aeration rate (for aerobic microorganisms), pH, p_{O_2}, which are usually controlled process variables, and the composition of the outlet gas (O_2 and CO_2 content), which sometimes is a controlled process variable (respiration quotient, RQ, CO_2 production rate, O_2 consumption rate), no other process variables are usually measured on-line in commercial equipment.

However, manufacturers have recently made great efforts to improve process analysis and control. In several laboratories, on-line systems for the analysis of the chemical medium composition are used to gain more information about the process and to control the concentrations of key components. Furthermore, mathematical models have been developed in order to describe the production proc-

ess and to effect process optimization and control. Therefore, the Series Editors decided to add to *Biotechnology* a separate volume with the title "Measuring, Modelling, and Control", and they asked the Volume Editor to organize it.

This volume consists of four main parts:

- instruments,
- measuring techniques,
- modelling, and
- control/automation.

Cell growth and product formation/substrate conversion is at the focus of attention, and here the measuring and control techniques are well-developed and generally applicable.

"Modelling, design, and control" of downstream processes are also considered because of the great importance of downstream processing. However, because of their broad scope, they are too heterogeneous and not yet sufficiently developed for treatment in the same way as processes for growth and product formation.

The instruments are subdivided into three groups:

- common instruments for medium analysis,

- instruments for gas analysis, and
- biosensors.

The last group of instruments is still being developed.

Only instruments that are (or can be) used for process control are considered in detail. However, modern off-line techniques (e.g., NMR) are also taken into account.

The measuring techniques are subdivided into four groups:

- physical techniques for the characterization of fluid dynamics,
- chemical methods for the analysis of broth composition,
- physical methods for the determination of cell concentration, and
- physical/biochemical methods for the characterization of the biological state of the cells.

Most of these are on-line techniques; others can only be carried out in a quasi on-line mode. All of them are used to characterize the reactor/medium/cell system.

There are interesting new developments in the characterization of such systems by modern mathematical methods, with optimization of sampling to gain maximal possible information. These new techniques are also included in this volume.

The models are subdivided into

- cell models,
- reactor models, and
- process models.

The process models cover

- product formation,
- aerobic wastewater treatment,
- anaerobic wastewater treatment, and
- models for recombinant microorganisms.

Models for animal and plant tissue cultures have not yet been included because no reliable kinetic data are available for mathematical modelling of these cultures.

Modern control techniques are increasingly applied to closed loop control of bioreactor systems. Therefore, different types of closed loop control techniques, including computer-aided control, are considered in detail.

Instrumental control is much more reliable than control by a human operator. Furthermore, long-range (many weeks or months) runs are only possible in the laboratories of research institutes and universities if automated equipment is used. Thus, automation of bioreactors is also taken into account.

Chapter 18 on modelling, design, and control of downstream processing covers only the most important downstream processes. Development in this field is still limited, so this chapter is not as extensive as the potential importance of downstream processing would warrant.

Expert systems are being developed in different aspects of technology, medicine, and the natural sciences. Their use in biotechnology is desirable, since they would permit the identification of equipment failures and their eventual elimination. Furthermore, by means of expert systems, large amounts of information from on-line and off-line measurements as well as from the literature and from heuristic knowledge can be used with high efficiency.

The reviews consider only the most important techniques and omit some detail because of limited space. Further information can be gained through the reference notations.

It is hoped that information given in this volume will help students, engineers, and scientists at universities, members of research institutes, and those in industry to increase their knowledge of this important and fast-growing field.

Hannover, March 1991   K. Schügerl

I. Instruments

1 Common Instruments for Process Analysis and Control

KARL SCHÜGERL

Hannover, Federal Republic of Germany

1 Introduction

Biological processes are influenced by several control variables: temperature, pH, dissolved oxygen partial pressure p_{O_2}, as well as by state variables such as redox potential, E_h, and dissolved CO_2 partial pressure, p_{CO_2}, which have a direct influence on cell metabolism (FORAGE et al., 1985).

Other control variables (power input, aeration rate) and state variables (liquid viscosity) have an indirect effect on cell growth and product formation. They influence gas dispersion (bubble size, gas holdup, and specific interfacial area) and the transport processes in the broth. The broth volume can be determined by means of the liquid level and the holdup. In continuous cultivation, the residence time of the broth or its dilution rate, which equals the specific growth rate of the cells, is determined by the liquid throughput and broth volume.

With highly foaming broth the cells are sometimes enriched in the foam by flotation. The diminution of the cell concentration in the broth reduces cell growth and product formation rates. Foam may be carried out of the reactor by the air flow and then may clog the gas analysis instruments and cause infection of the broth. Therefore, the foam detector belongs to the standard equipment of bioreactors. First, the use of these instruments for process analysis, optimization, and control is considered.

2 Use of the Variables Temperature, pH, p_{O_2}, E_h, and p_{CO_2} for Process Analysis, Optimization, and Control

2.1 Temperature and pH

Cells have an optimum temperature and pH for growth and frequently another optimum for product formation. Several authors have considered the calculation of the optimum temperature and pH profiles for product formation.

FAN and WAN (1963) used the discrete maximum principle to calculate the optimum temperature and pH profiles for a continuous multistage enzymatic reactor to maximize product concentration.

BOURDARD and FOULARD (1973) considered the optimization of yeast production in a batch process by means of optimum temperature and pH profiles using the continuous maximum principle.

SPITZER (1976) used a grid search method with subsequent steepest descent to maximize biomass productivity in a continuously operated bioreactor by optimizing pH and substrate profiles.

RAI and CONSTANTINIDES (1973) and CONSTANTINIDES and RAI (1974) investigated the production of gluconic acid with *Pseudomonas ovalis* and of penicillin G by *Penicillium chrysogenum* in batch operation and used the continuous maximum principle to maximize the productivity by means of optimal temperature and pH profiles.

CONSTANTINIDES et al. (1970) and KING et al. (1974) studied the production of penicillin G by *P. chrysogenum* in batch operation and used the continuous maximum principle and/or a specific optimal control to evaluate the optimum temperature profile for achieving maximum productivity. ANDREYEVA and BIRYUKOV (1973) also investigated the batch production of penicillin G and used the continuous maximum principle to find the optimum pH profile for maximum productivity.

BLANCH and ROGERS (1972) maximized profit in gramicidin S production by *Bacillus brevis* by evaluating the optimum temperature and pH as well as the number of stages using the discrete maximum principle.

Erythromycin biosynthesis in batch operation was maximized by CHERUY and DURAND (1979) by evaluating optimal temperature and pH profiles.

On the other hand, the pH variation can be used to control the production process. PAN et al. (1972) reported on penicillin production where carbohydrate and nitrogen source feed rates were controlled by measurement of the pH. The nitrogen source was metabolized to basic cations and the carbohydrate source to CO_2 and organic acids. The balance of the two ingredients provided a basis for the pH control.

ANDREYEVA and BIRYUKOV (1973) proposed a model for this pH effect and its use for calculating optimal fermentation conditions. CONSTANTINIDES (1979) reviewed these publications. SAN and STEPHANOPOULOS (1984) also proposed a relationship between the total rate of biomass growth and ammonia addition to the reactor for pH control.

ROSEN and SCHÜGERL (1984) used the consumption of sodium hydroxide solution at a constant pH to calculate the cell mass production rate of *Chaetomium cellulolyticum* and to control the substrate feed by means of a microprocessor in a fed-batch biomass production process. SHIOYA (1988) developed an advanced pH control system for the measurement of biological reaction rates.

In many microbial or cell culture systems the pH varies during growth. Acid or base must be added to the broth to keep the pH at the optimal value. In some enzymatic hydrolytic reactions acid or base must also be used to keep the pH constant by neutralizing the produced acid or base. From the amount of acid or base required for keeping the pH constant, the growth rate or enzyme reaction rate can be calculated.

If $\Delta pH(k+1)$ and $\Delta pH(k)$ are the differences of pH from the set point at times $k+1$ and k, $F(k)$ is the feeding rate of acid or base for pH control, and $R(k)$ is acid or base production or consumption rate, then Eq. (1) can be used to calculate $R(k)$:

$$\Delta pH(k+1) = \Delta pH(k) + aF(k) - bR(k) \qquad (1)$$

where a is a coefficient corresponding to the pH deviation caused by adding a unit amount of acid or base to the broth, and b is a coefficient that corresponds to the pH deviation caused by the formation of a unit amount of cell mass or product.

If the pH is kept constant, i.e., $\Delta pH(k+1) = \Delta pH(k) = 0$, then the acid or base production or consumption rate $R(k)$ can be evaluated from

$$R(k) = (a/b)F(k) \qquad (2)$$

This principle was applied to

- baker's yeast production using a disturbance predictive controller to determine the growth rate of the yeast,
- determination of the reaction rate of N-acetyltyrosine ethyl ester (ATEE) with α-chymotripsin to give ethanol and N-acetyltyrosine (AT) in a pH stat using a repeated feedforward/feedback controller (repeated PF system),
- determination of the overall production rate of (lactic) acid in hybridoma culture by an on-off controller.

The excretion of enzymes can sometimes be controlled by the pH value. For instance, LEE-LASART and BONALY (1988) reported on controlling the excretion of acid phosphatase by *Rhodotorula glutinis* by means of the pH of the medium. The enzyme was excreted only in the pH range 4.5 to 6.5.

Several microorganisms produce different metabolites depending on the pH. Thus, *Aspergillus niger* produces citric acid in the pH range from 2.5 to 3.5, whereas gluconic acid is produced at a higher pH and oxalic acid in the neutral pH range (SCHLEGEL, 1974).

2.2 Dissolved Oxygen Partial Pressure, p_{O_2}

Dissolved oxygen pressure or concentration is a state variable widely used to calculate the biomass concentration by O_2-balancing and to control the growth or production process of aerobic microorganisms.

Furthermore, p_{O_2}-electrodes are used as research tools for determining oxygen transfer rates (OTR) in bioreactors, biofilms, pellets, and cells immobilized in beads.

The use of oxygen balancing for real-time estimation of the biomass concentration was recommended first by HOSPODKA (1966). ZABRISKIE and HUMPHREY (1978) worked out this technique of observation in detail. Using the relationship between oxygen uptake rate (OUR), the yield coefficient of the cell growth with regard to the oxygen consumption, Y_{X/O_2}, the maintenance coefficient with regard to the oxygen consumption, $m_{O_2/X}$, and the growth rate dX/dt:

$$OUR = m_{O_2/X} X + \frac{1}{Y_{X/O_2}} \frac{dX}{dt} \qquad (3)$$

The cell mass concentration X can be calculated:

$$X(t) = \exp(-m_{O_2/X} Y_{X/O_2} t) \cdot \qquad (4)$$

$$\cdot \left[\int_0^t Y_{X/O_2} \exp(m_{O_2/X} Y_{X/O_2} t) \cdot OUR(t) dt + X_0 \right]$$

where X_0 is the initial biomass concentration.

Eq. (4) forms the basis for estimating the biomass concentration $X(t)$ from the oxygen uptake rate.

The growth rate may be approximated using Eqs. (3) and (4):

$$\frac{dX}{dt} = Y_{X/O_2} OUR(t) - m_{O_2/X} Y_{X/O_2} X(t) \qquad (5)$$

With this method the biomass concentration of *Saccharomyces cerevisiae* was estimated.

Several other authors used this technique in combination with the respiration coefficient, RQ, to estimate the biomass (e.g., COONEY et al., 1977; WANG et al., 1977; PERINGER and BLACHÈRE, 1979; TAKAMATSU et al., 1981). SQUIRES (1972) reported that p_{O_2} was used to control the sugar addition to the broth of *Penicillium chrysogenum* during penicillin production. The sugar feed was increased at a high p_{O_2} value; at a low p_{O_2} it was reduced. Since a close relationship exists between oxy-

gen and substrate uptake rates, this control of the substrate feed is very popular.

Under steady-state conditions, the oxygen uptake rate, OUR, and the oxygen transfer rate, OTR, are identical. Knowing the driving force for the oxygen transfer, $(O_2 - O_2^*)$, the volumetric mass transfer coefficient, $K_L a$, can be calculated:

$$K_L a = \frac{OTR}{(O_2^* - O_2)} \qquad (6)$$

where O_2 and O_2^* are the concentrations of the dissolved oxygen in the bulk and at the interface (in equilibrium with the gas phase). By measuring the oxygen balance during cell cultivation, the volumetric mass transfer coefficient can be calculated in real time.

In cell-free systems, $K_L a$ can be determined by non-stationary or stationary measurements. The non-stationary method is based on the relationship:

$$\frac{dO_2}{dt} = K_L a (O_2^* - O_2) \qquad (7)$$

in ideal systems.

By measuring the variation of the dissolved oxygen concentration in the bulk as a function of time, and calculating the dissolved oxygen concentration at the interface from the oxygen concentration in the gas phase, $K_L a$ can be evaluated from Eq. (7). However, the interrelationships between sorption rate and driving force are in practice more complex. Several relationships have been recommended for this calculation.

A good review of these methods is given in a 'Report of a Working Party on Mixing' of the European Federation of Chemical Engineering (LINEK and VACEK, 1986) and in the review article of LINEK et al. (1987).

Several papers consider the mass transfer of dissolved oxygen into biofilms, pellets, and cells immobilized in beads. The dissolved oxygen concentration profiles are determined by means of micro-oxygen electrodes (BUNGAY and HAROLD, 1971; CHEN and BUNGAY, 1981; BUNGAY and CHEN, 1981; BUNGAY et al., 1969, 1983; WITTLER et al., 1986).

2.3 Redox Potential, E_h, and Dissolved CO_2 Partial Pressure, p_{CO_2}

The oxidation and reduction of a compound is controlled by the redox potential of its environment.

The oxidation–reduction potential of a pair of reversible, oxidizable–reducible compounds is related to the equilibrium between the oxidized (ox) and reduced forms (red) and the number of electrons involved in the reaction (ne^-) (KJAERGAARD, 1977; KJAERGAARD and JOERGENSEN, 1979; THOMPSON and GERSON, in KJAERGAARD, 1977):

$$ox + ne^- \rightleftarrows red \qquad (8)$$

The redox potential of this reaction is given by the Nernst equation:

$$E_h = E_0 + \frac{RT}{nF} \ln \frac{\text{activity of ox}}{\text{activity of red}} \qquad (9)$$

where E_h is the redox potential referred to the normal hydrogen electrode,

E_0 is the standard potential of the system at 25 °C, when all activities of any reactants are at unity,

R the gas constant,

T the absolute temperature,

n the number of electrons involved in the reaction,

F the Faraday constant.

JOERGENSEN (1941) introduced a concept analogous to the pH, namely the rH, which is defined as

$$rH = -\log a_{H_2} \qquad (10)$$

where a_{H_2} is the activity of hydrogen in the hydrogen–hydrogen ion redox system according to Nernst. For hydrogen

$$E_h = \frac{RT}{2F} \ln a_{H_2} - \frac{RT}{F} 2.303 \, pH \qquad (11)$$

is obtained.

The combination of Eqs. (10) and (11) gives

$$rH = 2 \left(pH + \frac{F}{RT} \frac{1}{2.303} E_h \right) \qquad (12)$$

This rH value may vary from $rH = 0$, corresponding to a solution in which $p_{H_2} = 1$ bar and $pH = 0$, to $rH > 42$, corresponding to a solution with $p_{O_2} = 1$ bar and $pH = 0$. The rH value is a function of the pH value. Therefore, for the measurement of the redox potential, both rH *and* pH values are needed.

The redox potential is used in practice for microaerobic cultivations, i.e., at very low dissolved oxygen concentrations, which cannot be measured by standard oxygen electrodes. An example is the production of exoenzymes by *Bacillus amyloliquefaciens* in continuous culture at 0.5% oxygen saturation by means of redox-potential control (MEMMERT and WANDREY, 1987).

In small stirred tank reactors, the dissolved CO_2 concentration in the broth can be calculated from the gas composition by assuming an equilibrium between the phases. In tower reactors and large commercial units, no equilibrium distribution of CO_2 exists between the phases; therefore, the direct measurement of p_{CO_2} can be useful.

The driving force, $(p_{CO_2} - p_{CO_2}^*)$, can be evaluated from the calculated $p_{CO_2}^*$ at the interface and the measured p_{CO_2} in the bulk. The CO_2 production rate, *CPR*, can be determined from the evolved gas stream and the gas composition.

The volumetric mass transfer coefficient of the CO_2 desorption is given by

$$(K_L a)_{CO_2} = \frac{CPR}{(p_{CO_2} - p_{CO_2}^*)} \qquad (13)$$

CPR can also be used for the calculation of the cell mass concentration and the specific growth rate, μ. The instantaneous specific growth rate of *Penicillium chrysogenum* was calculated by MOU and COONEY (1983) by measuring the *CPR* during the growth phase.

By monitoring the O_2 and/or CO_2 concentrations in the outlet gas and its flow rate, O_2 and/or CO_2 balances can be calculated and used for state estimation of biochemical reactors (e.g., STEPHANOPOULOS and SAN, 1982). However, because this state estimation method is based on measurements of the gas composition, it will be discussed in Chapter 2.

3 Instruments for Determination of Physical System Properties

3.1 Temperature

Temperature is the most important control variable for most biotechnological processes, including sterilization as well as cell growth and product formation. In general, a precision of $\pm 0.5\,°C$ is necessary in the temperature range from $+20$ to $+130\,°C$. Only a few types of the various industrial thermometers are suitable because of this prerequisite (BÜSING and ARNOLD, 1980). Most popular are the Pt-100 (100 ohm at 0 °C and 123.2 ohm at 60 °C) resistance thermometers which are encased in a protective steel tube fixed with a sealing compound of high heat conductivity.

According to DIN 43760 (German Standard) the resistance of these instruments is guaranteed with the following precision: 100 ± 0.1 ohm at 0 °C, which corresponds to an error of $\pm 0.26\,°C$. Therefore, these resistance thermometers can be used without calibration. However, the resistances of all electrical connections must be controlled. These instruments are steam-sterilizable at 121 °C. Thermometers with short response times for fluid dynamical measurements are described in Chapter 4.

3.2 Pressure

The absolute pressure is measured with respect to zero pressure. Gauge pressure is measured with respect to that of the atmosphere. The SI unit of the pressure is Newton per square meter (N/m^2) called Pascal (Pa). (1 bar = 0.1 MPa = 10^5 N/m^2; 1 mbar = 100 Pa = 100 N/m^2.) Bar and millibar deviate with less than 2% from the technical and physical atmosphere.

Pressure measurements are necessary for the control of the sterilization and the state of the outlet gas filter as well as for the evaluation of the holdup and the partial pressures of the gaseous components in the gas and liquid phases. Membrane pressure gauges are commonly used in biotechnology, because they are particularly suited to aseptic operations. Numerous pressure gauges are used in the chemical industry (HIRTE, 1980; ANDREW and MILLER, 1979). In biotechnology, the commonly employed pressure gauges are based on strain and/or capacitance measurements. The capacitance pressure gauges can measure very small pressure differences; therefore, they are used for liquid level measurements. For the construction of the different pressure meters, see HIRTE (1980) and ANDREW and MILLER (1979).

3.3 Liquid Level and Holdup

Measurement of the liquid volume is important for filling bioreactors with nutrient solutions, for continuous and for fed-batch cultivations. It can be performed (OEDEKOVEN, 1980; ELFERS, 1964; ANDREW and RHEA, 1970) as follows:

● by measuring the hydrostatic pressure difference between the bottom of the reactor, p_b, and the head space, p_h, by means of pressure gauges. The pressure difference is proportional to the weight of the liquid in the reactor:

$$p_b - p_h = h \rho g \qquad (14)$$

where h is the liquid height above the bottom,
ρ the density of the broth, and
g the acceleration of gravity,

● by measuring the total weight of the reactor by load cells. The accuracy of the volume measurement is $\pm 0.2\%$ for large reactors and $\pm 1\%$ for laboratory reactors.

The measurement of the volume of an aerated broth is accomplished with a level controller. The common liquid level meters are based on the variation of the capacitance C of the sensor with the composition of the dielectricum. For plate condensers, the capacitance is given by

$$C = E^0 E_r \frac{A}{d} \tag{15}$$

where A is the area of the plates,
d the distance between the plates,
E^0 the absolute dielectric constant of vacuum, and
E_r the relative dielectric constant of the aerated broth between the plates.

The value of E_r of the broth and of the air differ by a factor of about 80. In the case of non-aerated broth the capacitance of the condenser, C, is given by

$$C = C_0 + \Delta C\, h \tag{16}$$

where C_0 is the capacity of the condenser with air,
ΔC the capacity difference due to broth per unit height, and
h the height of the liquid in the capacitor.

The capacity of the condenser with aerated broth is given by

$$C = C_0 + \Delta C h (1 - \varepsilon) \tag{17}$$

where ε is the gas holdup in the aerated broth. Analogous relationships hold true for cylindrical condensers (OEDEKOVEN, 1980).

The accuracy of the level control amounts to ± 2–4% depending on the uniformity of the liquid level. In the case of large reactors, the level variation can be extremely large. Therefore, only the level of the broth can be measured in the reactor, not that of the aerated broth, which is measured outside of the reactor, e.g., in a non-aerated section. Also in the case of foam formation, the measurement of the aerated broth level by capacitance instruments becomes difficult. Under these conditions floating bodies can be used as level controllers.

Since cultivation broths have adequate electrical conductivity, the liquid level can also be measured by inexpensive electrical conductivity probes. Their application is restricted to aqueous broths. In the presence of a second (organic) liquid phase, their application cannot be recommended.

For other level control instruments, see, e.g., OEDEKOVEN (1980), ELFERS (1964), ANDREW and RHEA (1970).

3.4 Liquid Throughput

Gas and liquid flow rates are important control variables for biotechnological processes; they must be known for reactor operation and for component balancing. In this chapter, only instruments for liquid throughput measurement are taken into account. Instruments for gas throughput measurements are considered in Chapter 2. Special techniques for measurements of local liquid velocities are treated in Chapter 4.

Of the large number of available instruments (SCHRÖDER, 1980; ANDREW et al., 1979; ERICSON, 1979) only three types are important in biotechnological practice:

- floating body flowmeters,
- differential pressure flowmeters, and
- magnetic-inductive flowmeters.

The floating body flowmeter or rotameter consists of a conical tube and a floating body with the upper diameter D_s, mass M_s, and density ρ_s (Fig. 1). In the upstreaming fluid, the lifting force, which is produced by the differential pressure across the slot between the tube wall and the floating body, is balanced by the weight of the floating body minus its buoyancy. The position of the float is a function of the flow rate and the density of the fluid, ρ. The volumetric throughput q_V is given by

$$q_V = \alpha D_s \frac{1}{\rho} \sqrt{g M_s \rho (1 - \rho/\rho_s)} \tag{18}$$

The flow coefficient α is a function of the Reynolds number and the diameter ratio D_k/D_s, where D_k is the diameter of the tube at the upper edge of the floating body.

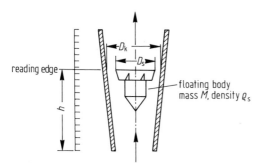

Fig. 1. Floating body flowmeter (SCHRÖDER, 1980).

Calibration of q_V is necessary because of the nonlinear relationship between the position of the floating body and the throughput. It can be carried out with water or air and recalculated for the nutrient medium with known density by means of the $\alpha - Ru$ diagram, where the Ruppel number

$$Ru = \frac{\eta}{\sqrt{g M_s \rho (1 - \rho/\rho_s)}} \qquad (19)$$

depends only on the instrument constants and fluid properties, but not on the throughput.

The accuracy of rotameters is between ± 1 and $\pm 3\%$ depending on the ratio $q_V/q_{V,\max}$ (SCHRÖDER, 1980).

Differential pressure flowmeters consist of a tube with a restriction (usually an orifice plate). The pressures p_1 and p_2 upstream and downstream of the orifice are measured. The throughput is (SCHRÖDER, 1980):

$$q_V = \alpha A_0 \sqrt{1/2\rho (p_1 - p_2)} \qquad (20)$$

where α is the flow coefficient, which is a function of D, η, ρ, w,
A_0 the cross-section of the orifice opening,
ρ the density of the fluid,
D the tube diameter,
η the dynamic viscosity of the fluid, and
w the mean flow rate of the fluid.

The flow coefficient α is usually given as a function of the orifice Reynolds number and the orifice-to-tube diameter ratio d/D for smooth tubes.

In practice, standardized orifices are used for which the flow coefficients are given in diagrams. The accuracy of calibrated orifice flowmeters is $\pm 0.5\%$ of $q_{V,\max}$.

According to the induction law of Faraday, an electrically conductive liquid passing a magnetic field induces a voltage between two electrodes positioned perpendicular to the direction of the flow. The voltage is proportional to the flow velocity:

$$U \sim BDW \qquad (21)$$

where U is the induced voltage,
B the magnetic induction,
D the tube diameter, and
w the mean liquid velocity.

The volumetric throughput q_V is given by (SCHRÖDER, 1980):

$$q_V \sim \frac{\pi}{4} \frac{D}{B} U \sim \frac{U}{B} \qquad (22)$$

Cultivation broths are electrically conductive, because they contain nutrient salts. Therefore, magnetic-inductive flowmeters can be employed for the measurement of nutrient medium throughputs. For the description of these instruments, see SCHRÖDER (1980).

Magnetic-inductive flowmeters are fairly expensive. However, they have important advantages:

- the voltage U is proportional to q_V,
- they are independent of the density and viscosity of the fluid as well as of the velocity profile of the fluid in tubes,
- they do not produce a pressure drop,
- they do not have moving parts,
- they can be used for suspensions,
- they can be steam-sterilized.

Their accuracy is $\pm 1\%$ at $q_{V,\max}$, and $\pm 1.5\%$ at $0.5\ q_{V,\max}$.

3.5 Power Input

In an agitated reactor, the power input, P, can be calculated by Eq. (23) by measuring the torque on the shaft, M_N, and the speed of rotation, N:

$$P = M_N 2\pi N \tag{23}$$

The torque is measured by torsion dynamometers or strain gauges and the impeller speed by an electronic tachometer. In large-scale reactors, the consumed electrical energy, as measured by the wattmeter, yields useful data on power input, if the mechanical losses in gear, seals, etc., are taken into account.

In small laboratory reactors, the mechanical losses are considerable in comparison with the power input into the broth. Therefore, power input measurements are inaccurate and are not recommended.

In bubble columns, P can be calculated by

$$P = M_G \left\{ RT \ln \frac{p_{in}}{p_{out}} - gH - \right.$$

$$\left. - \frac{1}{2} (w_{G,out}^2 - w_{G,in}^2) \right\} \tag{24}$$

where M_G is the gas mass flow,
      R the gas constant,
      T the absolute temperature,
      p_{in} the gas pressure at the column inlet,
      p_{out} the gas pressure at the column outlet,
      H the height of the bubbling layer,
      $w_{G,in}$ the linear gas velocity at the inlet,
      $w_{G,out}$ the linear gas velocity at the outlet, and
      g acceleration of gravity.

However, since the second and third terms together make up only 0.2% of the overall power input, the power input due to the gas expansion dominates:

$$P = M_G RT \ln \frac{p_{in}}{p_{out}} \tag{25}$$

In the power inputs, Eqs. (24) and (25), the energy losses due to mechanical energy (e.g., due to gas compression) are not considered.

3.6 Liquid Viscosity

The viscosity of the broth influences the operation of bioreactors considerably. At a high viscosity, a high specific power input is necessary to increase the intensity of transfer processes: oxygen transfer (into the broth of aerobic microorganisms) and mixing. Since at high specific power input the energy dissipation rate in the reactor is high, improvement of heat transfer is also important to keep the temperature constant.

High viscosity can be caused by high substrate concentration (e.g., starch), high product concentration (e.g., xanthan), high cell concentration (e.g., penicillin), high solid content (e.g., peanut flour), or by their combination. The most general description of the rheological properties of fluids is given by the relationship between the velocity gradient dv/dx and the stress, τ, the so-called flow equation:

$$\frac{dv}{dx} = f(\tau) \tag{26}$$

as long as viscoelastic behavior is not present or very slight. This flow equation can be calculated from the experimentally measured shear diagrams (shear rate versus shearing stress). It should be noted, however, that such a calculation is not always possible. In contrast to the shear diagram, the flow equation is independent of the experimental conditions (e.g., the type of viscosimeter) used for the determination of the viscosity.

There are many methods available to estimate the rheological behavior of fluids, but there are only a few that furnish true fluidity values. These include the capillary, the falling sphere, the Couette, the Searle, and the torsional pendulum methods. Until now, the evaluation of the flow equation from the shear diagram has only been possible for the capillary, Couette, and Searle methods (MUSCHELKNAUTZ and HECKENBACH, 1980).

The capillary viscosimeter cannot be employed for cultivation broths because of adverse wall effects in the capillary. As for the falling sphere and torsional pendulum viscosimeters, the flow equation cannot be calculated from the shear diagram (only partial solutions are known).

The Couette and Searle viscosimeters can only be used if the following conditions are fulfilled: the annular slit between inner and outer cylinders must be large enough to reduce the wall effects, and measurements must be made using different cylinder lengths to eliminate the end effects. In a Searle viscosimeter, the speed of rotation is limited by the occurrence of Taylor instabilities.

By measuring the torque M_N on the shaft of different types of stirrers at differing stirrer speeds, N is suited for the evaluation of the power input but not for the viscosity. These techniques, which are commonly used according to the literature, are not suitable for the evaluation of the shear diagram and the absolute viscosity.

Only the coaxial cylinder viscosimeters, Couette with rotation outer cylinder and Searle with rotation inner cylinder, are considered here, since they are the most popular ones.

The velocity gradient at distance r is

$$\frac{dv}{dx} = -\frac{d\omega}{dr} \qquad (27)$$

while the shear stress is

$$\tau = \frac{M_i}{2\pi r^2 L} \qquad (28)$$

where ω is the angular speed,
M_i the torque exerted on the inner cylinder, and
L the length of the inner cylinder.

From Eqs. (27) and (28) it follows that

$$d\omega = \frac{1}{2}f(\tau)\frac{d\tau}{\tau} \qquad (29)$$

Integration of Eq. (29) with $s^2 = R_i^2/R_a^2 = \tau_i/\tau_a$ gives:

$$\Omega = \frac{1}{2}\int_{\tau_a}^{\tau_i}\frac{f(\tau)}{\tau}d\tau \qquad (30)$$

where R_i and R_a are the radii of the inner and outer cylinders,
τ_i and τ_a are the shear stresses at the inner and outer cylinders.

The relationship between the angular velocity of the rotating cylinder Ω and τ is experimentally determined to obtain the shear diagram. The relationship $dv/dx = f(\tau)$ (flow equation) can be calculated from Eq. (30). For this evaluation, see MUSCHELKNAUTZ and HECKENBACH (1980) and DINSDALE and MOORE (1962).

For Newtonian fluids, the following relationship is valid:

$$\tau = -\eta\frac{dv}{dx} \qquad (31)$$

where η is the dynamical viscosity.

In practice, relative viscosities are frequently determined. The shear stress is measured for different shear rates with fluids of known (oils) and unknown (broth) viscosities, and the relative viscosity of the broth can be calculated from the ratio of their shear stresses at the same shear velocity, if the broth has Newtonian behavior.

On-line determination of the broth viscosity is sometimes useful for controlling a process. The on-line techniques only yield relative viscosities. The viscosity of the *Aspergillus niger* broth was measured on-line by means of a tube viscosimeter by BLAKEBROUGH et al. (1978). PERLEY et al. (1979) used an on-line capillary technique for the measurement of the viscosity of the *Hansenula polymorpha* broth. LANGER and WERNER (1981) and NEUHAUS et al. (1983) developed an on-line slot-type viscosimeter and measured the viscosity of the *Penicillium chrysogenum* broth. KEMBLOWSKI et al. (1985) used an on-line impeller type viscosimeter to determine the viscosity of the *Aureobasidium pullulans* broth.

3.7 Foaminess

Several cultivation broth components, especially a combination of different surfactants with proteins, may cause stable foams in aerated bioreactors. Foam control is necessary to avoid the loss of broth, the clogging of the gas analyzers, and infections caused by foam carry-out.

Foam can be suppressed by antifoam agents (BEROVIC and CIMERMAN, 1979; SIE and SCHÜGERL, 1983; SCHÜGERL, 1986; PRINS and VAN'T RIET, 1987; VIESTURS et al., 1982) or destroyed with mechanical foam breakers (VIESTURS et al., 1982). Foam can be detected by an electrical conductivity probe, capacitance probe, heat conductivity probe, or light scattering probe (HALL et al., 1973; VIESTURS et al., 1982). Antifoam and mechanical foam breakers are frequently combined, if the foam is very stable.

The presence of an antifoam agent in the broth may influence cell growth and product formation as well as downstream processing. Mechanical foam breaking may exert stress and selection pressure on the cells.

4 Instruments for Determination of Chemical System Properties

4.1 pH Value

The dissociation constant K_t of the purest water is very low ($10^{-15.74}$ at 25 °C). The concentration of water can be considered as constant because of the low K_t value. Thus, only the ion product K_w is taken into account:

$$[H^+] \cdot [OH^-] = K_w = 1.008 \cdot 10^{-14} \text{ at } 25\,°C \quad (32a)$$

Forming the logarithm of Eq. (32a)

$$\log[H^+] + \log[OH^-] = \log K_w \quad (32b)$$

and by multiplication with -1,

$$-\log[H^+] = pH$$
$$-\log[OH^-] = p_{OH}$$
$$-\log K_w = pK_w$$
and $pH + p_{OH} = pK_w$

are obtained.

At $[H^+] = [OH^-]$, the hydrogen ion concentration is $[H^+] = 10^{-7}$, therefore, at the neutral point $pH = p_{OH} = 7$.

The pH can be measured with a galvanic cell (chain). The potential E of the cell is given by the Nernst equation:

$$E = E_0 + 2.3 \frac{RT}{F} \log[H^+] \quad (33)$$

where E_0 is the standard potential and F the Faraday constant.

In this definition the thermodynamic activities of the ions were replaced by their concentrations since the activities cannot be measured. The absolute potential cannot be measured either, only the potential difference U between the indicator electrode and a reference electrode.

Silver–silver chloride electrodes are used in the galvanic chain for sterilizable electrodes. Fig. 2 shows the schematic assembly of a pH electrode (INGOLD I). In this figure E_1 is the potential on the outer surface of the glass membrane, which depends on the pH value of the sample solution. E_2 is the asymmetry (bias) potential, i.e., the potential of the glass membrane with the same solutions on both sides. E_3 is the potential on the inner surface of the glass membrane, which is a function of the pH value of the internal buffer solution. E_4 is the potential of the internal Ag/AgCl lead-out electrode, dependent on the KCl concentration in the internal buffer solution. E_5 is the potential of the reference AgCl/Ag electrode, which

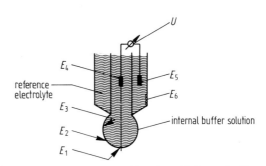

Fig. 2. Schematic assembly of a pH-electrode (Dr. W. Ingold AG, Brochure I, with permission). For details see text.

depends on the KCl concentration in the reference buffer solution, E_6 is the diaphragm or diffusion potential.

Since E_1 is the potential which we want to measure, the individual potentials E_2–E_6 should be kept constant. These are included in the standard potential U^0, which has to be determined by calibration.

In modern pH electrodes, U^0 varies only in a narrow range (e.g., Type U 402-K7 electrodes of Ingold AG have a potential of -10.4 ± 3.8 mV at pH 7.02 and 20 °C).

The potential difference between the indicator and reference electrodes U is also given by the Nernst equation:

$$U = U^0 + U_N \log [\mathrm{H}^+] \tag{34}$$

where the Nernst potential $U_N = \dfrac{2.3RT}{F} =$ 59.2 mV at 25 °C. However, in real pH electrodes, the Nernst potential is not attained, but only approached to 97.5% (in the case of new electrodes). Furthermore, U_N is reduced with increasing age of the electrode. The aging causes sluggish response, increasing electrical resistance, a smaller slope, and zero point (U^0) drift. During steam sterilization, a pressure difference builds up on both sides of the glass membrane. Therefore, a counter pressure is imposed to avoid the destruction of the electrode. Frequent steam sterilization has a considerable aging effect. Therefore, pH electrodes must be recalibrated frequently with buffer solutions.

During *in situ* steam sterilization a considerable, irreversible signal drift of the pH electrodes occurs. Therefore, it is advisable to measure the pH value of the broth in the reactor after each steam sterilization by an independent method and correct the reading of the pH meter.

Since the potential U depends on the temperature, pH-meters have a temperature compensation, which is usually calculated by the relationship

$$U_N(T) = U_N(25 \text{ °C}) (1 + \alpha t) \tag{35}$$

where $\alpha = 3.21 \pm 0.53 \cdot 10^{-3}/\text{°C}$,
$\quad\quad t = T$ – standard temperature (25 °C).

For more information on pH electrodes, their use, storage, aging, etc., see INGOLD I, PETERSEN (1980), MELZNER and JAENICKE (1980), and MCCULLOUGH and ANDREW (1979).

4.2 Dissolved Oxygen Partial Pressure, p_{O_2}

The dissolved oxygen concentration is also measured by electrochemical methods. Two types of electrodes are in use:

- polarographic electrodes
- galvanic electrodes.

In polarographic or amperometric electrodes the dissolved oxygen is reduced at the surface of the noble metal cathode in a neutral potassium chloride solution, provided it reaches 0.6–0.8 V negative with respect to a suitable reference electrode (calomel or Ag/AgCl). The current–voltage diagram is called the polarogram of the electrode (Fig. 3).

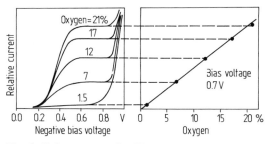

Fig. 3. Polarogram and calibration curve for a p_{O_2}-electrode (LEE and TSAO, 1979).

At the plateau of the polarogram, the reaction rate of oxygen at the cathode is limited by the diffusion of oxygen to the cathode. Above this voltage the water is electrolyzed into oxygen and hydrogen. In the plateau region (0.6–0.8 V), the current is proportional to the partial pressure of the dissolved oxygen (Fig. 3).

In this probe, the cathode, the anode, and the electrolyte are separated from the measuring liquid by a membrane which is permeable to gaseous oxygen. In the electrolyte, the following reactions occur:

cathodic reaction:
$$O_2 + 2H_2O + 2e^- \rightarrow H_2O_2 + 2OH^-$$
$$H_2O_2 + 2e^- \quad\quad \rightarrow 2OH^-$$

anodic reaction:
$$Ag + Cl^- \rightarrow AgCl + e^-$$

overall reaction:
$$4Ag + O_2 + 2H_2O + 4Cl^- \rightarrow 4AgCl + 4OH^-$$

Since hydroxyl ions are constantly being substituted for the chloride ions as reaction proceeds, KCl or NaCl must be used as an electrolyte. When the electrolyte becomes depleted of Cl^-, it has to be replenished.

The dissolved oxygen concentration is measured by the galvanic electrode which does not require an external voltage source for the reduction of oxygen at the cathode. Using a basic metal such as zinc or lead as anode and a nobler metal such as silver or gold as cathode, the voltage is generated by the electric pair and is sufficient for a spontaneous reduction of oxygen at the cathode surface. The reaction of the silver–lead galvanic electrode is given by:

cathodic reaction
$$O_2 + 2H_2O + 4e^- \rightarrow 4OH^-$$

anodic reaction
$$Pb \rightarrow Pb^{2+} + 2e^-$$

overall reaction
$$O_2 + 2Pb + 2H_2O \rightarrow 2Pb(OH)_2$$

During the reduction of oxygen, the anode surface is gradually oxidized. Therefore, occasional replacement of the anode is necessary.

The polarographic or amperometric electrode is in greater demand in biotechnological practice than the galvanic electrode. Fig. 4 shows a schematic view of a steam-sterilizable polarographic or amperometric oxygen electrode.

A constant voltage (ca. 650 mV) is applied between cathode (Pt) and anode (Ag/AgCl). A regular control of this voltage is necessary in order to avoid incorrect measurements. The

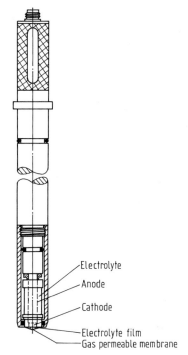

Electrolyte
Anode
Cathode
Electrolyte film
Gas permeable membrane

Fig. 4. Sterilizable p_{O_2}-electrode (Dr. W. Ingold AG, Brochure II, with permission).

control is carried out by measuring the polarogram and adjusting the bias voltage to maintain a voltage-independent current in the plateau region of the polarogram.

The current i_{p,O_2} is proportional to p_{O_2} only in the plateau region:

$$i_{p,O_2} = K \frac{A P p_{O_2}}{d} \tag{36}$$

where K is a constant,
  A the surface area of the cathode,
  P the membrane permeability,
  d the membrane thickness.

The response time is proportional to d^2/P. Therefore, thin membranes with high gas O_2-permeability are used. Two membranes are used for the p_{O_2} electrodes for sterile operation. The inner membrane consists of a 25 μm teflon foil, the outer one of a 150 μm silicone membrane reinforced by thin steel mesh. This type of electrode was developed by the Instrumentation Laboratory Inc., Lexington, Mass., USA, and also produced by Dr. W. Ingold AG, Urdorf, Switzerland (INGOLD II). This type of electrode has a fairly long response time (45 to 90 s to attain 98% of the final signal).

During steam sterilization, the membrane thickness and shape change irreversibly. An improved construction of BAUERMEISTER (1981) enables the electrodes to endure many (ca. 20) sterilizations without any change in the membranes.

The temperature of the calibrations and measurements must be controlled closely (± 0.1 °C) because of the temperature sensitivity of the signal (temperature coefficient 3%/°C). Since the electrode measures the partial pressure of oxygen, the signal is independent of the O_2-solubility in the broth. The calibration should be performed in the reactor under the same fluid dynamic conditions (stirrer speed) as those that prevail during cultivation to avoid errors due to differences in diffusion resistance at the surface of the membrane.

The calibration is carried out with nitrogen- and air-saturated broth by setting these values at 0 and 100%. The partial pressure of oxygen is expressed as follows:

$$p_{O_2} = [p_B - p(H_2O)] \times 0.2095 \tag{37}$$

where p_B is the temperature-corrected (barometric) pressure in the reactor,

$p(H_2O)$ the vapor pressure of the broth at the temperature of the calibration,

0.2095 the fraction of oxygen in atmospheric air.

The sources of error in the measurement of p_{O_2} are numerous: errors in reading of temperature and pressure, drift due to membrane fouling, change in membrane shape, variation of bias voltage and electrical resistance as well as capture of bubbles, etc. With sufficient accuracy of temperature and pressure measurements, and with bias voltage in the plateau region, the precision of the measurements is on the average $\pm 5\%$. Below 5% of the O_2-saturation, the error increases with decreasing p_{O_2}.

The dissolved oxygen concentration $[O_2]$ is calculated by the relationship:

$$[O_2] = p_{O_2}/H \tag{38}$$

where H is the Henry coefficient

$$H = \frac{22.4(760)}{1000\ \alpha}$$

and α is the Bunsen coefficient.

Bunsen coefficients α of oxygen for some simple aqueous solutions and a few cultivation broths have been given by SCHUMPE (1985). For more details, see MELZNER and JAENICKE (1980), INGOLD II, LEE and TSAO (1979), FRITZE (1980), BUEHLER and INGOLD (1976), and SCHINDLER and SCHINDLER (1983).

4.3 Redox Potential, E_h

The definition of the redox potential is given by Eq. (9). To determine E, the potential between the redox electrode and a standard reference electrode is measured. The universal reference reaction is the oxidation of hydrogen:

$$H_2 \rightarrow 2H^+ + 2e^-$$

$$E = E_0 + 2.3\ \frac{RT}{2F}\ \log\ \frac{[H+]}{[H_2]} \tag{39}$$

The standard potential $E_0(H^+/H_2)$ is by definition equal to zero at all temperatures. The universal reference electrode is known as the Standard Hydrogen Electrode (SHE), which consists of a platinum-coated platinum foil that is immersed in a solution containing 1 mol L^{-1} H^+, and over which flows hydrogen gas at a pressure of 1 bar. The reference electrodes (Hg/calomel/sat. KCl, or Ag/AgCl/KCl) used in practice are referred to the SHE:

$$E_h = E + E_{ref} \tag{40}$$

where E_h is the redox potential against the SHE,

E the redox potential against the reference electrode,

E_{ref} the standard potential of the reference electrode.

The sterilizable redox meter consists of a Pt electrode and an Ag/AgCl reference electrode. The electrodes are calibrated with redox buffers in the range of $E_h = +200$ mV to 600 mV (INGOLD III). Since the redox potentials have a high temperature coefficient, knowing the temperature of the broth is necessary for calculating the correct standard potential for the reference electrode. For example, standard potentials of Ingold reference electrodes are given in INGOLD III for different temperatures.

Redox potentials occur in a range of -1200 to $+1200$ mV. Measurement precision is ± 5 mV. A simple pH meter with a mV scale is an adequate measuring instrument. The redox potential depends on the p_{O_2} and pH in the broth. However, since both are measured in bioreactors, these effects can be taken into account. In aerobic cultivations the p_{O_2} and redox meters give nearly the same information at a constant pH value. In microaerobic and anaerobic cultivations the redox potential gives additional information about the state of the broth components. However, because of the complex composition of the broth this information is only qualitative.

For more information on the redox potential, see MELZNER and JAENICKE (1980), KJAERGAARD (1977), KJAERGAARD and JOERGENSEN (1979), INGOLD III, and FRITZE (1980a, b).

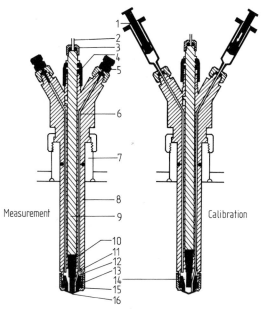

Fig. 5. Sterilizable p_{CO_2}-electrode (Dr. W. Ingold AG, Brochure IV, with permission).
Construction of a CO_2-sensor: (1) 20 mL syringe, (2) high-temperature coaxial cable, (3) cable screw connection, (4) adjustment nut, (5) locking plug, (6) supply duct, (7) welding socket, (8) bore hole conductor, (9) draw tube, (10) pH-electrode, (11) reference electrode, (12) CO_2-electrode, (13) membrane body, (14) calibration buffer, (15) glass membrane, (16) reinforced silicon membrane.

4.4 Dissolved CO_2 Partial Pressure, p_{CO_2}

The presence of dissolved CO_2 in the broth influences cell growth and product formation (HO et al., 1987). Therefore, the p_{CO_2} can be an important variable. The p_{CO_2} can be measured in-line using the p_{CO_2} meter of Dr. W. Ingold AG (INGOLD IV). The instrument consists of a pH meter and a hydrogen carbonate solution, which is separated from the broth by a gas-permeable membrane. Fig. 5 illustrates the main features of the electrode.

The dissolved CO_2 diffuses through the membrane into the hydrogen carbonate solution. The equilibrium of the reaction

$$CO_2 + H_2O \rightleftharpoons HCO_3^- + H^+$$

is determined by the dissociation constant K

$$K = \frac{[H^+][HCO_3^-]}{[CO_2]} \tag{41}$$

according to Henry's law

$$[CO_2] = p_{CO_2} H \tag{42}$$

where H is the Henry coefficient.

Since the hydrogen carbonate concentration in the electrolyte is high, it can be assumed to be constant. Thus, Eq. (41) can be simplified to

$$[H^+] = p_{CO_2} \text{ const.} \tag{43}$$

The potential of the inner pH electrode is a function of $[H^+]$:

$$E = E_0 + 2.3 \frac{RT}{F} \log [H^+] \qquad (44)$$

where $2.3\,RT/F = 59.16$ mV at 25 °C,
$\quad E$ is the measured potential, and
$\quad E_0$ is the standard potential.

From relationships (43) and (44)

$$E = E_0 + 2.3 \frac{RT}{F} \log (p_{CO_2}) \qquad (45)$$

is obtained.

The response time is fairly long (one to several minutes) and is influenced by the thickness of the membrane and by the electrolyte solution as well as by the response time of the pH electrode.

The measuring range of the electrode is 1 to 1000 mbar CO_2. The deviation is $\pm 2\%$, if the electrode is calibrated with gas mixtures. If the inner pH electrode is calibrated by buffer solutions, the deviation is $\pm 10\%$. To avoid errors due to the complex temperature dependence of the reading, the calibration should be carried out at broth temperature.

The electrode is sterilizable. The sterilization is performed after the reduction of the pressure of the pH electrode on the stainless-steel reinforced plastic membrane. The pH electrode is calibrated by buffer solutions after sterilization. Then the electrode is filled with the electrolyte and put into the measuring position. Fig. 5 shows the electrode during calibration and measurement. For more information, see INGOLD IV.

5 Performance of Instruments for Process Control

According to FLYNN (1982) the relevance, accuracy, and precision of the measured data and the reliability, accuracy, precision, resolution, specificity, response, sensitivity, availability, and costs of the sensors/instruments are important for their use in process control. All data which influence the productivity and the yield of the process and the quality of the product are relevant. Accuracy of the measured data is expressed as the difference between the observed value of the variable and its true value, which is usually determined by calibration.

The precision of the data relates to the probability that repeated measurements of the same system will produce the same values. The distribution of the values around their mean is usually characterized by the variance and/or standard deviation, or, e.g., the 95% confidence interval.

The most important property of a sensor is its reliability, which is made up of factors such as failure rate, failure mode, ease of preventive maintenance, ease of breakdown maintenance, physical robustness, and its credibility in the mind of process operators (FLYNN, 1982). The latter plays a role only if the data are used by the operator and not by an automated system.

Based on information from three chemical works, LEES (1976) published data on the reliability of the instruments important in the fermentation industry (Tab. 1).

One can observe that at the time of investigation (1970–1975) the pH meter and the O_2 and CO_2 analyzers were the least reliable instruments. During the last ten years, the reliability of these instruments has been improved considerably, provided an accurate flowmeter is used for the O_2 and CO_2 instruments.

FLYNN (1982) gave detailed results on the performance of the instruments in a 1 m^3 pilot plant bioreactor (Tab. 2).

One can see that the p_{O_2} measurement had the lowest accuracy, the p_{O_2} and air flow control the lowest precision, and the volume measurement the lowest resolution. In the meantime, the air-flow control should have attained a much higher accuracy and precision, provided the right instrument is used (e.g., mass flowmeter). In recent years, the accuracy of the p_{O_2} measurement has not been improved markedly, it can, however, be achieved by frequent calibration. The accuracy and precision of the p_{O_2} control is much better if one uses three sensors and parameter-adaptive control.

Tab. 1. Instrument Reliability (LEES, 1976)

Instrument	Number at Risk	Instrum. Years	Number of Failures	Failure Rate (faults/year)
Control valve	1531	747	447	0.60
Solenoid valve	252	113	48	0.42
Pressure	233	88	124	1.41
Flow	1942	943	1069	1.14
Level	421	193	327	1.70
Temperature	2579	1225	425	0.35
Controller	1192	575	164	0.29
pH Meter	34	16	93	5.88
O_2 Analyzer	12	6	32	5.65
CO_2 Analyzer	4	2	20	10.5

Tab. 2. Performance of Measurement and Control Instrumentation in a 1 m^3 Pilot Plant Bioreactor (FLYNN, 1982)

	Accuracy	Precision (stand. dev.)	Resolution
Temperature			
Measurement	±0.1 W		0.02 °C
Control	0.02	0.08	
pH			
Measurement	±0.1 4H		0.001
Control	0.1	0.01	
p_{O_2}			
Measurement	6.0 R		0.05 mbar
Control	0.52	9.95	
Pressure			
Measurement			0.004 psig
Control	0.41 R	0.07	
Volume			
Measurement			0.14 L
Control	0.4 R	3.4	
Air Flow			
Measurement			0.02 L/m
Control	−7.0 R	23.0	
Exit CO_2			
Measurement	±0.1 R		0.0005%
Exit O_2			
Measurement	±0.1 R		0.0005%

Notes: W, weekly; H, hourly; R, per run, which relates to the frequency with which the measuring instruments are recalibrated

6 Future Developments

In this chapter, only H^+-selective electrodes are considered. However, using ion-selective membranes, in principle the concentration of an arbitrary ion can be determined by means of the Nernst equation:

$$E = E_0 + 2.3 \frac{RT}{zF} \log [I_M] \qquad (46)$$

where z is the valence of the measured ion and

I_M the concentration of the measured ion in the broth.

The production of ion-selective carrier-membrane-electrodes is very easy. A standard pH-electrode is combined with an ion-selective membrane consisting of an ionophore in a polymer matrix.

The ionophore and the softener are usually dissolved in a PVC solution, put on the surface of the pH-electrode, and dried. Fig. 6 shows several constructions of such electrodes.

Ion-selective membranes may be prepared by ionophore antibiotics (valinomycin, nonactin, etc.) (SCHINDLER and SCHINDLER, 1983) and synthetic carriers (crown ethers and cryptates, cyclodextrins, cyclotriveratrylene, perhydrotriphenylene, etc.) (ATWOOD et al., 1984; VÖGTLE, 1975, 1981). Chiroselective transport molecules are particularly interesting for the more detailed analysis of broth components (LEHN, 1988).

At present, reliability and selectivity of ion-selective membranes are not always satisfactory. However, host–guest-complex chemistry is developing rapidly. Therefore, it is expected that some years from now reliable, ion-selective electrodes will be available.

The combination of pH, p_{O_2}-, p_{CO_2}-, and NH_4^+-selective transducers with biochemical receptors (enzymes, antibodies, lectins, etc.) is considered in Chapter 3 (Biosensors).

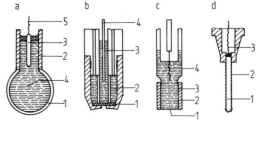

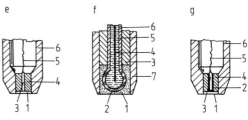

Fig. 6. Different types of ion-selective carrier membrane electrodes with liquid- and solid-lead-offs (SCHINDLER and SCHINDLER, 1983, with permission).

(a) Glass membrane electrode. (1) ion-selective glass membrane, (2) non-specific glass shaft, (3) Ag/AgCl-lead-off electrode, (4) lead-off electrode (liquid), (5) cable.

(b) Liquid membrane electrode with ion-exchanger reservoir. (1) porous membrane, (2) ion-exchanger reservoir, (3) lead-off electrolyte (liquid), (4) Ag/AgCl-lead-off electrode.

(c) PVC ion-exchanger membrane electrode. (1) PVC ion-exchanger membrane, (2) PVC tube, (3) lead-off electrolyte (liquid), (4) Ag/AgCl-lead-off electrode.

(d) Coated wire-electrode. (1) Pt-wire, (2) PVC ion-exchanger membrane, (3) cable.

(e) Disc electrode without O_2 reaction barrier. (1) carrier-PVC-membrane, (2) Pt-wire, (3) acryl/glass mantle, (4) PTFE-insulated, silver-coated Cu-wire, (5) PTFE or acryl glass.

(f) Coated glass electrode. (1) carrier-PVC-membrane, (2) ion-selective glass membrane, (3) acryl glass mantle, (4) Ag/AgCl-lead-off electrode, (5) inner electrolyte or cement lead-off, e.g., Ag/AgCl/Harvard cement, (6) non-specific glass shaft, (7) acryl glass or PTFE.

(g) Disc electrode with O_2 reaction barrier. (1) carrier-PVC-membrane, (2) Ag/AgCl (melt), (3) Pt-wire, (4) acryl glass mantle, (5) PTFE-insulated, silver-coated Cu-wire, (6) PTFE or acryl glass with Ag/AgCl/Pt-solid contact.

7 References

ANDREW, W. G., RHEA, K. G. (1970), Level measurement, in: *Applied Instrumentation in the Process Industries* (ANDREW, W. G., WILLIAMS, B., Eds.), 2nd Ed., Vol. 1, pp. 99–141. Houston: Gulf Publishing Co.

ANDREW, W. G., MILLER, W. F. (1979), Pressure measurement, in: *Applied Instrumentation in the Process Industries* (ANDREW, W. G., WILLIAMS, B., Eds.), 2nd Ed., Vol. 1, pp. 142–167. Houston: Gulf Publishing Co.

ANDREW, W. G., NORMAND, B. J., EDMONSON, F. E. (1979), Flow measurement, in *Applied Instrumentation in the Process Industries* (ANDREW, W. G., WILLIAMS, B., Eds.), 2nd Ed., Vol. 1, pp. 46–98. Houston: Gulf Publishing Co.

ANDREYEVA, L. N., BIRYUKOV, V. V. (1973), Analysis of mathematical models of the effect of pH on fermentation processes and their use for calculating optimal fermentation conditions, *Biotechnol. Bioeng. Symp. Ser.* **4**, 61–76.

ATWOOD, J. L., DAVIES, J. E., MACNICOL, D. D. (Eds.) (1984), *Inclusion Compounds,* Vol. 1–3. London: Academic Press.

BAUERMEISTER, G. D. (1981), *Elektrodenanordnung,* Ger. Patent 2801223, A. Z. P 2801 223.9.52G 01 N 27/28, Int. Cl. 3 (24. 4. 1980). *Electrode Device*, US Patent 4248712.

BEROVIC, M., CIMERMAN, A. (1979), Foaming in submerged citric acid fermentation on beet molasses, *Eur. J. Appl. Microbiol. Biotechnol.* **7**, 313–319.

BLAKEBROUGH, N., McMANAMEY, W. J., TART, K. R. (1978), Rheological measurements on *Aspergillus niger* fermentation systems, *J. Appl. Chem. Biotechnol.* **28**, 453–461.

BLANCH, H. W., ROGERS, P. L. (1972), Optimal conditions for gramicidin S production in continuous culture, *Biotechnol Bioeng.* **14**, 151–171.

BOURDARD, D., FOULARD, C. (1973), *Proc. First Eur. Conf. Computer Control in Fermentation,* Dijon, France.

BUEHLER, H., INGOLD, W. (1976), Measuring pH and oxygen in fermenters, *Process Biochem.,* 19–24.

BUNGAY, H. R., III, CHEN, Y. S. (1981), Dissolved oxygen profiles in photosynthetic microbial slimes, *Biotechnol. Bioeng.* **23**, 1893–1895.

BUNGAY, H. R., III, HAROLD, D. M., Jr. (1971), Simulation of oxygen transfer in microbial slimes, *Biotechnol. Bioeng.* **13**, 569–579.

BUNGAY, H. R., III, WHALEN, W. J., SANDERS, W. M. (1969), Microprobe techniques for determining diffusivities and respiration rates in microbial slime systems, *Biotechnol. Bioeng.* **11**, 765–772.

BUNGAY, H. R., III, PETTIT, P. M., DRISLANE, A.

M. (1983), Dissolved oxygen contours in *Pseudomonas ovalis* colonies, in: *Foundations of Biochemical Engineering: Kinetics and Thermodynamics of Biological Systems* (BLANCH, H. W., PAPOUTSAKIS, E. T., STEPHANOPOULOS, G., Eds.), *ACS Symp. Ser.* **207**, 395–401.

BÜSING, W., ARNOLD, J. U. (1980), *Temperaturmessung,* in: Messen, Steuern und Regeln in der Chemischen Technik (HENGSTENBERG, J., STURM, B., WINKLER, O., Eds.), 3rd Ed., Vol. 1, pp. 1–100. Berlin-Heidelberg-New York: Springer Verlag.

CHEN, Y. S., BUNGAY, H. R., III (1981), Microelectrode studies of oxygen transfer in trickling filter slimes, *Biotechnol. Bioeng.* **23**, 781–792.

CHERUY, A., DURAND, A. (1979), Optimization of erythromycin biosynthesis by controlling pH and temperature: Theoretical aspects and practical application, *Biotechnol. Bioeng. Symp.* **9**, 303–320.

CONSTANTINIDES, A. (1979), Application of rigorous optimization methods to the control and operation of fermentation processes, in: *Biochemical Engineering* (VIETH, R., VENKATASUBRAMANIAN, U., CONSTANTINIDES, A., Eds.). *Ann. N. Y. Acad. Sci.* **326**, 193–221.

CONSTANTINIDES, A., RAI, V. R. (1974), Application of the continuous maximum principle to fermentation processes, *Biotechnol. Bioeng. Symp.* **4**, 663–680.

CONSTANTINIDES, A., SPENCER, J. L., GADEN, E., Jr. (1970), Optimization of batch fermentation processes. II. Optimum temperature profiles for batch penicillin fermentations, *Biotechnol. Bioeng.* **12**, 1081–1098.

COONEY, C. L., WANG, H. Y., WANG, D. I. C. (1977), Computer-aided material balancing for prediction of fermentation parameters, *Biotechnol. Bioeng.* **19**, 55–67.

DINSDALE, A., MOORE, F. (1962), in: *Viscosity and its Measurement.* London – New York: Chapman & Hall; New York: Reinhold.

ELFERS, P. A. (1964), Liquid level, in: *Handbook of Applied Instrumentation* (CONSIDINE, D. M., ROSS, S. D., Eds.), Sect. 5, pp. 55–67. New York: McGraw-Hill.

ERICSON, F. A. (1979), Pumps and flow measurement, in: *Laboratory Engineering and Manipulations* (PERRY, E. S., WEISSBERGER, A., Eds.), 3rd Ed., pp. 182–220. New York: John Wiley & Sons.

FAN, L. T., WAN, C. G. (1963), An application of the discrete maximum principle to a stagewise biochemical reactor system, *Biotechnol. Bioeng.* **5**, 201–209.

FLYNN, D. S. (1982), Instrumentation for fermentation processes, in: *Modelling and Control of Bio-*

technological Processes, pp. 5–12. Helsinki: IFAC.

FORAGE, D. E., HARRISON, D. E. F., PITT, D. E. (1985), Effect of environment on microbial activity, in: *Comprehensive Biotechnology* (BULL, A. T., DALTON, H., Eds.), Vol. 1, pp. 251–280. Oxford–Toronto–Sydney–Frankfurt: Pergamon Press.

FRITZE, U. (1980a), Die Technische Messung der Redoxspannung, in: *Messen, Steuern und Regeln in der Chemischen Technik* (HENGSTENBERG, J., STURM, B., WINKLER, O., Eds.), 3rd Ed., Vol. 2, pp. 367–385. Berlin–Heidelberg–New York: Springer Verlag.

FRITZE, U. (1980b), Weitere potentiometrische und amperometrische Prozeßanalysenmethoden, in: *Messen, Steuern und Regeln in der Chemischen Technik* (HENGSTENBERG, J., STURM, B., WINKLER, O., Eds.), 3rd Ed., Vol. 2, pp. 383–402. Berlin–Heidelberg–New York: Springer Verlag.

HALL, M. J., DICKINSON, S. D., PRITCHARD, R., EVENS, J. I. (1973), Foams and foam control in fermentation processes, *Prog. Ind. Microbiol.* **12**, 170–234.

HIRTE, W. (1980), Druckmessung, in: *Messen, Steuern und Regeln in der Chemischen Technik* (HENGSTENBERG, J., STURM, B., WINKLER, O., Eds.), 3rd Ed., Vol. 1, pp. 101–190. Berlin–Heidelberg–New York: Springer Verlag.

HO, C. H. S., SMITH, M. D., SHANAHAN, J. F. (1987), Carbon dioxide transfer in biochemical reactors, *Adv. Biochem. Eng. Biotechnol.* **35**, 83–126.

HOSPODKA, J. (1966), Oxygen-absorption rate-controlled feeding of substrate into aerobic microbial cultures, *Biotechnol. Bioeng.* **8**, 117–134.

INGOLD I, Brochure I: *pH Electrodes.* Dr. W. Ingold AG, Urdorf, Switzerland.

INGOLD II, Brochure II: *O₂ Measurement Systems; pH and O₂ Measurement in Fermentors.* Dr. W. Ingold AG, Urdorf, Switzerland.

INGOLD III, Brochure III: *Redox Measurement.* Dr. W. Ingold AG, Urdorf, Switzerland.

INGOLD IV, Brochure IV: *Sterilizable CO₂ Probe.* Dr. W. Ingold AG, Urdorf, Switzerland.

JOERGENSEN, H. (1941), *Studies on the Nature of Bromate effect.* Thesis, Copenhagen, Munksgaard.

KEMBLOWSKI, Z., KRISTIANSEN, B., ALAYI, O. (1985), On-line rheometer for fermentation liquids, *Biotechnol. Lett.* **7**, 803–808.

KING, R. E., ARAGONA, J., CONSTANTINIDES, A. (1974), Special optimal control of batch fermentor, *Int. J. Control* **20**, 869–879.

KJAERGAARD, L. (1977), The redox potential: Its use and control in biotechnology, *Adv. Biochem. Eng.* **7**, 131–150.

KJAERGAARD, L., JOERGENSEN, B. B. (1979), Redox potential as a state variable in fermentation systems, *Biotechnol. Bioeng. Symp.* **9**, 85–94.

LANGER, G., WERNER, U. (1981), Measurements of viscosity of suspensions in different viscometer flows and stirring systems, *Ger. Chem. Eng.* **4**, 226–241.

LEE, Y. H., TSAO, G. T. (1979), Dissolved oxygen electrodes, *Adv. Biochem. Eng.* **13**, 35–86.

LEELASART, B., BONALY, R. (1988), Effect of control of pH on the excretion of acid phosphatase by *Rhodotorula glutinis, Appl. Microbiol. Biotechnol.* **29**, 579–585.

LEES, F. P. (1976), The reliability of instrumentation, *Chem. Ind.* (5), 195–205.

LEHN, J. M. (1988), Supramolekulare Chemie – Moleküle, Übermoleküle und molekulare Funktionseinheiten, *Angew. Chem.* **100**, 91–116.

LINEK, V., VACEK, V. (1986), Recommended procedure for the measurement of the volumetric mass transfer coefficient $k_L a$ in aerated agitated vessels. *Report of a Working Party on Mixing of the European Federation of Chemical Engineering,* Prague.

LINEK, V., VACEK, V., BENES, P. (1987), A critical review and experimental verification of the correct use of the dynamic method for the determination of oxygen transfer in aerated agitated vessels to water, electrolyte solutions and viscous liquids, *Chem. Eng. J.* **34**, 11–34.

McCULLOUGH, R. L., ANDREW, W. G. (1979), Analytical instruments, in: *Applied Instrumentation in the Process Industries* (ANDREW, W. G., WILLIAMS, B., Eds.), 2nd Ed., Vol. 1, p. 209. Houston: Gulf Publishing Co.

MELZNER, W., JAENICKE, D. (1980), Prozeßanalytik, in: *Ullmanns Encyklopädie der technischen Chemie,* 4th Ed., Vol. 5, pp. 891–944. Weinheim–Deerfield Beach, Florida–Basel: Verlag Chemie.

MEMMERT, K., WANDREY, C. (1987), Concept of redox-regulation for continuous production of *Bacillus* exoenzymes, *Proc. 4th Eur. Congr. Biotechnology* (NEISSEL, O. M., VAN DER MEER, R. R., LUYBEN, K. CH. A. M., Eds.), Vol. 3, pp. 153–154. Amsterdam: Elsevier Science Publisher B. V.

MOU, D. G., COONEY, C. L. (1983), Growth monitoring and control in complex medium: A case study employing fed-batch penicillin fermentation and computer-aided on-line mass balancing, *Biotechnol. Bioeng.* **25**, 257–269.

MUSCHELKNAUTZ, E., HECKENBACH, M. (1980), Rheologische Betriebsmeßverfahren, in: *Messen, Steuern und Regeln in der Chemischen Technik* (HENGSTENBERG, J., STURM, B., WINKLER, O., Eds.) 3rd Ed., Vol. 2, pp. 431–512. Berlin–Heidelberg–New York: Springer Verlag.

NEUHAUS, O., LANGER, G., WERNER, U. (1983), On-line Betriebsviskosimeter für Fermentationsprozesse, in: *Biotechnologie-Projektförderung. Projekte 1978-1984*, pp. 39-53. Bonn: BMFT (Bundesministerium für Forschung und Technologie).

OEDEKOVEN, G. (1980), Standmessung, in: *Messen, Steuern und Regeln in der Chemischen Technik* (HENGSTENBERG, J., STURM, B., WINKLER, O., Eds.), 3rd Ed., Vol. 1, pp. 589-640. Berlin-Heidelberg-New York: Springer Verlag.

PAN, C. H., HEPLER, L., ERLANDER, A. P. (1972), Control of pH and carbohydrate addition in the penicillin fermentation, *Dev. Ind. Microbiol.* **13**, 103-112.

PERINGER, P., BLACHÈRE, H. T. (1979), Modeling and optimal control of bakers' yeast production in repeated fed-batch culture, *Biotechnol. Bioeng. Symp.* **9**, 205-213.

PERLEY, C. R., SWARTZ, J. R., COONEY, C. L. (1979), Measurement of cell mass concentration with a continuous viscosimeter, *Biotechnol. Bioeng.* **21**, 519-523.

PETERSEN, O. (1980), pH-Messung in: *Messen, Steuern und Regeln in der Chemischen Technik* (HENGSTENBERG, J., STURM, B., WINKLER, O., Eds.), 3rd Ed., Vol. 2, pp. 305-333. Berlin-Heidelberg-New York: Springer Verlag.

PRINS, A., VAN'T RIET, K. (1987), Proteins and surface effects in fermentation: Foam, antifoam and mass transfer, *TIBTECH,* November 5, 296-301.

RAI, V. R., CONSTANTINIDES, A. (1973), Mathematical modeling and optimization of the gluconic acid fermentation, *AIChE Symp. Ser.* **132** (69), 114-122.

ROSEN, W., SCHÜGERL, K. (1984), Microprocessor controlled fed-batch cultivation of *Chaetomium cellulolyticum* on cellulose, *Third Eur. Congr. Biotechnology,* München, 10-14 Sept., Vol. II, pp. 617-619. Weinheim: Verlag Chemie.

SAN, K.-Y., STEPHANOPOULOS, G. (1984), Studies on on-line bioreactor identification. IV. Utilization of pH measurements for product estimation, *Biotechnol. Bioeng.* **26**, 1209-1218.

SCHINDLER, J. G., SCHINDLER, M. M. (1983), *Bioelektrochemische Membranelektroden,* Berlin-New York: Walter de Gruyter.

SCHLEGEL, H. G. (1974), *Allgemeine Mikrobiologie,* 3rd Ed. Stuttgart: Georg Thieme Verlag.

SCHRÖDER, A. (1980), Durchflußmeßtechnik, in: *Messen, Steuern und Regeln in der Chemischen Technik* (HENGSTENBERG, J., STURM, B., WINKLER, O., Eds.), 3rd Ed., Vol. 1, pp. 191-364. Berlin-Heidelberg-New York: Springer Verlag.

SCHÜGERL, K. (1986), Foam formation, foam suppression, and the effect of foam on growth, *Process Biochem.* (August) 122-123.

SCHUMPE, A. (1985), Gas solubilities in biomedia, in: *Biotechnology* (REHM, H.-J., REED, G., Eds.) Vol. 2, pp. 159-170. Weinheim-Deerfield Beach, Florida-Basel: VCH Verlagsgesellschaft.

SHIOYA, S. (1988), Measurement of biological reaction rates using advanced pH control, in: *Fourth Int. Congr. Computer Applications in Fermentation Technology,* IFAC, Sept. 25-29, University of Cambridge, UK (FISH, N. M., FOX, R. I., Eds.). Chichester: Society of Chemical Industry and Ellis Horwood Publ.

SIE, T.-L., SCHÜGERL, K. (1983), Foam behavior of biological media. IX. Efficiency of antifoam agents with regard to their foam suppression effect on BSA solutions, *Eur. J. Appl. Microbiol. Biotechnol.* **17**, 221-226.

SPITZER, D. W. (1976), Maximization of steady-state bacterial production in a chemostat with pH and substrate control, *Biotechnol. Bioeng.* **18**, 167-178.

SQUIRES, R. W. (1972), Regulation of the penicillin fermentation by means of a submerged oxygen-sensitive electrode, *Dev. Ind. Microbiol.* **13**, 128-135.

STEPHANOPOULOS, G., SAN, K.-Y. (1982), On-line estimation of the state of biochemical reactors, in: *Chemical Reaction Engineering,* Boston 1982 (WEI, J., GEORGAKIS, C., Eds.). *ACS Symp. Ser.* **196**, 155-164.

TAKAMATSU, T., SHIOYA, S., YOKOYAMA, K., IHARA, D. (1981), State estimation and control of a biochemical reaction process, in: *Biochemical Engineering II* (CONSTANTINIDES, A., VIETH, W. R., VENKATASUBRAMANIAN, K., Eds.). *Ann. N. Y. Acad. Sci.* **369**, 147-158.

VIESTURS, U. E., KRISTAPSONS, M. Z., LEVITANS, E. S. (1982), Foam in microbiological processes, *Adv. Biochem. Eng.* **21**, 169-224.

VÖGTLE, F., Ed. (1975), Host guest complex chemistry I, *Topics in Current Chemistry* 101. Berlin-Heidelberg-New York: Springer Verlag.

VÖGTLE, F., Ed. (1981), Host guest complex chemistry II, *Topics in Current Chemistry* 98. Berlin-Heidelberg-New York: Springer Verlag.

WANG, H. Y., COONEY, C. L., WANG, D. I. C. (1977), Computer aided baker's yeast fermentations, *Biotechnol. Bioeng.* **19**, 69-89.

WITTLER, R., BAUMGARTL, H., LÜBBERS, D. W., SCHÜGERL, K. (1986), Investigations of oxygen transfer into *Penicillium chrysogenum* pellets by microprobe measurements, *Biotechnol. Bioeng.* **28**, 1024-1036.

ZABRISKIE, D. W., HUMPHREY, A. E. (1978), Real-time estimation of aerobic batch fermentation biomass concentration by component balancing, *AIChE J.* **24**, 138-146.

2 Methods and Instruments in Fermentation Gas Analysis

ELMAR HEINZLE

IRVING J. DUNN

Zürich, Switzerland

List of Symbols and Abbreviations

C	concentration $(kg\ m^{-3})$
CPR	carbon dioxide production rate $(mol\ s^{-1})$
CTR	carbon dioxide transfer rate $(mol\ s^{-1})$
G	gas flow rate $(m^3\ s^{-1})$
H	Henry coefficient $(L\ bar\ mol^{-1})$
I	ion current (A)
K	equilibrium constant
$K_L a$	mass transfer coefficient (s^{-1})
L	liquid flow rate $(m^3\ s^{-1})$
m/z	mass-to-charge ratio
M	total mass flow $(kg\ s^{-1})$
n	number of moles
N	molar gas flow rate $(mol\ s^{-1})$
OUR	oxygen uptake rate $(mol\ s^{-1})$
OTR	oxygen transfer rate $(mol\ s^{-1})$
p	pressure (bar)
PHB	poly-β-hydroxybutyric acid
Q	specific reaction rate $(mol\ kg^{-1}\ s^{-1})$
r	reaction rate $(mol\ L^{-1}\ s^{-1})$
R	gas constant $(=0.08314\ bar\ L\ mol^{-1}\ K^{-1})$
RQ	respiratory quotient $(-)$
s	relative statistical error $(-)$
t	time (s)
T	temperature (K, °C)
v	velocity $(m\ s^{-1})$
V	volume (m^3, L)
X	biomass concentration $(g\ L^{-1})$
y	gas phase molar fraction $(-)$
Z	thickness (m)
τ	time constant (s)
ϕ, φ	molar flux $(=$ specific reaction rate$)$ $(mol\ kg^{-1}\ s^{-1})$

Subscripts and Superscripts

E	electrode
G	gas phase
i	index for component
inert	inert gas
L	liquid phase
rel	relative
X	biomass
0	input into the reactor
1	output from the reactor
*	refers to gas–liquid equilibrium
$'$	relative value $(-)$

1 Introduction

It is evident from Chapter 1 of this volume of "Biotechnology" that on-line fermentation analysis is of increasing importance because precise control of environmental variables is necessary to optimize process yield and selectivity. Most biological products are not volatile and are either dissolved in the fermentation fluid, precipitated, or enclosed within the cell membrane boundary. These products are usually difficult or presently impossible to measure on-line in a process environment. This is also true for the biocatalyst itself (cell or enzyme).

In industrial processes each sensor causes risks of infection, whether located in the sterile region or connected to the process with a liquid sampling device. This risk does not exist if measurements are made in the effluent gas stream outside the sterile region.

On-line gas analysis is of general interest because almost any biological process using living organisms involves consumption and production of gases and volatile compounds. Especially oxygen consumption and carbon dioxide production occur in any aerobic fermentation process. Measurement of these reaction rates gives direct information about the culture activity. Oxygen consumption rate usually is directly proportional to the heat evolution of any aerobic process (COONEY et al., 1969).

Historically, one of the first instruments for gas analysis was the Orsat apparatus (HERON and WILSON, 1959). In this apparatus CO_2 and O_2 are subsequently absorbed in sodium hydroxide and pyrogallol solutions, and volume changes are detected. Inert gases are determined by difference. CO_2 production was one of the first biological activities to be quantified in yeast alcohol production. Traditionally, the measurements were made using volumetric methods.

Under normal conditions, where the ideal gas law is valid, gas volume, pressure, and molar amount are directly linked with each other. This makes barometric and volumetric measurements very useful. In microbiology the Warburg apparatus is still a very popular method of measuring gas reaction rates. Oxygen and CO_2 production can be measured simultaneously by first measuring pressure change and subsequently absorbing CO_2 in an alkaline solution, then making a final pressure measurement. Volumetric and barometric methods were also further developed to give on-line readings of gas composition (VANA, 1982).

Historically, the results of on-line gas analysis have almost exclusively been used to monitor fermentations. Since more reliable analytical instruments and on-line data acquisition and computing hard- and software have been developed, it is now possible to use gas analysis data together with other measurements to quantitatively characterize fermentation kinetics. Cheap and reliable process computer systems, together with increasingly powerful and easy to use software, have dramatically improved capabilities.

Gas analysis usually involves measurement of gas flow rates and gas composition. Setting up appropriate mass balances allows evaluation of actual production and consumption rates. Today, gas flow rates can be measured with mass flow meters which directly give an electric signal. This facilitates automatic data evaluation using computers. A whole series of instruments to measure gas composition online has been developed. The instruments include paramagnetic oxygen analyzers, infrared absorption photometers, gas chromatographs (GC), mass spectrometers (MS), flame ionization detectors (FID), amperometric and potentiometric sensors, and semiconductor devices.

Generally speaking, excluding pH measurement, gas analysis is the most widespread and most reliable on-line analysis in industrial fermentation processes. It has been applied to the on-line analysis of bacterial, fungal, and higher cell culture systems. Its potential in animal and plant cell culture has not yet been fully exploited. This is clearly seen by the fact that in a recent review of on-line analysis of animal cell culture the possibility of oxygen uptake rate measurements has not even been mentioned (MERTEN et al., 1986).

2 Mass Balancing for Gas Analysis

2.1 Basic Gas Balance Equations

Balances which consider gas transfer and gas reaction rates are necessary to characterize the aeration efficiency and to follow biological activity. The same equations can be applied to any component. Well-mixed phases, whose concentrations can be assumed to be uniform, can be described simply, while situations with spatial variation require more complex models. The following general gas balance equations can be written for a well-mixed (tank geometry) system (Fig. 1):

Gas phase,

$$V_G \frac{dC_{G_1}}{dt} = G_0 C_{G_0} - G_1 C_{G_1} -$$
$$-K_L a (C_{L_1}^* - C_{L_1}) V_L \tag{1}$$

Liquid phase,

$$V_L \frac{dC_{L_1}}{dt} = L_0 C_{L_0} - L_1 C_{L_1} +$$
$$+ K_L a (C_{L_1}^* - C_{L_1}) V_L + \sum r V_L \tag{2}$$

The variables and their dimensions are as follows:

C_G and C_L ($\mathbf{M L^{-3}}$), gas and liquid phase concentrations; G and L ($\mathbf{L^3 T^{-1}}$), gas and liquid flow rates; $K_L a$ ($\mathbf{T^{-1}}$), mass transfer coef-

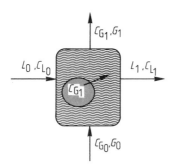

Fig. 1. Gas transfer in a stirred tank reactor with well mixed phases.

ficient; V_G and V_L ($\mathbf{L^3}$), gas and liquid volume; r ($\mathbf{M T^{-1} L^{-3}}$), reaction rate.

The above equations have been written to apply to any component (oxygen, carbon dioxide, ethanol, etc.). They include accumulation, convective flow, inter-phase transfer, and reaction terms. Usually there is only one biological reaction term, but a special exception is the case of CO_2 dissociation to yield bicarbonate. In a batch reactor the liquid flow terms are $L_1 = L_0 = 0$. In a fed-batch culture $L_0 \neq L_1$, and in a continuous culture $L_0 = L_1 > 0$.

Here $C_{L_1}^*$ is the liquid phase concentration in equilibrium with C_{G_1}, and it is calculated by Henry's law

$$C_{G_1} R T = C_{L_1}^* H \tag{3}$$

Henry coefficients for some important gases are: $H_{O_2} = 856.9$ L bar/mol; $H_{CO_2} = 34.01$ L bar/mol; $H_{N_2} = 1484$ L bar/mol. The solubility of pure gases in water can also be expressed in liters of gas per liter of water. At 30 °C the values are: N_2, 0.0134; O_2, 0.0261; CO_2, 0.665; H_2, 0.017; CH_4, 0.0276.

Pressure and Temperature Effects

It is often convenient to write gas balances in terms of partial pressures instead of concentrations. Using the ideal gas law

$$pV = nRT \tag{4}$$

where $R = 0.08205$ atm L/mol K = 0.08314 (bar L/mol K) = 8314 (Pa L/mol K) or its equivalent for a flowing system,

$$pG = NRT \tag{5}$$

where N is moles/time and G is volume/time. Thus, useful expressions are:

$$C_{G_i} = \frac{n_i}{V} = \frac{p_i}{RT} \tag{6}$$

and

$$n_i = C_{G_i} V = p_i \frac{V}{RT} \tag{7}$$

$$N_i = C_{G_i} G = p_i \frac{G}{RT} \qquad (8)$$

Mole fractions and partial pressure relationships are also useful (Dalton's law).

For constant V and T,

$$\frac{p_i}{p} = \frac{n_i}{n} = y_i \qquad (9)$$

Also temperature correction relations which allow correction to standard conditions at constant p and V,

$$\frac{T_1}{T_2} = \frac{n_2}{n_1} \qquad (10)$$

The above gas phase balance can be written in terms of partial pressures by replacing the (VC) terms with $(pV)/(RT)$ terms or the equivalent for (GC) terms.

Thus the accumulation terms become

$$\left(\frac{V_{G_1}}{RT_1}\right)\left(\frac{dp_{1,i}}{dt}\right)$$

or in terms of mole fractions at constant p_1,

$$\left(\frac{V_{G_1}p_1}{RT}\right)\left(\frac{dy_{1,i}}{dt}\right)$$

The flow terms have the form

$$\left(\frac{G}{RT}\right)p_i$$

or in terms of mole fractions,

$$\left(\frac{Gp}{RT}\right)y_i$$

If the pressures and temperatures are different at inlet and outlet the gas balance in terms of the mole fractions becomes,

$$\left(\frac{V_{G_1}p_1}{RT_1}\right)\left(\frac{dy_{1,i}}{dt}\right) =$$
$$= \left(\frac{G_0 p_0}{RT_0}\right)y_{0,i} - \left(\frac{G_1 p_1}{RT_1}\right)y_{1,i} -$$
$$- K_L a\left(\frac{p_1}{H}y_{1,i} - C_{L_1,i}\right)V_L \qquad (11)$$

If the total pressures and temperatures can be considered to be equal, then the balance equation simplifies,

$$V_{G_1}\frac{dy_i}{dt} = G_0 y_{0,i} - G_1 y_{1,i} -$$
$$- K_L a\left(\frac{RT}{H}y_{1,i} - \frac{RT}{p}C_{L_1,i}\right)V_L \qquad (12)$$

Correct application of the balance equation generally requires correcting for the inlet and outlet conditions. This is most easily done by calculating the moles of gas involved, in which case the balance equation would be

$$\frac{dn_{G_1,i}}{dt} = V_L\left(\frac{dC_{L_1,i}}{dt}\right) =$$
$$= N_0 - N_1 - K_L a\left(\frac{p_1}{H}y_{1,i} - C_{L_1,i}\right)V_L \qquad (13)$$

where $C_{L_1,i}$ is in mol/L and H in L bar/mol.

Calculation of the molar volumes is most easily done by applying the factor of 22.4 L, which is the volume of one mole of gas at standard conditions of 1 bar and 0°C. Thus,

$$n_G = \left(\frac{V_G}{22.4}\right)\left(\frac{p_1}{1}\right)\left(\frac{273}{T_1}\right) \qquad (14)$$

$$N_0 = \left(\frac{G_0}{22.4}\right)\left(\frac{p_0}{1}\right)\left(\frac{273}{T_0}\right) \qquad (15)$$

and

$$N_1 = \left(\frac{G_1}{22.4}\right)\left(\frac{p_1}{1}\right)\left(\frac{273}{T_0}\right) \qquad (16)$$

Oxygen Uptake and Heat Production Rates

The concept of yield allows a quick estimate of reactor heat loads. Thus,

$$r_Q = Y_{Q/O_2} r_{O_2} \qquad (17)$$

where r_Q is the heat production rate (kJ $L^{-1} s^{-1}$). Typical yield values for aerobic culture are given in the literature (ROELS, 1983): $Y_{Q/O_2} = 385$ to 490 kJ/mol O_2.

In the following, the general equations will be applied to special situations.

2.2 Inert Gas Balance to Calculate Flow Rates

Inert gases are not consumed or produced within the system ($r_i = 0$). Their mass streams must therefore be equal at the inlet and outlet of the reactor at steady state ($dC_{inert}/dt = 0$).

$$N_0 y_{0,\,inert} = N_1 y_{1,\,inert} \tag{18}$$

From this balance, an estimation of N_1 can be made on the basis of measurement of N_0 and inert gas partial pressures (y_{inert})

$$N_1 = N_0 \frac{y_{0,\,inert}}{y_{1,\,inert}} \tag{19}$$

2.3 Steady-State Gas Balance to Determine the Biological Reaction Rate

For oxygen the assumption of a liquid phase steady state ($dC_{L,O_2}/dt = 0$) usually will be fulfilled because of its low solubility. If reaction rates are not changing very dramatically, steady-state assumption in the gas phase ($dC_{G,O_2}/dt = 0$) is also valid. Then the oxygen uptake rate (*OUR*) can be estimated by simply measuring N_0 and molar fractions (y_{O_2}).

$$0 = N_0 y_{0,\,O_2} - N_1 y_{1,\,O_2} - r_{O_2} V_L \tag{20}$$

Applying Eq. (19) one gets

$$OUR = r_{O_2} V_L = -N_0 \left(y_{0,\,O_2} - y_{1,\,O_2} \frac{y_{0,\,inert}}{y_{1,\,inert}} \right) \tag{21}$$

Alternatively, this equation can be derived by balancing around the entire two-phase system. Using this equation, a knowledge of the gas flow rates and concentrations allows the calculation of the biological oxygen uptake

rate. This application is very important in fermentation technology, because it permits on-line monitoring of the rate of reaction.

Often specific activities are needed. These are defined as:

$$Q_{O_2} = \frac{r_{O_2}}{X} = \frac{OUR}{X V_L} \quad \text{and} \quad Q_{CO_2} = \frac{r_{CO_2}}{X} = \frac{CPR}{X V_L}$$

The units are mol/g s.

For CO_2 the situation often will be more complex because CO_2 is much more soluble than O_2: $H_{O_2} = 856.9$ L bar mol^{-1}; $H_{CO_2} = 34.01$ L bar mol^{-1}. Additionally CO_2 reacts with water to give HCO_3^-:

$$CO_2 + H_2O \rightleftharpoons HCO_3^- + H^+ \tag{22}$$

Only at low pH values when the concentration of HCO_3^- may be negligible can the equilibrium liquid phase reactions be neglected. In general the equilibrium needs to be considered. Otherwise, the sum of the two reaction rates will be measured. The equilibrium constant is

$$K = \frac{C_{HCO_3^-} C_{H^+}}{C_{L,CO_2}} \tag{23}$$

and has a value of $4.71 \cdot 10^{-7}$ mol L^{-1} at $30\,°C$. Assuming a well-mixed gas phase, the amount of CO_2 accumulated as HCO_3^- can be estimated from Eq. (23)

$$C_{HCO_3^+} = \frac{K C_{L,CO_2}}{C_{H^+}} = \frac{K y_{1,CO_2} p}{H_{CO_2} C_{H^+}} \tag{24}$$

Differentiation at constant pH gives

$$dC_{HCO_3^-} = \frac{K}{C_{H^+}} dC_{CO_2} \tag{25}$$

which makes it possible to determine the bicarbonate increase from an increase in dissolved carbon dioxide.

Dynamic errors can be significant if pH control is not adequate, because variations in pH will cause accumulation of bicarbonate. A pH increase of 0.2 units causes a 26% increase in $C_{HCO_3^-}$ if the gas phase concentration of CO_2 is kept constant. The higher the pH the more important this effect will be. At low pH values (≤ 6) and with good pH control, the

CO$_2$ production rate (*CPR*) is calculated analogously to Eq. (20)

$$CPR = r_{CO_2} V_L = -N_0 \left(y_{0,CO_2} - y_{1,CO_2} \frac{y_{0,inert}}{y_{1,inert}} \right)$$ (26)

In air y_{0,CO_2} is usually very small, leading to

$$r_{CO_2} V_L = N_0 y_{1,CO_2} \frac{y_{0,inert}}{y_{1,inert}}$$ (27)

2.4 Determination of *CPR* with Accumulation of CO$_2$ in the Liquid Phase

In the more general case where only gas phase accumulation is negligible, Eq. (2) has to be modified to include bicarbonate concentration

$$V_L = \frac{d(C_{L,CO_2} + C_{HCO_3^-})}{dt} = CTR - CPR$$ (28)

Combining with Eq. (25) yields

$$V \left(1 + \frac{K}{C_{H^+}} \right) \frac{dC_{CO_2}}{dt} = CPR - CTR$$ (29)

where *CTR* is the CO$_2$ transfer rate, which can be calculated from the gas balance at steady state:

$$CTR = N_1 y_{1,CO_2} - N_0 y_{0,CO_2}$$ (30)

Combining Eqs. (29), (30), and (19) and introducing the difference quotient instead of the differential quotient we get

$$CPR = N_0 \left(y_{1,CO_2} \frac{y_{0,inert}}{y_{1,inert}} - y_{0,CO_2} \right) -$$

$$- \frac{\Delta(C_{L,CO_2} + C_{HCO_3^-})}{\Delta t} V_L$$ (31)

Since C_{L,CO_2} and $C_{HCO_3^-}$ usually are not measured on-line, estimates have to be derived from gas analysis and a pH measurement. The assumption of equilibrium conditions

$(C_{L,CO_2} = C_{L,CO_2}^*)$ between gas and liquid phases provides an estimate of C_{L,CO_2} according to Eq. (3) and of $C_{HCO_3^-}$ using Eq. (23) giving

$$C_{L,CO_2} + C_{HCO_3^-} = \frac{y_{1,CO_2} p_1}{H_{CO_2}} \left(1 + \frac{K}{10^{-pH}} \right)$$ (32)

For non-equilibrium conditions, but still negligible gas phase CO$_2$ accumulation ($dC_G/dt \approx 0$), using Eq. (1) gives

$$(K_L a)_{CO_2} (C_{L,CO_2}^* - C_{L,CO_2}) V_L =$$

$$= N_0 \left(y_{1,CO_2} \frac{y_{0,inert}}{y_{1,inert}} - y_{0,CO_2} \right)$$ (33)

Assuming $(K_L a)_{CO_2} \approx (K_L a)_{O_2}$, which is reasonable according to SCHNEIDER and FRISCH-KNECHT (1977) and HEINZLE and LAFFERTY (1980), C_{L,CO_2} and $C_{HCO_3^-}$ can be calculated if $(K_L a)_{O_2}$ is known.

$$C_{L,CO_2} = \frac{N_0 \left(y_{1,CO_2} \frac{y_{0,inert}}{y_{1,inert}} - y_{0,CO_2} \right)}{V_L (K_L a)_{CO_2}} +$$

$$+ \frac{p_1 y_{1,CO_2}}{H_{CO_2}}$$ (34)

Methods to estimate the $K_L a$ value for stirred vessels using empirical correlations are available (JOSHI et al., 1982).

If $(K_L a)_{O_2}$ is known, an estimate of C_{L,O_2} can be obtained in exactly the same way using Eq. (34). For the measurement of dissolved oxygen concentrations, suitable electrodes are available. In those cases, however, where dissolved oxygen measurement is difficult or even impossible (non-Newtonian fluids with high viscosity) $K_L a$ is usually also not known. Therefore, this method of calculation tends to be of limited value for oxygen.

2.5 Determination of $K_L a$ by Steady-State Gas Balancing with Well-Mixed Gas and Liquid Phases

If C_{L,O_2} can be measured by an electrode, balancing for oxygen will give an estimate for $(K_L a)_{O_2}$ derived from Eq. (1)

$$(K_L a)_{O_2} = \frac{N_0 \left(y_{0,O_2} - y_{1,O_2} \dfrac{y_{0,\text{inert}}}{y_{1,\text{inert}}} \right)}{(C^*_{L,O_2} - C_{L,O_2}) V_L} \tag{35}$$

2.6 Determination of $K_L a$ by the Dynamic Method

If water is initially deoxygenated and then aerated, the dissolved oxygen concentration will increase with time. The exact form of the response curve depends on $K_L a$, the $C^*_{L_1} - C_{L_1}$ driving force, and the dynamics of the measurement. The liquid balance gives:

$$\frac{dC_{L_1}}{dt} = K_L a (C^*_{L_1} - C_{L_1}) \tag{36}$$

The classical dynamic $K_L a$ method assumes that $K_L a$ and C_{L_1} are constant. Then the differential equation can be integrated analytically to yield:

$$\ln \left(\frac{C^*_{L_1}}{C^*_{L_1} - C_{L_1}} \right) = K_L a t \tag{37}$$

Plotting the logarithmic function versus t gives $K_L a$ as the slope. This method is accurate at low rates of transfer, with $K_L a$ below $60 \, \text{h}^{-1}$.

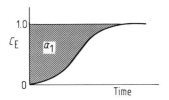

Fig. 2. Calculation of $K_L a$ from dynamic experiments. C_E, electrode signal; α_1, see Eq. (39).

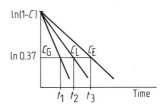

Fig. 3. Approximate method to determine $K_L a$ (RUCHTI et al., 1985).

$$t_1 = \tau_G; \; t_2 = \tau_G + \frac{1}{K_L a}; \; t_3 = \tau_G + \frac{1}{K_L a} + \tau_E$$

Usually deoxygenation is accomplished with nitrogen so that the initial gas phase consists of nitrogen, which is gradually displaced and mixed with air. Under these conditions C_L is not constant and the gas balance must be employed to calculate C_{G_1} versus t. Since C_{L_1} is measured with a membrane-covered oxygen electrode, the dynamics of the method of measurement cannot usually be neglected. Experimentally this is described approximately by a first order lag equation

$$\frac{dC_E}{dt} = \frac{C_{L_1} - C_E}{\tau_E} \tag{38}$$

The relative influence of the three processes depends on the values of the constants $\tau_G = V_G/G$, $1/K_L a$, and τ_E. Because of their linear nature, the differential equations can be solved analytically and it can be shown that the area above the response curve is equal to a simple sum of time constants (DANG et al., 1979):

$$\alpha_1 = \frac{1}{K_L a} + \frac{R T}{H} \frac{V_L}{V_G} \tau_G + \tau_G + \tau_E \tag{39}$$

The method of calculation is shown in Fig. 2.

As might be expected, very high $K_L a$ values cannot be measured accurately with relatively slow electrodes. Thus, when $1/K_L a \ll \tau_E$, this method is not accurate.

A simple method involves neglecting the transfer term in the gas balance and noting that the equations represent three lags in series whose output is approximately the sum of the individual time constants. This is shown in Fig. 3.

2.7 Determination of Oxygen Uptake Rates by a Dynamic Method

Low oxygen uptake rates which occur in slow growing systems (plant and animal cell cultures, aerobic sewage treatment processes), cannot easily be measured by a gas balance because the difference between inlet and outlet concentrations is small. Since the solubility of oxygen is low, even small uptake rates will cause measurably large rates of change in the dissolved oxygen. Thus it is possible either by sampling or by turning off the reactor air supply to measure $Q_{O_2}X$ by the liquid balance, Eq. (2):

$$\frac{dC_{L_1}}{dt} = r_{O_2} = Q_{O_2}X \qquad (40)$$

When the time for an appreciable decrease is short compared with the electrode time constant, no correction for the electrode dynamics is required.

An on-line measurement procedure for tank systems can be devised if the aeration is turned on and off at set values of minimum and maximum dissolved oxygen. A cyclic signal of dissolved oxygen will result. The slope of the dissolved oxygen curve with the air off, perhaps correcting for surface aeration, will be equal to *OUR*. This can be calculated directly from a plotted curve. Care must be taken that the electrode time constant is much smaller than the measurement time, which will be true for low *OUR*.

Respirometers based on oxygen electrodes involve the measurement of dissolved oxygen in a batch system without oxygen supply. The resulting slope is the *OUR* (mg/L min), provided the uptake rate is reasonably low (1 min from 100% to 0% saturation). Knowing the biomass concentration X, the specific uptake rate ($Q_{O_2} = OUR/X$) can be calculated. This method is accurate for low *OUR* in the range of 0.3 mmol/L h, which corresponds to complete consumption in 1 h. It requires sampling of the culture and off-line measurement.

2.8 Loop Reactors with External Aeration to Determine *OUR*

Reactors with a liquid circulation loop to supply dissolved oxygen from an external absorber are especially suited for measuring low *OUR*s. This is so because a difference measurement with two oxygen electrodes provides good accuracy at low *OUR*. But the biomass must be retained within the reaction zone, something that will not be possible for suspended cultures without the use of membranes. The method lends itself especially to immobilized systems. One convenient reactor configuration is shown in Fig. 4. The system consists of a reactor section in which the biocatalyst particles are held, either as a packed or fluidized bed. Variable rates of oxygen supply are obtained by varying the recycle rate. The oxygen electrodes can be calibrated with each other during operation using a bypass system (not shown), thereby obtaining accurate difference measurements.

Usually the reactor is operated at low residence times in the reaction zone (high flow, small volume) so that the conversion per pass is small. The difference in concentration must be large enough for accurate measurement but small enough to prevent oxygen limitation.

The recycle system may be operated batchwise without any flow entering or leaving the recycle loop and a small amount of conversion taking place per pass. Feeding the recycle sys-

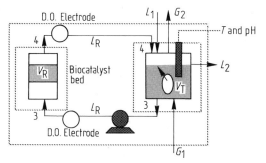

Fig. 4. Reactor oxygenator recycle loop system. L_R, recirculation liquid flow rate; $L_1 = L_2$, continuous liquid feed and effluent flow rate; G_1, G_2, gas feed and effluent flow rates; V_R, reactor volume; V_T, oxygenator tank volume.

tem converts it effectively into a continuous stirred tank. The well-stirred tank condition of homogeneity is fulfilled because there are only small differences in concentration between the reactor inlet and outlet.

Besides being flexible with respect to reactor mode (batch or continuous tank), the recycle reactor system provides the possibility of adding reactants (such as O_2) and controlling the temperature and pH in the recycle line.

The reactor-oxygenator recycle loop can be analyzed as a total system or broken down into its individual components as shown in Fig. 4.

The oxygen balance taken over the total system can be simplified by neglecting the accumulation terms and the liquid flow terms, which will be small compared to the gas rates and the consumption by reaction. Thus the oxygen balance becomes

$$0 = G_1 C_{G_1} - G_2 C_{G_2} + r_{O_2} V_R \qquad (41)$$

The absorption tank can be described by the oxygen balances for the liquid phase:

$$0 = L_R (C_{L_3} - C_{L_4}) + K_L a (C_L^* - C_{L_3}) V_T \qquad (42)$$

and for the gas phase:

$$0 = G_1 C_{G_1} - G_2 C_{G_2} - K_L a (C_L^* - C_{L_3}) V_T \qquad (43)$$

The oxygen balance of the liquid phase for the total system is

$$0 = K_L a (C_L^* - C_{L_3}) V_T + r_{O_2} V_R \qquad (44)$$

where r_{O_2} is the oxygen uptake rate of the reaction. These equations, which assume ideally mixed phases, are useful in designing the gas absorber according to the required oxygen transfer coefficient.

Balancing the oxygen around the reactor gives

$$0 = L_R (C_{L_3} - C_{L_4}) + r_{O_2} V_R \qquad (45)$$

Thus it is seen that either Eq. (41) can be used in a conventional gas balance form to calculate *OUR* ($= r_{O_2} V_R$) or the much more accurate liquid phase balance Eq. (45) can be used. In the range of 2 mg O_2 per min or 3.75 mmol/h reasonable accuracy can be expected. This cor-responds to about 20% difference in saturation with a 1 minute liquid residence time.

2.9 Methods to Measure Low Oxygen Uptake Rates

If oxygen uptake rates (*OUR*) are very low, there will be only a small difference between inlet and outlet oxygen concentrations in the gas phase. Examples are animal and plant cell culture, and sometimes wastewater treatment. The difference can be too small for error-free gas balancing, and in such cases other methods must be found for *OUR* measurement. The methods for low *OUR* can be separated into those applicable to all reactors and those applicable to special reactor types. General methods include the following:

1. Carbon dioxide analysis to obtain *OUR*
2. Recycling gas phase to increase oxygen utilization
3. Fed-batch oxygen control.

Methods involving other reactor types include:

- Circulation loop reactors with external aeration (Sect. 2.8)
- Intermittent aeration and dynamic measurement of dissolved oxygen in tank reactors (Sect. 2.7)
- Off-line batch oxygen uptake with electrodes (Sect. 2.7).

These methods will be discussed below.

Carbon Dioxide Balance

The ratio of the molar evolution rate of CO_2 to that of O_2 may often be equal to unity (*RQ* = 1) or equal to a constant (*RQ* = constant). Even if *OUR* is low, it may be possible to measure the concentration of CO_2 in the exit gas with sufficient accuracy, to allow accurate balancing for r_{CO_2}, which is proportional to r_{O_2}. This method is not applicable when CO_2 is used to control pH, as in cell culture. The solubility of CO_2 is much higher than that of O_2, and additionally CO_2 may be absorbed chemically as HCO_3^-. Therefore, this method is only useful at low pH (<5.5).

Circulation of Gas Phase

Recycling the gas phase in low rate systems with a gas circulation pump allows greater utilization of oxygen. In this case only small gas flow rates of air or oxygen would be fed to the circulating gas phase, and the same rate would be withdrawn as off-gas. By this means the concentration of O_2 in the exit gas will be greatly increased, thus permitting conventional analysis and balancing. Alternatively the gas phase can be enclosed in a membrane system, such as silicon tubing.

Fed-Batch Oxygen Control

If the gas phase is partially closed or circulated, a dissolved oxygen measurement and control system can be used to meter the required amount of oxygen, which would give the uptake rate directly. Here also a membrane system can be used.

2.10 Oxygen Transfer in Large-Scale Bioreactors

Large industrial fermenters can generally be expected to exhibit deviations from ideal mixing conditions. Thus, the assumptions of completely mixed gas and liquid phases may not be valid for large bioreactors, and some of the previously developed equations will not be applicable for the measurement of oxygen transfer and uptake. Little experimental information is available on concentration inhomogeneities or gradients within large bioreactors. Generally, information on the distribution of residence time is not available from which a physical and mathematical model could be established.

Dissolved oxygen must be considered a rapidly changing quantity because of the low solubility of oxygen, and it is, therefore, necessary to consider the possibility that differences in

the local oxygen transfer rates will result in local variations in the steady-state dissolved oxygen concentration. Consequently the dissolved oxygen concentration can vary appreciably from the top to the bottom of a deep tank.

The flow conditions of the gas phase in large-scale industrial fermentors fall between the extremes of plug flow and well-mixed states. Information on the distribution of residence time under operating conditions is necessary to characterize the gas phase flow. Unfortunately, little experimentation has been reported on industrial-scale equipment (JURECIC et al., 1984; OOSTERHUIS, 1984; GRIOT et al., 1988).

Hydrostatic pressure gradients in tall fermentors will cause large differences in oxygen solubility, C_L^*, from the liquid surface to the tank bottom. For example, in a 10 m tall reactor, the oxygen solubility for a given gas composition is twice as great at the bottom as at the top because the total pressure is doubled. It is shown by Henry's law, Eq. (3). This will double the maximum OTR ($=K_L a C_L^*$). Some compensation may be caused by smaller bubbles, due to compression, at the bottom of the column. If the number of bubbles remain constant (no coalescence), $K_L a$ should vary with $p^{-0.5}$.

It is possible to estimate the local dissolved oxygen concentration during normal fermentation conditions by equating the oxygen uptake rate to the oxygen transfer rate at any position in the vessel. Thus, this balance can be applied to an element of gas and liquid volume at any depth in the reactor.

Assuming that the biomass concentration and factors which influence Q_{O_2} are uniform within the vessel (no local substrate or oxygen limitation), the oxygen uptake rate may be taken as a constant throughout the vessel. The mass transfer coefficient, $K_L a$, will vary with turbulence level, bubble size, and gas holdup. Assuming it to be a constant, independent of position, the balance may be solved for C_{L, O_2} in exactly the same way as for CO_2, Eq. (34). Combination with Eq. (20) for the well-mixed case gives

$$C_{L, O_2} = \frac{r_{O_2}}{(K_L a)_{O_2}} + \frac{p y_{2, O_2}}{H_{O_2}} \tag{46}$$

The resulting equation shows C_{L,O_2} to be a linear function of p and, therefore, of depth. It predicts that when $K_L a = 300$ h^{-1} and $r_{O_2} = 900$ mg L^{-1} h^{-1}, the dissolved oxygen concentration will vary linearly from 5.4 mg L^{-1} at the surface to 12.1 mg L^{-1} at a depth of 10 m.

Eq. (46) contains assumptions which can only be applied with some uncertainty, and it should be tested against dissolved oxygen data from deep fermentation tanks.

The possibility that gas compositions, dissolved oxygen concentrations and oxygen solubilities, holdup, and transfer parameters may vary with depth in a tall fermentor introduces much additional complexity into the problem of modelling the reactor for the purpose of extracting oxygen transfer coefficients. Although it is impossible to give specific recommendations for particular situations, a discussion of the possible models and their underlying assumptions will help define the problem and clarify the issues. The factors of gas and liquid phase flow patterns, gas composition gradients, and hydrostatic pressure are incorporated into the models to be discussed below.

Steady-State Models

Equating *OUR* to *OTR* forms the basis for steady-state batch liquid methods. A modification is required to express variations in oxygen transfer rate with position. Thus,

$$r_{O_2} = \frac{1}{V_L} \int_0^{V_L} K_L a (C_L^* - C_L) \, dV_L \qquad (47)$$

where $K_L a$ and the driving force, $C_L^* - C_L$, may vary throughout the liquid volume. Because of the complexities of the gas–liquid flow it is not possible to evaluate the variation in $K_L a$. Therefore, it will be assumed to be constant. Gradients in gas composition and pressure will cause C_L^* to vary. The dissolved oxygen concentration, C_L, will be assumed to be uniform within the region of a single impeller.

Well-Mixed Gas and Liquid

The oxygen balance for homogeneous conditions in both phases and equal feed and effluent gas flow rates becomes

$$K_L a = \frac{r_{O_2}}{(C_{L_2}^* - C_{L_2}) V_L} = \frac{G(C_{G_1} - C_{G_2})}{(C_{L_2}^* - C_{L_2}) V_L} \qquad (48)$$

where r_{O_2} is obtained from the gas balance and $C_{L_2}^*$ is evaluated based on the exit gas composition and pressure.

The resulting equation is

$$K_L a = \frac{G(C_{G_1} - C_{G_2})}{\left(C_{G_2} \dfrac{RT}{H} - C_{L_2}\right) V_L} \qquad (49)$$

Well-Mixed Liquid and Plug Flow Gas

In many applications the gas may bubble up through the liquid without being circulated by a mechanical mixer. Under these conditions a gradient in oxygen gas composition exists from inlet to outlet. An approximate model can be obtained by using the well-mixed gas model and an average driving force. Thus,

$$C_L^* = \frac{C_{G_1} + C_{G_2}}{2} \frac{RT}{H} \qquad (50)$$

This approach is accurate when the differences between inlet and outlet composition are small.

A more exact formulation recognizes the plug flow nature of the gas flow patterns and can be derived by considering the balances around a segment of thickness ΔZ and volume ΔV. The steady-state oxygen balance of the gas phase can be formulated as:

Oxygen entering the segment by flow	Oxygen leaving − the segment by flow	Transfer of the = segment to the liquid

or

$$(G C_G)|_{z_1} - (G C_G)|_{z_2} = K_L a (C_L^* - C_L) \Delta V_L \qquad (51)$$

letting

$$G = v_G A_G$$

where $A_G = A(V_G/V_T)$; A, cross-sectional area; V_T, total volume; and $\Delta V_L = (V_L/V_T)\Delta V_T$ gives

$$C_G|_{z_1} - C_G|_{z_2} = \frac{K_L a}{v_G}(C_L^* - C_L)|_{z_2}\frac{V_L}{V_G}\Delta Z \quad (52)$$

Writing this in differential form gives

$$\frac{dC_G}{dZ} = -\frac{K_L a}{v_G}(C_L^* - C_L)\frac{V_L}{V_G} \quad (53)$$

Here, the quantity v_G represents the linear velocity of the gas phase. Eq. (53) can be integrated after substituting to give:

$$K_L a = \frac{V_G}{V_L}\frac{v_G}{L}\frac{H}{RT}\ln\left(\frac{C_{G_1} - C_{L_2}}{C_{G_2} - C_{L_2}}\right) \quad (54)$$

Eq. (54) accounts for the oxygen gas concentration gradients which are created by the oxygen transfer from the gas bubbles as they rise through the liquid phase. Inherent in Eq. (54) are the assumptions of constant $K_L a$, V_L/V_G, and v_G. The influence of pressure on C_L^* and v_G is not included.

Comparison of the Models

The selection of the appropriate model from among the two basic types: well-mixed and plug flow, and their approximations for a particular gas–liquid contacting situation may require experience or a certain amount of guesswork. Guidance can be obtained by comparing the use of the models for given situations and noting the errors which can arise by choosing a model which is not applicable. For example, by calculating the apparent $K_L a$, obtained from applying the well-mixed gas model to a plug flow gas situation, it can be seen that unless the difference between inlet and outlet oxygen gas concentrations is large, very little difference between the models can be detected.

By taking the ratio of $K_L a$'s as obtained by Eq. (48) for the well-mixed gas (WM), and

Eq. (54) for the plug flow gas (P), the following is obtained:

$$\frac{(K_L a)_{WM}}{(K_L a)_P} = \frac{C_{L_1}^* - C_{L_2}^*}{(C_{L_2}^* - C_{L_2})\ln\left(\dfrac{C_{L_1}^* - C_{L_2}}{C_{L_2}^* - C_{L_2}}\right)} \quad (55)$$

Rearranging and introducing $\Delta C_1 = C_{L_1}^* - C_{L_2}$, $\Delta C_2 = \Delta C_{WM} = C_{L_2}^* - C_{L_2}$, and defining the log mean (LM) concentration driving force as

$$\Delta C_{LM} = \frac{\Delta C_1 - \Delta C_2}{\ln\left(\dfrac{\Delta C_1}{\Delta C_2}\right)} \quad (56)$$

finally yields

$$(K_L a)_{WM}\Delta C_{WM} = (K_L a)_P \Delta C_{LM} \quad (57)$$

From Eq. (49) it can be seen that the total oxygen transfer rate per unit volume of liquid is equal to $(K_L a)_{WM}\Delta C_{WM}$. Clearly, from Eq. (57) the differences in $K_L a$ values, as obtained from well-mixed and plug flow models, depend on the differences in the respective driving forces ΔC_{WM} and ΔC_{LM}. It is also of interest to consider using an arithmetic mean (AM) driving force defined by

$$\Delta C_{AM} = \frac{C_{L_1}^* + C_{L_2}^*}{2} - C_{L_2} \quad (58)$$

such that

$$(K_L a)_{WM}\Delta C_{WM} = (K_L a)_{AM}\Delta C_{AM} \quad (59)$$

In Fig. 5 values of ΔC_{WM}, ΔC_{LM}, and ΔC_{AM} are compared as a function of the exit oxygen gas mole fraction, y_2. In this figure the driving forces, calculated from Eq. (56, 57, and 58), are shown for two different dissolved oxygen levels.

Particularly noteworthy in Fig. 5 is how close the arithmetic mean driving force, ΔC_{AM}, is to the logarithmic mean driving force, ΔC_{LM}, within the usual range of exit oxygen gas mole fractions, $0.15 > y_2 < 0.21$. Within this range, ΔC_{AM} can be used without error to calculate $K_L a$ values for large systems which exhibit oxygen gas gradients. Even for the situation in which $y_2 = 0.10$ and $C_L = 2$ mg/

L, $(K_La)_{AM}$ will be only 12% lower than $(K_La)_{LM}$. The use of ΔC_{AM} is a striking improvement over ΔC_{WM}. As can be calculated from Fig. 5, the errors in K_La at $y_2 = 0.15$ and 0.10 are 25% for $C_L = 2$ mg/L. If the exit oxygen mole fraction is below $y_2 = 0.18$ the errors using ΔC_{WM} will be less than 10%. As

3 Application of Gas Analysis Results to Elemental Balancing Methods

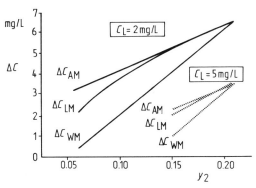

Fig. 5. Curves showing ΔC_{AM}, ΔC_{LM}, and ΔC_{WM} for $C_L = 2$ mg/L and 5 mg/L and $y_1 = 0.21$.

seen from the curves for $C_L = 5$ mg/L, the errors are rather sensitive to the dissolved oxygen concentration; at $y_2 = 0.18$ the error in using ΔC_{WM} is 20%. The main conclusion to be drawn is that the use of ΔC_{AM} will generally yield K_La values which are close to those obtained with ΔC_{LM}.

3.1 Systematics of Elemental Balancing

Consider the general case of a partially unknown stoichiometry written in C-moles (ROELS, 1983):

$$\underset{\text{[substrate]}}{CH_{b_2}O_{c_2}N_{d_2}} + \underset{\text{[nitrogen source]}}{\alpha CH_{b_4}O_{c_4}N_{d_4}} + \underset{\text{[oxygen]}}{\beta O_2} \rightarrow$$

$$\underset{\text{[biomass]}}{\gamma CH_{b_1}O_{c_1}N_{d_1}} + \underset{\text{[product]}}{\delta CH_{b_3}O_{c_3}N_{d_3}} +$$

$$+ \underset{\text{[water]}}{\varepsilon H_2O} + \underset{\substack{\text{[carbon} \\ \text{dioxide]}}}{\zeta CO_2} \qquad (60)$$

where c, d, and f are coefficients representing the fraction of carbon converted to biomass, product, and CO_2, respectively. Elemental balances give

$$\text{C} \quad 1 + a \quad\quad = \gamma + \delta + \zeta \qquad (61)$$

$$\text{H} \quad b_2 + b_4\alpha \quad = b_1\gamma + b_3\delta + 2\varepsilon \qquad (62)$$

$$\text{O} \quad c_2 + c_4\alpha + 2\beta = c_1\gamma + c_3\delta + \varepsilon + 2\zeta \qquad (63)$$

$$\text{N} \quad d_2 + d_4\alpha \quad = d_1\gamma + d_3\delta \qquad (64)$$

Additional information is required, since there are too many unknowns to permit a solution of this general problem. The elemental balances provide only 4 equations and hence can be solved for only 4 unknowns.

The whole system can be described in terms of molar fluxes (ϕ) (Fig. 6). The flow vector

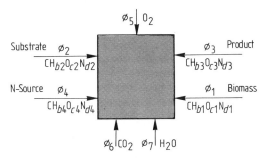

Fig. 6. Macroscopic fluxes (ϕ) in a fermentation system.

$$\phi = [\phi_1 \ldots \phi_7] \qquad (65)$$

at steady state

$$\phi \cdot E = 0 \qquad (66)$$

where E is the elemental composition matrix

$$E = \begin{array}{c} \\ \\ \\ \\ \\ \\ \\ \end{array} \begin{array}{cccc} C & H & O & N \\ \end{array}$$

$$E = \left| \begin{array}{cccc} 1 & b_1 & c_1 & d_1 \\ 1 & b_2 & c_2 & d_2 \\ 1 & b_3 & c_3 & d_3 \\ 1 & b_4 & c_4 & d_4 \\ 0 & 0 & 2 & 0 \\ 1 & 0 & 2 & 0 \\ 0 & 2 & 1 & 0 \end{array} \right| \begin{array}{l} \text{Biomass} \\ \text{Substrate} \\ \text{Product} \\ \text{N-Source} \\ O_2 \\ CO_2 \\ H_2O \end{array} \quad (67)$$

Three quantities are independent. Assuming ϕ_6 ($=r_{O_2}$), ϕ_5 ($=r_{CO_2}$), and ϕ_4 (nitrogen source) are known, equations for ϕ_1 (biomass), ϕ_2 (substrate), ϕ_3 (product), and ϕ_7 (water) can be obtained by methods of linear algebra, as detailed by ROELS (1983). A number of examples in the literature used this method:

- Systematics and examples (ROELS, 1983)
- Growth of baker's yeast on glucose (WANG et al., 1977)
- Penicillin fermentation (HEIJNEN et al., 1979; ROELS, 1983; MOU and COONEY, 1983)
- Tetracycline production (ROSS and SCHÜGERL, 1987)
- Gluconic acid production (REUSS et al., 1984)
- Production of poly-β-hydroxybutyric acid (HEINZLE and DETTWILER, 1986).

3.2 Elemental Balancing for Monitoring a Poly-β-Hydroxybutyric Acid (PHB) Producing Culture

The method of elemental balancing is discussed in more details using as example the synthesis of poly-β-hydroxybutyric acid (PHB) by *Alcaligenes latus* (HEINZLE and DETTWILER, 1986; HEINZLE et al., 1990). This example is later also used to analyze error propagation (Sect. 4).

Stoichiometry

The production of PHB ($CH_{1.5}O_{0.5}$) from sugar (CH_2O), in which CH_2O represents the

sum of all sugars (saccharose, glucose, and fructose), is defined by

$$CH_2O \rightarrow CH_{1.5}O_{0.5} + 0.125\,O_2 + 0.25\,H_2O \quad (68)$$

The production of residual biomass ($CH_{\beta_1}O_{\beta_2}N_{\beta_3}$) equals the total biomass minus PHB and is defined by

$$CH_2O + \alpha_1\,O_2 + \alpha_2\,H_2O + \alpha_3\,NH_3 \rightarrow \\ CH_{\beta_1}O_{\beta_2}N_{\beta_3} \quad (69)$$

From Eq. (69) it can be seen that $\alpha_3 = \beta_3$.

Oxidation of sugar for ATP-production is given by

$$CH_2O + O_2 \rightarrow CO_2 + H_2O \quad (70)$$

The composition of residual biomass $CH_{1.8}O_{0.5}N_{0.2}$ was taken from the literature (ROELS, 1983). This composition generally could be confirmed by elemental analysis. From this we get $\alpha_3 = 0.2$, $\alpha_2 = -0.4$, and $\alpha_1 = -0.05$. Fig. 7 shows the macroscopic principles of fluxes in PHB production according to ROELS (1983), where the flow vector

$$\phi = [\phi_1 \ldots \phi_7] \quad (71)$$

at steady state

$$\phi \cdot E = 0 \quad (72)$$

E is the elemental composition matrix

$$E = \left| \begin{array}{cccc} C & H & O & N \\ 1 & 1.8 & 0.5 & 0.2 \\ 1 & 2 & 1 & 0 \\ 1 & 1.5 & 0.5 & 0 \\ 0 & 3 & 0 & 1 \\ 0 & 0 & 2 & 0 \\ 1 & 0 & 2 & 0 \\ 0 & 2 & 1 & 0 \end{array} \right| \begin{array}{l} \text{Biomass} \\ \text{Glucose} \\ \text{PHB} \\ \text{Ammonia} \\ O_2 \\ CO_2 \\ H_2O \end{array} \quad (73)$$

The elemental balances are:

C: $\quad 0 = \phi_1 + \phi_2 + \phi_3 + \phi_6 \quad (74)$

H: $\quad 0 = 1.8\phi_1 + 2\phi_2 + 1.5\phi_3 + 3\phi_4 + 2\phi_7 \quad (75)$

O: $\quad 0 = 0.5\phi_1 + \phi_2 + 0.5\phi_3 + 2\phi_5 + 2\phi_6 + \phi_7 \quad (76)$

N: $\quad 0 = 0.2\phi_1 + \phi_4. \quad (77)$

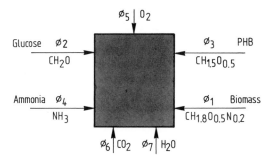

Fig. 7. Macroscopic fluxes (ϕ) in production of poly-β-hydroxybutyric acid (PHB) by *Alcaligenes latus*.

Multiplying each by an appropriate factor and adding all equations after rearrangement,

$$\phi_1(\lambda_C + 1.8\lambda_H + 0.5\lambda_O + 0.2\lambda_N)$$
$$+ \phi_2(\lambda_C + 2\lambda_H + \lambda_O)$$
$$+ \phi_3(\lambda_C + 1.5\lambda_H + 0.5\lambda_O)$$
$$+ \phi_4(3\lambda_H + \lambda_N)$$
$$+ \phi_5(2\lambda_O)$$
$$+ \phi_6(\lambda_C + 2\lambda_O)$$
$$+ \phi_7(2\lambda_H + \lambda_O) = 0 \tag{78}$$

Substituting $\lambda_H = 1$, then $\lambda_O = -2$, $\lambda_N = -3$, and $\lambda_C = 4$,

$$4.2\phi_R + 4\phi_S + 4.5\phi_P - 4\phi_{O_2} = 0 \tag{79}$$

From Eq. (77),

$$0.2\phi_R + \phi_{NH_3} = 0 \tag{80}$$

and from Eq. (74):

$$\phi_R + \phi_S + \phi_P + \phi_{CO_2} = 0 \tag{81}$$

Using Eq. (80) and combining Eqs. (79) and (81) yields the following expressions which can be used directly for the calculation of metabolic fluxes based on titration and gas analysis:

$$\phi_R = -5\phi_{NH_3} \tag{82}$$

$$\phi_P = 8(\phi_{O_2} + \phi_{CO_2} - 0.05\phi_R) \tag{83}$$

$$\phi_S = -\phi_R - \phi_P - \phi_{CO_2} \tag{84}$$

The above relations contain 6 unknowns and 3 equations; thus, measuring any 3 unknowns will determine the system. This means that if the fluxes of ammonia, oxygen, and carbon dioxide are measured, the calculation of the fluxes of substrate, cell biomass, and product is straightforward.

Determining CO_2 and the NH_3 Fluxes

Cells take up nitrogen, and the ammonia uptake rate can be related to biomass production using Eq. (69). The protons produced are directly related to the NH_3 consumed by the biomass according to Eq. (85):

$$NH_4^+ \Leftrightarrow NH_3 + H^+ \tag{85}$$

These can be determined by titrating with alkali, but must be corrected for other sources of protons. CO_2 produced by the cells also yields protons, Eqs. (22) and (23), which must be considered when using Eq. (85) to estimate biomass. To determine *CTR* the procedure described in Sect. 2.4 was applied.

On-Line Calculations

The concentration of dissolved CO_2 (C_{L,CO_2}) was estimated using Eq. (34). Differentiation with time allowed the calculation of the liquid phase accumulation. Combining Eqs. (25) and (31), ϕ_{CO_2} ($= CPR/V_L$) can be calculated:

$$\phi_{CO_2} = N_0 \left(\frac{Y_{1,CO_2} - Y_{0,CO_2} \dfrac{Y_{0,\text{inert}}}{Y_{1,\text{inert}}}}{V_L} \right) +$$
$$+ \left(1 + \frac{K}{C_{H^+}} \right) \frac{\Delta C_{L,CO_2}}{\Delta t} \tag{86}$$

At constant pH, ammonia consumption rate ($r_{NH_4^+}$) equals the difference between addition rate of alkali (r_{OH^-}) and the accumulation rate of bicarbonate:

$$\phi_{NH_4^+} = -r_{OH^-} + \frac{\Delta C_{HCO_3^-}}{\Delta t} \tag{87}$$

The molar fluxes of biomass, substrate, and product can be calculated using Eqs. (82) to (84).

Integration of Fluxes for the Batch Concentrations

Integration of the fluxes gives the concentrations of R, S, P, and NH_4^+ at any time t:

$$C_R = C_{R,0} + \int_0^t \phi_R \, dt \qquad (88)$$

$$C_P = C_{P,0} + \int_0^t \phi_P \, dt \qquad (89)$$

$$C_S = C_{S,0} + \int_0^t \phi_S \, dt \qquad (90)$$

$$C_{NH_4^+} = C_{NH_4^+,0} + \int_0^t \phi_{NH_4^+} \, dt \qquad (91)$$

It can be expected that due to integration over a period of time any errors in the fluxes would accumulate. This is analyzed below (Sect. 4).

Fig. 8 shows an example of on-line estimation of C_S and C_P compared with off-line results. Both concentration values continuously drifted away from the actual values. This may have been caused by an offset in the oxygen partial pressure measurement, whose effect would be accumulated by the integration of the fluxes. As shown, a simulated offset of $\Delta y_{2,O_2} = 0.0012$ would partially explain the deviations observed in this experiment.

The influence of errors in gas analysis on the estimation of gas reaction rates and fluxes of biomass, substrates, and product is discussed in more detail in the following Sect. 4.

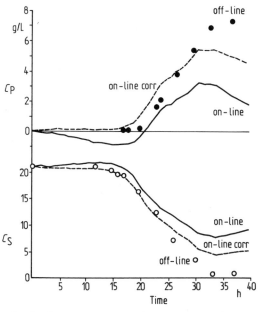

Fig. 8. Elemental balancing in production of PHB by *Alcaligenes latus*. Comparison of on-line (full line) and off-line measurement (symbols) results and possible cause of error due to incorrect calibration for the measurement of oxygen partial pressure at the reactor exit (y_{2,O_2}). C_P, product PHB concentration; C_S, substrate concentration (total sugar); pH = 7.3; $G_1 = 9.12$ vvm. Simulation with $y_{2,O_2,corr} = y_{2,O_2,true} - 0.0012$ (dashed line).

4 Error Analysis for Gas Balancing

In most cases the choice of a method for gas analysis and equipment depends on the required accuracy and reliability. These are usually difficult to define because of the complexity of error propagation. Higher quality instruments usually cost more. The requirements may differ tremendously from industrial large-scale process monitoring to research laboratory gas analysis.

Generally two types of errors have to be considered:

(1) Errors caused by simplification of gas balancing (Sect. 4.3)
(2) Measurement errors (Sect. 4.4).

Systematic balancing errors can easily be avoided by careful analysis of the system and proper setting up of balance equations. Measurement errors depend on a number of factors, some of which may directly influence the cost of equipment.

4.1 Objectives of On-Line Gas Analysis and Requirements for Accuracy and Reliability

Three different levels of complexity using the results of gas analysis may be identified:

- In industrial practice it is often of paramount importance to guarantee and prove the exact repeatability of batches. If on-line gas analysis is solely used for this purpose, good precision and reliability are of primary importance. The qualitative and quantitative traces of characteristic signals (e.g., signals of oxygen and CO_2 partial pressure meters, dissolved oxygen electrode signal) are used to trace the performance of a particular batch. In these cases medium composition and metabolic pathways are often very complex and only poorly understood. It is usually not necessary to know the true values of, e.g., oxygen uptake rates.

  In such cases, calibration has only to assure good reproducibility. These measurements are usually characteristic for a particular process and process analysis device and, therefore, cannot be applied directly to other processes.
- More specific information may be collected if "true" concentration and reaction rate values (e.g., *OTR*, *CPR*) and *RQ* (respiratory quotient) are determined. This information is of more general value and may be used to compare research results of different research groups. Calibration of instruments has to be done frequently using a good quality standard. The most critical measurement concerning precision and accuracy in gas analysis is oxygen partial pressure, because measured differences are small. This will be discussed in detail later.
- The third level of complexity is the use of the results of gas analysis, usually combined with other measurements, to indirectly estimate other concentration values and growth and production rates using elemental balancing techniques

(Sect. 3). In this case errors of measurement may be amplified dramatically and yield completely misleading results.

4.2 Definition of Measurement Requirements

Accuracy denotes the agreement with true values. In many cases absolute accuracy is not very important (e.g., dissolved oxygen measurement, traces of partial pressures of gases and volatiles). To obtain high accuracy, careful calibration procedures using good standards and careful analysis of the process are important to minimize systematic errors (e.g., total pressure or temperature changes in a partial pressure measurement apparatus – see Sect. 8).

Good precision is achieved if signals are not noisy and if there is no drift. Signal noise may be reduced using filters or statistical methods. Drift usually can be identified and corrected by frequent recalibration. Contrary to liquid phase on-line sensors, which are immersed into the sterile region of the fermenter, gas analysis equipment, which is usually located outside the sterile region, can be easily recalibrated without causing sterility problems. Constant measurement and process environments reduce drift.

Excellent stability over extended periods in a harsh process environment is required for process instruments. This requires the special design of instruments (power supply, housing, redundancy of critical instrument parts, e.g., filaments in mass spectrometers, and robustness).

In most cases the sensitivity of analytical apparatus is sufficient for gas analysis. Problems may occur when analyzing trace gas components. Examples of this are H_2 in an anaerobic reactor and less volatile compounds (e.g., acetoin and higher alcohols).

4.3 Errors Caused by Simplification of Balancing

Lists of possible and potentially useful simplifications in the balance equations are given in Tables 1 and 2. It is important to know the possible contribution of each simplification to the final error in measuring reaction rates.

For CO_2 the situation is more complex because of its higher solubility and its reaction to HCO_3^-. Estimation of r_{O_2} is also indirectly influenced by errors in CO_2 estimation.

The analysis of outlet gas concentrations is for most fermentations one of the most reliable on-line methods for reactor monitoring. In some cases an exact quantitative balance is not needed, and it is sufficient to note relative changes. This is most likely to be the case in production reactors. In such cases reliability of instrumentation is of primary importance. For research purposes specific respiration measurements (Q_{O_2}, Q_{CO_2}) are usually required, making a balance of the mass or molar rates necessary. A schematic drawing of the equipment for gas balancing is shown in Fig. 9.

4.3.1 Simplifications Concerning Pressure, Temperature, Humidity, and Gas Flow Rates

Starting from general balances for well-mixed gas and liquid phases Eqs. (1) and (2) and assuming steady state conditions ($dC_i/$

Tab. 1. Simplifications and their Consequences for Estimation of r_{O_2}

Section	Simplification	Acceptability for r_{O_2}
4.3.1	$G_1 = G_0$	Only if $RQ \approx 1$ (see Tab. 3)
4.3.1	$T_1 = T_2$ $p_1 = p_2$ $p_{H_2O} = 0$	No if G measured Yes if N_0 and $y_{1,i}$ measured, and Eq. (21) used
4.3.2	$dC_L/dt = 0$ $dC_G/dt = 0$	Usually o.k.; usually $\Delta r_{O_2} < 2\%$
4.3.2	$dC_{HCO_3^-}/dt = 0$	Usually o.k.
4.3.2	$C_{HCO_3^-} = 0$	Usually o.k.
For flow reactors		
	$L_0 C_{L_0} = 0$	o.k. with aeration in reaction zone

Tab. 2. Simplifications and their Consequences for Estimation of r_{CO_2}

Section	Simplification	Acceptability for r_{CO_2}
4.3.1	$G_1 = G_0$	Only if $RQ \approx 1$ (see Tab. 3)
4.3.1	$T_1 = T_2$ $p_1 = p_2$ $p_{H_2O} = 0$	No if G measured Yes if N_0 and $y_{1,i}$ measured, and Eq. (21) used
4.3.2	$dC_L/dt = 0$ $dC_G/dt = 0$	Usually o.k.; usually $\Delta r_{O_2} < 2\%$
4.3.2	$dC_{HCO_3^-}/dt = 0$ $C_{HCO_3^-} = 0$	o.k. at pH $\leqslant 5.5$ or $dC_L/dt \approx 0$ o.k. at pH $\leqslant 5.5$
For flow reactors		
	$L_0 C_{L_0} = 0$	o.k. for aeration in the reaction zone

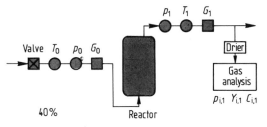

Fig. 9. Schematics of measurements in gas balancing.

$dt = 0$), the influence of p, T, humidity, and gas flow rate assumptions are discussed. Assuming steady state, the total mass balance for dry gas streams in terms of molar flow rates N (mol/s) can be written as

$$0 = N_0 - N_1 + (r_{O_2} + r_{CO_2}) V_L \qquad (92)$$

Assumption of Equal Flow Rates $N_1 = N_0$

Under the condition that $r_{O_2} = -r_{CO_2}$, then $N_0 = N_1$. Thus, the molar flow rates of dry gases are equal if the oxygen uptake rate (*OUR*) is equal to the CO_2 production rate (*CTR*). This condition is described by the respiration quotient equal to unity ($RQ = 1$). If, however, these rates are not equal, there will be a net production of gas during fermentation, and the molar flow rates will change.

If it is known that $N_0 \approx N_1$, only one volumetric flow rate need be measured at the inlet (N_0) along with the composition mole fraction at the outlet $y_{1,i}$. Calculating the molar flow from the known p and T values, Eq. (8), gives,

$$r_i = \frac{N_i (y_{0,i} - y_{1,i})}{V_L} \qquad (93)$$

For air: $y_{0,O_2} = 0.21$ and $y_{0,CO_2} = 0.0$, the results give r_{O_2} and r_{CO_2} in moles/volume · time.

Anything that causes N_0 to be unequal to N_1 would be a source of error if the above procedure is used. This is clearly illustrated in Tab. 3. For $RQ = 1$ (case I: upper part of the table)

$N_0 = N_1$ leads to correct results, whereas an error is obviously made if $N_0 \neq N_1$ (case II).

Unequal Flow Rates

Suppose there is a case in which $r_{O_2} \neq r_{CO_2}$ or $RQ \neq 1$. We then know that there will be a difference in molar flow, which could be measured by two flowmeters with suitable T and p corrections. Alternatively, information about the inert nitrogen or argon gas content can be used, Eq. (21). Using dry gas measurements and making an inert gas balance will always lead to correct results as is shown by column G in Tab. 3.

Calculation without T, p, and Humidity Correction, Assuming $G_0 = G_1$

Calculations following these assumptions must be incorrect, because the total volumetric flow rates as obtained from a gas flow meter cannot be correctly converted to molar flow rates without knowledge of pressure and temperature. Errors caused by these simplifications can be seen in Tab. 3, columns C, D, E, and G. In case I with $RQ = 1$ the errors are between 0 and 8%, whereas in case II the errors are up to 18.5%.

Not accounted for in the above calculations is the effect of humidity on the volumetric flow rate. The gas analysis would, except for mass spectrometry, always be made on a previously dried stream, and, therefore, the total flow rate must be expressed on a dry basis.

4.3.2 Errors Caused by Steady-State Assumption

As indicated in Tables 1 and 2, errors may also be caused by the steady-state assumption. Of course there will be no such errors in a true steady-state condition, e.g., in a continuous culture.

These errors are dynamic in nature and cannot be analyzed as simply as steady-state er-

Tab. 3. Examples of Calculations of Errors Caused by p, T, p_{H_2O} and, Gas Flow Assumptions
A, inlet stream; **B**, true values in outlet stream; **C**, assumption of equal volumetric gas streams (G) and pressures (p) at inlet and outlet and of dry air at outlet; **D**, same as **C**, but assumption of equal T instead of p; **E**, equal G, T and p and dry gas assumption; **F**, equal G, T, and p and dry gas assumption; **G**, equal T and p and dry gas assumption; **H**, equal T and p assumption. Upper part: $RQ=1$; lower part: $RQ=9.318$

	A	B	C	D	E	F	G	H
	Stream 0	Stream 1	$G_1=G_0$ $p_0=p_1$ $p_{H_2O}=0$	$G_1=G_0$ $T_0=T_1$ $p_{H_2O}=0$	$G_1=G_0$ $T_0=T_1$ $p_0=p_1$ $p_{H_2O}=0$	$G_1=G_0$ $p_{H_2O}=0.042$	$T_0=T_1$ $p_0=p_1$ $p_{H_2O}=0$	$T_0=T_1$ $p_0=p_1$
y_{O_2} (dry gas)	0.210	0.172	0.172	0.172	0.172	0.172	0.172	0.172
y_{O_2} (wet gas)		0.165				0.165		0.165
y_{CO_2} (dry gas)	0.0	0.038	0.038	0.038	0.038	0.038	0.038	0.038
G (L/min)	10.00	11.34	10.00	10.00	10.00	10.00	10.00	10.44
N (mol/min)	0.431	0.450	0.417	0.411	0.431	0.397	0.431	0.450
r_{O_2} (mol/min)		0.0162	0.0187	0.0198	0.0162	0.0250	0.0162	0.0162
r_{CO_2} (mol/min)		0.0162	0.0157	0.0155	0.0162	0.0149	0.0162	0.0169
RQ		1.0	0.840	0.782	1.000	0.598	1.000	1.045
$\Delta r_{O_2,rel}$		0.0	0.033	0.048	0.000	0.079	0.000	−0.044
$\Delta r_{CO_2,rel}$		0.0	−0.033	−0.048	0.000	−0.079	0.000	0.044
y_{O_2} (dry gas)	0.210	0.172	0.172	0.172	0.172	0.172	0.172	0.172
y_{O_2} (wet gas)		0.165				0.165		0.165
y_{CO_2} (dry gas)	0.0	0.128	0.128	0.128	0.128	0.128	0.128	0.128
G (L/min)	10.00	12.80	10.00	10.00	10.00	10.00	11.29	11.79
N (mol/min)	0.431	0.508	0.417	0.411	0.431	0.397	0.486	0.508
r_{O_2} (mol/min)		0.0067	0.0187	0.0198	0.0162	0.0250	0.0067	0.0067
r_{CO_2} (mol/min)		0.0621	0.0532	0.0524	0.0550	0.0507	0.0621	0.0649
RQ		9.318	2.850	2.652	3.392	2.027	9.318	9.730
$\Delta r_{O_2,rel}$		0.0	0.143	0.156	0.114	0.184	0.000	−0.044
$\Delta r_{CO_2,rel}$		0.0	−0.143	−0.156	−0.114	−0.184	0.000	0.044

rors. Generally speaking, they are significant if the changes in biological activity are large compared to the corresponding mass transfer terms. A typical case is the end of a batch culture, where a sudden depletion of the limiting substrate quickly decreases *OUR* and *CPR*. A similar case is diauxic growth, where one substrate is depleted and the culture has to adapt its metabolic activities to a new substrate. Another situation is the addition of an inhibitory or toxic substance to the culture, as in waste water treatment or during addition of a toxic precursor.

The dynamic behavior of complex systems can best be seen by simulation. Dynamic errors and their consequences for elemental balancing results are discussed with the example of an *Alcaligenes latus* fermentation producing PHB (see also Sect. 2).

4.4 Erroneous Estimation of Reaction Rates Caused by Measurement Errors

Three sources of errors in measurement may be identified:

- calibration offset
- instrument drift
- statistical noise.

There are different ways to avoid or minimize such errors. Calibration offset errors can be minimized by using accurate standards and by careful calibration procedure (pressure, temperature, humidity). For high accuracy, dynamic errors during calibration must be carefully avoided. It is usually difficult to prove whether such errors occur or not. In some cases inconsistent balancing results point to such errors. An error analysis for the estimation of gas reaction rates has been published by NEUBERT and MINKEVICH (1984).

Instrument drift can be minimized by purchasing high-quality instrumentation conforming to process equipment standards. Instruments should not be exposed to environmental conditions for which they are not built (temperature, humidity, vibrations, etc.). Measurement techniques using internal standards are less prone to such errors, since drift usually affects the intensity of the internal standard in the same way as that of the analyzed component. Examples are GC, MS, and analyzers employing dual-channel reference gas.

The first two types of error may be identified by recalibration or by the increasing inconsistency of balancing results. The application of estimators with dynamic filters (e.g., Kalman filter) including a measurement model or artificial intelligence systems may support the identification and eventually the compensation of such errors.

An example of systematic error is given in Fig. 8, where the on-line estimation resulted in negative concentrations for the product PHB. Since it was determined that the measurement of y_{O_2} was most sensitive to errors, a miscalibration for y_{O_2} may be assumed and compensated as illustrated in Fig. 8.

Statistical measurement noise leads to correspondingly noisy results. The noise may be amplified and cause results which cannot be interpreted without filtering properly. In most cases standard electrical filters (first order) will be sufficient to provide meaningful results. Such filters, however, introduce a delay which can lead to erroneous results.

4.4.1 Errors in the Measurement of Gas Flow

One kind of error which may enter into the determination of *OUR* or *CPR* by the gas balance method is caused by inaccuracies in the measurement of flow rate. If Eqs. (21) and (26) are used, the error in *OUR* or *CPR* will be proportional to the measured flow error ($\Delta OUR \propto \Delta N_0$; $\Delta CPR \propto \Delta N_0$). For example, a 10% error in the measurement of both gas flows will cause an error of equal percentage in the *OUR*.

4.4.2 Statistical Error Propagation

The situation is a little more complex if both G_0 and G_1 are measured and have their individual errors or if other errors are involved. In this case statistical error propagation has to be estimated according to methods described in standard statistics textbooks (e.g., KREYSZIG, 1970). For a large number of measurements with independent individual measurement errors having Gaussian distributions, the mean error of the estimated variable (s) of a function $z = h(x, y)$ is

$$s = \sqrt{\left(\frac{\delta h}{\delta x}\right)^2 s_x^2 + \left(\frac{\delta h}{\delta y}\right)^2 s_y^2} \tag{94}$$

Starting with Eqs. (1) and (2), assuming steady state conditions and $L_0 = L_1 = 0$, and also assuming $s_{C_0, O_2} = s_{C_1, O_2} = 0$, results in

$$s_{OUR} = \sqrt{\left(\frac{\delta(G_0 C_{0, O_2} - G_1, C_{1, O_2})}{\delta G_0}\right)^2 s_{G_0}^2 + \left(\frac{\delta(G_0 C_{0, O_2} - G_1 C_{1, O_2})}{\delta G_1}\right)^2 s_{G_1}^2} \tag{95}$$

and further

$$s_{OUR} = \sqrt{C_{0, O_2}^2 s_{G_0}^2 + C_{1, O_2}^2 s_{G_1}^2} \tag{96}$$

Having similar O_2 concentrations at inlet and outlet ($C_{0, O_2} \approx C_{1, O_2}$) and identical errors for the measurement of both gas flow rates ($s_{G_0} = s_{G_1}$), leads to the error

$$s_{OUR} = C_{O_2} s_G \sqrt{2} \tag{97}$$

With the above simplifications the mean error for *OUR* is proportional to the mean error in G. For unequal values of C_{O_2} and s_G the mean error in *OUR* is given by Eq. (96).

4.4.3 Errors in Oxygen Gas Analysis

An oxygen gas analysis which is accurate to 5% of a full scale measurement span of 18 to 21% will cause a 5% error in *OUR* when the exit gas is 18% oxygen. The importance of choosing a suitable measurement span for the oxygen analysis instrument can be appreciated by considering the error caused by a 5% full span inaccuracy when the span is 0 to 21% and the outlet gas is 18%. The resulting error in *OUR* would be approximately 30%. In general it should be kept in mind that *OUR* measurements involve rather small differences between inlet and outlet gas streams, and extreme care

must be exercised if accurate results are required.

HEINZLE et al. (1984) made an analysis of error propagation for *statistical errors* in estimating oxygen uptake rates. Their analysis applies only to situations in which inert internal standards are used (e.g., GC and MS). Small differences between inlet and outlet oxygen concentrations are the main reason for the amplification of measurement errors. Assuming that inlet concentrations are well known (air) and that gas flow measurements are accurate (mass flow meter), they obtained the equation

$$s_f' = \frac{d s_d'}{(c - d)} \tag{98}$$

where s_f' is the relative error for *OUR*, $d = (y_{1, O_2} / y_{1, \text{inert}})$, $c = (y_{0, O_2} / y_{0, \text{inert}})$, and s_d' is the relative error in measuring d. Consequently s_d' will also be equal to the error in the ratio of the corresponding ion currents. In Tab. 4 the relation between s_d' and s_f' is illustrated for a series of relative differences in oxygen concentrations at inlet and outlet $\{\Delta C_{O_2}' = 2[(y_{0, O_2} - y_{1, O_2})/(y_{0, O_2} + y_{1, O_2})]\}$.

Systematic errors in oxygen balancing may be treated as follows: Rearranging Eq. (21) gives

$$\frac{-OUR}{N_{\text{in}} y_{0, \text{inert}}} = \frac{y_{0, O_2}}{y_{0, \text{inert}}} - \frac{y_{1, O_2}}{y_{1, \text{inert}}} \tag{99}$$

Defining $f = (-OUR/N_{\text{in}} y_{0, \text{inert}})$, $c = (y_{0, O_2} / y_{0, \text{inert}})$, and $d = (y_{1, O_2} / y_{1, \text{inert}})$, we get

$$f = c - d \tag{100}$$

A deviation Δd in d and Δf in f gives relative errors $\Delta d' = \Delta d / d$ and $\Delta f' = \Delta f / f$, and thus Eq. (100) becomes

$$\Delta f' = \frac{\Delta d'}{\dfrac{c}{d} - 1} \tag{101}$$

Substituting for c/d gives

$$\Delta f' = \frac{\Delta d'}{\dfrac{y_{0,O_2}}{y_{1,O_2}} - 1} \tag{102}$$

This equation demonstrates the great sensitivity of the error in *OUR* to the relative gas composition $y_{0,O_2}/y_{1,O_2}$. As y_{1,O_2} approaches y_{0,O_2}, $\Delta f'$ increases rapidly. Some example values are listed in Tab. 4.

Tab. 4. Imprecision of Determination of *OUR* (d) According to Eqs. (98) and (102)

$\Delta C'_{O_2}$ (%)	50	10	5	1	Condition
s'_d	4.4	0.67	0.32	0.13	$s'_f = 5\%$
s'_f	1.1	3.5	16	40	$s'_d = 1\%$
$\Delta d'$	3.3	0.5	0.25	0.049	$\Delta f' = 5\%$
$\Delta f'$	1.5	9.5	19.4	98	$\Delta d' = 1\%$

$\Delta C'_{O_2}$, relative difference in concentration between inlet and outlet gas concentration
s'_d, relative statistical error in measuring the ratio of oxygen to nitrogen concentration
s'_f, relative statistical error in determination of oxygen uptake rate
$\Delta d'$, $\Delta f'$, corresponding systematic errors

From this table it can be seen that small relative differences in concentration ($\Delta C'_{O_2}$) will require very high precision in the determination of the relative peaks (ratios of ion currents for MS) in order to give sufficiently low values of s'_d and $\Delta d'$. If s'_d or $\Delta d'$ can be kept below 0.1%, *OUR* can be determined with sufficient accuracy (small enough values of s'_f and $\Delta f'$) in almost any realistic case.

For example, using Tab. 4, if the relative statistical error is to be smaller than 5% ($s'_f \leq 5\%$) and the relative difference between inlet and outlet oxygen concentration is 5% ($\Delta C'_{O_2} = 5\%$), then the relative statistical error in measuring the ratio of oxygen to nitrogen concentration must not exceed 0.32% ($s'_d \leq 0.32\%$).

It appears that systematic errors caused either by inaccurate calibration procedures, by lack of sufficiently precise gas mixtures or by slight drift in instrument signals may be more significant than statistical errors. The latter can be reduced by increased measurement time. Drift can be negligible with especially designed MS instruments or with analyzers equipped with automatic calibration.

If CO_2 and O_2 are measured by individual instruments which both have the same error, the resulting error automatically will be about double that of one instrument with an internal standard (HEINZLE et al., 1984).

4.4.4 Instantaneous Error Analysis for the Elemental Balancing Example PHB

Experiments showed that systematic errors because of instrument drift were the most significant. Especially the difference $(\phi_{O_2} + \phi_{CO_2})$ in Eq. (83) (ϕ_{O_2} is always negative and ϕ_{CO_2} positive) dominates because of an amplification factor of 8. Using Eq. (20) and assuming $N_0 = N_1$ results in a relative error in ϕ_{O_2} of

$$\Delta\phi_{O_2,\,rel} = -\frac{y_{0,O_2}\Delta y_{1,O_2,\,rel}}{y_{0,O_2} - y_{1,O_2}} \tag{103}$$

From Eq. (103) it is evident that excessively low measurements of y_{1,O_2} ($\Delta y_{1,O_2,\,rel} < 0$) result in excessively high values for estimated oxygen uptake rates.

An example calculation will illustrate this. Aeration with air ($y_0 = 0.21$) and a mole fraction of O_2 at the reactor exit of $y_{2,O_2} = 0.19$ and a relative error in the measurement of y_{1,O_2} of 1% ($\Delta y_{2,O_2,\,rel} = -0.01$) results in a relative error in the determination of ϕ_{O_2} of

$$\Delta\phi_{O_2,\,rel} = -\frac{0.21(-0.01)}{0.21 - 0.19} = 0.105$$

This error is propagated for the estimation of ϕ_P. If other measurements (ϕ_R and ϕ_{CO_2}) are correct, Eq. (83) gives

$$\Delta\phi_{P,\,rel} = 8\Delta\phi_{O_2,\,rel} \tag{104}$$

This gives an error of $\Delta\phi_{P,\,rel} = 0.84$ for the calculated example.

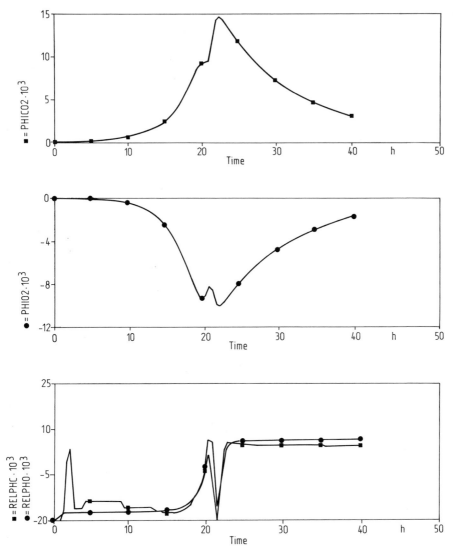

Fig. 10. Simulation for *OUR* (ϕ_{O_2}) and *CPR* (ϕ_{CO_2}) and relative errors. PHIO2, ϕ_{O_2} (mol h^{-1}); PHICO2, ϕ_{CO_2} (mol h^{-1}); RELPHO, RELPHC, relative errors for ϕ_{O_2} and for ϕ_{CO_2} (–).

Fig. 11. Comparison of true and measured values with error in the determination of y_{O_2} in the exit gas.
Left side: relative offset calibration error for y_{O_2} (yooffs = –0.01 = –1%)
Right side: relative measurement drift for y_{O_2} (yodrif = –0.0002 h^{-1} = –0.02% h^{-1})
PHIO2, PHICO2, true model values of ϕ_{O_2} and ϕ_{CO_2} (mol L^{-1} h^{-1})

MPHIO, MPHIC, measured values containing er- ▶ rors (mol L^{-1} h^{-1})
True concentration values:
R, residual biomass; S, substrate; P, product PHB all (c-mol L^{-1})
Values estimated from measurement containing error:
MR, residual biomass; MS, substrate; MP, product PHB all (c-mol L^{-1}).

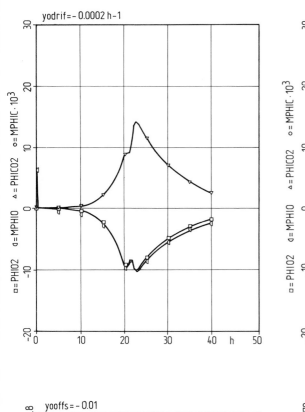

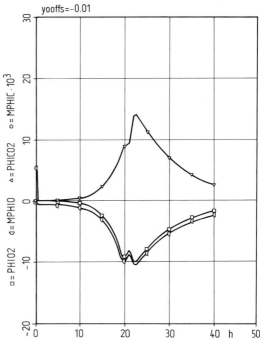

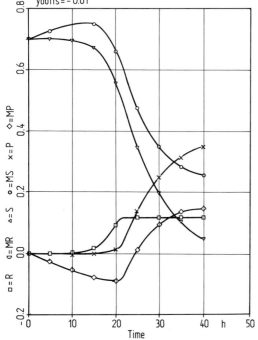

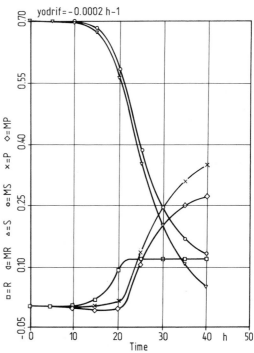

In a similar way the relative error in the determination of ϕ_{CO_2} at constant pH, accurate measurement of gas concentrations, and gas flows is calculated from Eq. (87):

$$\Delta\phi_{CO_2,rel} = \frac{\Delta\left(\left(1 + \dfrac{K}{C_{H^+}}\right)\dfrac{\Delta C_{CO_2}}{\Delta t}\right)}{\phi_{CO_2}} =$$

$$= \frac{K}{\phi_{CO_2}}\frac{\Delta C_{CO_2}}{\Delta t}\left(\frac{1}{C_{H^+}} - \frac{1}{C_{H^+} + \Delta C_{H^+}}\right) \quad (105)$$

An error in the pH measurement of 0.1 gives

$$\Delta\phi_{CO_2,rel} = \frac{3.16\cdot10^{-7}}{5\cdot10^{-6}}\frac{1\cdot10^{-3}}{3600}\cdot$$

$$\cdot\left(\frac{1}{10^{-7}} - \frac{1}{1.26\cdot10^{-7}}\right) = 0.036 \quad (106)$$

If ϕ_{O_2} and ϕ_R are measured correctly, using Eq. (83) gives a relative error in estimation of ϕ_P of

$$\Delta\phi_{P,rel} = 8\,\Delta\phi_{CO_2,rel} = 0.29 \quad (107)$$

4.4.5 Dynamic Error Analysis for Reaction Rates

The instantaneous errors in the estimation of the fluxes as explained above (Sect. 4.4.4) will vary in a batch process with time. As discussed in Sect. 3.2 the estimation of the concentration values requires integration over time, Eqs. (88) to (91), and this integration will lead to error accumulation. The best way to understand this is to consider a detailed dynamic error propagation analysis by simulation.

This was again done using the PHB example with the kinetic model of HEINZLE and LAFFERTY (1980) and following the data of HEINZLE and DETTWILER (1986). Well-mixed gas and liquid phases were assumed ($V_{G_1} = V_L \cdot 0.1$). A slight modification was to couple another well-mixed gas element ($V_{G_2} = V_L \cdot 0.3$) representing the fermenter head space. Concentration–time relations, gas reaction rates, acid production rates, and gas

flow values were created by the model. Based on these data, sugar and ammonia consumption as well as biomass and PHB production were estimated using the measured gas partial pressures, flow rates, and pH values following the procedure described in Sect. 3.2. The error model for the measurements was either that of calibration offset or of measurement drift.

Errors Caused by Steady-State Assumption

Fig. 10 gives results of simulations for the estimation of ϕ_{O_2} and ϕ_{CO_2} and corresponding relative errors (relpho, relphc) caused by steady-state assumptions for the gas phase concentrations and for dissolved oxygen. It is evident that these assumptions cause only relative errors of $<2\%$. Resulting errors vary during the process according to the process dynamics.

Errors Caused by Oxygen Measurement Offset

Fig. 11 shows true and measured values of ϕ_{O_2}, ϕ_{CO_2}, as well as true values (from simulation) of residual biomass (R = total biomass-PHB), substrate (S), and product PHB (P), and values estimated from measurements containing only errors in the determination of y_{O_2} (yooffs and yodrif). It can be seen that, in the case of calibration offset, estimation of ϕ_{O_2} also shows basically a negative offset, whereas estimation of ϕ_{CO_2} is scarcely affected. In the case of a measurement drift (right side), the

---

Fig. 12. Estimation errors caused by error in determination of y_{O_2} in the exit gas.
Left side: relative offset calibration error for y_{O_2} ($-0.01 \le$ YOOFFS ≤ 0.01)
Right side: relative measurement drift for y_{O_2} ($0.0004 \le$ YODRIF ≤ 0.0002 h^{-1})
MEPHIO, Absolute estimation error for ϕ_{O_2} (mol L^{-1} h^{-1})
REPHIO, Relative estimation error for ϕ_{O_2} $(-)$
FINERP, Estimation error for product PHB (c-mol L^{-1})
TFIN, time (h).

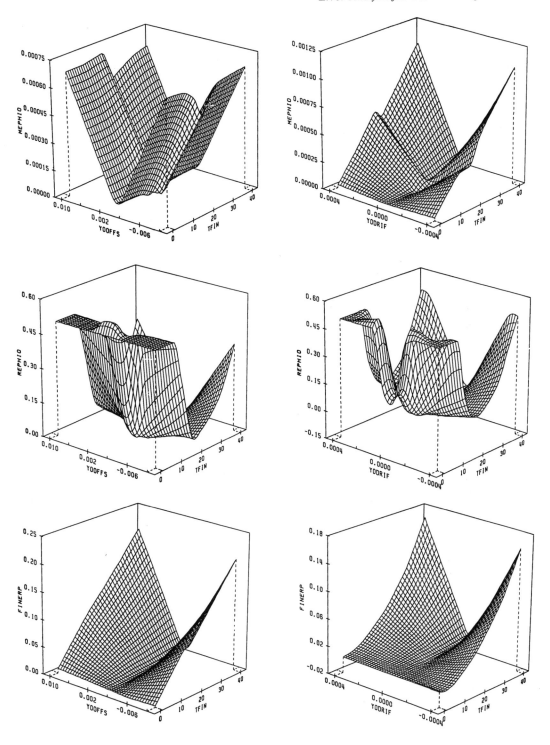

measured value for ϕ_{O_2} gradually drifts away from the true value.

The estimation of R (MR) is not affected, since this is mainly calculated from titration results and CO_2 measurements. Measured values of concentrations S and P (MS, MP) are greatly affected by errors in y_{O_2}. Comparing these results with results shown earlier (Fig. 8), it is again clear that very small errors in the estimation of y_{O_2} (<1% relative) can be amplified dramatically when elemental balancing methods are applied to estimate concentration values.

Fig. 12 gives a more general overview of the effect of an error of the measurement of y_{O_2} for offset error (left side) and for drift error (right side). In the upper row errors of measurement for ϕ_{O_2} are drawn. In the center relative values of this error are shown. In the bottom part the final error in the estimation of the product P is given. The largest values of this error are about 50% as can easily be seen in Fig. 11 for a constant offset and drift error in the measurement of y_{1,O_2}.

5 Sample Pretreatment and Multiplexing

5.1 System without Removal of Condensable Volatiles

Analyzers which are not disturbed by water vapor (e.g., MS) need a minimum of sample pretreatment. Fermentation liquid must be prevented from entering any of the gas analyzers discussed later. Because of the high gas flow rates, gas streams leaving bioreactors usually contain aerosols, which are created when gas bubbles burst at the liquid surface of the reactor. These aerosols will usually disturb the measurement. In many cases foaming is a problem, and it is not always certain that foam will remain in the reactor. Simple methods are usually sufficient to prevent foam from entering gas analyzers. Simple goose necks or surge or foam traps with an electrical contact may be used to detect liquid leaving the reactor (Fig.

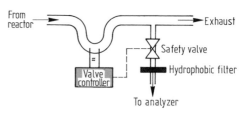

Fig. 13. Goose neck with electric contact to prevent liquid entering the analyzer.

13). As all biological fluids have high electrical conductivity, there will immediately be good electrical contact between the two electrodes. This signal can be used to shut off gas lines leading to the analyzers. It is always advisable to put a hydrophobic filter into the sampling line to guarantee that no liquids, even aerosols, will enter the analyzers. To avoid condensation of volatiles, all the lines of such a sampling system have to be heated.

5.2 Application of Paramagnetic and Infrared Analyzers to the Measurement of Oxygen and Carbon Dioxide

The most common methods of gas analysis for fermentations use paramagnetic and infrared analyzers. As seen in Fig. 14, the analysis has the following aspects:

1. The selection of the gas stream from multiple fermenters using a valve manifold
2. Gas drying
3. Gas filtration
4. Transport to the analyzer
5. Analysis with facility for frequent or automatic calibration
6. Data processing using information on flow, humidity, pressure, and temperature

The valve system for selecting the fermenter exit gas is often controlled by the data processing computer. Care must be taken to allow sufficient time between samples to remove previous samples from the lines, normally 4 times

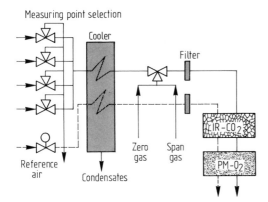

Fig. 14. Sampling and calibration system for analysis of O_2 using a paramagnetic analyzer (PM) and of CO_2 using an infrared analyzer (IR).

the residence time of the gas in the lines. This can be speeded up by evacuating the lines if necessary.

Water, especially condensate, is detrimental to the operation of analyzers, and humidity influences the calibration. It must be removed in a drying step. This usually involves a cooler with dewpoint control and condensate removal.

Gas flow to the analyzers is achieved either with overpressure control on each fermenter line or with a gas membrane pump. Since the flow rate through the analyzer will influence the calibration, care must be taken to hold this constant.

Air is usually used for calibration of the paramagnetic analyzer, and in some instruments it is compared directly with the gas sample. Carbon dioxide gas mixtures are used to calibrate the infrared analyzer. It is useful to automate this calibration procedure.

The data from the analyzers are usually reduced by the computer to rates of uptake and production, expressed in mmol/L h. As explained earlier, this requires careful attention to the measurement of flow rates and their conversion to molar quantities. Plotting the reduced data as a function of time greatly aids in process control and interpretation. The data can be stored for later use. Feedback control of the process can be programmed, for example, to control feeding rates based on oxygen uptake.

5.3 Special Valve Manifolds for Mass Spectrometers

Valve manifolds, as described before, are not ideal for fast and very accurate analysis of a number of gas streams, because it is very difficult to avoid dead zones. Therefore, a number of sample valves have recently been developed especially for fermentation gas analysis.

Balzers AG (MÜLLER and RETTINGHAUS, 1988) has recently developed a gas inlet system shown in Fig. 15. Special solenoid valves having practically no dead zones are used. The whole valve system is located in a box within which the temperature is kept usually constant at 110 °C to avoid condensation and to speed up adsorption equilibration.

It is very useful to keep the total pressure within the MS constant by controlling the flow through the capillary (Cap). This can be done by controlling the pressure at the capillary entrance using a piezo pressure sensor (p) (Fig. 15).

The sampling stream valves are under permanent flow to guarantee instant sample availability. Connections to the capillary inlet are usually closed. Switching between gas streams is done as follows: close V1 and V2, open V3 to evacuate sampling system, close V3, open V2, open one single sample stream valve (SV) or calibration gas valve (CV), open V1 to allow gas to enter the MS. The pressure at V1 is

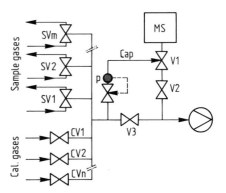

Fig. 15. Control of MS vacuum by controlling inlet gas flow. V1, MS inlet valve; V2, V3, switching valves; SV1 to SVm, sample stream valves; CV1 to CVn, calibration gas valves (see text).

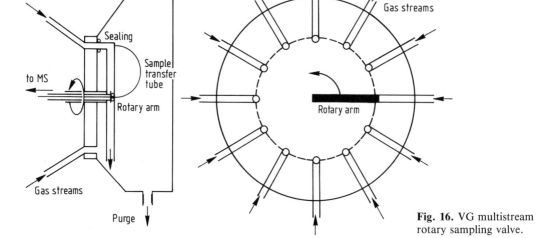

Fig. 16. VG multistream rotary sampling valve.

approximately 1 mbar. For very accurate analysis repeated evacuation and flushing with analysis gas can be done.

One problem when using solenoid valves is cross-contamination by leakage of neighboring sample stream valves. In the system shown above such leakage can quite easily be detected by monitoring pressure reduction with time when evacuating the sample system.

Conventional rotary valves, usually with flat or tapered sealing, provide a much faster sample response, but prolonged use results in wear of the sealing faces and causes leakage and innerstream contamination. This problem led VG GAS ANALYSIS SYSTEMS LTD. (1988) to develop a completely new rotary valve system shown in Fig. 16.

In this valve the sample streams are presented on the periphery of a circular flat distribution plate and vent into a cylindrical housing and from there to a common exhaust.

To sample a selected stream, the rotary arm is positioned to the sample stream. Ingress of other gas at this point is inhibited by a face sealing sleeve, which has a light spring loading. The sample flows directly into the sample transfer tube and is transferred to the static sample outlet tube. The sample bypass exhaust tube inhibits back migration of other stream exhaust into the bypass seal chamber.

6 Gas Flow Measurement

Estimation of *OUR* and *CPR* and generally of reaction rates using gas analysis requires the measurement of gas flow rates as has been described previously. In research pilot-plants and production reactors, data are usually treated automatically using computer data acquisition systems. For the on-line acquisition of flow rate data, electrical signals are required. There is an extensive literature on flow rate measurement (e.g., BAKER and POUCHOT, 1983a, b; GINESI and GREBE, 1985, 1987; CHEREMISINOFF and CHEREMISINOFF, 1988) and numerous devices are available. Three methods are discussed briefly in the following.

6.1 Positive Displacement Devices

These instruments measure the volume of gas entering the system and causing rotation of a measurement chamber or bulb, either dry or wet. The error is usually small (< 1% rel.) and within the specified measurement flow rate region (e.g., 10 to 100% of maximum flow rate). Dry versions are less prone to be affected by impurities that would change surface tension of the liquid in a wet gasometer. An electric signal is created by counting revolutions.

Measurements obviously are directly influenced by pressure and temperature. This can be compensated by using the ideal gas law. Gas composition does not influence the measurement except when gaseous components change the surface tension of a liquid gasometer. With CO_2, accumulation in the liquid gasometer fluid has to be taken into account. Such devices are not very useful at fermentation reactor inlets because of pressure fluctuations.

6.2 Rotameters

A measurement body is floated in a vertical conical tube by an upstream gas flow. The drag coefficient is a function of the Reynolds number and hence depends on the fluid viscosity. Rotameters are available in a variety of materials and can be used for any gas and flow rate. The measurement is influenced by pressure, temperature, and less by gas composition. The location of the floating body can be converted into an electric signal either by electromagnetic or optical transmission. The simple construction and low costs make rotameters very popular.

6.3 Thermal Mass Flow Monitors (MFM)

Fig. 17 illustrates two types of thermal MFMs, the immersible type and the capillary tube type (GINESI and GREBE, 1987).

Immersible thermal MFMs have two sensors immersed in the flow: one is self-heated and monitors mass flow, and the other monitors gas temperature and automatically corrects for temperature changes. Typically, each sensor is a resistance temperature detector. The sensor is driven as a constant-temperature or constant-current anemometer. The instrument has to be calibrated with the gas to be used. In fermentations this is usually dry air. This instrument has a very fast response (≈ 100 ms) and reasonable accuracy ($\approx 2\%$).

In a *capillary* MFM the total mass flow (M) is divided into two paths, one through the by-

pass (M_2) and one through the sensor tube (M_1). This tube is designed to have laminar flow conditions. Two coils surround the sensor capillary. The coils direct a constant amount of heat through the thin walls of the sensor tube into the gas. The coils sense changes in temperature through changes in their resistance. During operation, heat is transported from the upstream coil to the downstream one,

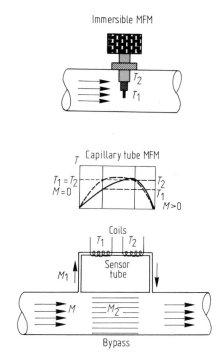

Fig. 17. Types of thermal mass flow meters (MFM): immersible MFM and capillary tube MFM. For details see text.

thus changing coil resistances. A bridge circuit with a constant current input gives an output voltage in direct proportion to the difference in resistance. This gives ΔT. The other two parameters, heat input and specific heat (C_P), are both constant. C_P is constant over a wide temperature and pressure range. The ΔT output is almost linear with M. The design of the sensor tube and the bypass is critical. The sensors are calibrated with the gas used. C_P is a function of gas composition.

7 Instruments for Analysis of Gas Composition

Most important and usually also most expensive in gas analysis are the instruments for the measurement of gas composition. From the previous chapters it is evident that by far the most important variables are oxygen and carbon dioxide. In Sect. 2 it was shown why the analysis of inert gas components (usually nitrogen and argon) and inert gas balancing greatly simplifies and improves gas analytical results.

In some cases other gases are important reactants. In anaerobic processes methane is the main product, and its analysis may be required for control of substrate feeding in order to avoid toxic overload. There is probably no current industrial process in which molecular nitrogen (N_2) is an important substrate. It is an important waste product in waste water denitrification processes, but its analysis would be quite difficult because of the high background of air nitrogen.

There is a whole group of volatile substances which are either used as substrates or are major products (Tab. 5). These products may be measured directly using liquid samples. This is, however, not ideal for online analysis. The advantages and disadvantages of direct liquid phase analysis and gas phase analysis of such volatile compounds are compared in Tab. 6.

Volatile compounds may also be interesting as by-products indicating the physiological state of a biological culture. Empirical correlations between formation of by-product and actual product may be useful for process on-line analysis (HEINZLE et al., 1985).

Aerosols are also usually present in effluent streams of bioreactors. They are created by bursting bubbles at the liquid surface and usually disturb the measurement of gas composition.

7.1 Paramagnetic Oxygen Analyzers

The principles of operation and application have been treated in several reviews (NICHOLS,

Tab. 5. Volatile Substrates and Products in Fermentation

Compounds	Importance	Reference
Ethanol	Fuel and beverages	KOSARIC et al. (1983)
Methanol	Cheap carbon substrate	FAUST and PRÄVE (1983)
Butanol	Acetone-butanol fermentation	BAHL and GOTTSCHALK (1988)
Acetone	Acetone-butanol fermentation	MCLAUGHLIN et al. (1985)
Acetoin		SHARPELL (1985)
Higher *n*-alkanes	Biomass production	EINSELE (1982)
Butanediol	Potential organic feedstock	MAGEE and KOSARIC (1987), MADDOX (1988)
Volatile acids (Acetic, propionic)	Anaerobic digestion	ERICKSON and FUNG (1988)
Acetic acid	Vinegar production	EBNER and FOLLMANN (1983)
Fragrances	Natural perfumes	SHARPELL (1985)
Water	Solid state fermentation	ZADRAZIL and GRABBE (1983)

Water vapor is present in any gas stream leaving a bioreactor system. Its measurement and control is of great importance in solid state fermentation processes since active organisms require a minimum amount of water activity.

1988; VANA, 1982; CARLEYSMITH and FOX, 1984). Oxygen has a high *magnetic susceptibility* which changes with temperature. It is paramagnetic and is thus attracted by a magnetic field at temperatures below its Curie point of 80 °C. Only nitric oxide (NO) and nitrogen

Tab. 6. Comparison of Analysis of Volatile Compounds in the Liquid and Gas Phases

Liquid phase analysis	
Advantages	Original sample
Disadvantages	Risk of infection
	Separation of solid particles required
	Complex matrix
	Imhomogeneities important because of local sampling
Gas phase analysis	
Advantages	Only minor sample treatment required
	No additional risk of infection
	Integral sample of the gassed reactor region
Disadvantages	Non-equilibrium between gas and liquid likely
	Delay due to gas–liquid transfer

dioxide have appreciable magnetic susceptibilities which could interfere.

Thermomagnetic Instruments

Heated above 80 °C, oxygen becomes diamagnetic and is repelled by a magnetic field.

Certain oxygen analyzers exploit this *thermomagnetic* property to create a convective flow or "magnetic wind" within the measurement cell, which is used to cool a heated coil; this cooling effect is detected by a Wheatstone bridge. Fluctuations in pressure will disturb the flow and cause errors. Only hydrogen has a significantly different thermal conductivity that would require special calibration of the instrument. In some instruments the flow of a stream of nitrogen across the heated filament is induced by the magnetic properties of the test gas. This avoids contact of the test gas with the hot filament and eliminates thermal conductivity interference. Other instruments utilize air as a reference gas; this also has the result of eliminating the influence of thermal conductivity. The characteristics of the individual instruments are summarized in Tab. 7.

Magnetomechanical Instruments

Some instruments use the paramagnetic properties of oxygen to create the *mechanical* deflection of a rigid test body, often a nitrogen-filled glass dumbbell, which is suspended from a fine fiber in a non-uniform magnetic field. The force on the body depends on the field intensity, the intensity gradient, and the

Tab. 7. Characteristics of Paramagnetic O_2 Analyzers

Model Manufacturer	Ranges	Repro. (% span)	Flow (L/min)	Response (s)	Drift (% week)	Air Reference
Magnos 4 G Hartmann & Braun	4	0.5	0.01–1.6	2.5	2.25	yes
1100 A Servomex	11	0.5	<0.2	5	–	no
Oxynos 100 Leybold-Heraeus	programmable	2	0.01–1.0	5	2	no
Oxygor Maihak	2	0.5	0.2–2.5	13	2	yes
Oxymat 5 Siemens	4	0.5	0.3–1	1	–	yes

Addresses of manufacturers:
Hartmann & Braun AG, Gräfstr. 97, D-6000 Frankfurt 90, FRG
Servomex Ltd., Crowborough, Sussex TN6 3DU, GB
Leybold AG, D-6450 Hanau 1, FRG
Maihak AG, Semperstr. 38, D-2000 Hamburg, FRG
Siemens AG, D-8000 München 2, Wittelsbacher Platz 2, FRG

difference between the magnetic susceptibilities of the body and the surrounding gas. The exerted force is balanced by the torque on the suspending fiber so that the angle of deflection, which is measured optically with a mirror and light beam, is proportional to the oxygen concentration in the measurement chamber. Alternatively, in a compensation-type instrument the light beam can be used to produce a compensating force with an electrical field, which results in zero deflection. Gas pressure and mechanical vibrations will influence the measurement with these devices. Tab. 7 lists the manufacturers of paramagnetic oxygen analyzers and characteristics of the devices.

Practical Problems with Paramagnetic Oxygen Analyzers

Oxygen analyzers are often pressure sensitive, except for special dual channel systems or those with automatic pressure correction. Frequent or automatic calibration is usually necessary to compensate for drift. Many instruments are sample-flow sensitive. Drying of the sample gas is required, and calibration should be made at the same humidity. In practice, the difference between inlet and outlet gas concentration is often so small as to create the need for unusually high measurement accuracy. This is often only attainable with great care in calibration and error analysis.

The actual response times of all instruments depend on the volume of gas to be replaced and not on the instrument itself. In the tables only the nominal response times for the instruments are given. Three of the oxygen analyzers provide a differential method with air as a reference; this would allow, for example, a 19–21% range. A number of instruments are fitted with microprocessors which provide for automatic calibration, automatic correction for pressure, programmable ranges, linearity correction, compensation for the interference of known gases, computer interface, and other features previously unavailable.

7.2 Infrared Analyzers

Most chemical compounds absorb infrared light. Infrared spectrometers are very important tools in analyzing the structure of chemical compounds. Infrared analyzers are presently used in a number of gas analysis instruments. Multicomponent analysis is possible, though most instruments are specifically designed for one particular component (NICHOLS, 1988; VANA, 1982; CARLEYSMITH and FOX, 1984).

Carbon dioxide is often analyzed using instruments based on infrared absorption (IR). Conveniently, compounds containing only one kind of atom, such as oxygen and nitrogen gas, do not absorb IR radiation. However, care must be taken for all hydrocarbons absorb in the infrared region. Analyzers consist of a radiation source, cuvettes for sample and reference gases, and a detector. Selectivity is obtained by use of monochromatic radiation and selective detectors. Filters, often consisting of cuvettes filled with the interfering gases, are used to obtain further selectivity. Two popular instruments use a discontinuous source, created by a chopper. This discontinuity creates cyclic pressure variations due to heating, which cause cyclic deformations of a membrane. The amplitude is proportional to the sample gas concentration and can be measured.

Infrared analyzers are available for monitoring two components and can be used in biological systems, e.g., for carbon dioxide and methane. Tab. 8 summarizes the available analyzers and their characteristics.

IR analyzers must be purchased for a specific purpose, especially when a particular range or application with complex mixtures is required. Only two IR instruments listed are capable of analyzing for two components, which would be useful for CO_2 and CH_4.

The information given in the tables is not intended as a buyer's guide, but only as an introduction to the leading manufacturers. Choice of an instrument will depend on a more detailed examination of the models available with the assistance of the detailed description from the manufacturer. Also in some cases local representation and service will make a difference.

Tab. 8. Characteristics of Infrared CO_2 Analyzers

Model Manufacturer	Ranges	Repro. (% span)	Flow L/min	Response (s)	Drift (% week)	Component
Uras 3 G Hartmann & Braun	4	0.5	0.5 to 1	8	1	1
PSA 402 Servomex	2	1	0.1–2	3	1	1
Binos 1000 Leybold-Heraeus	2	2	0–2.5	–	2	2
Unor Maihak	2	0.5	0.2–2.5	–	1	1
Ultramat 5 Siemens	4	0.5	0.5–2	–	–	2

Manufacturer addresses see Tab. 7

7.3 Mass Spectrometers

The application of mass spectrometers (MS) to fermentation gas analysis has been treated in several review articles (HEINZLE, 1987, 1988; BUCKLAND et al., 1985; WINTER, 1987; HEINZLE and REUSS, 1987). Many details contained in these articles will not be discussed here.

Sample introduction of gases is usually via a pumped capillary inlet (≈ 1 mL s^{-1}), but it also occurs over membranes connected to the high vacuum region. The latter interface type consumes much less gas and allows selective enrichment of gases of interest. It is, therefore, especially interesting for monitoring of dissolved gas and trace gas components.

The principles of operation of MS involve ionization of the molecules into fragments of various mass to charge ratios, their separation and detection, and analysis of the resulting spectra.

Ionization

The most stable and most widely used ionization technique is electron impact ionization. Because of its high energy input it creates a number of fragment peaks. The good stability and reliability come at the expense of a complex fragmentation pattern, which limits the analysis of mixtures. In the ion source, stable species M are ionized by fast electrons giving

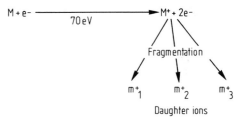

Fig. 18. Ionization of molecules (M) yielding fragment or 'daughter' ions (m_1^+, m_2^+, m_3^+).

parent and fragment or 'daughter' ions (Fig. 18).

Since each species has its own unique mass spectrum, all components of a mixture may be quantitatively determined. This feature makes mass spectrometry a universal detector.

For the analysis of mixtures it would be ideal to use a soft ionization technique, which would create only a few peaks per substance (BIER and COOKS, 1987). Recently soft ionization process instruments have been developed by two companies (ALCATEL, 1987; IONENTECHNIK, 1987).

Ion Separation

Ion separation can be done according to several principles. Magnetic sector and quadrupole instruments are both well established and for most analyses will give the same results (BARTMAN, 1987).

Quadrupole versus Magnetic Sector Mass Analyzer

In the quadrupole analyzer, mass selection is achieved through suitable adjustment of radio-frequency and d.c. electric fields. Magnetic sector analyzers depend upon the deflection of an ion beam by a magnetic field. In the single detector instrument any m/z peak of interest can be selected by suitable adjustment of the magnetic field.

In the multi-collector arrangement, each ion is exposed to the same fixed magnetic field and thus travels a circular orbit with a radius according to its mass to charge ratio. Each ion fragment is measured by small Faraday bucket collector assemblies, located along the focal plane according to which particular signals are of interest.

Multi-collector arrangements permit only the measurement of the compounds specified when the instrument is ordered. High sensitivity measurements with SEM (see below) are not possible due to the limited space.

Opinions on possible accuracies and comparisons between different instrument configurations are somewhat controversial. It has been claimed (WINTER, 1987) that for the highest precision and accuracy magnetic sector instruments with single detection capability are required. Other researchers and manufacturers claim that attainable precisions are comparable. It is quite likely that the most critical feature in reaching maximum precision and accuracy is the gas inlet system, including transfer from the reactor to the instrument, multiplexing, gas pretreatment, and pressure reduction.

Detectors

The ions can be detected either by a Faraday cup and electrometer amplifier or, for faster operation and higher sensitivity, with a secondary-electron-multiplier (SEM) and subsequent electrometer amplification. The Faraday cup is rugged and very reliable; it can be regarded as an absolute detector of ions and is typically capable of measurements down to 1–10 ppm concentration levels. The SEM detector is a fast-response, high-gain device by which sub-ppm sensitivities may be achieved.

SEM has poorer long-term stability and is generally less accurate. If whole spectra have to be analyzed, it may be necessary to use SEM to obtain sufficient speed of analysis. Because of the detection limits of the Faraday cup detector, the use of an SEM is inevitable for trace component analysis.

Using a single amplifier with automatic gain switching controlled from the data system, the single collector instrument maintains the highest precision over the widest dynamic range. In the fixed collector arrangement, each signal amplifier may show different offset or gain drift; this results in instability and drifting. In the single detector arrangement, any such drift is equal for all components and can therefore be compensated for by using any internal standard.

In the single collector arrangement (both magnetic sector and quadrupole, WINTER, 1987; MÜLLER and RETTINGHAUS, 1988), remotely switchable Faraday cup/SEM assemblies can be employed. These allow all species to be monitored – from 100% down to sub-ppm levels, the appropriate detector being chosen automatically through the software.

The limited number of detectors in the multi-collector design greatly restricts the amount of mass spectral information. This may prevent a proper analysis of complex mixtures.

To make full use of the great potential of MS, it is necessary to control it by a microprocessor and to reduce the amount of primary data (i.e., series of spectra) in order to allow its interpretation by the operator.

Spectral Analysis

Following mass analysis and ion detection, the spectral data must be reduced to the concentrations of the various components of a potentially very complex mixture. This process is known as spectral analysis.

The recorded mass spectral peaks can be represented as a column vector of various ion current signals, S, occurring at p different m/z values, due to q different components of a mixture whose concentrations are c_1–c_q:

$$s_1 = c_1 m_{11} + c_2 m_{12} + \ldots + c_q m_{1q}$$
$$s_2 = c_1 m_{21} + c_2 m_{22} + \ldots + c_q m_{2q}$$
$$\vdots \qquad \vdots \qquad \vdots \qquad \ldots \qquad \vdots$$
$$s_p = c_1 m_{p1} + c_2 m_{p2} + \ldots + c_q m_{pq} \qquad (108)$$

or

$$S = MC \qquad (109)$$

This matrix equation represents the observed spectrum as the product of the fragmentation pattern or sensitivity factor matrix M with the concentration columns vector C. Rearranging Eq. (109) gives:

$$C = M^{-1} S \qquad (110)$$

and allows the concentrations, C, to be uniquely determined.

However, since most chemical species show several mass spectral peaks, more ion current signals are generally available than there are compounds whose concentrations are to be determined ($p > q$). This requires a multi-dimensional analysis of least squares.

Interfaces for Gases

Continuous gas analysis is preferably carried out using a capillary gas inlet (Fig. 19), which provides the sample to the ion source. The dynamic response of this system is very fast. For gases such as oxygen and nitrogen the response time is much less than 1 s, for ethanol and acetone it is about 1 s. The response time may increase to several seconds when measuring less volatile compounds. Fast transients of a single gas stream can therefore be measured. In industrial practice it is more interesting to measure a number of gas streams (e.g., several gas outlets of reactors).

Calibration of the MS involves only a few calibration points using a number of known

gas mixtures, since the response is usually linear over very wide ranges of partial pressures. Analysis and calibration for all species of interest, N_2, O_2, Ar, CO_2, H_2, CH_4, C_2H_6, C_2H_4, methanol, ethanol, H_2S, NH_3, and others are easily adjusted by use of software. Calibration is necessary because of atmospheric pressure changes and normal drift; this can easily be automated.

Gas analysis using such a capillary inlet is very much dependent on the total pressure at the capillary, but as there is no enrichment in the inlet, all partial pressures are affected in the same way. This allows correction to 100% if all gases are analyzed. It is also possible to use an internal standard, e.g., an inert gas. An even better method is to control the pressure at the inlet of the capillary (e.g., 500 mbar) to keep the pressure within the high-vacuum chamber constant. This improves performance of measurement (SCHAEFER and SCHULTIS, 1987; MÜLLER and RETTINGHAUS, 1988).

Most striking is the very high precision of the results – the standard deviations for the nitrogen and oxygen concentrations are less than 0.01% (WINTER, 1987; HEINZLE, 1987). Clearly this allows the uptake of oxygen and the respiratory quotient to be determined with the highest accuracy. At its ambient level (of 330–400 ppm) carbon dioxide is determined with a precision of just a few ppm. This allows the onset of the fermentation process to be very precisely determined, for instance, as a consistent 5 ppm rise in measured CO_2 levels, etc. (WINTER, 1987).

Another possible gas inlet involves a membrane interface. This is especially useful if very small gas streams have to be measured. Another useful application is trace gas analysis with membrane enrichment of the desired compounds. The application of such systems has been discussed in a number of review articles (WEAVER and ABRAMS, 1979; BOHATKA, 1985; LLOYD et al., 1985; HEINZLE, 1987).

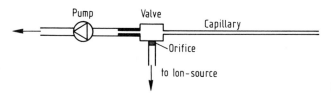

Fig. 19. Capillary process–MS-interface.

Comparison with Other Gas Analysis Instruments

MS has the following advantages over the conventional techniques in fermentation off-gas analysis (IR absorption measurement for CO_2 and paramagnetic analysis for O_2):

- Multi-component analysis is standard
- No susceptibility to moisture and other interferences
- Very large dynamic range
- Direct measurement of inert gases
- Low gas consumption (≈ 1 mL/s at 1 bar)
- Large numbers of fermentors can be monitored.

Practical Problems with Mass Spectrometers

Magnetic instruments with simultaneous multi-component detection (fixed magnetic field and acceleration voltage) must be purchased with a particular analysis problem in mind. Due to the multi-detectors, the instrument can be used only for preset mass/charge ratios whose numbers are limited to just 4 to 8 fixed mass-to-charge ratio signals. Any adjustment or extension of the analysis demands physical movement of existing detectors or acquisition of new ones. For commercial instruments this requires a return of the analyzer to the manufacturer. For process instruments this may not be a disadvantage. In a research environment the requirements constantly change. This gives an edge to the more flexible single detector scanning magnetic sector and quadrupole instruments. The higher technology of the mass spectrometers may require personnel who are not afraid of complexity, including use of computers to analyze the spectra. If the only requirement is for oxygen and carbon dioxide in a few fermentors, probably the conventional paramagnetic and IR analyzers are preferable. The well-equipped research laboratory can hardly do without mass spectrometry because of its wide variety of possible uses.

Mass spectrometers are now available with a range of capabilities and prices. Some, although perhaps too small to be considered process instruments, may exceed the capabilities and total price of the gas analysis instruments which they replace. Tab. 9 summarizes the available equipment. Only part of the listed equipment represents fully assembled process analytical instrumentation, including sample treatment, multiplexing, analysis, data treatment, and link to process computer systems.

7.4 Gas Chromatography

Gas chromatography (GC) is a separation process that employs the absorption equilibrium between an immobile liquid phase of large surface area and the components of the gaseous sample, which are transported by a carrier gas. The immobile phase is usually coated on small solid particles or within a fused silica capillary and is chosen to have an affinity for the gas sample and to exhibit the necessary differences in equilibrium to achieve separation of the sample components (GUIOCHON and GUILLEMIX, 1988).

GC equipment consists basically of a source of carrier gas, a sample injection system, a column in a temperature-controlled oven, a suitable detector for the component peaks, and a data handling system. The sample is injected in small quantities, either manually or with a sampling valve, to the GC and is separated into components peaks. A component is identified by its retention time in the column. The peak area, by integration, or sometimes peak height is calibrated to concentration, possibly with the aid of an internal standard component. Due to this non-continuous nature, the sampling and analysis time must be considered when using the GC in a control loop.

The thermal conductivity detector (TCD) and flame ionization detector (FID) are the most common detectors employed in gas chromatography. In the following these detectors and their applications will be described (GUIOCHON and GUILLEMIN, 1988).

Tab. 9. Mass Spectrometers for Gas Analysis. Precision for Single Stream Analysis of Oxygen in Air (% absolute); for Switching Time no Accuracy Value is Specific (Specified Absolute Accuracy). Calibration Interval Recommended. M – magnetic sector, Q – quadrupole, FB – Faraday bucket, SEM – secondary electron multiplier

Model Manufacturer	Ion Sep.	Ion Detection	Precision (% abs.)	Switching Time (min)	Cal. Interv. (h)	Reference No.
MGA 1200 Perkin-Elmer	M	Multi-FB	0.5 to 1	0.1 (<1%)	⩾24	1
PGM 407 Balzers	Q	Single-FB/ SEM	<0.02	0.5 (<0.02%)	–	2
MM 8-80, Prima 600 VG Gas Analysis	M	Single-FB/ SEM	<0.01	0.5	24	3
Micromass PC VG Quadrupoles	Q	Single-FB/ SEM	–	–	–	4
Quester Extrel	Q	Single-FB	<0.5 (% rel.)	0.3	–	5
MSQ 1003 Leybold-Heraeus	Q	–	–	–	–	6
Atomki	Q	Single-FB/ SEM	–	–	–	7
Bioquad Spectramass	Q	Single-FB/ SEM	–	–	–	8
Unigas 325 TT UTI	Q	Single-FB	–	–	–	9
ANAGAZ 200 Kenos (Nermag)	Q	Single-FB	–	–	–	10
DSMS Hiden	Q	Single-FB/ SEM	<0.5 (% rel.)	–	–	11
Nuclide 3-60-G Nuclide	M	Single-FB/ SEM	–	–	–	12
MAT 271/45 Finnigan MAT	M	Single-FB/ SEM	≪0.01	–	–	13

References: (1) MGA 1200 – Perkin-Elmer, Norwalk, Connecticut, USA; BUCKLAND et al. (1985); SCHAEFER and SCHULTIS (1987). (2) PGM 407 – Balzers AG, Balzers, Liechtenstein; MÜLLER and RETTINGHAUS (1988); HEINZLE et al. (1990). (3) MM 8-80 – VG Gas Analysis System Ltd., Aston Way, Middlewich, Cheshire CW10 0HT, England; WINTER (1987); Prima 600 – VG Gas Analysis Systems Ltd. (4) Micromass PC – VG Quadrupoles, Aston Way, Middlewich, Cheshire CW10 0HT, England. (5) Questor – Extrel Corporation, 240 Alpha Drive, Box 11512, Pittsburgh, PA 15238, USA; BARTMAN (1987). (6) MSQ 1003 – Leybold Heraeus GmbH, Bonner Strasse 498, PO-Box 510760, D-5000 Köln 51, FRG. (7) BOHATKA et al. (1987); Atomki, Bem ter 18/C, H-4001 Debrecen, Hungary. (8) Bioquad – Spectramass Ltd., Radnor Park Industrial Estate, Back Lane, Congleton, Cheshire CW12 4XR, England. (9) Unigas 325 TT – UTI Instruments Comp., 325 N. Mathilda Av., PO-Box 519, Sunnyvale, CA 94088-3519, USA. (10) ANAGAZ 200 – Kenos Analyse, 199, avenue du Marechal-Foch – Elisabethville, 78410 Aubergenville, France. (11) DSMS – Hiden Analytical Ltd., 231 Europe Boulevard, Warrington, Cheshire WA5 5TN, England. (12) Nuclide 3-60-G – MAAS, Inc., Nuclide Div., 1155 Zion Road, Bellefonte, PA 16823, USA. (13) MAT 271/45 – Finnigan MAT, 355 River Oaks Parkway, San José, CA 95134, USA

Thermal Conductivity Detector (TCD)

The TCD measures the transfer of heat from an electrically heated element (resistance wire or thermistor) located in a gas cavity within a metallic block. The rate of heat transfer depends on the thermal conductivity of the gas. Usually a differential method is used in which the heat transfer of the sample gas is compared with the transfer of the carrier gas. Thus, a Wheatstone bridge is used to balance the current through the sample element to restore its temperature to that of the reference. Resistance wires (tungsten) are used for ambient and low temperatures, while thermistors (metallic oxides) are used for high temperatures. Following the response of the eluted peaks requires detectors of small internal volume. The gas flow must be kept constant, but operation over a wide range is possible.

Calibration gives a generally linear relation between concentration and peak area. A detection threshold of 30 ppm can generally be obtained. The most important sources of error are caused by temperature fluctuations, excessive gas flows, and mechanical vibrations. Its most important application is for gases to which the FID does not respond (O_2, CO_2, N_2, H_2, Ar, NH_3, H_2O).

Flame Ionization Detector (FID)

This detector involves mixing the detector gas with hydrogen and burning in air or oxygen. Ionization of the sample causes a current to flow between two electrodes. Very high linearity and sensitivity (few ppb) are achieved between the measured voltage and the concentration of gas. The peak area is calibrated to the concentration. The FID is also sometimes used without GC, for example, to detect alcohol in an off-gas. There is no response to O_2, CO_2, N_2, H_2, Ar, NH_3, H_2O, and simple organic compounds such as formic acid and formaldehyde. It is advantageous that inorganic compounds, such as water, air, and ammonia, have no response. The FID is simple and can be considered the most universal detector for GC, detecting almost any organic compound.

GC for Gas Analysis in Fermentation Processes

The use of GC for the analysis of oxygen and carbon dioxide in fermentation off-gas is not very common, but reliable results can be obtained (example in Fig. 20). In this case a Hewlett-Packard GC, Model 5890, was used.

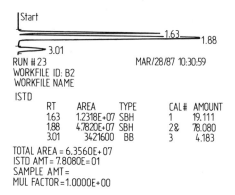

Fig. 20. Gas chromatogram of *Bacillus subtilis* culture gas sample. HP 5890A – Poropak Q and molecular sieve columns. Dried gas stream 0.2 mL/min. Automatic gas injection valve. $T = 30\,°C$, $F_{He} = 25$ mL/min. The peaks are: 1.63, O_2; 1.88, N_2; 3.01, CO_2.

It was equipped with two packed columns (Poropak Q and a molecular sieve), a gas sampling valve, and a TCD detector. The dried gas sample passed first through the Poropak column. The high retention time of CO_2 in the column allowed this component to be separated from the remaining N_2 and O_2, which were separated in the molecular sieve column. The system was programmed to switch the outlet of the first column to prevent CO_2 from entering the second column (Fig. 21).

Most of the companies producing GC equipment offer suitable systems for fermentation off-gas analysis. Process gas chromatography has found application in fermentation off-gas analysis, but the long analysis time (several minutes) and relative expense prevent optimum on-line measurement and control. GC analysis has been used to control glucose feed to a continuous culture of yeast (Spruy-

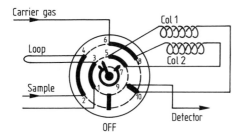

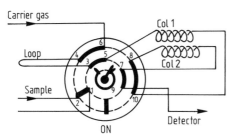

Fig. 21. Column switching for the gas chromatographic analysis of air containing CO_2. Col 1, Poropak Q; Col 2, molecular sieve.

TENBURG et al., 1979) and to monitor volatile compounds on-line (PONS and ENGASSER, 1988).

Recent developments in applying silicon micromachining technology to the design of GC have resulted in a fast analyzing instrument (MICROSENSOR TECHNOLOGY, 1988). The analysis is usually completed within 30 s. This is also true for a number of volatile compounds. The specified repeatability is $\pm 2\%$.

7.5 Flame Ionization Detector

This sensitive analyzer is based on the ionization of the sample in a hydrogen flame. It is most frequently used as a detector in GC and has therefore been described in some detail in the previous section on GC. The positively and negatively charged ions migrate to their respective electrodes, thus producing a current proportional to their concentration and to the number of carbon atoms in the molecule.

It has been used as a stand-alone instrument without chromatographic separation to deter-

mine ethanol content of the off-gas in fermentations or in the liquid by using a carrier gas-tubing method (DAIRAKU and YAMANE, 1979; HEINZLE et al., 1981). Oxygen and carbon dioxide do not interfere. Since it is also insensitive to H_2S and CO_2, it could be used to monitor methane in anaerobic fermentations.

7.6 Electrochemical Analyzers

Electrochemical analyzers are especially popular for measurement of dissolved O_2 (HITCHMAN, 1978) and CO_2 (PUHAR et al., 1980). These electrodes can be used to measure gas partial pressures, but their accuracy is low.

Based on similar principles, two types of electrochemical analyzers are also available for gas analysis (NICHOLS, 1988). A fuel cell amperometric device using cathodic reduction of O_2 to OH^- and anodic oxidation of Pb to Pb^{2+} is offered (Systech, Wellington Street, Thame, Oxfordshire, England; Teledyne Analytical Instruments, 16830 Chestnut Street, City of Industry, CA 91749, USA). This instrument has a wide dynamic range detecting as low as 0.01 ppm and up to 30% of oxygen. The accuracy is 0.1% O_2.

Another instrument using a zirconia ceramic electrolyte is of the potentiometric type. An electrochemical potential is generated which is proportional to the logarithm of the oxygen partial pressure in the gas stream. The measuring cell must be operated at 600 to 800 °C. It is especially useful for low concentrations of oxygen. The attainable accuracies would be sufficient for many applications ($\pm 0.1\%$ in the %-range).

7.7 Semiconductor Devices

Special semiconducting devices are influenced by chemical environmental variables. Semiconductors may respond to oxygen (e.g., Senox 1050, Häfele, Umweltverfahrenstechnik, Durlacher Allee, D-7500 Karlsruhe, FRG, accuracy $\pm 2\%$ relative), but usually the required accuracy cannot be obtained.

More interesting developments have occurred for monitoring volatile organic compounds such as ethanol. Such sensors have been used for control (VORLOP et al., 1984; AXELSSON, 1988). Vogelbusch (PO Box 52, A-1051, Vienna) offers complete equipment for control of molasses feed to ethanol-producing fermentation and for vinegar production based on such semiconductor sensors.

Palladium-MOSFETs are sensitive to hydrogen (DANIELSSON et al., 1984) and can, therefore, be used for monitoring this gas in corresponding biochemical reactions. This sensor could be useful in monitoring the key intermediate hydrogen in anaerobic digestion. Unfortunately, H_2S, which usually is present in anaerobic digestion, interferes with the measurement.

Gas phase biosensors (GUILBAULT and LUONG, 1988) have an interesting potential in high sensitivity monitoring of trace gases. This may be a future method for sensing fermentation components with very low vapor pressure.

8 Examples of Application

Numerous examples in the literature describe monitoring of traces of off-gases during fermentations. Many examples involve the estimation of oxygen uptake rate (*OUR*), carbon dioxide production rate (*CPR*), and usually also the respiratory quotient (*RQ*), but in almost all cases neglecting any deviation from steady state and also neglecting CO_2 accumulation as HCO_3^- in the liquid phase. Rather few examples exist in which the gas analysis results have been used for further process analysis and eventual process control.

Stoichiometric relations have especially been used for primary metabolite production, where metabolic pathways usually are well established.

Examples are:

8.1 Ethanol Production

If no analytic method of direct on-line measurement for ethanol in yeast cultures is available, gas balancing can be used to get an esti-

mation of ethanol production. The metabolic basis is that complete oxidation of sugar gives a respiratory quotient of nearly 1.0 according to the following reaction:

$$C_6H_{12}O_6 + O_2 \rightarrow 6\,CO_2 + 6\,H_2O \tag{111}$$

Equimolar amounts of oxygen and CO_2 are consumed and produced. Ethanol is produced according to

$$C_6H_{12}O_6 \rightarrow 2\,C_2H_5OH + 2\,CO_2 \tag{112}$$

For this reaction the *RQ* would be infinitely high. Summing these two equations, we can see that the amount of ethanol produced is proportional to the difference of CO_2 production and oxygen consumption and that the rate of ethanol production correspondingly would be proportional to the difference of CO_2-production rate minus O_2 consumption rate

$$Q_E = Q_{CO_2} - Q_{O_2} \tag{113}$$

The respiratory quotient $(RQ = CPR/OUR = Q_{CO_2}/Q_{O_2})$ can be used to control sugar feed to a glucose-repressed baker's yeast culture (WANG et al., 1979; SPRUYTENBURG et al., 1979) to avoid accumulation of ethanol.

8.2 Mass Balancing in a Penicillin Fed-Batch Fermentation by CO_2 Measurements
(MOU and COONEY, 1983)

In this example biomass production was found to be proportional to CO_2 production rate (*CPR*)

$$X_t = Y_{X/CO_2} \int_0^t CPR\,dt \tag{114}$$

where Y_{X,CO_2} is the yield coefficient (g cell/mmol CO_2) and X_t is the biomass concentration at time t.

By making a carbon balance for all significant components (substrate, precursor, CO_2, penicillin, and cell mass) and using an empirical correlation for penicillin production, the following equation was obtained:

$$X_t = \frac{C_S + C_{PAA} - C_{CO_2} - C_{pen}}{C_X} \qquad (115)$$

where C_S is the carbon in the substrate consumed; C_{PAA} is the carbon in the precurser fed; C_{CO_2} is the carbon in CO_2; C_{pen} is the carbon in penicillin, and C_X is the carbon content in cells, which was found to remain constant during fermentation.

Penicillin contributed only slightly to the carbon balance, and therefore an experimentally found correlation could be used with sufficient accuracy:

$$P_t = q_p \int_{t_1}^{t} X \, dt \qquad (116)$$

where t_1 is the time of transition from fast to slow growth, when synthesis of penicillin starts. q_p is an experimentally determined constant.

The biomass concentration is obtained from Eq. (114). This estimation procedure was used to estimate biomass continuously online and further, to control glucose feed to the culture.

8.3 Further Examples

Other examples of using indirect measurements for control are:

1. Using the CO_2 production rate to control the feeding of carbon substrate for cellulase production by *Trichoderma reesei* cultures. (ALLEN and MORTENSEN, 1981)
2. Using the liquid phase ethanol concentration as measured with porous teflon tubing and carrier gas with an FID detector to control glucose addition to a baker's yeast fed-batch culture. (DAIRAKU et al., 1983)
3. Monitoring product concentration by fingerprinting volatiles in fermentations (HEINZLE et al., 1985). Monitoring of volatiles can be achieved (besides using an experienced nose) by automatic head space gas analysis using gas chromatography or by using mass spectrometry (MS) either with membrane probes in

the gas or liquid phase or using a direct gas inlet. Measurements with a direct MS-membrane inlet were made in a series of bioprocesses. Spectra of volatiles showed correlation with product formation.

Reviews have appeared recently covering various aspects of the literature on the direct and indirect measurement of fermentation variables, together with their use in process control (ZABRISKIE, 1985; PARULEKAR and LIM, 1985; YAMANE and SHIMIZU, 1984; ROELS, 1983).

9 References

ALCATEL (1987), *Gas-SIMS Massenspektrometrie*, D-6980 Wertheim a.M., FRG: Alcatel Hochvakuum GmbH.

ALLEN, A. L., MORTENSEN, R. E. (1981), Production of cellulase from *Trichoderma reesei* in fed-batch fermentation from soluble carbon sources, *Biotechnol. Bioeng.* **23**, 2641–2645.

AXELSSON, J. P. (1988), Experimental techniques and data treatment for studying the dynamics of ethanol production/consumption in baker's yeast, *Anal. Chim. Acta* **213**, 151–163.

BAHL, H., GOTTSCHALK, G. (1988), Microbial production of butanol/acetone, in: *Biotechnology*, 1st Ed., Vol. 6b, pp. 1–30 (REHM, H.-J., REED, G., Eds.). Weinheim: VCH.

BAKER, W. C., POUCHOT, J. F. (1983a), The measurement of gas flow, Part 1, *J. Air Pollut. Control Assoc* **33**, 66–72.

BAKER, W. C., POUCHOT J. F. (1983b), The measurement of gas flow, Part 2, *J. Air Pollut. Control Assoc.* **33**, 156–162.

BARTMAN, C. D. (1987), The application of quadrupole mass spectrometer to biotechnology process control, in: *Mass Spectrometry in Biotechnological Process Analysis and Control*, pp. 49–62 (HEINZLE, E., REUSS, M., Eds.). New York: Plenum Press.

BIER, M. E., COOKS, R. G. (1987), Membrane interface for selective introduction of volatile compounds directly into the ionization chamber of a mass spectrometer, *Anal. Chem.* **59**, 597–601.

BOHATKA, S. (1985), Quadrupole mass spectrometric measurement of dissolved and free gases, in: *Gas Enzymology*, pp. 1–16. (DEGN, H., et al., Eds.). Dordrecht: D. Reidel Publ. Company.

BOHATKA, S., SZILAGYI, J., LANGER, G. (1987),

Application of MS to industrial fermentation, in: *Mass Spectrometry in Biotechnological Process Analysis and Control*, pp. 115–124 (HEINZLE, E., REUSS, M., Eds.). New York: Plenum Press.

BUCKLAND, B., BRIX, T., FASTERT, H., CIBEWO-NYO, K., HUNT, G., DEEPAK, J. (1985), Fermentation exhaust gas analysis using mass spectrometry, *Biotechnology* **3**, 982–988.

CARLEYSMITH, S. W., FOX, R. I. (1984), Fermenter instrumentation and control, *Adv. Biotechnol. Processes* **3**, 1–51.

CHEREMISINOFF, N. P., CHEREMISINOFF, P. N. (1988), *Flow Measurement for Engineers and Scientists*. New York: Marcel Dekker.

COONEY, C. L., WANG, D. I. C., MATELES, R. I. (1969), Measurement of heat evolution and correlation with oxygen consumption during microbial growth, *Biotechnol. Bioeng.* **11**, 269.

DAIRAKU, K., YAMANE, T. (1979), Use of a porous Teflon tubing method to measure gaseous or volatile substances dissolved in fermentation liquids, *Biotechnol. Bioeng.* **21**, 1671–1676.

DAIRAKU, K., IZUMOTO, E., MORIKAWA, H., SHIOYA, S., TAKAMATSU, T. (1983), An advanced micro-computer coupled control system in a baker's yeast fed-batch culture using a tubing method, *J. Ferment. Technol.* **61**, 189–196.

DANG, N. D. P., KARRER, D. A., DUNN, I. J. (1979), Oxygen transfer coefficients by dynamic model moment analysis, *Biotechnol. Bioeng.* **19**, 853–860.

DANIELSSON, B., WINQUIST, F., LUNDSTRÖM, I., SPETZ, A., MOSBACH, K. (1984), Process monitoring and bioanalysis with hydrogen- and ammonia-sensitive semiconductors, in: *Third European Congress on Biotechnology*, Vol. 3, pp. 621–626. Weinheim: Verlag Chemie.

EBNER, H., FOLLMANN, H. (1983), *Acetic acid*, in: *Biotechnology*, 1st Ed., Vol. 3, pp. 387–408 (REHM, H.-J., REED, G., Eds.). Weinheim: Verlag Chemie.

ERICKSON, L. E. (1988), Bioenergetics and yields for anaerobic digestion, in: *Handbook on Anaerobic Fermentations*, pp. 325–344 (ERICKSON, L. E., FUNG, D. Y.-C., Eds.). New York: Marcel Dekker.

FAUST, U., PRÄVE, P. (1983), Biomass from methane and methanol, in: *Biotechnology*, 1st Ed., Vol. 3, pp. 83–108 (REHM, H.-J., REED, G. Eds.). Weinheim: Verlag Chemie.

GINESI, D., GREBE, G. (1985), Flowmeter selection. A comparison of performance features vs. economic costs, *Adv. Instrum.* **40**, 1173–1190.

GINESI, D., GREBE, G. (1987), Flowmeters, a performance review, *Chem. Eng. (N. Y.)* **94**, 102–118.

GRIOT, M., SANER, U., HEINZLE, E., DUNN, I. J., BOURNE, J. R. (1988), Fermenter scale up using an oxygen-sensitive culture, *Chem. Eng. Sci.* **43**, 1903–1908.

GUILBAULT, G. G., LUONG, J. H. (1988), Gas phase biosensors, *J. Biotechnol.* **9**, 1–10.

GUIOCHON, G., GUILLEMIX, C. L. (1988), *Quantitative Gas Chromatography*. Amsterdam: Elsevier.

HEIJNEN, I., ROELS, A. J., STOUTHAMER, A. H. (1979), Application of balancing methods in modeling the penicillin fermentation, *Biotechnol. Bioeng.* **21**, 2175–2201.

HEINZLE, E. (1987), Mass spectrometry for on-line monitoring of biotechnological processes, *Adv. Biochem. Eng.* **35**, 2–45.

HEINZLE, E. (1988), Estimation of metabolites in bioprocesses using mass spectrometry, in: *Proc. 8th Int. Biotechnology Symp.*, Vol. 1, pp. 503–512 (DURAND, G. et al., Eds.). Paris: Societé Française de Microbiologie.

HEINZLE, E., DETTWILER, B. (1986), Einsatz von Elementbilanzen zur on-line Analyse von Fermentationen: Produktion von Poly-β-Hydroxybuttersäure, in: *4. DECHEMA Jahrestagung der Biotechnologen*. Frankfurt a.M.: DECHEMA.

HEINZLE, E., LAFFERTY, R. M. (1980), A kinetic model for growth and synthesis of poly-β-hydroxybutyric acid (PHB) in *Alcaligenes eutrophus* strain H 16, *Eur. J. Appl. Microbiol. Biotechnol.* **11**, 17–22.

HEINZLE, E., REUSS, M. (1987), *Mass Spectrometry in Biotechnological Process Analysis and Control*. New York: Plenum Press.

HEINZLE, E., BOLZERN, O., DUNN, I. J., BOURNE, J. R. (1981), A porous membrane-carrier gas measurement system for dissolved gases and volatiles in fermentation systems, in: *Advances in Biotechnology*, Vol. 1, pp. 439–444 (MOO-YOUNG, M., ROBINSON, C. W., VEZINA, C., Eds.). Toronto: Pergamon Press.

HEINZLE, E., MOES, J., GRIOT, M., KRAMER, H., DUNN, I. J., BOURNE, J. R. (1984), On-line mass spectrometry in fermentation, *Anal. Chim. Acta* **163**, 219–229.

HEINZLE, E., KRAMER, H., DUNN, I. J. (1985), State analysis of fermentation using a mass spectrometer with membrane probe, *Biotechnol. Bioeng.* **27**, 238–246.

HEINZLE, E., OEGGERLI, A., DETTWILER, B. (1990), On-line fermentation gas analysis: Error analysis and application of mass spectrometry, *Anal. Chim. Acta,* in press.

HERON, A. E., WILSON, H. N. (1959), Gas analysis, *Comprehensive Anal. Chem.* **1A**, 236–327.

HITCHMAN, M. L. (1978), *Measurement of Dissolved Oxygen*. New York: Wiley.

IONENTECHNIK GmbH (1987), *GASP Abgas-Analysengerät*. Innsbruck, Austria.

JOSHI, J. B., PANDIT, A. B., SHARMA, M. M. (1982), Mechanically agitated gas-liquid reactors, *Chem. Eng. Sci.* **37**, 813–844.

JURECIC, R., BEROVIC, M., STEINER, W., KOLOINI, T. (1984), Mass transfer in aerated fermentation broths in a stirred tank reactor, *Can J. Chem. Eng.* **62**, 334–339.

KOSARIC, N., WIECZOREK, A., COSENTINO, G. P., MAGEE, R. J., PRENOSIL, J. E. (1983), *Ethanol fermentation*, in: Biotechnology, 1st Ed., Vol. 3, pp. 257–386 (REHM, H.-J., REED, G., Eds.). Weinheim: Verlag Chemie.

KREYSZIG, E. (1970), *Introductory Mathematical Statistics. Principles and Methods*. New York: Wiley.

LLOYD, D., BOHATKA, S., SZILAGYI, J. (1985), Quadrupole mass spectrometry in the monitoring and control of fermentations, *Biosensors 1,* 179–212.

MADDOX, I. S. (1988), Microbial production of 2,3-butanediol, in: *Biotechnology*, 1st Ed., Vol. 6 b, pp. 31–50 (REHM, H.-J., REED, G., Eds.). Weinheim: VCH.

MERTEN, O.-W., PALFI, G. E., STEINER, J. (1986), On-line determination of biochemical/physiological parameters in the fermentation of animal cells in a continuous or discontinuous mode, *Adv. Biotechnol. Processes* **6**, 111–178.

MICROSENSOR TECHNOLOGY Inc. (1988), *M200 Microsensor Gas Analyzer*. 41762 Christy Street, Fremont, CA 94538, USA.

MOU, D. G., COONEY, C. L. (1983), Growth monitoring and control through computer-aided on-line mass balancing in a fed-batch penicillin fermentation, *Biotechnol. Bioeng.* **25**, 225–255.

MÜLLER, N., RETTINGHAUS, G. (1988), Measures for precise process gas analysis with a quadrupole mass spectrometer, *11th International Mass Spectrometry Conference*, Bordeaux.

NEUBERT, M., MINKEVICH, I. G. (1984), Microbial gas balance measurements: basic interrelations and error estimation, *Acta Biotechnol.* **4**, 313–322.

NICHOLS, G. D. (1988), *On-line Process Analyzers*. New York: J. Wiley & Sons.

OOSTERHUIS, N. M. G. (1984), *Scale-up of Bioreactors, a Scale-down Approach*, Ph. D. Thesis, Delft University of Technology, The Netherlands.

PARULEKAR, S. J., LIM, H. C. (1985), Modeling, optimization and control of semi-batch bioreactors, *Adv. Biochem. Eng. Biotechnol.* **32**, 207–258.

PONS, M. N., ENGASSER, J. M. (1988), Monitoring of alcoholic fed-batch cultures by gas chromatography via a gas-permeable membrane, *Anal. Chim. Acta.* **213**, 231–236.

PUHAR, E., EINSELE, A., BUEHLER, H., INGOLD, W. (1980), Steam sterilizable pCO_2 electrode, *Biotechnol. Bioeng.* **22**, 2411–2416.

REUSS, M., FRÖHLICH, S., KRAMER, B., MESSERSCHMIDT, K., NIEBELSCHÜTZ, H. (1984), Mathematical modelling of microbial kinetics and oxygen transfer for gluconic acid production with *Aspergillus niger*, in: *Third European Congress on Biotechnology*, Vol. 2, pp. 455–460. Weinheim: Verlag Chemie.

ROELS, J. A. (1983), *Energetics and Kinetics in Biotechnology*. Amsterdam: Elsevier Biomedical Press.

ROSS, A., SCHÜGERL, K. (1987), Application of elemental balances to complex systems, in: *Proc. 4th Eur. Congr. Biotechnology*, Vol. 3, pp. 94–95 (NEIJSSEL, O. M., VAN DER MEER, R. R., LUYBEN, K. C. A. M., Eds.). Amsterdam: Elsevier.

RUCHTI, G., DUNN, I. J., BOURNE, J. R., V. STOCKAR, U. (1985), Practical guidelines for determining oxygen transfer coefficients with the sulfite oxidation method, *Chem. Eng. J.* **30**, 29–38.

SCHAEFER, K., SCHULTIS, M. (1987), A microprocessor controlled multiple inlet system for mass spectrometers in biotechnology, in: *Mass Spectrometry in Biotechnological Process Analysis*, pp. 39–48 (HEINZLE, E., REUSS, M., Eds.). New York: Plenum Press.

SCHNEIDER, K., FRISCHKNECHT, K. (1977), Determination of the O_2 and CO_2 $K_L a$ values in fermenters with dynamic method measuring the step responses in the gas phase, *J. Appl. Chem. Biotechnol.* **17**, 631–642.

SPRUYTENBURG, R., DUNN, I. J., BOURNE, J. R. (1979), Computer control of glucose feed to a continuous culture of *Saccharomyces cerevisiae* using the respiratory quotient, *Biotechnol. Bioeng. Symp.* **9**, 359.

VANA, J. (1982), *Gas and liquid analysis, Comprehensive Anal. Chem.* **17**, 1–742.

VG GAS ANALYSIS SYSTEMS Ltd. (1988), *MM8-80 Process Gas Analysis Mass Spectrometer*. Middlewich, Cheshire CW10OHT, England.

VORLOP, K. D., BECKE, J. W., STOCK, J., KLEIN, J. (1984), Semiconductor gas sensors in bioreactor control, *Anal. Chim. Acta.* **163**, 287–291.

WANG, H. Y., COONEY, C. L., WANG, D. I. C. (1979), Computer control of baker's yeast production, *Biotechnol. Bioeng.* **21**, 975–995.

WEAVER, J. C. (1977), Possible biomedical applications of the volatile enzyme product method, in: *Biomedical Applications of Immobilized Enzymes and Proteins*, Vol. 2, pp. 207–225

(CHANG, T. M. S., Ed.). New York: Plenum Press.

WEAVER, J. C., ABRAMS, J. H. (1979), Use of a variable pH interface to a mass spectrometer for the measurement of dissolved volatile compounds, *Rev. Sci. Instrum.* **50**, 478–481.

WINTER, M. J. (1987), The application of single detector sector mass spectrometer systems in fermentation off-gas analysis, in: *Mass Spectrometry in Biotechnological Process Analysis*, pp. 17–38 (HEINZLE, E., REUSS, M., Eds.). New York: Plenum Press.

YAMANE, T., SHIMIZU, S. (1984), Fed-batch techniques in microbial processes, *Adv. Biochem. Eng. Biotechnol.* **30**, 147–194.

ZABRISKIE, D. W. (1985), Data analysis, in: *Comprehensive Biotechnology*, Vol. 2, pp. 175–190 (MOO-YOUNG, M., Ed.). Oxford: Pergamon Press.

ZADRAZIL, F., GRABBE, K. (1983), Edible mushrooms, in: *Biotechnology*, 1st Ed., Vol. 3, pp. 145–189 (REHM, H.-J., REED, G., Eds.). Weinheim: Verlag Chemie.

3 Biosensors

BO MATTIASSON

Lund, Sweden

1 Biosensors
– a Definition

Biosensors are man-made sensing devices constructed by combining the specificity of biomolecules with the signal transducing and processing capability of components of electrical and/or optical origin. The biomolecules used in this context all offer unique properties in terms of specificity and, in some cases, binding strength. In nature, many of these are involved in signal registration and transmission. Biosensors can thus be regarded as man-made imitations of principles that have been long applied in nature and are now being used in various analytical contexts. A schematic presentation of the biosensor concept is given in Fig. 1.

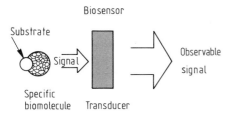

Fig. 1. Schematic presentation of a biosensor, with the biocomponent imparting biospecificity and a transducer registering the signal and transforming it into an observable signal.

It should be mentioned that there are also other definitions of biosensors. The broadest definition involves any kind of measurement made on a biological system. Since in that case almost any kind of analytical device can be regarded as a biosensor, as long as it is used in studying biological systems, the first, more restricted definition will be employed.

2 The Biocomponent
of the Biosensor

As seen from Fig. 1 the specificity lies in the biocomponent of the sensor. When recognition has taken place the signal may be transferred either as it is or after proper processing. It is thus important to investigate what kind of biological entities are at hand and what their characteristics are. The biochemical literature contains a rather impressive spectrum of various types of molecular recognition pairs (Tab. 1). A common theme when discussing these entities is their relative lability. In relation to many other chemical structures these biological macromolecules are labile and might easily denature, e.g., by unfolding or by proteolytic attack. These constraints on the molecular level have formed the basis for the very special handling of the biocomponent in the biosensor concept.

2.1 Biocatalysts

2.1.1 Enzymes

The most commonly used binding pair is enzyme-substrate. Large numbers of enzymes are available. So far more than 2000 enzymes have been characterized and listed in the enzyme nomenclature handbook (IUB, 1979). There are many more that have been studied, but so far not isolated in pure form and characterized. In other words, there are many enzyme-substrate interactions that can be used for the monitoring of many different substrates. Enzymes can also be used to detect and even quantify inhibitors. Such molecules influence the catalytic process. By exposing the enzyme to substrate under controlled conditions, the influence of an inhibitor may be recognized as an altered enzyme activity. Substances activating the enzyme may likewise be monitored in this way. In general such interactions can be used for quantitation down to 10^{-6} mol/L for substrates and competitive inhibitors. The measurement of noncompetitive inhibitors is more sensitive by a few orders of magnitude.

Tab. 1. Association Constants for Some Naturally Occurring Reactant Pairs

Reactant Pair	K_{ass} (mol/L)	Application	Reference
Avidin–biotin	10^{15}	Amplifier in immunoassays	GREEN, 1963
Antibody–hapten	10^5–10^{11}	Immunoassays	PARKER, 1976
Protein A–Fc Region of IgG	10^6	Immunoassays	LANCET et al., 1978
Lectin–carbohydrate			HUBBARD and GOHN, 1976
Simple sugars	10^3–10^4		
Multipoint attachment	10^6–10^7		

Tab. 2. Biosensors Based on Coupled Enzyme Sequences

Target	Enzyme Sequence	Product	Transducer	Reference
Sucrose	Invertase + mutarotase + glucose oxidase	H_2O_2	Polarographic oxygen electrode	OLSSON et al., 1986
Lactose	β-Galactosidase + glucose oxidase	Heat	Thermistor	DANIELSSON et al., 1979
Lactose	β-Galactosidase + glucose oxidase	H_2O_2	Electrochem. sensor	PILLOTON et al., 1987
Cellobiose	β-Glucosidase + glucose oxidase	H_2O_2	Thermistor	DANIELSSON et al., 1981b
Xylose + xylulose	Xylose isomerase + mutarotase + glucose dehydrogenase	NADH	Chemically modified electrode	DOMINGUEZ et al., 1988
Starch	α-Amylase + β-amylase + mutarotase + glucose oxidase	H_2O_2	Amperometric cell	APPELQVIST et al., 1986

In some special cases it has been possible to push the sensitivity towards substrate even further by the use of enzyme-catalyzed cycling of substrate (SCHELLER et al., 1985).

It is, of course, important that the enzyme-catalyzed process yields a product that can be monitored by a suitable transducer, or that the substrate *per se* can be monitored. Since there are many different transducers, it may be possible either to find a suitable transducer for the reaction or to add a second enzyme that will catalyze a subsequent reaction by converting the product of the first enzyme reaction into a second product that may be recognized by the transducer. Tab. 2 lists some examples of such arrangements.

2.1.2 Cells

It is not necessary to operate with purified enzymes. This may be especially convenient when reaction sequences are needed to carry out the catalytic process. In such cases, cell homogenates or even whole cells may be successfully employed (ARNOLD and RECHNITZ, 1980; MATTIASSON, 1983). However, one must bear in mind that when operating with crude enzyme preparations there is a risk of side reactions competing for some of the substrate, thereby becoming a source of error in the analysis. By using cells this risk is even more pronounced, and this is actually one of the reasons why cell-based sensors have not yet become more popular.

The physiological status of the cells employed may vary; Tab. 3 lists some different stages that have been used. Fully viable cells have all the characteristics of the native cell

Tab. 3. Conditions of Cells Used for Constructing Biosensors

Only one or a few enzymes involved
Resting cells
Permeabilized cells
Living cells with an intact surface
Actively growing and dividing cells
Mixed cultures of living cells, or mixtures of enzymes and cells
Cells co-immobilized with a specific enzyme or organelle
Immobilized cells or organelles with a substantial part of the metabolism intact

and may be very suitable when used to monitor events that are coupled to cell growth and cell division. In other cases the cell may only be regarded as a package of the appropriate enzyme for the catalytic step to be used in the analysis. In addition, one may exploit the membrane properties of the intact cell in the sense that membrane-modifying activities can be monitored by following the metabolism of intact cells.

2.1.3 Organelles

The same approach used for cells also applies to organelles, but their higher fragility and resulting tendency to lose activity has hampered development.

An electrode based on rat liver microsomes was used for the determination of thyroxin (MEYERHOF and RECHNITZ, 1979), and a glutamine sensor was obtained by immobilizing the mitochondrial fraction from porcine kidney cortex cells on an ammonia gas sensor (ARNOLD and RECHNITZ, 1980). Microsomes may either be used separately or co-immobilized with other biocatalysts (SCHUBERT and SCHELLER, 1988).

2.2 Binding Assays

There is a whole range of interactions that can be summarized as binding reactions. Immunochemical binding reactions are the best studied and also the most successful on the market. Receptor-based analysis has been predicted to have a bright future, but so far there has been no real breakthrough for this kind of assay. Lectin–carbohydrate interactions are still another group of interactions.

Generally speaking, these assays have the potential for high sensitivity since the formation of affinity complexes between a macromolecule and a ligand is very efficient even at low concentrations. In most of the reported cases the binding assay has been set up as a competitive assay where the native ligand in the sample to be analyzed competes with a labeled ligand that is added in a fixed concentration. The result of the competition is then read by evaluating how much of the label is bound.

Tab. 4. Comparison between Enzymatic and Immunochemical Analyses

	Enzyme/ Substrate	Antibody/ Antigen
Operational range (mol/L)	10^{-6}	10^{-12}
Specificity	+	+ +
Availability for new analyses	−	+
Ease of production	−	+
Usefulness in quantifying macromolecular structures	−	+
Time needed	+	−

In more recent developments, the trend is towards direct monitoring of the affinity interaction between the binder and the ligand. If this can be done directly, easier and perhaps faster assays can be set up.

Binding assays can be set up according to the dipstick method, in which the binder is used as a disposable chemical; otherwise, the binder must be regenerated, which then puts some constraints on the stability of the binding molecule.

In many cases binding assays are useful in concentration ranges down to 10^{-10} mol/L, and they are also useful for quantifying macromolecules. Since most of the reported work is focused on immunoassays, the discussion in this chapter will mainly deal with such assays. Tab. 4 shows a comparison between the properties of enzyme-based assays and immunoassays. At the present state of development the two types complement each other.

2.2.1 Immunochemicals

The basis for immunochemical binding assays is the specific binding of antigens to antibodies. When immunizing an animal by introducing a foreign macromolecule, a protective reaction is initiated that leads to formation of antibodies. The mechanism involves activation of lymphocytes to produce antibodies. Each such activated lymphocyte produces only one type of antibody, but since many lymphocytes respond to the immunization, the result is a polyclonal antiserum. This means that the antibodies may differ in their molecular properties, e.g., binding strength, specificity, pH-dependence, and stability. However, because of the number of different strains, it often seems as if the overall characteristics of the antiserum are stability and good binding properties.

The binding constants are reported to be in the region of 10^{-5} to 10^{-11} mol/L. By tradition, all selections of antisera have been for high-affinity antibodies, since these allow higher sensitivity in the subsequent binding assay.

The introduction of the hybridoma technology for production of monoclonal antibodies made it possible to obtain large amounts of individual clones of antibodies. This led to the standardization of some assays, but also helped to unravel some points that had been taken for granted when dealing with polyclonal antibodies. The different clones had to be screened for good binding constants and for the desired specificity. Furthermore, it soon turned out that stability varied drastically between different clones. Selection therefore had also to include tests for stability, which is especially important when repeated use is planned.

For discrete assays, when the antibody is a disposable reagent, one must screen for the strongest binder with the highest affinity. This may, however, not be ideal when constructing biosensors. As will be discussed later in this review, repeated use of the antibody involves dissociation of immunocomplexes. When dealing with very strong complexes, harsh conditions must be applied and partial denaturation of the antibody may occur.

When immobilizing antibodies it is desirable to avoid binding to their specific binding sites and instead to couple via the Fc-parts. This can either be done by using covalent coupling to the carbohydrate chains on the Fc-fragment (O'SHANNESSY and HOFFMAN, 1987) or by biospecific reversible immobilization via interaction with protein A or protein G (LANCET et al., 1978).

2.2.2 Receptors

Receptor-based analysis is a great challenge, since our knowledge of receptors and their properties is rather poor. Only recently have receptors become available in pure form, and it is still far from standard technology to prepare such entities. Furthermore, since the receptors are membrane bound, the handling of the intact molecule is quite different from the treatment that is given traditional proteins. This shows that at present we have no technology to successfully integrate such molecules into applications such as biosensors. One easy and probably viable way would be to use very crude preparations, consisting of more or less whole cells or cell fragments, since by using such an approach the purification and stability problem of the receptor could be circumvented.

3 Methods to Integrate the Biocomponent with the Transducer

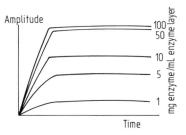

Fig. 2. Schematic presentation of the response of an enzyme electrode as a function of the load of enzyme.

It has been said that biosensors are based on proximity of the biocomponent and the transducer. In order to make this technically feasible, the biocomponent must be rendered more easily handled than an enzyme in solution. Furthermore, if a stable signal is expected from the biosensor, precautions must be taken to compensate for denaturation, etc. One way to do that is by operating at high concentrations, i. e., with an excess of the biocomponent, so that the reaction becomes diffusion controlled.

Immobilization of biomolecules is a technology that is handy and permits operation with high concentrations. By immobilizing the biocomponent one can apply it in situations and under circumstances that would otherwise have been practically impossible. Furthermore, immobilized biochemicals can in principle be reused. This leads to conditions suitable for continuous measurements or repetitive assays. However, under these circumstances the operational stability of the preparation is important, since a reliable signal is of utmost importance. Fig. 2 shows that using a large excess of the biomolecule in enzyme-based assays has turned out to be successful.

There are many different ways to carry out immobilization. This has been well described in recent reviews and books (MOSBACH, 1976, 1988; MATTIASSON, 1983) and, therefore, it shall only be treated as required for the discussion later on in the chapter.

The most popular way of immobilizing a protein is by covalent coupling (MOSBACH, 1976). A solid support is used with chemical groups that can be activated in such a way that the activated form will react with specific groups on the protein surface. There are many different coupling methods from which to choose. The groups available on the support are of course important, and often that is the first consideration in the choice of a coupling method. However, it is even more important that the coupled protein maintains its biological activity and that the coupling is stable, i. e., the biomolecule sticks to the support and does not leak away. In the case of immobilization of enzymes, it has sometimes turned out that an essential amino acid residue in the active site is very reactive and thus reacts first in the coupling procedure. In this case, other chemical approaches must be considered to avoid utilizing these amino acid residues. An alternative is to try to protect the enzyme during coupling, e.g., by immobilizing it in the presence of a competitive inhibitor that will block the active site (JULLIARD et al., 1971; MATTIASSON et al., 1974; BÜLOW and MOSBACH, 1982). In some cases substrates have been used.

Another immobilization method that has become very popular, especially for cells and cell organelles, is entrapment (MATTIASSON, 1983). The particulate matter is mixed with a solution of either a water-soluble polymer or of monomers. The water-soluble polymers are then crosslinked to form a three-dimensional network in which the particulate matter is entrapped. In the case of monomer solutions, polymerization is initiated and a similar three-dimensional network is formed.

Still another way of carrying out immobilization has been developed especially for labile and/or expensive enzymes. In these cases it is not ideal to immobilize a large excess of the enzyme, since from an economical point of view it may be impossible or ill advised. If the enzyme is very labile, a large excess of enzyme molecules will only marginally improve the operational stability. In such cases it is important to be able to immobilize the biocomponent shortly before it is used in the assay. Further-more, since the time between immobilization and utilization is short, one can operate at a fairly low concentration of the enzyme. When the activity decreases a new immobilization is carried out. One way to achieve these characteristics is to use reversible immobilization (MATTIASSON, 1981, 1988). The principle for reversible immobilization is presented in Fig. 3.

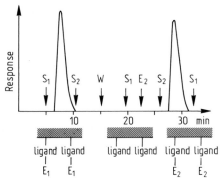

Fig. 3. Schematic presentation of an assay cycle when dealing with reversible immobilization. The arrows indicate changes in the perfusion medium. At S_1 substrate for enzyme E_1 is introduced, and at S_2 for enzyme E_2. At arrow W a washing procedure is shown for rinsing the column and making it possible to either reload fresh enzyme E_1 or to change to another enzyme with affinity for the sorbent, in this case enzyme E_2.

Most work on biosensors has so far been conducted on bioelectrodes. Fig. 4 schematically illustrates different ways to apply the biocomponent to the tranducer.

3.1 Immobilized Layer Covering the Transducer

The traditional way of constructing a biosensor is to cover the transducer with a layer of immobilized biochemicals. The biocomponent was originally immobilized by crosslinking on the tip, e.g., of an oxygen electrode. In order to keep it in place the whole unit was covered by a semipermeable membrane. There are

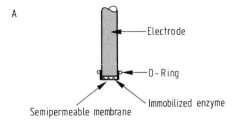

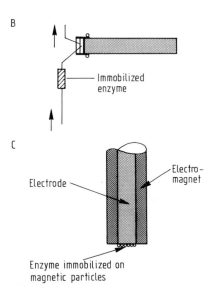

Fig. 4. Schematic presentation of different ways to arrange the biocomponent in relation to an electrode. A: immobilized enzyme layer covering the transducer; B: pre-column with the biocomponent; C: reversible immobilization of enzyme-carrying magnetic particles.

also examples where crosslinking did not take place, so that only a concentrated solution of enzyme was deposited on the sensor, which was then covered. Other alternatives involve the entrapment of the biocomponent in a membrane that is then used to cover the transducer tip. The technique has been used for cell-based electrodes (MATTIASSON et al., 1982). This kind of biosensor is well suited for dipping into a solution and registering the concentration of the analyte. However, when a faster response is desired, a small flow cell is placed on top of the sensor so that the sample can be administered as a short pulse followed by washing (Fig. 3).

3.2 Pre-Column of Immobilized Biocomponent

An even more convenient way to construct an analytical device may be to use the biocomponent in immobilized form as a pre-column placed prior to the transducer in a flow system. By this arrangement the sample must be passed over the biocomponent column, and during this passage biocatalysis or binding takes place in a much more efficient way than with the membrane-covered sensor. In this case it becomes a semantic question of what is a biosensor and what is not. The proximity of the pre-column and the transducer is not *per se* important for the analytical performance and, therefore, the two entities of the "biosensor" may well be separated in space.

3.4 Reversible Immobilization by Means of Mechanical Arrangement

The ultimate goal for many scientists is the development of a biosensor in which the biocomponent is directly bound to the surface of the transducer. This binding can either be covalent or by adsorption (WINGARD, 1984; APPELQVIST et al., 1985, 1986; MARKO-VARGA et al., 1985). Most of the applications so far have dealt with cofactor-dependent enzymes. In such cases the electrode surface may be doped with suitable mediators. Conducting polymers comprise another interesting area for immobilization. So far, only a minor fraction of the work on biosensors has been carried out on the basis of this principle (FOULDS and LOWE, 1986, 1988; PANDEY, 1988; UMANA and WALLER, 1986), but for the utilization of microelectronics this approach seems to be most interesting (SHINOHARA et al., 1988).

3.4 Reversible Immobilization by Means of Mechanical Arrangement

It may be attractive to replace an enzyme layer covering an electrode tip. To achieve this without having to dismount the whole configuration, some special arrangements have been developed with the idea of constructing a biosensor that can be autoclaved. ENFORS and NILSSON (1979) presented a stainless steel fitting covered with a semipermeable membrane. This unit was mounted into a fermenter. A polarographic electrode was placed close to the membrane. After autoclaving, fresh enzyme was introduced through specially designed capillary ports in the steel holder. The enzyme was kept in the space between the electrode and the membrane, creating a stationary enzyme layer until replaced. Enzyme in soluble form as well as immobilized to small particles was introduced. One problem with this device is that the thickness of the enzyme layer may vary, and thus the response time will also vary.

One alternative to achieve a reversible coverage of an electrode surface by immobilized enzyme is to use magnetic particles and an electromagnetic field over the electrode surface (MATTIASSON and MIYABAYASHI, 1988). With this approach it is possible to change the enzyme layer by turning off the electricity to the magnet, washing out the particles, and turning on the field before new magnetic particles are introduced.

4 Transducers

Most biosensor work so far reported has been done with electrochemical transducers. Amperometric as well as potentiometric electrodes have been the most utilized transducers over the years. The main types of transducers used will be discussed below.

4.1 Electrochemical Detectors

Electrochemical detectors are commonly used to construct biosensors. Two types dominate: amperometric and potentiometric detectors. Besides these groups one may also mention fuel cells.

4.1.1 Amperometric Detectors

Amperometric detection is based on measuring the current obtained when the analyte is either electrochemically oxidized or reduced at a given solution–electrode interface. The measured current is given by

$$i = nF \, dN/dt$$

Proportionality between the current i and the concentration of the analyte N can be obtained, provided some prerequisites are met.

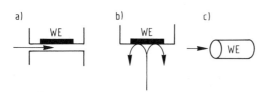

Fig. 5. Geometries for amperometric detectors. a) Thin-layer, b) wall-jet, c) tubular. WE, working electrode.

Amperometric detectors are classified according to their cell geometries: thin-layer, wall-jet, and tubular (Fig. 5). The thin layer type is popular in flow systems such as HPLC and also when combined with a preceding enzyme column. The wall-jet may give a faster response, since the solution flows perpendicular to the electrode surface. The result is a more efficient mass transfer to the electrode surface. Tubular electrodes have not gained the same popularity as those with the other two geometries. The design and construction is simple. The difficulties in polishing the electrode surface and in having a controlled potential all along the tube also affect the performance of the electrode.

Polarographic oxygen electrodes are well established in bioanalysis (FATT, 1976). The very first enzyme electrode was built around such a polarographic oxygen electrode (CLARK and LYONS, 1962). Today a wide variety of applications are available using both purified oxidases and whole cells as biocomponents.

Miniaturization of amperometric sensors has begun, and there are already examples of new microbiosensors (KARUBE et al., 1988;

SUZUKI et al., 1988). There is still a long way to go before these new miniaturized sensors are comparable with the more traditional ones as far as performance, reliability, and stability are concerned. Thus, the microglucose sensor constructed along these lines was reported to have a lower stability than the traditional sensors. While the biocomponent had the same properties as usual, the transducer was less stable (KARUBE et al., 1988). The small dimensions, however, hold such interesting possibilities that it is still of great interest to continue this development.

A limiting factor for a wider application of amperometric enzyme assays has been a shortage of suitable oxidases. As new enzymes are discovered and isolated, the area of amperometric enzyme sensors will broaden.

4.1.2 Potentiometric Biosensors

Potentiometric biosensors may be regarded as variations of ion selective electrodes (ISE). When combined with a suitable reference electrode, an extremely useful analytical tool is obtained. The most commonly used ISE is the one sensitive to H^+. It is used for measuring pH. Other ISEs are useful for monitoring other ionic species, e.g., NH_4^+, CN^-, S^{2-}. When combining these ion-selective electrodes with a suitable biocomponent, new biosensors are developed. The analytically useful range of these sensors is generally from 10^{-1} to 10^{-5} mol/L and some sensors may even be useful at lower concentrations (KUAN and GUILBAULT, 1987).

4.1.3 Fuel Cells

Most biosensors are based on an indirect measurement in the sense that the transducer registers any concentration change due to the action of the biocomponent of the biosensor. However, with the biofuel cell, the biocomponent gives a direct electrical signal. An example discussed elsewhere in this chapter is the BOD sensor based on the action of *Clostridium butyricum*. The activity was registered as produced hydrogen which was converted in the fuel cell (KARUBE et al., 1977b).

More recent development has focused on tapping the electron flow from microorganisms. By introduction of suitable redox mediators the efficiency of this process has been improved. One interesting application of these sensors is the cell counter, where the amplitude of the signal reflects the number of active cells in the analyzed sample. Such sensors are applicable both to bacterial and mammalian cells (MATSUNAGA et al., 1980; MIYABAYASHI et al., 1987).

Biological fuel cells attract great interest because of the potential that one may be able to generate electricity directly from the biological process.

4.2 Thermometric Detectors

Most biocatalytic reactions produce/consume some heat, and thus in theory most reactions could also be monitored by means of a thermometrical detector. Because of the lack of selectivity of the transducer it has been important to compensate for various nonspecific reactions. This has mainly been carried out by the use of split-flow devices with one sample and one reference transducer.

The most commonly used transducer is the thermistor, a semiconductor with a resistance that is temperature-dependent. With a Wheatstone bridge it is possible to monitor the change in resistance as a biocatalytic process proceeds. Thermistors measure temperature changes down to 10^{-2} °C. This has been converted into sensitivities in the enzyme thermistor concept of 10^{-6} mol/L, and for the enzyme thermal probe it is less sensitive by a few orders of magnitude (WEAVER et al., 1976; FULTON et al., 1980). Thermocouples have also been used, as have Peltier elements (PENNINGTON, 1976).

4.3 Optical Transducers in Biosensor Construction

Measurement of light absorption/transmission is a standard technique in a biochemical laboratory. Application of small flow cells has made it possible to combine immobilized enzyme technology, flow injection analysis, and spectrophotometric detection. This area is relatively well developed and some devices have also reached the market. These devices were mainly developed for clinical analysis and have recently been applied for on-line analysis of bioprocesses (SCHÜGERL, 1988). The sensitivity is determined by Lambert–Beer's law. The light path used is often 1 cm or shorter. A higher sensitivity would be attainable if a longer light path were used. There are several advantages in using spectrophotometric detection. A huge amount of reference literature from conventional biochemistry is available.

Besides monitoring absorbance/transmittance and fluorescence, it is also of interest to register light scattering, since in that case the amount of particulate matter may be analyzed. This is valuable when quantifying cells and also flocs created by, e.g., immunochemical binding.

The instrumentation required for the above analyses has often been a conventional laboratory unit equipped with a flow cell. However, there has been a need for making the optical reading outside the instrument and close to the bioreaction. This has been achieved by the introduction of fiber optics.

An optical sensor configuration is schematically shown in Fig. 6 A–C. The basic construction of an optical fiber sensor is shown in Fig. 7 where the two mechanisms of light exchange are both illustrated. The first involves an illumination cone from the end of the fiber. The alternative is an evanescent wave from the naked portion of the fiber. In the general configuration the fiber is covered with an opaque covering, but when evanescent wave measurements are used, the wall of the fiber must be directly exposed to the sample solution.

Today many different applications are based on this technology. The pH, for example, is monitored using an optical fiber with a pH indicator immobilized at its tip (PETERSON et al., 1980). When the pH changes, the absorption spectrum of the indicator changes, and thus the pH at the tip of the fiber can be read.

Optical sensor configurations for monitoring pCO_2, pO_2 or dissolved oxygen, metal ions, etc. have also been described (SEITZ, 1987).

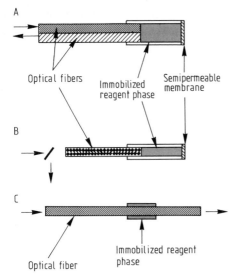

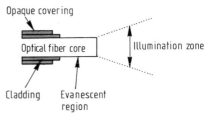

Fig. 7. An optical fiber illustrating the two different principles used for constructing optical biosensors. At the tip of the sensor is an illumination zone. If an immobilized reagent phase is placed in this zone, any changes, e.g., by binding to the zone may be detected. In the evanescent region the fiber is naked after removal of the cladding layers; here evanescent wave measurements may be performed. In that case a solution similar to that in Fig. 6C is achieved.

Fig. 6. Optical sensor configurations. Example A is based on separate fibers carrying the light to and from the immobilized reagent phase. In B the same fiber operates with light to and from the reagent. A beam splitter is used to capture the emerging light. C: a sensor with the immobilized reagent phase coated on the outside of the fiber. As light passes through the fiber it will come into contact with the immobilized reagent phase.

5 Biocatalytically Active Biosensors

5.1 Enzyme-Based Sensors: Examples and Applications

Enzyme-based optical sensors are starting to appear. Recently a glucose sensor has been reported based on the action of glucose oxidase trapped at the outer tip of the fiber. The reduced enzyme has a different fluorescence spectrum than the oxidized form. This difference was exploited when setting up the glucose sensor (WOLFBEIS, 1989).

Combinations of optical fiber technology with biorecognition has been studied (SCHULTZ and SIMS, 1979; SCHULTZ et al., 1982).

An attractive characteristic of optical fiber-based sensors compared to electrochemical sensors is the fact that no electrical signals are needed until the light signals are read. This has been regarded as a major advantage when dealing with *in vivo* measurements. The optical signals are also less susceptible to electrical disturbances, and this may be of great advantage in certain environments. Optical biosensors will become very important in the future.

The most popular sensor studied has been the glucose sensor (CLARK and LYONS, 1962; DANIELSSON et al., 1977; CLELAND and ENFORS, 1984; MARKO-VARGA et al., 1985; D'COSTA et al., 1986; BROOKS et al., 1987; HOLST et al., 1988; HÅKANSON, 1988; WOLFBEIS, 1989). Therefore, when describing the different solutions applied in glucose monitoring, a broad spectrum of the different developments in biosensor research will be presented. The reasons for this development are several: glucose is an important metabolite in clinical chemistry. It is an interesting metabolite to monitor for intensive care and is also an interesting target in diabetes. Glucose is also a frequently used carbon source in fermentation technology. Upon degradation of polymeric carbohydrates such as starch and cellulose, glucose is formed, and it is therefore a compound of interest in food and feed technology.

Still another area in which glucose monitoring may be useful is contamination control.

When determining the degree of freshness of meat, a measure of the glucose content is inversely proportional to the number of bacteria present (J. HIGGINS, Cranfield Institute, UK, personal communication).

Much of the biosensor development has been conducted by clinical chemists. In order to standardize and simplify the handling of the analyses, application of immobilized enzymes in analysis became attractive. The interest focused on analysis of discrete samples (DANIELSSON et al., 1977), whereas in intensive care and in fermentation control the desire has been for continuous monitoring (ENFORS, 1981; CLARK et al., 1988), or at least frequent intermittent analyses (HOLST et al., 1988; HÅKANSON, 1989).

The enzyme used for setting up these assays has in most cases been glucose oxidase (E.C. 1.1.3.4). This enzyme, produced by *Aspergillus niger*, is a glycoprotein with FAD as prosthetic group. It does not depend upon any soluble cofactor, and this has facilitated its development. A drawback is that hydrogen peroxide is formed during the reaction:

$$glucose + O_2 \rightarrow gluconolactone + H_2O_2$$

Hydrogen peroxide is harmful to many enzymes. During the initial experiments with glucose electrodes these harmful effects were not observed, and the enzyme was regarded as extremly stable. It turned out later that the intermittent exposure of the enzyme preparation to the hydrogen peroxide did not cause much harm, whereas when monitoring continuously and thus exposing the enzyme continuously to hydrogen peroxide, a rapid decrease in activity was observed. The harmful effects are ascribed to the oxidative activity of the peroxide as well as to the radicals formed. In order to stabilize the enzymes, co-immobilization has been applied. These stability problems were first encountered when immobilized glucose oxidase was used in enzyme technological process applications (MESSING, 1974; BOUIN et al., 1976 a, b). Experience from that field was applied in the biosensor work. By coupling catalase with glucose oxidase, instantaneous degradation of the peroxide was achieved. When superoxide dismutase was also included, any radicals formed were eliminated.

Recently, an alternative way to avoid the negative effects of hydrogen peroxide resulting from the oxidase activity has been presented. The enzyme performs the oxidation of the substrate with the concomitant reduction of FAD to FADH$_2$. The latter is regenerated via mediators replacing oxygen. The mediators are then oxidized by the electrode.

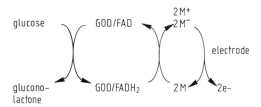

Ferricyanide may be used as mediator, but the most popular one belongs to the ferrocene/ferrocinium group (CASS et al., 1984).

One step further would be to directly branch off the electrons from FADH$_2$ without use of mediators. Efforts in this direction have involved direct coupling of FAD to the electrode surface, thereby facilitating the reoxidation of FADH$_2$ (WINGARD 1984; ALBERY et al., 1985).

An alternative arrangement is to covalently attach mediator molecules to the enzyme, thereby making it easier to branch off the electrons from the interior of the protein over to the electrode surface (DEGANI and HELLER, 1987, 1988; BARTLETT et al., 1987).

This development further leads to the use of conducting polymers to tap the electron flux to or from the biocomponent of the biosensor. By this way of linking the reaction before formation of hydrogen peroxide, two things may be achieved: avoidance of hydrogen peroxide formation and elimination of oxygen dependence, thereby making it possible to operate over extended ranges of concentration.

When monitoring the glucose oxidase catalyzed process, several options are at hand. The most common option is to use a polarographic oxygen sensor to register the oxygen tension. When glucose is present, this will be reflected in a decreased oxygen tension. An alternative is to use a different polarization current and register the hydrogen peroxide formed. This makes it easier to avoid interference by other substances in the sample, e.g., by ascorbic acid. This sounds like a minor problem in fermentation control, but in clinical analysis and also in medical applications it is a definite problem. A third alternative in monitoring the glucose oxidase catalyzed process is detection of the protons formed upon hydrolysis of the resulting gluconolactone. The gluconic acid thus formed is dissociated and it lowers the pH. However, it should be stated that pH-sensitive electrodes as transducers in biosensor work have not been very successful. The buffering capacity rarely remains constant for long periods of time, and thus the sensitivity in detecting the acid produced varies with time (NILSSON et al., 1973; ENFORS and NILSSON, 1979; RUSLING et al., 1976). To make this kind of sensor useful, eleborate standardization and calibration has to be carried out.

Another detection principle that has been applied to monitor glucose oxidase activity is thermal registration by thermistors (DANIELSSON et al., 1977). Here, the heat of reaction is registered. The glucose-containing sample is pumped through a bed of immobilized glucose oxidase, and at the outlet of the bed a thermistor registers the temperature of the effluent. When glucose is present in the sample, enzyme catalysis takes place during transit through the column, and the reaction heat is transported with the effluent. The heat is proportional to the amount of glucose converted. Since the system operates with reproducible volumes, a clear relation exists between concentration of glucose and heat generated. In this system as well as in the others, oxygen concentration is limiting. The solubility of oxygen in water is only 0.25 mmol/L at 25 °C. This means that one can only expect a linear correlation between the concentration of glucose and the response up to approx. 0.5 mmol/L. However, when catalase is present some oxygen is produced as the hydrogen peroxide formed is degraded. In this manner an expansion of the concentration range is achieved.

By operating with thick enzyme-membranes it has been possible to shift the dynamic range for glucose electrodes toward higher concentrations.

There are different ways to improve oxygen supply to such sensing systems (ADLERCREUTZ and MATTIASSON, 1982). One elegant system was reported by ENFORS and coworkers (CLE-LAND and ENFORS, 1984). Oxygen was produced at the tip of an oxygen electrode by electrolysis of water. The oxygen tension was set at a constant value, and thus the amount of water that needed to be hydrolyzed was proportional to the amount of glucose present. The energy needed to perform this hydrolysis was used to read the glucose concentration. This approach provided an oxygen electrode that was linear up to at least 20 mmol L^{-1}.

Other enzymes are available for constructing an enzyme based glucose assay. Hexokinase catalyzes the phosphorylation of glucose at the expense of ATP, and the glucose-6-phosphate formed may in a subsequent step be oxidized by glucose-6-phosphate dehydrogenase with a concomitant reduction of $NADP^+$. This can be monitored by a fluorometer or spectrophotometer. Electrochemical alternatives are also available (SCHELTER-GRAF et al., 1984). However, in this case there is a requirement for a free cofactor, and therefore such a method is only suitable for analyzing discrete samples under well-controlled laboratory conditions.

A more recent alternative is based on a new group of enzymes, the PQQ-enzymes utilizing pyrrolo quinoline quinone as cofactor. It has been found that many enzymes that carry out oxidation reactions do not operate with any of the traditional biochemical cofactors, but instead use PQQ (MATSUSHITA et al., 1982; NEIJSSEL et al., 1983). This cofactor is spontaneously regenerated *in situ* and such systems therefore require no external addition of a cofactor (D'COSTA et al., 1986).

Glucose Analyzer for Blood Monitoring or Fermentation Control

In addition to the mutual arrangement of the immobilized enzyme and the transducer in certain positions, there is a need for a rational way of treating the sample. It does not matter how well a sensor functions if the sample handling technology is poor. Below, the sampling unit, the sample treatment system, and the analytical part are described in detail. These three units belong together when setting up successful monitoring systems (Fig. 8) (HOLST et al., 1988). The preparation of immobilized

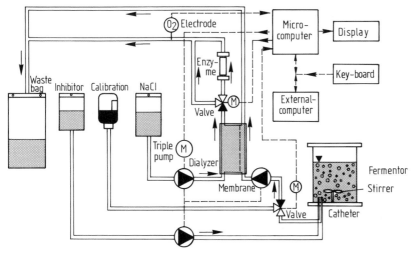

Fig. 8. Flow scheme for the glucose monitor used for fermentation control. (From HOLST et al., 1988, with permission)

enzyme is designed to last in the clinical situation for five days, and for the industrial application to last for one fermentation. Even if it would be possible to use one and the same enzyme unit for two or more fermentations, it is better to replace it after each fermentation. For glucose monitoring in clinical samples a different operational range is needed than for fermentations. The level of operation may well be controlled by designing the dialysis step properly, either by shifting to a more suitable membrane, or by varying the flows on either side of the membrane. Blood glucose levels are normally around 5 mmol/L; therefore, the system is optimized for that concentration range. Since the unit is based on a polarographic oxygen sensor, any oxidative process may be monitored. By applying lactate oxidase, a useful lactate sensor for continuous monitoring was achieved. However, due to stability problems of the enzyme, a sampling frequency has to be used that protects the enzyme from too long an exposure to hydrogen peroxide. For applications in cultivation of microbial cells (lactic acid bacteria) and for cultivation of mammalian cells these modifications are fully acceptable. Fig. 9 shows the results of one such cultivation where both glucose and lactate were monitored using two separate monitors (HÅKANSON et al., 1990). In this ap-

plication it is not the sensitivity that raises problems; it is merely the sampling and the sample handling.

Thermistor Application

Thermometric registration of the enzymatic reaction is the most general of the different ways to follow enzymatic reactions discussed here. Since it may also be used in optically very dense solutions, it offers certain potential

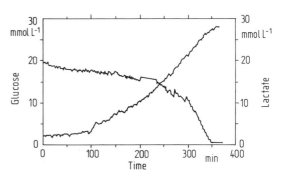

Fig. 9. Tracings of glucose and lactic acid using two analytical devices as shown in Fig. 8, one equipped with immobilized glucose oxidase and the other with immobilized lactate oxidase.

advantages. When setting up a flow-calorimetric assay it is important to insulate the biosensor so that any heat generated during the enzymatic reaction will be detected by the transducers used. The device that is most commonly described is the enzyme thermistor (DANIELSSON et al., 1981a). Here, the immobilized enzyme is packed into a small column placed in a continuous flow. During baseline conditions plain buffer is used for perfusion, and the sample is then introduced into the continuous stream. When passing through the enzyme column, any substrate present is converted by the enzyme, and the products and any heat generated are transported towards the outlet where a sensitive thermistor accurately registers the temperature of the liquid. The sample can either be introduced in short pulses or on a continuous basis. The peak height in the former case and the plateau level in the latter reflect the concentration in the solution. The signal registered can conveniently be used for process control (DANIELSSON et al., 1979; MANDENIUS et al., 1980, 1981, 1985). The time needed for one sample is 1–5 minutes, but this can of course be varied by changing the size of the enzyme column, length of tubing, etc. The applications using this instrument have mainly focused on analyzing discrete samples (MANDENIUS et al., 1985) whenever needed and on continuous measurements (SCHÜGERL, 1988).

FIA System for Sucrose

Flow-injection analysis (FIA) has in essence already been described for enzyme electrode analysis and for the enzyme thermistor. However, FIA in its pure form is more optimized and is an attractive technique for analysis of liquid samples. When combined with small columns of immobilized enzymes, such devices are very attractive because they combine the specificity of the enzyme with the speed of analysis of the flow system. By using microcolumns with immobilized enzymes it is possible to introduce many enzymes at the same time into the flow system, thereby making it realistic to analyze one metabolite in samples that also contain competing substrates. This is achieved by arranging the different enzyme columns in the flow system so that the un-

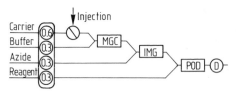

Fig. 10. Schematic presentation of the flow system used for analyzing sucrose in the presence of glucose. The following enzyme columns used were: MGC contained immobilized mutarotase, glucose oxidase, and catalase in order to remove all glucose present in the sample when injected into the flow stream. IMG contained immobilized invertase, mutarotase, and glucose oxidase that converted sucrose to glucose, converting it to the proper configuration for glucose oxidase to work, and finally oxidizing it to gluconic acid with the concomitant production of hydrogen peroxide. In the final column containing POD, horse radish peroxidase immobilized to porous glass, the peroxide reacted and a colored product was formed. This was registered using the detector D. (After OLSSON et al., 1986, with permission)

wanted substrate is first removed before the desired substrate has reacted and has been analyzed. As seen in Fig. 10, it is entirely possible to analyze sucrose in beet juice in the presence of some glucose. The time for each analysis of the flow systems is very short (<1 min/sample) which permits high frequencies. This fact has made this type of analysis also attractive for process control.

Much of the discussion above has focused on polarographic oxygen sensors. Glucose is a metabolite of major interest, and the development of glucose sensors was therefore used to illustrate the biosensor development. But other systems could also have been discussed in some detail.

The ammonium electrode as well as the ammonia gas-sensitive electrode have been used for quantifying amino acids (GUILBAULT and SHU, 1971; JOHANSSON et al., 1976), urea (GUILBAULT and MONTALVO, 1970; TRANMINH and BROUN, 1975) and other substances that produce ammonia when enzymatically modified.

That ammonia/ammonium coupled assays are important in biochemistry, biotechnology, clinical analysis, and biomedicine is well illustrated by the number of publications in this area.

These analyses are useful in fermentation technology to monitor nitrogen-containing substrates that are often limiting in a cultivation. For mammalian cell cultivations it is important to monitor glutamine as a nitrogen source and ammonia as a toxic waste product. Urea is another waste product of great interest in biomedicine and clinical analysis (GUILBAULT and MONTALVO, 1970; TRAN-MINH and BROUN, 1975). An elevated ammonia level may indicate a reduced function of kidney filtration. Creatinine is an alternative indicator molecule of kidney function that can also be monitored (THOMPSON and RECHNITZ, 1974; CHEN et al., 1980).

Analysis of amino acids may be performed by deaminating enzymes, and in this case ammonia/ammonium transducers are applicable. Amino acids may also be reacted in an oxidative deamination, which opens the possibility for both the detection of ammonia and the monitoring of the oxygen level (JOHANSSON et al., 1976). Other transducing systems that have been applied in this type of assay involve conductivity and calorimetry.

In general terms most enzyme-based biosensors operate in the concentration range down to 1–10 μmol/L; in some cases with slightly higher, in other cases with lower sensitivities. The response time is dependent upon the arrangement. When the sensor is dipped into the sample to be analyzed, a fairly long time is needed to reach equilibrium (5–10 min). When operating in continuous flow systems, faster responses are achieved, mainly because equilibrium is no longer required for the assay (1–2 min). Furthermore, by reducing the thickness of the layer of immobilized enzymes, diffusion is facilitated and thus a quicker response may be achieved. The extreme arrangement from that point of view is direct coupling of the enzyme to the surface of the transducer (response time 5–10 s).

Over the last ten or twenty years quite a long list of electrodes with various specificities have been described. Many of these electrodes have been combined with enzymes to form specific enzyme electrodes. The electrode is thus specific for registering the product of the enzymatic reaction (GUILBAULT, 1984).

5.2 Cell-Based Sensors: Examples and Applications

The intact cell offers optimal conditions for the maintenance of complex biological reactions (MATTIASSON, 1983). Stability of enzymes is usually rather high within the cell. This is also true for enzymes present in crude preparations, provided that proteolytic activity is removed. This is probably due to the interaction of enzymes with a variety of functional groups which stabilize the enzyme. When approaching a higher degree of purity the enzyme is more exposed to hostile conditions in the environment and is thus less stable.

Immobilized cells have been used in technical applications and in analysis. By replacing enzymes with whole cells, biosensors have been constructed that respond to specific substances as well as to more general stimuli (MATTIASSON, 1983). Cell-based sensors often exploit metabolic sequences before a readable signal can be generated from the cells; coenzyme recycling presents no problem in these systems.

5.2.1 Analysis of Specific Compounds, e.g., Vitamins

Bioassays have long been used for analysis of vitamins. Most often a mutant of an organism is used that requires the vitamin for growth. In the biosensor assay a dense cell population is immobilized and kept close to the transducer. When the cellular metabolism starts to prepare for cell division it is possible to monitor the increased activity long before the first cell division takes place.

Thiamin was quantified using *Saccharomyces cerevisiae* entrapped in "armed" alginate membranes (MATTIASSON et al., 1982). Small sheets of membrane were cut out and exposed to the vitamin-containing sample. This could either be done when mounted in the sensor configuration or in a separate pretreatment step. The latter procedure gives a higher throughput for the analysis and is therefore preferred. In this type of analysis the cells are activated such that they cannot be used for additional analyses. Therefore, after each analysis the cell preparation has to be replaced. In order to make comparisons between different experiments, it is important to mount the membrane on the transducer in a very reproducible way. This was achieved by using an electromagnetic holder. Several analyses of vitamins have been described (MATSUNAGA et al., 1978).

The introduction of biosensors has speeded up bioassays to such an extent that they may become a realistic alternative to binding assays.

5.2.2 Analysis of BOD

Another extreme example of using cells in analysis is the determination of BOD (Biological Oxygen Demand). The traditional way of carrying out these assays is very time-consuming and the result of the analysis is not available until five to seven days after the sample has been taken. At that time the waste water tested is already in the recipient. In their imaginative work, KARUBE and coworkers utilized a mixed culture to monitor the biological oxygen demand in specific industrial waste waters. In the examples presented the waste water has been fairly well defined in composition and has contained substances that are readily degraded. Using the immobilized mixed culture in combination with different transducers, it was possible to achieve a response within a few hours instead of the days required by the conventional approach (HIKUMA et al., 1979; KARUBE et al., 1977a, b, c). In this case, it is possible to reuse the cell preparation after metabolic activity has ceased, a process that takes substantially longer than is normal with enzyme-based assays.

If macromolecular substrates are to be quantified, there is a need for a pretreatment to degrade them to molecules that can easily be processed. This can, of course, be carried out using a pre-column with suitable immobilized hydrolases. As an alternative to the use of aerobic cells, it was assumed that a mixed anaerobic culture would be suitable when measuring the total content of biologically degradable material in a waste water. By monitoring the gas volume produced it was possible

to use a dense population of immobilized cells and to obtain a quick answer (DISSING et al., 1984).

5.2.3 Analysis of Inhibiting Substances

The effects of inhibitory substances may be analyzed by exposure of cells to a surplus of substrate and a suspected inhibiting factor. Since whole cell metabolism is involved, one can monitor both specific toxic compounds interacting with one specific site in the cell and more generally acting compounds.

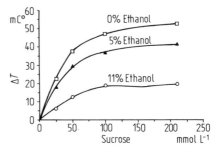

Fig. 11. Registered heat from immobilized *Saccharomyces cerevisiae* fed with an excess of sucrose in the presence of different concentrations of ethanol.

As an example of the latter case, a biosensor based on immobilized yeast cells was utilized to monitor the inhibiting effects of the medium from an ethanol fermentation using the same yeast. A cell-free medium was mixed with an excess of sucrose before being fed to the preparation of immobilized cells. The degree of inhibition was detected as a reduced efficiency in converting sucrose (Fig. 11).

5.2.4 Analysis of Specific Metabolites

Cells may often be used as convenient packages of certain enzymes. When applying cell-based biosensors for analysis of specific molecules, this is often the case. Depending on the complexity of the medium to be analyzed, it is important to make sure that there are no competing or alternative reactions going on. To achieve this, partial denaturation of the cells has sometimes to be carried out. The analyzing technique and the problems encountered are very much the same as for enzyme-based biosensors.

6 Binding Assays: Examples and Applications

Biochemical recognition assays involving no catalytic reaction are often grouped as binding assays. A broad spectrum of reactant pairs can be utilized to set up such assays. Tab. 5 lists some of the reactant pairs and their analytical applications. Enzymes are not included in the table even if one, at least in theory, can use an enzyme for binding without catalysis. This may be carried out in pH-regions where the enzyme is capable of binding, but where catalysis does not occur. The most commonly studied reactant pair is antigen/antibody. It forms the

Tab. 5. Reactant Pairs with Biospecific Binding and Examples of Their Use in Analysis

Entity 1	Entity 2	Application
Antibody	Antigen	Immunoassay
Antibody	Hapten	Immunoassay
Avidin	Biotin	Immuno/histo-chemistry
Protein A	Fc on IgG	Immunoassay
Protein G	Fc on IgG	Immunoassay
Lectin	Carbohydrate Glycoprotein Glycolipid	Carbohydrate assay
DNA	DNA-probe	Specific DNA analysis
RNA	RNA-probe	Specific RNA analysis
Receptor	Ligand	Receptor assay Ligand assay
Enzyme	Inhibitor	Enzyme/inhibitor
Enzyme	Activator	Enzyme/activator

basis for immunoassays that are widely used in clinical chemistry and are now expanding into veterinary medicine, agricultural analysis, environmental analysis, and food analysis. The assays available on the market are mainly based on establishing equilibrium in the binding reaction between the two reactants. This leads to long incubation times and thus laborious handling. When designing biosensors for the same reactant pairs, two main types of immunosensors can be identified: the dip-stick type and the reversible type.

The other reactant pairs described in Tab. 5 can all be utilized for biosensor construction. So far very few sensors have been described. Receptor-based sensors will be discussed in this paragraph. The other potential biosensors, e.g., lectin sensors, may very well become important in the future. Today, however, very few publications are available (BORREBAECK and MATTIASSON, 1980).

6.1 Immunosensors

6.1.1 Dip-Stick Immunosensor

By tradition, antisera have been selected according to the binding constant. The stronger the binding the better. This has formed the basis for sensitive binding assays. With strong binding, reversibility in the binding has been sacrificed. This means that immunoassays set up according to traditional principles are sensitive, but that it is difficult to split the complex and reuse the immobilized entity. When designing biosensors in this way, a disposable sensor is the result, i.e., an immuno-dip-stick. There may be reasons for using such devices, especially if they can be designed to avoid wet chemistry. However, it is doubtful whether they can be regarded as biosensors in the strict sense.

A variation on this theme is to dissociate the bound compound and to reuse the immobilized entity. This is the basis for flow injection ELISA (MATTIASSON et al., 1977; MATTIASSON and LARSSON, 1987; MATTIASSON et al., 1989). Here, the antibody is immobilized to a support with very low nonspecific adsorption, e.g., Sepharose. It is then placed in a small column in a continuous flow of buffer. The sample to be analyzed is mixed with labelled antigen prior to injection into the flow system. The sample will pass the column of immobilized antibodies as a short pulse. During passage there is competition for binding between labelled and native antigen. In a subsequent step a pulse of substrate is administered to the system, and the amount of enzyme-labelled antigen is read. From this it is possible to deduce the amount of native antigen present in the sample. The assay cycle is terminated by splitting the immunocomplex, for instance, at a low pH (glycine pH 2.2), and after reconditioning the system is ready for another cycle. Upon repeated use one can expect inactivation of the antibodies and, therefore, a decreased capacity of the immunosorbent. However, on passage through the column, the native and the labelled antigen will compete under identical conditions even if the absolute number of

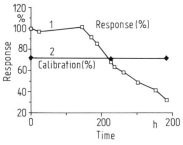

Fig. 12. Response stability of an antibody–Sepharose CL 4B-column. Curve 1: peak height when passing a 100% sample, i.e., a sample with only enzyme-labelled antigen. Curve 2: peak height in percent of a preceding pure aggregate pulse obtained for reference samples containing a constant amount of native antigen, in this case human serum albumin. (From BORREBAECK et al., 1978, with permission)

binding sites decreases. Therefore, the relative amount of the bound labelled antigen will remain constant, even if the absolute quantity of both labelled and native antigen is reduced, as can be seen in Fig. 12 (BORREBAECK et al., 1978).

This concept of fast competitive binding operates under conditions far from equilibrium.

The shorter the time of exposure, the faster the analysis. However, to achieve speed one has to sacrifice sensitivity. In most cases it is possible to carry out analyses in the region of 10^{-8} to 10^{-9} mol/L. If higher sensitivity is needed longer times of exposure can be used. The experimental procedure involves very few manual steps, and, therefore, it is possible to substantially reduce experimental error in comparison with what is normal for ELISA.

Flow injection ELISA has been performed with several different sensors such as polarographic oxygen sensors (MATTIASSON and NILSSON, 1977), spectrophotometers (MATTIASSON and BORREBAECK, 1978), and flow calorimeters (MATTIASSON et al., 1977).

More recently a sandwich type of assay using the same experimental setup has been published (DE ALWIS and WILSON, 1985).

As indicated in Sect. 3, several arrangements of the biocomponent in relation to the transducer are known. By covering the tip of the electrodes with a layer of the immunoactive component, true immunoelectrodes may be achieved. In most cases they are based on competitive binding assays as described for the flow-ELISA-procedure. When such an electrode is placed in a flow system, a flow-ELISA-system is obtained (MATTIASSON et al., 1977; MATTIASSON and LARSSON, 1987). The alternative is to dip the immunoelectrode into the solution (AIZAWA et al., 1979; DOYLE et al., 1984).

There may be many further alternative configurations (e.g., immuno-FET) but the basic chemistry and performance is the same.

6.1.2 Reversible Immunoassays

If immunoassays are to be used in fermentation control or continuous monitoring in medical biotechnology, it facilitates the handling substantially if a direct binding assay is set up. In such a system direct interaction between the native antigen in the sample and the immobilized antibody results in a readable signal. Several potential solutions have been presented to meet these requirements (Tab. 6). Some of these are based on expensive equipment and others are very sensitive to external influences.

Tab. 6. Direct Reading of Interactions During an Assay

Measuring Principle	Analyte Studied	Reference
Ellipsometry	NAD–lactate DH	MANDENIUS et al., 1984
	Mannan–Con A	HORISBERGER, 1980
	Biotinyl–Con A	
	Avidin	HORISBERGER, 1980
	Fibrinogen–anti-fibrinogen	CUYPERS et al., 1978
Reflectometry	Albumin	IWAMOTO et al., 1982
	IgG	WELIN et al., 1984
ImmunoFET	Antisyphilis	COLLINS and JANATA, 1982
Piezoelectric systems	Human IgG	ROEDERER and BASTIAANS, 1983
Streaming potential	Con A–glucosides	GLAD et al., 1986
	mcAb–anti-Ig	MIYABAYASHI and MATTIASSON, 1989

It is at present too early to state which of the techniques, if any, will be successful.

Both electrochemical and optical transducers have been used (NGO, 1987).

6.1.2.1 Streaming Potential

When a liquid is pumped over a bed of charged particles it is possible to monitor a streaming potential across the bed. The amplitude is dependent upon many things, among them the flow rate, the ionic strength and the charge distribution in the bed. If the flow rate and the ionic strength are kept constant, it is possible to register the signal as a function of a changing charge distribution on the solid support. By placing an affinity sorbent in the flow and monitoring the streaming potential over the bed, it is possible to register affinity binding in a direct way. This kind of assay is, of course, very general and also very sensitive to interference. By operating with a reference bed (Fig. 13) without suitable ligands it is possible to subtract any nonspecific interaction that otherwise might contribute to the amplitude of the signal (GLAD et al., 1986). In some recent publications the streaming potential system was operated continuously, at least for the time period needed to saturate the affinity column (MATTIASSON and MIYABAYASHI, 1988; MIYABAYASHI and MATTIASSON, 1990). The

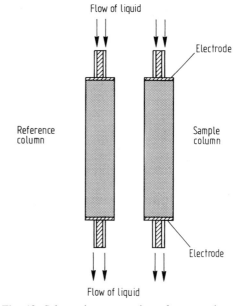

Fig. 13. Schematic presentation of a streaming potential dual-flow cell. The liquid is pumped over two identical columns, except that one column contains a specific antibody and the other is blank. The streaming potential between the electrodes at the top and the bottom of the column is registered. By subtracting the reference reading from that of the sample column, a value for the specific binding is obtained.

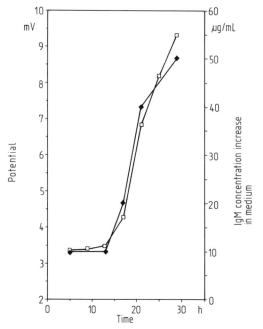

Fig. 14. Time course for production of IgM from a hybridoma cell cultivation. Filled symbols show the results from a RIA, and open symbols are the potentiometric readings from the streaming potential.

sensitivity varies with the charge density of the compound bound to the support, but in general terms it is possible to achieve signals separated from background noise at concentration levels above 10^{-9} mol/L. It is still too early to state what the potential of this technique will be. At present it is possible to monitor the concentration of monoclonal antibodies during cultivation (Fig. 14).

6.1.2.2 Reflectometry

When plane-polarized light is reflected from a solid surface there is a minimum reflectance at a certain angle of incidence, the pseudo-Brewster angle. When organic material is adsorbed onto the surface an increase in reflectance is observed. The more material is adsorbed, the larger the change. High sensitivity is obtained in immunoassays because of the large difference in refractive index between silicon and organic material.

Using reflectometry it was possible first to monitor the adsorption of an antigen to the surface, and, in a subsequent step when antibodies were bound, to detect and quantify them (WELIN et al., 1984).

6.2 Receptor-Based Sensors

Pheromones and insect antennae demonstrate clearly what can be achieved when constructing biosensors based on receptor–ligand interactions. To set up such an assay one needs to be able either to isolate the receptor or to include the whole cell or organ where the receptor is located.

One preliminary sensor was constructed by using whole crab antennae. The sensor showed very low stability and may be regarded only as an academic demonstration that it is possible to use receptors for constructing biosensors.

The construction of a receptor-based assay for nerve gases was described in an interesting presentation (TAYLOR et al., 1988). The sensitivity was high as expected, but the stability was far better than what had been achieved before. It still remains a task for chemists to disclose the mechanism of immobilization before one can decide whether this is a general technique for stabilizing receptors.

6.3 Analytical Performance of Affinity Biosensors

From the examples discussed above it is obvious that immunoassays can be designed to be very sensitive, fast, and even operative in a continuous manner. At the present state of development these characteristics cannot all be obtained using one set of operating conditions. One must, therefore, make compromises and optimize the assay according to set requirements. However, there is still a very small data base available from biotechnological processes regarding variation of concentrations of macromolecules with time.

The concentration range can easily be defined by carrying out conventional off-line analyses. In most cases if not all, the concentrations of the target molecule will be far above the detection limits of equilibrium binding assays. This opens the possibility for applying faster, non-equilibrium binding assays or looking for continuous measurements. It is, however, still far too early to formulate a general strategy when designing and utilizing immunobased biosensors.

7 Biosensors in Extreme Environments

Efforts have mainly been directed towards applying biosensors in aqueous environments under conditions a biochemist would call gentle. However, there are many other needs for bioanalysis that raise strong challenges to biosensor technology.

7.1 Biosensors in Biological Liquids

A challenge encountered by every biosensor in fermentation broth or body fluids is fouling by protein and, eventually, by cells. Such secondary layers may change the response characteristics of the biosensor and may lead to totally erroneous signals.

Furthermore, for *in vivo* measurements on patients such protein-depositing activities may activate the coagulation system. It is therefore of utmost importance for future medical applications as well as for use in bioreactors that biosensors be made with biocompatible surfaces.

7.2 Biosensors in Organic Solvents

In organic chemistry it would be desirable in many cases to follow a process more accurately. Today, samples are removed from the reaction vessel and analyzed outside, either using gas chromatography or HPLC. However, if it were possible to apply the biosensor concept under these conditions, one could exploit the unique selectivity and regioselectivity of enzymes for monitoring various synthetic

processes. Is this science fiction, or is there a chance that the dream will become a reality?

Bioorganic synthesis, i.e., the technique of using enzymes in organic solvents for organic synthesis, has undergone a development that few even dreamt of ten years ago. Today it is fully possible to operate with enzymes in pure organic solvents, e.g., in chloroform, toluene, or hexane (TRAMPER et al., 1985; LAANE et al., 1987). There is a requirement for an extremely small amount of water for hydration of essential regions on the protein molecule. In several studies it has been shown that the enzymes under these very non-physiological conditions are more stable than in aqueous solution or immobilized in contact with an aqueous environment (KLIBANOV, 1986; RESLOW et al., 1987).

Very few examples of successful utilization of the enzyme-biosensor concept with organic solvents have as yet been reported. In two of these, oxidases (mainly cholesterol and phenol oxidase) have been studied, since both the substrate and oxygen have higher solubilities in the organic phase than in water (DANIELSSON et al., 1989; HALL et al., 1988; FLYGARE et al., 1988; KAZANDIJAN et al., 1986). We have tried to monitor, e.g., ester synthesis in organic solvents using a potentiometric electrode. It was clearly shown that the signals obtained correlated very well with the substrate concentrations. When the enzyme was omitted or any other factor was changed, no enzymatic activity was noted and no potentiometric signal was registered (MIYABAYASHI et al., 1989). The performance of the enzyme electrode correlated very well with earlier observations on the dependence on the water activity in the medium (RESLOW et al., 1987).

From the studies carried out so far it is absolutely clear that one can operate with enzyme-based sensors in organic solvents and generate signals that reflect the process in the medium. However, to apply such sensors to traditional organic processes may require much time and effort. Temperature has been carefully controlled and many of the aggressive reagents often used in organic synthetic work are avoided. The basic strategy may be to draw off a small stream and use that for monitoring. Only in the longer perspective might it be realistic to insert the sensors directly into the reactors.

7.3 Biosensors Operated in Air

It is a great challenge to analyze air using biosensors. One early report in this field was the nerve gas sensor developed for the US army (GOODSON and JACOBS, 1976). However, in this application the air was passed through a scrubber and any nerve gas present was thus transferred into aqueous solution (BAUMAN et al., 1965). Since it has been shown that one can operate with enzymes with a very thin hydration layer in organic solvents, it seems reasonable to assume that the same conditions should hold for measurements in air. One area of application is the analysis of various volatile compounds such as nerve gases, small organic alcohols, and aldehydes.

8 Future Developments

Biosensor research has led to the development of a broad spectrum of biosensors, at least as judged by the academic literature. However, commercialization has not been as successful as expected. Though many types of biosensors have been described, there are few that work reliably under realistic conditions outside the well-controlled laboratory environment. This lack of commercial success has led to the situation that only the true enthusiasts use biosensors for process control. The more common applications will have to wait until reliable and robust instruments are available. Therefore, in the future much work will be done to make biosensors more adapted to the field so that at least some of this enormous potential can be tapped.

Implantable biosensors have great potential importance in medicine and in fermentation control. By tailoring the surface properties of the sensors so that no growth or adsorption takes place, it will be possible to apply such sensors in many new environments.

Miniaturization of sensors may be attractive, especially for implantable sensors, and if multifunctional sensors are to be developed. Much work is already under way in this direction. Using such small sensors it will be possible to include sensors for several essential compounds in the same sensor so that one exposure of the sensing device can provide information on many different essential components. To reach this goal, the integration of microelectronics and biosensor work needs to advance further. Current reports in this field are very promising, which is why it seems realistic to forecast that this development will take place in the near future.

Protein stability is one of the bottlenecks when constructing stable biosensors. By applying protein engineering it is possible, e.g., to eliminate sensitive SH groups or to establish $S-S$ bonds (PERRY and WETZEL, 1984). It may thus be possible to tailor the enzymes to the applications. Another alternative is to apply more stable enzymes from thermophilic organisms when constructing the sensor. The introduction of doped electrodes and the emerging awareness of how to integrate electrically conducting polymers show promise for the future.

Some recent developments on applications of biosensors in extreme environments are mentioned in Section 7. These developments will continue, and further developments, especially in the area of bioorganic sensors, can be expected.

9 References

ADLERCREUTZ, P., MATTIASSON, B. (1982), Oxygen supply to immobilized biocatalysts. A model study, *Acta Chem. Scand.* **B36**, 651–653.

AIZAWA, M., MORIOKA, A., SUZUKI, S., NAGAMURA, Y. (1979), Enzyme immunosensor. III. Amperometric determination of human chorionic gonadotropin by membrane bound antibody, *Anal. Biochem.* **94**, 22–28.

ALBERY, W. J., BARLETT, P. N., CRASTON, D. H. (1985), Amperometric enzyme electrodes. Part II. Conducting organic salts as electrode materials for the oxidation of glucose oxidase, *J. Electroanal. Chem.* **194**, 223–235.

ALWIS, DE, W. U., WILSON, G. S. (1985), Rapid sub-picomole electrochemical enzyme immunosensor for immunoglobulins, *Anal. Chem.* **57**, 2754–2756.

APPELQVIST, R., MARKO-VARGA, G., GORTON, L., TORSTENSSON, A., JOHANSSON, G. (1985), Enzymatic determination of glucose in a flow system by catalytic oxidation of the nicotinamide coenzyme at a modified electrode, *Anal. Chim. Acta* **169**, 237–247.

APPELQVIST, R., MARKO-VARGA, G., GORTON, L., JOHANSSON, G. (1986), Amperometric determination of starch, glycogen and amylopectin in a FIA system using enzyme reactors and a modified electrode as detecting device, *Proc. 2nd Int. Meeting Chemical Sensors*, Bordeaux 1986, G-25, 603.

ARNOLD, M. A., RECHNITZ, G. A. (1980), Comparison of bacterial, mitochondrial, tissue and enzyme biocatalysts for glutamine selective membrane electrodes, *Anal. Chem.* **52**, 1170–1174.

BARLETT, P. N., WHITAKER, R. G., GREEN, M. J., FREW, J. (1987), Covalent binding of electron relays to glucose oxidase, *J. Chem. Soc. Chem. Commun.*, 1603–1604.

BAUMAN, G., GOODSON, L., GUILBAULT, G. G., KRAMER, D. (1965), Preparation of immobilized cholinesterase for use in analytical chemistry, *Anal. Chem.* **37**, 1378–1381.

BORREBAECK, C., MATTIASSON, B (1980), Lectin-carbohydrate interactions studied by a competitive enzyme inhibition assay, *Anal. Biochem.* **107**, 446–450.

BORREBAECK, C., BÖRJESSON, J., MATTIASSON, B. (1978), Thermometric enzyme linked immunosorbent assay in continuous flow system: optimization and evaluation using human serum albumin as a model system, *Clin. Chim. Acta* **86**, 267–278.

BOUIN, J. C., ATALLAH, M. T., HULTIN, O. H. (1976a), The glucose oxidase-catalase system, *Methods Enzymol.* **44**, 478–488.

BOUIN, J. C., ATALLAH, M. T., HULTIN, O. H. (1976b), Parameters in the construction of an immobilized dual enzyme catalyst, *Biotechnol. Bioeng.* **18**, 179–187.

BROOKS, S. L., ASHBY, R. E., TURNER, A. P. F., CALDER, M. R., CLARKE, D. J. (1987), Development of an on-line glucose sensor for fermentation monitoring. *Biosensors* **3**, 45–56.

BÜLOW, L., MOSBACH, K. (1982), Ligation of restriction endonuclease-generated DNA fragments using immobilized T4 DNA ligase, *Biochem. Biophys. Res. Commun.* **107**, 458–465.

CASS, A. E. G., DAVIS, G., FRANCIS, G. D., HILL, H. A. O., ASTON, W. J., HIGGINS, I, J., PLOTKIN, E. V., SCOTT, L. D. L., TURNER, A. F. P. (1984), Ferrocene mediated enzyme electrode for amperometric determination of glucose, *Anal. Chem.* **56**, 667–671.

CHEN, B., KUAN, S., GUILBAULT, G. G. (1980), A creatinine specific enzyme electrode, *Anal. Lett.* **13**, 1607–1624.

CLARK, L. C., JR., LYONS, C. (1962), *Ann. N. Y. Acad. Sci.* **102**, 29.

CLARK, L. C., NOYES, L. K., SPOKANE, R. B., SUDAN, R., MILLER, M. L. (1988), Long-term implantation of voltammetric oxidase/peroxide glucose sensor in the rat peritoneum, *Methods Enzymol.* **137**, 68–89.

CLELAND, N., ENFORS, S.-O. (1984), Monitoring of glucose consumption in an *Escherichia coli* cultivation with an enzyme electrode, *Anal. Chim. Acta* **163**, 281–285.

COLLINS, S., JANATA, J. (1982), A critical evaluation of the mechanism of potential response of antigen polymer membranes to the corresponding antisera, *Anal. Chim. Acta* **136**, 93–99.

CORCORAN, C. A., RECHNITZ, G. A. (1985), Cell based biosensors, *Trends Biotechnol.* **3**, 92–96.

CUYPERS, P. A., HERMENS, W. T., HEMKER, H. C. (1978), Elipsometry as a tool to study protein films at liquid-solid interfaces, *Anal. Biochem.* **84**, 56–57.

DANIELSSON, B., GADD, K., MATTIASSON, B., MOSBACH, K. (1977), Enzyme thermistor determination of glucose in serum using immobilized glucose oxidase, *Clin. Chim. Acta* **81**, 163–175.

DANIELSSON, B., FLYGARE, L., VELEV, T. (1989), Biothermal analysis performed in organic solvents, *Anal. Lett.* **22** (6), 1417–1428.

DANIELSSON, B., MATTIASSON, B., KARLSSON, R., WINQVIST, F. (1979), Use of enzyme thermistor in continuous measurements and enzyme reactor control, *Biotechnol. Bioeng.* **21**, 1749–1766.

DANIELSSON, B., MATTIASSON, B., MOSBACH, K. (1981a), Enzyme thermistor devices and their analytical applications, *Appl. Biochem. Bioeng.* **3**, 97–143.

DANIELSSON, B., RIEKE, E., MATTIASSON, B., WINQUIST, F., MOSBACH, K. (1981b), Determination by the enzyme thermistor of cellobiose formed on degradation of cellulose, *Appl. Biochem. Biotechnol.* **6**, 207–222.

D'COSTA, E. J., HIGGINS, I. J., TURNER, A. P. F. (1986), Quinoprotein glucose dehydrogenase and its application in an amperometric glucose sensor, *Biosensors* **2**, 71–87.

DEGANI, Y., HELLER, A. (1987), Direct electrical communication between chemically modified enzymes and metal electrodes. 1. Electron transfer from glucose oxidase to metal electrodes via electron relays, bound covalently to the enzyme, *J. Phys. Chem.* **91**, 1285–1289.

DEGANI, Y., HELLER, A. (1988), Direct electrical communication between chemically modified enzymes and metal electrodes. 2. Methods for bonding electron-transfer relays to glucose oxidase and D-amino-acid oxidase, *J. Am. Chem. Soc.* **110**, 2615–2620.

DISSING, U., LING, T. G. I., MATTIASSON, B. (1984), Monitoring of methanogenic processes with an immobilized mixed culture in combination with a gas-flow meter, *Anal. Chim. Acta* **163**, 127–133.

DOMININGUEZ, E., MARKO-VARGA, G., HAHN-HÄGERDAL, B., GORTON, L. (1988), A flow injection analysis system for the amperometric determination of xylose and xylulose using coimmobilized enzymes and a modified electrode, *Anal. Chim. Acta* **213**, 139–150.

DOYLE, M. J., HALSALL, H. B., HEINEMAN, W. R. (1984), Enzyme-linked immunoadsorbent assay with electrochemical detection of α_1-acid glycoprotein, *Anal. Chem.* **56**, 2318–2322.

ENFORS, S. O. (1981), Oxygen stabilized enzyme electrode for glucose analysis in fermentation broths, *Enzyme Microb. Technol.* **3**, 29–32.

ENFORS, S. O., NILSSON, H. (1979), Design and response characteristics of an enzyme electrode for measurement of penicillin in fermentation broth, *Enzyme Microb. Technol* **1**, 260–264.

FATT, I. (1976), *Polarographic Oxygen Sensor. Its Theory of Operation and its Application in Biology, Medicine and Technology.* Cleveland, Ohio: CRC Press.

FLYGARE, L., DANIELSSON, B., MOSBACH, K. (1988), *Abstr. B 29,* 8th International Biotechnology Symposium, Paris, July 1988.

FOULDS, C. N., LOWE, C. R. (1986), Enzyme entrapment in electrically conducting polymers. Immobilization of glucose oxidase in polypyrrole and its application in amperometric glucose sensors, *J. Chem. Soc. Faraday Trans. I.* **82**, 1259–1264.

FOULDS, C. N., LOWE, C. R. (1988), Immobilization of glucose oxidase in ferrocene-modified pyrrole polymers, *Anal. Chem.* **60**, 2473–2478.

FULTON, S. P., COONEY, C. L., WEAVER, J. C. (1980), Thermal enzyme probe with differential temperature measurements in a laminar flow-through cell, *Anal. Chem.* **52**, 505–508.

GLAD, C., SJÖDIN, K., MATTIASSON, B. (1986), Streaming potential – a general affinity sensor, *Biosensors* **2**, 89–100.

GOODSON, L. H., JACOBS, W. B. (1976), Monitoring of air and water for enzyme inhibitors, *Methods Enzymol.* **44**, 647–658.

GREEN, N. M. (1963), Avidin. I. The use of ^{14}C biotin for kinetic studies and for assay, *Biochem. J.* **89**, 585–591.

GUILBAULT, G. G. (1984), *Analytical Uses of Immobilized Enzymes.* New York: Marcel Dekker.

GUILBAULT, G. G., MONTALVO, J. G. (1970), An enzyme electrode for the substrate urea, *J. Am. Chem. Soc.* **92**, 2533–2538.

GUILBAULT, G. G., SHU, F. R. (1971), An electrode for the determination of glutamine, *Anal. Chim. Acta* **56**, 333–338.

HALL, G. F., BEST, D. J., TURNER, A. P. F. (1988), Amperometric enzyme electrode for the determination of phenols in chloroform, *Enzyme Microb. Technol.* **10**, 543–546.

HIKUMA, M., SUZUKI, H., YASUDA, T., KARUBE, I., SUZUKI, S. (1979), Amperometric estimation of BOD by using living immobilized yeast, *Eur. J. Appl. Microbiol. Biotechnol.* **8**, 289–297.

HOLST, O., HÅKANSON, H., MIYABAYASHI, A., MATTIASSON, B. (1988), Monitoring of glucose in fermentation processes using a commercial glucose analyser, *Appl. Microbiol. Biotechnol.* **28**, 32–36.

HORISBERGER, M. (1980), An application of elipsometry. Assessment of polysaccharide and glycoprotein interactions with lectin at a liquid/solid interface, *Biochim. Biophys. Acta* **632**, 298–309.

HUBBARD, A. L., GOHN, Z. A. (1976), in: *Biochemical Analysis of Membrane* (MADDE, A. H., Ed.), pp. 427–501. New York: Wiley.

HÅKANSON, H. (1988), Portable continuous blood glucose analyzer, *Methods Enzymol.* **137**, 319–326.

HÅKANSON, H., HOLST, O., WEHTJE, E., MATTIASSON, B. (1990), Monitoring of glucose and lactic acid in cultivations of lactic acid bacteria using an automated biosensor-based analyzer, in preparation.

IUB (1979), *Enzyme Nomenclature 1978.* New York: Academic Press.

IWAMOTO, G. K., VAN WAGENEN, R. A., ANDRADE, J. D. (1982), Insulin adsorption: intrinsic tyrosine interfacial fluorescence, *J. Colloid Interface Sci.* **86**, 581–585.

JOHANSSON, G., EDSTRÖM, K., ÖGREN, L. (1976), An enzyme reactor electrode for determination of amino acids, *Anal. Chim. Acta* **85**, 55–60.

JULLIARD, J. H., GOLDINOT, G., GAUTHERON, D. C. (1971), Some modifications of the kinetic properties of bovine liver glutamate dehydrogenase (NAD(P)) covalently bound to a solid matrix of collagen, *FEBS Lett.* **14**, 185–188.

KARUBE, I., MATSUNAGA, T., MITSUDA, S., SUZUKI, S. (1977a), Microbial electrode BOD sensor, *Biotechnol. Bioeng.* **19**, 1535–1547.

KARUBE, I., MATSUNAGA, T., SUZUKI, S. (1977b), A new microbial electrode for BOD estimation, *J. Solid-Phase Biochem.* **2**, 97–104.

KARUBE, I., MITSUDA, S., MATSUNAGA, T., SUZUKI, S., KADA, T. (1977c), A rapid method for the estimation of BOD using immobilized microbial cells, *J. Ferment. Technol.* **55**, 243–248.

KARUBE, I., SODA, K., TAMIYA, E., GOTOH, M., KITAGAWA, Y., SUZUKI, H. (1988), New microbiosensor for estimation of fermentation parameters. *Proc. 8th Int. Biotechnol. Symp.*, Vol. I, pp. 537–546 (DURAND, G., BOBICHAN, L., FLORENT, J., Eds.). Paris: Société Française de Microbiologie.

KAZANDIJAN, R. Z., DORDICK, J. S., KLIBANOV, A. M. (1986), Enzymatic analysis in organic solvents, *Biotechnol. Bioeng.* **28**, 417–421.

KLIBANOV, A. M. (1986), Enzymes that work in organic solvents, *CHEMTECH* **16**, 354–359.

KUAN, S. S., GUILBAULT, G. G. (1987), Ion-selective electrodes and biosensors based on IESs, in: *Biosensors – Fundamentals and Applications*, pp. 135–152 (TURNER, A. P. F., KARUBE, I., WILSON, G. S., Eds.). Oxford: Oxford University Press.

LAANE, C., TRAMPER, J., LILLY, M. D. (Eds.) (1987), *Biocatalysis in Organic Media*. Amsterdam: Elsevier.

LANCET, D., ISENMAN, D., SJÖDAHL, J., SJÖQUIST, J., PECHT, I. (1978), Interaction between staphylococcal protein A and immunoglobulin domains, *Biochem. Biophys. Res. Commun.* **85**, 608–614.

MANDENIUS, C. F., DANIELSSON, B., MATTIASSON, B. (1980), Enzyme thermistor control of the sucrose concentration of a fermentation with immobilized yeast, *Acta Chem. Scand.* **34B**, 463–465.

MANDENIUS, C. F., DANIELSSON, B., MATTIASSON, B. (1981), Process control of an ethanol fermentation with an enzyme thermistor as a sucrose sensor, *Biotechnol. Lett.* **3**, 629–634.

MANDENIUS, C. F., WELIN, S., DANIELSSON, B., LUNDSTRÖM, I., MOSBACH, K. (1984), The interaction of proteins and cells with affinity ligands, covalently coupled to silicon surfaces as monitored by elipsometry, *Anal. Biochem.* **137**, 106–114.

MANDENIUS, C. F., BÜLOW, L., DANIELSSON, B., MOSBACH, K. (1985), Monitoring and control of enzymic sucrose hydrolysis using on-line biosensors, *Appl. Microbiol. Biotechnol.* **21**, 135–142.

MARKO-VARGA, G., APPELQVIST, R., GORTON, L. (1985), A glucose sensor based on glucose dehydrogenase adsorbed on a modified carbon electrode, *Anal. Chim. Acta* **179**, 371–379.

MATSUNAGA, T., KARUBE, I., SUZUKI, S. (1978), Electrochemical microbioassay of vitamin B₁, *Anal. Chim. Acta* **98**, 25–30.

MATSUNAGA, T., KARUBE, I., SUZUKI, S. (1980), Electrochemical determination of cell populations, *Eur. J. Appl. Microbiol. Biotechnol.* **10**, 125–132.

MATSUSHITA, K., OHNO, O., SHINAGAWA, E., ADACHI, O., AMEYAMA, M. (1982), Membrane bound, electron transport linked D-glucose dehydrogenase of *Pseudomonas fluorescens*. Interactions of the purified enzyme with ubiquinone or phospholipid, *Agric. Biol. Chem.* **46**, 1007–1011.

MATTIASSON, B. (1981), Reversible immobilization of enzymes with special reference to analytical applications, *J. Appl. Biochem.* **3**, 183–194.

MATTIASSON, B. (1983), Analytical applications of immobilized cells, in: *Immobilized Cells and Organelles* Vol. II, pp. 95–113 (MATTIASSON, B., Ed.). Boca Raton, Florida: CRC Press.

MATTIASSON, B. (1988), Affinity immobilization, *Methods Enzymol.* **137**, 647–656.

MATTIASSON, B., BORREBAECK, C. (1978), Nonequilibrium, isokinetic enzyme immunoassay of insulin using reversibly immobilized antibodies, in: *Enzyme Labelled Immunoassay of Hormons and Drugs* (PAL, S. B., Ed.), pp. 91–105. Berlin: Walter de Gruyter.

MATTIASSON, B., LARSSON, K. (1987), Flow injection enzyme immunoassay – a quick and convenient binding assay, in: *Proc. 4th Eur. Congr. Biotechnology*, Amsterdam, June 14–19, 1987, Vol. 4, pp. 517–522 (NEIJSSEL, O. M., VAN DER MEER, R. R., VAN LUYBEN, K. Ch. A. M., Eds.).

MATTIASSON, B., MIYABAYASHI, A. (1988), Registration of the streaming potential over an affinity column in continuous flow system as a means to biospecifically quantify proteins, *Anal. Chim. Acta* **213**, 79–89.

MATTIASSON, B., NILSSON, H. (1977), An enzyme immunoelectrode. Assay of human serum albumin and insulin, *FEBS Lett.* **78**, 251–254.

MATTIASSON, B., GESTRELIUS, S., MOSBACH, K. (1974), Some observations on the behaviour of an immobilized allosteric enzyme: phosphofructokinase, in: *Enzyme Engineering Vol. 2*, pp. 181–182 (PYE, E. K., WINGARD, L. B., Jr., Eds.). New York: Plenum Press.

MATTIASSON, B., BORREBAECK, C., SANFRIDSSON, B., MOSBACH, K. (1977), Thermometric enzyme linked immunosorbent assay: TELISA, *Biochim. Biophys. Acta* **483**, 221–227.

MATTIASSON, B., DANIELSSON, B., WINQUIST, F., NILSSON, H., MOSBACH, K. (1981), Enzyme thermistor analysis of penicillin in standard solutions and in fermentation broth, *Appl. Environ. Microbiol.* **41**, 903–908.

MATTIASSON, B., LARSSON, P.-O., SAHLIN, P. (1982), Vitamin analysis with use of a yeast electrode, *Enzyme Microb. Technol.* **4**, 251–254.

MATTIASSON, B., BERDÉN, P., LING, T. G. I. (1989), Flow-injection binding assays. A way to improve the speed in binding analyses, *Anal. Biochem.* **181**, 379–382.

MESSING, R. A. (1974), Simultaneously immobilized glucose oxidase and catalase in controlled-pore titania, *Biotechnol. Bioeng.* **16**, 897–908.

MEYERHOF, M. E., RECHNITZ, G. A. (1979), Microsomal thyroxine measurements with iodide selective membrane electrode, *Anal. Lett.* **12**, 1336–1346.

MIYABAYASHI, A., MATTIASSON, B. (1990), A dual streaming potential device used as an affinity sensor for monitoring hybridoma cell cultivations, *Anal. Biochem.* **184**, 165–171.

MIYABAYASHI, A., DANIELSSON, B., MATTIASSON, B. (1987), Development of a flow-cell system with dual fuel-cell electrodes for continuous monitoring of microbial populations, *Biotechnol. Tech.* **1**, 219–224.

MIYABAYASHI, A., RESLOW, M., ADLERCREUTZ, P., MATTIASSON, B. (1989), A potentiometric enzyme electrode for monitoring in organic solvents, *Anal. Chim. Acta* **219**, 27–36.

MOSBACH, K. (Ed.) (1976), *Methods Enzymol.* **44**. New York: Academic Press.

MOSBACH, K. (Ed.) (1988), *Methods Enzymol.* **135–137**. New York: Academic Press.

NEISSEL, O. M., TEMPEST, D. W., POSTMA, P. W., DUINE, J. A., JZN, J. F. (1983), Glucose metabolism by K$^+$-limited *Klebsiella aerogenes*: evidence for the involvement of a quinoprotein glucose dehydrogenase, *FEMS Microbiol. Lett.* **20**, 35–39.

NGO, T. T. (Ed.) (1987), *Electrochemical Sensors in Immunological Analysis.* New York: Plenum Press.

NILSSON, H., ÅKERLUND, A.-C., MOSBACH, K. (1973), Determination of glucose, urea and penicillin using an enzyme-pH-electrode, *Biochim. Biophys. Acta* **320**, 529–534.

OLSSON, B., STÅLBOM, B., JOHANSSON, G. (1986), Determination of sucrose in the presence of glucose in a flow-injection system with immobilized multi-enzyme reactors, *Anal. Chim. Acta* **179**, 203–208.

O'SHANNESSY, D. J., HOFFMAN, W. L. (1987), Immobilization of glycoproteins to hydrazide containing supports, *Hoppe-Seyler's Z. Biol. Chem.* **368**, 767.

PANDEY, P. C. (1988), A new conducting polymer-coated glucose sensor, *J. Chem. Soc. Faraday Trans. I* **84**, 2259–2265.

PARKER, C. W. (1976), *Radioimmunoassay of Biologically Active Compounds.* Englewood Cliffs, New Jersey: Prentice Hall.

PENNINGTON, S. N. (1976), A small volume microcalorimeter for analytical determinations, *Anal. Biochem.* **72**, 230–237.

PERRY, J. L., WETZEL, R. (1984), Disulfide bond engineering into T4 lysozyme. Stabilization of the protein toward thermal inactivation, *Science* **226**, 555–557.

PETERSON, J. I., GOLDSTEIN, S. R., FITZGERALD, R. V., BUCKHOLD, D. K. (1980), Fiber optic pH-probe for physiological use, *Anal. Chem.* **52**, 864–869.

PILLOTON, R., MASCINI, M., CASELLA, I. G., FESTA, M. R., BOTTARI, E. (1987), Lactose determination in raw milk with a two-enzyme based electrochemical sensor, *Anal. Lett.* **20**, 1803–1814.

RESLOW, M., ADLERCREUTZ, P., MATTIASSON, B. (1987), Organic solvents for bioorganic synthesis. I. Optimization of parameters for a chymotrypsin catalyzed process, *Appl. Microbiol. Biotechnol.* **26**, 1–8.

ROEDERER, J. E., BASTIAANS, G. J. (1983), Microgravity immunoassay with piezoelectric crystals, *Anal. Chem.* **55**, 2333–2336.

RUSLING, J. F., LUTTRELL, G. H., CULLEN, L. F., PAPARIELLO, G. J. (1976), Immobilized enzyme-based flowing stream analyzer for measurement of penicillin in fermentation broths, *Anal. Chem.* **48**, 1211–1215.

SCHELLER, F., SIEGBAHN, N., DANIELSSON, B., MOSBACH, K. (1985), High sensitivity enzyme thermistor determination of L-lactate by substrate recycling, *Anal. Chem.* **57**, 1740–1743.

SCHELTER-GRAF, A., SCHMIDT, H.-L., HUCK, H. (1984), Determination of substrates of dehydrogenases in biological material in flow-injection systems with electrostatic NADH-regeneration, *Anal. Chim. Acta* **163**, 299–303.

SCHUBERT, F., SCHELLER, F. W. (1988), Organelle electrodes, *Methods Enzymol.* **137**, 152–160.

SCHULTZ, J. S., SIMS, G. (1979), Affinity sensors for individual metabolites, *Biotechnol. Bioeng. Symp.* **9**, 65–71.

SCHULTZ, J. S., MANSOURI, S., GOLDSTEIN, I. J. (1982), Affinity sensor: a new technique for developing implantable sensors for glucose and other metabolites, *Diabetes Care* **5**, 245–253.

SCHÜGERL, K. (1988), On-line analysis and control of production of antibiotics, *Anal. Chim. Acta* **213**, 1–9.

SEITZ, W. R. (1987), Optical sensors based on immobilized reagents, in: *Biosensors. Fundamentals and Applications*, pp. 599–616 (TURNER, A. F. P., KARUBE, I., WILSON, G. S., Eds.). Oxford: Oxford University Press.

SHINOHARA, H., CHIBA, T., AIZAWA, M. (1988), Enzyme microsensor for glucose with an electrochemically synthesized enzyme polyaniline film, *Sens. Actuators* **13**, 79–86.

SUZUKI, H., TAMIYA, E., KARUBE, I. (1988), Fabrication of an oxygen electrode using semiconductor technology, *Anal. Chem.* **66**, 1078–1080.

TAYLOR, R. F., MARENCHIC, I. G., COOK, E. J.

(1988), An acetylcholine receptor-based biosensor for the detection of cholinergic agents, *Anal. Chim. Acta* **213**, 131–138.

THOMPSON, H., RECHNITZ, G. (1974), Ion electrode based enzymatic analysis of creatinine, *Anal. Chem.* **46**, 246–249.

TRAMPER, J., VAN DER PLAS, H. C., LINKO, P. (Eds.) (1985), *Biocatalysis in Organic Synthesis*. Amsterdam: Elsevier.

TRAN-MINH, C., BROUN, G. (1975), Construction and study of electrodes using cross-linked enzymes, *Anal. Chem.* **47**, 1359–1364.

UMANA, M., WALLER, J. (1986), Protein-modified electrodes. The glucose oxidase/polypyrrole system, *Anal. Chem.* **58**, 2979–2983.

WEAVER, J. C., COONEY, C. L., FULTIN, S. P., SCHULER, P., TANNENBAUM, S. R. (1976), Experiments and calculations concerning a thermal enzyme probe, *Biochim. Biophys. Acta* **452**, 285–291.

WELIN, S., ELWING, H., ARWIN, H., LUNDSTRÖM, I. (1984), Reflectometry in kinetic studies of immunological and enzymatic reactions on solid surfaces, *Anal. Chim. Acta* **163**, 263–267.

WINGARD, L. B., JR. (1984), Cofactor modified electrodes, *Trends Anal. Chem.* **3**, 235–238.

WOLFBEIS, O. S. (1989), Fiber-optic biosensor based on the intrinsic fluorescence of enzymes. Poster presented at the Symposium *"Forward Look into Detection and Characterization of Chemical and Biological Species"*, Salamanca, Spain, April 1989.

II. Measurements in Bioreactor Systems

4 Characterization of Bioreactors

ANDREAS LÜBBERT

Hannover, Federal Republic of Germany

1 Introduction

Bioreactors are generally regarded as containers which are used to synthesize products by means of biochemical reactions on a commercial scale. In most instances, the metabolism of living cells is exploited to convert substrates into desired products. Organisms, however, often grow and produce only under optimized environmental conditions.

In this respect, the bioreactor is a tool to solve a logistic problem: to supply the microorganisms with the correct amounts of substrates just in time to prevent their branching into unwanted metabolic pathways. Simultaneously, all inhibitors or repressors, whether they are the main products or unwanted by-products, must be removed from the cells. Moreover, the organisms should be processed at their optimal temperature, which must be kept constant by means of a controller, since most conversion processes consume or produce heat. All related transport problems are solved by inducing appropriate fluid motions within the reactors.

As DAMKÖHLER has already pointed out in 1937, the biochemical conversion process generally cannot be separated from the physical transport process maybe sometimes academically, but not in industrial practice. The reverse is also true. No transport system can be discussed independently of what is to be transported in what quantities, at what velocities, and especially across what distances. The amount of material which is consumed by the cells in a bioreactor and that must be transported is determined by microbial (enzyme) kinetics. These tell us about the biochemical reaction pathways and the relative amounts of the different components which can be converted by the organisms. What is really converted, essentially depends on the reactor's ability to supply the cells appropriately. In this way, the bioreactor controls the effective macroscopic reaction rates and, hence, the performance of the system: If there is insufficient transport, the organisms cannot produce at their highest rate, or cannot synthesize the desired products.

In terms of bioreactor design, this interdependency of transport and biochemical kinetics leads to some problems, which result from the fact that biochemical kinetics is scale-independent, whereas transport of heat and matter is not. The amount of energy that must be expended for the transport to match the needs of the cells drastically increases with the reactor volume (SWEERE et al., 1987, 1988b). Eventually, different transport systems must be used in bioreactors of different sizes. In any case, however, the transport system must be optimized according to the requirements of a specific biotechnological system.

The current discussion in biotechnology is mainly focused on the performance of cells, and, hence, most interest is directed toward enhancing their metabolic activity. This is because it is believed that the highest innovation rate is there; but one has overlooked the fact that every increase in cell activity and active cell density leads to a higher conversion rate per volume unit of the bioreactor and hence necessitates a more efficient transport. Thus, the interest in high-performance reactors will increase with the development of improved organisms.

In the literature, performance of bioreactors is mainly discussed in terms of the efficiency of the most critical transport processes, the transport of components and heat within the continuous liquid bulk phase, and, especially in aerobic processes, the mass transfer between the dispersed gas phase and the continuous liquid phase across interphase boundaries (MOSER, 1987). The characterization of bioreactors, consequently, describes their performance in transporting mass and heat.

From the engineering point of view, the relevant bioreactors are very similar to chemical reactors. Hence, most problems can be approached in analogy to the experience gained in chemical engineering (DRAHOS and CERMAK, 1989). There are mainly three reasons for an interest in bioreactor characterization. The first is to gain a general understanding, i.e., to appreciate the profound influence of hydrodynamics on the final result of the production process. The second is to provide a general base for a comparison of the performance of different reactors for a given biochemical system. Finally, and this is the most ambitious reason, characterization is a prerequisite for a scientifically based reactor scale-up. In all cases, appropriate measuring techniques are indispensible to base these activities on reliable data. Therefore, the main interest of this chapter must be directed toward methods which can actually be used to obtain data from actually existing biological systems.

2 Characteristic Properties

The transport of mass and heat in bioreactors can be divided as follows. First of all, substrates, fed into a real system locally, must be distributed all over the reactor. This homogenization procedure is referred to as bulk mixing property. Another task is to transfer components from one phase into another. The oxygen supply of aerobically growing cells is a cardinal example. Another specific transport step, which has not been investigated as thoroughly, is the transport from the liquid phase into microparticles, especially microorganisms which, as ARMENANTE and KIRWAN (1989) claimed, may be significantly different from the transport into the larger particles usually encountered in chemical engineering. Furthermore, since the conversion is an exothermic process in most cases, the heat produced must be removed to keep the environment of the cells at optimal temperatures. All these transports are performed or at least supported by flows within the reactors.

The characteristics of bioreactors are thus well described in terms of the physical and hydrodynamic properties of the flow generated within the reactors. Flows encountered in bioreactors are multiphase flows, and it must be stressed at the beginning of the discussion that they cannot be treated as single phase flows (JONES, 1983). Hence, descriptions, and particularly measuring methods, are completely different.

Since most bioreactors are submerse reactors, we restrict the discussion to reactors whose continuous bulk phase is a liquid in which the cells are dispersed. As a central example, for which all aspects can be demonstrated, we use reactors for aerobic cultures, e.g., reactors for antibiotics or yeast produc-

tion. It should be mentioned, however, that the different aspects discussed cannot be put into a universal order of importance, because the order of importance depends upon the particular application.

2.1 Mass Transfer Across Gas/Liquid Interfacial Areas

The oxygen supply of microorganisms is one of the most essential limiting transport processes in aerobic fermentations as, for example, in the mycelial culture broths of antibiotic productions. Since oxygen is only insufficiently soluble in water, the oxygen capacity of most fermentation media is so low that it must continuously be supplied to keep up with demand.

2.1.1 Mechanism

Oxygen is transported through a succession of steps from gas bubbles, which are dispersed in the continuous liquid phase, via the liquid bulk to the submersed microorganisms. All steps can be related to mass transfer resistances, and, as in an electrical circuit, the total resistance to the concentration current in such a serial line is mainly determined by the largest resistance. In practice, the limiting step is assumed to be oxygen transfer across the liquid boundary layer around the physical interfacial area between the continuous liquid and the bubbles. Therefore, in the literature, much attention is devoted to the absorption of oxygen from gas bubbles into a liquid bulk phase.

The dominating mechanism of mass transport across this boundary layer is molecular diffusion. According to basic physics, the current of n molecules, dn/dt, through the interfacial area A, driven by a concentration gradient dC/dx (Fick's first law), is

$$dn/dt = -DA\,dC/dx \tag{1}$$

D is the diffusion constant. If the film of strength δ is assumed to be stagnant with constant concentrations of oxygen on both sides, then Fick's law can be linearized, and, within some reference volume V, rewritten as representation that shows the essential properties of the gas absorption more clearly:

$$dC/dt = DA(C_i - C_b)/(\delta V)$$
$$= (DA/\delta V)(C_i - C_b) \qquad (2)$$

C_i denotes the concentrations at the interfacial area and C_b those within the bulk fluid. With the definition of the specific interfacial area a

$$a = A/V \qquad (3)$$

it is convenient to rewrite the equation according to the film model to

$$dC/dt = (D/\delta)a(C_i - C_b) = k_L a(C_i - C_b) \qquad (4)$$

The constant k_L is called the mass transfer coefficient. Normally, the driving concentration difference $C_i - C_b$ is calculated by the difference between the concentration C_i at the bubble–liquid interface side of the film and the measurable bulk concentration C_b of the gas. C_i is calculated by the partial pressure within the gas feed and the Henry constant.

Eq. (4) can be interpreted as an analog of Ohm's law: The current of concentration dC/dt is drawn by the potential difference represented by the concentration difference $(C_i - C_b)$ across the resistance $1/(k_L a)$. If C_b is assumed to be variable, $C_b = C$, then $1/(k_L a)$ assumes the meaning of relaxation time, which can be taken as the characteristic time of mass transfer across the interfacial area.

It must be kept in mind that this is an extremely rough model. This must be considered, if one uses the model as a base for characterizing the mass transfer in real systems. The basic variables, however, the specific interfacial area a, the mass transfer coefficient k_L, and the driving force $(C_i - C_b)$, which control the mass transfer, are readily contained.

The mass transfer is often assumed to be homogeneously distributed across the reactor, and the key quantities k_L and a are thus taken as parameters which are constant throughout the reactor, or at least can be taken as (integral) mean values. However, there is enough evidence from experimental studies to show that there are considerable gradients of nearly all physical quantities across a real reactor

(OOSTERHUIS, 1984). This directly influences the mass transfer; thus, the true transport rates are drastically different at different places in the vessels, especially in large production units. In order to characterize a bioreactor in detail, there is a need for spatially resolved data.

2.1.2 Characteristics of Mass Transfer

In order to determine the performance of a given bioreactor, one is interested in estimating all quantities relevant to mass transfer at different places in its flow, especially the mass transfer resistance $1/(k_L a)$ and its factors k_L and a.

The specific interfacial area a is that parameter which can be most directly influenced by engineering. One is interested in a large transport cross-section in order to obtain the desired absorption rate, since molecular diffusion is known to be a slow process. a depends on the gas holdup ε and the bubble size distribution. At a given holdup ε, small bubbles are required to obtain a high interfacial area.

At the same time, the rate of mass transfer can also be improved by increasing the mass transfer coefficient k_L. As can be seen from the simple film model mentioned above, this is primarily a function of the film strength δ. It can be understood immediately that δ is dependent on the shear experienced by the bubbles. This shear directly depends on the relative velocity between the bubbles and the surrounding liquid phase. Since the bubble-rise velocity in stagnant fluids primarily increases with the bubble size, one is interested in larger bubbles to reduce the film strength. Thus, there are competing requirements concerning the bubble size to reduce mass transfer resistance.

The mass transfer rate not only depends on the coefficient $k_L a$, but also on the driving force and the time the bubbles spend in the reactor as long as they release oxygen. The utilization of the oxygen, thus, depends on the distribution of bubble residence time, which is mainly influenced by the reactor geometry, especially by the height of the reactor.

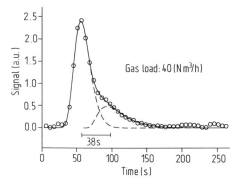

Fig. 1. Example of gas residence time data (symbols) measured in a pilot-scale bioreactor of 4 m³ total volume, obtained during yeast production (LÜBBERT et al., 1990). The structure of the data can be explained by fitting a model to the data that allows extraction of the mean residence time, the circulation time, and the amount of recirculated gas, as explained in the text.

However, efficient mass transfer from the bubbles to the bulk liquid means that their oxygen concentration decreases. Hence, the driving force for the mass transfer is reduced. Since the liquid phase that stays inside the reactor moves globally in circulatory motion, small bubbles, which almost completely follow the liquid phase motion, are likely to stay inside the reactor for much longer periods of time than they efficiently take part in the oxygen absorption. Bubbles that do not take part in the mass transfer can be regarded as dead volumes, which decrease the productivity of the process. Consequently, because of competitive requirements, the residence time distribution of the bubbles must be controlled at its optimum.

Fig. 1 shows a result obtained in an airlift loop reactor, which is an example of a reactor with a highly deterministic liquid phase flow. The structure of this residence distribution reflects the circulatory motion of the gas bubbles. Thus, additional information can be gained from the gas residence time distribution. If an appropriate model is fitted to the data, one can elicit not only the mean residence time and its variance in the statistical sense, but also from the appearance of the first peak, the time of the passage of the gas through the riser, and from the distance of ad-

jacent peaks, the mean circulation time of the gas phase. Moreover, the amount of the gas circulated once around the loop can be calculated from the relationship of the integrals of the resolved peaks. The circulatory flow of bubbles in loop reactors leads to a feedback of bubbles into the gas input and bubble formation region, where they may coalesce with

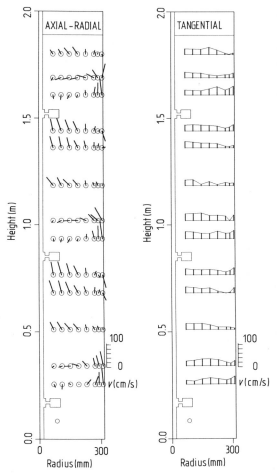

Fig. 2. Velocity profile of the bubbles within a stirred tank reactor operated with a Rushton turbine (BRÖRING et al., 1990). The measurement was made with the ultrasound pulsed Doppler method in an air-in-water model medium (BRÖRING et al., 1990). The velocity profile depends on the impeller type, but in all stirrer types investigated a feed back of the bubbles into the stirrer region was observed at realistic stirrer speeds.

fresh bubbles, reducing their mean oxygen concentration and, hence, reducing the driving force for mass transfer.

This is a drawback not only of loop reactors. Even in stirred tank reactors a considerable part of the gas is recirculated back into the stirrer region, where it is caught by the ventilated cavities formed behind the stirrer blades and, hence, in all probability coalesces. It is immediately clear that the amount of gas recirculated drastically depends on the bubble size distribution and, hence, also on the coalescence/redispersion behavior in the bulk of the dispersion. Smaller bubbles follow the liquid phase motion much better and contribute to backmixing much more. In this way, small bubbles, which do not tend to coalesce in the bulk phase, paradoxically contribute more to backmixing than the coalescing ones.

Experimental evidence on such bubble recirculation can be obtained from the bubble velocity field within a common stirred tank reactor, as shown in Fig. 2. This figure clearly demonstrates that the bubble motion is complicated and is completely different from the flow structure of the continuous liquid flow. As opposed to the gross liquid flow structure, which is essentially independent of the rheological properties of the media, Fig. 3 shows that a change in the viscous properties of the continuous liquid phase leads to completely different bubble paths. Since such bubble motion maps cannot be calculated, they must be obtained experimentally. Such detailed data are also necessary to put forward the theoretical investigation of bioreactor hydrodynamics (TRÄGARDH, 1988).

This recirculation leads to another essential hydrodynamic effect. It is known from various experiments that the power absorbed by a dispersion at constant stirrer speed decreases with the gas flow rate. The main reason for this power loss is thought to be the formation of previously mentioned ventilated gas cavities behind the stirrer blades, which lower the hydrodynamic resistance (drag) of the impeller. The gas flow through these cavities determines their size. This gas flow cannot be calculated, since one does not know the amount of gas which is recirculated into the stirrer. It is assumed that about the same amount of recirculated gas as of freshly fed gas is recirculated

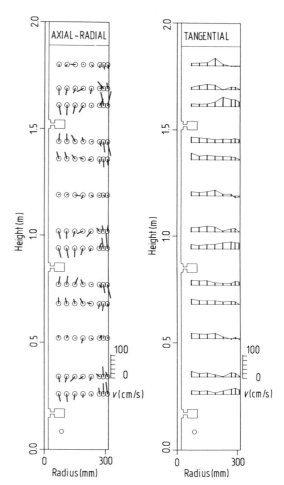

Fig. 3. Velocity profile of the bubbles within a stirred tank reactor operated with a Rushton turbine (BRÖRING et al., 1990). As compared to Fig. 2 the measurement was performed in a model fermentation medium. One percent CMC in water was chosen. The bubble flow profiles are then totally different from the results shown in Fig. 2.

(MIDDLETON, 1985). This is another argument for measuring and controlling bubble sizes and velocities.

Thus, the coalescence/redispersion process of gas bubbles is a general problem of highest importance, not only for mass transfer, but also for most other aspects of multiphase flow hydrodynamics. Coalescence of bubbles is greatly reduced in solutions compared to pure solvents. Thus, many important parameters determining the fluid dynamics and the mass transport, e.g., bubble size distribution, interfacial area, and holdup, are changing during the fermentation. Immediate consequences are grossly different residence times and, especially in airlift reactors, different liquid circulation times. Even a minor amount of a surfactant component (ppm amount) can drastically change the coalescence and redispersion behavior of the contactor (BUCKLAND et al., 1988). This will result in extreme changes of $k_L a$. As a consequence, such systems cannot be modelled from first principles. It is necessary to rely on experimental determination of the mentioned quantities. For the same reasons, measurements must be performed in exactly the same material system as the original, preferably under the same fluid dynamical conditions. This fact has not been taken due notice of in the past. It is not possible to predict $k_L a$ for other systems with confidence (MIDDLETON, 1985). The coalescence behavior also influences the power uptake of the liquid from a stirrer, because this influences the formation and the stability of the cavities behind the stirrer blades. The relevant mechanisms and the possibilities of influencing them are not completely understood at this date.

2.2 Bulk Mixing Properties of Bioreactors

Another essential task of bioreactors is the distribution of substrates or other components which are introduced at one place in the reactor or, more generally, the reduction of the concentration and temperature gradients in a sufficiently fast and efficient way. The success of the different mixing mechanisms, which take part in this homogenization, depends on the spatial resolution of interest. For biochemical reactions, homogeneities, down to the molecular scale, are required.

2.2.1 Relevant Dispersion Mechanisms

The problem to be solved is analogous to supplying the population of a country with a highly perishable food, e.g., fruit that must be imported. Several transport systems are required for an efficient distribution. First of all, a fast transport of larger amounts of this food covering all states of the nation is necessary. This may be done by big trucks covering long distances on interstate highways across the whole country. More locally, regional wholesale dealers transport the goods on regional roads in order to distribute them within states to the stores in different cities and villages. These stores must present the food to as many families as possible, who buy it and distribute it directly to the smallest units, the family members, where it is consumed.

2.2.1.1 Convective Large-Scale Flows

Geometrically, any material that is initially fed into a bioreactor at one place appears as a separate volume inside the reactor. The aim of the flow inside the reactor is to break this volume element into parts and to distribute them homogeneously over the whole reactor. Analogous to the trucks driving on the highways, large-scale convectional flows are guided by appropriate devices in order to obtain a fast but coarse distribution of matter all over the bioreactor. The first aim of bioreactor construction is thus to guide these deterministic flows so that they reach every corner of the bioreactor within a short time.

2.2.1.2 Statistical Turbulence

Wholesale business must correspond to the induction of several additional random dispersive flows. In practice, one tries to apply turbulence, which is known to be the most efficient regional dispersion mechanism. The advantage of turbulence in aiding mixing can best be understood in terms of the statistical turbulence theory. The basis of this theory is that turbulent flows consist of a wide variety of eddies of different sizes. Larger eddies stochastically decay into smaller ones of approximately half the initial diameter. These also decay, and so on, so that there is a cascade of decaying eddies transferring kinetic energy from a large-scale to a small-scale motion. The largest eddies are generated by the deterministic flow induced by the agitation system. Their sizes are determined by the scale of the reactor and the stirring device. There is also a lower limit, which, by Kolmogoroff's theory, appears at scales where the kinetic energy of the eddy motion is balanced by viscous energy dissipation, so that the eddy motion dies out. Compared to the molecular scale, however, the Kolmogoroff scale is orders of magnitude larger. Maintaining turbulence requires continuous eddy generation, and thus a continuous energy input at the scale where the largest eddies are generated.

In recent years, much effort has been put into studies of isotropic turbulence (KAWASE and MOO-YOUNG, 1989) to elucidate mixing in bioreactors, in the hope that at least locally the isotropy of the eddy motion can be reasonably assumed. Then the relationship

$$l \approx (v^3/\varepsilon)^{1/4} \tag{5}$$

can provide a rough idea as to how far down in size turbulence can disperse liquid fluid elements (microscale size l, scale of the smallest eddies) in a fluid of viscosity v, if an energy dissipation density ε is assumed. This estimation is very rough, since, as is well-known, the local ε can vary widely across the reactor. It can be more than 100 times larger around the impeller than at places far away from the stirrer.

The Kolmogoroff theory, which has been developed for single-phase flows of large-scale flow systems, cannot be applied quantitatively in chemical reactors (MAHOUAST et al., 1989). Hence, it is not surprising that it is difficult to apply in multiphase reactors (LÜBBERT and LARSON, 1990). Qualitatively, however, it is perceptible that more energy in random motion will lead to a smaller scale of the segregated fluid elements, whether the flow is turbulent or not.

2.2.1.3 Molecular Diffusion

Eddies can be understood as fluid elements, which have some identity. Although they can exchange matter with their environment across their boundaries, their primary dispersion activity is due to their decay. During the decay cascade, these fluid elements are broken into parts and reshuffled.

The exchange of matter between liquid-phase fluid elements and their environment takes place by molecular diffusion. Essentially the same mechanism as in the previously mentioned mass transfer across interphase boundaries is responsible for this type of mass transfer. Hence, the rate of the diffusional mass transfer is largest at the low end of the eddy-size spectrum. In this way, turbulence that generates small eddies supports molecular diffusion. Another argument is that each breakage of fluid elements leads to fresh interfaces and, thus, to relatively large concentration gradients. The existence of some of the smallest eddy sizes, however, limits this support. All finer dispersion can only be obtained by molecular diffusion.

Turbulence is known to appear at a higher Reynolds number Re, which is defined by

$$Re = dv/v \tag{6}$$

Apart from single-phase tube flow, however, it is not always simple to find appropriate characteristic dimensions d and velocities v to calculate a meaningful Reynolds number. But it can be seen immediately from this simple characteristic number that the different mechanisms are not independent of each other. Fast global transport supports local mixing mechanisms, since it directly increases the Reynolds number by the mean velocity v. It can also be seen that larger reactors, driven at the same mean velocity v, are more susceptible to turbulence, since the characteristic dimensions become larger.

2.2.1.4 Viscous Shear

In highly viscous systems, the Reynolds number, necessary for turbulence, may not be obtainable. Then, larger transport cross-sections for molecular diffusion must be obtained with streaking fluid elements by applying appropriate shear stresses. This can be done, e.g., by some special agitation devices which deform the fluid elements so that their interfacial areas become larger until the streaks eventually break into small parts leading to even larger transport cross-sections for molecular diffusion (GODFREY, 1985; EDWARDS, 1985). Optimally, the striation thickness becomes so slight that particles can penetrate by diffusion within a short time. Concentration differences can be kept relatively high by folding such streaks, thereby building layers of different concentrations.

2.2.1.5 Dispersion by Rising Bubbles

Since we are dealing with multiphase flows in bioreactors, there are additional mixing mechanisms which result from the interplay of the different phases in the reactor. The most interesting effects are due to the displacements necessary for a bubble to rise in a liquid phase and those due to transport within the wakes of particles (WEBER and BHAGA, 1982; RIETEMA, 1982; WASOWSKI and BLASS, 1989; LÜBBERT and LARSON, 1990). This effect is significant if the rising or settling velocities of the particles relative to the continuous phase are sufficiently high. Mixing occurs, since a rising bubble pushes away the liquid ahead in an irreversible way so that the fluid elements behind it are rearranged. Furthermore, liquid is carried along inside the bubble wakes. This leads to a continuous exchange of matter with the surrounding bulk phase.

2.2.1.6 Interaction of the Different Mechanisms

Generally, as in the mentioned analogy, all fluid dispersion mechanisms are active simultaneously. Their relative efficiencies, however, differ widely at different scales. While turbulence is far the most effective means to homogenize fluids at intermediate scales, it cannot

compete with ordered, forced convective flows in transporting matter over distances of the order of the reactor dimension. On the other end of the spectrum, at the microscale, molecular diffusion, which is neglible at large scales, is the only effective mechanism.

Since, ultimately, molecular diffusion is indispensible, one must concentrate on maximizing the overall rate of diffusion in every possible way. In a somewhat oversimplified formulation, one can state that all other dispersion mechanisms must be directed to support molecular diffusion. At any rate, there is a synergism in the interplay of the different mechanisms, which must be exploited by bioreactor engineering.

In sum, however, it is not necessarily the diffusion process which is limiting. As NAUMAN (1989) showed, bulk circulatory flow may be the rate-limiting transport process in batch reactors. This seems to be confirmed by numerous experimental results obtained from larger bioreactors, e.g., from the results of HANSFOLD and HUMPHREY (1966) to those of BUCKLAND et al. (1988). Thus, detailed data on mixing and circulation times in real fermentation systems are a prerequisite to gaining more insight into the rate-limiting processes. This is also necessary to choose the optimal agitator system for a specific application.

2.2.2 Characteristics of Mixing in Bioreactors

In order to bring the different dispersion mechanisms effectively into play, the proper flows should be induced in bioreactors by means of an appropriate input of mechanical energy. In stirred tank reactors this is done by using an optimal impeller system, in airlift reactors this is obtained by introducing the gas phase appropriately. It has been known for a long time that in industrially important aerobic fermentations, critical performance parameters, e.g., the oxygen uptake rate by the microorganisms, are highly dependent on the physical parameters, both operational and geometrical (WANG and FEWKES, 1977). FIELDS and SLATER (1984) showed that the local liquid mixing behavior in airlift loop reactors affected the respiration of the microorganisms.

2.2.2.1 Deterministic Circulatory Flow Modes

In order to achieve fast global transport, some kind of forced convective flow is induced in all bioreactors. In most cases, the resulting large-scale circulatory motion is characteristic for a given bioreactor. Form and size of the flow structure depends (apart from the rheology of the fluid) mainly on boundary conditions, i.e., on the geometry of the vessel which confines the flow, and on the properties of the power input. The liquid in a given cylindrical vessel, for example, is capable of flowing in several different global flow modes (LÜBBERT, 1987). Different modes are supported or suppressed by the way in which they are supplied with the required power. These modes, characteristic- or eigen-motions, form the intimate basis for the similarity approach, which can be used as a first approach in design and scale-up of bioreactors.

A complete description of this circulatory motion would require knowledge of the whole velocity field. However, measurements for its comprehensive construction would be virtually impossible. Therefore, a model assumption is necessary. This is examined by measuring profiles of the flow along critical cross-sections or by point measurements. Detailed information on the velocity structure is needed to detect dead volumes which are insufficiently supplied by the flow. In highly viscous media, it is, e.g., possible to detect the flow volumes which are not agitated by simply looking for the places where there is no mean bubble motion. This can be done by ultrasound Doppler measurement. An example of measuring curves obtained in a xanthan bioreactor is shown in Fig. 4. This permits a determination of whether the bubbles are moving or only fluctuating around a fixed position. More globally, the flow can also be characterized by the size of the circulation loops and by the mean time particles need to circulate once around the reactor. Since this is a random variable, a more complete characterization is given by a circulation time distribution. Circulation velocities influence the cooling capacity of the bioreactors (OUYOUNG et al., 1989) and are, furthermore, necessary to estimate the power which must be applied to prevent sedimentation in respective applications.

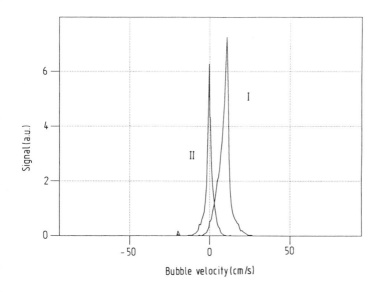

Fig. 4. Two velocity distributions of small bubbles measured at different points within a highly viscous xanthan fermentation broth during a production run. The measurements were performed with an ultrasound pulsed Doppler instrument within a 3 m^3 stirred tank bioreactor. With such measurements, which take only two minutes for one velocity distribution, the flow can be probed for dead regions. Curve I indicates a moving dispersion, whereas curve II shows that the bubbles are only vibrating around a zero mean velocity.

2.2.2.2 Random Dispersive Motions

In order to obtain an efficient dispersion on a smaller scale, stochastic fluid motions are applied. In real bioreactors, higher mechanical energies are supplied to obtain the high flow velocities, and, thus, the higher Reynolds numbers, which are necessary to obtain turbulence, known to be most effective in fluid dispersion, at least in low-viscosity single-phase flows. Turbulent fluid motions are most often characterized by flow velocity fluctuations measured at one fixed point in the flow (Eulerian representation), in most cases represented by its spectral density function. Here, low frequency fluctuations represent large-scale eddies, and high frequency fluctuations are mainly brought about by small eddies.

The most direct representation of mixing is the Lagrangian representation, in which one looks at the fate of particles within a flow. If one examines particles which were at some initial time t_0 close together in the flow, one will find that they will increase their mutual mean distance with time through the influence of different dispersion mechanisms. It is obvious that the speed at which their mean distance increases is characteristic for the mixing efficiency of the flow. Hence, one is interested in the speed at which a set of particles fed into the reactor at a confined region is spread out until it eventually becomes homogeneously distributed inside the reactor.

Whether or not turbulence can be used to increase the dispersion rate in a bioreactor primarily depends on the viscosity of the cultivation broth. At high viscosities, the energy required to establish the stochastical eddy cascade may become uneconomically high. Consequently, only some kind of laminar flow stretching and folding can be made available to support diffusion. Hence, viscosity becomes the critical fluid property which most significantly influences the performance of a given bioreactor and, thus, is the dominating factor in the choice of the reactor vessel and agitation system, one of the most important parts of the bioreactor (MIDDLETON 1985; COOKE et al., 1988).

2.2.2.3 Balancing Random and Deterministic Motions

Since all dispersion activities cost some energy, a well-balanced application of the different transport mechanisms is required to obtain an optimal result. More concretely stated, the applied energy must be channeled into the different mechanisms in an optimal way.

This idea has been taken up by the designers of bioreactors by introducing constructive measures to control the fraction of mechanical energy introduced that is channeled into deterministic and random routes. In bubble column reactors, it can be done by means of inserting draft tubes. In stirred tank reactors, it is more often done by using special new impellers. This development started with the Intermig impellers. The recently developed fluidfoil impellers, designed to give higher flow per unit of power, can produce a very uniform axial flow pattern leaving the impeller zone, which is comparable to the application of a solid draft tube (OLDSHUE, 1989). This can also be regarded as a way of maximizing the pumping capacity of the impeller in gas–liquid flows. At a given power level, this must be done at the expense of fluid shear rates and microscale mixing. The reason for this development is the realization that in larger reactors most processes demand high flow and require less fluid shear and micromixing than has usually been assumed (BUCKLAND et al., 1988; OLDSHUE, 1989).

Since the pumping capacity cannot be measured directly as in a normal pump, one must understand the mode in which the flow in the bioreactor tends to rotate. Based on this flow pattern, one must perform point measurements of the mean flow velocities. Thus, local measurements are necessary for its estimation. In such estimates, however, it is necessary to pay special attention to the fact that not all of the circulated fluid passes the stirrer region. Thus, the total circulation flow may be much higher than the primary flow from the impeller through its discharge region.

2.2.2.4 Characteristics of Heat Transfer in Bioreactors

The energy introduced into the reactor mechanically via the agitation system or as exothermic heat of the metabolic activities of the organisms must be removed in order to keep the temperature at an optimal value.

The heat produced in a fermentation is of the order of 1–6 kW/m^3, while the energy of the agitation can range up to 15 kW/m^3. This excess heat must be removed via cooled interfaces. In bubble columns and airlift reactors, the heat transfer coefficient depends mainly on the gas holdup. The heat transfer coefficient is of the order of 6 kW/m^2 (DECKWER, 1980; ZEHNER, 1986; VERMA, 1989).

2.3 Conclusions about Practice of Characterization

If one considers the interplay of the two rate processes, the relative influence of one or the other is determined by the relationship of their time constants. The faster one normally controls the overall behavior. Since bioreactions are known to be slow by most chemical rate standards, it is not apparent at first sight why fluid dynamics are practically relevant. The reason is that, as compared to normal chemical conversion processes, the limits set for the different concentrations and temperatures are much narrower in biochemical kinetics. Hence, the transport problem is directed toward maintaining highly homogeneous environmental conditions in the cells.

As experience shows, it is possible even in slow bioconversion processes that highly nonuniform concentration profiles will develop in industrial-scale fermenters, no matter how intensive the agitation appears to be (OOSTERHUIS, 1984). This indicates that transport is a significant problem in such situations. On account of the complex fluid dynamics, most transport parameters cannot be calculated with *a priori* knowledge; instead, they must be determined experimentally. Thus, as in chemical engineering practice (HOFMANN and EMIG, 1983), experimental investigation of heat, mass

transport, and kinetics builds the basis for reactor characterization and layout. Unfortunately, generally usable scale-up rules for bioreactors based on global data do not exist (OLDSHUE, 1989). Larger bioreactors must be investigated by structurized techniques, which require spatially resolved measurements (RIKMANIS et al., 1987).

3 Global Measuring Techniques

In this section, measuring methods which allow a characterization of the bioreactor as a whole are reviewed. Only those techniques are considered which can be used during real cultivations, or at least in media which sufficiently simulate actual fermentation broths. Such a medium may be a CMC-solution, as shown by ALLEN and ROBINSON (1989).

3.1 Integral Gas Holdup

The most important integral parameters in multiphase systems are the integral holdups of the different phases. A holdup is defined as the spatial fraction of the phase or component in question, integrated over the whole volume of the multiphase dispersion. It is not integrated over the whole geometric reactor volume, as one could also assume.

3.1.1 Volume Expansion Method

If it is possible to measure the volume of the dispersion and, in a separate run, the volume of the continuous phase alone, then the holdup of the dispersed particles is easily determined by the difference between the two. This principle is often used to measure the gas holdup ε in gas–liquid reactors (PHILIP et al., 1990). In cylindrical reactor vessels, the volumes can be measured by the liquid level heights with (L_G) and without (L_N) as gas input. If, in the latter case, the fluid is completely free of bubbles, the height difference can be used to calculate the gas holdup:

$$\varepsilon = (L_G - L_N)/L_G \tag{7}$$

In many practical cases, it would be difficult to measure the gas holdup during the fermentation by this technique, except in batch fermentations, where the liquid-phase volume is known initially, and one can assume that it will be nearly constant throughout the cultivation. In any other case, the liquid volume can be estimated from the fermenter weight, since it is impossible to stop the aeration until all bubbles are eliminated from the dispersion. Furthermore, it may be difficult to read the height of the aerated broth, since its surface in strongly stirred and aerated systems is not necessarily planar. Most often, it fluctuates and is covered by more or less thick foam layers.

If it is possible to determine the dispersion level height within short time periods, one can get additional information on the bubble sizes which influence the gas holdup in bubble columns by means of a dynamic gas disengagement measurement technique, as developed by SRIRAM and MANN (1977). The general procedure is to stop the gas supply to the reactor suddenly and, at the same time, to start recording the level height of the dispersion as a function of time. Since the bubble-rise velocities depend on the size of the bubbles, the larger bubbles will leave the reactor sooner than the smaller ones. If one assumes the liquid to be stagnant, one can estimate a bubble-size distribution from the data, but if the liquid motion is not known during the disengagement of the bubbles, the technique only gives a rough estimate of the size distribution. Another serious source of error is the bubble-bubble interaction, which is known to significantly influence the bubble-rise velocity and, hence, the disengagement time. Moreover, the bubble-rise velocities depend on the mobilities of interfacial areas. The same error sources affect the fractional gas holdups and the bubble-rise velocities estimated from this method. Nevertheless, it is a simple method which can be used in model media, which degas within a short time (SCHUMPE and GRUND, 1986).

3.1.2 Pressure Differences

In column reactors it is possible to determine the mean holdup between two horizontal planes by a simple pressure-difference measurement, provided the pressure can be assumed to be constant across these two planes and the density ρ of the medium without gas bubbles is known. If there are no gas bubbles, the pressure difference along the height difference h can be calculated as $h\rho$. The gas holdup leads to a reduction in the density of the fluid, hence to a lower pressure difference Δp. It is easy to see that the relative pressure is simply the mean gas holdup

$$\varepsilon = (h\rho - \Delta p)/(h\rho) \tag{8}$$

In this way, it is possible to measure holdups in different reactor parts, e.g., in the riser and in the downcomer section of an airlift tower loop reactor. Usually such measurements are carried out with U-tube manometers filled with special indicating fluids.

3.2 Integral Specific Interfacial Areas

Mean specific interfacial areas, integrated over the whole reactor, can be measured by chemical methods (WESTERTERP et al., 1963). The method is based on the measuring technique for the gas–liquid mass transfer coefficient $k_L a$, as originally applied by COOPER et al. (1944). The standard model reaction system is the cobaltous ion-catalyzed reaction of oxygen with sodium sulfite. Other reactions have been compiled by SHAH et al. (1982).

The chemically determined specific interfacial area a_{chem} deviates from the true geometrical area a_{geo} because the overall conversion of the gas phase reactant represents an incorrect average, if bubble sizes and residence time distributions are not uniform. As OYEVAAR and WESTERTERP (1989) pointed out, the deviations become larger the broader are the distributions and the higher is the overall conversion of the reactant in the gas phase. In biotechnology, the major drawback of the chemical method is its restriction to specific gas–

liquid systems and their particular physico-chemical properties, e.g., the coalescence behavior of the two-phase system. The latter is essential to all fluid-dynamical aspects of bioreactors. This limitation led to various alternative physical methods, which — in contrast — are local measuring techniques. SRIDHAR and POTTER (1978) compared the chemical and the physical light transmission method of CALDERBANK (1958) and found discrepancies. They found that the chemical method consistently yielded higher values of the interfacial areas in stirred vessels and supposed that this was due to the much larger mass transfer coefficients within the impeller area. This essentially says that at a constant specific interfacial area the results of the chemical method are dependent on the hydrodynamics in the multiphase flow.

3.3 Residence Time Distributions and Related Measurements

3.3.1 Tracer Techniques

3.3.1.1 General Methodology

Residence time distribution (RTD) measurements are techniques taken over from chemical engineering (DANCKWERTS, 1953; NAOR and SHINNAR, 1963; NAUMAN and BUFFHAM, 1983), where an extensive literature exists on the different aspects of the measuring techniques. All techniques are based on a stimulus/response approach. A tracer is used to mark fluid elements in the input stream of the reactor and, viewed as the system's response to this marking, its concentration is observed at the outlet of the reactor as a function of time. Although there may be situations in multiphase systems in which the interpretation of residence time distributions measured by tracer techniques is difficult (SINNAR and RUMSCHITZKI, 1989), the methods can provide much insight into complex multiphase flows in biotechnology (BRYANT, 1977; LÜBBERT et al., 1987a; SWAINE and DAUGULIS, 1988, 1989).

The main problems of these simple measuring techniques are due to the kind of marker

and its proper detection in the output flow. It is not easy to mark the fluid elements at the desired rate without changing their properties. Traced fluid elements must be assumed not to differ from all other elements of the fluid component. Otherwise one cannot use them to estimate the flow behavior of this component.

3.3.1.2 Gas Residence Times

A gaseous component which is steadily fed into the bioreactor, e.g., the air in aerobic fermentations, can easily be labelled by addition of small amounts of a tracer gas, injected by pulses through small steel tubes directly into the gas distributor. The tracer pulses can simply be controlled by fast valves. As tracers, noble gases, e.g., helium, are most often used in practice. Small amounts can be detected in the gas outlet by mass spectrometers. A considerable experimental problem is the sampling of the tracer leaving the dispersion, since the surface is strongly fluctuating and covered by foam layers. There are two possible ways of solving the problem. One is to use membrane sampling devices (HEINZLE et al., 1983), and the other is to mechanically destroy the foam entering the detection system by means of a small foam destroyer, as shown in Fig. 5. The mechanical technique (LÜBBERT et al., 1987a) can be used when one cannot make sure that the membrane will be of constant transparency for the probe gas (e.g., in technical-grade systems) over longer periods of time.

In order to enhance the signal-to-noise ratio of the results, it is advantageous to use signal forms $s(t)$ which differ from the conventionally used Dirac delta functions in order to mark the gas flow dispersed into the reactor. Gains in the signal/noise ratio of up to two orders of magnitude can be obtained using pseudostochastically distributed pulse trains of tracers. The system's response to such pseudorandom signals then also becomes a random signal. Therefore, the weighting function, in this case the time-of-flow distribution, must be calculated from $s(t)$ and the system's response $a(t)$ measured at the reactor exit by means of the cross-correlation function $R_{sa}(\tau)$ of both signals (Fig. 6). $R_{sa}(\tau)$ is related to the weighting function $W(t)$ by the convolution integral

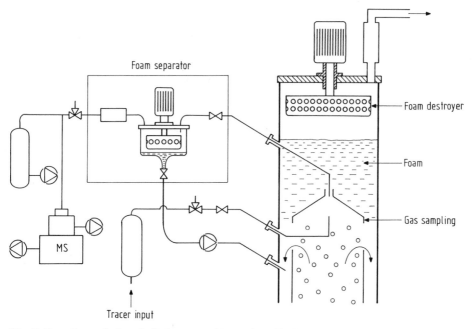

Fig. 5. Sampling technique built for gas residence time distribution measurements in pilot scale fermenters which are operated with technical substrates (FRÖHLICH, 1986). It consists of a bell-shaped sampling funnel, a mechanical foam destroying unit, and a mass spectrometer connected by a transmission line.

(BENDAT and PIERSOL, 1986):

$$R_{sa}(\tau) = \int_{-\infty}^{\infty} R_{ss}(t-\tau)\,W(t)\,\mathrm{d}t \qquad (9)$$

If the stimulus function $s(t)$ is a pseudorandom function, as assumed here, its autocorrelation function is simply a Dirac function, $R_{ss}(\tau) = \delta(\tau)$. Then, by the convolution theorem of the Dirac functions, the cross-correlation function is

$$R_{sa}(\tau) = W(\tau) \qquad (10)$$

Hence, the time-of-flow distribution W can be obtained by numerical calculation of the cross-correlation function between the known signal $s(t)$ used to control the tracer gas valve, and signal $a(t)$, measured at the detector (Fig. 6).

Such a measurement has been automated with the use of a simple 8-bit microprocessor (LÜBBERT et al., 1987a). The processor can be

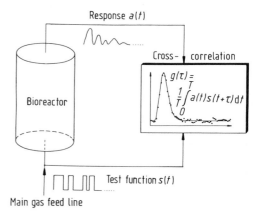

Fig. 6. Reconstruction of the residence time distribution from the data if the tracer was added pseudostochastically. The pseudostochastical test function $s(t)$ must be cross-correlated with the tracer concentration signal $a(t)$, which is measured at the reactor gas outlet. After subtraction of the amplitude offset, this cross-correlation function $g(t)$ is proportional to the residence time distribution.

used simultaneously to control the stimulus signal to read the system's response from the detector, to calculate the correlation function in real time, and to perform an ensemble average of the signals obtained from successive repetitions of the measuring cycle.

3.3.1.3 Liquid Residence Times

In chemical engineering, one often uses salt solutions or acids to label liquid-phase fluid elements. The marked fluid elements are then measured by their conductivity or their pH. In many model systems, the detection at the exit of the reactor is an easy task (KHANG and FITZGERALD, 1975). Unfortunately, this is not as simple in bioreactors, since the electrolyte content is already high, and the addition of small amounts of a salt would not be as readily detectable at the outflow. Larger amounts of salt are no alternatives to overcome this difficulty, since they would change the flow behavior and/or the metabolism of the organisms.

JIMENEZ et al. (1988) proposed to use dyes for tracer studies in bioreactors. They found some dyes which are not adsorbed by the biomass, which are stable over time, have good solubility, and do not change their colors within the pH interval of 6.5 to 8.5. An example is bromocresol green. An alternative for bioreactors is to use small amounts of fluorescent substances to mark liquid flow elements. These can be detected by means of fluorescence detectors at the exit of the bioreactor. Coumarin dyes can be used, which, at low concentrations, are known not to influence the flow behavior or the microorganisms (BEYELER et al., 1981; GSCHWEND et al., 1983; MEYER et al., 1984; SCHNECKENBURGER et al., 1985; SCHEPER, 1985; CEVEY and VON STOCKAR, 1985; SCHEPER and SCHÜGERL, 1986).

Measuring the mixing behavior of a bioreactor by means of its residence time distribution is a model-supported measurement. The parameters characterizing the mixing behavior are defined by the residence time distribution model used. They are determined by numerically fitting the model to the data. If one uses the axial dispersion model, the characteristic numbers are the mean residence time τ and the Bodenstein number Bo.

3.3.1.4 Circulation and Mixing Times

Residence time distributions are only defined if there is a net throughput of the component in question. Similar measuring methods can be used to investigate the global mixing behavior of a continuous liquid phase, if the fermentation is operated batchwise. Blending is dominated by the more or less stable large-scale circulatory fluid motion inside the reactor. The most straightforward characteristics of such a circulatory flow are the circulation time distribution with its moments, the mean circulation time, and the variance. These quantities can never be neglected in reactor characterization (OLDSHUE, 1989).

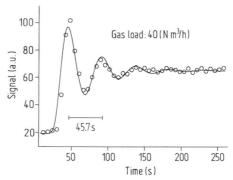

Fig. 7. Data (symbols) of a tracer measurement to investigate the circulatory liquid-phase motion in an airlift tower loop bioreactor of total volume 4 m³ (LÜBBERT et al., 1990). A single coumarin pulse was introduced into the downcomer and detected at an adjacent point upstream with a fluorescence probe. The solid line is due to a model function fitted to the data to obtain the parameters mean circulation time and mixing time.

Using tracer experiments to investigate these circulatory flows gives rise to nearly the same problems as with RTD. Some fluid elements must be marked at one properly chosen place within the flow, and the tracer concentration must be measured at another. Both points must be selected carefully to obtain response curves which can be used to extract the desired characteristic parameters. The system response functions are often measured at two locations, as pointed out by EINSELE (1976) and SITTIG and RAMSPECK (1979).

If one uses a single Dirac pulse of the tracer as the signal, then the tracer concentration at the detector will be a damped oscillation (Fig. 7). The period τ_p of this oscillation is the exact circulation time. The damping of the tracer concentration is due to liquid mixing, which smears out the marked fluid elements in space and reduces the tracer intensity at the detection point when the cloud passes. From the envelope of this damped oscillation one can, thus, obtain the time constant of the decay of the tracer cloud, which is one way of characterizing the time necessary for mixing (EINSELE, 1976). In most cases, however, a simpler definition of the characteristic mixing time is used. This is the time after which the decaying concentration fluctuations stay within predefined limits around the final concentration value c_f. Normally, one uses the interval $[c_f - 5\%, c_f + 5\%]$.

The choice of an appropriate tracer is a difficult task and depends upon the particular process. As tracers one can use different dyes (EINSELE, 1976; BEYELER et al., 1981; EINSELE et al., 1978; LAINE and KUOPPAMÄKL, 1979; SCHEPER, 1985; SCHÜGERL et al., 1987), alcohols (SWAINE and DAUGULIS, 1988, 1989), or radioactive tracers (PROKOP et al., 1969; SEHER and SCHUHMACHER, 1979).

3.3.2 Flow Follower Techniques

There are several practical difficulties with tracers in measuring liquid circulation time distributions in bioreactors. One is due to tracer accumulation inside the bioreactor, if one repeats the addition of tracer pulses to obtain more reliable results. Since, furthermore, it has not been simple to find proper liquid tracers to mark fluid elements, some investigators have replaced tracers by macroscopic, inert, solid particles adjusted in density to the dispersion so that they are of neutral buoyancy. Thus, it can be assumed that they follow the global liquid flow circulating through the reactor, at least in the time average. From the sequence of the times at which these flow followers pass a defined control region, the liquid

circulation time in batch fermentations has been estimated.

SYKES (1965) was the first to report use of this method in investigating the circulation paths traced by fluid particles inside biochemical reactors. He used strips of plastics, which were followed visually, and the number of circulations seen was written down. Such visual experiments, which are restricted to clear liquids, have been performed by several investigators. For opaque media, more complicated flow followers have been developed. In table-tennis-ball-like bodies, small radio frequency transmitters were installed. The transmitted radiation was then detected by antennas which, appropriately mounted, could detect the passage of such bodies (BRYANT, 1969, 1977; MANN et al., 1981; OOSTERHUIS, 1984). BRYANT (1977) reported use of radio pills, 20 mm in diameter, each consisting of a plastic sphere with a transmitter (10 MHz) and a battery for its power supply. The even more miniaturized pills of SCHMIDT and BLENKE (1983) and BLENKE (1988), 13 mm in diameter, worked for several months. In their experiments in airlift tower loop reactors, they used two antennas. By recording the time difference distribution between the passage times of the probe body through both control volumes, they additionally calculated a mean velocity of the dispersion.

There are several other ways to construct flow followers. SCHMIDT and BLENKE (1983) reported on pills which only contained an electric resonance circuit. Their advantage is that no battery is needed, which is the most voluminous component of radio pills. The antenna within such a flow follower absorbs power, radiated from the transmitting antenna at the reactor wall, every time the pill crosses through a control volume. This method has been implemented in tower jet loop reactors.

Another possibility which may not be as attractive for bioreactors is the use of radioactive pills in conjunction with radiation detectors.

MUKAKATA et al. (1976, 1980) developed a permanent magnetic flow follower to study the circulation time distribution in stirred tank bioreactors. They twisted a wire around the reactor several times to build an inductance coil in the plane defined by the impeller, as

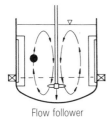

Flow follower

Fig. 8. Principle of an experimental arrangement to measure circulation times by means of magnetic flow-followers. The flow-follower pill is symbolized by the full circle, its mean path projected on the plane of the drawing by the arrows, and the induction coil in the plane perpendicular to the stirrer axis at the height of the impeller by the symbols outside the vessel.

shown in Fig. 8. On passing through the coil, the magnetic pill induces current pulses, which can easily be detected and counted. They used the method in *Penicillium chrysogenum* cultivations and in paper pulp suspensions. FUNA-HASHI et al. (1987) simplified this technique and used it to investigate the circulating flow of high viscosity xanthan gum solutions inside a stirred tank reactor.

Primarily, the data from this measuring technique give the transmission frequency distribution of the pill through the detection area. But without a proper model one cannot relate this passage time distribution to a realistic flow pattern inside the reactor. Such models have been proposed by MANN et al. (1981) for an aerated stirred tank reactor and extended by OOSTERHUIS (1984). In this way, one can take into account the fact that the flow follower does not follow the mean circulation path. CLARK and FLEMMER (1985), on the other hand, argue that secondary circulations disturb the path of the probe so much that the determined circulation rates are not reliable. An estimate of the pumping rate of the impeller, thus, cannot be made without additional assumptions about the flow patterns, which must be examined by special and more detailed measurements. It should be noticed at this point that the term "circulation time" within a stirred tank reactor is not well-defined. However, the technique is a direct measuring method for an essential physical quantity of direct

interest to aerobic bioprocesses: Adjusted correctly, the method gives valuable information about the frequency at which the organisms pass through the regions optimally supplied with oxygen.

The main conceptual difficulties in using these techniques can be summarized as follows:

- Densities of the cultivation broths in fermenters change during the production run (OOSTERHUIS, 1984).
- The macroscopic solid probes behave quite differently from flexibly deformable fluid elements of the same size.
- Long measuring times are necessary to obtain good statistics.
- The accuracy of the result depends on a reliable model of the circulatory flow inside the bioreactor.

3.3.3 Biological Test Systems

In order to compare the overall performances of different bioreactor constructions, it is advantageous to cultivate a standard biological system in all reactors under consideration, as proposed by FIECHTER's group at the ETH Zürich (ADLER and FIECHTER, 1983, 1986; JARAMILLO, 1985; JARAMILLO et al., 1986). Comparative characterization of bioreactors by means of standard biological test systems can be regarded as a global test of the success of a more device-oriented bioreactor construction based on physical, i.e., fluid-dynamical principles (ADLER and FIECHTER, 1983).

One can use a strictly aerobic culture if the oxygen-transfer rate capacity of the reactor is the main aspect of interest. KREBSER et al. (1988) proposed using the yeast *Trichosporum cuteanum* to characterize several bioreactors. GRIOT (1987) took *Bacillus subtilis K (AJ 1992)* cultures, which are sensitive to the oxygen tension of the culture broth, to characterize geometrically similar stirred tank bioreactors of different sizes. He used the product distribution as an actual sensing effect to characterize the oxygen transport inside the reactor.

It should be kept in mind, however, that the relative performance of bioreactors depends on the special biological system involved and on the rheological conditions. Thus, the results must be assessed accordingly.

3.4 Power Input

Since nearly all fluid-dynamical parameters in chemical or biochemical reactors depend on the mechanical power put into the medium, most of the transport mechanisms acting in bioreactors are controlled by it. Hence, its actual value is of primary importance in characterizing reactors.

In stirred tank bioreactors, the power is introduced primarily by the agitation system, although in high, slender aerobic bioreactors a considerable part is due to the air compressed into the liquid. A simple measurement of the electrical power, drawn by the agitation system, is too inaccurate. It is very difficult to estimate the efficiency of the motor and especially the frictional power losses due to the gear and agitator shaft seal. The best method of determining the effective power input by a stirrer is to measure the torque put onto the impellers by the stirrer wheel.

Torque T, which the impellers transfer into the liquid, leads to a mean strain shear rate τ at the wall of the reactor (KAI and SHENGYAO, 1989):

$$T = k_1 V \tau \quad (\text{N m}) \tag{11}$$

where k_1 is a constant of proportionality and V the reactor volume.

With stirrer speed N, general mechanics lead to the power input P

$$P = 2\pi N T \quad (\text{N m/s}) \tag{12}$$

In chemical engineering, one often uses the shaft strain gauge technique to measure torque T. This requires the insertion of torsion elements into the wheel of the agitator. The measurement is then done by means of a calibrated strain gauge mounted onto the shaft of the torsion element. These elements are covered by protective coatings to prevent the penetration of the fluid into the gauge. Such strain gauges are essentially electric metal layer resistances (ca. 1 kΩ) which change their resist-

ance when a strain is placed on them. In this application they measure the power delivered to the shaft at the impeller end. The problem that occurs in transmitting the measuring signal from the moving wheel outside the reactor is frequently solved by telemetry readouts, which are available commercially. Strain gauges must be calibrated. This can be done statically using weights and a lever arm. The accuracy of a power measurement is of the order of a few percent (KUBOI et al., 1983).

Unfortunately, the strain gauges cannot be sterilized, therefore, measurements done during real fermentation runs in bioreactors cannot be found in the literature. As EINSELE and FIECHTER (1974) proposed, a strain gauge can be mounted on the stirrer wheel between the gear and the shaft seal. In that way, the seal is a mechanical resistance, and it must be considered accordingly.

A method, which, however, is only applicable to smaller reactors, is the dynamometer technique proposed by RUSHTON et al. (1950). Here it is assumed that the torque transferred into the liquid can be measured by means of the torque experienced by the reactor vessel. A motor driven at variable speeds at a constant torque is mounted on a revolving platform that can be turned coaxially to the stirrer axis. The frictional loss of the table is reduced as much as possible by means of a roller-bearing. At the periphery of the table, the force necessary to compensate for the torque introduced by the action of the stirrer via the dispersion is measured. Very small vessels can also be placed on an air-bearing to reduce friction losses (CALDERBANK, 1967). These techniques have been used in vessels stirred with a Rushton turbine, a radial conveying device. Stirrers which promote axial flow components may not give correct results with such a measuring setup.

One may ask, why not measure the electrical power applied to the motor of the agitation system? This will yield only very approximate results, since such a measurement is affected by large bias errors due to the efficiency of the motor and the power losses by the gear and the agitator shaft seals. Since these sources of power loss are neither easy to determine nor constant, a proper correction cannot be made with sufficient accuracy.

In airlift reactors, the power for the agitation is introduced pneumatically with the gas phase. It can be measured in a much easier way. Since the gas input is usually assumed to take place by isothermal expansion of the gas, it can be calculated by

$$P_G = \delta_L g Q_G H_L = W w_{sg} \tag{13}$$

where δ_L is the density of the liquid phase, g the gravitational acceleration, Q_G the gas feed rate, and H_L the height of the ungassed liquid. Thus, in a cylindrical air-lift reactor, the power input P_G can simply be determined by the weight W of the dispersion times and the superficial gas velocity w_{sg}, which can be measured via commercially available flowmeters.

The maximal shear rate will appear at the tips of the impellers. It has been proposed to measure it electrochemically, e.g., by inserting a platinum wire into one blade of the turbine with only one end of the wire exposed to the forward-facing surface of the blade. The exposed end must be located near the outermost edge of the blade to measure the shear directly at its tip (WICHTERLE et al., 1984; ROBERTSON and ULBRECHT, 1987). The signal transmission must be done by special techniques, e.g., by mercury contact bearings.

There are several methods for measuring wall shear stresses. A review of these techniques was given by HANRATTY and CAMPBELL (1983). An actual measuring device, developed for bioreactors, was published by PAULI et al. (1989). They showed how to measure the shear rate exerted on the wall of a small probe, which can be moved through the reactor. The authors used a limiting-current electrodiffusion technique. They proposed using the special three-segment probe patented by SOBOLIK et al. (1986).

A well-known effect in gassed, stirred tanks, such as bioreactors, is the decrease in power uptake on aeration, P_g, as compared to power P_u, absorbed under ungassed conditions. The decrease in power is assumed to be due to the formation of gas cavities behind the stirrer blades (BRUIJN et al., 1974). Its magnitude, however, is not fully predictable. The power drawn can be linked to the flow pattern of the gas phase (NIENOW et al., 1978), especially to the manner in which bubbles are recirculated into the stirrer region. The latter depends on the coalescence/redispersion properties of the dispersion, which cannot be predicted with sufficient accuracy. Only very recently have such patterns been measured experimentally (BRÖRING et al., 1990).

3.5 Flow Visualization

To get some idea of the global characteristics of the flow, it is believed by many investigators that its visualization must be the first key step. An impressive example is the visualization of the cavities behind impeller blades, which strongly influence the mechanical power input into aerated stirred tanks. The phenomenon has been investigated by photography (BRUIJN et al., 1974; VAN'T RIET, 1975; KUBOI et al., 1983).

DENK and STERN (1979, 1980) proposed a technique to investigate the global flow pattern in cylindro-conical tank bioreactors used in breweries. With their light sheet technique, which can be used in model vessels with transparent walls, they illuminated special dispersed tracer particles (of the same density as the liquid) in a stroboscopic manner. If the illumination is restricted to a thin layer parallel to a particular plane through the tank, the motion of the particles along their stream lines in this plane can be visualized and recorded by movies. In this way, they could resolve the fluid structure in their tanks. Such knowledge is a prerequisite for the design of beer fermentation tanks of this type, especially with respect to their cooling capacity.

WALTER and BLANCH (1986) investigated the bubble breakup in turbulent flow fields inside bioreactors by means of a high speed movie camera and found by visual inspection of successive photographs that most breakup processes proceed in three steps, an oscillation, a dumb-bell stretching, and a pinching-off. Such a sequence must be observed with individual bubbles. Visualization seems to be the only way to keep track of that complicated process.

The general aim of flow visualization techniques is to obtain spatially resolved information on the flow structure. This is often necessary to obtain an overall picture of the flow.

These techniques provide quantitative local data on the geometrical form of the flow. Hence, they may also be regarded as local measuring techniques. For the dynamic properties of the flow, they usually build a basis for more detailed investigations using the local measuring techniques discussed in the next sections, since local measurements can only be performed at a limited number of selected points that seem to be interesting with regard to the whole flow structure.

4 Local Measuring Techniques

Local measuring techniques are designed to deliver values integrated over a more or less restricted measuring volume. This volume may merely be a point or a larger, but confined, volume. Local measurements are necessary to verify structured models and to obtain parameters for them. Here we concentrate on measuring techniques for the parameters which are

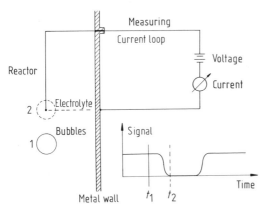

Fig. 9. Principle of the conductivity probe technique. An insulated wire is put into the flow of a dispersion with a conductive continuous liquid phase. At the end of the wire, a point-like metal surface is exposed to the electrolyte. If a voltage is applied between the metal wall of the reactor and the wire, a current is drawn via the electrolyte. This is interrupted, if the point probe is surrounded by an (insulating!) bubble.

most significant for the characterization of the performance of bioreactors. These are mainly the properties of the continuous liquid and the dispersed gas phase. The developments in this field have been stimulated by investigations in many other fields (WILD et al., 1986), e.g., by measuring techniques for multiphase flows in modern power-generation systems (JONES, 1983) or blood-flow measuring techniques in medicine (WEBSTER, 1978; PAYNE, 1985).

Sometimes it may be interesting to calculate global quantities from local measuring values, e.g., the mean specific interfacial area. This is not the main goal of local measuring techniques. To obtain values of quantities spatially averaged over the whole reactor, many local values must be measured and integrated. These may be very difficult to obtain in practice.

4.1 Bubble Parameters

The most interesting bubble parameters are the local gas holdup, the bubble size distribution, the specific interfacial area, and the velocity distribution of the bubbles.

4.1.1 Bubble Probes

4.1.1.1 Conductivity Probes

The conductivity probe technique can be used to measure bubble properties locally inside bioreactors. The main parameters that can be obtained are: local gas holdup, bubble velocity, and size distributions.

Measurements with conductivity probes require a continuous liquid phase that is electrically conductive. Since the conductivity of tap water is generally sufficient to apply the method, this condition is fulfilled in bioreactors. A conductivity probe that can be used to measure gas holdups consists of a point-like electrode and a second larger electrode, which must be so large that it cannot be covered by single bubbles (Fig. 9). It can be positioned elsewhere in the dispersion where it does not disturb the flow.

If one applies a voltage between these two electrodes, one observes an electrical current flowing through the electrolyte. This current, however, is interrupted if the point electrode is covered by an (insulating) gas bubble. Then, by means of an electronic signal-conditioning unit, one can obtain a binary signal, which takes the value 1, if the point electrode is within a bubble and 0 if it is not.

1. Single-point probes

Local gas holdup measurement

If the current is recorded as a function of time, then, at a sufficiently long measuring interval T, the time-integral of the dimensionless signal can be interpreted as the integral time T_B the probe spent within bubbles. Divided by T, this is equal to the probability of meeting a bubble at the probe tip, i.e., to the local gas holdup there.

$$\varepsilon = \lim_{t \to \infty} T_B/T \qquad (14)$$

In modern implementations, these times are determined in simple digital microprocessor-controlled devices. On account of the binary signal obtained from the analog electronics, one can omit an analog-to-digital converter. The data can simply be sampled using one bit of a parallel digital input port of a microprocessor card. The necessary sampling rate of about 100 kHz can then be reached with simple 8-bit microprocessors.

Although the principle of this method is extremely simple, there are some important technical problems with the signal-conditioning electronics. First of all, one cannot assume the current to be a smooth, stationary signal if the probe is operated within real liquids. On account of changes in the conductivity of technical broths and electrochemical effects at the point electrode, one observes significant drift in the current. The long-term conductivity drift must be compensated by appropriate controllers and by using alternating voltages at sufficiently high frequencies to reduce electrochemical effects. Generally, a few kHz would be enough to avoid these effects; but if one takes into account that the signal must be sampled at a rate of at least tens of kHz, one normally needs very different frequencies for the voltages which draw the current through the liquid phase. LEWIS and DAVIDSON (1983)

could only reach frequencies of up to 10 kHz because of electronic difficulties. YASUNISHI et al. (1986), however, reported 95 kHz. The difficulties arise mainly with small point probes.

A necessary condition for the performance of these probes is that the liquid film must disappear at a sufficiently high rate from the probe surface after the detection point has entered a bubble to interrupt the current between the two electrodes. This limits the viscosity of the liquid phase of the dispersions that can be investigated. Another severe limitation relates to the bubble sizes that can be detected. These must be large as compared with the characteristic probe diameter, otherwise the probe will not pass across the bubble. The smaller the bubbles, the more they tend to follow the motion of the continuous liquid phase around the probe. Keeping the probe dimensions small is necessary to reduce the interference between probe and flow; normally this can only be accomplished at the price of reduced mechanical stability.

The measuring system must be calibrated because of the numerous effects which influence the signals of the probes. Such a calibration in real flow conditions is not simple, since a measuring standard does not exist. Thus, the probes are calibrated in slug flows within small tubes, where one can measure and control the gas flow and the mean bubble volume by sampling a predefined number of bubbles.

Bubble probes proved to be applicable to fermentation broths, provided they are not too viscous (CZECH, 1986; FRÖHLICH, 1987; LARSON et al., 1990). They are a simple means of measuring profiles of the local gas holdups which develop in all larger fermenters.

2. Two-point probes

Bubble velocity distributions

Bubble probes can be constructed based upon the same principles as discussed above. They can be used to measure bubble velocity distributions. Such bubble probes must contain at least two point-like bubble detectors, which are posed at adjacent points along the mean bubble path. These two detectors will

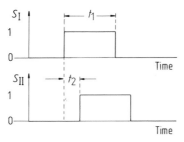

Fig. 10. Two signals of a two-point bubble probe. The points are assumed to be placed on a single stream line of the bubble at a distance d_P apart. Thus the first signal S_I is the first to detect the bubble, S_{II} follows with a time delay t_2.

ranged so that the straight lines connecting the lower with the three upper electrodes are linearly independent, it is possible, at least in principle, to calculate a complete set of vector components of a single bubble-rise velocity (BURGESS and CALDERBANK, 1975). There are, however, some practical problems concerning the interaction between the probes and the bubbles in a real dispersion, which limit the application of this method. The more detection points are used, the more extended the probe becomes geometrically, and the more often bubbles are carried around the probe and, hence, are not detectable.

give output signals (Fig. 10) with a mutual time delay t_2, since the bubble surface needs some time to proceed from one point to the other. Since the distance d_P between the two detection points can be measured quite accurately by means of a measuring microscope, the velocity component v_x of the bubble surface along the direction defined by the two points can be calculated by

$$v_x = d_P/t_2 \qquad (15)$$

Since the electrodes must be at small distances of less than a millimeter, problems may arise with cross-talk and capacity effects, which require special electronic measures in order to obtain sufficiently high alternating currents between the point electrodes and the common counter-electrode at frequencies up to 100 kHz (LARSON et al., 1990).

An interesting alternative was used by ZUN and SAJE (1982), who obtained the mean time delay between the two signals by calculating the cross-correlation function of both signals. The advantage is that one can eliminate the efforts to adjust the trigger levels which determine the location of the edges of the conditioned signals. However, as can be seen later, with this correlation method one loses the information about bubble diameters that is also contained in the bubble signals.

3. Multi-point probes

If one constructs a 4-point bubble detector (Fig. 11), in which the detection points are ar-

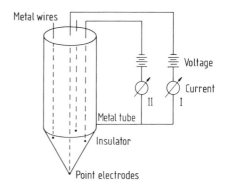

Fig. 11. Principle of a four-point conductivity probe. The point electrodes are the end points of thin wires placed parallel into a glass rod which end on the cone ground at the end of the rod. The signals obtained from two of the point sensors for a time interval within which a bubble was hit by the probe are shown on the right-hand side of the figure. From the times indicated, the bubble velocity and chord length can be estimated as explained in the text.

BUCHHOLZ and coworkers (BUCHHOLZ et al., 1983; STEINEMANN and BUCHHOLZ, 1984, 1985; STEINEMANN, 1985) optimized the technique by constructing extremely small multipoint conductivity probes. The authors used probe diameters down to about 0.5 mm, and they claimed that bubbles down to two millimeters (STEINEMANN and BUCHHOLZ, 1984) were detectable. Such extremely thin detectors are, however, very susceptible to mechanical damage, which restricts their use to scientific work.

Bubble size distributions

Bubble probes having two or more point-like detectors (Fig. 11), which are positioned closer together than the mean bubble diameter, can also be used to estimate the bubble diameter (BUCHHOLZ and SCHÜGERL, 1979). In this case, one takes the interval lengths t_1 (Fig. 10) in one of the signals corresponding to individual bubbles. If one multiplies this time interval by the velocity v_B of this bubble, calculated from the time delays between the signals of the different detection points, as explained in the last section, one obtains the bubble chord lengths s_B:

$$s_B = t_1 v_B \qquad (16)$$

The diameter of a spheric bubble is somewhat larger. With statistical arguments, the bubble diameter d_B distribution can be assumed (FUKUMA et al., 1987) to be calculated from these chord lengths by

$$d_B = 1.5 s_B \qquad (17)$$

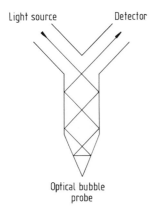

Light source   Detector

Optical bubble
probe

Fig. 12. Principle of an optical bubble probe. A light beam is guided through an optical fiber. At the end of the fiber the reflecting coating material is removed so that the light can be transmitted into the bubble dispersion. If the tip of the probe is surrounded by a bubble, the low refractive index outside the fiber gives rise to total reflection at the inner surface of the fiber and more light is reflected back into the fiber. This light is guided onto a detector by a light switch.

Since the bubble surface can be thought to behave similarly to an elastic membrane, the surface of the bubble will deform before it breaks, and the probe penetrates through the bubble. Such interferences have been observed by FRIJLINK (1987) with comparable probe geometries. This effect will influence the accuracy of the results about the direction of the bubble velocity from the experiments with multi-point bubble probes. The signals of the upper-point detectors, however, can be used to distinguish between bubbles which deviate too much from a vertical path and those which rise properly. This can be done by setting time gates within which the leading edge of all upper-point signals must be detected. If the gate width is small enough, one can be sure to detect only bubbles that are hidden centrally, and the calculation of the bubble chord length of spherical bubbles should approach the bubble diameter. In this elimination procedure, however, the accuracy is increased at the cost of the total measuring time necessary to obtain a statistically significant sample size of the bubbles. Further limitations of the technique are described by RAPER et al. (1978) and CALDERBANK (1978).

Thus, if bubble probes are applicable, they deliver simultaneous information about the bubble size and velocities, so that one can construct two-dimensional probability density functions. Such probes have been used by several authors to characterize bioreactors (STEINEMANN and BUCHHOLZ, 1984; CZECH, 1986; LÜBBERT et al., 1987a; SUN and FURUSAKI, 1988).

4.1.1.2 Optical Fiber Probes

Point-like bubble sensors, which are the essential components of conductivity probes, can also be built up with fiber-optical techniques (DE LASA et al., 1984; LEE et al., 1984; LEE and DE LASA, 1987, 1988; SPINDLER et al., 1987a, b, 1988; YU and KIM, 1988). Such probes deliver the same signals as the conductivity probes. Once again, these point sensors are built to give a binary signal, which takes the value 1 if there is a bubble at the measuring point and 0 if not. Hence, the signals are analyzed in the same way. The principle underlying the optical fiber probe is a measurement of the light intensity reflected at the specially prepared tips of the fiber.

In a fiber-optical probe, a light beam is guided through a thin glass fiber (Fig. 12). At the tip of the fiber, where there is no coating, most light is transmitted into the liquid provided the probe is immersed in a fluid with a larger refraction index than the glass. Only a small part of the light is reflected back into the fiber. This reflected intensity is measured. However, if the fiber tip is within a bubble, which usually has a lower refractive index than the fiber, total reflection appears at the interface, and consequently much more light arrives at the detector. Although an optical technique, it is not necessary that the liquid phase be optically transparent. The time resolution of the detectors is usually high.

There are some advantages to the optical technique as compared to conductivity probes. Thin liquid films on the sensitive fiber tip only slightly decrease the signal level of the optical sensor. Thus, one of the main disadvantages of the conductivity probes, e.g., the time constant of the liquid film flowing off the point electrode, is essentially removed. Another advantage, common to all fiber-optical probes, is that the signal is not susceptible to electromagnetic cross-talk effects or other effects of electric or magnetic fields that influence electric signal lines. The requirement for conductive liquids obviously disappears. This, however, is not relevant to biotechnological applications.

FRIJLINK (1987) investigated several probes thoroughly. Some probe constructions suffer from difficulties with contamination of the probe surfaces, especially probes with sharp edges due to grinding. Moreover, sharp polished tips are expensive and more fragile than round shapes. At elevated viscosities, interference effects between probe and bubbles caused by bubble deformation were observed by FRIJLINK (1987).

MILLER and MITCHIE (1969, 1970) were the first to publish the idea of fiber-optical probes in the chemical engineering literature. Instead of fibers, they used single glass rods of several millimeters in diameter in their pioneering experiments. DANEL and DELHAYE (1971) were the first to use U-shaped optical fibers to construct bubble probes. GALAUP (1975) em-

ployed two such sensors to measure bubble velocities and sizes. CALDERBANK and PEREIRA (1977) constructed a multiple-fiber arrangement similar to the conductivity probe of BURGESS and CALDERBANK (1975). ABUAF et al. (1978) improved the technique to feed back the reflected light. These authors connected two thin 0.125 mm fibers to one sensor. DELHAYE (1981) took even thinner fibers, 0.04 mm mono-mode ones. LASA et al. (1984) used a U-shaped probe in a three-phase fluidized bed. VAN DER LANS (1985) applied a probe consisting of five-step index quartz fibers. FRIJLINK (1987) improved the shape of the probe tip of the step index fiber after a careful investigation of the influence of the shape.

4.1.1.3 Isokinetic Sampling Probe Method

The isokinetic sampling probe is a bubble probe which differs from the point probes discussed so far. A small stream of the dispersion is sucked out of the reactor at a constant flow velocity through a glass capillary (Fig. 13). In the capillary, the bubbles are elongated to form slugs filling the whole cross-section. At

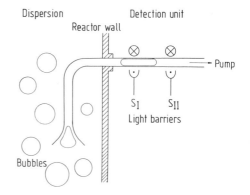

Fig. 13. Principle of the isokinetic sampling probe. A small representative sample flow of the dispersion is sucked through a capillary by means of a pump. Bubbles are stretched to elongated slugs within the capillary, filling its whole cross-section. The length of these slugs, which is proportional to the bubble volume, is measured by means of two adjacent light barriers.

two adjacent points along the capillary, detectors are installed which can distinguish between gas or liquid inside the capillary at the detection points. Usually the detection is done electro-optically. A specific bubble thus sets off two signals from the two detectors, which are similar to those of the two-point bubble probes. They are employed to determine the time the bubble front requires to pass from the first to the second probe. With the known distance between the detectors and the known cross-sectional area of the capillary, one can determine the volume of the bubble. With the assumption of a spherical bubble, one can calculate a diameter d for that bubble.

The main problem that makes this method difficult to use is that the flow rate has to be adjusted so that the bubbles are withdrawn as they arrive at the entrance of the capillary. If the flow rate is too low, a backup is likely to occur and the bubbles will coalesce in the entrance; if it is too high, they may be dispersed into individual parts. These so-called isokinetic conditions can be obtained in laminar flows, but are difficult to obtain in turbulent flows.

Isokinetic sampling probes have been developed since the early 1960s (ADORNI et al., 1961); a review of the early developments can be found in JONES (1983). TODTENHAUPT (1971) and PILHOFER et al. (1972, 1974) used the method in chemical engineering. BRENTRUP et al. (1980) and WEILAND et al. (1980) demonstrated by measurements in *Candida utilis* cultures that the method can be applied in fermentation broths as well. The data analysis was refined later by GREAVES and KOBBACY (1984). They also optimized the sampling technique in order to take a representative sample of the dispersion from which it was possible to determine the local gas holdup. The method is still in use in some laboratories (HO et al., 1987). GREAVES and BARIGOU (1988) constructed a map of bubble size distributions by measuring at numerous places within a 1 m^3 gassed stirred tank reactor.

4.1.1.4 Restrictions of Bubble Probes

Bubble probes have the conceptual advantage of being able to measure most physical quantities necessary to characterize the bubble phase inside a bioreactor, but they suffer from severe restrictions regarding their applicability in real fermentation broths. Bubble probes obviously require the bubbles to move along the probe axis to obtain valid results; this at least poses a significant restriction on the use of this method, since it excludes measurements in turbulently stirred tanks.

Moreover, it is virtually impossible to measure small bubbles, e.g., bubbles less than 2 mm in diameter, since these bubbles are conducted around the probes by the continuous liquid flow. In real fermentation systems, such bubble sizes are common. Non-spherical bubbles are often underestimated in their Sauter diameters. Furthermore, it is often doubtful whether the bubbles that are detected are unaffected by the probe. Nonetheless, bubble probes are in some situations the only means of obtaining information about the bubbles in a real system.

4.1.2 Radiation Techniques

4.1.2.1 Optical Methods

Light scattering methods of determining gas–liquid interfacial areas have been used in chemical engineering since the original work of CALDERBANK (1958). He developed a theory of scattering by monodisperse particles. This theory was extended by McLAUGHLIN and RUSHTON (1973) to polydisperse systems. They developed a mathematical model based on the probability theory to simulate light transmission through dispersions and showed that the fraction f of parallel (!) light which passes through is a unique function of

$$f = \exp(-\varepsilon L/d_s) \qquad (18)$$

regardless of the particle size distribution. ε is the dispersed particle holdup, L the characteristic length of the optical path, and d_s the Sauter diameter, defined by

$$d_s = \Sigma d_i^3 / \Sigma d_i^2 \qquad (19)$$

With the assumption of the presence of merely spherical bubbles, geometrical relationships between the gas holdup, the specific interfacial area, and the Sauter bubble diameter lead to

$$a = 6\varepsilon/d_s \qquad (20)$$

This, introduced into the above equation for f, leads to

$$f = \exp(-aL/6) \qquad (21)$$

which is nearly the result of CALDERBANK.

McLAUGHLIN and RUSHTON (1973) verified their results in light scattering in polydisperse liquid–liquid dispersions by determining the specific interfacial area. The authors stated that the results were also applicable to other dispersions, because no requirements were made on physical parameters other than the geometrical ones. The method has been improved with regard to the influence of multiple

scattering effects by LANDAU et al. (1977a,b), AL TAWEEL et al. (1984), and URUA and DEL-CERRO (1987).

Optical measuring techniques usually require sufficient light transmission through the investigated part of the medium. In most practical cases of biotechnological broths, this condition is not fulfilled.

In former years, many investigators tried to measure bubble-size distributions or specific interfacial areas by means of photographic observations. These measurements cannot lead to representative results, since the bubble-size and density distributions change significantly across the radius of a fermenter. Photographically, however, one only observes bubbles in a layer adjacent to the reactor wall. Thus, the observations are not necessarily representative of the overall values in the reactor. They can lead to contradictory conclusions, as OYE-VAAR et al. (1988) have recently stated.

4.1.2.2 Ultrasound Techniques

1. Ultrasound Doppler technique

The ultrasound Doppler technique is a method for measuring bubble velocity distributions. The measuring device is very robust and does not suffer from unknown interactions between the probes and the flow of the dispersion within the measuring volume. As opposed to bubble probes, this technique can also be applied to turbulent flows.

As schematically shown in Fig. 14, ultrasound pulses are transmitted into the dispersion. Bubbles act as excellent reflectors, hence, part of the ultrasound power is reflected into a detector. In the case of moving bubbles, the detected signal is shifted in frequency according to Doppler's principle. This frequency shift is proportional to the bubble's velocity component in the direction along half angles between the input and the reflected beam. With electronic circuits it is possible to construct a signal, which fluctuates at the Doppler shift frequencies, from the signal employed to drive the transmitter and the detected reflection signal. A spectral analysis of that signal then results in the distribution of the bubble velocities along that direction in space.

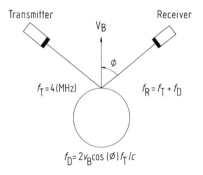

Fig. 14. Principle of the ultrasound Doppler technique to measure bubble velocities. An ultrasound wave is transmitted from the probe into the dispersion and reflected at all bubble surfaces within the region seen by the ultrasound detector. If the reflecting bubble is moving with the velocity v_B, the reflected ultrasound is shifted in frequency by an amount f_D. This shift is proportional to the bubble velocity component along half angles between input and output beam. Furthermore, it depends on the sound velocity c in the liquid phase.

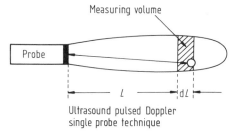

Measuring volume

Probe

L ⟷ dL

Ultrasound pulsed Doppler
single probe technique

Fig. 15. Single-probe ultrasound reflection technique. An ultrasound pulse is transmitted from the probe into the dispersion. The pulse is reflected by bubbles within its path. A part of the reflected ultrasound power is then detected by the same ultrasound probe. By means of an electronic gating technique, only those reflections are recorded, which arrive at the probe within a predefined time window. Together with the known sound velocity a defined measuring volume can be obtained.

It is not necessary to use a separate detector because the same probes can be used as transmitters and receivers as well. If one employs ultrasound pulses instead of a continuous ultrasound wave, it suffices to use only a single probe, which is alternatively switched to be the transmitter or the receiver (Fig. 15). Using electronic gating techniques, the measuring volume can then be shifted within several centimeters apart from the probe. In that case, the bubble velocities along the direction of the ultrasound beam are probed.

Ultrasound probes have been constructed that can be used in fermentations (SOLLIN-

ger 1989). They are steam-sterilizable and mounted on top of a steel tube, and they can be installed in reactors of any size.

Typically, an ultrasound transmission frequency of 4 MHz is used in bioreactors (KORTE and LÜBBERT, 1985). The band width of the Doppler shift signal is then less than 5 kHz in most cases. Such signals can easily be digitized. Using modern microelectronic components, e.g., signal processors, the digital signal analysis can be done in real-time. Unfortunately, a simple calculation of the power spectral density function is not sufficient for an analysis of the data as long as some bubbles approach the ultrasound probe while others disappear. If such bubbles have the same absolute value for their velocity component along the axis, defined by the probe alignment, they can not be distinguished, because, physically, Doppler frequencies are always positive.

Approaching and departing bubbles can, however, be distinguished if one also takes advantage of the phase information in the signal. This can be exploited by means of a quadrature detection technique, as described by KORTE (1986). LÜBBERT et al. (1987b) sampled and digitized Doppler signals at a rate of 20 kHz and showed how to calculate the velocity distribution of the bubbles on-line. To cope with the large data rate, they performed the analysis by means of special electronic devices based on signal processors (Texas Instruments TMS320). Such processors allow for calculating the bubble velocity distribution simultaneously with the data sampling procedure.

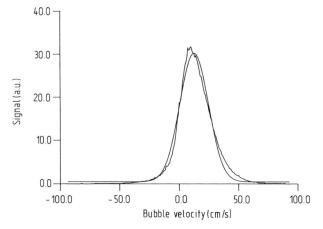

Fig. 16. Typical bubble velocity distribution as measured with the ultrasound pulsed Doppler technique in a stirred tank bioreactor of 1 m^3 operating volume during a penicillin fermentation. The turbulent flow within this reactor normally shows negative as well as positive velocity values, corresponding to bubbles approaching and disappearing from the probe. The noiseless curve is the fit of a normal distribution to the result, which shows that the velocity distribution is slightly non-symmetric.

Thus, the resulting velocity distribution is directly available at the end of the data acquisition period.

A typical result obtained in a stirred tank reactor during a penicillin production is shown in Fig. 16. Due to turbulent motions, one observes positive as well as negative velocity values along most flow directions in such flows. The total measuring time required to obtain a result of the quality shown in Fig. 16 is about 10 minutes. The moments of the distributions measured at different points in the reactor can be used to characterize the global bubble motion. The mean values along the coordinate axis lead to contour maps like that shown in Fig. 2. The variances give some information about the randomness of the bubble motion, i.e., on the local mixing behavior.

2. Ultrasound transmission methods

Since the work of SAUTER (1928), the transmission of radiation through dispersions has been a familiar technique to determine properties of a dispersed phase. Unfortunately, it is not possible to use the elegant light techniques, since nearly all fermentation media are optically opaque. Thus, some radiation must be chosen which can be transmitted through cultivation broths. One possibility is ultrasound.

If one transmits an ultrasound wave through a dispersion containing ultrasound reflectors such as gas bubbles, one observes a reduction in the transmitted power, as compared to the transmission through the continuous medium alone. This attenuation can be described by the general physical equation of absorption of radiation, where the intensity I decreases exponentially with the path length L:

$$I/I_0 = \exp(-qL) \quad \text{or} \quad q = -\ln(I/I_0)/L \quad (22)$$

The absorption coefficient q is dependent on the wavelength of the transmitted radiation. If the scattering particles appear at the ultrasound beam as a projection area per unit volume, a_p, then q is equal to a_p, as CALDERBANK (1958, 1967) showed for light scattering, in which the wavelength is small compared to the scattering centers. This is intuitively clear, since q should only depend on the number density n of the reflecting bubbles and their

mean reflecting surface, which, in turn, is characterized by their mean projected area A. Since q is independently proportional to both physical quantities, it is proportional to their product, which turns out to be the projected area a_p per unit volume.

For particles of random shapes that do not have concave surfaces, it was found that

$$q = a/4 \quad (23)$$

where a is the specific interfacial area. The validity of this relationship was tested with light scattering in air-in-water dispersions, where the scatterers were bubbles.

As STRAVS and VON STOCKAR (1985) pointed out, this relationship can also be used in ultrasound transmission experiments, if a is constant along the radiation path and the frequency is far away from the resonance frequencies of the bubbles, which are capable of compression or elastic form oscillations. Hence, if one measures the ultrasound absorption I/I_0 along a predefined path length L, one can determine the specific interfacial area a.

One avoids the effects of perpetual wave interferences, if one uses an ultrasound pulse instead of continuous waves. By means of proper gating, one reduces the influence of multiple scattering. A closer look at this topic shows that the scattering of ultrasound waves by particles is a function of the bubble radius and the frequency of the wave.

STRAVS and VON STOCKAR applied this simple transmission technique successfully to *Candida utilis* cultivations. RIEBEL and LÖFFLER (1989) reported on initial investigations exploiting the wavelength dependency of ultrasound scattering. They used a broadband ultrasonic radiation source and measured the transmission at different frequencies simultaneously. If the ultrasonic extinction cross-sections were known (!), they showed how to set up a linear equation system that can be solved to obtain particle-size distributions and particle concentrations. With this ultrasound spectrometric technique, they demonstrated in model media that solid particles ranging from 20 to 1000 μm in diameter can be analyzed, if frequencies in the range from 1.7–81 MHz are used.

3. Ultrasound pulse reflection methods

Specific interfacial areas can also be measured by sound reflection at the surface of the dispersed particles. The ultrasound power reflected back into the detector depends on two competing effects. On the one hand, the power received at the detector increases with the number density and with the size of the reflecting bubbles. On the other hand, it decreases due to extinction along the path from the transmitter via the measuring volume back into the detector, as discussed above for the ultrasound transmission method.

The increase is proportional to the specific interfacial area a, whereas, as shown above, the attenuation effect is roughly proportional to

$$\exp(-La/4) \qquad (24)$$

L being the length of the ultrasound path, which can be adjusted electronically by shifting the time gate within which the reflected signal is recorded.

Since both effects are statistically independent, the overall signal power P is proportional to the product of both:

$$P(a) = \text{const.} \, a \exp(-La/4) \qquad (25)$$

Fig. 17 shows a plot of this function. The constant can be obtained experimentally by simply reducing the gas flow through the reactor until the detected ultrasound power reaches its maximum value. The coordinates of this maximum are sufficient to determine the calibration constants of both axes of the coordinate system. Thus, the relation $P(a)$ can then be used to determine the interfacial area from the measured ultrasound power, provided one can make sure that the specific interfacial area a is definitely larger or smaller than $a_{\max}$. This decision is easy to make in normal cultivation practice.

The difficulty of making special calibration measurements can be avoided by measuring the reflected power at two different but known positions along the probe axis. Then one can determine the specific interfacial area from the two observed ultrasound intensity values. If these two measuring points are close together,

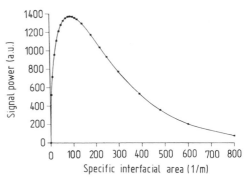

Fig. 17. Plot of the dependency of the ultrasound power, reflected back into the ultrasound detector, as a function of the specific interfacial area a. The normalization constants of both axes must be determined by a simple calibration procedure.

one can assume that practically the same a is found at both places. BRÖRING (1990) was able to enhance the signal-to-noise ratio of the a measurements in bioreactors with this technique.

4.1.2.3 Further Radiation Techniques

Since ultrasound transmission through biological culture media is also limited to a maximum of ten centimeters, it leads one to think about radiation of other wavelengths or of other qualities. Electromagnetic radiation, for example, does not suffer from the transmission limitation, if a much lower wavelength is chosen than that of light. In the literature one can find examples in which light is replaced by gamma radiation. YOUNG et al. (1987) used a translating gamma densitometer to measure the cross-sectionally averaged gas holdup profile along the riser of an airlift reactor. The void fraction can be calculated directly from the attenuation of the gamma beam. This technique has the advantage that it can be used non-intrusively, i.e., one can irradiate through the reactor walls. It is also possible to use X-ray radiation. The technique is well established in nuclear reactor thermal hydraulics, where it is very important to measure void fractions (GINOUX, 1978). BEINHAUER (1971) and PIKE et al. (1965) measured gas holdups in two-phase flows by X-ray attenuation. As a matter of principle, however, measuring techniques using ionizing radiation do not find much acceptance in biotechnology.

4.2 Liquid Flow Properties

4.2.1 General Anemometers

Many different instruments have been developed in hydrodynamics. As mentioned before, most of them, especially the elegant optical devices, cannot be used in bioreactors, since the broths are optically opaque. Even if the continuous liquid phase is transparent, the great density caused by the large number of small bubbles prevents the transmission of visible light through the whole reactor. There may be some very special cases where photographic methods, laser-Doppler, phase-Doppler, and the many others can be applied, but in normal cases of practical interest they cannot be used. Thus, optical techniques are not discussed in much detail here.

4.2.1.1 Laser Doppler Technique

Laser Doppler techniques are the most elegant optical measuring methods to investigate single-phase flows. Sophisticated measurements have been made to investigate the liquid flow inside chemical reactors by several authors (LAUFHÜTTE and MERSMANN, 1985; BAUCKHAGE et al., 1987; WEETMAN and OLDSHUE, 1988; COSTES and COUDERC, 1988; WU and PATTERSON, 1989). Especially, local investigations of velocity profiles and Reynolds stresses in the surroundings of impellers can give much insight into the pumping characteristics of mixing impellers under ungassed conditions.

One way to reduce the limitations in light transmission is to reduce the light beam paths in the flow to only a few centimeters by means of optical fibers (KYUMA et al., 1981). If the ungassed liquid is optically transparent, there is a chance that one can then measure liquid velocities.

There is, however, another problem that limits the application of laser-Doppler anemometry (LDA) in multiphase flows. Since usually the measuring volume is small compared to the dispersed particles, e.g., bubbles or solid particles, the signal, measured at one particular point within the multiphase flow, is often interrupted by signal segments which contain invalid information. These correspond to those time intervals during which the measuring point falls inside a bubble or a particle. At these places, the liquid velocity is undefined. To extract an exact mean velocity from such a signal would be a formidable task, since it would require detection and elimination of the invalid signal segments. Thus, the method is practically restricted to multiphase flows, either with extremely small dispersed particles (CARTELLIER and ACHARD, 1985) or with only a few large bubbles, the effect of which on the signal can easily be recognized and eliminated.

4.2.1.2 Hot Film Anemometers and Other General Techniques

In some bioreactors, e.g., in wastewater treatment reactors at low gas loads, the medium is so similar to water that one can use hot film anemometry (HFA) to investigate the local flow field within the reactors. The HFA technique has been applied to investigate flows in bubble columns as well as stirred tanks (LU and JU, 1987). Modern versions of anemometer probes are developed to make the technique direction sensitive. Dual split-film probes were applied by FRANZ (1983) and triple-split-film probes by MENZEL et al. (1989). The arguments regarding signal interruptions, however, raised in the discussion of point-like measuring volumes in multiphase flows in connection with laser-Doppler anemometry, also apply to HFA. Therefore, HFA is not recommended in normal biotechnological applications.

Other methods used in single-phase flows, e.g., Pitot tubes (TANAKA et al., 1989) or fan velocimeters (HÖNTZSCH GmbH, 1986), did not give reliable results in multiphase flows (HILLS, 1983).

4.2.2 Special Techniques

4.2.2.1 Mechanical Fluctuation Probes

An estimate of the fluctuations in the continuous liquid phase can be made by very simple means. One example of practical importance is a flexible mechanical vane sensor developed by SMITH et al. (1987). The vane tries to follow the fluctuating motion of the dispersion to which it is exposed. Like a forced oscillator, it will follow the exciting frequencies, but with an amplitude which depends on its own mechanical properties. In a turbulently fluctuating flow it can only follow the slower motions due to the larger-scale motions.

With such a simple probe it is possible to obtain evidence for the ventilated cavities behind impeller blades, if the sensor is installed adjacent to the rotating blades (WARMOES-KERKEN and SMITH, 1988). The power transferred into the dispersion depends on the structure of the cavities, since the streamlining actions of different cavities lead to different impeller drag coefficients. Alternative cavity sizes can be detected by correlation and spectral analysis of the fluctuations of the vane sensor.

4.2.2.2 Heat-Pulse-TOF Technique

The time-of-flow (TOF) measurement technique is a variant of the residence-time distribution measurement. Primarily it measures the distribution of passage times of particles, which start at one point in the flow and happen to meet a specified second point later on. The technique has been proposed by LÜBBERT and LARSON (1986) to determine the mean flow velocity and local mixing characteristics of the continuous liquid phase in complex multiphase systems. They used heat to label the fluid elements instead of using material tracers.

The technique can be implemented in arbitrarily large bioreactors. Fig. 18 shows the simple outline of the measuring setup: At one point in the cultivation broth, a small heating element is used to label passing fluid elements.

Adjacently downstream on the same trajectory of the mean flow, a fast and sensitive temperature detector registers the arrival of the marked fluid elements.

It is sufficient in many practical applications to use simple platinum metal layer resistance thermometers as detectors. If higher sensitivities are necessary, one can also use hot film anemometer probes switched as temperature sensors (LARSON, 1988). The latest development is an optical detector on a semiconductor basis capable of following temperature fluctuations at more than 10 kHz (SCHMIDT, 1990).

If the temperature pulses are released to the liquid pseudorandomly, the signal-to-noise ratio can be kept high enough to keep the required temperature increases within the labelled fluid elements below 1 K. Even in turbulent flows in bioreactors during real fermentations it was possible to obtain sufficiently high

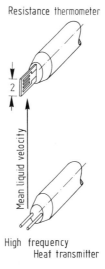

Fig. 18. Probe arrangement used in the temperature pulse technique to measure time-of-flow distributions. These can be used to estimate the liquid flow velocities and local mixing parameters. The lower probe is the heater, which uses the voltage applied to the two short stainless steel wire ends to draw a current through the conductive medium. In this way one can heat the liquid between the wires sufficiently rapidly. The upper probe is a temperature sensor on PT20 base. It is able to detect the heat-labelled fluid elements, if the mean liquid velocity between the two probes is greater than zero.

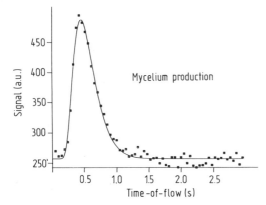

Fig. 19. Time-of-flow result obtained within a laboratory-scale airlift loop reactor during a production of cephalosporin. Solid peanut meal at a volume fraction of about 30% was used as substrate. The probe distance was 13 cm.

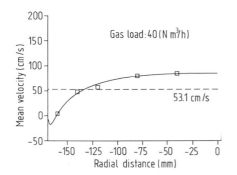

Fig. 20. Half profile of the axial component of the mean liquid velocity along the radius of the riser of a pilot-scale airlift tower loop reactor (4 m³). The data were obtained with the heat pulse technique at higher gas throughputs corresponding to 25 cm/s superficial gas velocity.

temperature signal intensities at probe distances of more than 10 cm.

The fluid elements can be heat-labelled by several simple techniques. The simplest one is to use a heated wire (LÜBBERT and LARSON, 1986). A faster method is to use high frequency currents directly through the electrically conductive biosuspension (LARSON and LÜBBERT, 1990). As demonstrated, the method can be implemented very inexpensively.

The heat pulse technique is especially suited for measurements in dispersions with high

void fractions flowing at higher local velocities. Even in biotechnological cultivations at higher viscosities, which are inaccessible by other methods, measurements are possible. Fig. 19 should serve as an example of a time-of-flow curve measured during a rheologically extremely complicated production of cephalosporin in a tower loop reactor. This cultivation was operated with an additional solid phase in the form of peanut meal at a volumetric solid-phase holdup of about 20%. It could be shown that this technique can also be used in larger airlift reactors to measure flow profiles at higher gas loads. Fig. 20 shows an example obtained at 40 N m³/h, corresponding to 25 cm/s superficial gas velocity. Even three times that gas load does not lead to problems.

Time-of-flow distributions contain information not only about the mean liquid velocity, but also about the mixing behavior of the flow. To understand this, consider the fluid elements, which are marked by one pulse of the heater, as a particle cloud. This cloud will become larger with time through the influence of the different mixing mechanisms. Thus, the time dependence of the cloud width directly reflects the active mechanisms. LÜBBERT and LARSON (1990) showed that the standard deviations *s* of the time-of-flow curves can be related to the mixing mechanisms. They analyzed the measured data by means of a nonlinear least square fit of an appropriate model. In the risers of airlift tower loop reactors, they found that most mixing was done by the wakes of the rising bubbles.

The pseudorandom time-of-flow technique involving heat pulses to tag the fluid elements has also been used to measure very low volumetric flow rates of simple single-phase flows in narrow tubes, e.g., in oil feed lines (WITTE and BAIER, 1985), or even non-invasively in biotechnological and medical applications (KOHLRUSCH, 1988; NEUGEBAUER, 1989).

Heat as a possible tracer in local flow experiments has been discussed several times in the literature. KRIZAN and PILHOFER (1981) described a correlation method, in which they correlated two temperature signals measured at two adjacent points along a trajectory in the flow. Originally, an attempt was made to use the correlations between natural temperature fluctuations within the dispersion. However, measurable effects turned out to require the use of a third probe which dispersed heat to the flow continuously.

Methods, which are based on a correlation between fluctuating signals, e.g., velocity fluctuations obtained from two measuring points within an extended flow, cannot generally be applied in the same way as those known from tube flows. The distortions, leading to fluctuations, may be influenced by a wave front, which propagates in a direction different from the line defined by the two probes or which may have a propagation velocity different from the fluid flow. In both cases, the correlation peak will not give correct information about the time-of-flow of real fluid elements. Therefore, the pseudostochastical heat pulse technique is to be preferred.

4.2.2.3 Ultrasound Techniques

POPOVIC and ROBINSON (1988, 1989) also used an ultrasound Doppler device, but they tried to determine the circulating liquid velocity in an external loop airlift reactor. In such an application 'no slip assumptions' must be made, i.e., the bubbles must be assumed to move at the same velocity as the liquid phase. Such an assumption may be justified at the end of the downcomer of an airlift reactor, since there are predominantly small bubbles there, which travel practically at the liquid velocity. The authors do not report applications of this method during real cultivations, but their measurements in cultivation broths of *Chaetomium cellulolyticum* transferred into a loop reactor, especially for test measurements, give enough assurance that the technique can also be used during cultivation.

5 Conclusions and Future Trends

Fluid-dynamical characterization of bioreactors during their normal operation cannot be done with sufficient accuracy using the well-known measuring techniques developed in sin-gle-phase fluid dynamics. New methods have to be developed to provide insight into the complex multiphase flows in real cultivation broths. The techniques must continue to be developed in the future, since we have by no means as yet attained a satisfactory state-of-the-art method.

Better measuring techniques are a prerequisite to obtaining data necessary to improve our knowledge of bioreactors. Global measuring techniques, such as residence and circulation time distribution measurements, well-known in chemical engineering, can yield much valuable information about the global flow structure. By new implementations, e.g., using pseudo-stochastical instead of single-pulse or step excitation, the decisive signal-to-noise ratio can be increased so much that more details can be extracted from the measuring data.

Many techniques developed so far utilize probes in direct contact with the fluid elements. In most cases, this leads to an interaction between the two, which may change the sensors and/or the fluid flow. Although one tries to minimize the interactions, it is better to avoid direct contact. This means that, as far as possible, methods must be developed which are based on radiation effects. Unfortunately, elegant optical methods, which have been developed in single-phase or low-density fluid dynamics, cannot be used in most bioreactors. Thus, new instruments using different wavelengths or other types of radiation, e.g., ultrasound, must be developed. As shown in this chapter, there are some promising first steps in this direction.

Even from another point of view, the state of development of multiphase flow measuring techniques leaves much to be desired. Measuring devices, if they exist for a nontrivial physical quantity in question, cannot be applied like a simple thermometer. In most cases, they must be adapted laboriously to the rheological system under investigation.

Furthermore, the measurement techniques do not provide data with spectroscopic accuracy. Whenever possible one should build in redundancy, using more than one method, to measure a quantity accurately and to ensure that the devices are working correctly.

Dealing with many sensors simultaneously using model-aided measuring methods or sophisticated measuring schemes requires some routine to handle the data, to calculate an indirect measuring quantity, or to control a complex measuring procedure. Nowadays, this is usually done by means of microcomputers. They may be part of a larger process control system, a single computer, or a PC. The trend is to use microprocessors and to install them as close as possible to the sensors. This has several advantages, the most obvious being that a user is mostly interested in the results of the measurement and not in the obstacles that had to be overcome to obtain them. Thus, he prefers to separate all activities required for the measurement and put them into a separate system. Everyone wishes to obtain from measuring instruments only their output in the form of the final result, including information about its reliability. In large automated systems with many measuring devices, this is indispensible in order to keep control over the whole system.

6 References

ABUAF, N., JONES, O. C., ZIMMER, G. A. (1978), Optical probe for local void fraction and interface velocity measurement, *Rev. Sci. Instrum.* **49**, 1090.

ADLER, I., FIECHTER, A. (1983), Charakterisierung von Bioreaktoren mit biologischen Testsystemen, *Chem. Ing. Tech.* **55**, 322–323, Synopse 1093; *Swiss Biotech.* **1**, 17–24.

ADLER, I., FIECHTER, A. (1986), Valuation of bioreactors for low viscous media and high oxygen transfer demand, *Bioprocess Eng.* **1**, 51.

ADORNI, N., CASAGRANDE, I., CRAVAROLO, L., HASSID, A., SILVESTRI, M. (1961), Experimental data on two-phase adiabatic flow, liquid film thickness, phase and velocity distributions, pressure drops in vertical gas-liquid flows, *CISE-R-35* (EUREAC-150), Milan, Italy: Centro Informazioni Studi Esperanzi.

ALLEN, D. G., ROBINSON, C. W. (1989), Hydrodynamics and mass transfer in *Aspergillus niger* fermentations in bubble column and loop bioreactors, *Biotechnol. Bioeng.* **34**, 731–740.

ARMENANTE, P. M., KIRWAN, D. J. (1989), Mass transfer to microparticles in agitated systems, *Chem. Eng. Sci.* **44**, 2781–2796.

BAUCKHAGE, K., FLÖGEL, H. H., SCHULTE, G. (1987), Partikelgröße und Geschwindigkeit gleichzeitig, *Chem. Ind.* **4**, 96–101.

BEINHAUER, R. (1971), *Dissertation*, Technische Universität, Berlin.

BENDAT, J. S., PIERSOL, A. G. (1986), *Random Data*, 2nd Ed., New York: Wiley-Interscience.

BEYELER, W., EINSELE, A., FIECHTER, A. (1981), On-line measurements of culture fluorescence: Method and application, *Eur. J. App. Microbiol. Biotechnol.* **13**, 10–14.

BLENKE, H. (1988), Verfahrenstechnische Beiträge zur Entwicklung von Bioreaktoren, *BTF Biotech-Forum* **5**, 6–23.

BRENTRUP, L., WEILAND, P., ONKEN, U. (1980), Messung von Blasengrößenverteilungen in Fermentationsmedien mittels einer photoelektrischen Sondenmethode, *Chem. Ing. Tech.* **52**, 72–73, Synopse 758.

BRÖRING, S. (1990), Messung der spezifischen Phasengrenzflächen in Mehrphasenreaktoren mit einem neuen Ultraschall-Reflexionsverfahren, *Ongoing Dissertation,* Universität Hannover, FRG.

BRÖRING, S., FISCHER, J., KORTE, T., SOLLINGER, S., LÜBBERT, A. (1990), Flow structure of dispersed gas flows in multiphase reactors investigated by a new ultrasound Doppler technique, *Can. J. Chem. Eng.,* submitted.

BRUIJN, W., RIET, K. VAN'T, SMITH, J. M. (1974), Power consumption with aerated Rushton turbines, *Trans. Inst. Chem. Eng.* **52**, 88–104.

BRYANT, J. (1966), *PhD Thesis*, Cambridge University, Cambridge, UK.

BRYANT, J. (1977), The characterization of mixing in fermenters, *Adv. Biochem. Eng.* **5**, 101–123.

BUCHHOLZ, R., SCHÜGERL, K. (1979), Bubble column bioreactors: Methods for measuring the bubble size, *Eur. J. Appl. Microbiol. Biotechnol.* **6**, 301–315.

BUCHHOLZ, R., TSEPOTONIDES, I., STEINEMANN, J., ONKEN, U. (1983), Influence of gas distribution on the interfacial area and mass transfer in bubble columns, *Ger. Chem. Eng.* **6**, 105–113.

BUCKLAND, B. C., GBEWONYO, K., JAIN, D., GLAZOMITSKY, K., HUNT, G., DREW, S. W. (1988), Oxygen transfer efficiency of hydrofoil impellers in both 800 l and 19000 l fermentors, *Proc. 2nd Int. Conf. Bioreactor Fluid Dynamics* (KING, R. Ed.) London: Elsevier Applied Science Publisher.

BURGESS, J. M., CALDERBANK, P. H. (1975), The measurement of bubble parameters in two-phase dispersions, I: The development of an improved probe technique, *Chem. Eng. Sci.* **30**, 743–750.

CALDERBANK, P. G. (1958), Physical rate processes in industrial fermentation, Part 1: The interfacial area in gas/liquid contacting with mechanical agitation, *Trans. Inst. Chem. Eng.* **36**, 443–463.

CALDERBANK, P. H. (1967), *Chem. Eng. London* **212**, CE209.

CALDERBANK, P. H. (1978), Limitations of the Burgess-Calderbank probe technique for characterization of gas liquid dispersions on sieve trays, *Chem. Eng. Sci.* **33**, 1407.

CALDERBANK, P. H., PEREIRA, J. (1977), The prediction of distillation plate efficiencies from froth properties, *Chem. Eng. Sci.* **32**, 1427–1433.

CARTELLIER, A., ACHARD, J. L. (1985), Limitations of the classical L.D.A. formula for velocity measurements for large particles in two-phase suspension flows, *Physicochem. Hydrodyn.* **6**, 463–481.

CEVEY, P. F., STOCKAR, U., VON (1985), A tracer system based on a photochromic dye and on fibre optics for measuring axial dispersion of organic liquids in pilot-scale packed columns, *Chem. Eng. J.* **31**, 7–13.

CLARK, N. N., FLEMMER, R. L. (1985), Turbulent circulation in bubble columns, *AIChE J.* **31**, 500.

COOKE, M., MIDDLETON, J. C., BUSH, J. R. (1988), Mixing and mass transfer in filamentous fermentations, pp. 37–64, in *Proc. 2nd Int. Conf. Bioreactor Fluid Dynamics* (KING, R., Ed.), London: Elsevier Applied Science Publisher.

COOPER, C. M., FERNSTROM G. A., MILLER, S. A. (1944), Performance of agitated gas-liquid contactors, *Ind. Eng. Chem.* **36**, 504–509.

COSTES, J., COUDERC, J. P. (1988), Study by laser Doppler anemometry of the turbulent flow induced by a Rushton turbine in a stirred tank: Influence of the size of the units – I. Mean flow and turbulence, *Chem. Eng. Sci.* **43**, 2751–2764.

CZECH, K. (1986), On-line Charakterisierung eines Airliftbioreaktors, *Dissertation*, Universität Hannover, FRG.

DAMKÖHLER, G. (1937), Einfluß von Diffusion, Strömung und Wärmetransport auf die Ausbeute bei chemisch-technischen Reaktionen, in: *Der Chemie-Ingenieur* (EUCKEN, A., et al., Eds.), Vol. III, Part 1, p. 359.

DANCKWERTS, P. V. (1953), Continuous flow systems: Distributions of residence times, *Chem. Eng. Sci.* **2**, 1.

DANEL, F., DELHAYE, J. M. (1971), Sonde optique pour mesure du taux de presence local en ecoulement diphasique, *Mes. Reg. Autom.* **36**, 99–101.

DECKWER, W. D. (1980), On the mechanism of heat transfer in bubble column reactors, *Chem. Eng. Sci.* **35**, 1341–1346.

DELHAYE, J. M. (1981), *Entropie* **99**, 2–25.

DENK, V., STERN, R. (1979), Beitrag zur Kenntnis der Bewegungsvorgänge während der Gärung in zyklonischen Gärtanks, *Brauwissenschaft* **32**, 254.

DENK, V., STERN, R. (1980), Examples of flow visualization in food technology, *Proc. 2nd Int. Symp. Flow Visualization* (MERZKIRCH, W., Ed.), Bochum.

DRAHOS, J., CERMAK, J. (1989), Diagnosis of gas-liquid flow patterns in chemical engineering systems, *Chem. Eng. Process.* **26**, 147–164.

EDWARDS, M. F. (1985), Mixing in low viscosity liquids in stirred tanks, pp. 131–144, in: *Mixing in the Process Industries* (HARNBEY, N., EDWARDS, M. F., NIENOW, A. W., Eds.), London: Butterworths.

EINSELE, A. (1976), Charakterisierung von Bioreaktoren durch Mischzeiten, *Chem. Rundsch.* **29**, 53–55.

EINSELE, A., FIECHTER, A. (1974), Anwendung der Dehnungsmeßstreifentechnik in der Fermentationsindustrie, *Chem. Ing. Techn.* **46**, 701, Synopse 121.

EINSELE, A., RISTROPH, D. L., HUMPHREY, A. E. (1978), Mixing times and glucose uptake measured with a fluorometer, *Biotechnol. Bioeng.* **20**, 1487–1492.

FIELDS, P. R., SLATER, N. K. H. (1984), The influence of fluid mixing upon respiratory patterns for extended growth of a methylotroph in an airlift fermentor, *Biotechnol. Bioeng.* **26**, 719–726.

FRANZ, K. (1983), Untersuchungen zur Turbulenzstruktur in Blasensäulen mittels einer neu entwikkelten richtungsspezifischen Heißfilm-Anemometertechnik, *Doctoral Dissertation* Universität Dortmund, FRG.

FRIJLINK, J. J. (1987), Physical aspects of gassed suspension reactors, *Doctoral Thesis*, Technical University Delft, The Netherlands.

FRÖHLICH, S. (1987), Charakterisierung von Turmreaktoren im Labor- und Pilotmaßstab, *Doctoral Dissertation*, Universität Hannover, *VDI Fortschrittberichte*, Reihe 3: Verfahrenstechnik, Nr. 131, VDI Verlag, Düsseldorf.

FUKUMA, M., MUROYAMA, K., YASUNISHI, A. (1987), Properties of bubble swarm in a slurry bubble column, *J. Chem. Eng. Jpn.* **20**, 28–33.

FUNAHASHI, H., HARADA, H., TAGUSHI, H., YOSHIDA, T. (1987), Circulation time distribution and volume of mixing regions in highly viscous xanthan gum solution in a stirred vessel, *J. Chem. Eng. Jpn.* **20**, 277–282.

GALAUP, J. P. (1975), Contribution a l'etude des methodes de mesure en encoulement diphasique, *Doctoral Thesis*, Université de Grenoble, France.

GINOUX, J. J. (1978), *Two-phase Flows and Heat Transfer*, McGraw-Hill, New York.

GODFREY, J. C. (1985), Mixing of high-viscosity fluids, pp. 185–201, in: *Mixing in the Process Industries* (HARNBEY, N., EDWARDS, M. F., NIENOW, A. W., Eds.), London: Butterworths.

GREAVES, M., BARIGOU, M. (1988), The internal structure of gas-liquid dispersions in a stirred reactor, pp. 313–320, in: *Proc. 6th Eur. Conf. Mixing*, Pavia, Italy.

GREAVES, M., KOBBACY, K. A. H. (1984), Measurement of bubble size distribution in turbulent gas-liquid dispersions, *Chem. Eng. Res. Des.* **62**, 3–12.

GRIOT, M. U. (1987), Maßstabsvergrößerung von Bioreaktoren mit einer sauerstoffempfindlichen Testkultur, *Doctoral Dissertation*, ETH Zürich, Nr. 8412.

GSCHWEND, K., BEYELER, W., FIECHTER, A. (1983), Detection of reactor nonhomogeneities by measuring culture, *Biotechnol. Bioeng.* **25**, 2789–2793.

HANRATTY, T. J., CAMPBELL, J. A. (1983), Measurement of wall shear stress, pp. 559–615, in: *Fluid Mechanics Measurements* (GOLDSTEIN, R. J., Ed.), Washington: Hemisphere.

HANSFOLD, G. S., HUMPHREY, A. E. (1966), Effect of equipment scale and degree of mixing on continuous fermentation yield at low dilution rates, *Biotechnol. Bioeng.* **8**, 85–96.

HEINZLE, E., K. FURUKAWA, I. J. DUNN, J. R. BOURNE (1983), Experimental methods for on-line mass spectrometry in fermentation technology, *Biotechnology* **1**, 181–188.

HILLS, J. H. (1983), Investigations into the suitability of a transverse pilot tube for two-phase flow measurements, *Chem. Eng. Res. Des. 61,* 371.

HO, C. S., STALKER, M. J., BADDOUR, R. F. (1987), The oxygen transfer coefficient in aerated stirred reactors and its correlation with oxygen diffusion coefficients, in: *Biotechnological Processes: Scale-up and Mixing* (HO, C. S., OLDSHUE, J. Y., Eds.), New York: American Institute of Chemical Engineers.

HOFMANN, H., EMIG, G. (1983), Systematik und Prinzipien der Auslegung chemischer Reaktoren, *DECHEMA Monograph.* **94**, 1–14.

HÖNTZSCH GMBH (1986), *Flügelradanemometer*, personal communication.

JARAMILLO, A. (1985), Fluid dynamics and oxygen transfer during the cultivation of *Trichosporum cuteanum* in jet loop bioreactors, *Doctoral Dissertation*, ETH Zürich.

JARAMILLO, A., ADLER, I., FIECHTER, A. (1986), Fluiddynamik und Sauerstoff-Transport in Strahlschlaufenreaktoren mit dem biologischen Modellsystem *Trichosporon cutaneum*, Chem. Ing. Tech. **58**, 914–195, Synopse 1542.

JIMENEZ, B., NOYOLA, A., CAPDEVILLE, B. (1988), Selected dyes for residence time distribution evaluation in bioreactors, *Biotechnol. Techn.* **2**, 77–82.

JONES, O. C. (1983), Two-phase flow measuring techniques in gas-liquid systems, pp. 479–558, in: *Fluid Mechanics Measurements* (GOLDSTEIN, R. J., Ed.), Washington: Hemisphere.

KAI, W., SHENGYAO, Y. (1989), Heat transfer and power consumption of non-Newtonian fluids in agitated vessels, *Chem. Eng. Sci.* **44**, 33–40.

KAWASE, Y., MOO-YOUNG, M. (1989), Mixing time in bioreactors, *J. Chem. Technol. Biotechnol.* **44**, 63–75.

KHANG, S. J., FITZGERALD, T. J. (1975), A new probe and circuit for measuring electrolyte conductivity, *Ind. Eng. Chem. Fundam.* **14**, 208–213.

KOHLRUSCH, J. (1988), Berührungslose Messung geringer Volumenströme in der geschlossenen Schlauchleitung eines Infusionsapparates, *Diploma Thesis*, Universität Hannover, FRG.

KORTE, T. (1986), Meßtechniken zur Charakterisierung von Blasensäulenreaktoren, Doctoral Dissertation, Universität Hannover; *Fortschritt-Berichte*, Reihe 8, Nr. 110, Düsseldorf: VDI-Verlag.

KORTE, T., LÜBBERT, A. (1985), Messung von Blasengeschwindigkeiten in Bioreaktoren mit Hilfe eines Ultraschallimpuls-Doppler Anemometers, *Chem. Ing. Tech.* **57**, 1114, Synopse 1428.

KREBSER, U., MEYER, H. P., FIECHTER, A. (1988), A comparison between the performance of continuously stirred-tank bioreactors and a torus bioreactor with respect to highly viscous culture broths, *J. Chem. Technol. Biotechnol.* **43**, 107–116.

KRIZAN, P., PILHOFER, T. (1981), Entwicklung einer Sonde zur Messung örtlicher Geschwindigkeiten der Flüssigphase in Blasensäulen, *Chem. Ing. Tech.* **53**, 105.

KUBOI, R., NIENOW, A. W., ALLSFORD, K. (1983), A multipurpose stirred tank facility for flow visualisation and dual impeller power measurement, *Chem. Eng. Commun.* **22**, 29–39.

KYUMA, K., TAI, S., HAMANAKA, K., NUNOSHITA, M. (1981), Laser Doppler velocimeter with a novel optical fiber, *Appl. Opt.* **20**, 2424.

LAINE, J., KUOPPAMÄKL, R. (1979), Development of the design of large-scale fermentors, *Ind. Chem. Eng. Proc. Des. Dev.* **18**, 501–506.

LANDAU, J., BOYLE, J., GOMAA, H. G., TAWEEL, A. M., AL (1977a), Comparison of methods for measuring interfacial areas in gas-liquid dispersions, *Can J. Chem. Eng.* **55**, 13–18.

LANDAU, J., GOMAA, H. G., TAWEEL, A. M., AL (1977b), Measurement of large interfacial areas by light attenuation, *Trans. Inst. Chem. Eng.* **55**, 212–215.

LANS, R. G. J. M., VAN DER (1985), Hydrodynamics of a bubble column loop reactor, *Doctoral Thesis*, Technical University, Delft, The Netherlands.

LARSON, B. (1988), Eine Methode zur Charakterisierung der Flüssigphasenströmung in Mehrphasensystemen mit Hilfe pseudostochastischer Wärmeimpulsfolgen, *Dissertation*, Universität Hannover, *VDI Fortschrittsberichte*, Reihe 8: Meß-, Steuerungs- und Regelungstechnik, Nr. 169, Düsseldorf: VDI Verlag.

LARSON, B., MUTHESIUS, M., LÜBBERT, A. (1990), Local gas hold-up as measured with conductivity probes during the production of baker's yeast in a pilot scale fermentor, to be published.

LASA, H., DE, LEE, S. L. P., BERGOUGNOU, M. A. (1984), *Can. J. Chem. Eng.* **62**, 165–169.

LAUFHÜTTE, H. D., MERSMANN, A. (1985), Dissipation of power in stirred vessels, *Proc. 5th Eur. Conf. Mixing*, Würzburg, Cranfield: BHRA-Press.

LEE, S. A., LASA, H. I., DE (1987), Phase holdups in three-phase fluidized beds, *AIChE J.* **33**, 1359–1370.

LEE, S. L. P., LASA, H. I., DE (1988), Radial dispersion model for bubble phenomena in three phase fluidized beds, *Chem. Eng. Sci.* **43**, 2445–2449.

LEE, S. A., LASA, H. I., DE, BERGOUGNOU, M. A. (1984), Bubble phenomena in three-phase fluidized beds viewed by a U-shaped fiber optic probe, *AIChE Symp. Ser.* **80**, 110.

LEWIS, D. A., DAVIDSON, J. F. (1983), Bubble sizes produced by shear and turbulence in a bubble column, *Chem. Eng. Sci.* **38**, 161–167.

LU, W. M., JU, S. J. (1987), Local gas holdup, mean liquid velocity and turbulence in an aerated stirred tank using hot-film anemometry, *Chem. Eng. J.* **35**, 9–17.

LÜBBERT, A. (1987), Grundlegende Eigenschaften der Zweiphasenströmung in Blasensäulen-Reaktoren, *Chem. Ing. Techn.* **59**, 513, Synopse 1604.

LÜBBERT, A., KORTE, T. (1987), Fortschritte beim Ultraschall-Doppler-Verfahren, *Chem. Ing. Tech.* **59**, 84, Synopse 1561.

LÜBBERT, A., LARSON, B. (1986), A new method for measuring local velocities of the continuous liquid-phase in strongly aerated gas-liquid multiphase reactors, *Chem. Eng. Technol.* **10**, 27–32.

LÜBBERT, A., LARSON B. (1990), Detailed investigation of the multiphase flow in airlift tower loop reactors, *Chem. Eng. Sci.*, in press.

LÜBBERT, A., FRÖHLICH, S., SCHÜGERL, K. (1987a), Characterization of bioreactors by mass spectrometry analysis, pp. 125–142, in: *Mass Spectrometry in Biotechnological Process Analysis and Control* (HEINZLE, E., REUSS, M., Eds.), New York: Plenum Press.

LÜBBERT, A., KORTE, T., LARSON, B. (1987b), Simple measuring techniques for the determination of bubble- and bulkphase velocities in bioreactors, *Appl. Biochem. Biotechnol.* **14**, 207–219.

LÜBBERT, A., LARSON, B., WAN, L. W., BRÖRING, S. (1990), Local mixing behaviour of airlift multiphase chemical reactors, *AIChE Symp. Ser.*, submitted.

MAHOUAST, M., COGNET, G., DAVID, R. (1989), Two-component LDV measurements in a stirred tank, *AIChE J.* **35**, 1770–1778.

MANN, R., MAVROS, P. P., MIDDLETON, J. C. (1981), A structured stochastic flow model for interpreting flow-follower data from a stirred vessel, *Trans. Inst. Chem. Eng.* **59**, 271–278.

MCLAUGHLIN, C. M., RUSHTON, J. H. (1973), Interfacial areas of liquid-liquid dispersions from light transmission measurements, *AIChE J.* **19**, 817–822.

MENZEL, T., JÄGER, W. R., EWALD, G., ONKEN, U. (1989), Lokale fluiddynamische Parameter in Blasensäulenreaktoren bei höheren Flüssigkeitsviskositäten, *Chem. Ing. Tech.* **61**, 70–71, Synopse 1722.

MEYER, H. P., BEYELER, W., FIECHTER, A. (1984), Experiences with the online measurements of culture fluorescence during cultivation of *Bacillus subtilis, Escherichia coli,* and *Sporotrichum thermophile, J. Biotechnol.* **1**, 341.

MIDDLETON, J. C. (1985), Gas-liquid dispersion and mixing, pp. 322–355, in: *Mixing in the Process Industries* (HARNBY, N., EDWARDS, M. F., NIENOW, A. W., Eds.), London: Butterworths.

MILLER, N., MITCHIE, R. E. (1969), The development of a universal probe for measurement of local voidage in liquid/gas two-phase flow systems, pp. 82–88, in: *Two-phase Flow Instrumentation* (TOURNEAU, B. W., LE, BERGLES, A. E., Eds.), New York: ASME.

MILLER, N., MITCHIE, R. E. (1970), Measurement of local voidage in liquid/gas two phase flow system using a universal probe, *J. Br. Nucl. Energy Soc.* **9**, 94–100.

MOSER, A. (1987), Mixing in bioreactors – Quantification and modelling, pp. 51–92, in: *Physical Aspects of Bioreactor Performance* (CRUEGER, W., Ed.), Frankfurt: DECHEMA.

MUKATAKA, S., KATAOKA, H., TAKAHASHI, J. (1976), *Kagaku Kogaku Ronbunshu*, **2**, 628.

MUKATAKA, S., KATAOKA, H., TAKAHASHI, J. (1980), Effects of vessel size and rheological properties of suspensions on the distribution of circulation times in stirred vessels, *J. Ferment. Technol.* **58**, 155–161.

NOAR, P., SHINNAR, R. (1963), Representation and evaluation of residence time distributions, *Ind. Chem. Eng. Fundam.* **2**, 278.

NAUMANN, E. B. (1989), A circulation time model for batch reactors, *Chem. Eng. J.* **40**, 101–109.

NAUMANN, E. B., BUFFHAM, B. A. (1983), *Mixing in Continuous Flow Systems*, New York: John Wiley.

NEUGEBAUER, T. (1989), Komponenten zur Mikrowellenerwärmung von Flüssigkeiten, *Diploma Thesis*, Universität Hannover, FRG.

NIENOW, A. W., WISDOM, D. J., MIDDLETON, J. C. (1978), The effect of scale and geometry on flooding, recirculation and power in gassed stirred vessels, paper F1, 1–16, *Proc. 2nd Eur. Conf. Mixing*, Cranfield: BHRA.

OLDSHUE, J. Y. (1989), Fluid mixing in 1989, *Chem. Eng. Progr.*, May 1989, 33–42.

OOSTERHUIS, N. M. G. (1984), Scale-up of bioreactors – A scale-down approach, *Doctoral Dissertation*, Technical University, Delft, The Netherlands.

OUYOUNG, P. K., CHISTI, M. Y., MOO-YOUNG, M. (1989), Heat transfer in airlift reactors, *Chem. Eng. Res. Des.* **67**, 451–456.

OYEVAAR, M., WESTERTERP, R. (1989), The use of the chemical method for the determination of interfacial areas in gas-liquid contactors, *Chem. Eng. Sci.* **44**, 2691–2701.

OYEVAAR, M., ZIJL, A., WESTERTERP, R. (1988), Interfacial areas and gas hold-ups at elevated pressures in a mechanically agitated gas-liquid reactor, *Chem. Eng. Technol.* **11**, 1–10.

PAULI, J., MENZEL, T., ONKEN, U. (1989), Directional specific shear-rate measurements in gas-liquid two-phase flow, *Chem. Eng. Technol.* **12**, 374–378.

PAYNE, P. A. (1985), Medical and industrial applications of high resolution ultrasound, *J. Phys. E.* **18**, 465.

PHILIP, J., PROCTOR, J. M., NIRANJAN, K., DAVIDSON, J. F. (1990), Gas-holdup and liquid circulation in internal loop reactors containing highly viscous Newtonian and non-Newtonian liquids, *Chem. Eng. Sci.* **45**, 651–664.

RIKE, R. W., WILKING, B., WARD, H. C. (1965), Measurement of void fraction in two-phase flow by X-ray attenuation, *AIChE J.* **11**, 794–800.

PILHOFER, T., MILLER, H. D. (1972), Photoelektrische Sondenmethode zur Bestimmung der Blasengrößenverteilung mitteldisperser Tropfen in einem nichtmischbaren flüssigen Zweistoffsystem, *Chem. Ing. Tech.* **44**, 295.

PILHOFER, T., JEKAT, H., MILLER, H. D. (1974), *Chem. Ing. Tech.* **46**, 913.

POPOVIC, M., ROBINSON, C. W. (1988), External-circulation-loop airlift bioreactors: study of the liquid circulating velocity in highly viscous non-Newtonian liquids, *Biotechnol. Bioeng.* **32**, 301–312.

POPOVIC, M., ROBINSON, C. W. (1989), Mass transfer studies of external-loop airlifts and a bubble column, *AIChE J.* **35**, 393–405.

PROKOP, A., ERICKSON, L. E., FERNANDEZ, J., HUMPHREY, A. E. (1969), Design and physical characteristics of a multistage, continuous tower fermentor, *Biotechnol. Bioeng.* **11**, 945–966.

RAPER, J. A., DIXON, D. C., FELL, C. J. D., BURGESS, J. M. (1978), Limitations of the Burgess-Calderbank probe technique for characterization of gas liquid dispersions on sieve trays, *Chem. Eng. Sci.* **33**, 1405.

RIEBEL, U., LÖFFLER, F. (1989), On-line measurement of particle size distribution and particle concentration in suspensions by ultrasonic spectrometry, *Chem. Eng. Technol.* **12**, 433–438.

RIET, K., VAN'T (1975), Turbine agitator hydrodynamics and dispersion performance, *PhD Thesis*, Technical University, Delft, The Netherlands.

RIETEMA, K. (1982), Science and technology of dispersed two-phase systems, *Chem. Eng. Sci.* **37**, 1125–1150.

RIKMANIS, M., VANAGS, J., VIESTURS, U. (1987), Determination of the characteristic parameters of mixing of media in bioreactors (in Russian), *Biotechnologija* **3**, 70–78.

ROBERTSON, B., ULBRECHT, J. J. (1987), Measurement of shear rate on an agitator in a fermentation broth, pp. 31–35, in: *Biotechnological Processes: Scale-up and Mixing* (HO, C. S., OLDSHUE, J. Y., Eds.), New York: American Institute of Chemical Engineers.

RUSHTON, J. H., COSTRICH, E. W., EVERETT, H. J. (1950), Power characteristics of mixing impellers, *Chem. Eng. Progr.* **46**, 395 and 467.

SAUTER, J. (1928), *Forsch. Arb. Geb. Ing.* **2/8**, 312.

SCHEPER, T. (1985), Messung zellinterner und zellexterner Parameter zur Fermentationskontrolle, *Doctoral Dissertation*, Universität Hannover, FRG.

SCHEPER, T., SCHÜGERL, K. (1986), Bioreaktor Charakterisierung mittels in-situ Fluorimetrie, *Chem. Ing. Tech.* **58**, 433.

SCHMIDT, J. (1990), Mixing mechanisms in multiphase flows with a continuous liquid phase, *Ongoing Dissertation*, Universität Hannover, FRG.

SCHMIDT, K. G., BLENKE, H. (1983), Meßmethode zur Bestimmung der Flüssigkeitsgeschwindigkeit von heterogenen Gas-Liquid-Systemen in Schlaufenreaktoren, *vt Verfahrenstechnik* **17**, 593–597.

SCHNECKENBURGER, H., REUTER, B. W., SCHOBERTH, S. M. (1985), Fluorescence techniques in biotechnology, *Trends Biotechnol.* **3**, 257.

SCHÜGERL, K., LÜBBERT, A., SCHEPER, T. (1987), On-line Prozeßanalyse in Bioreaktoren, *Chem. Ing. Tech.* **59**, 701–714.

SCHUMPE, A., GRUND, G. (1986), The gas disengagement technique for studying gas holdup structure in bubble columns, *Can. J. Chem. Eng.* **64**, 891-896.

SEHER, A., SCHUHMACHER, V. (1979), Determination of residence times of liquid and gas phases in large bubble columns with the aid of radioactive tracers, *Ger. Chem. Eng.* **2**, 117-122.

SHAH, Y. T., KELKAR, B. G., GODBOLE, S. P., DECKWER, W. D. (1982), Design parameters estimations for bubble column reactors, *AIChE J.* **28**, 353-379.

SHINNAR, R., RUMSCHITZKI, D. (1989), Tracer experiments and RTD's in heterogeneous reactor analysis and design, *AIChE J.* **35**, 1651-1658.

SITTIG, W., RAMSPECK, W. (1979), Methode zur Mischzeitbestimmung durch pH-Messung in Gas-Flüssig-Reaktoren, *Hoechst Forschungsbericht* Nr. **76**.

SMITH, J. M., WARMOESKERKEN, M. M. C. G., ZEEF, E. (1987), Flow conditions in vessels dispersing gases in liquids, pp. 107-115, in: *Biotechnology Progress*, New York: AIChE.

SOBOLIK, V., MITSCHKA, P., MENZEL, T. (1986), CSSR-Patent PV727886.

SOLLINGER, S. (1989), Entwicklung einer Ultraschall-Doppler-Sonde, Diploma Thesis, Universität Hannover, FRG.

SPINDLER, K., LORENZ, G., HAHNE, E. (1987a), Faseroptischer Sensor zur Messung des örtlichen Gasgehalts in Flüssigkeiten, *Technisches Messen* **2**, 50-55.

SPINDLER, K., LORENZ, G., HAHNE, E. (1987b), Messungen des lokalen Gasvolumenanteils mit Hilfe einer faseroptischen Sonde, *Chem. Ing. Tech.* **59**, 73-737.

SPINDLER, K., LORENZ, G., ERHARD, A., HAHNE, E. (1988), Faseroptischer Sensor zur lokalen Gasgehaltmessung in Zweiphasenströmungen, *VDI-Berichte* **677**, 237-240.

SRIDHAR, T., POTTER, O. E. (1978), Interfacial area measurements in gas-liquid agitated vessels, *Chem. Eng. Sci.* **33**, 1347-1353.

SRIRAM, K., MANN, R. (1977), Dynamic gas disengagement: A new technique for assessing the behaviour of bubble columns, *Chem. Eng. Sci.* **32**, 571-580.

STEINEMANN, J. (1985), Charakterisierung des Gasphasenverhaltens in Blasensäulen mittels einer neuentwickelten elektrischen Mehrpunkt-Leitfähigkeitssonde, *Doctoral Dissertation*, Universität Dortmund, FRG.

STEINEMANN, J., BUCHHOLZ, R. (1984), Application of an electrical conductivity microprobe for the characterization of bubble behavior in gas-liquid bubble flow, *Part. Charact.* **1**, 102-107.

STEINEMANN, J., BUCHHOLZ, R. (1985), Leitfähigkeits-Mikrosonden zur Charakterisierung von Gas-/Flüssigströmungen, *Chem. Techn.* **14**, 146-153.

STRAVS, A. A., STOCKAR, U., VON (1985), Measurements of interfacial areas in gas-liquid dispersions by ultrasonic pulse transmission, *Chem. Eng. Sci.* **40**, 1169-1175.

SUN, Y., NOZAWA, T., FURUSAKI, S. (1988), Gas holdup and volumetric oxygen transfer coefficient in a three-phase fluidized bed bioreactor, *J. Chem. Eng. Jpn.* **21**, 20-24.

SWAINE, D. E., DAUGULIS, A. J. (1988), Review of liquid mixing in packed bed biological reactors, *Biotechnol. Progr.* **4**, 134-148.

SWAINE, D. E., DAUGULIS, A. J. (1989), Liquid residence time distributions in immobilized cell bioreactors, *Biotechnol. Bioeng.* **33**, 604-612.

SWEERE, A. P. J., MESTERS, J. R., JANSE, L., LUYBEN, K. C. A. M., KOSSEN, N. W. F. (1988a), Experimental simulation of oxygen profiles and their influence on baker's yeast production: I. One fermentor system, *Biotechnol. Bioeng.* **31**, 567-578.

SWEERE, A. P. J., JANSE, L., LUYBEN, K. C. A. M., KOSSEN, N. W. F. (1988b), Experimental simulation of oxygen profiles and their influence on baker's yeast production: II. Two fermentor system, *Biotechnol. Bioeng.* **31**, 579-586.

SWEERE, A. P. J., LUYBEN, K. C. A. M., KOSSEN, N. W. F. (1987), Regime analysis and scaledown: tools to investigate the performance of bioreactors, *Enzyme Microbiol. Technol.* **9**, 386-398.

SYKES, J. (1965), cited by BRYANT (1966).

TANAKA, M., SENDAI, T., HOSOGAI, K. (1989), Flowing characteristics in a circular loop reactor, *Chem. Eng. Res. Des.* **67**, 423-427.

TAWEEL, A. M., AL, LANDAU, J. (1977), Turbulence modulation in two-phase jets, *Int. J. Multiphase Flow* **3**, 341-351.

TAWEEL, A. M., AL, DIVARKARLA, R., GOMMA, H. G. (1984), Measurement of large gas-liquid interfacial areas, *Can J. Chem. Eng.* **62**, 73-77.

TAWEEL, A. M., AL, DIVARKARLA, R., GOMMA, H. G. (1987), Erratum: Measurement of large gas-liquid interfacial areas, *Can. J. Chem. Eng.* **65**, 570.

TODTENHAUPT, E. K. (1971), Blasengrößenverteilung in technischen Begasungsapparaten, *Chem. Ing. Tech.* **43**, 336.

TRÄGARDH, C. (1988), A hydrodynamic model for the simulation of an aerated agitated fed-batch fermentor, pp. 117-134, in: *Proc. 2nd Int. Conf. Bioreactor Fluid Dynamics* (KING, R., Ed.), London: Elsevier Applied Science Publisher.

URUA, I. J., DELCERRO, M. C. G. (1987), Measurement of large gas-liquid interfacial areas by the light transmission method, *Can. J. Chem. Eng.* **65**, 565-569.

VERMA, A. K. (1989), Heat transfer in bubble columns, *Chem. Eng. J.* **42**, 205–208.

WALTER, J. F., BLANCH, H. W. (1986), Bubble break-up in gas-liquid bioreactors: Breakup in turbulent flows, *Chem. Eng. J.* **32**, B7–B17.

WANG, D. I. C., FEWKES, R. C. J. (1977), Effect of operating and geometric parameters on the behavior of non-Newtonian mycelial antibiotic fermentations, *Dev. Ind. Microbiol.* **18**, 39–57.

WARMOESKERKEN, M. M. C. G., SMITH, J. M. (1988), Impeller loading in multi-turbine vessels, 179–197, in: *Proc. 2nd Int. Conf. Bioreactor Fluid Dynamics* (KING, R., Ed.), London: Elsevier Applied Science Publisher.

WASOWSKI, T., BLASS, E. (1989), Neue Aspekte zum Einfluß des Blasenwakes auf die Hydrodynamik und den Stoffaustausch in einer Blasensäule, *Chem. Eng. Sci.* **61**, 519–530.

WEBER, M. E., BHAGA, D. (1982), Fluid drift caused by a rising bubble, *Chem. Eng. Sci.* **37**, 113–126.

WEBSTER, J. G. (1978), Measurement of flow and volume of blood, Chap. 8 in: *Medical Instrumentation* (WEBSTER, J. G., Ed.), Houston: Houghton Mifflin Co.

WEETMAN, R. J., OLDSHUE, J. Y. (1988), Power, flow and shear characteristics of mixing impellers, *Proc. 6th Eur. Conf. Mixing*, Pavia, Italy.

WEILAND, P., BRENTRUP, L., ONKEN, U. (1980), Measurement of bubble size distribution in fermentation media using a photo-electric probe, *Ger. Chem. Eng.* **3**, 296–302.

WESTERTERP, K. R., VAN DIERENDONCK, L. L.,

DE KRAA, J. (1963), Interfacial areas in agitated gas-liquid contactors, *Chem. Eng. Sci.* **18**, 157–176.

WICHTERLE, K., KADLEC, M., ZAK, L., MITSCHKA, P. (1984), *Chem. Eng. Commun.* **26**, 25–32.

WILD, G., ANDRE, J. C., MIDOUX, N. (1986), Mehr Licht, photophysikalische Meßverfahren zur Untersuchung von Gas/Flüssig-Reaktoren, *Chem. Ing. Tech.* **58**, 142–143.

WITTE, W., BAIER, P. W. (1983), Korrelative Volumenstrom-Messung von Mineralölen mit pseudozufälligen Wärmeimpulsen, *Chem. Ing. Tech.* **55**, 795.

WU, H., PATTERSON, G. K. (1989), Laser-Doppler measurements of turbulent-flow parameters in a stirred mixer, *Chem. Eng. Sci.* **44**, 2207–2221.

YASUNISHI, A., FUKUMA, M., MUROYAMA, K. (1986), Measurement of behavior of gas bubbles and gas holdup in a slurry bubble column by a dual electroresistivity probe method, *J. Chem. Eng. Jpn.* **19**, 444–449.

YOUNG, M. A., CARBONELL, R. G., OLLIS, D. F. (1987), Non-Newtonian broths in airlift bioreactors, pp. 45–49 in: *Biotechnology Processes: Scale-up and Mixing* (HO, C. S., Oldshue, J. Y., Eds.), New York: American Institute of Chemical Engineers.

YU, Y. H., KIM, S. D. (1988), Bubble characteristics in the radial direction of three-phase fluidized beds, *AIChE J.* **34**, 2069–2072.

ZEHNER, P. (1986), *Int. Chem. Eng.* **26**, 29.

ZUN, I., SAJE, F. (1982), Statistical characteristics of bubble flow, p. 112, in: *Proc. 3rd Austrian-Italian-Jugoslavian Chem. Eng. Conf.*, Vol. 2, Graz.

5 On-Line Analysis of Broth

Karl Schügerl

Hannover, Federal Republic of Germany

1 Introduction

A coordinated cooperation of several enzymes is necessary for the biosynthesis of primary and secondary metabolites (BRITZ and DEMAIN, 1985).

The activity of a particular enzyme depends upon the concentration of several medium components (substrates, products, inductors, repressors, inhibitors, and activators). Some of these are known, while others have not yet been discovered. In order to achieve the full biological potential of the cells, the optimal production conditions must be maintained, at least with regard to the most important key parameters.

The first step towards this goal is the identification of the key components. This is usually done by medium optimization. However, this method does not give information about the regulation of the biosynthesis. Acquisition of this information requires special investigations combined with on-line analysis of the medium and cellular components. In addition, knowledge of the dynamic behavior of the microorganisms is required for process control and for coping with operational disturbances.

2 Need for On-Line Process Analysis and Control

2.1 Carbon Catabolite Regulation

Catabolite repression, transient repression, and catabolite inhibition regulate the utilization of many carbohydrates. *Catabolite repression* is a reduction in the rate of synthesis of certain enzymes — particularly those of degradative metabolism — in the presence of glucose or other readily metabolized carbon sources during steady-state growth (PAIGEN and WILLIAMS, 1970).

A period of more intense repression occurs immediately after the cells are exposed to glucose. This is called *transient repression*.

Catabolite inhibition is a control exerted by glucose on enzyme activity rather than on enzyme formation, analogous to feedback inhibition in biosynthesis pathways (PAIGEN and WILLIAMS, 1970).

PAIGEN and WILLIAMS (1970) compiled a large number of examples of catabolite repression and some examples of catabolite inhibition. Best investigated is the regulation of glucose metabolism in the growing yeast cell (FIECHTER et al., 1981; GANCEDO and GANCEDO, 1987; GANCEDO, 1988).

Baker's yeast production requires "oxidative" growth, i.e., subcritical substrate flux and pure carbon limitation. "Oxidoreductive" or "reductive" conditions are needed for ethanol production with *Saccharomyces cerevisiae*. Growth is "oxidoreductive" under aerobic conditions and with critical or supracritical glucose flux. Growth is purely "reductive" only under anaerobic conditions (SONNLEITNER and KÄPPELI, 1986). p_{O_2}- and S(C-substrate concentration) measurements and control are necessary for aerobic oxidative growth. For ethanol production under oxidoreductive conditions, p_{O_2} and S measurements and control, and under anaerobic conditions, E_h (redox potential) and S measurements and control are imperative.

Carbon catabolite regulation is also involved in controlling the initiation of secondary metabolite synthesis. Usually, glucose acts as a co-repressor or an inhibitor, repressing formation or inhibiting action of, e.g., antibiotic synthetases. Therefore, the co-repressor or inhibitor must be depleted before antibiotic synthesis can occur (MARTÍN and DEMAIN, 1980). A transition of the growth phase to the production phase is initiated by the reduction of the glucose concentration below its critical value. Thus, according to the investigations of REVILLA and MARTÍN (MARTÍN and DEMAIN, 1980), glucose represses the incorporation of valine into penicillin during penicillin biosynthesis (HERSBACH et al., 1984).

SCHEIDEGGER et al. (1988) proved that the glucose consumption rate is the essential factor for maximal cephalosporin production, not the actual glucose concentration in the broth. A decrease of the glucose consumption rate between the 20th and 30th hour of cultivation was found to induce the early onset of antibiotic synthesis. Oxygen starvation after the onset of cephalosporin C production led to a pronounced increase in penicillin N formation, which indicates that the ring expansion enzyme is repressed by oxygen limitation (QUEENER et al., 1984; SCHEIDEGGER et al., 1988).

Several other carbon substrates (lactose, galactose, mannan) do not influence the biosynthesis of antibiotics, e.g., penicillin, actinomycin, and streptomycin. The C-substrate consumption rate must be kept within a well-defined range by on-line measurement and control of the substrate concentration in the broth during the production phase of secondary metabolites because of the carbon catabolite regulation.

2.2 Nitrogen Metabolite Regulation

Nitrogen metabolite regulation is well-documented in bacteria, yeasts, and molds (MARTÍN and DEMAIN, 1980; MARTÍN and LIRAS, 1981). Ammonia (or some other readily used nitrogen source) represses enzymes involved in the use of other nitrogen sources. Penicillin (HERSBACH et al., 1984) and cephalosporin (MARTÍN and DEMAIN, 1980; QUEENER et al., 1984) production is reduced by the inhibition of the activity of glutamate synthetase by increasing the ammonium concentration in the broth. This holds true also for the synthesis of other antibiotics (AHARONOWITZ, 1980).

The ammonium (or any other quickly consumable nitrogen substrate) concentration must be measured and controlled in the broth in order to achieve a high productivity because of this nitrogen metabolite regulation.

2.3 Phosphate Regulation

Phosphate is the crucial growth-limiting nutrient in many fermentations. It must be exhausted before the onset of the syntheses of the various antibiotics (MARTÍN and DEMAIN, 1980; MARTÍN and LIRAS, 1981; MARTÍN, 1977; ROSS and SCHÜGERL, 1988; RHODES, 1984).

MARTÍN (1977) compiled a long list of antibiotic processes which are inhibited by inorganic phosphate. Phosphate concentrations ranging from 0.3 to 300 mmol/L generally support extensive cell growth, but concentrations of 10 mmol/L and above suppress the biosynthesis of many antibiotics (MARTÍN and DEMAIN, 1980; MARTÍN, 1977). Therefore, by on-line measurement and control of the inorganic phosphate concentration in the broth, a high growth rate in the growth phase and a high productivity during the production phase can be attained.

2.4 Enzyme Induction and Regulation by Other Components

Several enzymes are formed by induction (BÖING, 1982), e.g., amylase formation is induced by starch, invertase by sucrose, β-galactosidase by lactose, etc. Some of these inducers are used as substrates. Therefore, substrate fed-batch is practiced to keep the enzyme formation rate high. By measuring the substrate concentration and keeping it in the optimal range, a high enzyme productivity can be attained. The product of the enzymatic reaction often represses enzyme formation. By keeping the product concentration below the critical level, the productivity can be improved.

The synthesis of some secondary metabolites can be enhanced by a precursor (e.g., the production of penicillin G by phenylacetic acid) or by a stimulator (cephalosporin C production by methionine or sulfate, depending on the type of the strain) (MATSUMURA et al., 1978, 1980; SMITH, 1985).

Productivity can be increased considerably by keeping the concentration of the precursor, inducer, or stimulator in the optimal range.

2.5 Realization of the Concentration Control of Key Components

Several possibilities exist to control the concentration of key components: often components with low solubility (e.g., peanut flour for cephalosporin and tetracycline production) are used, or substrates are used that can only be consumed slowly by the microorganisms (e.g., lactose for penicillin production). However, with strain improvement the optimal concentrations of the key component may vary. Such flexible operation can only be realized by on-line process analysis and control. Generally speaking, high concentrations cause repression or inhibition and low concentrations reduce the growth and/or production rate. Sometimes the optimal concentration range is very narrow.

Control of the concentration of the limiting component is particularly difficult because of its low concentration. In such cases, the concentration of the key component sometimes is evaluated from the data of measurable components, e.g., by means of an observer.

3 On-Line Analysis in Non-Sterile Areas

Except for temperature, pH, p_{O_2}, E_h, p_{CO_2}, and perhaps certain ion concentrations measured *in situ*, the concentrations of all other medium components are measured outside the reactor in a non-sterile area. For on-line analysis, continuous aseptic sampling is imperative.

3.1 Aseptic Sampling

Four types of aseptic sampling systems are known:

> *Sampling systems integrated into a recirculation loop* (1) with flat membrane and stirrer for cross-flow filtration or without stirrer for cake filtration, and (2) with tubular membrane; *in situ sampling systems* (3) with discus-shaped body and flat membrane, and (4) with rod-shaped body and tubular membrane.

3.1.1 Sampling Systems Integrated into a Recirculation Loop

3.1.1.1 Systems with a Flat Membrane

A steam-sterilizable filtration cell made of stainless steel has been developed by the Gesellschaft für Biotechnologische Forschung (GBF) and B. Braun Melsungen AG (FRG) and has been commercialized as Biopem® by B. Braun Melsungen AG (Fig. 1) (KRONER et al., 1986; RECKTENWALD et al., 1985a,b). The cell is operated in a cross-flow mode. The Biopem sampling module has the following properties: cell capacity: 200 mL; membrane area: 45 cm^2; membrane diameter: 90 mm; cross-flow rate: 10–60 L/h; and the following operating data: stirrer speed: 300–600 rpm; filtrate rate: 0.1–1.5 mL/min; operating overpressure: 0.1–0.3 bar (Biopem) with Durapore HVLP membrane 0.45 μm (Millipore).

A similar sampling module has been developed by MANDELIUS et al. (1984). Ultrafiltration or dialysis membranes — 25 mm in diameter — can be used in this sterilizable cross-flow filtration module. No further data have been published.

A disc-shaped microfiltration module has been developed by Millipore, but has never been commercialized (Fig. 2) (GARN et al., 1989). The module consists of stainless-steel half-shells, with an inserted Durapore membrane (0.2 μm pores, diameter: 47 mm, membrane area: 13 cm^2). The broth is circulated at 30–42 L/h on the spiral side (GARN et al., 1989). This module has also been tested for the sampling of animal tissue culture broths (GRAF, 1989).

The Bio-filtrator of WAZAU CO., developed by REUSS (LENZ et al., 1985), forms a cake of the mold during sampling. The permeate is used for on-line analysis of the broth components by HPLC, and the height and the pressure drop across the cake are used to determine the concentration of the mold in the broth (LENZ et al., 1985). Replaceable filters are used because of the cake formation. The filter medium is a band of filter paper or membrane filter that is advanced after each cake formation and successive cleaning process.

a)

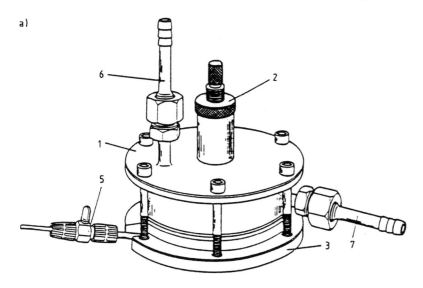

b)

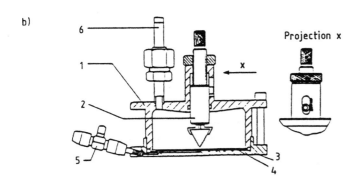

Projection x

Fig. 1. Biopem. Aseptic on-line sampling device with disc-shaped ultrafiltration membrane unit integrated in a loop, a) sampling module, b) schematic drawing with adjustable stirrer assembly.
1 Top plate, 2 adjustable magnetic stirrer assembly, 3 bottom plate, 4 channels, 5 shut-off valve, 6 outlet line, 7 inlet line (not shown in b).
(B. Braun + Diessel Biotech GmbH, P.O. Box 120, D-3508 Melsungen).

A similar system has been developed by EN-GASSER (GHOUL et al., 1986). The filter medium consists of a continuous band of 10 cm wide ultrafiltration membrane, which is transported after each filtration process to avoid reduction of the permeation rate by membrane fouling.

3.1.1.2 Systems with a Tubular Membrane

A rod-shaped, stainless-steel module, 5.5 mm in inner diameter, and a 200 mm long, tubular polypropylene microfiltration membrane were used for on-line sampling of fermentation broths in a cross-flow mode (LORENZ et al., 1987; SCHMIDT et al., 1986). The dead volume of the module was 2.56 mL. However, the response time also depends on the volume of the loop and on the broth circulation rate. In the special case of a laboratory fermentor and a short loop, the response time approached 5 min or more at a broth flow rate in the tube of 1 m/s.

The permeation rate decreased from about 1 mL/min to 0.6 mL/min with nylon and Durapore membranes and from 0.6 to 0.2 mL/min with polysulfon 100000 membranes within several days, provided broth of a high solid content (peanut flour) was sampled. With broth of a low viscosity (e.g., broth of *Zymomonas mobilis*), the permeation rate remained constant (about 1 mL/min). At a high broth circulation rate — necessary to avoid fouling — marprene tubing must be used in the peristaltic pump to

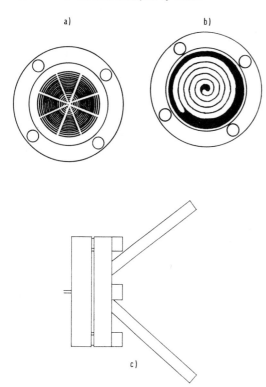

a)   b)

c)

Fig. 2. Aseptic on-line sampling device with disc-shaped ultrafiltration membrane unit integrated in a loop, a) permeate side, b) filter side, c) side view. (Millipore; GARN et al., 1989). (With permission of John Wiley & Sons)

avoid destruction of the tubing (LORENZ et al., 1987).

ROSS (1986) used this membrane module for sampling *Streptomyces aureofaciens* broth with 80 g/L peanut flour. It was possible to

maintain permeation rates from 0.2 to 0.4 mL/min for 10 days.

MIZUTANI et al. (1987) used a ceramic filter of 0.5 μm pore size from the Toshiba Ceramics Co. for broth sampling. Sampling through an external loop has the advantage that — in case of membrane clogging — the membrane can be replaced during the fermentation.

3.1.2 *In situ* Sampling Systems

3.1.2.1 Systems with a Discus-Shaped Body and a Flat Membrane

External sampling systems which are integrated into a loop have several disadvantages:

- a pump is needed for maintaining a high broth circulation rate across the loop. Solids in the broth erode the tubing of the peristaltic pump and cause leakage and infection of the culture;
- mechanical pumps are the least reliable parts of a continuously operated reactor system;
- with a high solid content, the probability for clogging of the tubing is high;
- in pellet- or floc-containing broths, the sampling changes the broth properties;
- when sampling animal tissue culture broths, the mechanical stress exerted on the cells at a high pumping rate may affect cell viability;
- cells may suffer from oxygen limitation and may change their metabolism. For

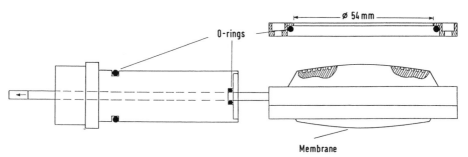

ø 54 mm

O-rings

Membrane

Fig. 3. Aseptic *in situ* on-line sampling device with discus-shaped ultrafiltration membrane unit (NIEHOFF et al., 1986).

example, during the sampling of *Saccharomyces cerevisiae* broth through an external loop, ethanol was produced, which falsified the analysis of the broth composition. This is the reason why *in situ* sampling systems have been developed.

Different types of discus-shaped sampling systems have been developed. A microfiltration device was placed near the stirrer to avoid fouling (LORENZ et al., 1987). Better performance was attained with a discus-shaped ultrafiltration membrane module (Fig. 3) (LORENZ et al., 1987; SCHMIDT et al., 1986) and the discus-shaped module with a porous plate membrane support (Fig. 4), (SCHÜGERL, 1988).

These modules with a 100000 Dalton polysulfon-membrane have the following properties:

	discus-shaped module (Fig. 3)	disc-shaped module (Fig. 4)
free filtration surface area (cm^2)	13.5	47.5
dead volume (mL)	1.35	2.5
membrane diameter (mm)	62	62
response time (min)	3.5	9

Both were used for the control of antibiotic production (SCHÜGERL, 1988). They were used with microfiltration as well as with ultrafiltration membranes. It was possible to steam-sterilize them several times, and they were stable for long periods of time (1000 h) (NIEHOFF et al., 1986).

The disadvantage of these types of modules is that they can only be mounted inside the reactor. Therefore, they are unsuitable for large commercial reactors. This drawback is avoided by the use of rod-shaped modules which can be mounted outside the reactor.

3.1.2.2 Systems with a Rod-Shaped Body and a Tubular Membrane

WAARVIK (1985) developed a rod-shaped module with a microporous Accurel polypropylene membrane of 0.1 μm mean pore diameter. The membrane tubing had an inner diameter of 5.5 mm, an outer diameter of 8.6 mm, and a surface roughness of 125 μm. It was possible to maintain a filtration rate of 1.6 mL/min during the growth of *Escherichia coli* (approx. 12 h) and 0.8 mL/min during the production of penicillin V by *Penicillium chrysogenum* (approx. 250 h). Also, a 25 cm long Ceraflo layered ceramic tubing with an outer diameter of about 5 mm and inner diameter of about 3 mm, a nominal surface pore size of 0.001 μm (ultrafiltration membrane) to 0.1 μm (microfiltration membrane) and a surface roughness of 0.12 to 1.25 μm, was installed and used for sampling the penicillin broth. It was possible to maintain a filtration rate of 0.5 mL/min during the production.

A prototype of a nylon dialysis sampling probe was described by GIBSON and WOODWARD (1988) without information about its performance.

The ABC Company (ABC, 1988) offers a rod-shaped, sterilizable stainless steel module with polypropylene microfiltration tubing, developed at the University of Hannover (Fig. 5) (GRAF, 1989; WENTZ, 1989). This module has the following properties:

free filtration area (cm^2)	34.0
dead volume (mL)	7.0
response time (min)	9.0
pore radius (μm)	0.2

This module was used for on-line measurement and control of several fermentation proc-

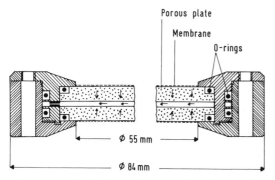

Porous plate

Membrane

O-rings

⌀ 55 mm

⌀ 84 mm

Fig. 4. Aseptic *in situ* on-line sampling device with disc-shaped ultrafiltration membrane unit (MÖLLER et al., 1986).

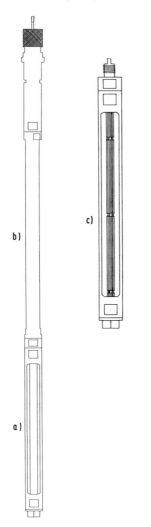

Fig. 5. Aseptic *in situ* on-line sampling device with rod-shaped microfiltration membrane unit, a) filter, b) extension, c) filter without membrane.
(ABC Biotechnologie/Bioverfahrenstechnik GmbH, P.O. Box 1151, D-8039 Puchheim).

esses. It is the only one which was also used for the analysis and control of the cultivation of animal cell cultures (BHK, hybridoma cells). The on-line analysis of these cultures is especially difficult because of their high protein content. Depending on the position of the module in the reactor, the filtration rate decreased from 1.0 mL/min to 0.2 mL/min after

1000 h of continuous operation, provided it was positioned in the dead water region in the reactor. It only diminished from 1.0 mL/min to 0.8 mL/min if the module was close to the impeller (GRAF, 1989; WENTZ, 1989).

A rod-shaped module with ceramic ultrafiltration tubing was also developed (WENTZ, 1989; HÜBNER, 1990). This module was employed for measurements during the cultivation of microorganisms and animal cells.

3.2 Analysis Techniques

3.2.1 Continuous Air-Segmented Analyzer

Typical continuous air-segmented automatic analyzer systems consist of multichannel peristaltic pumps with air-injectors, sample-reagent mixing units, thermostated reaction coils, in which the wet chemical reactions are performed, and detectors with a bubble separator for measuring the concentration of the reaction product. Back diffusion and possible carry-over should be avoided by air-segmentation (BRAN + LÜBBE GmbH; SKALAR ANALYTIKA).

Autoanalyzer systems are well established in various biotechnological laboratories (BAYER et al., 1986; BORDONS, 1986; HOLST et al., 1988; IMMING et al., 1982; KUHLMANN et al., 1984; MEINERS and SCHALLER, 1986; NIEHOFF et al., 1986; SCHMIDT et al., 1984, 1985; SCHÜGERL, 1988; SZIGETI et al., 1986; TAWFIK and MARDON, 1985; VOSER, 1966).

The electrical signal produced by the detector must bear a defined relationship to the concentration of the product. Many well-known manual analytical methods are transferable to automatic analyzers. The automatic methods are often more precise, because the reaction times and reaction conditions are well reproducible.

Frequently the concentration of the analyte happens to be too high, therefore, a dilution step is necessary. The accuracy of the determination is considerably influenced by the dilution process. For high dilutions of the sample, the accuracy of the measurement is controlled by the accuracy of the dilution.

It is necessary to calibrate the channels in 2–12 h cycles because of the variation of the effectiveness of the reagents with age, or because of the change of the electrodes' sensitivities with time. Enzyme-based procedures must be calibrated frequently because of the decrease of the enzyme activity with time.

When using spectrophotometric methods, blank determinations are also necessary because of the possible variation of the transparency of the light source- and detector-windows due to protein adsorption on the glass surface and solids precipitation in the detectors, or to change of the light absorbance of the medium. For blank determinations, the analyte must be removed from the sample, e.g., by means of an enzymatic reaction.

Regular cleaning of the tubing–coil–detector system with acid, alkali, or other solutions is necessary for the removal of adsorbed layers or sediments or for the regeneration of the detectors (e.g., gas-sensitive electrodes) with buffer solutions.

The growth of microorganisms in the tubing–coil–detector system can be suppressed by occasional flushing with an azide solution (NIEHOFF et al., 1986).

The dilution procedures, the calibration and blank measurements, and the cleaning and regeneration procedures should be performed automatically (see Chapter 17).

In Fig. 6 six air-segmented analyzer channels for the determination of medium components in *Cephalosporium acremonium* cultivation broths are shown (BAYER et al., 1986), and in Tab. 1, analyzer channels for *Penicillium chrysogenum* and *Cephalosporium acremonium* cultivation broths are depicted (SCHÜGERL, 1988).

For chemical reactions, it is very important to determine the selectivity of the reagent for the investigated component; e.g., when monitoring the concentration of reducing sugars with *p*-HBAH, it is not possible to distinguish between different sugars if the broth contains sugar mixtures. Tab. 2 shows the relative molar absorbances of different reducing sugars and sugar derivatives. Only in the presence of a single sugar or with mixtures of sugars in which only one of them exhibits absorbance is it possible to monitor and control this component.

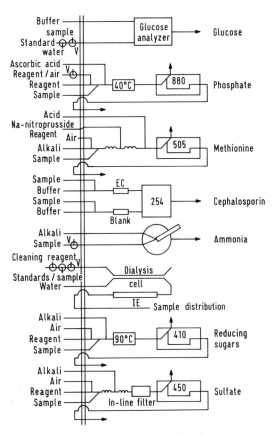

Fig. 6. Continuous air-segmented analyzer system for production of cephalosporin C (BAYER et al., 1986).

Also in the case of enzymatic reactions, the specificity is often not sufficient to monitor the concentration of a single medium component (e.g., a particular amino acid).

The continuous air-segmented automatic analyzer technique is popular in laboratories, however, it is, in general, not suitable for monitoring broth components for extended cultivation times. Very often cell growth occurs after 3–4 days in spite of the frequent cleaning of the system with growth-suppressing chemicals, because the analysis is performed in a non-sterile region, and the sample contains nutrients. Furthermore, protein precipitation often occurs after the sample has been mixed with reagents.

Tab. 1. Air-Segmented Automatic Analyzer Channels (SCHÜGERL, 1988)

Species	Method	Detector
Phosphate	Ammonium molybdate	Photometer
Sulfate	Ba-methylthymol blue/ion-exchanger	Photometer
Ammonium	NaOH/EDTA, membrane	Ion-sensitive electrode
Urea	Diacetylmonoxime, thiosemicarbazide	Photometer
Penicillin	Hydroxylammonium chloride, nickel chloride	Photometer
Dissolved organic carbon	K-persulfate/UV-light membrane	Micro-CO_2 gas sensor
Reducing sugar	p-Hydroxybenzoic acid hydrazide (p-HBAH)	Photometer
Methionine	Na-nitroprusside[a]	Photometer
Cephalosporins	Absorbance of cepham chromophore[b]	Photometer

[a] no cross-sensitivity exists with other amino acids and medium components except for histidine (37%) and cephalosporin C (5%) (BAYER et al., 1986)
[b] with blank after enzymatic cleavage of β-lactam ring

Tab. 2. Relative Molar Absorbances of Different Reducing Sugars and Sugar Derivatives with p-HBAH (p-hydroxybenzoic acid hydrazide) (SCHMIDT et al., 1985)

Substance Conc. (0.4 g L^{-1})	Rel. Absorbance (%)	Rel. Mol. Absorbance (%)
Glucose	100	100
Mannose	95	95
Galactose	78	78
Fructose	104	104
Xylose	73	61
Ribose	74	62
Lactose · H_2O	39	77
Maltose · H_2O	57	213
Cellobiose	60	114
Sucrose	0	0
Raffinose · 5 H_2O	1	3
Trehalose · 2 H_2O	0	0
Glucosamine · HCl	58	69
Glucuronic acid (gamma lactone)	67	65
N-acetylgalactosamine	26	32
N-acetylglucosamine	26	32

Selectivity, detection limit, and accuracy of this technique depend on several factors, such as the analyte and its concentration range (dilution), composition of the broth, and the reaction used for detection. Therefore, it is not possible to give generally valid data about the reliability of the analysis. Selectivity, detection limit, and accuracy can only be given for a definite analyte in a particular broth and the analyte concentration range using a specific dilution and reaction.

When using glucose oxidase (GOD) for glucose analysis, H_2O_2 is formed, which reacts with an oxygen acceptor and forms a colored product. The sensitivity and detection limit of glucose depends on the chromogene selected. With four different chromogenes (sulfonated 2,4-dichlorophenol and 4-aminophenazone; 3-methyl-2-benzothiazolinone (MBTH) and formaldehyde azine of MBTH; 3,3′,5,5′-tetramethylbenzidine; 3-dimethylamino benzoic acid (DMAB) and MBTH), the sensitivities vary in *Saccharomyces cerevisiae* cultivation broths by a factor of 12 (0.0033, 0.0021, 0.030, and 0.042 (mg/mL) (BROWN et al., 1987).

Calibration curves are shown in Fig. 7 for NH_4^+, phosphate, sulfate, urea, DOC, glucose, galactose, and lactose in penicillin broth (NIEHOFF, 1983). The concentration ranges covered by the analyzer system are depicted. The maximum error is less than 2%.

Large amounts of chemicals are needed for long-term cultivations, because the analysis is carried out continuously. Since the detection is often based on equilibrium reactions, these

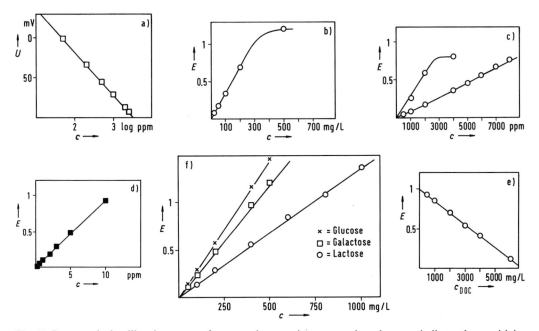

Fig. 7. Some typical calibration curves for a continuous air-segmented analyzer to indicate the sensitivity and precision of measurements (NIEHOFF, 1983).
a) NH_4^+, b) urea, c) sulfate, d) phosphate, e) dissolved organic carbon (DOC), f) glucose, galactose, lactose.

Tab. 3. Advantages and Disadvantages of Continuous Air-Segmented Automatic Analyzers

Advantages
● Continuous signal (easy to use for control)
● Simple automation
● Relatively inexpensive ($10000/channel)

Disadvantages
● Protein precipitation and cell growth in tubing, coils, and detector
● Large amounts of chemicals are needed
● Long response time (the determinations are often based on the equilibrium of the reactions)

analyzers usually have long response times. On the other hand, process control can be handled easily by employing a continuous signal.

In Tab. 3 the advantages and disadvantages of the continuous air-segmented automatic analyzers are compiled.

This technique will gradually be replaced by on-line flow-injection analyzer and on-line HPLC systems for industrial process control because of the above disadvantages.

3.2.2 On-Line Flow-Injection Analyzers, FIA

Flow-injection analysis (FIA) has become very popular recently (RUZICKA and HANSEN, 1988). This technique is also based on the classical wet chemical analysis.

The principle of the flow injection analysis is simple, being based on the injection of a definite volume of a liquid sample solution into a moving, non-segmented continuous carrier stream of a suitable liquid (Fig. 8) (HANSEN, 1985). The injected sample forms a zone which begins to disperse and reacts with the carrier stream as it is transported toward a detector, in which the concentration of the reaction product and/or substrate is measured. A typical recorder output is shown in Fig. 8. The peak height (*H*) or the surface below this peak

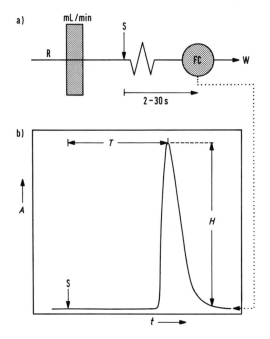

a)

mL/min

2 - 30 s

b)

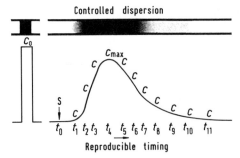

Controlled dispersion

C_0

C_{max}

t_0 t_1 t_2 t_3 t_4 t_5 t_6 t_7 t_8 t_9 t_{10} t_{11}

Reproducible timing

Fig. 9. Dispersed sample zone of original concentration C_0-injected at position S and the corresponding recorder output. To each concentration C corresponds a specific dispersion value. Each specific dispersion value can be related to a fixed delay time t, which is at its minimum at C_{max} and increases to large values along the continuous gradient (HANSEN, 1985).

Fig. 8. a) Single-line FIA manifold. R carrier stream of reagent, b) typical recorder output (HANSEN 1985). S sample injection, FC flow-through detector, W waste, H peak height, T residence time.

is related to the concentration of the analyte. Since the residence time (T) of the sample in the FIA system is usually less than 30 s, at least two samples can be analyzed per minute. The injected volume is small (1 to 200 µL, typically 25 µL). This usually requires no more than 0.5 mL of reagent per analysis and makes FIA an

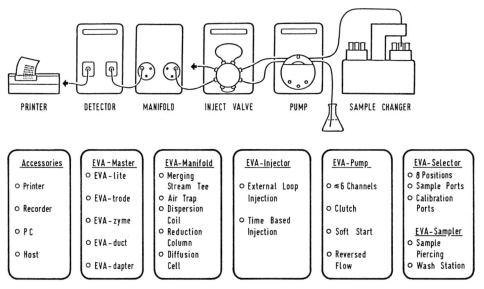

PRINTER   DETECTOR   MANIFOLD   INJECT VALVE   PUMP   SAMPLE CHANGER

Accessories	EVA – Master	EVA–Manifold	EVA–Injector	EVA–Pump	EVA–Selector
	○ EVA–lite	○ Merging			○ 8 Positions
○ Printer		Stream Tee	○ External Loop	○ ≤6 Channels	○ Sample Ports
	○ EVA–trode	○ Air Trap	Injection		○ Calibration
○ Recorder		○ Dispersion		○ Clutch	Ports
	○ EVA–zyme	Coil	○ Time Based		
○ PC		○ Reduction	Injection	○ Soft Start	EVA–Sampler
	○ EVA–duct	Column			○ Sample
○ Host		○ Diffusion		○ Reversed	Piercing
	○ EVA–dapter	Cell		Flow	○ Wash Station

Fig. 10. Typical flow injection analyzer system with sample changer: EVA (Eppendorf Variables Analysersystem).
(EPPENDORF GERÄTEBAU, Netheler & Hinz, Barkhausenweg 1, D-2000 Hamburg).

Tab. 4. Some Analytes Measured Off-line by FIA which are Important for Biomedical Technique and Biotechnology

Analyte	Immob. Enzyme/Antibody	Detector	Reference
Glucose	Glucose oxidase (GOD) on membrane	H_2O_2 electrode	[1], [2], [3], [13]
Lactate	Lactate oxidase (LO) on membrane	H_2O_2 electrode	[1], [2], [3], [13]
Glutathione	Glutathione sulfhydryl oxidase, cartridge	H_2O_2 electrode	[10]
Glucose	GOD, cartridge	H_2O_2 with luminol chemiluminescence	[6]
Glucose	GOD, cartridge horseradish peroxidase, 4-aminonantipyrine, 3,5-dichloro-2-hydroxyphenyl sulfonate	H_2O_2 with photometer at 510 nm	[12]
Ammonia	NaOH/EDTA, membrane	NH_3 gas sensor	[2]
Urea	Urease/cartridge	NH_3 gas sensor	[2], [7]
Glucose	GOD/cartridge	O_2 electrode	[2]
Lactate	LO/cartridge	O_2 electrode	[2]
Lactose	β-Galactosidase/GOD cartridge	O_2 electrode	[2]
Maltose	α-Glucosidase/GOD cartridge	O_2 electrode	[2]
Glucose	GOD/catalase	Enzyme thermistor (ET)	[2]
Penicillin G	Penicillin G-amidase	ET	[2]
Penicillin G	Penicillinase	ET	[2]
Penicillin V	Penicillin G-amidase	ET	[2]
Penicillin V	Penicillinase	ET	[2]
Ampicillin	Penicillin G-amidase	ET	[2]
L-Lactate	Lactate dehydrogenase (LDH) cartridge	Amperometric determination of NADH	[4]
D-Lactate	LDH cartridge		[4]
L-Isocitrate	Isocitrate dehydrogenase cartridge		[4]
Ethanol	Alcohol dehydrogenase cartridge		[4]
L-Malate	Malate dehydrogenase cartridge		[4]
Formate	Formate dehydrogenase cartridge		[4]
β-D-Glucose	Glucose dehydrogenase (GDH) cartridge		[4]
L-Alanine	Alanine dehydrogenase		[4]
L-Glutamate	Glutamate dehydrogenase		[4]
L-Leucine	Leucine dehydrogenase		[4]
Xylose	Xylose isomerase mutarotase, glucose dehydrogenase, cartridge	Amperometric NADH determination	[8]
L-Lactate	LDH, glutamic pyruvic transaminase (GPT) cartridge	Amperometric NADH determination	[9]
Glucose	Hexokinase, glucose-6-phosphate dehydrogenase, cartridge	Fluorometric NADH determination	[5]
Fructose	Hexokinase, fructose-6-phosphate dehydrogenase, cartridge	Fluorometric NADH determination	[5]

Tab. 4. Continued

Analyte	Immob. Enzyme/Antibody	Detector	Reference
Tetracycline	Bromine	Chemiluminescence	[11]
Transferrin	Antitransferrrin/horseradish cartridge, 2,2-azinodi-3-ethylbenzthiazoline sulfonic acid (ABTS) + H_2O_2	Photometer	[14], [15]
IgG Antigen	F(ab′)$_2$ and Fc fragments of IgG, cartridge, antigen sensitized carboxyfluorescein containing liposomes (LCA), unbound LCA lysed by surfactant	Fluorescence detector	[16]

References: [1] YELLOW SPRINGS INSTRUMENTS, [2] SCHÜGERL, 1988, [3] PETERSSON, 1988a, [4] SCHELTER-GRAF et al., (1984), [5] LINARES et al., 1987, [6] SWINDLEHURST and NIEMAN, 1988, [7] PETERSSON, 1988b, [8] DOMINGUEZ et al., 1988, [9] GORTON and HEDLUND, 1988, [10] SATOH et al., 1988, [11] ALWARTHAN and TOWSHEND, 1988, [12] STULTS et al., 1987, [13] THOMPSON et al., 1987, [14] LARSSON et al., 1987, [15] MATTIASSON and LARSSON, 1987, [16] PLANT et al., 1988

automated microchemical technique capable of a sampling rate of at least 100 determinations per hour (HANSEN, 1985).

At a peak maximum, the analyte concentration is at its highest; to its left and right, this concentration gradually diminishes to zero. Since the mixing is a first-order (linear) process, each concentration can be associated with a definite time (Fig. 9) (HANSEN, 1985). This also holds true for linear chemical reactions. By measuring the signal height at different delay times $t_1, t_2, t_3, \ldots, t_n$, the signal-to-analyte

concentration relationship can be evaluated without dilution of the sample (gradient dilution), and the calibration of the FIA-system can be performed in this manner (gradient calibration). The dilution of the sample can be avoided within a not-too-large concentration range.

A variant of the FIA technique is the FIA stopped flow approach, which increases the sensitivity of the measurement and permits evaluation of the kinetics of the reaction. Knowledge of the kinetics is necessary for a

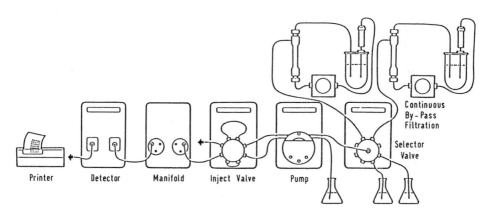

Fig. 11. Typical flow injection analyzer system with on-line sampling unit: EVA (Eppendorf Variables Analysersystem).
(EPPENDORF GERÄTEBAU, Netheler & Hinz, Barkhausenweg 1, D-2000 Hamburg).

fine adjustment of the reagent-to-sample ratio by choosing the appropriate delay time (RU-ZICKA and HANSEN, 1988).

The FIA is now also an established off-line technique for chemical analysis in biomedicine and biotechnology (Tab. 4). Only a small amount of this kind of equipment is on the market (TECATOR AB; MÖLLER, 1982; KARL-BERG, 1983; FIAtron; EPPENDORF GERÄTE-BAU, 1989).

Figs. 10 and 11 show typical modular systems with sample changer and bypass filtration of the FIAchem-system of the FIAtron- and EVA-systems of Eppendorf. The first one consists of the sample changer, pump, injector valve, manifold and detector, the second one of the continuous bypass filter selector valve, pump, injector valve, manifold, and detector. In the manifold, the analyte reacts with the carrier and the reactants to form a product. The concentrations of product or consumed substrate are measured by a detector. Depending on the complexity of the reaction, manifolds consist of different elements. Fig. 12 shows some manifold types for the determination of glucose and ethanol, phosphate, and ammonia (GARN, 1989). The manufacturers offer various detectors: conductivity-, pH-, ion-selective-, and amperometric electrodes, which are combined with different enzyme membranes or small reactors with immobilized enzymes (enzyme cartridge) as well as photometers, fluorometers, etc. Tab. 4 lists some analytes whose concentrations have been measured by FIA off-line.

HANSEN (1989) has reviewed flow-injection enzymatic assays.

Different possibilities for obtaining several peaks per injection by using a single detector and for improving the selectivity, sensitivity, and frequencey of the analysis have been discussed by VALCÁRCEL et al. (1989). Besides the large number of papers on the use of FIA for off-line analysis, some reports have also been published on on-line FIA-systems (Tab. 5).

With respect to long-range (many weeks) analysis with FIA, only experience with low-molecular weight analytes has been reported (SCHÜGERL, 1988). With high-molecular weight analytes, only operations of several hours up to two days are known (RECKTEN-

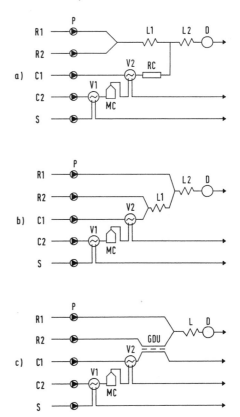

Fig. 12. Typical FIA manifolds (GARN et al., 1989). (With permission of John Wiley & Sons)
a) For determination of ethanol and glucose: C_1 phosphate buffer (1 mL/min), C_2 phosphate buffer (4.6 mL/min), D spectrophotometer, L_1, L_2 mixing coils (both 100 cm long, 0.8 mm id), MC stirred mixing chamber, P piston pumps; 1) Trinder reaction: R_1 sulfonated phenol (0.75 mL/min), R_2 4-AP+POD (0.5 mL/min); 2) ABTS reaction: R_1 ABTS+POPD (0.5 mL/min), RC enzyme reactor, S conditioned sample, V_1, V_2 injection valves (both 75 µL).
b) For determination of phosphate: C_1 water (1 mL/min), C_2 water (4.6 mL/min), D spectrophotometer, L_1, L_2 mixing coils (10 and 200 cm long, 0.8 mm id), MC stirred mixing chamber, P piston pumps, R_1 heptamolybdate (0.25 mL/min), R_2 ascorbic acid (0.25 mL/min), S conditioned sample, V_1, V_2 injection valves (both 75 µL).
c) For determination of ammonia: C_1 borax buffer (0.5 mL/min), C_2 water (4.6 mL/min), D spectrophotometer, GDU gas diffusion unit, L mixing coil (100 cm long), MC stirred mixing chamber, P piston pumps, R_1 NaOCl (0.5 mL/min), R_2 phenol (0.5 mL/min), S conditioned sample, V_1, V_2 injection valves (both 75 µL).

Tab. 5. On-line FIA-Systems in Biotechnology

Analyte	Detection	Detector	Reference
Glucose	GOD/membrane	H_2O_2 electrode	[1]
Lactate	Lactate oxidase (LO)/membrane	H_2O_2 electrode	[1]
Glucose	GOD cartridge	H_2O_2 detection with luminol (chemiluminescence)	[7]
Lactate	Lactate oxidase cartridge	H_2O_2 detection with luminol (chemiluminescence)	[7]
Ammonia	NaOH/EDTA/membrane	NH_3 gas sensor	[1]
Urea	Urease/cartridge	NH_3 gas sensor	[1]
Lactate	LO cartridge	O_2 electrode	[1]
Lactose	β-Galactosidase/GOD cartridge	O_2 electrode	[1]
Glucose	GOD/membrane	O_2 electrode	[2]
Glucose	GOD/catalase	ET	[1]
Penicillin G	Penicillinase cartridge	ET	[1]
Penicillin G	Penicillin G-amidase cartridge	ET	[1]
Penicillin V	Penicillinase cartridge	ET	[1]
Penicillin V	Penicillin G-amidase cartridge	ET	[1]
Urease	Urea in solution	ET	[1]
Penicillin G-amidase	Penicillin G in solution	ET	[1]
Glucose	Peroxidase, 4-amino-phenazone, phenol	Photometer	[13]
Glucose and ethanol	Peroxidase cartridge, sulfurated 2,4-dichlorophenol, 4-amino-2,3-dimethyl 1-phenyl-3-pryrazolin-5-one	Photometer (510 nm)	[3]
Glucose	Glucose dehydrogenase/mutarotase/ NADA redox indicator	Photometer (560 nm)	[4]
Fructose	H_2SO_4/cysteine/carbazole	Photometer (560 nm)	[4]
Phosphate	Ammonium heptamolybdate in 1 N H_2SO_4, ascorbic acid	Photometer (660 nm)	[3]
Phosphate	Ammonium molybdate H_2SO_4, stannous chloride, hydroxylammonium chloride	Photometer (660 nm)	[12]
Ammonia	NaOH/EDTA/NaOCl, phenol	Photometer (655 nm)	[3]
Ammonia	Hypochlorite, H_2SO_4, Na-nitro-prusside, phenol, methanol, NaOH (Berthelot r.)	Photometer (625 nm)	[12]
Penicillin G	Penicillinase cartridge	pH electrode	[5]
Amino acids	o-Phthaldialdehyde/mercapto-propionic acid/methanol/ borate/NaOH buffer	Fluorescence detector	[6]
Protein	Biuret reagent	Photometer (565 nm)	[7], [11]
	Bradford reagent	Photometer (550 nm)	[11]
Formate dehydrogenase	Formate/NAD$^+$	NADH measured by fluorometer at 340 nm	[8]
Leucine dehydrogenase	Ketoleucine/NADH, NH_4^+	NADH measured by fluorometer at 340 nm	[8]
Protease			[9], [10]
Pullulanase			[9], [10]

References: [1] SCHÜGERL, 1988, [2] MIZUTANI et al., 1987, [3] GARN et al., 1989, [4] GRAM et al., 1987, [5] OLSSON, 1988, [6] NALBACH et al., 1988, [7] NIKOLAJSEN et al., 1988, [8] RECKTENWALD et al., 1985a, [9] KRONER and KULA, 1984, [10] HUSTEDT et al., 1985, [11] RECKTENWALD et al., 1985b, [12] PEDERSEN et al., 1985, [13] VALERO et al., 1988

Tab. 6. Advantages and Disadvantages of On-Line Flow-Injection Analyzers

Advantages
- No extensive protein precipitation and cell growth in the tubing, coils, or detector (the carrier stream does not contain medium components)
- Short response time (kinetic regime is used for the determination)
- A small amount of chemicals is needed

Disadvantages
- Discrete data (difficult to use for control)
- High expenditure for process automation
- Expensive ($30000–40000/channel)

WALD et al., 1985a, b; KRONER and KULA, 1984; HUSTEDT et al., 1985).

Selectivity, detection limit, and accuracy of the FIA again depend on several factors. They can only be given for a definite analyte in a particular broth and analyte concentration range using a specific dilution and reaction. For instance, in *Saccharomyces cerevisiae* cultivation broths the detection limits given by GARN et al. (1989) are: for glucose and ethanol (with Trinders reaction) 5 mg/L, for phosphate (with ammonium heptamolybdate) 1 mg/L, and for ammonia (with NaOCl and phenol) 50 mg/L. The low sensitivity for the latter analysis is due to the additional separation and dilution steps in the gas diffusion unit. The errors of the analysis are usually less than 1–2%.

In Tab. 6, advantages and disadvantages of the on-line flow injection analysis are compiled.

FIA is also becoming popular in the on-line analysis of long-range cultivation processes because of the advantages given in Tab. 6.

3.2.3 On-Line High Performance Liquid Chromatography, HPLC

HPLC is a standard technique for the analysis of complex liquid mixtures, where the separation is based on adsorption-, reversed phase-, liquid/liquid partition-, ion exchange-, ion pair-, gel-, or affinity chromatography (SNY-DER and KIRKLAND, 1979; HORVATH, 1980). It is also used in biotechnology (KENNEDY et al., 1989).

The equipment usually consists of a solvent storage vessel, porous metal filter, high pressure pump, pre-column, injection syringe, injection valve, guard column, column, thermostat, detector and integrator (MEYER, 1988; ENGELHARDT, 1977).

The various components present in the sample interact to different extents with the chromatographic material, which is optimized for the separation of medium components with a possible high resolution. Each compound elutes in a peak from the bed after a particular volume of the liquid has been passed through the bed. The presence and amount of the eluted compounds in the medium leaving the column are usually detected by some non-specific detectors sensitive to most components in the mixture or by specific biosensors (immobilized enzymes or antibodies) detecting only a particular component. Typical non-specific detectors are refractive index detectors, photometers, fluorometers, and polarimeters. The record of the signal from the instrument as a function of time or volume of the liquid passed through the column is known as a chromatogram.

Peaks of the chromatogram can be identified by determining the retention time of the particular components of the mixture by means of calibration with the pure components or with their mixtures, provided that a peak is completely resolved from its neighbors. Integration of the peak and reference to calibration data yields the amount of the component present in the analyzed sample (CHASE, 1986). In the case of incompletely resolved peaks or irregular base-lines, specific programs have been developed to determine the peak surface areas of the components.

In several cases the components have to be derivatized to permit their detection, e.g., by fluorometry. The reaction is usually carried out by pre-column derivatization. For amino acids, the most popular derivatization reagents are: *ortho*-phthaldialdehyde, fluorenyl-methyloxycarbonyl chloride, ninhydrine and phenylisothiocyanate (SCHNEIDER and FÖLDI, 1986; BRUTON, 1986). Sometimes post-column derivatization is used (HUEN, 1984).

Tab. 7. Analytes Monitored with On-Line HPLC in Biotechnology

Analyte	Method	Detector	References
Phenoxyacetic acid K-Penicillin V p-Hydroxypenicillin V Penicilloic acid Penicilloic acid	Tetrabutyl ammonium hydrogen sulfate/ methanol on nucleosil C_{18} 5 μm (reversed phase)	UV-detector	[1], [2], [3]
Methionine Penicillin N 2-Hydroxy-4-methyl-mercaptobutyric acid Deacetylcephalo-sporin C Diacetoxycephalo-sporin C Cephalosporin C	Tetrabutylammonium hydrogen sulfate on nucleosil C_{18} 10 μm	UV-detector	[1], [4], [5], [6]
Ethanol		Differential refractometer	[7]
Ethanol Glucose Glycerol		Differential refractometer	[8]
Erythromycin		UV-detector	[8]
D- and L-Amino acids		D- and L-Amino oxidases, immobilized on Pt-electrode, amperometric detection of H_2O_2	[9]
Amino acids, especially tryptophan	Phosphate/methanol/ THF on nucleosil ODS	4-Fluoro-7-nitrobenzo-2-oxa-1,3-diazole (NBD-F) adducts, fluorometer	[11]
Lactose Glucose Galactose Lactic acid Acetic acid Ethanol Gluconic acid	Bio-Rad HPX-87-H	UV-detector	[10]

References: [1] SCHÜGERL, 1988, [2] MÖLLER, 1987, [3] MÖLLER et al., 1986, [4] BAYER, 1987, [5] BAYER et al., 1988, [6] HOLZHAUER, 1987, [7] MATHERS et al., 1986, [8] DINCER et al., 1984, [9] YAO and WASA, 1988, [10] MONSEUR and MOTTE, 1988, [11] IMAI et al., 1988.

The separation of amino acids and peptides from protein hydrolysates and their quantitative determination by HPLC is a fairly difficult problem (HEARN et al., 1983). Recently a combination of HPLC and mass spectrometer has been used for this purpose (LEE and HENION, 1989).

In contrast to the large number of publications on HPLC, very few papers have been published about the use of this technique in an on-line mode (Tab. 7).

In liquid chromatography very high purity and care is required, since the smallest dust particles can stick to the capillary wall or in the pores of the metal filter and cause clogging. The eluents are prepared with twice-distilled water, cleaned through a microfiltration (e.g., 0.45 μm) membrane and degassed in an ultrasound bath. To avoid gas absorption they are gassed with helium during the analysis. The samples have to be freed of compounds (e.g., proteins) which can precipitate in the analyzer or adsorb on the column material. Therefore, the samples are usually treated with methanol (1:1) and deproteinated by ultrafiltration, if low molecular weight components are to be analyzed. For more details see the original papers.

Usually a single calibration of the components for all of the samples from different production phases is adequate. A comparison of the off-line and on-line analysis results is recommended.

If the separation of the peaks is satisfactory, and the concentration of the components is in the intermediate concentration range, the error of the analysis is usually less than 1–2%. Usually 20–30 minutes are necessary for a complete analysis of a multicomponent system. Therefore, the analysis frequency is low.

In Tab. 8 the advantages and disadvantages of the on-line HPLC analysis are shown.

On account of its high flexibility and excellent performance, it is expected that on-line HPLC will gain in importance in the future.

Tab. 8. Advantages and Disadvantages of On-Line HPLC Analysis

---

Advantages
- Analyzes several components at the same time
- Easy automation
- Relatively inexpensive (about $7500/component)

Disadvantages
- Very sensitive to impurities
- Discrete data (difficult to use for control)
- Only 2–3 analyses per hour (low scanning rates and large dead times)

3.2.4 Fast Protein Liquid Chromatography, FPLC

Protein liquid chromatography is an established technique in biochemical laboratories (RICHEY, 1983; HORVATH, 1980; BUSSOLO, 1984). However, it is rarely used for the monitoring of particular protein concentrations in fermentation broths. The main reasons are the difficulties connected with analytical techniques (reliability), the technological problems (sampling), and the working conditions (GRESSIN, 1988). For the wider acceptance of this technique, its selectivity, sensitivity, and reliability must be improved.

GRESSIN (1988) reported on such a system used for monitoring of the protein concentration in fermentation broth, which was developed at Rhone Poulenc Santé and based on an FPLC manufactured by Applied Automation. Reversed liquid chromatography was used in an isocratic mode on Spherosil P_{35}-600 C18 as stationary phase at a moderate pressure (< 100 bar), together with a UV detector with a fixed wavelength, and a pre-column for increasing the lifetime of the analytical column. The analyzer was coupled to a data-acquisition system. Adaptive feeding of the substrate was possible by means of this system.

The analysis is more difficult if intracellular protein monitoring is desired. Such a case was described by GUSTAFSSON et al. (1986). The cells (*Escherichia coli*) were separated from the broth by centrifugation at 4900 g for 10 min to eliminate the extracellular proteases. After sonification of the cells, the cell debris was separated again by centrifugation and the supernatant filtered through a microporous (0.45 µm) membrane. The filtrate was injected into the FPLC anion exchanger column (Mono Q HR 5/5, Pharmacia AB), which was equipped with a gradient programmer. Again, high-purity water and analytical grade chemicals are needed, and buffer solutions must be degassed and filtered by microfiltration (e.g., 0.22 µm) membranes, as in the HPLC analysis. With β-galactosidase as the model analyte, the elution time was reduced to 9 minutes.

It is expected that the use of FPLC for the monitoring of protein formation will gain in importance.

3.2.5 Ion Chromatography, IC

For the separation of ions of strong acids and bases (e.g., Cl^-, NO_3^-, Na^+, K^+), ion chromatography has been developed by SMALL et al. (1987) (MEYER, 1988; SMITH and CHANG, 1983; GJERDE and FRITZ, 1987). It is based on three different separation techniques: ion-exchange (high performance ion chromatography, HPIC, for larger ions), ion exclusion (high performance ion chromatography exclusion, HPICE), and ion pair formation (mobile phase ion chromatography, MPIC) (WEISS, 1985).

The equipment consists of a dispository for eluents, a pump, an injection valve, a column for exchange reaction, a suppressor column, a conductivity detector or photometer, and a recorder/integrator (DIONEX CORP.).

In contrast to HPLC columns, in which the carrier is silica gel, the carrier in ion chromatography columns often consists of polystyrene/divinylbenzene (PS/DVB) resins. The latter have a much better pH stability (they are stable in the pH range 0 to 14) than silica (pH 1 to 9). Eluents with extreme pH values must be applied to bring compounds such as sugars and alcohols to an ionogenic state. The degree of the cross-linking of the resin ranges between 2 and 5% to achieve the optimal porosity of the carrier (WEISS, 1985).

The anion exchanger of Dionex consists of PS/DVB particle cores (10–25 μm) with a sulfonated surface and aminated porous latex particles (0.1 μm), which adhere to the surface of the ion-exchanger particles by electrostatic and van der Waals forces. The surface sulfonation hinders the diffusion of other anionic species into the inner stationary phase by Donnan exclusion, therefore, the groups on the surface of the latex particles dominate (MEYER, 1988).

A so-called suppressor column is used for the sensitive detection of ions by means of their conductivity. In this column, e.g., after the separation of Cl^- and Br^- in the ion exchange column, the following reactions occur:

$$resin\text{-}SO_3^- \, H^+ + NaHCO_3 \rightarrow$$
$$resin\text{-}SO_3^- \, Na^+ \, (+ CO_2 + H_2O)$$

and

$$resin\text{-}SO_3^- \, H^+ + NaCl \rightarrow$$
$$resin\text{-}SO_3^- \, Na^+ + HCl$$

$$resin\text{-}SO_3^- \, H^+ + NaBr \rightarrow$$
$$resin\text{-}SO_3^- \, Na^+ + HBr$$

With the exchange of Na^+ for the H^+ of the cation exchanger, the highly conductive $NaHCO_3$ is converted into H_2CO_3, which has a low conductivity. On the other hand, NaCl and NaBr are converted into their highly conductive acids. The highly conductive acids in the weak conductivity H_2CO_3 solution permit highly sensitive detection by the electrical conductivity detector (especially suitable for tracer analysis).

Besides the potentiometric detectors, UV and amperometric detectors are also used. However, the potentiometric detector remains the standard. The use of the universal potentiometric detector based on electrical conductivity is hampered by the high-ion-strength eluents. Therefore, either other detectors or low capacity resins are used for the separation (HADDAD et al., 1986; HORVAI et al., 1988).

Organic acids can also be separated by the anion exchange column. The separation of carbohydrates is difficult. They can only be separated by means of strong alkaline eluents on a strong basic anion exchanger.

Cation exchange chromatography columns also consist of a surface-sulfonated PS/DVB resin core. On the surface of the core between the $SO_3^- \, H^+$ groups and the cation M^+, the following exchange occurs:

$$resin\text{-}SO_3^- \, H^+ + M^+ A^- \rightleftharpoons$$
$$resin\text{-}SO_3^- \, M^+ + H^+ A^-$$

Diffusion of the strongly dissociated ions into the pores of the core can be neglected because of the strong hydrophobic character of the core. The alkaline metal cations are eluted with dilute HCl and the divalent cations with complex-forming agents. Suppressor columns are also used to improve the sensitivity of detection with electrical conductivity.

Ion exclusion chromatography is used for the separation of organic acids with low acid strength, amino acids, and alcohols (WEISS, 1985). Due to Donnan exclusion, the strong inorganic acids pass the stationary phase rapidly. Only non-dissociated compounds can diffuse into the resin pores, since they are not subject to Donnan exclusion. In this case, the separation is based on polar and van der Waals forces between solutes and the stationary phase.

Again, suitable suppressor systems and electrical conductivity cells are used as detectors. Inorganic and organic anions can be separated in a single run within 30 min in combination with ion-exchange chromatography.

Alternatively to ion exchange chromatography, ion pair chromatography is gaining importance, since anions as well as cations can be separated and analyzed by this process. An organic ionic component, which forms an ion pair with the counter-charged analyte of the sample, is added to the mobile phase. This ion pair is a salt, but it behaves like a non-ionic organic molecule. It can be separated by reversed-phase chromatography.

One uses, e.g., alkylsulfonate for the analysis of cationic analytes and, e.g., tetrabutylammonium phosphate for the analysis of anionic analytes.

The advantages of ion pair chromatography are:

- separation in a reversed-phase system is possible,
- mixtures of acids, bases, and neutral components as well as amphoteric molecules can be separated,
- the selectivity can be controlled by means of the counter-ion (TOMLINSON et al., 1978; BIDLINGMEYER, 1980; GLOOR and JOHNSON, 1977).

For more details see WEISS (1985) and MEYER (1988).

Ion chromatography is mainly used in industry for water analysis (WEISS, 1985). Up to the present, no application for the analysis of fermentation broth is known. This lack is probably due to the complex matrix and high ion strength in the broths.

It is expected that these difficulties will be overcome and ion chromatography will become a standard method for broth analysis.

3.2.6 Chemically Specific Detectors

Each analyzer system needs a detector that indicates the variation of the analyte concentration after separation (e.g., by HPLC) or in a complex matrix (e.g., by FIA). When the analytes are separated during the analysis, the ideal detector should have the same sensitivity for each analyte. If the analytes are not separated, the ideal detector should detect only a single analyte. It should be selective, i.e., it should have a high chemical specificity.

Furthermore, the ideal detector should have a short time constant and should be very sensitive to the analyte, but it should not be sensitive to variations in temperature, pressure, flow rate, or chemical composition of the matrix (i.e., it should be noise-free).

Chromatographic techniques require nonspecific detectors based, e.g., on the refractive index (*RI*), ultraviolet (UV) absorbance, fluorescence emission, electrochemical reaction, or electrical conductivity. Continuous air-segmented and flow injection analyzers need chemically specific detectors.

Measurement of the concentration of several analytes is based on specific wet chemical reactions that form easily detectable compounds. However, inorganic ions can be directly detected by ion-selective electrodes (SCHINDLER and SCHINDLER, 1983), low molecular weight organic compounds by enzymatic reaction (BOWERS and CARR, 1980; SCHINDLER and SCHINDLER, 1983), and proteins by means of immunoassays (SCHARPÉ et al., 1976; VORLAENDER, 1980; SCHINDLER and SCHINDLER, 1983).

For ion-selective detectors see Chapter 1, Section 6.

The combination of transducers and biochemical receptor membranes and their *in situ* use as biosensors are considered in Chapter 3 (Biosensors). However, this receptor–transducer combination is often applied as a detector for continuous air-segmented analyzers or FIA-systems. In such a case, the biochemical receptor is not necessarily used as a membrane directly connected to the transducer.

The following combinations are used:

a) The receptor membrane is connected directly to the transducer,
b) the receptor is immobilized on carrier particles, applied in a cartridge, and combined with a separate transducer,
c) the receptor is immobilized on the tubing wall and combined with a separate transducer,

d) the receptor is immobilized on a membrane and combined with a separate transducer.

The signal quality depends on the fractional conversion of the enzymatic reaction (which is related to the amount of enzyme present and the time during which the analysis is run) as well as on the mass dispersion during the analysis. A larger amount of enzyme can be immobilized on the surface of carrier particles than on a tubing surface or a membrane. Furthermore, the reaction time in a cartridge or tubing is longer than the contact time with a membrane.

The mass dispersion in a small cartridge filled with carrier particles is very low (close to the plug flow) as long as the cartridge is uniformly filled. Because of the parabolic laminar velocity profile of the sample flow in the tubing, the mass dispersion in it is higher than in the cartridge.

On the other hand, a comparison of dispersion in the membrane and cartridge systems indicates that the mass dispersion is only lower in the cartridge than in the tubing if the flow across the membrane is completely uniform.

Generally speaking, when using a cartridge the highest signal can be expected with the highest fractional conversion and the smallest mass dispersion. In addition, the enzyme activity of the cartridge changes only slowly, thus its utilization time is much longer than that of a receptor immobilized on tubing or membranes.

Transducers are based on the variation of pH [potentiometric pH electrodes, IS-FETs (pH-sensitive field effect transistors)], p_{O_2} (amperometric p_{O_2} electrodes), p_{CO_2} (potentiometric pH-electrodes), p_{NH_3} (potentiometric NH$_3$-sensitive electrodes), H_2O_2 (amperometric electrodes) values, or the change of the local temperature (thermistor, a minicalorimeter; due to reaction enthalpy), fluorescence intensity (fluorometer, e.g., due to cofactor-NADH conversion), and color or luminescence intensity (photometer) (BOWERS and CARR, 1980; KRICKA and THORPE, 1986; TURNER et al., 1987; RECHNITZ, 1988; SCHMID et al., 1987).

The performance of the detector also depends on the transducer properties (time constant, sensitivity, selectivity, stability), as well

as on the volume and time constant of the receptor–transducer system and the mass dispersion therein. Detectors consisting of thin receptor membranes with direct contact to the transducer surface have the lowest time constant.

The behavior of the electrochemical transducer is discussed in Chapter 1, that of the transistor in Chapter 3, and that of the thermistor and fluorometer in Chapter 6.

4 Prerequisites for *in situ* Analysis

On-line aseptic sampling is required for on-line analysis of the medium composition outside the reactor. Depending on the molecular weight of the analytes, ultrafiltration or microfiltration membrane filters are used. The sample must be prepared for the analysis with continuous air-segmented, FIA, or chromatographic techniques (removal of proteins, mixing with reagents, derivatization, etc.). Furthermore, frequent calibrations and blank determinations are required to compensate for drifts (caused by variations of the activity of the chemically-specific receptor, of the transparency of the optical detector windows, precipitations, etc.).

Sample preparation, regular calibrations, and blank determinations are not possible for *in situ* analysis by sensors. Furthermore, when using monocultures for product formation, the analyzer system must be sterilized. When using, e.g., an enzyme electrode, the signal is influenced by the variation of the enzyme activity due to enzyme loss, its deactivation (denaturation, decomposition by proteases or nonspecific protein adsorption) or inhibition, as well as by the response of the transducer to the local and instantaneous variations of the properties of the medium (pH, p_{O_2}, p_{CO_2}, turbidity, absorbance, fluorescence emission). These are the reasons why biosensors cannot be used *in situ* for process control.

By measuring the local and instantaneous variation of broth properties by a second and perhaps a third transducer of the same type close to the biosensor, and using this information to correct the signal, the direct transducer response can be eliminated. However, the problems with *in situ* sterilization and enzyme deactivation can only be avoided, if the chemically-specific receptor (that is, the enzyme) can be withdrawn before sterilization, and through occasional calibration and blank determination. This means that immobilized receptors cannot be applied in membranes. The receptor must be used in solution, and immobilization must be performed by a separate membrane.

The sensors have to be small to be able to measure the medium composition and the transducer response at the same position. Only miniaturized sensors fulfill these prerequisites; but receptor withdrawal is especially difficult with such miniaturized sensors. Probably a new type of sensor will need to be developed for this purpose.

5 Future Developments

Future developments in the analysis of fermentation broths will probably run along the following main lines:

- more general use of the spectroscopic techniques,
- adaptation of the immunoassays for on-line measurements,
- broader use of supercritical chromatography,
- improvement of the reliability of the applied techniques by means of expert systems.

5.1 General Use of Spectroscopic Methods

The spectroscopic techniques (in the UV-, VIS-, and IR-ranges, as well as fluorescence, electron spin resonance, nuclear magnetic resonance, and optical rotary dispersion and circular dichroism) have a broad application in biochemistry (GALLA, 1988). They can yield a large amount of information about biological processes. However, for on-line analysis of the

Tab. 9. Important Aspects of FT-IR Technology (FINK and CHITTUR, 1986)

Advantages	Constraints
Rapid data acquisition, up to 10 spectra per second	Instrumental instability, especially in high-vibration environment
Capability of acquiring spectra in aqueous media by background subtraction	Similarity of spectral features of biological compounds
Potential for multicomponent analysis of complex spectra	Sensitivity of molecular vibrations to environmental factors
Observation of macromolecular conformation by deconvolution of spectral features	Subjectivity of water subtraction algorithms
Availability of several sample/optic modes — transmission, attenuated total reflectance (ATR), and diffuse reflectance (DRIFT)	Labor-intensive learning curve and data work-up

broth, at present only UV-, VIS-, and IR-spectroscopy can be applied. Especially IR spectroscopy is developing rapidly. Infrared analyzers for gas analysis have been available for many years, but only with the availability of Fourier transform infrared (FT-IR) spectrometers in the past decade has the rapid analysis of aqueous-phase analytes by infrared methods been feasible. Some important aspects of FT-IR technology are summarized in Tab. 9.

These techniques have been applied to investigations of protein structure/function, protein adsorption, nucleic acid/sugar structure/function, lipid transformations, protein-lipid interactions, and cell membrane properties as well as cell characterization (FINK and CHITTUR, 1986; MANTSCH and CASAL, 1986). Especially interesting is the use of the FT-IR technique for investigations of adherent Chinese hamster ovary (CHO) cells on a germanium IRE crystal to elucidate the mechanism of attached cell growth (FINK and CHITTUR, 1986).

However, only few applications in process monitoring are known: determination of the concentration of glucose in biological fluids (BAUER and FLOYD, 1987), β-lactam antibiotics in fermentation broths (WONG et al., 1984), pyruvate consumption rate in *Escherichia coli* cultivation (WHITE et al., 1985), concentration of glucose, ethanol, glycerol in the broth of *Saccharomyces cerevisiae* (ALBERTI et al., 1986), and concentration of glucose and excreted metabolites during the cultivation of a recombinant *E. coli* strain (ALLEN and LULI, 1985).

These investigations indicate that IR-techniques have a high potential for monitoring broth components. A new generation of smaller, less expensive instruments with better computer equipment and algorithms is needed for on-line process analysis.

5.2 Adaptation of Immunoassays for On-Line Monitoring of Proteins

Immuno-techniques have a broad use in clinical diagnostics (VORLAENDER, 1980). They are based on the specific binding reaction that occurs between antigens and their antibodies. The agglutinated adduct of antigen/antibody reaction can be measured by visual inspection, turbidometry, nephelometry, or weighing after centrifugation. When the amount is too small, special techniques can be used: radio-immunoassay (RIA), enzyme-immunoassay (EIA), enzyme-linked immunosorbentassay (ELISA), enzyme multiplied immunoassay technique (EMIT), and fluorescence immunoassay (FIA), all permit the determination of proteins, hormones, and drugs in the human body at low concentrations (SCHARPÉ et al., 1976; ENGVALL and PERLMANN, 1971; SCHUURS and VAN WEMEN, 1978).

Recently some of these immunoassays have been adapted for biotechnological applications (KARUBE and SUZUKI, 1986; LARSSON et al., 1987; OH and KIM, 1986). Especially the flow injection ELISA technique appears to be a quick, reproducible, and reliable method that is easy to automate (LARSSON et al., 1987).

It is expected that during the next few years on-line techniques will be developed for protein analysis based on turbidometry, nephelometry, and on a different type of immunoassay.

5.3 Supercritical Fluid Chromatography

With regard to its behavior and performance, supercritical fluid chromatography (SFC) is positioned between gas chromatography (GC) and liquid chromatography (LC) and represents a continuous transition from GC to LC. The number of applicable mobile phases for SFC is large in comparison with GC, but small compared to LC. The most important criterion for the application of a mobile phase is its critical temperature, followed by its ability to dissolve the components and its thermal stability. Mixtures of mobile phases consisting of a basic component and a second component (modifier) are important for SFC, for isocratic as well as for programmed composition of the mobile phase.

The use of capillary columns has been investigated and worked out in recent years. Their resolution is lower than that of GC, however, they can also be applied to analytes with low vapor pressure. One of the most interesting aspects of capillary supercritical fluid chromatography SFC is its potential for interfacing with GC as well as with LC detectors (LATER et al., 1987). The most common systems are the: SFC-UV spectrophotometer (LATER et al., 1987), SFC-mass spectrometer (RANDALL and WAHRHAFTIG, 1978), and SFC-Fourier-transform infrared spectrophotometer (SHAFER and GRIFFITS, 1983). However, flame ionization detectors (FJELDSTED et al., 1983a), fluorescence detectors (FJELDSTED et al., 1983b), flame photometric detectors (MARKIDES et al., 1986), etc., are also used.

SFC is especially popular in food technology, e.g., determination of caffeine in beverages (GERE, 1983a), or of carotenoids in plants (GERE, 1983b). It is well suited for the analysis of hydrophobic analytes such as fatty acids and their esters (LATER et al., 1987). However, no biotechnological application of SFC in on-line analysis is yet known. It is expected that it will first be used for off-line analysis and later also for on-line analysis, especially for on-line multidimensional chromatography in combination with HPLC-techniques (LURIE, 1988).

It should be mentioned here that the combination of different chromatographic systems (e.g., size exclusion chromatography with reversed-phase chromatography, HPLC and capillary GC, capillary GC and capillary SFC, HPLC and capillary SFC) is an excellent method for obtaining increased resolving power for multicomponent mixtures, especially if completely independent techniques are used (LURIE, 1988).

5.4 Improvement of the Reliability of the Applied Techniques by Means of Expert Systems

When using process analysis data for process control, reliability is extremely important. Several possibilities exist for malfunction of the sterile sampling system as well as of the analyzer system. In all analyzers, sample pretreatment (dilution, neutralization, filtration) is necessary. The filtration, due to variation in the transmembrane permeation rate caused by membrane fouling (usually protein adsorption) or drift of the pumping rates on the different channels can cause serious errors in the dilution of the analyte as well as in the pH of the reaction mixture and the rate of the reaction needed for formation of the compound to be detected.

With the continuous air-segmented analyzer, the most common failures are caused by incomplete phase separation, base-line drift due to protein precipitation, and cell growth in the coils and in the detectors.

In FIA systems, the most common failures are due to poor reproducibility of repeated injections, sluggish peak response in reaching the base-line, base-line drift, air bubbles, abrupt noise on peaks, noisy signal, double peaks, and negative peaks (RUZICKA and HANSEN, 1988).

In addition to the possible failures of FIA systems, base-line drift in HPLC systems may occur due to the precipitation of proteins.

If such data are to be used for process control, an automated system is necessary to recalibrate the analyzer system, evaluate the blank values, clean the channels periodically, control the flow rates and the transparency of the windows, etc. (see Chapter 17). However, the automated system is not able to diagnose errors or subsequently to correct data or eliminate the failures.

For error diagnosis and failure correction, a special closed-loop control expert system is necessary. The first steps in this direction have already been taken (FISH and FOX, 1988; especially LÜBBERT et al., 1988).

It is expected that in the near future several expert systems will be developed for error diagnosis in on-line process analysis. This is a prerequisite for the application of complex analyzer systems in process control.

6 References

ABC (1988), *Sampling System*. ABC Biotechnologie/Bioverfahrenstechnik GmbH, P.O. Box 1151, D-8039 Puchheim.

AHARONOWITZ, Y. (1980), Nitrogen metabolite regulation of antibiotic biosynthesis, *Annu. Rev. Microbiol.* **34**, 209–233.

ALBERTI, J. C., PHILLIPS, J. A., FINK, D. J., WASACZ, F. M. (1986), Off-line monitoring of fermentation samples by FTIR/ATR: A feasibility study for real-time process control, *Biotechnol. Bioeng. Symp.* **15**, 689–722.

ALLEN, B. R., LULI, G. W. (1985), A gradient-feed process for obtaining high cell densities for recombinant *E. coli*, Paper presented at the *National American Chemical Society Meeting, Chicago*, September, 1985.

ALWARTHAN, A., TOWNSHEND, A. (1988), Determination of tetracycline by flow injection with chemiluminescence detection, *Anal. Chim. Acta* **205**, 261–265.

BAUER, B., FLOYD, T. A. (1987), Monitoring of glucose in biological fluids by Fourier-transform infrared spectrometry with a cylindrical internal reflectance cell, *Anal. Chim. Acta* **197**, 295–301.

BAYER, T. (1987) Reaktionstechnische Untersuchungen zur Produktion von Cephalosporin C in Mammutschlaufenreaktor, *Dissertation*, Universität Hannover, FRG.

BAYER, T., HEROLD, T., HIDDESSEN, R., SCHÜGERL, K. (1986), On-line monitoring of media components during the production of cephalosporin C, *Anal. Chim. Acta* **190**, 213–219.

BAYER, T., ZHOU, W., HOLZHAUER, K., SCHÜ-

GERL, K. (1989), Investigations of cephalosporin C production in an airlift tower loop reactor, *Appl. Microbiol. Biotechnol.* **30**, 26–33.

BIDLINGMEYER, B. A. (1980), Separation of ionic compounds by reversed phase liquid chromatography: An up-to-date of ion-pairing techniques, *J. Chromatogr. Sci.* **18**, 525–539.

BÖING, J. T. P. (1982), Enzyme production in: *Prescott & Dunn's Industrial Microbiology*, 4th Ed. (REED, G., Ed.). Westport, Connecticut: AVI Publishing Co., Inc.

BORDONS, A. (1986), Automated determination of lysine by colorimetric method with ninhydrin, *Biotechnol. Lett.* **6**, 411–414.

BOWERS, L. D., CARR, P. W. (1980), Immobilized enzymes in analytical chemistry, *Adv. Biochem. Eng.* **15**, 89–129.

BRAN + LÜBBE GmbH, *Analysis Technique (Autoanalyser® II and TRAACS 800™)*, *Brochure*. Bran + Lübbe GmbH, Postfach 1360, D-2000 Norderstedt/Hamburg.

BRITZ, M. L., DEMAIN, A. L. (1985), Regulation of metabolite synthesis, in: *Comprehensive Biotechnology*, Vol. 1 (BULL, A. L., DALTON, H., Eds.), pp. 617–636. Oxford: Pergamon Press.

BROWN, C., BERRY, D. R., LARSEN, V. F. (1987), Improved enzymatic analysis of low level glucose concentrations in batch and continuous fermentations, in: *Proc. 4th Eur. Congr. Biotechnology*, Vol. 3, pp. 179–181 (NEISSEL, O. M., VAN DER MEER, R. R., LUYBEN, K. Ch. A. M., Eds.). Amsterdam: Elsevier Science Publisher B.V.

BRUTON, C. (1986), Fully automated amino acid analysis using AminoTag pre-column derivatization, IBL June, 48–57.

BUSSOLO, J. M. (1984), A practical introduction to reversed-phase liquid chromatography of proteins and peptides, IBL Sept./Oct., 14–31.

CHASE, H. A. (1986), Rapid chromatographic monitoring of bioprocesses, *Biosensors* **2**, 269–286.

DINCER, A. K., KALYANPUR, M., SKEA, W., RYAN, M., KIERSTED, T. (1984), Continuous on-line monitoring of fermentation processes, *Dev. Ind. Microbiol.* **25**, 603–611.

DIONEX CORP. Sunnyvale, CA, USA, *Brochures*, and/or DIONEX GMBH, Einstein-Str. 1, D-6108 Weiterstadt, *Applikationsberichte*.

DOMINGUEZ, E., HAHN-HÄGERDAL, B., MARKO-VARGA, G., GORTON, L. (1988), A flow injection system for the amperometric determination of xylose and xylulose with coimmobilized enzymes and modified electrode, *Anal. Chim. Acta* **213**, 139–150.

ENGELHARDT, H. (1977), *Hochdruck-Flüssigkeits-Chromatographie*, 2nd Ed. Berlin–Heidelberg: Springer Verlag.

ENGVALL, E., PERLMANN, P. (1971), Enzyme-linked immunosorbent assay. Quantitative assay of immunoglobulin G, *Immunochemistry* **8**, 871–880.

EPPENDORF GERÄTEBAU (1989), *Brochure: EVA (Eppendorf Variable Analysing) System*. Netheler & Hinz, Barkhausenweg 1, D-2000 Hamburg.

FIAtron, FIAchem, *Brochure*, 510 S Worthington St., Oconomowoc, WI 53066, USA.

FIECHTER, A., FUHRMANN, G. F., KÄPPELI, O. (1981), Regulation of glucose metabolism in growing yeast cells, *Adv. Microb. Physiol.* **22**, 122–183.

FINK, D., CHITTUR, K. K. (1986), Monitoring biological processes by Fourier transform infrared spectroscopy, *Enzyme Microb. Technol.* **9**, 568–572.

FISH, N. M., FOX, R. I. (Eds.) (1988), *Fourth Int. Congr. Computer Applications in Fermentation Technology: Modelling and Control of Biotechnical Processes*, 25–29 September. University of Cambridge, UK, Society of Chemical Industry.

FJELDSTED, J. C., KONG, R. C., LEE, M. L. (1983a), Capillary supercritical fluid chromatography with conventional flame detectors, *J. Chromatogr.* **279**, 449–455.

FJELDSTED, J. C., RICHTER, B. E., JACKSON, W. P., LEE, M. L. (1983b), Scanning fluorescence detection in capillary supercritical fluid chromatography, *J. Chromatogr.* **279**, 423–430.

GALLA, H. J. (1988), *Spektroskopische Methoden in der Biochemie*. Stuttgart: Georg Thieme Verlag.

GANCEDO, J. M. (1988), Approaches to the study of catabolite repression in yeast, *Chim. Oggi*, May, 41–44.

GANCEDO, C., GANCEDO, J. M. (1987), General catabolite control in yeast, in: *Proc. 4th Eur. Congr. Biotechnology*, Vol. 4, pp. 485–490 (NEISSEL, O. M., VAN DER MEER, R. R., LUYBEN, K. Ch. A. M., Eds.). Amsterdam: Elsevier Science Publishers B.V.

GARN, M., GISIN, M., THOMMEN, C., CEVEY, P. (1989), A flow injection analysis system for fermentation monitoring and control, *Biotechnol. Bioeng.* **34**, 423–428.

GERE, D. R. (1983a), Assay of caffeine in beverages by supercritical fluid chromatography, *Application Note 800-6*, Hewlett-Packard Co, Avondale, PA, USA.

GERE, D. R. (1983b), Separation of paprika oleoresins and associated carotenoids by supercritical fluid chromatography, *Application Note 800-5*, Hewlett-Packard Co, Avondale, PA, USA.

GHOUL, M., RONAT, E., ENGASSER, J. M. (1986), An automatic and sterilizable sampler for laboratory fermentors: Application to the on-line control of glucose concentration, *Biotechnol. Bioeng.* **28**, 119–121.

GIBSON, T. D., WOODWARD, J. R. (1988), Continuous, reliable on-line analysis of fermentation media by simple enzymatic/spectrophotometric assays, *Anal. Chim. Acta* **213**, 61–68.

GJERDE, D. T., FRITZ, J. S. (1987), *Ion Chromatography*, 2nd Ed. Heidelberg: Hüthig.

GLOOR, R., JOHNSON, E. L. (1977), Practical aspects of reversed phase ion pair chromatography, *J. Chromatogr. Sci.* **15**, 413–423.

GORTON, L., HEDLUND, A. (1988), A flow-injection method for the amperometric determination of L-lactate with immobilized enzymes and a chemically modified electrode, *Anal. Chim. Acta* **213**, 91–100.

GRAF, H. (1989), *Dissertation*, Universität Hannover, FRG.

GRAM, J., DE BANG, M., VILLADSEN, J. (1987), Flow injection analysers for on-line monitoring of glucose isomerisation process, in: *Proc. 4th Eur. Congr. Biotechnology*, Vol. 4, pp. 409–415 (NEISSEL, O. M., VAN DER MEER, R. R., LUYBEN, K. Ch. A. M., Eds.). Amsterdam: Elsevier Science Publishers B.V.

GRESSIN, J. C. (1988), On-line process liquid chromatography, a useful tool in biochemical processes, *BTF-Biotech-Forum* **5**, 38–43.

GUSTAFSSON, J. G., FREJ, A. K., HEDMAN, P. (1986), Monitoring of protein product formation during fermentation with fast protein liquid chromatography, *Biotechnol. Bioeng.* **28**, 16–20.

HADDAD, P. R., ALEXANDER, P. W., TROJANOWICZ, M. (1986), The application of a metallic copper electrode to potentiometric detection of reducing species in ion chromatography, *J. Liq. Chromatogr.* **9**, 777–789.

HANSEN, E. H. (1985), Recent advances in flow injection analysis, *Int. Lab.*, October, 14–23.

HANSEN, E. H. (1989), Flow-injection enzymatic assays, *Anal. Chim. Acta* **216**, 257–273.

HEARN, T. W., REGNIER, F. E., WEHR, C. T. (1983), HPLC of peptide and proteins, *Int. Lab.*, Jan./Febr., 16–35.

HERSBACH, G. J. M., VAN DER BEEK, C. P., VAN DIJCK, P. W. M. (1984), The penicillins: Properties, bio-synthesis, fermentation, in: *Biotechnology of Industrial Antibiotics*, pp. 45–140 (VANdamme, E. J., Ed.). New York: Marcel Dekker.

HOLST, O., HAKANSON, H., MIYABASHI, A., MATTIASSON, B. (1988), Monitoring of glucose fermentation processes using a commercial glucose analyser, *Appl. Microbiol. Biotechnol.* **28**, 32–36.

HOLZHAUER, K. (1987), Entwicklung und Anpassung chromatographischer Verfahren zur Verbesserung der Antibiotikaproduktion, *Diploma Thesis*, Universität Hannover, FRG.

HORVAI, G., PÁL, H., NIEGREISZ, T., TOTH, K., PUNGOR, E. (1988), Electrochemical detectors in ion chromatography, *LC-GC-International* **2** (1), 32–38.

HORVATH, C. (Ed.) (1980), *High Performance Liquid Chromatography, Advances and Perspectives*. New York: Academic Press.

HUEN, J. M. (1984), HPLC analysis of amino acids in beverages, IBL May/June, 50–53.

HUSTEDT, H., KRONER, K. H., KULA, M. R. (1985), On-line-Bestimmungen von Enzymen und Proteinen zur Prozeßkontrolle in der Enzymgewinnung, *BTF-Biotech-Forum* **2**, 57–63.

HÜBNER, U. (1990), *Dissertation*, Universität Hannover, FRG.

IMAI, K., UEDA, E., TOYOKA, T. (1988), High-performance liquid chromatography with photochemical fluorometric detection of tryptophan based on 4-fluoro-7-nitrobenzo-2-oxa-1,3-diazole, *Anal. Chim. Acta* **205**, 7–14.

IMMING, G., SCHALLER, K., MEINERS, M. (1982), The use of automated analyser to control microbial processes, in: *Proc. First IFAC Workshop on Modelling and Control in Biotechnical Processes, Helsinki* (HALME, A., Ed.). Oxford–New York: Pergamon Press.

KARLBERG, B. I. (1983), Automation of wet chemical procedures using FIA, *Int. Lab.*, Jan./Febr. 82–90.

KARUBE, I., SUZUKI, M. (1986), Novel immunosensors, *Biosensors* **2**, 343–362.

KENNEDY, J. F., RIVERA, Z. S., WHITE, C. A. (1989), The use of HPLC in biotechnology, *J. Biotechnol.* **9**, 83–106.

KRICKA, L. J., THORPE, G. H. G. (1986), Immobilized enzymes in analysis, *Trends Biotechnol.* **4** (10), 253–258.

KRONER, K. H., KULA, M. R. (1984), On-line measurement of extracellular enzymes during fermentation by using membrane techniques, *Anal. Chim. Acta* **163**, 3–15.

KRONER, K. H., STACH, W., KUHLMANN, W. (1986), Kontinuierliches Probeentnahmeverfahren für die Bioprozeßanalytik, *Chem. Tech.* **15**, 74–77.

KUHLMANN, W., MEYER, H. D., BELLGARDT, K. H., SCHÜGERL, K. (1984), On-line analysis of yeast growth and alcohol production, *J. Biotechnol.* **1**, 171–185.

LARSSON, K. M., OLSSON, B., MATTIASSON, B. (1987), Flow injection enzyme immuno assay, in: *Proc. 4th Eur. Congr. Biotechnology, Amsterdam*, Vol. 3, pp. 196–199 (NEISSEL, O. M., VAN DER MEER R. R., LUYBEN, K. Ch. A. M., Eds.). Amsterdam: Elsevier Science Publishers BV.

LATER, D. W., BORNHOP, D. J., LEE, E. D., HENION, J. D., WIRBOLDT, R. C. (1987), Detection techniques for capillary supercritical fluid chromatography, *LC-GC International* **1**, 36–42.

LEE, E. D., HENION, J. D. (1989), Micropore high performance liquid chromatography–ion spray mass spectrometry for the determination of peptides, *J. Microcolumn* **1**, Jan./Febr., 14–18.

LENZ, R., BOELCKE, C., PECKMANN, U., REUSS, M. (1985), A new automatic sampling device for the determination of filtration characteristics and the coupling of an HPLC to fermentors, in: *Modelling and Control of Biotechnical Processes*, pp. 55–60 (HALME, A., Ed.), *1st IFAC* (International Federation of Automatic Control) *Symposium, Noordwijkerhout*, Dec. 11–13.

LINARES, P., LUQUE DE CASTRO, M. D., VALCÁRCEL, M. (1987), Sequential determination of glucose and fructose in foods by flow-injection analysis with immobilized enzymes, *Anal. Chim. Acta* **202**, 199–205.

LORENZ, T., SCHMIDT, W., SCHÜGERL, K. (1987), Sampling devices in fermentation technology: a review, *Chem. Eng. J.* **35**, B15–B22.

LURIE, I. S. (1988), On-line coupled HPLC-capillary SFC, *LC-GC-International* **2** (1), 39–40.

LÜBBERT, A., HITZMANN, B., KRACKE-HELM, H.-A., FRÜH, K., SCHÜGERL, K. (1988), On experiences with expert systems in the control of bioreactors, in: *Fourth Int. Congr. Computer Applications in Fermentation Technology: Modelling and Control of Biotechnical Processes*, Sept. 25–29. University of Cambridge, UK, Society of Chemical Industry.

MANDELIUS, C. F.. DANIELSSON, B., MATTIASSON, B. (1984), Evaluation of a dialysis probe for continuous sampling in fermentors and in complex media, *Anal. Chim. Acta* **163**, 135–141.

MANTSCH, H., CASAL, H. L. (1986), Biological applications of infrared spectrometry, *Fresenius Z. Anal. Chem.* **324**, 655–661.

MARKIDES, K. E., LEE, E. D., BOLICK, R., LEWE, M. L. (1986), *Anal. Chem.* **58**, 740–743.

MARTÍN, J. F. (1977), Control of antibiotic synthesis by phosphate, *Adv. Biochem. Eng.* **6**, 105–127.

MARTÍN, J. F., DEMAIN, A. L. (1980), Control of antibiotic biosynthesis, *Microb. Rev.*, June, 230–251.

MARTÍN, J. F., LIRAS, P. (1981), Biosynthetic pathways of secondary metabolites in industrial microorganisms, in: *Biotechnology*, Vol. 1, pp. 211–277 (REHM, H. J., REED, G., Eds.), Weinheim-Deerfield Beach, FL – Basel: Verlag Chemie.

MATHERS, J. J., DINWOODIE, R. C., TALAROWICH, M., MEHNERT, D. W. (1986), Computer-linked HPLC system for feedback control of fermentation substrates, *Biotechnol. Lett.* **8**, 311–314.

MATSUMURA, M., IMANAKA, T., YOSHIDA, T., TAGUCHI, H. (1978), Effect of glucose and methionine consumption rates on cephalosporin C production by *Cephalosporium acremonium, J. Ferment. Technol.* **56**, 345–353.

MATSUMURA, M., IMANAKA, I., YOSHIDA, T., TAGUCHI, H. (1980), Regulation of cephalosporin C production by endogenous methionine in *Cephalosporium acremonium, J. Ferment. Technol.* **58**, 205–214.

MATTIASSON, B., LARSSON, K. M. (1987), Flow injection enzyme immuno assay – a quick and convenient binding assay, in: *Proc. 4th Eur. Congr. Biotechnology, Amsterdam*, Vol. 4, pp. 517–522 (NEISSEL, O. M., VAN DER MEER, R. R., KUYBEN, K. Ch. A. M., Eds.). Amsterdam: Elsevier Science Publishers BV.

MEINERS, M., SCHALLER, K. (1986), Einsatz von Auto-Analysatoren zur Prozeßanalytik in der Biotechnologie, *BTF-Biotech-Forum* **3**, 193–202.

MEYER, V. (1988), *Praxis der Hochleistungs-Flüssigchromatographie*, 5th Ed. Frankfurt: Verlag Moritz Diesterweg; Aarau–Frankfurt–Salzburg: Verlag Sauerländer.

MIZUTANI, S., IIJIMA, S., MORIKAVA, M., SHIMIZU, K., MATSUBARA, M., OGAWA, Y., IZUMI, R., MATSUMOTO, K., KOBAYASHI, T. (1987), On-line control of glucose concentration using an automatic glucose analyser, *J. Ferment. Technol.* **65**, 325–331.

MONSEUR, X., MOTTE, J. C. (1988), On-line monitoring during fermentation of whey based on ultrafiltration and liquid chromatography, *Anal. Chim. Acta* **204**, 127–134.

MÜLLER, J. (1982), FIA eine neue Analysenmethode, *Labor Praxis* **6**, Heft 4.

MÖLLER, J. (1987), Penicillin-Produktion mit einem Hochleistungsstamm von *Penicillium chrysogenum* – chromatographische Methoden zur Prozeßkontrolle, *Dissertation*, Universität Hannover, FRG.

MÖLLER, J., HIDDESSEN, R., NIEHOFF, J., SCHÜGERL, K. (1986), On-line high-performance liquid chromatography for monitoring fermentation processes for penicillin production, *Anal. Chim. Acta* **190**, 195–203.

NALBACH, U., SCHIEMENZ, H., STAMM, W. W., HUMMEL, W., KULA, M. R. (1988), On-line flow-injection monitoring of enzyme inductor L-phenylalanine in the continuous cultivation of *Rhodococcus* sp. M4, *Anal. Chim. Acta* **213**, 55–60.

NIEHOFF, J. (1983), Erweiterung der Analytik und Messung der Sauerstoffverbrauchsrate bei der Fermentation von *Penicillium chrysogenum, Diploma Thesis*, Universität Hannover, FRG.

NIEHOFF, J., MÖLLER, J., HIDDESSEN, R., SCHÜ-

GERL, K. (1986), The use of an automatic on-line system for monitoring penicillin cultivation in a bubble-column loop reactor, *Anal. Chim. Acta* **190**, 205–212.

NIKOLAJSEN, K., NIELSEN, J., VILLADSEN, J. (1988), In-line flow injection analysis for monitoring lactic acid fermentations, *Anal. Chim. Acta* **209**, 139–145.

OH, T. K., KIM, S. H. (1986), Determination of cellobiohydrolase from culture filtrate of *Trichoderma viride* by the method of single immunodiffusion and enzyme-linked immunosorbent assay, *Biotechnol. Lett.* **8**, 403–406.

OLSSON, B. (1988), A flow-injection system based on immobilized penicillinase and a linear pH-buffer for potentiometric determination of penicillin, *Anal. Chim. Acta* **209**, 123–133.

PAIGEN, K., WILLIAMS, B. (1970), Catabolite repression and other control mechanisms in carbohydrate utilization, *Adv. Microb. Physiol.* **4**, 251.

PEDERSEN, K. M., HAAGENSEN, P., KÜMMEL, M., SOEBERG, H. (1985), Automated measuring system for biological wastewater plant with biological removal of phosphate, in: *Modelling and Control of Biotechnological Processes, 1st IFAC Symposium, Noordwijkerhout*, Dec. 11–13, pp. 101–106 (JOHNSON, A., Ed.). Oxford–New York–Toronto–Sydney–Frankfurt: Pergamon Press.

PETERSSON, B. A. (1988a), Amperometric assay of glucose and lactic acid by flow injection analysis, *Anal. Chim. Acta* **209**, 231–237.

PETERSSON, B. A. (1988b), Enzymatic determination of urea in undiluted whole blood by flow injection analysis using an ammonium ion-selective electrode, *Anal. Chim. Acta* **209**, 239–248.

PLANT, A. L., LOCASCIO-BROWN, L., BRIZGYS, M. V., DURST, R. A. (1988), Liposome enhanced flow injection immunoanalysis, *Bio/Technology*, March, 266–269.

QUEENER, S. W., WILKERSON, S., TUNIN, D. R., MCDERMOTT, J. P., CHAPMAN, J. L., NASH, C., PLATT, C., WESPHELING, J. (1984), Cephalosporin C fermentation: Biochemical and regulatory aspects of sulfur metabolism, in: *Biotechnology of Industrial Antibiotics*, pp. 141–170 (VANDAMME, E. J., Ed.). New York: Marcel Dekker, Inc.

RANDALL, L. G., WAHRHAFTIG, A. L. (1978), Dense gas chromatography/mass spectrometer interface, *Anal. Chem.* **50**, 1703–1705.

RECHNITZ, G. A. (1988), Biosensors, *Chem. Eng. News*, Sept. 5, 24–36.

RECKTENWALD, A., KRONER, K. H., KULA, M. R. (1985a), On-line monitoring in downstream processing by flow injection analysis (FIA), *Enzyme Microbiol. Technol.* **7**, 607–612.

RECKTENWALD, A., KRONER, K. H., KULA, M. R. (1985b), Rapid on-line protein detection in biotechnological processes by flow injection analysis (FIA), *Enzyme Microbiol. Technol.* **7**, 146–149.

RHODES, P. M. (1984), The production of oxytetracycline in chemostat culture, *Biotechnol. Bioeng.* **26**, 382–385.

RICHEY, J. (1983), FPLC: A comprehensive separation technique for biopolymers, *Int. Lab.*, Jan./Febr., 50–75.

ROSS, A. (1986), Probeentnahmesystem zur Ankupplung eines Autoanalysers an einen Fermenter, in: *Technische Membranen in der Biotechnologie* (KULA, M. R., SCHÜGERL, K., WANDREY, C., Eds.), *GBF Monographien*, Vol. 9, pp. 221–224. Weinheim: VCH Verlagsgesellschaft.

ROSS, A., SCHÜGERL, K. (1988), Tetracycline production by *Streptomyces aureofaciens:* the time lag of production, *Appl. Microbiol. Biotechnol.* **29**, 174–180.

RUZICKA, J., HANSEN, E. H. (1988), *Flow Injection Analysis,* 2nd Ed., New York: John Wiley & Sons.

SATOH, I., ARAKAWA, S., OKAMOTO, A. (1988), Flow-injection determination of glutathione with amperometric monitoring of the enzymatic reaction, *Anal. Chim. Acta* **214**, 415–419.

SCHARPÉ, S. L., COOREMAN, W. M., BLOMME, W. J., LAEKEMAN, G. M. (1976), Quantitative enzyme immunoassay: Current status, *Clin. Chem.* **22** (6), 733–738.

SCHEIDEGGER, A., KÜENZI, M. T., FIECHTER, A., NÜESCH, J. (1988), Effect of glucose and oxygen on β-lactam biosynthesis by *Cephalosporium acremonium, J. Biotechnol.* **7**, 131–140.

SCHELTER-GRAF, A., SCHMIDT, H. L., HUCK, H. (1984), Determination of the substrates of dehydrogenases in biological material in flow-injection systems with electrocatalytic NADH oxidation, *Anal. Chim. Acta* **163**, 299–303.

SCHINDLER, J. G., SCHINDLER, M. M. (1983), *Bioelektrochemische Membran-Elektroden.* Berlin: Walter de Gruyter.

SCHMID, R. D., GUIBAULT, G. G., KARUBE, I., SCHMIDT, H. L., WINGARD, L. B. (Eds.) (1987), *Biosensors. International Workshop 1987.* Weinheim: VCH Verlagsgesellschaft.

SCHMIDT, W. J., MEYER, H. D., SCHÜGERL, K., KUHLMANN, W., BELLGARDT, K. H. (1984), On-line analysis of fermentation media, *Anal. Chim. Acta* **163**, 101–109.

SCHMIDT, W. J., KUHLMANN, W., SCHÜGERL, K. (1985), Automated determination of glucose in fermentation broths with *p*-hydroxy-benzoic-acid hydrazide (*p*-HBAH), *Appl. Microbiol. Biotechnol.* **21**, 78–84.

SCHMIDT, W. J., AZZOPARDI, G., GRABOSCH, M., HEROLD, T., SCHÜGERL, K. (1986), On-line analysis and control of production of antibiotics, *Anal. Chim. Acta* **213**, 1–9.

SCHNEIDER, H. J., FÖLDI, P. (1986), Aminosäuren-Analyse, I und II, *GIT Fachz. Lab.* **8**, 783, 873–889.

SCHUURS, A. H. W. M., VAN WEMEN, B. K. (1978), Immunoassay using antigen–enzyme conjugates, *FEBS Lett* **15**, 232–236.

SCHÜGERL, K. (1988), On-line analysis and control of production of antibiotics, *Anal. Chim. Acta* **213**, 1–9.

SHAFER, K. H., GRIFFITS, P. R. (1983), On-line supercritical fluid chromatography/Fourier transform infrared spectrometry, *Anal. Chem.* **55**, 1939–1942.

SKALAR ANALYTIKA, *Brochure*, Breda, The Netherlands.

SMALL, H., STEVENS, T. S., BAUMANN, W. C. (1975), Novel ion-exchange chromatographic method using conductometric detection, *Anal. Chem.* **47**, 1801.

SMITH, A. (1985), Cephalosporins, in: *Comprehensive Biotechnology. The Practice of Biotechnology. Current Commodity Products*, Vol. 3, pp. 163–185 (BLANCH, H. W., DŘEW, S., WANG, D. I. C., Eds.). Oxford: Pergamon Press.

SMITH, Jr., F. C., CHANG, R. C. (1983), *The Practice of Ion Chromatography*, New York: Wiley-Interscience.

SNYDER, L. R., KIRKLAND, J. J. (1979), *Introduction to Modern Liquid Chromatography*, 2nd Ed. New York: John Wiley & Sons.

SONNLEITNER, B., KÄPPELI, O. (1986), Growth of *Saccharomyces cerevisiae* is controlled by its limited respiratory capacity: Formulation and verification of a hypothesis, *Biotechnol. Bioeng.* **28**, 927–937.

STULTS, C. L. M., WADE, A. P., CROUCH, S. R. (1987), Computer-assisted optimization of an immobilized enzyme flow-injection system for the determination of glucose, *Anal. Chim. Acta* **192**, 155–163.

SWINDLEHURST (Koerner), C. A., NIEMAN, T. A. (1988), Flow-injection determination of sugars with immobilized enzyme reactors and chemiluminescence detection, *Anal. Chim. Acta* **205**, 195–205.

SZIGETI, L., PÉCS, M., LOVREZ, G., PUNGOR, Jr., E., NYESTE, L., HOLLÓ, J. (1986), Measurement and control of phosphate concentration of fermentation processes, *Hung. Sci. Instrum.* **61**, 57–64.

TAWFIK, A. M., MARDON, C. J. (1985), Automated analytical method for the determination of individual sugars in mixtures of glucose, fructose and sucrose, *J. Sci. Agric.* **35**, 621–627.

TECATOR AB, *Brochure: FIAstar for Flow Injection Analysis*. P.O. Box 70, S-263 01 Höganäs, Sweden.

THOMPSON, R. Q., KIM, H., MILLER, C. E. (1987), Comparison of immobilized enzyme reactors for flow injection systems, *Anal. Chim. Acta* **192**, 165–172.

TOMLINSON, E., JEFFERIS, T. M., RILEY, C. M. (1978), Ion pair high-performance liquid chromatography, *J. Chromatogr.* **159**, 315–358.

TURNER, A. P. F., KARUBE, I., WILSIN G. S. (Eds.) (1987), *Biosensors: Fundamentals and Applications*. Oxford: Oxford University Press.

VALCARCEL, M., LUQUE DE CASTRO, M. D., LAZARO, F., RIOS, A. (1989), Multiple peak recordings in flow injection analysis, *Anal. Chim. Acta* **216**, 275–288.

VALERO, F., POCH, M., SERRA, A., BARTROLI, J., MEDINA, M. J. (1988), On-line monitoring of glucose fermentation by flow injection analysis, in: *DECHEMA Biotechnology Conferences*, Vol. 2, pp. 103–115. *Lectures at ACHEMA '88*, Weinheim–Basel–Cambridge–New York: VCH.

VORLAENDER, K. O. (1980), *Diagnostik unter Verwendung immunologischer Methoden*. Stuttgart: Georg Thieme Verlag.

VOSER, W. (1966), Verwendung flüssiger Ionenaustauscher in der automatischen Analyse organischer Verbindungen mit dem Technicon Auto-Analyser, in: *Automation in the Analytical Chemistry, Paris*, 2.–4. Nov., pp. 47–52.

WARWICK, T. L. (1985), Culture monitoring system, *Eur. Patent Application*, Appl. No. 85308834.2, Publ. No. 0 184 441, Date filing: 04.12.85 Appl. Eli Lilly & Company, Indianapolis, IN, USA.

WAZAU CO. *Bio-filtrator, Bauart TU Berlin, Brochure*. Dr.-Ing. Georg Wazau, Keplerstr. 12, D-1000 Berlin.

WEISS, J. (1985), *Handbuch der Ionenchromatographie*. Weiterstadt, FRG: Dionex GmbH.

WENTZ, D. (1989), *Dissertation*, Universität Hannover, FRG.

WHITE, R. L., ROBERTS, D. E., ATTRIGE, M. C. (1985), Fourier transform infrared detection of pyruvic acid assimilation by *E. coli*, *Anal. Chem.* **57**, 2487–2491.

WONG, J. S., REIN, A. J., WILKS, D., WILKS, Jr., P. (1984), Infrared spectroscopy of aqueous antibiotic solutions, *Appl. Spectrosc.* **38**, 32–35.

YAO, T., WASA, T. (1988), High-performance liquid chromatographic detection of L- and D-amino acids by use of immobilized enzyme electrodes as detectors, *Anal. Chim. Acta* **209**, 259–264.

YELLOW SPRINGS INSTRUMENTS, YSI 23A *Brochure*.

6 Determination of Cell Concentration and Characterization of Cells

KENNETH F. REARDON

Fort Collins, Colorado 80523, U.S.A.

THOMAS H. SCHEPER

Hannover, Federal Republic of Germany

1 Introduction

Cells can be regarded as the real bioreactors in biotechnological processes. For this reason, an accurate description and analysis of the cell concentration and the state of microorganisms is one of the main interests in biotechnological analysis. With as detailed a knowledge as possible of physiology, metabolism, viability, and morphology, better process optimization can be performed. In particular, on-line and *in situ* sensors or analytical systems for biomass concentration and analysis of cell components and characteristics are urgently needed. In this chapter, several methods for these measurements during cultivation processes will be reviewed to show that the first attempts to devise such on-line analysis systems have been made. Although their application is still in its initial stages, their potential is clear.

2 Biomass Concentration

One of the most important parameters in biotechnology is biomass concentration. Biomass is usually defined as the number or mass of viable cells or as total microbial biomass, although the exact quantity of interest varies with the application. For example, information on the content of viable biomass is valuable for the production of chemicals, while in mixed cultures it is the analysis of the various populations that is important. The exact, continuous, and on-line determination of biomass concentration is one of the "main dreams" in biotechnology.

The most common method to determine biomass concentration is to withdraw samples for gravimetric (cell dry weight) or spectrophotometric (optical density) tests during a cultivation. Despite the fact that these tests are time-consuming and often not very accurate, every biomass sensor will be compared with these methods. Since gravimetric and optical methods are often falsified by non-biomass particles or substances, modifications of these off-line tests are necessary. In particular, chemical methods can be used to determine the viable biomass in the presence of solid particles. These methods include measurement of proteins, DNA, RNA, ATP, or other key cell components such as phospholipids and muramic acid (HUANG et al., 1971; FORSBERG and LAM, 1977; CHAPMAN and ATKINSON, 1977; MOREIRA et al., 1978; HENDY and GRAY, 1979; HYSERT et al., 1979; SOLOMON et al., 1983; KANG et al., 1983; KOLIANDER et al., 1984; TAYA et al., 1986). Most of these methods are time-consuming and cannot be easily automated for quasi on-line process analysis. Spectrophotometric methods are affected by particle size and shape, as well as by background effects (e.g., solid particles, light absorbing products), but can be carried out accurately after proper sample preparation (HONG et al., 1987) or calibration (FINGUERUT et al., 1978). Spectrophotometric techniques can be used for on-line biomass sensors but have the same problems as the related off-line versions.

Coulter counters and flow cytometers are used for the counting and sizing of microorganisms (KUBITSCHEK, 1969; DRAKE and TSUCHIYA, 1973; HATCH et al., 1979). In addition, flow cytometry offers interesting opportunities for the rapid, single-cell determination of cell components and cell viability, as well as the analysis of mixed cultures (SHAPIRO, 1988; HATCH et al., 1979). Exact cell concentration measurements can be performed by analyzing defined sample volumes in a flow cytometer (BATTYE et al., 1985), but problems arise when the whole instrument must be run automatically. Further discussion of the uses of flow cytometry can be found in Sect. 3.

Several new techniques for biomass determination in biotechnology have recently been applied and will have to demonstrate their potential for on-line biomass estimation in the coming years. One of these is nuclear magnetic resonance spectroscopy, a powerful technique for the non-invasive monitoring of many cellular parameters (e.g., intra- and extracellular metabolites, and intracellular pH) (GONZALES-MENDEZ et al., 1982; FERNANDEZ and CLARK, 1987; GILLIES et al., 1989). Since the signals are affected by biomass concentration (GONZALES-MENDEZ et al., 1982), it may be possible to use this technique for biomass determination in special circumstances. Unfortunately, costs for instrumentation, complexity of analytical procedure, and limitation to very small bioreactor systems still limit a broad application of this promising method.

Biomass determination is not only of interest in microbiology and biotechnology. For example, reviews have been published for biomass estimation in the fields of medicine (HEDÉN and Illéni, 1975; COONROD et al., 1983), water and soil technology (JENKINSON and LADD, 1981; KARL, 1986), and the food industry, microbiology, and, of course, biotechnology (FLEISCHAKER et al., 1981; COONEY, 1981; WANG and STEPHANOPOULOS, 1984; HARRIS and KELL, 1985; CLARKE et al., 1986; MERTEN et al., 1986). In this section, primarily on-line sensors and some practical biotechnological applications will be presented. The applications, problems, and limitations of these biomass sensors will be discussed.

2.1 Optical Sensors

Optical sensors appear to be very promising for biomass estimation. Nephelometric and spectrofluorometric methods have produced a wide variety of sensor types for application in biotechnology. These sensors are non-invasive and the response times are nearly instantaneous. The use of glass fiber technology makes these sensors small, robust, and reduces their cost. These advantages have led to the development of a large number of commercially available optical biomass sensors.

2.1.1 Nephelometric Methods

Nephelometric methods monitor the response of turbid samples to light signals. In general, turbidity results from suspended solid particles in the liquid and is largely dependent on the physical size and concentration of the particles. In biological systems, suspended cells, solid particles, and air bubbles are the major causes of turbidity. It is therefore not possible to distinguish between viable and non-viable biomass with nephelometric methods.

Fig. 1 shows the scattering behavior of particles that are of sizes relevant for biotechnolo-

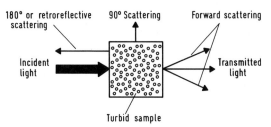

Fig. 1. Scattering behavior of particles with sizes relevant in biotechnology (above 1 μm) (adapted from VANOUS, 1978)

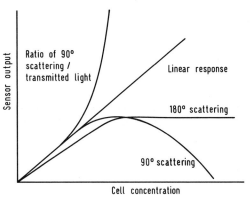

Fig. 2. Different methods for measuring turbidity.

Fig. 3. Sensor signals as a function of cell mass concentration for three different turbidity measurement methods.

gy (VANOUS, 1978). Most of the incident light is scattered in the forward direction. Different nephelometer designs can be used to measure turbidity (Fig. 2). Although it is not the most sensitive for concentration variations, the 90° scattered light is usually measured because it is less sensitive to variations in particle size. The measurement of light scattering at other angles is generally used for particle-size analysis.

Most optical sensors in biotechnology measure transmitted light. The amount of transmitted light is reduced by the light scattering of the particles, and thus the turbidity can be measured at low optical densities. At very high cell densities, both transmitted and scattered light are absorbed inside the measuring volume (Fig. 3). This "inner filter" effect makes all detection designs other than retroreflective monitoring of scattered light ineffective for high biomass concentrations. The accurate use of nephelometric techniques requires the correct application of light scattering analysis (including the limitations of the technique) as well as proper interpretation of the data obtained.

As previously mentioned, the measurement of transmitted light can be used for cell mass estimation at low cell concentrations (below 1 g/L wet weight) (HANCHER et al., 1974). In the region of these low particle concentrations, the Lambert–Beer law is valid:

$$\log I_0/I = k \cdot l \cdot c \tag{1}$$

where l is the length of the light path through the sample and k is the extinction coefficient. The ratio between the incident light (I_0) and the transmitted light (I) is commonly known as the optical density (OD).

The ratio of scattered to transmitted light is also measured occasionally. As shown in Fig. 3, this method becomes ineffective at higher biomass concentrations because the transmitted light intensity decreases rapidly. The measurement of transmitted light is only accurate up to extinction values of about 0.5 (KOCH, 1970). As shown in Fig. 1, most of the incident light is scattered in the forward direction. At higher concentrations, part of the scattered light will be scattered again by other cells and reoriented so that it reaches the detector unit. These effects are responsible for deviations from the Lambert-Beer law (KOCH, 1970). Higher biomass concentrations can be monitored by changing the length of the light path or by diluting the sample. For continuous monitoring, sample dilution would require a waste of fermentation broth or a recycle of diluted broth. In order to avoid this, a side stream from the cultivation can be pumped through a flow cell with a variable light path and recycled to the fermentor without dilution. Fig. 4a shows a schematic diagram of such a flow cell (LEE and LIM, 1980) in which the light path is adjusted by varying the diameter of the water-filled inner tube. With this device (outer diameter (D) 12 mm; inner diameter (d) 8 mm), extinction measurements were directly proportional to dry weight masses up to 2.0 g/L. This also shows the limitations of nephelometric methods to relatively low biomass concentrations.

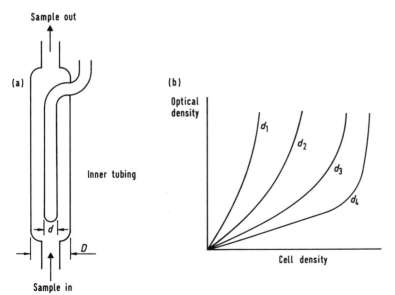

Fig. 4. (a) Schematic diagram of a flow-through cell for measurement of transmitted light at various light paths ($D-d$). (b) Sensor readings as a function of cell density at various diameters ($d_1 < d_2 < d_3 < d_4$) of the inner tubing (adapted from LEE and LIM, 1980).

A considerable disadvantage of transmitted light measurements is that a flow cell designed for high cell concentrations is relatively insensitive to lower biomass concentrations at the beginning of the fermentation (Fig. 4b). This problem can be overcome by varying the light path during the process monitoring.

More sophisticated *in situ* sensors have been constructed for light transmission measurements directly inside the fermentor. Small light sources (e.g., high intensity light-emitting diodes (LED) (LEE, 1981; LIMA FILHO and LEDINGHAM, 1987) or tungsten lamps (OHASHI et al., 1979) were placed into sterilizable probes close to a photodiode. The medium to be analyzed can circulate through the defined light path between light source and detector. Often, fiberoptics are used to guide the incident or transmitted light inside the probe (LEE, 1981).

OHASHI et al. (1979) designed their sensor in such a way that the medium broth flows through the probe as a result of agitation by the fermentor stirrer. On its way through the probe (residence time approximately 1 to 3 min) the medium is degassed before it enters the measuring chamber. With this technique, interference due to air bubbles is minimized. Since the probe construction required to establish the directed flow is complicated, the non-variable light path is relatively large and thus the linear detection range is below 5×10^7 cells/mL.

The probes constructed by LEE (1981) and LIMA FILHO and LEDINGHAM (1987) have smaller light paths that are adjustable below 0.5 cm. Their biomass detection range is linear up to 6 g/L. By chopping the LED light signal, interference from external light sources could be excluded. Problems caused by air bubbles were decreased by placing the sensor at a relatively gas-free position in the fermentor, while problems caused by wall growth were diminished by covering the top of the photodetector unit with thin teflon membranes to provide a permanently non-stick surface.

All transmission sensors can be used in cultivations with higher biomass concentrations, although the sensor readings might then no longer be linear with cell concentration. Suitable calibration measurements are then necessary.

Several companies supply optical sensor systems based on the measurement of light transmission (e.g., Bonnier Technology Group, Monitek, Guided Wavelength, and Hach). Most of these sensors were originally constructed for use in waste water treatment. For use in biotechnological applications, most sensors were reconstructed to be sterilizable and to fit into fermentors via standard electrode ports.

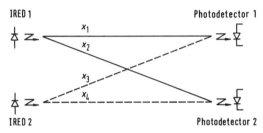

Fig. 5. Four-beam infrared measuring principle (Bonnier Technology off while both photodetectors are operated continuously) (reprinted with permission of Bonnier Technology Group, POB 102068, D-4630 Bochum).

Fig. 5 shows the principle of a four-beam infrared sensor from BTG (Bonnier Technology Group). Two infrared-emitting diodes are used as light sources. They are alternately turned off and on while the two photodetectors are operated continuously. Two signals with different light paths are received by the detectors. These signals are passed through a logarithmic converter to linearize the absorption function. Through the alternating measurement of the signals, several analysis errors derived from wall growth or ageing of the light source or detector can be avoided. METZ (1981) reported applications of this type of sensor to cultivations of *Saccharomyces cerevisiae* and *Tritirachium album*. Both organisms showed different absorption/scattering behavior at the same biomass concentrations because cell size strongly influences scattering behavior. Clearly, all optical sensors must be calibrated to each type of microorganism.

The combination of scattering and transmission measurement achieved in a sensor from Monitek is shown schematically in Fig. 6.

Here, forward-scattered light and transmitted light are measured at the same time. This flow-through sensor must be placed in a fermentor bypass loop. By using the ratio of transmitted to scattered light, the linearity of signal and cell concentration should be better for this class of sensors (VANOUS, 1978).

Another turbidometric sensor is described in detail by HANCHER et al. (1974). In this case, a retroreflective turibidometer unit was used in a flow-through cell coupled to a fermentor. By measuring the back-scattered light, wet biomass concentrations in the range of 1–60 g/L for *Escherichia coli* cultivations could be monitored. The principle of a commercial retroreflective sensor (Aquasant) is shown in Fig. 7.

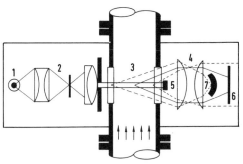

Fig. 6. Monitek turbidity probe in which forward-scattered and transmitted light are measured simultaneously. 1 Light source, 2 optical system, 3 medium, 4 optical system, 5 transmission light detector, 6 scattering light detector, 7 light trap (reprinted with permission of Monitek, Mörsenbroicher Weg 200, D-2000 Düsseldorf 30).

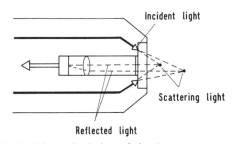

Fig. 7. Schematic design of the Aquasant retroreflective turbidity probe (reprinted with permission of Aquasant Meßtechnik AG, Hauptstraße 20, CH-4416 Bubendorf).

Fiberoptic bundles are used to conduct the light signals, resulting in a small sensor that fits into standard electrode ports of fermentors (MERTEN et al., 1987).

2.1.2 Fluorescence Sensors

Another important group of optical sensors for biomass concentration is on-line, *in situ* fluorosensors. During the last few years, these sensors have gained importance for biotechnological purposes not only for biomass estimation, but also for reactor characterization (EINSELE et al., 1978, BEYELER et al., 1983, GSCHWEND et al., 1983, SCHEPER and SCHÜGERL, 1986a), process monitoring (e.g., metabolic state of the cells, see Sec. 3.6.1), and process control (MEYER and BEYELER, 1984).

The principle of these sensors is based on the fluorescence studies of DUYSENS and AMESZ (1957). They demonstrated that the reduced adenine dinucleotides (NADH and NADPH) in living cells fluoresce at a wavelength of about 460 nm when irradiated with 340–360 nm (UV) light. By measuring this emitted light, the NAD(P)H pool inside living microorganisms can be monitored. The intensity of fluorescence is affected by the amount of viable biomass, the metabolic state of the cells, and also by abiotic factors (e.g., bubbles and fluorescent medium components).

Originally, large, complicated fluorometer devices were interfaced with fermentors. Incident as well as emitted light passed through the fermentor walls via quartz windows (HARRISON and CHANCE, 1970; ZABRISKIE and HUMPHREY, 1978). During the next several years, studies were performed with this type of fluorometer equipment. For example, HARRISON and CHANCE (1970) studied aerobic-anaerobic transitions of cultivations of *Klebsiella aerogenes,* and ZABRISKIE and HUMPHREY (1978) showed that biomass estimation is possible on the basis of culture fluorescence. Studies on feeding strategies for fed-batch cultures of *Candida utilis* were performed with this device by RISTROPH et al. (1977), while EINSELE et al. (1978) used it for substrate uptake and mixing time studies in bioreactors. All of these studies clearly showed the potential of this method.

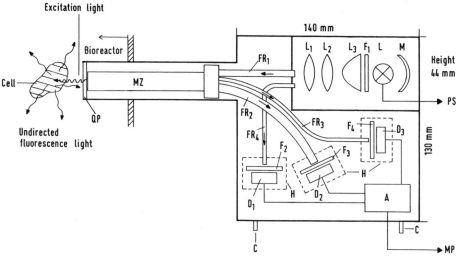

Fig. 8. Schematic design of a fluorescence probe for the simultaneous detection of two different wavelengths. M mirror, L lamp, L_{1-3} lenses, F_{1-4} filters, FR_{1-4} fiber cables, D_{1-3} photodetectors, H housing, MZ mixing zone of fibers, QP quartz plate, A amplifier, C cooling, MP microprocessor unit, PS power supply.

During the next few years, smaller fluorosensors were developed that could be interfaced with a fermentor via a standard electrode port (BEYELER et al., 1981). These sensors are arranged as an open-ended detector to measure the fluorescent light in the backward direction. A schematic design of such a fluorosensor is given in Fig. 8 (SCHEPER and SCHÜGERL, 1986a). A low-pressure mercury lamp is used as UV source, optical filters select the excitation light of 360 nm, and lenses focus this light on a fiberoptic bundle in which the light is guided into the reactor. The 180° fluorescence of the intracellular reduced adenine dinucleotides (460 nm) is collected by the same fiberoptic system. The collecting bundle is split into two fiber bundles that guide the fluorescent light to two different photodiodes. This type of fluorosensor allows the simultaneous measurement of two different wavelengths (e.g., NAD(P)H fluorescence and that from tracers). Two fluorosensor devices are commercially available: BioChem Technology, Malvern, USA (MACBRIDE et al., 1986) and Ingold Meßtechnik AG, Urdorf, Switzerland. Both can only measure the NAD(P)H-dependent culture fluorescence. It is also possible to couple a spectrofluorometer to a fermentor via

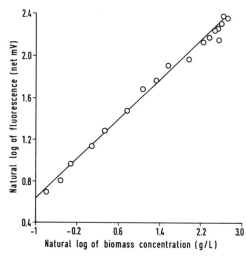

Fig. 9. Linearized biomass and fluorescence data from a cultivation of *Saccharomyces cerevisiae* (adapted from ZABRISKIE and HUMPHREY, 1978).

a fiberoptic cable and collect in-line fluorescence spectra during cultivation processes.

The first application of culture fluorescence monitoring in biotechnology was reported by HARRISON and CHANCE (1970). Although their experimental equipment was complicated and adjusted to one single type of fermentor, they were able to show that the on-line monitoring of biotechnological processes on the basis of the redox state of cells was possible. ZABRISKIE and HUMPHREY (1978) used culture fluorescence as an on-line estimator of viable biomass during cultivation of *Saccharomyces cerevisiae,* a species of *Streptomyces,* and a species of *Thermoactinomyces.* The culture fluorescence and biomass data could be linearized by the following equation (see Fig. 9) (ZABRISKIE and HUMPHREY, 1978):

$$X = [e^{-b} \cdot I_{net}]^{1/a} \qquad (2)$$

During the next few years, several other authors reported on the estimation of biomass concentration from culture fluorescence data (Tab. 1). A strict linear relation between biomass and culture fluorescence was found for the growth of *Zymomonas mobilis* under nonlimited conditions (SCHEPER et al., 1987b). Fig. 10 show the results for cultivation at different glucose levels. An increase of biomass of 1 g/L was followed by an increase of fluorescence intensity of about 21 mV. However, the intracellular metabolism changed completely at glucose concentrations of 15%. Under these conditions, the NAD(P)H values differed totally from those at lower glucose concentrations. A linear increase of culture fluorescence with biomass was also found by LUONG and CARRIER (1986) for cultivations of *Methylomonas mucosa* and by BOYER and HUMPHREY (1988) for cultivations of *Pseudomonas putida.*

Most NAD(P)H-fluorescence studies have been performed in suspended cell culture. DORAN and BAILEY (1987), REARDON et al. (1986), and MÜLLER et al. (1988) showed that the monitoring of culture fluorescence could also be applied to immobilized cell systems. The growth of *Clostridium acetobutylicum* and *Saccharomyces cerevisiae* immobilized in calcium alginate was observed by these investigators. Although no accurate calibration of fluorescence signal and biomass content could be performed, the authors showed that the in-

Tab. 1. Application of Fluorescence Monitoring for Biomass Concentration Estimation

Organism	References
Pseudomonas putida	SAMSON et al. (1987), BOYER and HUMPHREY (1988)
Zymomonas mobilis	SCHEPER et al. (1986b, 1987a)
Sporotrichum thermophile	MEYER et al. (1984)
Bacillus subtilis	MEYER et al. (1984), HEINZLE et al. (1986)
Alcaligenes eutrophus	GROOM et al. (1988)
Escherichia coli	MEYER et al. (1984), SCHEPER et al. (1984, 1987c), GEBAUER et al. (1987), WALKER and DHURJATI (1989)
Clostridum acetobutylicum	SRINIVAS and MUTHARASAN (1987a)
Pediococcus sp.	ARMIGER et al. (1985)
Streptomyces sp.	ZABRISKIE et al. (1975), ZABRISKIE and HUMPHREY (1978)
Thermoactinomyces sp.	ZABRISKIE et al. (1975)
Saccharomyces cerevisiae	ZABRISKIE and HUMPHREY (1978), BEYELER et al. (1981), MEYER and BEYELER (1984), SCHEPER and SCHÜGERL (1986b), ARMIGER et al. (1985), SRINIAVAS and MUTHARASAN (1987b)
Candida tropicalis	BEYELER et al. (1981)
Candida utilis	WATTEEUW et al. (1979), ZABRISKIE (1979)
Penicillium chrysogenum	SCHEPER et al. (1986)
Cephalosporium acremonium	SCHEPER et al. (1987b)
Mucor mucosa	LUONG and CARRIER (1986)
Spodoptera frugiperda	SCHEPER et al. (1987b)
Plant cells	FORRO et al. (1984)
Human melanoma cells	LEIST et al. (1986)
Mouse hybridoma cells	ARMIGER et al. (1985)
Hybridoma cells ATCC HB32	MACMICHAEL et al. (1987)

crease of viable immobilized biomass was followed by an increase of fluorescence of the culture due to NAD(P)H.

The monitoring of viable cell mass in technical media is also possible with this technique (SCHEPER et al., 1987a). Signal disturbances

2.2 Calorimetric Methods

When microorganisms are cultivated, heat is generated. This heat production is a unique parameter that is closely related to metabolic activity, biomass concentration, and substrate

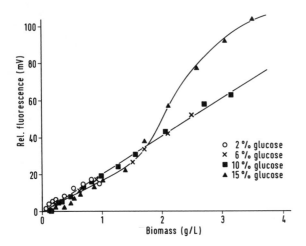

Fig. 10. Culture fluorescence signals as a function of biomass concentration at various glucose concentrations during cultivation of *Zymomonas mobilis*.

can be caused by air bubbles, which lower the sensor readings while displacing cell-containing medium in front of the sensor's observation window. In addition, fluorescent medium components, especially in fed-batch cultivations, affect the signal, as do drastic changes in the metabolic state of the cells. Wall growth on the sensor's observation window has not been found by any author. In general, accurate biomass estimation should only be possible when the NAD(P)H pool per cell is constant during the whole cultivation process. However, under controlled conditions an accurate estimation is still possible even when metabolic changes occur, e.g., for aerobic yeast cultivations, in which the metabolism changes from oxidative-reductive to purely oxidative (ZABRISKIE and HUMPHREY, 1978; see Fig. 9).

The non-invasive monitoring of culture fluorescence offers interesting insights into biotechnological processes. The on-line estimation of viable biomass is only one of them. Although limitations for general usage as an on-line biomass sensor do exist, several applications to cultivation monitoring have shown the high analytical potential of this technique.

consumption via different catabolic pathways (LUONG and VOLESKY, 1983). Different calorimetric devices (external-flow microcalorimeter, twin-type, and heat-flux calorimeter) and different calorimetric techniques (dynamic or continuous calorimetry) have been used to monitor heat evolution during cultivation processes.

Most problems with this technique arise when the heat produced by the microorganisms must be separated from other heat-producing or heat-consuming processes (e.g., aeration, stirring, and addition of nutrients or alkali/acid). An overall heat balance of the whole fermentor/analysis system is necessary (LUONG and VOLESKY, 1983).

In earlier calorimetric studies in biotechnology, small (5–50 mL volume), but well-defined calorimetric devices were used (DERMOUN and BELAICH, 1979; ISHIKAWA et al., 1981; ISHIKAWA and SHODA, 1983). These twin-type heat conduction calorimeters utilize two similar vessels. One is run as a cultivation vessel (with microorganisms) and the other functions as the reference under the same conditions (without microorganisms). This technique makes it possible to measure the heat evolved by the cells without the influence of the aforementioned disturbances. However, sampling is nearly impossible for batch processes and thus large fermentors must be run in parallel as reference.

Calorimetric techniques can be applied to real fermentation processes by recirculating fermentation broth through an external flow microcalorimeter where the evolved heat is monitored (DJAVAN and JAMES, 1980; NICOLS and JAMES, 1981; ORIOL et al., 1987; SAMSON et al., 1987; ROY and SAMSON, 1988). Problems with the flow calorimeter might arise from changes in cellular metabolism while the medium is pumped through the loop, especially if oxygen or nutrients are exhausted during transport. In this case, the broth inside the flow cell might not represent the actual state of cultivation inside the fer-

mentor. Heat losses during pumping and wall growth can also become problems.

Dynamic calorimetry is another technique for performing calorimetric measurements on bench-scale fermentors (MOU and COONEY, 1976; WANG et al., 1978). Here, the heat that is released during cellular metabolism is measured by the increase in temperature of the fermentation broth when the temperature control system is turned off. The heat accumulation is corrected for heat loss and gain in the fermentor system. Although real-time, on-line monitoring is not possible and the fermentation conditions are not strictly isothermic, this simple method has been successfully applied to various bench-scale cultivations (MOU and COONEY, 1976; WANG et al., 1978).

LUONG and VOLESKY (1980, 1982) have described the use of continuous calorimetry to study heat evolution in cultivations in larger

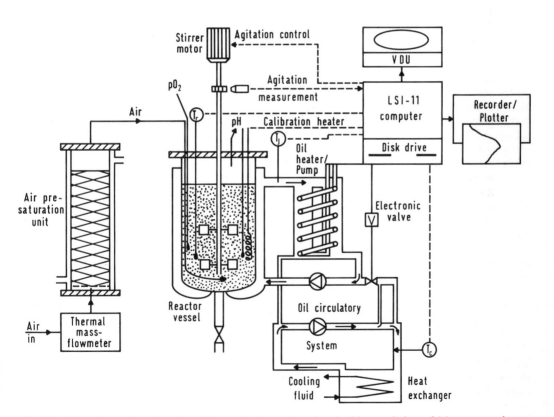

Fig. 11. Schematic design of a "heat flux" calorimeter (reprinted with permission of MARISON and VON STOCKAR, 1986).

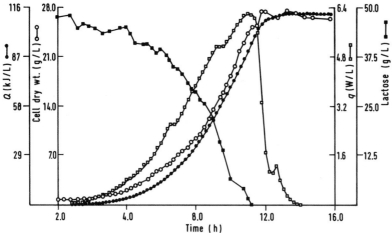

Fig. 12. Thermogram monitored during a cultivation of *Kluyveromyces fragilis* (reprinted with permission of MARISON and VON STOCKAR, 1987).

fermentors (up to 14 L). Here, the fermentation broth was overcooled using a defined cooling water flow rate in the fermentor jacket. Heat produced by the microorganisms and by a separately-controlled electrical immersion heater was responsible for the temperature of the fermentor liquid. The temperature was kept constant by controlling the immersion heater. An overall heat balance was employed to determine the amount of heat evolved by the microorganisms (LUONG and VOLESKY, 1980, 1982; LUONG et al., 1983).

The system shown in Fig. 11 is called a "heat-flux calorimeter" (MARISON and VON STOCKAR, 1986). A 1.8 L fermentor is temperature controlled by a silicone-oil circulation system composed of two different subunits. These are connected via an electronic valve that is controlled by a computer. One subunit is run at a slightly higher temperature (electrically heated) than desired for the cultivation process, while the other subunit is run at a lower temperature. The reactor temperature is monitored continuously and kept constant by computer-controlled mixing of both flows. A rise in the reactor temperature, brought about by heat evolution of the microorganisms, is followed by a temperature decrease in the jacket oil. The temperature difference that is necessary to overcome heat-transfer resistance between the temperature controlling fluid in the jacket and the cultivation broth is used as the measuring signal. Thus, insulation of the complete system and a heat balance are no longer necessary.

Fig. 12 shows that the monitoring of heat evolution can be used for the on-line measurement of biomass concentration (MARISON and VON STOCKAR, 1986, 1987; BIROU et al., 1987; BIROU and VON STOCKAR, 1989). Evolved heat corresponded closely with biomass concentration. However, changes in the metabolic state of the cells caused by medium limitations or diauxies resulted in drastic changes in the thermograms (MARISON and VON STOCKAR, 1987). In such cases, the relationship between biomass and evolved heat is quite different.

2.3 Filtration Methods

It is well known that cells can be separated from the medium by filtration. Thus, biomass concentration can be estimated from the filtration cake volume, the filtation flow rate, the filtrate volume, and the pressure difference across the filter system during the filtration process (SILVENNOINEN and KOIVO, 1982). NESTAAS and WANG (1981) demonstrated that the biomass concentration can be calculated from:

$$X = \frac{1}{v} \cdot \frac{V_c}{V_f + V_c} \tag{3}$$

in which X (g/L) is the biomass concentration, v (L/g) is the specific cake volume, V_c (L) is the cake volume, and V_f (L) is the filtrate volume. The authors showed that an automatically-operated filtration unit could be coupled to a *Penicillium chrysogenum* cultivation for semicontinuous estimates of biomass concentration (NESTAAS and WANG, 1981; NESTAAS et al., 1981). Samples of 50–70 mL were withdrawn from the fermentor, loaded into the filtration probe, and filtered under constant pressure. The filtrate volume could be measured gravimetrically using a load cell, while the cake volume was determined optically (light transmission). The filter cake volume can also be measured by the pressure drop across the system (NESTAAS et al., 1983). The measurement of the filter cake volume as a function of filtration time is therefore possible with this device. When the specific filter cake volume is nearly constant during cultivation, biomass can be measured accurately with this filtration probe. The filtration time for one sample was

105 s (NESTAAS et al., 1981, NESTAAS and WANG, 1983). After filtration, the filtrate and the filter cake volume were recycled to the fermentor in order to minimize the loss of fermentation broth. The probe had to be cleaned after each filtration, resulting in a sampling cycle time of 30 min. A similar device was used by THOMAS et al. (1985) for biomass calculation during cultivation of *P. chrysogenum*. The correlation of off-line data and the automatic filtration unit was very good in the range of 2–40 g/L biomass; however, problems arose at lower biomass concentrations.

In order to overcome difficulties with the backflushing of filter media and with filter contamination, LENZ et al. (1985) constructed a computer-controlled filtration probe with renewable filters. The principle of one analysis cycle is shown in Fig. 13 (REUSS et al., 1987). During sampling, the filtration unit was filled with fermentation broth (approximately 100 mL). The measuring cycle was then started by applying a constant filtration pressure. Five in-

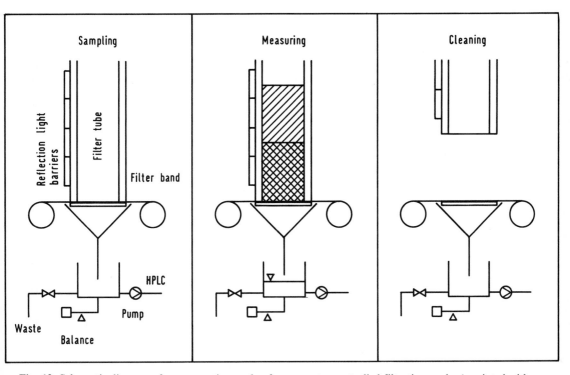

Fig. 13. Schematic diagram of one operation cycle of a computer-controlled filtration probe (reprinted with permission of REUSS et al., 1987a).

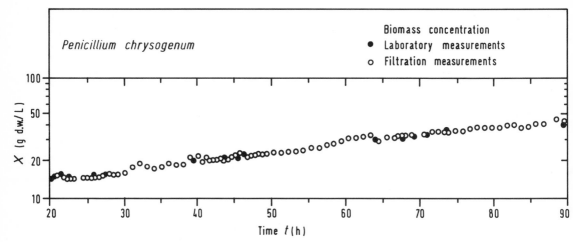

Fig. 14. Comparison of on-line filtration and off-line measured biomass data during cultivation of *Penicillium chrysogenum* (reprinted with permission of REUSS et al., 1987b).

dependent retroflective light barriers were used to monitor the growth of the filter cake, thus increasing the analytical accuracy. After the measuring procedure, the filtration unit was cleaned with water and dried with air while the filter was renewed. Fig. 14 shows that the off-line gravimetric data correlated very well with data of the filtration probe (REUSS, 1987). In addition, the filtrate could be used for analysis of the cell-free medium. Although filtration probes are relatively complex devices and exact on-line monitoring of biomass concentration is not possible with them, the technique could be successfully applied to bioprocess monitoring during cultivations of *P. chrysogenum* (NESTAAS and WANG, 1981, 1983; NESTAAS et al., 1981; LENZ et al., 1985; THOMAS et al., 1985; REUSS, 1987; REUSS et al., 1987) and to the Pekilo process (SILVENNOINEN and KOIVO, 1982).

2.4 Viscosity Methods

The viscosity of a fermentation broth is affected by biomass concentration, among other factors. Therefore, the cell concentration can be derived from viscosity measurements when no other influence such as extracellular material, cell morphology changes, or temperature variations complicate this process (SHIMMONS

et al., 1976). PERLEY et al. (1979) showed that a capillary-type viscosimeter in a continuous sample flow from the fermentor could be used for monitoring broth viscosity. The viscosity of cell suspensions from cultivations of *Hansenula polymorpha* increased in a non-linear manner with cell mass concentration. The effect of cell mass on broth viscosity was small, and therefore biomass estimates from viscosity measurements were rather inaccurate. The application of an in-line rotational viscometer for baker's yeast (SHIMMONS et al., 1976) showed better results, although these experiments were performed using yeast cells suspended in buffer solution. No real data on growth have been supplied.

2.5 Electrochemical Methods

Different electrochemical methods can be used for the estimation of biomass concentration (RAMSAY et al., 1985). Here, four of these methods for on-line analysis will be reviewed.

2.5.1 Impedimetric Methods

When a sinusoidal electrical field is applied to two electrodes between which a cell-containing sample is placed, the impedance of the biological sample can be measured. Two parameters, ε, the electrical permittivity (for capacitance) and σ, the conductivity (for conductance) are necessary to describe the electrical properties of this sample. Both are affected by the biomass concentration. In a carefully temperature-controlled system under defined growth conditions (e.g., medium composition), the measurement of the conductivity can be used for biomass monitoring. Since instrumentation (e.g., from Malthus Instruments Ltd, Bactomatic, or Bactobridge) is often quite expensive, these methods are used in the routine control of microbial contamination for large numbers of off-line samples for clinical or food industry analyses (CADY, 1975; RAMSAY et al., 1985). Only a small number of reports on the analysis of higher biomass concentrations in these commercially available instruments have been published. One of these reports (MUNOZ and SILVERMAN, 1979) contains data on measurements of *Escherichia coli* concentrations in sewage effluents in the range of 10^6 to 10^7 cells per mL.

Routine conductivity meters can also be used for the determination of biomass. An example is given by TAYA et al. (1989) for plant tissue cultures. A linear relation between conductivity and cell mass was found for biomass concentrations up to 10 g/L for *Coffea arabica, Nicotiana tabacum, Withania sonnifera,* and *Catharanthus roseus.* In addition, BLUTE et al. (1988) used conductivity measurements for on-line cell mass estimation in hollow-fiber bioreactors.

The measurement of the frequency-dependent permittivity seems to be more promising for biotechnological purposes. Simple devices for off-line analysis have shown the potential of this technique for monitoring biotechnological processes (GENCER and MUTHARASAN, 1979). HARRIS et al. (1987) clearly showed that the permittivity at low radio frequencies (below 1 GHz) has a linear correlation with biomass concentrations in cell suspensions. A small probe suitable for use in standard electrode ports was used to measure the permittivity and conductivity in fermentation broths. KELL (1987) and HARRIS et al. (1987) reported a good linear correlation between capacitance and biomass in yeast cultures up to 100 g/L (at

300 kHz frequency). A version of this type of *in situ*, real-time biomass sensor is commercially available (Aber Instruments Ltd); this device measures the radio frequency (100 kHz–10 MHz), conductance and capacitance during cultivation processes.

2.5.2 Potentiometric Sensors

This kind of sensor measures the potential that is developed at an electrode in a cell containing a sample with respect to a reference electrode. The growth of microorganisms and the associated production of electroactive substances cause changes in the potential. WILKINS et al. (1974) described such a biomass sensor based on the electrochemical detection of hydrogen evolved by the microorganisms. This test system consisted of a platinum electrode and a reference electrode. The hydrogen evolved was detected by a voltage increase in the cathodic direction. WILKINS (1978) and WILKINS et al. (1978) showed that a wide variety of microorganisms could be measured using this method. It is very sensitive: one cell per mL could be detected (with a detection time of 9 h) (WILKINS, 1978). However, these sensors appear to be restricted to relatively low biomass concentrations, and a long analysis time is necessary (e.g., for Gram-negative cells, 10^6 cells/mL could be measured in 2 h) (WILKINS, 1978; WILKINS et al., 1978; JUNTER et al., 1980). Microorganisms as well as electroactive substances affect the sensor readings.

2.5.3 Fuel Cell Type Sensors

The principle of this method is shown in Fig. 15 (MATSUNGA et al., 1979; BENNETTO et al., 1987). Two fuel cells are inserted into the cell containing sample. The anodes of both cells are in contact with the sample. The current produced in probe I results from the oxidation of microorganisms and electroactive substances, while the current of reference probe II results only from the oxidation of electroactive substances, because microorganisms cannot permeate through the dialysis membrane (MATSUNGA et al., 1979). Thus,

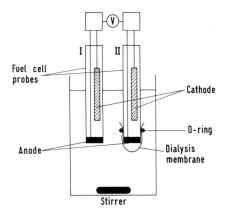

Fig. 15. Schematic design of a fuel-cell type biomass sensor (adapted from MATSUNGA et al., 1979).

the current difference between both probes is a function of biomass concentration. MATSUNGA et al. (1979) reported detection times for this sensor in the range of 15 to 20 min, limited by diffusion through the dialysis membrane. They applied their sensor successfully to batch cultivations of *Saccharomyces cerevisiae* and *Lactobacillus fermentum*. The linear detection range was 10^7 to 4×10^8 cells/mL for *S. cerevisiae* and 10^8 to 4×10^9 cells/mL for *L. fermentum*.

The signals can be drastically enhanced by the addition of redox mediators. These substances mediate electron transfer between electron donors in the cell membrane and the platinum anode. Mediators such as 2,6-dichlorophenolindophenol (Fig. 16; NISHIKAWA et al., 1982) or phenazine ethosulfate (TURNER et al., 1982, RAMSAY et al., 1985) were used. The sensitivity of this sensor system differs for various types of microorganisms.

2.5.4 Amperometric Sensors

These sensors monitor the current produced during the reaction of substances to be analyzed (e.g., electroactive substances, microorganisms). Such systems can be made up of two electrodes (working and counter electrodes) or three electrodes (working, reference, and counter electrodes). In both cases, the working electrode is controlled at a pre-set potential (MATSUNGA et al., 1980). In the simpler two-electrode system, it is difficult to control the potential at the working electrode. However, several applications (RAMSAY et al., 1985, RAMSAY and TURNER, 1988) have shown that these sensors can accurately measure *E. coli* concentrations in a range of 5×10^6 to 9×10^7 cells/mL.

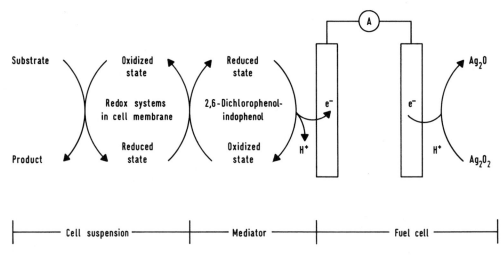

Fig. 16. Principle of a mediated electron transfer between the redox systems in cell membranes and a fuel-cell type sensor (adapted from NISHIKAWA et al., 1982).

MATSUNGA et al. (1980) presented a three-electrode system for estimating biomass concentration in cultivation of *Bacillus subtilis;* SAKATO et al. (1981) used a similar device to monitor the growth of *Micromonospora olivosterospora.* Both authors used two sets of probes (each containing a working, counter, and reference electrode). One set was covered with a dialysis membrane to measure the cell-free medium as a reference. The measuring times were in the range of 5 to 10 min. Again, mediators increased the sensitivity of this amperometric method (RAMSAY et al., 1985, RAMSAY and TURNER, 1988).

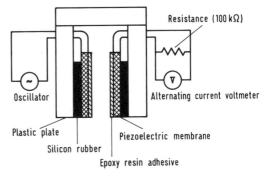

Fig. 17. Schematic design of a piezoelectric biomass sensor device (adapted from ISHIMORI et al., 1981).

2.6 Acoustic Methods

The estimation of biomass concentrations by measurements of acoustic propagation in cell-containing media is another non-invasive technique (CLARKE et al., 1987). ISHIMORI et al. (1981) reported the use of piezoelectric sensor devices for biomass estimation in populations of *S. cerevisiae, Bacillus subtilis,* and *Klebsiella* sp. The device used in these experiments is shown schematically in Fig. 17. The sensor was run with an ultrasonic frequency of 40 kHz and the output voltage was a function of cell concentration. The effect of buffer concentration, pH, and the medium composition on the sensor signal was low. Only changes in the compressibility of the sample affected the voltage output strongly; this parameter is a function of cell growth. Continuous monitoring of the biomass concentration during growth of *S. cerevisiae* in a molasses-containing medium was possible (up to 2×10^8 cells/mL).

BLAKE-COLEMAN et al. (1984, 1986) described the principles of acoustic resonance densitometry (ARD) for estimating biomass concentration. In this method, the specific gravity or relative density of the cell-containing samples is determined with high resolution by a stable, direct-measuring technique. Mathematically, the sample is regarded as a spring-mass system. It is injected into a measuring cell in a closed oscillatory circuit. Since the spring coefficient is considered to be constant, the frequency of a controlled, sustained resonance behavior of the sample is mass dependent. Thus, this frequency can be used to monitor the specific mass of the sample.

BLAKE-COLEMAN et al. (1984, 1986) applied this technique to monitor cultivations of *Erwinia chrysanthemi* and *Escherichia coli* and found that the relative density was linear up to optical densities of 540 (measured at 600 nm). Gas bubbles and foam affect the ARD measurements, so on-line biomass estimation must be done in an external flow loop. The on-line monitoring by this technique of cell mass in mammalian cell cultures is reported by KILBURN et al. (1989). A good correlation between cell mass and ARD data was found, except when changes in the average cell size occurred during growth.

3 Cell Components and Characteristics

Nearly all currently available bioreactor instruments measure physical (e.g., temperature and pH) or chemical (e.g., dissolved oxygen and carbon source concentrations) properties of the culture medium. While these parameters certainly reflect the presence and activity of the cells in the bioreactor, they provide no direct information on the state of the microorganisms. Such information is crucial for several reasons, the most important of which are process development and control. These considerations are especially significant for cultivations of genetically modified cells, in which the product often accumulates inside the organism. Although off-line techniques are, in many cases, readily available, they are far less desirable for process development because the measured quantities are not necessarily those that would be found *in situ*. The metabolism and physiology of microorganisms have been observed to change very rapidly in response to alterations in the cellular environment. The use of off-line measurements is nearly worthless for process control, since the time scale of the changes occurring in the bioreactor, and hence the required time scale of the control response, is often much shorter than the time needed for analysis.

Unfortunately, very few techniques or sensors for truly on-line measurement exist, and even fewer are commercially available. Thus, the scope of this section has been expanded to include instruments or methods that are quasi-on-line (i.e., directly interfaced with the bioreactor but providing non-continuous data) or that seem applicable to on-line analysis. In addition, the focus is on cultivations of suspended cells, primarily because most work has involved such systems.

As a rapid perusal of the following pages reveals, three instrument systems are capable of measuring many cellular characteristics of interest: flow cytometry, nuclear magnetic resonance spectroscopy, and systems based on sample-flow analysis (flow injection analytical devices and autoanalyzers). Therefore, it is useful to give a brief overview of each of these techniques before discussing the measurement of individual cellular quantities.

Flow Cytometry

Flow cytometry is a technique by which various characteristics of individual cells can be analyzed at an extremely high rate, up to 10 000 cells per second. Thus, one can rapidly obtain information on the distribution of a particular cell parameter in a population. These distributions reveal much more than data of average values and are especially useful in understanding how different subpopulations

contribute to the overall behavior of a cultivation. In addition, it is possible to sort the sample on the basis of the measured parameter to acquire relatively homogeneous cell populations.

The sample stream that enters the flow cytometer must be a suspension of single cells that contains as little debris as possible. Both of these requirements make the direct interface of a flow cytometer with a fermentor somewhat problematic, although suitable pre-treatment steps can be envisioned. In a typical cytometer (Fig. 18), the cell suspension is introduced into the center of a flow of sheath fluid in such a way that the cells flow one at a time past a focused light source, usually a laser or mercury arc lamp. The light beam that hits a cell is scattered and absorbed, and, if the cell has been stained, it is re-emitted as fluorescent light. The resulting light signals are collected by various lenses and measured with photomultipliers or photodiodes, from which the output can be analyzed and processed.

Flow cytometry has been used for many years in medicine and biology, and many different cellular parameters have been measured (Tab. 2). Although most of these assays require treatment with dyes and other reagents and add some complications to automated bioreactor analysis, the wide spectrum of detailed measurements possible makes flow cytometry a potentially powerful quasi-on-line technique. Details on instruments, methods, and data analysis have been presented in a number of reports (e.g., HORAN and WHEELESS, 1977; KRUTH, 1982; STEINKAMP, 1984) and books, including an excellent text by SHAPIRO (1988).

Nuclear Magnetic Resonance Spectroscopy

The basis of nuclear magnetic resonance (NMR) spectroscopy is the resonance of certain atomic nuclei when placed in a high-strength magnetic field and exposed to high-frequency electromagnetic radiation. Only those nuclei possessing an inherent magnetic moment can be studied. The resonance frequency is a function of the particular nucleus

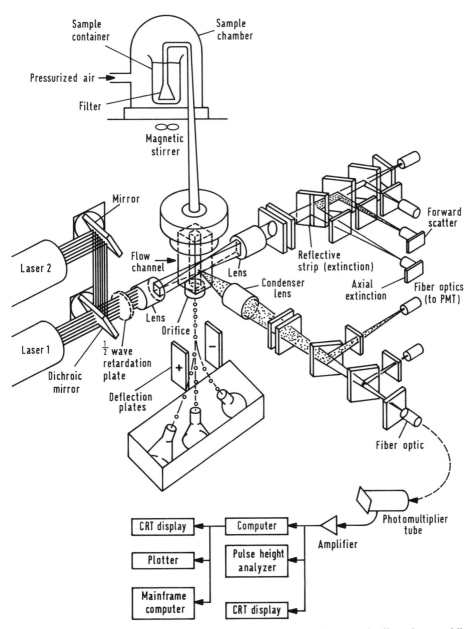

Fig. 18. Schematic diagram of a typical flow cytometer with two lasers and cell-sorting capability (reprinted with permission from BAILEY and OLLIS, 1986).

Tab. 2. Some Cellular Parameters Measurable by Flow Cytometry (adapted from SHAPIRO, 1988)

Components	Characteristics
DNA content	Cell size
DNA base ratio	Cell shape
RNA content	Cytoplasmic granularity
Total protein	Membrane integrity
Basic protein	Membrane fluidity
Pigment content	Membrane potential
Antigens	Surface charge
Surface sugars	Intracellular pH
Surface receptors	Intracellular redox state
Intracellular receptors	
Enzyme (activity)	
Lipid content	

Tab. 3. Some Cellular Parameters Measurable by NMR (adapted from HARRIS, 1986)

Nucleus	Natural Abundance (%)	Relative Receptivity	Applications
^{1}H	99.985	1.00	Metabolic intermediates
^{13}C	1.108	1.76×10^{-4}	Metabolic intermediates
^{14}N	99.63	0.001	NH_3; amino acids; urea, protein turnover; nitrogen assimilation; intracellular pH
^{15}N	0.37	3.85×10^{-6}	
^{19}F	100	0.834	Intracellular Ca^{2+}, Zn^{2+}, and pH
^{23}Na	100	0.0927	Intracellular Na^+
^{31}P	100	0.0665	ATP; ADP; NAD(P)H; intracellular pH and Mg^{2+}; phosphorylated intermediates
^{35}Cl	75.53	3.56×10^{-3}	Intracellular Cl^-
^{39}K	93.1	4.75×10^{-4}	Intracellular K^+

as well as the local magnetic field to which the nucleus is exposed. Thus, the phosphorus in ATP can be distinguished from that in NADH. The frequency spectra are usually expressed relative to the frequency of a reference compound by a quantity called the chemical shift. NMR spectra contain a great deal of information, and the technique has been used with success for many years on solids, including tissue samples.

Tab. 3 presents some of the nuclei that have been used in NMR experiments on biological systems. From all of the listed components, ^{31}P is most commonly used for *in vivo* investigations, because its spectra are simpler and yet still able to provide information on a number of important phosphate compounds (Fig. 19). In addition, ^{31}P has the advantage of being the natural isotope.

Since NMR spectroscopy is a non-invasive, non-destructive technique, it is a powerful method to study fundamental physiological and metabolic phenomena *in vivo*. The major drawback of NMR analysis of microorganisms is the low sensitivity of the technique, requiring the use of high-density samples. This can lead to further problems with maintenance of cell viability. However, the power of this tool

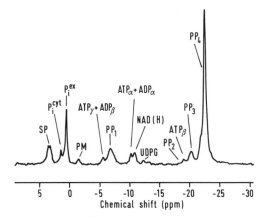

Fig. 19. Typical ^{31}P-NMR spectrum. SP sugar phosphates, P_i inorganic phosphate, PM phosphomannan, PP polyphosphate compounds, UDPG uridine-diphosphoglucose (reprinted with permission from SHANKS and BAILEY, 1988).

has inspired a great deal of work in recent years, covering both applications (Tab. 3) and suitable perfusion and immobilized-cell reactors (DRURY, 1988; FERNANDEZ et al., 1988). A number of reviews of NMR applications and techniques have been published (ROBERTS and JARDETZKY, 1981; GADIAN, 1982; KANAMORI and ROBERTS, 1984; GUPTA et al., 1984).

Sample-Flow Analysis

This term refers not to an instrument but rather to a type of analytical system. Many detectors cannot be utilized directly in a fermentor vessel because of probe fouling, chemical interference of medium components, electrical interference, calibration requirements, and because they cannot be sterilized. Other analytical techniques may require the addition of reagents to the sample. All of these problems can, in theory, be overcome by analyzing a sample flow withdrawn from the bioreactor.

This type of system offers several advantages. Not only are the previously mentioned problems avoided, but calibration and probe maintenance are greatly simplified. There are also some potential problems associated with the use of these systems, among them a higher risk of culture contamination and the possibility that the properties of the sample stream may change if too much time is required for transport from the fermentor to the sensor. These problems are by no means without solution, since it should be possible to design systems that could overcome them.

There are several types of flow-sampling techniques, including continuously sampling flow systems (autoanalyzers) and flow-injection analysis (FIA) systems. In the former, a virtually undiluted sample stream flows continuously past the desired sensor. In this case, the sensor is exposed to the same, possibly interfering, chemical environment that it would face in the fermentor. In addition, there could be problems with the volume of the sample stream, since it most likely could not be returned to the fermentor. In contrast, flow injection analysis involves the periodic injection of small quantities of culture medium into a capillary flow stream (Fig. 20). Reagents are usually added to this stream, and thus the en-

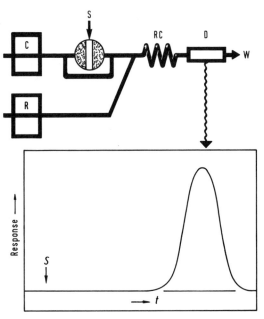

Fig. 20. Top: Scheme of a two-line injection system. C carrier, S sample, R reagent, RC reaction coil, D detector, W waste.
Bottom: Typical detector readout. "S" denotes the sample injection (adapted from RUZICKA and HANSEN, 1988).

vironment to which the sensor is exposed can be manipulated as desired. The main disadvantage of this method is the requirement for rapid and sensitive sensor response and the discontinuous output of data. Still, the large number of applications to the analysis of culture broth reported in the literature (see Sect. 2.2) attest to the immense usefulness of these quasi-on-line FIA systems. To date, relatively few reports of the analysis of cell constituents or characteristics have been published, but there are no significant obstacles to such applications. Further details of FIA and other flow-sampling systems are provided in several review articles (CLARKE et al., 1985; HANSEN, 1988; RUZICKA and HANSEN, 1988).

In the following sections, reported applications of these and other techniques for the on-line, quasi-on-line, and potentially (quasi-)on-line analysis of a variety of cellular parameters are summarized. The parameters that are discussed are clearly not the only measurable ones (see, for example, Tabs. 2 and 3), but are those of greatest general interest.

3.1 Cell Morphology

The morphological characteristic most often of interest in cell cultivation is the cell size, but cell shape determination has also proven useful. Although these parameters may be investigated with a microscope, several instruments are available for rapid analysis, and cell size distribution can also be obtained by some of them. The two mostly used instruments are the Coulter counter and the flow cytometer.

3.1.1 Coulter Counter

The ability of the Coulter counter to measure cell size is based on the disturbance of an electric field as it is traversed by small particles, in this case microorganisms. The cells must be suspended in an electrically conducting fluid; most cultivation media would suffice, although dilution may be required. This suspension is then drawn through a small orifice, across which an electric field has been applied. The resistance of the orifice increases when a (relative to the liquid) low-conductivity

cell crosses the field, and thus the voltage drop across the aperture also increases. The increase of the voltage drop is generally proportional to the fraction of the volume of the orifice occupied by the cell. The detection limit of the commercially available instrument is about 0.5 µm, but modifications have led to detection limits as low as 0.09 µm (DeBLOIS and BEAU, 1970). Reviews of the instrument and technique are available (KUBITSCHEK, 1969; HARRIS and KELL, 1985).

The Coulter counter technique suffers from several limitations, one of which is that cells of all shapes are sized as spheres. Other problems include false measurements caused by the simultaneous passage of two or more cells through the orifice and the inability to discriminate between viable cells, non-viable cells, and contaminating particulates. A more subtle problem that is often overlooked concerns the magnitude of the applied electric field. Above a certain critical value, dielectric breakdown of the cell membrane takes place, with the result that the measured cell size increases with increasing field strength (JELTSCH and ZIMMERMANN, 1979). Since most Coulter counter instruments are capable of exceeding this critical field strength, the actual cell size can easily be underestimated. However, when adequate caution is used, these problems can usually be minimized or eliminated.

There have been many reports on the use of the Coulter counter, but most of them involve counting rather than sizing of particles. Size determination is of interest because this parameter can be related to the cell cycle or to an age distribution, both of which are needed to understand the biological aspects of the culture. For example, ALBERGHINA et al. (1983) used Coulter counter measurements to study the cell cycle of the budding yeast *S. cerevisiae*. In another investigation, the size-discriminating capability of the instrument was used to distinguish and enumerate the different populations in a cultivation of *Tetrahymena pyriformis* growing on *E. coli* and *Azobacter vinelandii* (DRAKE and TSUCHIYA, 1973).

To date, no report of the quasi-on-line use of a Coulter counter for monitoring of cultures has been published, although it seems possible to use a modified Coulter instrument for this purpose.

3.1.2 Flow Cytometry

Two types of flow cytometers have been utilized for cell-size measurements. In a single-beam instrument, the intensity of narrow-angle forward-scattered light resulting from the interaction of a cell with the focused light beam can be related to the size of the cell, although the difference between the refractive indices of the cell and the surrounding fluid also plays a role. Typically, the refractive indices remain constant, and the value of the forward-scatter intensity can be used as a measure of cell size. There can be significant problems with particulate matter in the sample fluid, and careful tuning is necessary to measure size distributions in samples of small microorganisms such as bacteria.

The second type of instrument utilizes a split beam configuration. The two laser beams are tightly focused at separate spots, allowing calculation of the cell flow velocity. This velocity measurement can be combined with measurements of absorption (for cells larger than 2 µm) or 90° light scatter (for bacteria or other cells smaller than 2 µm).

The sensitivity of both types of flow cytometer can be improved by labelling the cells with a fluorescent dye, either a general label or one which labels only viable cells (see Sect. 3.2).

Flow cytometry has been used to study cell size for several applications. Batch and continuous cultivations of *S. cerevisiae* have been investigated with this technique to study the effects of dilution rate on cell size and protein content (RANZI et al., 1986). SCHEPER et al. (1987e) studied a different strain of the same species in oscillating continuous cultures and in cultivations employing cell recycle.

Size measurements have also been performed on cultivations of recombinant *E. coli* (DENNIS et al., 1983, 1985; SCHEPER et al., 1984, 1987d). These studies utilized the antibiotic resistance of the genetically modified cells to estimate the fraction of plasmid-containing cells. The addition of antibiotic to the medium blocked cell wall formation in plasmid-free cells and caused them to elongate, but had no effect on plasmid-containing cells. The size differences between the two populations could clearly be seen in the flow cytometer results.

WITTRUP et al. (1988) used flow cytometry to investigate the morphological changes occurring in several strains of *E. coli* that formed inclusion bodies of plasmid-encoded proteins. Both forward-angle (cell size) and right-angle (internal structure) scatter intensities increased with foreign protein production and inclusion body formation. Thus, quasi-on-line monitoring of light scatter during growth of recombinant cells would be both feasible and informative.

At this time, there have been no such reports on the quasi-on-line use of flow cytometers, although automatic sample injection and calibration have been achieved. No pretreatment steps are necessary for cell size determination, but the presence in the medium of particulates that are of the size range of the cells causes significant error.

3.1.3 Other Instruments

Nephelometric instruments are also able to measure cell-size distribution; these devices have already been discussed with regard to the monitoring of biomass concentration. An example of one of these nephelometers is given by PREIKSCHAT (1987). The instrument utilizes a scanning laser beam and measures the resulting scattered light pulse, the width of which is proportional to the cell size. The maximum counting rate is reported to be 300 000 particles per second, with a size range of 0.5 to 250 µm. The published results were measured off-line, but the author suggests that a flow-through cuvet could be used and that he intends to develop an *in situ* probe.

3.2 Viability

It is clearly important to measure the concentration of viable cells in a fermentor, as these are the microorganisms that actually perform the desired bioconversion. As mentioned in the discussion of biomass sensors, most methods in use today measure only the total biomass concentration (and some also include the mass of particulate matter).

A major problem with the design of viability assays is the vagueness of the concept of a viable cell. The meaning of "viability" is not clear – is a cell "dead" when it can no longer reproduce? Or when it no longer shows metabolic activity? The best functional definition may vary with different types of cultivations.

Currently, the fraction of viable cells is determined off-line by three methods: vital staining, cell replication, and metabolic activity (JONES, 1987). The most common of these is probably the use of plate counting as a measure of reproductive ability. It is known, however, that plate counting underestimates the number of viable cells (relative to vital stain methods) even in healthy cultures; stressed populations yield even lower plate-count estimates. All of these off-line methods are time-consuming.

3.2.1 Flow Cytometry

Although it is not possible to assay for reproductive ability in a straightforward manner by flow cytometry, many of the methods for vital staining and metabolic activity can be adapted for use in a flow cytometer. Vital stains usually measure the ability of a microorganism to exclude a dye from its cytoplasm, although a few require the uptake of a dye as an indication of viability. Methods based on metabolic activity usually assay the ability of the cell to enzymatically convert a non-fluorescent dye to a fluorescent form; this conversion requires a functional energy metabolism.

Many researchers have used dyes such as trypan blue, propidium iodide, and fluorescein diacetate to measure viability (by vital staining) with flow cytometers (off line) (SHAPIRO, 1988). For example, fluorescein diacetate (FDA) is a non-fluorescent dye that is cleaved by intracellular esterases to form fluorescein, which is retained by viable cells. This stain has been used to select viable erythroleukemia cells (HAMORI et al., 1980), and both FDA and 4-methyl-umbelliferone cleaved by phosphatases were used to detect viable leukocytes (MALIN-BERDEL and VALET, 1980). Recently, two new viability stains were introduced: vita blue dibutyrate (LEE et al., 1989) and calcofluor white M2R (BERGLUND et al., 1987).

As previously mentioned, the use of flow cytometry for quasi-on-line analysis has not yet been achieved. The application of this technique for quantifying viable cells requires the use of reagents, and is thus somewhat more complex than cell-size measurements. An automated system for adding reagents has recently been developed (PENNINGS et al., 1987), although it was not interfaced with a reactor.

3.2.2 Sample-Flow Analysis

Although no such systems have been described, the development of a sample-flow analysis technique for cell viability should be possible. Since the general scheme of vital staining and metabolic activity measurements involves reagent addition and brief incubation, followed by photometric or fluorometric detection of the dye, a FIA-type system might be used. The best system would involve a dye that changes its characteristics once inside the cell, so that unused dye would not interfere with the measurement. In addition, the incubation period should be very brief.

3.3 Intracellular pH

Many cellular processes, if not the majority, are pH-dependent, and the pH value of interest is thus not that of the medium, but rather the intracellular pH (usually written pH_i). Although the pH of the cytoplasm is the most common pH_i, the internal pH of various organelles (e.g., mitochondria or vacuoles) may also be relevant. Examples of important cellular activities that are pH-dependent include metabolic reactions such as glycolysis, as well as the transport of small molecules across the cell membrane. In addition, the usual strategy of controlling the pH of the medium is often an ineffective means of maintaining a constant pH_i, especially for microorganisms that produce weak acids. Thus, on-line measurements of pH_i would provide valuable information on the progress of the cultivation.

A number of off-line techniques are available for the determination of intracellular pH values (NUCCITELLI and DEAMER, 1982). Methods suitable for on-line or quasi-on-line adaptation include NMR spectroscopy and the use of pH-sensitive dyes (e.g., in a flow cytometer).

3.3.1 Flow Cytometry

There are many dyes available that change color with changes in pH, and some of these have been utilized in flow cytometric pH_i measurements (MUSGROVE et al., 1986; SHAPIRO, 1988). More accurate pH_i values can be obtained from ratiometric measurements, i.e., the ratio of two wavelengths at which absorption, excitation, or emission intensities change differently with pH. Examples of the off-line use of these dyes include fluorescein for rat bone marrow cells (VISSER et al., 1979), $2',7'$-bis(carboxyethyl)-5,6-carboxy fluorescein (BCECF) for human leukemic T-cells (MUSGROVE et al., 1986), and 2,3-dicyanohydroquinone (DCH) for mouse Ehrlich ascites tumor cells (VALET et al., 1981). DCH and BCECF have pH-dependent fluorescence emission profiles and are thus easier to use than fluorescein, which has a pH-dependent fluorescence excitation spectrum (SHAPIRO, 1988).

Measurements can also be made with distributional pH-sensitive dyes, which are weak acids and bases that are distributed between the cytoplasm and the medium depending on the transmembrane pH gradient. In general, however, the ratiometric measurements discussed above are considered to be more accurate for use with a flow cytometer because the ratio measurement is independent of fluorophore concentration.

3.3.2 Nuclear Magnetic Resonance Spectroscopy

Several different nuclei can be used to measure intracellular pH by NMR, including ^1H and ^{13}C (ROBERTS and JARDETZKY, 1981), ^{14}N and ^{15}N (KANAMORI and ROBERTS, 1983), ^{19}F (OKERLUND and GILLIES, 1988), and ^{31}P (GADIAN, 1982). However, nearly all investigations have used ^{31}P for this measurement. As mentioned in the general comments on NMR, this nucleus has the advantage of being

the naturally abundant isotope, so no special substrates are required. In addition, several other phosphate-containing species of interest can be studied at the same time. Using ^{31}P, the pH_i is measured by determining the pH-dependent chemical shift of intracellular inorganic phosphate and relating it to a titration curve (GADIAN, 1982).

Intracellular pH measurements by *in vivo* ^{31}P NMR have become relatively common and have been used to study metabolic processes in many different organisms. For example, the pH dependence of glycine–proton symport in *S. cerevisiae* has been reported (BALLARIN-DENTI et al., 1984); pH_i values were also measured in a study of the effects of oxygen on glycolysis in the same organism (DEN HOLLANDER et al., 1981). An NMR investigation showed that the pH_i response to glucose pulses of *E. coli* strains with a high plasmid copy number was different from that of plasmid free cells (AXE and BAILEY, 1987). Both cytoplasmic and vacuolar pH were measured in *Catharanthus roseus* and *Daucus carota* plant cells (BRODELIUS and VOGEL, 1985). Intracellular pH has also been measured in immobilized cells. Whereas the pH_i of agarose- and alginate-entrapped *C. roseus* cells was identical to that of free cells (VOGEL and BRODELIUS, 1984), GALAZZO et al. (1987) found that *S. cerevisiae* cells immobilized in calcium alginate had a lower pH_i than otherwise identical suspended cells. This was correlated with higher rates of glucose uptake and ethanol production.

Since NMR measurements are not instantaneous, it was necessary to provide the cells in such investigations with nutrients, including oxygen, making them technically on-line measurements. In most of these reports, the experiments lasted only a few hours, but in the study involving CHO cells (GONZALEZ-MENDEZ et al., 1982), the cells were maintained in the NMR reactor for many days. Reactors for NMR studies will be discussed in Sect. 3.7.

3.3.3 Sample-Flow Analysis

Several of the dyes discussed in Sect. 3.3.1 were first used in non-flow cytometric applications, including the use of fluorescein diacetate to measure the pH_i of *Bacillus acidocaldarius* (THOMAS et al., 1976) and mouse Ehrlich ascites tumor cells (THOMAS et al., 1979), and of BCECF in pig and mouse lymphocytes (RINK et al., 1982). This suggests that these dyes could be used in a sample-flow analysis system in which the cells would be exposed to the dye for a short period of time before fluorescence emission (BCECF) or excitation (fluorescein diacetate) intensities were measured in a flow-through fluorometer.

3.4 DNA and RNA

On-line or quasi-on-line measurement of intracellular DNA and RNA concentrations is of interest in the cultivation of many microorganisms and it is particularly important in genetically modified cells (e.g., those containing plasmids). In that case, the ability to monitor DNA levels would be a valuable aid in controlling the cultivation, especially if plasmid instability were important. Information on the levels of RNA would be useful in assessing the extent of transcriptional activity.

No system for these on-line measurements has been developed, but a number of studies have measured intracellular DNA and RNA levels off-line, mainly by using flow cytometry.

3.4.1 Flow Cytometry

One of the major applications of flow cytometry in recent years has been the measurement of cellular DNA levels, and to a lesser extent, levels of RNA. As with other flow cytometric measurements, single-cell concentration distributions are obtained in addition to population-average data, thus providing more information on the state of the cultivation. Another advantage of flow cytometry is the possibility of measuring more than one parameter at the same time, e.g., DNA/RNA or DNA/protein.

Measurement of DNA and RNA with a flow cytometer requires that the cells be stained. Often, other treatment steps are also necessary, depending on the dye that is used. These dyes are fluorescent when bound to the nucleic acid, resulting in greater sensitivity than that obtainable with an absorbance method. Tab. 4 lists some of these fluorochromes along with some comments on their use.

As can be seen, some of the dyes (ethidium bromide and propidium iodide) stain both double-stranded (DS) DNA and DS RNA. If the distribution of one of these nucleic acids is desired, the other must be destroyed by the appropriate nuclease, especially for DNA determinations. However, the RNA content of many cells is much higher than that of DNA,

Tab. 4. Some Stains for Use in Flow Cytometric Measurements of DNA and RNA (based on SHAPIRO, 1988)

Stain	Substrate	Comments
Ethidium bromide (EtBr)	DS DNA; DS RNA	Measurement of RNA or DNA individually requires use of DNAse or RNAse (resp.)
Propidium iodide (PI)	DS DNA; DS RNA	As for EtBr; sharper peaks than EtBr; relatively short incubation times possible
Mithramycin Chromomycin A_3 Olivomycin	DS DNA	Simple protocols; relatively short incubation times
Mithramycin/EtBr	DS DNA	Less interference from RNA; higher fluorescence intensity
Hoechst 33342, 33258	DS DNA	No permeabilization required
4',6'-Diamidino-2-phenylindole (DAPI)	DS DNA	Very sharp peaks
Acridine orange (AO)	DS DNA; SS RNA	RNA and DNA complexes fluoresce at different λs; some difficulties with method
Hoechst 33342/ Pyronin Y	DS DNA; DS RNA	RNA and DNA complexes fluoresce at different λs; no permeabilization required

and thus treatment with DNAse may be unnecessary. It is possible to obtain distributions of both DNA and RNA intracellular concentration simultaneously by using stain combinations such as Hoechst 33342 and pyronin Y (SHAPIRO, 1981). Other fluorochromes such as DAPI are useful when only the DNA levels are to be monitored (STÖHR et al., 1977). Although the acridine orange method requires more pretreatment and has been known to be difficult to use (SHAPIRO, 1988), it is the only stain to provide information on total RNA content (DS RNA is converted to SS (single-stranded) RNA before staining). In general, DS and SS RNA levels in cells parallel one another, so DS RNA content measurements can be useful.

A number of the dyes in Tab. 4 have been used in off-line studies of cultivations and are briefly mentioned here as examples of the manner in which such information can be utilized.

The analysis of bacterial nucleic acid content with a flow cytometer is more difficult than that of mammalian cells, chiefly because of the size difference. Nonetheless, several reports providing intracellular distributions of the nucleic acid population have been published in recent years. Examples include a series of studies on batch cultivations of *Bacillus subtilis* by BAILEY and coworkers (BAILEY et al., 1978; FAZEL-MADJLESSI and BAILEY, 1979, 1980), in which propidium iodide (PI) was used to measure total nucleic acids, and an investigation of the effects of several antibiotics on the *E. coli* cell cycle using mithramycin (STEEN et al., 1982). Mithramycin DNA staining has also been combined with a mathematical model in an investigation of the cell cycle of a plasmid-containing *E. coli* strain (SEO and BAILEY, 1987).

Yeast strains have also been the subject of DNA analyses by flow cytometry. Since yeast cells are larger than bacterial cells, the acquisition of the histograms is simpler. SCHEPER et al. (1987e) have reported the distribution of DNA and RNA content for several different batch and continuous cultivations of *S. cerevisiae;* PI was used to stain both of the nucleic acids (RNAse was added prior to the DNA measurements). AGAR and BAILEY (1981, 1982) used a similar staining procedure to study continuous cultivations of *Schizosaccharomyces pombe*. In one of these studies, the distribution of DNA and RNA of a population undergoing synchronous growth was measured (AGAR and BAILEY, 1982), and in another, changes in the DNA frequency as a function of different dilution rates were measured (AGAR and BAILEY, 1981).

3.4.2 Sample-Flow Analysis

To date, no system for the analysis of nucleic acids in a bioreactor sample stream has been developed. Although most chemical protocols for nucleic acid quantification are too time-consuming and complex for use in such a system, an analytical scheme involving the fluorochromes used in flow cytometry might be envisioned. Alternatively, a scheme suggested by KOLIANDER et al. (1984) might be automated for quasi-on-line RNA analysis. In this method, RNA is hydrolyzed rapidly in 5 mol/L NaOH, leaving guanylic acid, which is then measured by HPLC.

3.5 Protein

The on-line measurement of cellular protein content, like that of DNA and RNA, is of interest because it provides detailed insights into growth and metabolism during a bioprocess. However, the monitoring of intracellular protein levels is especially important in cultivations of genetically modified cells, in which the product is a protein that is retained in the cell or only partially secreted. In such cases, measurement of the level of a specific protein would be desired. Among other uses, on-line or quasi-on-line quantification of the concentration of the protein product is valuable in determining the effects of cultivation conditions for process development and to determine the end point of a batch cultivation. It has been observed that the maximum intracellular product concentration is often reached before the maximum cell density (LOW, 1986).

Various methods are available for these measurements (e.g., LOWRY et al., 1951; BRADFORD, 1976). The following discussion will focus on the monitoring of intracellular protein content, but will include (quasi) on-line techniques for extracellular protein measurement as well as off-line methods for intracellular proteins, since few on-line systems for intracellular protein monitoring have been reported. Hopefully, other systems will be adapted for intracellular proteins and on-line use.

A more thorough review of methods for measuring extracellular proteins in cultivations is provided in Chapter 5 of this volume.

3.5.1 Flow Cytometry

A number of fluorescent, protein-binding stains have been utilized for the flow cytometric measurement of the distributions of total protein content. These include fluorescein isothiocyanate (FITC), sulfaflavine, and several rhodamine compounds such as rhodamine 640 (also known as rhodamine 101), sulforhodamine (SR101), tetramethylrhodamine isothiocyanate (TRITC), substituted rhodamine isothiocyanate (XRITC), and Texas Red (a derivative of SR101). Of these, FITC is the most commonly used; it binds covalently to proteins and has good spectral characteristics that allow it to be combined with DNA stains for dual-parameter analysis. However, the combination of rhodamine 640 and Hoechst 33342 (for DNA) would be preferable for quasi-on-line application of flow cytometry because no permeabilization step is required (CRISSMAN and STEINKAMP, 1982). The spectral properties of XRITC and Texas Red allow them to be used with FITC in double protein-labelling studies (TITUS et al., 1982). In some cases, the total protein content of cells that have been fixed (e.g., with glutaraldehyde) has been found to be proportional to measurements of 90° light scattering (SHAPIRO, 1988), a relationship that could be exploited to shorten the pretreatment procedure in the application of flow cytometry to process monitoring.

The distribution of a specific protein in a population of cells can also be ascertained with flow cytometry. This permits one to detect differences in the content of a specific protein among various subpopulations in a cultivation, a phenomenon not measurable by other

methods. Typically, the protein to be studied is an enzyme, the activity of which is measured with the use of substrates that are converted to fluorophores. These substrates are often derivatives of fluorescein, 4-methylumbelliferone, resorufin, and 7-bromo-3-hydroxy-2-naphtho-s-anisidine (naphthol AS-BI) (KRUTH, 1982). Specific protein assays can be difficult to develop because they require that the substrate enter the cell and react to a fluorescent product that remains within the cell. Often, cell membranes that have been permeabilized to allow the substrate to enter the cell are also permeable to the product. In such cases, a trapping reagent can be added to convert the product into an insoluble compound.

Many studies have used flow cytometry (off-line) to characterize the total protein distribution in cell populations, especially for medical applications. Examples of published studies that are more directly related to cultivation processes include a series of papers by BAILEY and coworkers on batch fermentations of *Bacillus subtilis* (FAZEL-MADJLESSI and BAILEY, 1979, 1980; FAZEL-MADJLESSI et al., 1980), an investigation of the diauxic batch growth of *Saccharomyces cerevisiae* (GILBERT et al., 1978), and the measurement of changes of total protein content during synchronous growth of *Schizosaccharomyces pombe* (AGAR and BAILEY, 1982). Additional studies with these two yeasts focused on the variations of protein distributions of a population with changes in the chemostat dilution rate and pulsewise addition of substrates. Protein content was found to be very sensitive to such changes (AGAR and BAILEY, 1981; ALBERGHINA et al., 1983; RANZI et al., 1986). In an example of the applicability of flow cytometry to the monitoring of recombinant bacterial cultivations, SCHEPER et al. (1987d) measured the distribution of total cell protein in a genetically modified *E. coli* strain before and after thermal induction of plasmid transcription. In all of these examples, proteins were stained with fluorescein isothiocyanate.

Procedures have also been published for the measurement of the distribution of particular cloned-gene proteins in yeasts and bacteria. WITTRUP and BAILEY (1988) reported an assay for β-galactosidase in *S. cerevisiae* that utilizes resorufin-β-D-galactopyranoside as the substrate. In this protocol, the staining procedure was extremely rapid, the influence of released resorufin was minimized by the addition of bovine serum albumin, and a quantitative relationship between measured fluorescence and enzyme activity was obtained. An assay for penicillin-G acylase has also been reported (SCHEPER et al. 1987d). The substrate in this method is 7-phenylacetic-4-(trifluoromethyl)-coumarinylamide, which is converted to a fluorescent coumarin product. This assay was used to study the penicillin-G acylase distribution in continuous cultivations of *E. coli* 5K(pHM12) cells at different dilution rates.

Both of these assays could be used to provide information on the distribution of the plasmid copy number during cultivations of genetically modified cells. The measurement of total cell protein could be used to optimize and control many types of fermentations.

3.5.2 Sample-Flow Analysis

The development of flow-analysis systems for the monitoring of protein in bioprocesses is becoming more widespread. Most authors measure the concentration of one or more extracellular proteins. The key steps that need to be added to these systems are the aseptic sampling of cultivation broth containing the cells and the permeabilization or disruption of the cells. In the only truly on-line protein analysis system reported to date, AHLMANN et al. (1986) investigated, individually and in combination, three chemicals (EDTA, chloroform, and toluene), lysozyme, and ultrasonic treatment for use in cell disintegration. Ultrasonic disruption yielded the best results as judged by the amount of protein released from the cell containing sample, but operational limitations prohibited continuous use of this method. Instead, a mixture of lysozyme and EDTA was used; the combination of chemical and enzymatic agents resulted in more protein release than single-component methods. This type of treatment could certainly be integrated into other, similar analytical systems.

Both total and specific protein assays have been adapted for (quasi) on-line use for extracellular proteins. Examples of the former include flow injection analysis (FIA) systems

based on the biuret assay (SHIDELER et al., 1980), the autoanalyzer (KUSOV and KALINCHUK, 1978) and FIA (SALERNO et al., 1985) versions of the Lowry assay, a Bradford-FIA method (RECKTENWALD et al., 1985a), and modifications of the bicinchoninic acid-based assay for use in FIA systems (DAVIS and RADKE, 1987). Not all of these systems were developed for use with cultivation broth samples, the turbidity and medium composition of which might prevent accurate measurements. In addition, each system retained the characteristics of the assay upon which it was based, e.g., protein-to-protein variations, sensitivity, and linearity.

Monitoring of a specific protein obviously requires different types of assays than those listed above. A variety of automated protein measurement schemes have been developed, including the autoanalyzer system of AHLMANN et al. (1986) for the monitoring of intracellular penicillin-G acylase produced during cultivations of a genetically modified *E. coli*

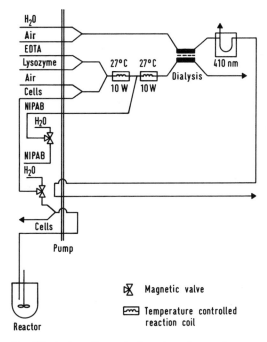

Fig. 21. Automatic analysis system for continuous monitoring of intracellular penicillin-G acylase (from AHLMANN et al., 1986).

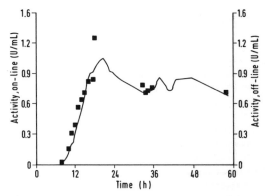

Fig. 22. Penicillin-G acylase production by *Escherichia coli* 5K (pHM12) during a fed-batch fermentation. The line indicates the data from the on-line system in Fig. 22, and the points represent off-line measurements (from AHLMANN et al., 1986).

strain (Figs. 21 and 22). A thermistor-based system was also used to measure the amount of this enzyme present in the cultivation broth (SCHEPER et al., 1984). The thermistor device consisted of a nylon tube through which substrate and buffer solutions were pumped continuously. Small amounts of sample were injected into this stream, and the temperature change was monitored by two thermistors. RECKTENWALD et al. (1985b) modified two manual assay procedures for use in FIA systems and were able to monitor formate dehydrogenase and leucine dehydrogenase levels in sample streams from cell disintegration processes.

With the exception of the above-mentioned thermistor-based system, all of the automated protein analysis systems developed to date utilize photometric detectors. Many other types of detectors are available for use with FIA or autoanalyzer systems, e.g., fluorometers, luminescence detectors, and various biosensors. Of these, biosensors have the most potential for future development; advances in all areas of biosensor development occur rapidly.

A biosensor can be thought of as an analytical device in which a biological component is coupled to a transducer to detect a chemical species. A very wide spectrum of biological components (e.g., enzymes, antibodies, organelles, and whole cells) and transducer types (e.g., ion-selective electrodes, field-effect transistors, thermistors, optoelectronic devices, and piezoelectric crystals) have been utilized in the design of these sensors. The designs and applications of biosensors are discussed in detail in Chapter 3 of this volume.

In addition to the thermistor device already mentioned, other biosensors that have been reported for protein detection include an electrode system for cholinesterase (GUILBAULT and IWASE, 1976), an optoelectronic sensor for serum albumin (GOLDFINCH and LOWE, 1980), and immunosensors for human serum albumin (KARUBE, 1988) and immunoglobulin G (WEHMEYER et al., 1985). At present, however, most of these sensors have several drawbacks that preclude their use for on-line or quasi-on-line monitoring of cultures, including the limited functional stability of the biological components and problems with mechanical stability. It is hoped that these difficulties will be overcome in the near future.

Liquid chromatography, especially fast protein liquid chromatography (FPLC), has also been modified for use in the monitoring of protein levels in cultivations, although no report of an on-line application has been published. One problem is the requirement for a particulate-free sample. One such FPLC system has been developed and used to measure concentrations of intracellular β-galactosidase (LOW, 1986; GUSTAFSSON et al., 1986), and a system for rapid affinity chromatography for immunoglobulin G has also been described (CHASE, 1986). Although these methods can be designed to detect specific proteins, their response times are often rather long (up to one hour), which limits their usefulness in the monitoring of quasi-on-line cell components.

3.6 NAD(P)H

Nicotinamide adenine dinucleotide (NAD$^+$) and the related compounds NADH, NADP$^+$, and NADPH are the major cofactors involved with the transfer of reducing equivalents in metabolic reactions. As such, their intracellular concentrations are closely related to the oxidation/reduction state of the cell, a parameter that appears to affect several cellular processes. Since the intracellular redox state is very sensitive to changes in the environment of the cell, the on-line monitoring of this parameter by measurements of NAD(P)H levels provides valuable information that is not available by other methods.

Manual, wet-chemistry assays for NADH and related compounds are not straightforward, and accurate measurements are made more difficult because of the rapid response of the redox state to environmental perturbations. However, NADH and NADPH are fluorescent while NAD^+ and $NADP^+$ are not, and this characteristic has been used to develop fluorescence probes for use in bioreactors. In addition, it is possible to measure NAD^+ and NADH with ^{31}P NMR, but the two compounds show the same peak. Because the desired quantity is actually the ratio of the oxidized and reduced forms of the couple, this technique is not useful for these measurements. Finally, a few biosensors for NADH have been reported.

3.6.1 Fluorescence Sensors

The background, design, and application of fluorescence sensors have been presented earlier in this chapter, when their use for the indirect measure of biomass concentration was discussed. For that application, the intracellular level of NAD(P)H must remain constant throughout the cultivation. Here, we will focus on the use of these probes to monitor the level of NAD(P)H in cells. Variations in this level, observed as variations in the fluorescence signal, can be used to detect environmental and metabolic changes.

Tab. 5 presents a number of published studies in which the intracellular NAD(P)H-dependent fluorescence levels were monitored in cultures that underwent metabolic changes during growth (e.g., glycolytic oscillations in yeast and the metabolic shift of *Clostridium*

Tab. 5. Some Phenomena Studied with NAD(P)H-Dependent Fluorescence Monitoring

Phenomenon Studied	Organism	References
Aerobic-anaerobic transition	*Klebsiella aerogenes*	HARRISON and CHANCE (1970)
	Saccharomyces cerevisiae	ZABRISKIE and HUMPHREY (1978), ARMIGER et al. (1986), SCHEPER et al. (1987a), MÜLLER et al. (1988)
	Candida tropicalis	BEYELER et al. (1981)
	Escherichia coli	MEYER et al. (1984), GEBAUER et al. (1987)
	C. guilliermondii	MANESHIN and AREVSHATYAN (1972)
Aeration rate	*Penicillium chrysogenum*	SCHEPER et al. (1986)
Addition of carbon source to starved cells	*S. cerevisiae*	ZABRISKIE and HUMPHREY (1978), BEYELER et al. (1981), ARMIGER et al. (1986), SCHEPER and SCHÜGERL (1986b), MÜLLER et al. (1988)
	C. tropicalis	EINSELE et al. (1979)
	E. coli	MEYER et al. (1984)
Diauxic growth	*S. cerevisiae*	MÜLLER et al. (1988)
Dilution rate changes	*S. cerevisiae*	SCHEPER and SCHÜGERL (1986b)
	E. coli	SCHEPER et al. (1987c)
	Pseudomonas putida	LI and HUMPHREY (1989)
Culture synchrony	*S. cerevisiae*	SCHEPER et al. (1987a, d)
Glycolytic oscillations	*S. cerevisiae*	BETZ and CHANCE (1965), CHANCE et al. (1964), KUCHENBECKER et al. (1981), DORAN and BAILEY (1987)
Metabolic shifts	*Thermoactinomyces* sp.	ZABRISKIE and HUMPHREY (1978)
	Clostridium acetobutylicum	REARDON et al. (1986, 1987), SRINIVAS and MUTHARASAN (1987a), RAO and MUTHARASAN (1989)
Additon of metabolic uncoupler	*S. cerevisiae*	BETZ and CHANCE (1965), MÜLLER et al. (1988)

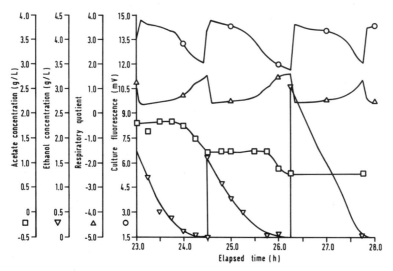

Fig. 23. NAD(P)H fluorescence, respiratory quotient, and acetate and ethanol concentrations during a fed-batch fermentation of *Candida utilis*. The fluorescence signal was used to control ethanol (substrate) additions (reprinted with permission from RISTROPH et al., 1977).

acetobutylicum), or were subjected to various environmental perturbations (e.g., changes of dilution rate in a chemostat). In all of these cases, the changes in the intracellular redox state were clearly measurable as changes in the fluorescence signal.

In several studies the rapid, sensitive response of the fluorescence signal to environmental changes has been used to control certain aspects of bioreactor operation. In one case (Fig. 23), the addition of ethanol to a fed-batch cultivation of *Candida utilis* (for single cell protein) was activated when the fluorescence signal decreased below a certain value because it had been observed that such a decrease indicated low levels of ethanol (RISTROPH et al., 1977). In a second report, decreases in the culture fluorescence were used to control the cellobiose addition to a fed-batch cultivation of a *Thermomonospora* species (MOREIRA et al., 1981). In both of these examples, fluorescence measurements provided a basis for control that would otherwise have been difficult to obtain, since on-line ethanol and cellobiose sensors are not commercially available. MEYER and BEYELER (1984) developed the concept of reaction rate control and combined information on carbon dioxide pro-

duction rate with NADH fluorescence measurements to form a sensitive control system that was applied to yeast fermentations.

Fluorescence measurements can be combined with those of other intracellular parameters to obtain more detailed information on the state of the cells. An experiment of this type has been reported by SCHEPER et al. (1987e) in which off-line flow cytometric measurements of DNA, RNA, protein, and cell-size distributions were augmented by on-line fluorosensor monitoring of a self-synchronized *S. cerevisiae* culture. A more complete image of the behavior of the cells was provided by the composite data set than could be obtained from any individual quantity.

There are some problems with the interpretation of data measured with fluorescence sensors, because the observed fluorescence signal can be influenced by a variety of phenomena such as inner filter effects and fluorescent medium components. These have been mentioned previously in the discussion of the use of fluorescence probes for biomass estimation. Because of these interfering factors one cannot easily correlate a measured fluorescence intensity with a certain intracellular NAD(P)H concentration. However, this difficulty need not detract at all from the usefulness of fluorosensor monitoring. Rather, the user must merely exercise caution when interpreting the fluorescence signal.

3.6.2 Biosensors

At, least two types of biosensors for NAD(P)H measurements have been developed. One of these is a system based on fiber optics, in which the tip of the optical fiber has been coated with bacterial luciferase. NADH in the test solution reacts at the fiber tip, and the resulting bioluminescence is measured (ARNOLD and MEYERHOFF, 1988). The other type of sensor is based on an enzyme-coated electrode (SCHELLER et al., 1985). The use of either type could be somewhat problematic, as the cells would have to be disrupted before the measurement, a process that would most likely change the levels of NAD(P)H very rapidly.

3.7 Other Biomolecules and Metabolic Intermediates

Naturally, it may be of interest to monitor the intracellular levels of compounds other than those previously mentioned. Some of these measurements, including those of ATP, lipids, and various metabolites, will be discussed in this section.

3.7.1 Adenosine Triphosphate

The intracellular concentration of adenosine triphosphate (ATP) is an important measure of the energetic state of microorganisms, since this compound functions as the cellular energy carrier. Actually, since adenosine diphosphate (ADP) can also donate energy to a reaction, it is more appropriate to quantify the bioenergetic state in terms of the energy charge:

$$\text{energy charge} = \frac{[ATP] + 1/2\,[ADP]}{[ATP] + [ADP] + [AMP]} \quad (4)$$

It would be particularly interesting and valuable to monitor both the energy charge and the redox state (by NADH measurements) of the cells simultaneously.

Unfortunately, ATP and energy charge measurements are extremely difficult to perform, since ATP levels can change very rapidly, especially when the cells are stressed. It is therefore necessary to measure this quantity on-line or to use a sampling protocol in which the cells are immediately treated to stop all reactions involving ATP. To date, the available techniques for ATP/energy charge measurements include bioluminescence assays, the use of biosensors, and ^{31}P NMR. The first two of these methods require extraction and stabilization of ATP, as well as ADP and AMP, if the energy charge is to be measured.

Several studies have utilized various rapid extraction procedures and bioluminescence assays (the luciferin–luciferase reaction) to determine the level of ATP in different types of cells, although most of these measurements were made off-line. An exception is the report of an ATP autoanalyzer system that uses trichloroacetic acid or benzalkonium extraction (SIRO et al., 1982). Using this analyzer, monitoring of the ATP concentration in batch cultivations of *S. cerevisiae* showed that intracellular ATP levels peaked in the lag phase before increasing again during exponential growth.

Several types of ATP-sensitive biosensors have been reported, including an optoelectronic sensor based on bioluminescence (SCHELLER et al., 1985) and an enzyme field effect transistor (ENFET) (GOTOH et al., 1986). These sensors have only been tested in solutions of ATP. As is the case with bioluminescence assays, it is necessary to employ a rapid extraction technique if these sensors are to be used in whole-cell measurements. This requires the sensors to be unaffected by medium components and any extraction chemicals.

It is also possible to use ^{31}P nuclear magnetic resonance spectroscopy to monitor intracellular ATP levels as well as the ratio [ATP]/[ADP]. These measurements have been performed on-line in modified NMR tubes containing perfused, suspended, or immobilized cells. Examples of such studies can be found in Tab. 6 of Sect. 3.7.3, which also includes a brief discussion of these test reactors.

3.7.2 Lipids

For some cultivations, monitoring of the intracellular lipid content may be important. For example, a variety of molds, yeasts, and algae are able to accumulate various lipid products in high concentrations. In such cases, process optimization and control would be facilitated by on-line measurements.

Two types of measurements have been reported; both have only been used as off-line techniques. One of these involves photometric or fluorometric detection of stained lipid molecules. However, the most rapid staining protocol reported to date requires extensive rinsing of the excess Sudan Black B stain before measurement (THAKUR et al., 1989), and thus this method would require substantial modifications before it could be used as a sample-flow analysis method.

Flow cytometry can also be used to quantify intracellular lipid content; in fact, distributions of the lipid concentration of a population have been obtained. A fluorescent dye, Nile red, has been used for these measurements (GREENSPAN et al., 1985). When using this stain, it is not necessary to fix or permeabilize the cells, and the flow cytometric analysis can be made immediately.

3.7.3 Metabolic Intermediates

The on-line monitoring of certain metabolic intermediates can provide valuable insights into the metabolism and energetics of microorganisms, information that would certainly be of use in the optimization of a cultivation process. While various assays and sensors that can measure metabolites exist, few, if any, are capable of performing on-line intracellular determinations. However, these on-line measurements can be made with nuclear magnetic resonance spectroscopy. Many studies employing *in vivo* NMR have recently been reported; a few of these are outlined in Tab. 6.

Perhaps the major problem with these on-line NMR measurements is the limited type of bioreactor that can be used in a spectrometer. In order to achieve good signal-to-noise ratios, it is necessary to use cultures of high cell density during the measurements. Most experiments have been performed in perfused NMR tubes, in which a highly concentrated cell suspension is flushed with nutrients and oxygen if required. Several modifications have been used to solve difficulties associated with these perfusion systems, especially the problem of keeping cells in the tube while allowing perfusate to escape. Solutions to this problem include the circulation of oxygenated medium through the small-diameter hollow-fiber dialysis tubing (KARCZMAR et al., 1983) as well as cell immobilization in agarose threads (FOXALL and COHEN, 1983), alginate beads (GALAZZO and BAILEY, 1989), or dialysis tubing (FERNANDEZ et al., 1988). Using a different approach, SANTOS and TURNER (1986) incorporated an airlift device into an NMR tube to solve the problems of insufficient oxygenation and mixing in perfused cultures. Another drawback to many bioreactor designs is their inability to support cell growth, even for relatively short times. In those cases, cells are first grown in batch culture, harvested, concentrated, and then placed in the NMR culture tube. Bioreactor systems that allow growth over a period of several days by supplying nutrients at a sufficient rate are clearly preferable for more realistic monitoring of the culture. Although some of the NMR bioreactor designs described above may be suitable for such studies, small hollow-fiber reactors that have been utilized in special wide-bore NMR probes (Fig. 24) appear to have ad-

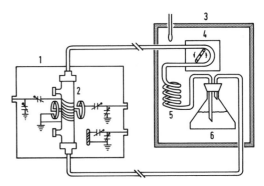

Fig. 24. Schematic diagram of a NMR reactor/ probe for long-term measurements. 1 NMR probe, 2 hollow fiber module, 3 incubator, 4 pump, 5 oxygen-permeable tubing, 6 nutrient reservoir (reprinted with permission from GONZALEZ-MENDEZ et al., 1982).

Tab. 6. Some *in vivo* NMR Studies of Intracellular Processes

Organism	Nucleus	Species Measured	Process Studied	References
Chinese hamster	^{31}P	Sugar phosphates, ATP	Metabolic responses to starvation and low temperature	GONZALEZ-MENDEZ et al. (1982)
Catharanthus roseus Daucus carota	^{31}P	Sugar phosphates, ATP + ADP,	Batch growth	BRODELIUS and VOGEL (1985)
Clostridium thermocellum	^{31}P	ATP	Effects of ethanol on glycolysis	HERRERO et al. (1985)
Escherichia coli	^{31}P	Sugar phosphates, ATP, UDPG	Glycolysis in plasmid-containing cells	AXE and BAILEY (1987)
Saccharomyces cerevisiae	^{31}P	G6P, F6P, FDP, 3PG, UDPG, ATP, phosphomannan	Glycolysis	SHANKS and BAILEY (1988)
S. cerevisiae	^{31}P	G6P, F6P, FDP, 3PG, Gal1P, ATP	Glycolysis in wild-type and *reg1* mutants	SHANKS and BAILEY (1990)
S. cerevisiae	^{31}P	G6P, F6P, FDP, 3PG, ATP + ADP, UDPG	Effects of immobilization	GALAZZO and BAILEY (1989)
	^{13}C	Glucose, trehalose, glycogen		
S. cerevisiae	^{31}P	ATP, FDP, G6P,	Pasteur effect	DEN HOLLANDER et al. (1986)
	^{13}C	αGP glucose, ethanol, glycerol, aspartate, glutamate, DHAP, αGP		
Hybridoma cells	^{31}P	ATP	Metabolic response to dissolved oxygen and glucose changes	FERNANDEZ et al. (1988)
	^{13}C	glucose, lactate, alanine		
	^{15}N	glutamine, alanine		
Bacillus lactofermentum	^{15}N	Glutamate, glutamine, alanine, acetylglutamine, asparagine, aspartate, and others	Nitrogen assimilation and glutamate production	HARAN et al. (1983)
Pseudomonas putida	^{19}F	2,5- and 3,5-difluorobenzoate	Difluorobenzoate metabolism	CASS et al. (1987)

Abbreviations: G6P glucose-6-phosphate, F6P fructose-6-phosphate, FDP fructose-1,6-diphosphate, 3PG 3-phosphoglycerate, UDPG uridine diphosphoglucose, Gal1P galactose-1-phosphate, DHAP dihydroxyacetone phosphate, αGP α-glycerol phospate

vantages for longer-term cultivations (GONZALEZ-MENDEZ et al., 1982; DRURY et al., 1988).

In the majority of the reports presented in Tab. 6, the authors utilized ^{31}P NMR in their experiments. This nucleus is useful because it is the sole naturally occurring isotope (see Tab. 3) and provides relatively simple, yet informative spectra. With ^{31}P NMR, it is possible to obtain information on the levels of phosphorylated metabolic intermediates, cellular energetics (ATP and ADP levels), pH$_i$, and inorganic phosphate transport. The various phosphorylated sugar compounds of the glycolytic pathway and others appear as one large peak in ^{31}P spectra and have usually been reported as a combined "sugar phosphate" concentration. But a method recently described by SHANKS and BAILEY (1988) allows this peak to be deconvoluted into the peaks of the individual compounds, providing the investigator with more information.

Other nuclei are also useful in NMR studies, including ^{13}C and ^{15}N. Since these nuclei are not the naturally abundant isotopes, they are useful for metabolic pathway studies, in which the conversion of a small amount of labelled substrate is followed with time.

4 Future Developments

The need for bioreactor monitoring for process research, development, optimization, and control is clear, and many analytical devices have been developed to fill this need. However, the overwhelming majority of these instruments quantify extracellular chemical and physical quantities. Information on the biological components of the cultivation is of vital importance.

Although the preceding discussion of sensors and analytical devices for on-line and quasi-on-line monitoring of biomass and cell characteristics during cultivations shows that very few such systems are available at this time, it should be clear from the descriptions of other currently used off-line measurement methods that many of these could be developed for bioreactor monitoring. For example, many manual assays can readily be modified for use in sample-flow analytical systems (FIA or autoanalyzer), and the addition of an efficient, automated cell disruption step. The powerful technique of flow cytometry could potentially be applied to quasi-on-line monitoring by utilizing a FIA system for sample preparation. The resulting output of the population-distribution would offer far more information than the average-value output of other instruments. The high cost of this instrument could be lowered by simplifying its design for a particular application.

On-line NMR measurements are currently feasible, although the expense of the instrument and the restrictions on the design of the bioreactor tend to limit the use of this technique to process research, perhaps the most likely area for flow cytometer use as well.

The development of biosensors is proceeding at a high rate, but here again most sensors are not designed to measure intracellular compounds. Those that are could be incorporated into a sample-flow analytical system (preceded by a cell-disruption step), as long as they are unaffected by medium components and are relatively stable with time.

The most basic biological parameter in a bioreactor is the cell density. Although many systems can perform this measurement, few are capable of determining the concentration of viable cells, and all have drawbacks.

Further research is clearly needed to develop all of these systems and to design new sensors and analytical devices for monitoring of the cultivation.

Acknowledgements

The authors would like to thank K. Dane Wittrup, Jacqueline Vanni Shanks, and Gerd Wehnert for their valuable assistance.

5 References

AGAR, D. W., BAILEY, J. E. (1981), Continuous cultivation of fission yeast: classical and flow microfluorometry observations, *Biotechnol. Bioeng.* **23**, 2217–2229.

AGAR, D. W., BAILEY, J. E. (1982), Measurements and models of synchronous growth of fission yeast in induced by temperature oscillations, *Biotechnol. Bioeng.* **24**, 217–236.

AHLMANN, N., NIEHOFF, A., RINAS, U., SCHEPER, T., SCHÜGERL, K. (1986), Continuous on-line monitoring of intracellular enzyme activity, *Anal. Chim. Acta* **190**, 221–226.

ALBERGHINA, L., MARIANI, L., MARTEGANI, E., VANONI, M. (1983), Analysis of protein distribution in budding yeast, *Biotechnol. Bioeng.* **25**, 1295–1310.

ARMIGER, W. B., LEE, J. F., MONTALVO, L. M., FORRO, J. R. (1985), Fed batch control based upon the measurement of intracellular NADH, *Proc. 190th ACS Meeting, Chicago,* MBTD 40.

ARMIGER, W. B., FORRO, J. R., MONTALVO, L. M., LEE, J. F. (1986), The interpretation of on-line process measurements of intracellular NADH in fermentation processes, *Chem. Eng. Commun.* **45**, 197–206.

ARNOLD, M. A., MEYERHOFF, M. E. (1988), Recent advances in the development and analytical applications of biosensing probes, *CRC Crit. Rev. Anal. Chem.* **20** (3), 149–196.

AXE, D. D., BAILEY, J. E. (1987), Application of ^{31}P nuclear magnetic resonance spectroscopy to investigate plasmid effects on *Escherichia coli* metabolism, *Biotechnol. Lett.* **9** (2), 83–88.

BAILEY, J. E., FAZEL-MADJLESSI, J., McQUITTY, D. N., LEE, L. Y., ORO, J. A. (1978), Measurement of structured microbial population dynamics by flow microfluorometry, *AIChE J.* **24** (4), 570–577.

BAILEY, J. E., OLLIS, D. F. (1986), *Biochemical Engineering Fundamentals,* 2nd Ed., New York: McGraw Hill.

BALLARIN-DENTI, A., DEN HOLLANDER, J. A., SANDERS, D., SLAYMAN, C.W. (1984), Kinetics and pH-dependence of glycine-proton symport in *Saccharomyces cerevisiae, Biochim. Biophys. Acta* **778**, 1–16.

BATTYE, F. L., DARLING, W., BEALL, J. (1985), A fast cell sampler for flow cytometry, *Cytometry* **6**, 492–495.

BENNETTO, H. P., BOX, J., DELANEY, G. M., MASON, J. R., ROLLER, S. D., STIRLING, J. L., THURSTON, C. F. (1987), Redox-mediated electrochemistry of whole micro-organisms: from fuel cells to biosensors, in: *Biosensors* (TURNER, A. P. F., KARUBE, I., WILSON, G. S., Eds.) pp. 291–314, Oxford: Oxford Science Publications.

BERGLUND, D. L., TAFFS, R. E., ROBERTSON, N. P. (1987), A rapid analytical technique for flow cytometric analysis of cell viability using calcofluor white M2R, *Cytometry* **8**, 421–426.

BETZ, A., CHANCE, B. (1965), Influence of inhibitors on the oscillation of reduced pyridine nucleotides in yeast cells, *Arch. Biochem. Biophys.* **109**, 759–584.

BEYELER, W., EINSELE, A., FIECHTER, A. (1981), On-line measurements of culture fluorescence: method and application, *Eur. J. Appl. Microbiol. Biotechnol.* **13**, 10–14.

BEYELER, W., GSCHWEND, K., FIECHTER, A. (1983), In-situ Fluorometrie: Eine neue Methode zur Charakterisierung von Bioreaktoren, *Chem. Ing. Tech.* **55** (5), 869–871.

BLAKE-COLEMAN, B. C., CALDER, M. R., CARR, R. J. G., MOODY, S. C., CLARKE, D.J. (1984), Direct monitoring of reactor biomass in fermentation control, *Trends Anal. Chem.* **3** (9), 229–235.

BLAKE-COLEMAN, B. C., CALDER, M. R., CARR, R. J. G., MOODY, S. C. (1986), Determination of

reactor biomass by acoustic resonance densitometry, *Biotechnol. Bioeng.* **28**, 1241–1249.

BLUTE, T., GILLIES, R. J., DALE, B. E. (1988), Cell density measurements in hollow fiber bioreactors, *Biotechnol. Prog.* **4** (4), 202–209.

BIROU, B., MARISON, I. W., VON STOCKAR, U. (1987), Calorimetric investigation of aerobic fermentations, *Biotechnol. Bioeng.* **30**, 650–660.

BIROU, B., VON STOCKAR, U. (1989), Application of bench-scale calorimetry to chemostat cultures, *Enzyme Microb. Technol.* **11**, 12–16.

BOYER, P. M., HUMPHREY, A. E. (1988), Fluorometric behaviour of phenol fermentation, *Biotechnol. Lett.* **2**(3), 193–198.

BRADFORD, M. M. (1976), A rapid and sensitive method for the quantitation of microgram quantities of protein utilizing the principle of protein-dye binding, *Anal. Biochem.* **72**, 248–254.

BRODELIUS, P., VOGEL, H. (1985), A Phosphorus-31 Nuclear Magnetic Resonance Study of Phosphate Uptake and Storage in Cultured *Catharanthus roseus* and *Daucus carota* plant cells, *J. Biol. Chem.* **260** (6), 3556–3560.

CADY, P. (1975), Rapid automated bacterial identification by impedance measurements, in: *New Approaches to Identification of Microorganisms* (HEDÉN, C.-G., ILLÉNI, T., Eds.), pp. 73–99, New York: John Wiley & Sons.

CASS, A. E. G., RIBBONS, D. W., ROSSITER, J. T., WILLIAMS, S. R. (1987), Biotransformation of aromatic compounds, *FEBS Lett.* **220** (2), 353–357.

CHANCE, B., ESTABROOK, R. W., GOSH, A. (1964), Damped Sinusoidal Oscillations of cytoplasmic reduced pyridine nculeotide in yeast cells, *Proc. Natl. Acad. Sci. USA* **51**, 1244–1251.

CHAPMAN, A. G., ATKINSON, D. E. (1977), Adenine nucleotide concentrations and turnover rates. Their correlation with biological activity in bacteria and yeast, *Adv. Microb. Physiol.* **15**, 253–306.

CHASE, H. A. (1986), Rapid chromatographic monitoring of bioprocesses, *Biosensors* **2**, 269–286.

CLARKE, D. J., CALDER, M. R., CARR, R. J. G., BLAKE-COLEMAN, B. C., MOODY, S. C., COLLINGE, T. A. (1985), The development and application of biosensing devices for bioreactor monitoring and control, *Biosensors* **1**, 231–320.

CLARKE, D. J., BLAKE-COLEMAN, B. C., CARR, R. J. G., CALDER, M. R., ATKINSON, T. (1986), Monitoring reactor biomass, *Trends Biotechnol.* **4** (7), 173–178.

CLARKE, D. J., BLAKE-COLEMAN, B. C., CARR, R. J. G., CALDER, M. R., ATKINSON, T. (1987), Principles and potential of piezo-electric transducers and acoustic techniques, in: *Biosensors* (TURNER, A. P. F., KARUBE, I., WILSON, G. S.,

Eds.), pp 551–571, Oxford: Oxford Science Publications.

COONROD, J. D., KUNZ, L. J., FERRARO, M. J., (Eds.) (1983), *The Direct Detection of Microorganisms in Clinical Samples,* London: Academic Press.

COONEY, C. L. (1981), Growth of microorganisms, in: *Biotechnology* (REHM, H. J., REED, G., Eds.), Vol. 1, pp 73–112, Weinheim–Deerfield Beach/Florida–Basel: Verlag Chemie.

CRISSMAN, H. A., STEINKAMP, J. A. (1982), Rapid, one-step staining procedures for analysis of cellular DNA and protein by single and dual laser flow cytometry, *Cytometry* 3 (2), 84–90.

DAVIS, L. C., RADKE, G. A. (1987), Measurement of protein using flow injection analysis with Bicinchoninic Acid, *Anal. Biochem.* 161, 152–156.

DEBLOIS, R. W., BEAU, C. P. (1970), Counting and sizing of submicron particles by the resistive pulse technique, *Rev. Sci. Instrum.* 41 (7), 909–915.

DEN HOLLANDER, J. A., UGURBIL, K., BROWN, T. R., SHULMAN, R. G. (1981), Phosphorus-31 nuclear magnetic resonance studies of the effect of oxygen upon glycolysis in yeast, *Biochemistry* 20, 5871–5880.

DEN HOLLANDER, J. A., UGURBIL, K., SHULMAN, R. G. (1986), ^{31}P and ^{13}C NMR studies of intermediates of aerobic and anaerobic glycolysis in *Saccharomyces cerevisiae, Biochemistry* 25, 212–219.

DENNIS, K., SRIENC, F., BAILEY, J. E. (1983), Flow cytometry analysis of plasmid heterogeneity in *Escherichia coli* populations, *Biotechnol. Bioeng.* 25, 2485–2490.

DENNIS, K., SRIENC, F., BAILEY, J. E. (1985), Ampicillin effects on five recombinant *Escherichia coli* strains: Implications for selection pressure design, *Biotechnol. Bioeng.* 27, 1490–1494.

DERMOUN, Z., BELAICH, J. P. (1979), Microcalorimetric study of *Escherichia coli* aerobic growth: kinetics and experimental enthalpy associated with growth on succinic acid, *J. Bacteriol.* 140 (2), 377–380.

DJAVAN, A., JAMES, A. M. (1980), Determination of the maintenance energy of *Klebsiella aerogenes* growing in continuous culture, *Biotechnol. Lett.* 2, 303–308.

DORAN, P. M., BAILEY, J. E. (1987), Effects of immobilization on the nature of glycolytic oscillations in yeast, *Biotechnol. Bioeng.* 29, 892–897.

DRAKE, J. F., TSUCHIYA, H. M. (1973), Differential counting in mixed cultures with Coulter counters, *Apl. Microbiol.* 26 (1), 9–13.

DRURY, D. D. (1988), Oxygen transfer properties of a bioreactor for use within a nuclear magnetic re-

sonance spectrometer, *Biotechnol. Bioeng.* 32, 966–974.

DUYSENS, L. N. M., AMESZ, J. (1957), Fluorescence spectrophotometry of reduced phosphopyridine nucleotide in intact cells in the near-ultraviolet and visible region, *Biochim. Biophys. Acta* 24, 19–26.

EINSELE, A., RISTROPH, D. L., HUMPHREY, A. E. (1978), Mixing times and glucose uptake measured with a fluorometer, *Biotechnol. Bioeng.* 20, 1487–1492.

EINSELE, A., RISTROPH, D. L., HUMPHREY, A. E. (1979), Substrate uptake mechanisms for yeast cells. A new approach utilizing a fluorometer, *Eur. J. Appl. Microbiol. Biotechnol.* 6, 335–339.

FAZEL-MADJLESSI, J., BAILEY, J. E. (1979), Analysis of fermentation processes using flow microfluorometry: single parameter observations of batch bacterial growth, *Biotechnol. Bioeng.* 21, 1995–2010.

FAZEL-MADJLESSI, J., BAILEY, J. E. (1980), Analysis of fermentation processes using flow microfluorometry: Amylase and protease activities in *Bacillus subtilis* batch cultures, *Biotechnol. Bioeng.* 22, 1657–1669.

FAZEL-MADJLESSI, J., BAILEY, J. E., McQUITTY, D. N. (1980), Flow microfluorometry measurements of multicomponent cell composition during batch bacterial growth, *Biotechnol. Bioeng.* 22, 457–462.

FERNANDEZ, E. J., MANCUSO, A., CLARK, D. S. (1988), NMR spectroscopy studies of hybridoma metabolism in a simple membrane reactor, *Biotechnol. Prog.* 4 (3), 173–183.

FERNANDEZ, E. J., CLARK, D. S. (1987), N.M.R. Spectroscopy: a non-invasive tool for studying intracellular processes, *Enzyme Microb. Technol.* 9, 259–271.

FINGUERUT, J., GUARDA, E. T. L., CAMARGO, E. (1978), Influence of the growth on the spectrophotometric determination of the yeast concentration in liquid hydrocarbon fermentations, *Biotechnol. Bioeng.* 20, 1285–1286.

FLEISCHAKER, R. J., WEAVER, J. C. SINSKEY, A. J. (1981), Instrumentation for process control in cell culture, *Adv. Appl. Microbiol.* 27, 137–167.

FORRO, J. R., MAENNER, G. F., ARMIGER, W. B. (1984), Monitoring cell activity by use of the culture fluorescence, *Proc. 188th ACS National Meeting, Philadelphia,* MBTD 79.

FORSBERG, C. W., LAM, K. (1977), Use of adenosine-5′-triphosphate as an indicator of microbiota biomass in rumen contents, *Appl. Environ. Microbiol.* 33 (3), 528–537.

FOXALL, D. L., COHEN, J. S. (1983), NMR studies of perfused cells, *J. Mag. Res.* 52, 346–349.

GADIAN, D. G. (1982), *Nuclear Magnetic Reson-*

ance and Its Applications to Living Systems. London: Oxford Unversity Press.

GALAZZO, J. L., SHANKS, J. V., BAILEY, J. E. (1987), Comparison of suspended and immobilized yeast metabolism using ^{31}P nuclear magnetic resonance spectroscopy, *Biotechnol. Tech.* **1** (1), 1–6.

GALLAZZO, J. L., BAILEY, J. E. (1989), *In vivo* nuclear magnetic resonance analysis of immobilization effects on glucose metabolism of yeast *Saccharomyces cerevisiae, Biotechnol. Bioeng.* **33**, 1283–1289.

GEBAUER, A., SCHEPER, T., SCHÜGERL, K. (1987), Growth of *E. coli* in a stirred tank in an airlift tower reactor with an outer loop, *Bioprocess Eng.* **2**, 13–23.

GENCER, M. A., MUTHARASAN, R. (1979), Determination of biomass concentration by capacity measurement, *Biotechnol. Bioeng.* **21**, 1097–1103.

GILBERT, M. F., MCQUITTY, D. N., BAILEY, J. E. (1978), Flow microfluorometry study of diauxic batch growth of *Saccharomyces cerevisiae, Appl. Environ. Microbiol.* **36** (4), 615–617.

GILLIES, R. J., MACKENZIE, N. E., DALE, B. E. (1989), Analysis of bioreactor performance by nuclear magnetic resonance spectroscopy, *Bio/Technology* **7**, 50–54.

GOLDFINCH, M. J., LOWE, C. R. (1980), A solid-phase optoelectric sensor for serum albumin, *Anal. Biochem.* **109**, 216–221.

GONZALEZ-MENDEZ, R., WEMMER, G., HAHN, G., WADE-JARDETZKY, N., JARDETZKY, O. (1982), Continuous-flow NMR culture system for mammalian cells, *Biochim. Biophys. Acta* **720**, 274–280.

GOTOH, M., TAMIYA, E., KARUBE, I., KAGAWA, Y. (1986), A microsensor for adenosine-5'-triphosphate pH sensitive field effect transistors, *Anal. Chim. Acta* **187**, 287–291.

GREENSPAN, P., MAYER, E. P., FOWLER, S. D. (1985), Nile red: a selective fluorescent stain for intracellular lipid droplets, *J. Cell Biol.* **100**, 965–973.

GROOM, C. A., LUONG, J. H. T., MULCHANDANI, A. (1988), On-line culture fluorescence measuring during the batch cultivation of poly-hydroxybutyrate producing *Alcaligenes eutrophus, J. Biotechnol.* **8**, 271–278.

GSCHWEND, K., BEYELER, W., FIECHTER, A. (1983), Detection of reactor nonhomogeneities by measuring culture fluorescence, *Biotechnol. Bioeng.* **25**, 2789–2793.

GUILBAULT, G. G., IWASE, A. (1976), Assay of cholinesterase in an electrode system with an immobilized substrate, *Anal. Chim. Acta* **85**, 295–300.

GUPTA, R. K., GUPTA, P., MOORE, R. D. (1984), NMR studies of intracellular metal ions in intact cells and tissues, *Annu. Rev. Biophys. Bioeng.* **13**, 221–246.

GUSTAFSSON, J.-G., FREJ, A.-K., HEDMAN, P. (1986), Monitoring of protein product formation during fermentation with fast protein liquid chromatography, *Biotechnol. Bioeng.* **28**, 16–20.

HAMORI, E., ARNDT-JOVIN, D. J., GRIMWADE, B. G., JOVIN, T. M. (1980), Selection of viable cells with known DNA content, *Cytometry* **1** (2), 132–135.

HANCHER, C. W., THACKER, L. H., PHARES, E. F. (1974), A fiber-optic retroreflective turbidimeter for continuously monitoring cell concentration during fermentation, *Biotechnol. Bioeng.* **16**, 475–484.

HANSEN, E. H. (1988), Flow injection analysis – recent developments and future trends, in: *Analytical Uses of Immobilized Biological compounds for Detection, Medical and Industrial Uses* (GUILBAULT, G. G., MASCINI, M., Eds.), pp. 291–308, Norwell, MA: D. Reidel Publ. Co.

HARAN, N., KAHANA, Z. E., LAPIDOT, A. (1983), *In vivo* ^{15}N NMR studies of regulation of nitrogen assimilation and amino acid production by *Brevibacterium lactofermentum, J. Biol. Chem.* **258** (21), 12929–12933.

HARRIS, C. M., TODD, R. W., BUNGARD, S. J., LOVITT, R. W., MORRIS, J. G., KELL, D. B. (1987), Dielectric permittivity of microbial suspensions at radio frequencies: a novel method for the real-time estimation of microbial biomass, *Enzyme Microb. Technol.* **9**, 181–186.

HARRIS, C. M., KELL, D. B. (1985), The estimation of microbial biomass, *Biosensors* **1**, 17–84.

HARRIS, R. K. (1986), *Nuclear Magnetic Resonance: A Physiochemical View,* Harlow, England: Longman House.

HARRISON, D. E. F., CHANCE, B. (1970), Fluorometric technique for monitoring changes in the level of reduced nicotinamide nucleotides in continuous cultures of microorganisms, *Appl. Microbiol.* **19** (3), 446–450.

HATCH, R. T., WILDER, C., CADMAN, T. W. (1979), Analysis of control of mixed cultures, *Biotechnol. Bioeng. Symp.* **9**, 25–37.

HEDÉN, C. G., ILLÉNY, T. (1975) (Eds.) *New Approaches to the Identification of Microorganisms,* New York: John Wiley & Sons.

HEINZLE, E., GOLDSCHMIDT, B., MOES, J., DUNN, I. J. (1986), Modelling of mixing phenomena observed in *Bacillus subtilis* batch cultivations, *Proc. 5th Yugoslavian-Austrian-Italian Chem. Eng. Conf., Portoroz, Yugoslavia,* pp. 525–534.

HENDY, N. A., GRAY, P. P. (1979), Use of ATP as an indicator of biomass concentration in *Tri-*

choderma viride Fermentation, *Biotechnol. Bioeng.* **21**, 153–156.

HERRERO, A. A., GOMEZ, R. F., ROBERTS, M. F. (1985), ^{31}P NMR studies of *Clostridium thermocellum*, *J. Biol. Chem.* **260** (12), 7442–7451.

HONG, K., TANNER, R. D., MALANEY, G. W., WILSON, D. J. (1987), A spectrophotometric method for estimating the yeast cell concentration in a semi-solid state fermentation, *Process Biochem.*, 149–153.

HORAN, P. K., WHEELESS Jr., L. L. (1977), Quantitative single cell analysis and sorting, *Science* **198**, 149–157.

HUANG, T.-L., HAN, Y. W., CALLIHAN, C. D. (1971), Application of the Lowry method for determination of cell concentration in fermentation of waste cellulosics, *J. Ferment. Technol.* **49** (6), 574–576.

HYSERT, D. W., KNUDSEN, F. B., MORRISON, N. N. M., VAN GHELUWE, G., LOM, T. (1979), Application of a bioluminescence ATP assay in brewery wastewater treatment studies, *Biotechnol. Bioeng.* **21**, 1301–1314.

ISHIKAWA, Y., SHODA, M,. MARUYAMA, H. (1981), Design and performance of a new microcalorimetric system for aerobic cultivation of microorgnisms, *Biotechnol. Bioeng.* **23**, 2692–2640.

ISHIKAWA, Y., SHODA, M. (1983), Calorimetric analysis of *Eschericha coli* on continuous culture, *Biotechnol. Bioeng.* **25**, 1817–1827.

ISHIMORI, Y., KARUBE, I., SUZUKI, S. (1981), Determination of microbial populations with piezoelectric membranes, *Appl. Environ. Microbiol.* **42** (4), 632–637.

JELTSCH, A., ZIMMERMANN, U. (1979), Particles in a homogeneous electric field: A model for the dielectric breakdown of living cells in a Coulter counter, *Bioelectrochem. Bioenerg.* **6**, 349–384.

JENKINSON, D. S., LADD, J. N. (1981), Microbiol biomass in soil: Measurement and turnover, in: *Soil Biochemistry* (PAUL, E. A., LADD, J. N., Eds.), Vol. 5, pp. 415–471, New York: Marcel Dekker.

JONES, R. P. (1987), Measures of yeast death and deactivation and their meaning, *Process Biochem.* **22**, 118–133.

JUNTER, G. A., LEMELAND, J. F., SELEGNY, E. (1980), Electrochemical detection and counting of *E. coli* in the presence of a reducible coenzyme, lipoic acid, *Appl. Environ. Microbiol.* **39**, 307–316.

KANAMORI, K., ROBERTS, J. D. (1983), N-15 NMR studies of biological systems, *Acc. Chem. Res.* **16**, 35–41.

KANG, S. J., PUGH, L. B., BORCHARDT, J. A.

(1983), ATP as a measure of active biomass concentration and inhibition in biological wastewater treatment processes, *Proc. Ind. Waste Conf.,* pp. 751–759.

KARL, D. M. (1986), Determination of *in situ* microbial biomass, viability, metabolism, and growth, in: *Bacteria in Nature* (POINDEXTER, J. S., LEADBETTER, E. R., Eds.), Vol. 2, pp. 85–176, New York: Plenum Press.

KARUBE, I. (1988), Immunosensors, in: *Analytical Uses of Immobilized Biological Compounds for Detection, Medical and Industrial Uses* (GUILBAULT, G. G., MASCINI, M., Eds.), pp. 267–279. Norwell, MA: D. Reidel Publ. Co.

KARCZMAR, G. S., KORTESKY, A. P., BISSELL, M. J., KLEIN, M. P., WIENER, M. W. (1983), A device for maintaining viable cells at high densities for NMR studies, *J. Mag. Res.* **53**, 123–128.

KELL, D. B. (1987), Forces, fluxes and the control of microbial growth and metabolism, *J. Gen. Microbiol.* **133**, 1651–1665.

KILBURN, D. G., FITZPATRICK, P., BLAKE-COLEMAN, B. C., CLARKE, D. J., GRIFFITS, J. B. (1989), On-line monitoring of cell mass in mammalian cell cultures by acoustic densiometry, *Biotechnol. Bioeng.* **33**, 1379–1384.

KOCH, A. L. (1970), Turbidity measurements of bacterial cultures in some available commercial instruments, *Anal. Biochem.,* 252–259.

KOLIANDER, B., HAMPEL, W., ROEHR, M. (1984), Indirect estimation of biomass by rapid ribonucleic acid determination, *Appl. Microbiol. Biotechnol.* **19**, 272–276.

KRUTH, H. S. (1982), Flow cytometry: rapid biochemical analysis of single cells, *Anal. Biochem.* **125**, 225–242.

KUBITSCHEK, H. E. (1969), Counting and sizing micro-organisms with the Coulter counter, in: *Methods in Microbiology* (NORRIS, J. R., RIBBONS, D. W., Eds.), pp. 593–604, London–New York: Academic Press.

KUCHENBECKER, D., BLEY, T., SCHMIDT, A. (1981), Short-periodic oscillations of culture fluorescence in yeast cell populations, *Stud. Biophys.* **86** (2), 92–96.

KUSOV, Y. Y., KALINCHUK, N. A. (1978), Automated Lowry method for microdetermination of protein in samples and effluents containing nonionic detergents, *Anal. Biochem.* **88**, 256–262.

LEE, Y. H. (1981), Pulsed light probe for cell density measurement, *Biotechnol. Bioeng.* **23**, 1903–1906.

LEE, C., LIM, H. (1980), New device for continuously monitoring the optical density of concentrated microbial cultures, *Biotechnol. Bioeng.* **22**, 639–642.

LEE, L. G., BERRY, G. M., CHEN, C.-H. (1989),

Vita Blue: A new 633-nm excitable fluorescent dye for cell analysis, *Cytometry* **10**, 151–164.

LEIST, C., MEYER, H. P., FIECHTER, A. (1986), Process control during the suspension culture of a human melanoma cell line in a mechanically stirred loop bioreactor, *J. Biotechnol.* **4**, 235–246.

LENZ, R., BOELECKE, C., PECKMANN, U., REUSS, M. (1985), A new automatic sampling device for determination of filtration characteristics and the coupling of an HPLC to fermentors, *Proc. Modelling and Control of Biotechnological Processes, 1st IFAC Symposium, Noordwijkerhout*, 11–13 December (JOHNSON, A., Ed.), IFAC Publications, New York: Pergamon Press.

LI, J., HUMPHREY, A. E. (1989), Kinetic and fluorometric behavior of a phenol fermentation, *Biotechnol. Lett.* **11** (3), 177–182.

LIMA FILHO, J. L., LEDINGHAM, W. M. (1987), Continuous measurement of biomass concentration in laboratory-scale fermenters using a LED-electrode system, *Biotechnol. Tech.* **1** (3), 145–150.

Low, D. K. R. (1986), The use of f.p.l.c.^R system in method development and process monitoring for industrial protein chromatography, *J. Chem. Tech. Biotechnol.* **36**, 345–350.

LOWRY, O. H., ROSEBROUGH, N. J., FARR, A. L., RANDALL, R. J. (1951), Protein measurement with the Folinphenol reagent, *J. Biol. Chem.* **193**, 265–275.

LUONG, J. H. T., VOLESKY, B. (1980), Determination of the heat of some aerobic fermentations, *Can. J. Chem. Eng.* **58**, 497–504.

LUONG, J. H. T., VOLESKY, B. (1982), Indirect determination of biomass concentration in fermentation processes, *Can. J. Chem. Eng.* **60**, 163–167.

LUONG, J. H. T., VOLESKY, B. (1983), Heat evolution during the microbial process-estimation, measurement and applications, *Adv. Biochem. Eng. Biotechnol.* **28**, 1–40.

LUONG, J. H. T., CARRIER, D. J. (1986), On-line measurement of culture fluorescence during cultivation of *Methylomonas mucosa*, *Appl. Microbiol. Biotechnol.* **24**, 65–70.

LUONG, J. H. T., YERUSHALMI, L., VOLESKY, B. (1983), Estimating the maintenance energy and biomass concentration of *Saccharomyces cerevisiae* by continuous calorimetry, *Enzyme Microb. Technol.* **5**, 291–297.

MacBRIDE, W. R., MAGAE, J. A., ARMIGER, W. B., ZABRISKIE, D. W. (1986), Optical Apparatus and Method for Measuring the Characteristics of Materials by their Fluorescence, *U. S. Patent*, 4 577 110.

MacMICHAEL, G., ARMIGER, W. B., LEE, J. F.,

MUTHARASAN, R. (1987), On-line measurement of hybridoma growth by culture fluorescence, *Biotechnol. Techn.* **1** (4), 213–218.

MALIN-BERDEL, J., VALET, C. (1980), Flow cytometric determination of esterase and phosphatase activities and kinetics in hematopoietic cells with fluorogenic substrates, *Cytometry* **1** (3), 222–228.

MANESHIN, S. K., AREVSHATYAN, A. A. (1972), Change in the fluorescence intensity of $NADH_2$ in *Candida guilliermondii* in the transition from an anaerobic to an aerobic state, *Appl. Biochem. Microbiol.* **8**, 273–275.

MARISON, I. W., VON STOCKAR, U. (1986), The application of a novel heat flux calorimeter for studying growth of *Escherichia coli* W in aerobic batch culture, *Biotechnol. Bioeng.* **28**, 1780–1793.

MARISON, I., VON STOCKAR, U. (1987), A calorimetric investigation of the aerobic cultivation of *Kluyveromyces fragilis* on various substrates, *Enzyme Microb. Technol.* **9**, 33–43.

MATSUNAGA, T., KARUBE, I., SUZUKI, S. (1979), Electrode system for the determination of microbial populations, *Appl. Environ. Microbiol.* **37** (1), 117–121.

MATSUNAGA, T., KARUBE, I., SUZUKI, S. (1980), Electrochemical determination of cell populations, *Eur. J. Appl. Microbiol. Biotechnol.* **10**, 125–132.

MERTEN, O.-W., PALFI, G. E., STEINER, J. (1986), On-line determination of biochemical/physiological parameters in the fermentation of animal cells in a continuous or discontinuous mode, *Adv. Biotechnol.* **6**, 111–178.

MERTEN, O.-W., PALFI, G. E., STÄHELI, J., STEINER, J. (1987), Invasive infrared sensor for the determination of cell number in a continous fermentation of hybridomas, *Dev. Biol. Stand.* **66**, 357–360.

METZ, H. (1981), Kontinuierliche Trübungsmessung in Bioreaktoren, *Chem. Tech.* **10**, 691–696.

MEYER, C., BEYELER, W. (1984), Control strategies for continuous bioprocess based on biological activities, *Biotechnol. Bioeng.* **26**, 916–925.

MEYER, H.-P., BEYELER, W., FIECHTER, A. (1984), Experiences with the on-line measurement of culture fluorescence during cultivation of *Bacillus subtilis, Escherichia coli, Sporotrichum thermophile*, *J. Biotechnol.* **1**, 341–349.

MOREIRA, A. R., PHILLIPS, J. A., HUMPHREY, A. E. (1978), Method for determining the concentration of adsorbed protein and cell biomass in cellulose fermentations, *Biotechnol. Bioeng.* **20**, 1501–1505.

MOREIRA, A. R., PHILLIPS, J. A. HUMPHREY, A. E. (1981), Utilization of carbohydrates by *Ther-*

momonospora sp. grown on glucose, cellobiose, and cellulose, *Biotechnol. Bioeng.* **23** (6), 1325–1338.

MOU, D. G., COONEY, C. L. (1976), Application of dynamic calorimetry for monitoring fermentation processes, *Biotechnol. Bioeng.* **18**, 1371–1392.

MÜLLER, W., WEHNERT, G., SCHEPER, T. (1988), Fluorescence monitoring of immobilized microorganisms in cultures, *Anal. Chim. Acta* **213**, 47–53.

MUNOZ, E. F., SILVERMAN, M. P. (1979), Rapid, single-step most-probable-number method for enumerating fecal coliforms in effluents from sewage treatment plants, *Appl. Environ. Microbiol.* **37** (3), 527–530.

MUSGROVE, E., RUGG, C., HEDLEY, D. (1986), Flow cytometric measurement of cytoplasmic pH: A critical evaluation of available fluorochromes, *Cytometry* **7**, 347–355.

NESTAAS, E., WANG, D. I. C. (1981), A new sensor, the "filtration probe", for quantitative characterization of the penicillin fermentation. I. Mycelial morphology and culture activity, *Biotechnol. Bioeng.* **23**, 2803–2813.

NESTAAS, E., WANG, D. I. C. (1983), A new sensor, the "filtration probe", for quantitative characterization of the penicillin fermentation. III. An automatically operating probe, *Biotechnol. Bioeng.* **25**, 1981–1987.

NESTAAS, E., WANG, D. I. C., SUZUKI, H., EVANS, L. B. (1981), A new sensor, the "filtration probe", for quantitative characterization of the penicillin fermentation. II. The monitor of mycelial growth, *Biotechnol. Bioeng.* **23**, 2815–2824.

NICOLS, S. L., JAMES, A. M. (1981), A calorimetric determination of maintenance energy requirements during the aerobic growth of *Klebsiella aerogenes, Biotechnol. Lett.* **3**, 119–124.

NISHIKAWA, S., SAKAI, S., KARUBE, I., MATSUNGA, T., SUZUKI, S. (1982), Dye-coupled electrode system for the rapid determination of cell populations in polluted water, *Appl. Environ. Microbiol.* **43** (4), 814–818.

NUCCITELLI, R., DEAMER, D. W. (1982), *Intracellular pH: Its Measurement, Regulation and Utilization in Cellular Functions.* New York: Alan R. Liss., Inc.

OHASHI, M., WATABE, T., ISHIKAWA, T., WATANABE, Y., MIWA, K., SHODE, M., ISHIKAWA, Y., ANDO, T., SHIBATA, T., KITSUNAI, T., KAMIYAMA, N., OIKAWA, Y. (1979), Sensors and instrumentation: steam-sterilizable dissolved oxygen sensors and cell mass sensor for on-line fermentation system control, *Biotechnol. Bioeng. Symp.* **9**, 103–116.

OKERLUND, R. S., GILLIES, R. J. (1987), Determination of Ca^{++} in whole cells by ^{19}F NMR spectroscopy, in: *NMR Spectroscopy of Cells and Organells* (GUPTA, R. K., Ed.), Boca Raton, FL: CRC Press.

ORIOL, E., CONTRERAS, R., RAIMBAULT, M. (1987), Use of microcalorimetry for monitoring the solid state culture of *Aspergillus niger, Biotechnol. Tech.* **1** (2), 79–84.

PENNINGS, A., SPETH, P., WESSELS, H., HAANEN, C. (1987), Improved flow cytometry of cellular DNA and RNA by on-line reagent addition, *Cytometry* **8**, 335–338.

PERLEY, C. R., SWARTZ, J. R., COONEY, C. L. (1979), Measurement of cell mass concentration with a continuous-flow viscosimeter, *Biotechnol. Bioeng.* **21**, 519–523.

PREIKSCHAT, E. (1987), New inline method to measure cell count and cell size in fermentors using a focussed laser beam, *Proc. 4th Eur. Congr. Biotechnol.,* Vol. 3, pp. 122–125.

RAMSEY, G., TURNER, A. P. F., FRANKLIN, A., HIGGINS, I. J. (1985), Rapid bioelectrochemical methods for the detection of living microorganisms, *Proc. Modelling and Control of Biotechnologica Processes, 1st IFAC Symposium, Noordwijkerhout,* 11–13 December, New York: Pergamon Press.

RAMSAY, G., TURNER, A. P. F. (1988), Development of an electrochemical method for the rapid determination of microbial concentration and evidence for the reaction mechanism, *Anal. Chim. Acta* **215**, 61–69.

RANZI, B. M., COMPAGNO, C., MARTEGANI, E. (1986), Analysis of protein and cell volume distribution in glucose-limited continuous cultures by budding yeast, *Biotechnol. Bioeng.* **28**, 185–190.

RAO, G., MUTHARASAN, R. (1989), NADH levels and solventogenesis in *Clostridium acetobutylicum:* New insights through culture fluorescence, *Appl. Microbiol. Biotechnol.* **30**, 59–66.

REARDON, K. F., SCHEPER, T., BAILEY, J. E. (1986), *In situ* fluorescence monitoring of immobilized *Clostridium acetobutylicum, Biotechnol. Lett.* **8** (11), 817–822.

REARDON, K. F., SCHEPER, T., BAILEY, J.E. (1987), Metabolic pathway rates and culture fluorescence in batch fermentations of *Clostridium acetobutylicum, Biotechnol. Progr.* **3** (3), 153–167.

RECKTENWALD, A., KRONER, K.-H., KULA, M.-R. (1985a), Rapid on-line protein detection in biotechnological processes by flow injection analysis (FIA), *Enzyme Microb. Technol.* **7**, 146–149.

RECKTENWALD, A., KRONER, K.-H., KULA, M.-R. (1985b), On-line monitoring of enzymes in downstream processing by flow injection analysis (FIA), *Enzyme Microb. Technol.* **7**, 607–612.

REUSS, M. (1987) Process computer coupled substrate feeding for fermentation processes, in: *Biochemical Engineering* (CHMIEL, HAMMES, BAILEY, Eds.), pp. 149–168, Stuttgart–New York: Gustav Fischer Verlag.

REUSS, M., BOELCKE, C., LENZ, R., PECKMANN, U. (1987), A new automatic sampling device for determination of filtration characteristics of biosuspensions and coupling of analyzers with industrial fermentation processes, *BTF-Biotech-Forum 4*, 2–12.

RINK, T. J., TSIEN, R. Y., POZZAN, T. (1982), Cytoplasmic pH and free Mg^{2+} in lymphocytes, *J. Cell Biol.* **95**, 189–196.

RISTROPH, D. L., WATTEEUW, C. M., ARMIGER, W. B., HUMPHREY, A. E. (1977), Experiences in the use of culture fluorescence for monitoring fermentations, *J. Ferment. Technol.* **55**, 599–

ROBERTS, J. K. M., JARDETZKY, O. (1981), Monitoring of cellular metabolism by NMR, *Biochim. Biophys. Acta* **639**, 53–76.

ROY, D., SAMSON, R. (1988), Investigation of growth and metabolism of *Saccharomyces cerevisiae* (baker's yeast) using microcalorimetry and bioluminometry, *J. Biotechnol.* **8**, 193–206.

RUZICKA, J., HANSEN, E. H. (1988), Homogeneous and heterogeneous systems, flow injection analysis today and tomorrow, *Anal. Chim. Acta* **214**, 1–27.

SAKATO, K., TANAKA, H., SAMEJIMA, H. (1981), Electrochemical measurement of cell populations, *Ann. NY Acad. Sci.* **369**, 321–334.

SALERNO, R. A., ODELL, C., CYANOVICH, N., BUBNIS, B. P., MORGES, W., GRAY, A. (1985), Lowry protein determination by automated flow injection analysis for bovine serum albumin and hepatitis B surface antigen, *Anal. Biochem.* **151**, 309–314.

SAMSON, R., BEAUMIER, D., BEAULIEU, C. (1987), Simultaneous evaluation of on-line microcalorimetry and fluorometry during batch culture of *Pseudomonas putida* ATCC 11172 and *Saccharomyces cerevisiae* ATCC 18824, *J. Biotechnol.* **6**, 175–190.

SANTOS, H., TURNER, D. L. (1986), Characterization of the improved sensitivity obtained using a flow method for oxygenating and mixing cell suspensions in NMR, *J. Mag. Res.* **68**, 345–349.

SCHELLER, F. W., SCHUBERT, F., RENNEBERG, R., MÜLLER, H.-G. (1985), Biosensors: Trends and commercialization, *Biosensors* **1**, 135–160.

SCHEPER, T., SCHÜGERL, K. (1986a), Characterization of bioreactors by in-situ fluorometry, *J. Biotechnol.* **3**, 221–229.

SCHEPER, T., SCHÜGERL, K. (1986b), Culture fluo-

rescence studies on aerobic continuous cultures of *C. cerevisiae, Appl. Microbiol. Biotechnol.* **23**, 440–444.

SCHEPER, T., GEBAUER, A., SAUERBREI, A., NIEHOFF, A., SCHÜGERL, K. (1984), Measurement of biological parameters during fermentation processes, *Anal. Chim. Acta* **163**, 111–118.

SCHEPER, T., LORENZ, T., SCHMIDT, W., SCHÜGERL, K. (1986), Measurement of culture fluorescence during the cultivation of *P. chrysogenum* and *Z. mobilis, J. Biotechnol.* **3**, 231–238.

SCHEPER, T., LORENZ, T., SCHMIDT, W., SCHÜGERL, K. (1987a), On-line measurement of culture fluorescence for process control of biotechnological processes, *Ann. NY Acad. Sci.* **506**, 431–445.

SCHEPER, T., WEHNERT, G., SCHÜGERL, K. (1987b), in: *DECHEMA Biotechnology Conferences,* Vol. 1, pp. 63–66, Weinheim–Deerfield Beach/Florida–Basel: VCH.

SCHEPER, T., GEBAUER, A., SCHÜGERL, K. (1987c), Monitoring of NADH-dependent culture fluorescence during cultivation of *Escherichia coli, Chem. Eng. J.* **34**, B7–B12.

SCHEPER, T., HITZMANN, B., RINAS, U., SCHÜGERL, K. (1987d), Flow cytometry of *Escherichia coli* for process monitoring, *J. Biotechnol.* **5**, 139–148.

SCHEPER, T., HOFFMANN, H., SCHÜGERL, K. (1987e), Flow cytometric studies during culture of *Saccharomyces cerevisiae, Enzyme Microb. Technol.* **9**, 399–405.

SEO, J.-H., BAILEY, J. E. (1987), Cell cycle analysis of plasmid-containing *Escherichia coli* HB 101 populations with flow cytometry, *Biotechnol. Bioeng.* **30**, 297–305.

SHANKS, J. V., BAILEY, J. E. (1988), Estimation of intracellular sugar phosphate concentrations in *Saccharomyces cerevisiae* using ^{31}P nuclear magnetic resonance spectroscopy, *Biotechnol. Bioeng.* **32**, 1138–1152.

SHANKS, J. V., BAILEY, J. E. (1990), Comparison of wild type and *reg1* mutant *Saccharomyces cerevisiae* metabolic levels during glucose and galactose metabolism using ^{31}P NMR, submitted.

SHAPIRO, H. M. (1981), Flow cytometric estimation of DNA and RNA content in intact cells stained with Hoechst 33342 and pyronin Y, *Cytometry* **2** (3), 143–150.

SHAPIRO, H. M. (1988), *Practical Flow Cytometry,* New York: Alan R. Liss, Inc.

SHIDELER, C. E., STEWART, K. K., CRUMP, J., WILLS, M. R., SAVORY, J., RENCE, B. W. (1980), Automated multiple flow injection analysis in clinical chemistry: determination of total protein with biuret reagent, *Clin. Chem.* **26** (10), 1454–1458.

SHIMMONS, B. W., SVRCEK, W. Y., ZAJIC, J. E. (1976), Cell concentration by viscosity, *Biotechnol. Bioeng.* **18**, 1793–1805.

SILVENNOINEN, E., KOIVO, H. N. (1982), Estimation of the biomass concentration in the Pekilo process from the filtrate flow rate of the concentrator, in: *Modelling and Control of Biotechnical Processes* (HALME, A., ed.), pp. 219–224, IFAC-Publications.

SIRO, M.-R., ROMAR, H., LÖVGREN, T. (1982), Continuous flow method for extraction and bioluminescence assay of ATP in baker's yeast, *Eur. J. Appl. Microbiol. Biotechnol.* **15**, 258–264.

SOLOMON, B. O., ERICKSON, L. E., YANG, S. S. (1983), Estimation of biomass concentration in the presence of solids for the purpose of parameter estimation, *Biotechnol Bioeng.* **25**, 2469–2477.

SRINIVAS, S. P., MUTHARASAN, R. (1987a), Culture fluorescence characteristics and its metabolic significance in batch cultures of *Clostridium acetobutylicum, Biotechnol. Lett.* **9** (2), 139–142.

SRINIVAS, S. P., MUTHARASAN, R. (1987b), Inner filter effects and their interferences in the interpretation of culture fluorescence, *Biotechnol. Bioeng.* **30** (6), 769–774.

STEEN, H. B., BOYE, E., SKARSTAD, K., BLOOM, B., GODAL, T., MUSTAFA, S. (1982), Applications of flow cytometry on bacteria: cell cycle kinetics, drug effects, and quantitation of antibody binding, *Cytrometry* **2** (4), 249–257.

STEINKAMP, J. A. (1984), Flow cytometry, *Rev. Sci. Instrum.* **55** (9), 1375–1400.

STÖHR, M., EIPEL, H., GOERTTLER, K., VOGT-SCHADEN, M. (1988), Extended application of flow microfluorometry by means of a dual laser excitation, *Histochemistry* **51**, 305–313.

TAYA, M., AOKI, N., KOBAYASHI, T. (1986), Estimation of microbial mass concentration based on fluorometric measurement of cell protein, *J. Ferment. Technol.* **64** (5), 411–417.

TAYA, M., HEGGLIN, M., PRENOSIL, J. E., BOURNE, J. R. (1989), On-line monitoring of cell growth in plant tissue cultures by conductometry, *Enzyme Microb. Technol.* **11**, 170–176.

THAKUR, M. S., PRAPULLA, S. G., KARANTH, N. G. (1989), Estimation of intracellular lipids by the measurement of absorbance of yeast cells stained with Sudan black B, *Enzyme Microb. Technol.* **11**, 252–254.

THOMAS, J.A., COLE, R. E., LANGWORTHY, T. A. (1976), Intracellular pH measurements with a spectroscopic probe generated *in situ, Fed. Proc., Fed. Am. Soc., Exp. Biol.* **35**, 1455.

THOMAS, J. A., BUCHSBAUM, R. N., ZIMNIAK, A., RACKER, E. (1979), Intracellular pH measurements in Ehrlich ascites tumor cells utilizing spectroscopic probes generated *in situ, Biochemistry* **18** (11), 2210–2218.

THOMAS, D. C., CHITTUR, V. K., CAGNEY, J. W., LIM, H. C. (1985), On-line estimation of mycelial cell mass concentrations with a computer-interfaced filtration probe, *Biotechnol. Bioeng.* **27**, 729–742.

TITUS, J. A., HAUGLAND, R., SHARROW, S. O., SEGAL, D. M. (1982), Texas Red, a hydrophilic, red-emitting fluorophore for use with fluorescein in dual parameter flow microfluorometric and fluorescence microscopic studies, *J. Immunol. Methods* **50**, 193–204.

TURNER, A. P. F., RAMSAY, G., HIGGINS, I. J. (1982), Applications of electron transfer between biological systems and electrodes, *Biochem. Soc. Trans.* **11**, 445–448.

VALET, G., RAFFAEL, A., MORODER, L., WÜNSCH, E., RUHENSTROTH-BAUER, G. (1981), Fast Intracellular pH determination in single cells by flow-cytometry, *Naturwissenschaften* **68**, 265–266.

VANOUS, R. D. (1978), Understanding nephelometric instrumentation, *Am. Lab.* **6**, 38–46.

VISSER, J. W. M., JONGELING, A. A. M., TANKE, H. J. (1979), Intracellular pH-determination by fluorescence measurements, *J. Histochem. Cytochem.* **27**, 32–35.

VOGEL, H. J., BRODELIUS, P. (1984), An *in vivo* ^{31}P NMR comparison of freely suspended and immobilized *Catharantus roseus* plant cells, *J. Biotechnol.* **1**, 159–170.

WALKER, C. C., DHURJATI, P. (1989), Use of culture fluorescence as a sensor for on-line discrimination of host and overproducing recombinant *Escherichia coli, Biotechnol. Bioeng.* **33**, 500–505.

WATTEEUW, C., ARMIGER, W. B., RISTROPH, D., HUMPHREY, A. E. (1979), Production of single cell protein from ethanol by fed-batch process, *Biotechnol. Bioeng.* **21**, 1221–1237.

WANG, N. S., STEPHANOPOULOS, G. N. (1984), Computer applications to fermentation processes, *CRC Crit. Rev. Biotechnol.* **2** (1), 1–103.

WANG, H., WANG, D. I. C., COONEY, C. L. (1978), The application of dynamic calorimetry for monitoring growth of *Saccharomyces cerevisiae, Eur. J. Appl. Microbiol. Biotechnol.* **5**, 207–214.

WEHMEYER, K. R., HALSALL, H. B., HEINEMANN, W. R. (1985), Heterogeneous enzyme immunoassay with electrochemical detection: competitive and "sandwich"-type immunoassays, *Clin. Chem.* **31** (9), 1546–1549.

WILKINS, J. R. (1978), Use of platinum electrodes for the electrochemical detection of bacteria, *Appl. Env. Microbiol.* **36** (5), 683–687.

WILKINS, J. R., STONER, G. E., BOYKIN, E. H.

(1974), Microbial detection method based on sensing molecular hydrogen, *Appl. Microbiol.* **27**, 949–952.

WILKINS, J. R., YOUNG, R. N., BOYKIN, E. H. (1978), Multichannel electrochemical microbial detection unit, *Appl. Environ. Microbiol.* **35** (1), 214–215.

WITTRUP, K. D., BAILEY, J. E. (1988), A single-cell assay of β-galactosidase activity in *Saccharomyces cerevisiae, Cytometry* **9**, 394–404.

WITTRUP, K. D., MANN, M. B., FENTON, D. M., TSAI, L. B., BAILEY, J. E. (1988), Single-cell light scatter as a probe of refractile body forma-

tion in recombinant *Escherichia coli, Bio/Technology* **6**, 423–426.

ZABRISKIE, D. W. (1979), Use of culture fluorescence for monitoring of fermentation systems, *Biotechnol. Bioeng. Symp.* **9**, 117–123.

ZABRISKIE, D. W., HUMPHREY, A. E. (1978), Estimation of fermentation biomass concentration by measuring culture fluorescence, *Eur. J. Appl. Microbiol.* **35** (2), 337–343.

ZABRISKIE, D. W., ARMIGER, W. B., HUMPHREY, A. E. (1975), Estimation of cell biomass and growth rate by measurement of culture fluorescence, *Proc. ASM Meeting, New York,* p. 195.

7 Bioreactor State Estimation

GREGORY STEPHANOPOULOS

SEUJEUNG PARK

Cambridge, Massachusetts 02139, U.S.A.

1 Scope of Bioreactor Identification

The objective of bioreactor identification is to determine what has happened, is happening, and probably will happen in the bioreactor, based on all available knowledge about the bioreactor and the process. General knowledge available for this purpose consists of process measurements and genetic/biochemical information about the specific cellular biocatalyst. The former are usually sufficient in those cases where an estimate of the bioreactor state is sought in terms of culture parameters, either directly measurable or otherwise observable through correlation with the measurements. If, on the other hand, a more mechanistic description of past history and present state of the process is needed, then measurements must be complemented by cause–effect relationships, which provide a more fundamental basis for the system observations. Such relationships, and models derived from them, are indispensable in those cases where past history and present observations are combined in an attempt to extrapolate into the future.

A problem exists here due to the different levels of noise present in the process measurements; furthermore, when state estimation is involved, the level of confidence in the underlying relationships varies from the almost certain to the more phenomenological ones. The scope of bioreactor identification is then defined by the extent to which such identification is attempted: if the system can be adequately portrayed in terms of its measurable parameters, identification simply involves noise filtering of the raw measurements. If additional parameters correlated with the measurements are desired for complete identification, then, in addition to the noise elimination, a parameter estimation component is also involved. The latter, based on some least-square schemes, yields parameter estimates which are optimally balanced in terms of noise content and dynamic response. Finally, if one wishes to include more fundamental parameters as part of the identification process, relevant fundamental mechanisms must also be considered and the identification becomes more involved.

System identification theory applied to bioreactors offers a systematic method of integrating disparate pieces and types of information for the purpose of efficiently and reliably identifying bioreactor events. This becomes all the more important if one considers that measurements in the fermentors can be of different nature, such as of on- or off-line and of varying frequency. Furthermore, knowledge of fundamental processes and mechanisms is often incomplete and uncertain, which impairs the direct utilization of such information in identifying and reconstructing critical events. The above functions of bioreactor identification are shown schematically in Fig. 1, which also shows the needed inputs. Genetic/biochemical knowledge is included in the system model representation. Measurements include both on-line and off-line data, and uncertainties are present in all inputs. In the framework of bioreactor identification, the different inputs complement one another and mutually compensate for incompleteness and uncertainties of the other.

Another issue of importance is carrying out the identification process on-line. The latter is obviously unnecessary for those cases in which the objective of identification is post-mortem analysis by enumerating state variables and events that occur during the course of the fermentation. If, on the other hand, system identification is connected to fermentor control, the control time dynamics is of importance. There are some cases in which the characteristic time of control activation is of the order of hours, allowing the co-processing of off-line system identification. Many applications, however, involve high productivity, high cell density cultures and exhibit rapid dynamics and strong interaction between the dense cell population and the abiotic culture environment. This rapid dynamics is further complicated by the need to maintain the concentrations of limiting substrates or inhibitory metabolites at or close to their threshold values. This is the case, for example, for sparingly soluble substances (such as oxygen), or substances whose concentrations need to be regulated at low levels to optimally balance opposing trends (such as glucose in baker's yeast fermentation, where a low concentration ensures an optimal balance between yield and productivity). In these

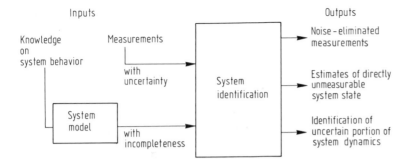

Fig. 1. Schematic description of system identification.

cases, fermentor identification should be implemented on-line to allow optimal operation through timely control action. Improvements in process control by incorporating on-line system identification have been demonstrated in several systems to date, including chemical processes (YDSTIE, 1986).

There have been reports on the use of on-line system identification for fault detection with other systems (ISERMANN, 1984). This presents interesting possibilities for on-line, model-based or black-box, system identification applications to fermentation diagnosis. Due to the relatively poor control of key intracellular events, it is highly desirable that the state of the fermentation be accurately assessed at all points during the course of the process. Fermentor estimates so obtained could provide the basis for detecting departures from the standard course of operation and the activation of prescribed controls.

In summary, on-line fermentor identification is a systematic method to utilize information of varying forms and accuracy for the purpose of identifying the state of fermentation. As indicated in Fig. 1, the methodology utilizes general knowledge as well as that contained in the measurements of the process in order to produce noise-free, corrected measurements, to estimate nonmeasurable parameters, and to identify unknown or uncertain elements of the system dynamics. These outputs can be instrumental for control and culture diagnosis, thereby supporting successful bioreactor operations.

2 Mathematical Formalism

Two questions arise naturally in relation to the implementation of the identification scheme of Fig. 1. The first addresses the modelling requirements that distill our knowledge of system behavior into a set of mathematical equations. Once a model is available, the second question concerns the methodology that yields the optimal system identification from an incomplete set of inputs. The modelling issue is first discussed in the context of fermentor identification. Following this, the underlying concept of identification theory is presented, first for general linear systems, and subsequently for nonlinear fermentation processes.

2.1 Models for Bioreactor Identification

Although one would like to confine the modelling scope to the aspect of metabolism relevant to the process, the latter is usually under global cellular controls which make dissociation from other cellular functions a very difficult task. Model structure simplification leads to lack of biological reality. On the other hand, attempts to account for increasing cellular complexity result in complicated models unsuitable for practical applications. The approach taken here is for simple descriptive models even at the expense of increased phenomenology. Furthermore, in composing the state vector, i.e., the array of biomass and abiotic variables that define the situation in the fermentor, one needs to be concerned with the issue of observability. The latter is, of course, related to the number and type of available measurements and implies that only such variables should be included in the state vector that affect the measurements directly or indirectly but always in a distinguishable manner, and that are, in this context, observable. The requirement for observability limits the use of structured models for identification applications. Compared to chemical reactors, fermentations are supported by considerably more instrumentation for on-line analysis and off-line assays. Yet, fermentations are significantly less reproducible, controllable, and understood due to the complexities of the cell biocatalyst.

In order to illustrate the procedure of model formulation, consider a simple bacterial fermentation, in a defined limiting medium, producing extracellular products. Measurements typically include biomass, substrate, and product concentrations, off-gas analysis, dissolved oxygen, pH, and pH-control-related parameters. Since no intracellular components are measured, the use of structured models is not justified. The overall fermentation model consists rather of material balances based either on fully known growth models or models with unknown growth and yield parameters that need to be identified. The state vector only contains those variables that can be observed from the measurements. Including additional measurements, such as the generated heat or rate of nitrogen addition, does not justify increasing the model structure, but it may allow the identification of additional parameters. Fermentation identification in this case consists of estimating the concentrations of biomass and other metabolites, as well as their rates and yields, during the course of fermentation.

It may be argued that the above task can be achieved by simple algebraic calculations and without any special identification method. The growth rate, for example, may be obtained from the difference between two consecutive biomass measurements, and the specific growth rate can be determined by dividing the total growth rate by the average biomass concentration between the two measurements. This approach, however, does not account for the noise which is present in all measurements.

Taking the average of consecutive measurements is an *ad hoc* procedure and is a suboptimal solution to this problem. Such averaging discards much valuable information about the process, yields unstable estimates, and is equivalent to the arbitrary filtering of process measurements. Identification theory accomplishes the bioreactor estimation objective in an optimally balanced manner. Furthermore, it produces optimal estimates of an array of additional variables and parameters which are not directly measurable but are observable through their indirect effect on the measured variables.

After the state variable vector, x, has been selected, the next step is to derive the model equations for the system dynamics. As mentioned earlier, these models are basically material balances of the state variables, and as such they include other models of growth and metabolic rates. In those cases where reliable model structures and parameter values are available, they are incorporated directly in the balances. If, however, as is often the case, such models are not available, or are known with a limited degree of confidence, provisions need to be made for their determination in the formulation of the model. This is accomplished by explicitly introducing a parameter vector θ as unknown in the model equations. This vector θ contains all unknown kinetic parameters appearing in growth and production rates, yield expression, stoichiometric coefficients, transport coefficients, and other quantities that enter into the balances. Sometimes, in the absence of such model structures, identification of the quantities (i.e., specific growth rate, etc.) themselves can be accommodated by including them directly in the vector θ.

Under these conditions, a suitable general representation of the system dynamics is

$$\dot{x}(t) = f[x(t), \theta(t), u(t), t] + w(t) \qquad (1)$$

Besides vectors x and θ, the model of Eq. (1) also contains a control vector u encompassing the substrate feed, oxygen supply, agitation, and other control variables of the fermentor. Vector w represents random noise sequences to account for all random variations in the balances as written. These variations are introduced by unanticipated factors and also include uncertainties in the structure of the models and the values of the parameters.

Of the various measurements made in a fermentation, some are direct measurements of the state variables, but others are only related to one or more of the state variables. Off-gas analysis, for example, or measurements of heat generation or ammonia addition for pH control, do not directly address any of the components of the state. However, they contain useful information about a number of state variables, and the latter influence these measurements. The relationship between measurements $z(t)$ and state variables is expressed in general terms as

$$z(t) = h[x(t), \theta(t), t] + v(t) \qquad (2)$$

where the vector v again represents a sequence of random noise to account for measurement errors and uncertainties in the relationships between measurements and the state.

Eqs. (1) and (2) are the starting point for the identification process. In a later section, it will be shown how fermentor balances and measurements can be cast in this form. Before this, however, the second question, namely how to optimally identify the system, needs to be addressed.

2.2 Rudiments of System Identification

The objective of system identification is to produce the most reliable estimate of the state and parameter vectors, denoted as $\hat{x}$ and $\hat{\theta}$, from the available measurements and system model dynamics. This objective is accomplished by maximizing

$$f_{x(t_i), \theta(t_i) | Z(t_i), \text{model}}$$

which represents the conditional probability distribution of the system state and parameters conditioned upon the inputs, namely the model and the measurements, $Z(t_i)$. The latter is the collection of all measurements up to time t_i, $[z(t_i), z(t_{i-1}), \ldots, z(t_0)]$. The optimal estimate is the one that is most likely to be true. Therefore, if the conditional probability is

known, the optimal state is chosen as the one that maximizes f. The spread, or variance, of the density function is a measure of the reliability of the estimate.

In on-line system identification, one would like to update the distribution f as new measurements become available. This can be accomplished by applying the following recursive rule (Bayes theorem):

$$f_{x(t_i), \theta(t_i) | Z(t_i), \text{model}} =$$
$$= \frac{f_{z(t_i) | x(t_i), \theta(t_i), Z(t_{i-1}), \text{model}}}{f_{z(t_i) | Z(t_{i-1}), \text{model}}} \cdot$$
$$\cdot f_{x(t_i), \theta(t_i) | Z(t_{i-1}), \text{model}} \qquad (3)$$

This describes the on-line update of the density function. The rightmost density function indicates the distribution of x and θ based on information accumulated up to time t_{i-1}. The fraction of density functions is a measure of the information contained in the measurements made at time t_i. The density function on the left-hand side is the updated distribution of x and θ after the measurement at time t_i is made.

To demonstrate the implementation of this mathematical formalism, the special case of a linear model without unknown parameters is considered:

$$\dot{x}(t) = F(t)x(t) + w(t) \qquad (4)$$

$$z(t_i) = H(t_i)x(t_i) + v(t_i) \qquad (5)$$

Then, the on-line density function update is

$$f_{x(t_i) | Z(t_i), \text{model}} =$$
$$= \frac{f_{z(t_i) | x(t_i), Z(t_{i-1}), \text{model}}}{f_{z(t_i) | Z(t_{i-1}), \text{model}}} f_{x(t_i) | Z(t_{i-1}), \text{model}} \qquad (6)$$

If one assumes that the noise vectors are independent random sequences of Gaussian distribution, then the distribution of state and measurement vectors is also Gaussian. This density function can be defined by the mean vector and covariance matrix as

$$N_x(x_m; Q_x) \equiv [(2\pi)^{-n/2} |Q_x|^{-1/2}] \cdot$$
$$\cdot \exp\{-\tfrac{1}{2}(x - x_m)^T Q_x^{-1}(x - x_m)\} \qquad (7)$$

where x_m and Q_x are the mean and covariance

of the x-vector distribution, respectively, and n is the dimension of the vector. A similar equation can be written for the density of the parameter vector. If the estimate at time t_{i-1}, conditioned by the model and the measurements up to t_{i-1}, is designated as $\hat{x}(t_{i-1})$ and its covariance as $P(t_{i-1})$, then the density functions on the right-hand side of Eq. (6) can be expressed as follows:

$$f_{x(t_i)\mid Z(t_{i-1}),\text{model}} = N_{x(t_i)}(\tilde{x}(t_i); \tilde{P}(t_i)) \qquad (8)$$

In Eq. (8), $\tilde{x}(t_i)$ is the predicted estimate of $x(t_i)$ based on the measurements up to t_{i-1}, and $\tilde{P}(t_i)$ is its covariance matrix. These are obtained by integrating, from t_{i-1} to t_i, the following two equations which describe the time propagation of the mean and covariance based on Eq. (4):

$$\dot{\tilde{x}}(t) = F(t)\tilde{x}(t) \qquad \tilde{x}(t_{i-1}) = \hat{x}(t_{i-1}) \quad (9)$$

$$\dot{\tilde{P}}(t) = F(t)\tilde{P}(t) + \tilde{P}(t)F^{\mathrm{T}}(t) + Q$$
$$\tilde{P}(t_{i-1}) = P(t_{i-1}) \quad (10)$$

where Q is the covariance matrix of $w(t)$.

The other density functions appearing in Eq. (6) are similarly expressed as

$$f_{z(t_i)\mid x(t_i),Z(t_{i-1}),\text{model}} = N_{z(t_i)}(Hx(t_i); R) \qquad (11)$$

$$f_{z(t_i)\mid Z(t_{i-1}),\text{model}} =$$
$$= N_{z(t_i)}(H\tilde{x}(t_i); H\tilde{P}(t_i)H^{\mathrm{T}} + R) \qquad (12)$$

$$f_{x(t_i)\mid Z(t_i),\text{model}} = N_{x(t_i)}(\hat{x}(t_i); P(t_i)) \qquad (13)$$

where R is the covariance matrix of the measurement noise vector $v(t)$. Upon substitution of Eq. (8), Eq. (11), and Eq. (13) into Eq. (6), one obtains the following expressions for the state estimate and its variance:

$$\hat{x}(t_i) = [\tilde{P}(t_i)^{-1} + H^{\mathrm{T}}R^{-1}H]^{-1}$$
$$[\tilde{P}(t_i)^{-1}\tilde{x}(t_i) + H^{\mathrm{T}}R^{-1}z(t_i)] \qquad (14)$$

$$P(t_i)^{-1} = \tilde{P}(t_i)^{-1} + H^{\mathrm{T}}R^{-1}H \qquad (15)$$

Eq. (14) can be equivalently expressed in a more familiar form using the gain matrix as

$$\hat{x}(t_i) = \tilde{x}(t_i) + K(t_i)[z(t_i) - H\tilde{x}(t_i)] \qquad (16)$$

$$K(t_i) = \tilde{P}(t_i)H^{\mathrm{T}}[H\tilde{P}(t_i)H^{\mathrm{T}} + R]^{-1} \qquad (17)$$

The above equations along with Eq. (9) and Eq. (10) form the celebrated Kalman filter, the optimal linear on-line data processing algorithm. In identifying the system state vector at time t_i, it uses both the $\tilde{x}(t_i)$, which is the predicted state estimate based on the known system dynamics, and $(H^{\mathrm{T}}H)^{-1}H^{\mathrm{T}}z(t_i)$, which is the estimate of the state according to the measurements. The above knowledge and information are used to an extent proportional to their reliability, defined by $\tilde{P}(t_i)^{-1}$ and $H^{\mathrm{T}}R^{-1}H$, respectively. This is a rather special case of system identification in which the system behavior is known accurately enough. In the following sections, the more general situations are considered in which the system model contains uncertain portions parametrized for on-line identification, as shown in Fig. 1. The Kalman filter derived in this section offers a good starting point for this generalization.

It has often been the practice to determine state variables directly from the measurements, without any prior filtering or data preprocessing. The simultaneous measurement, for example, of nitrogen uptake rate and off-gas analysis in combination with elemental balances can yield estimates of the biomass concentration. This approach leaves the obtained estimates entirely exposed to measurement errors and process disturbances. This can be seen in Eq. (14), which, assuming that measurements alone are relied upon for the estimation of the state, yields

$$\hat{x}(t_i) = (H^{\mathrm{T}}H)^{-1}H^{\mathrm{T}}z(t_i) \qquad (18)$$

The above equation shows that the state estimates are completely vulnerable to any error that may affect the value of $z(t_i)$ that is used in this estimation. Furthermore, the current state value has no bearing on the prediction of the next time instant, and any additional knowledge about the process is simply disregarded. As a result, estimates obtained this way will be noisy, unstable, and inaccurate. The average of several consecutive measurements can yield smooth estimates, but this imposes arbitrary rules on measurement processing solely for the sake of smoothing. Besides conflicts that may arise if this averaging is incompatible with the process, valuable information contained in the measurements will be lost in the smoothing process.

3 Extended Kalman Filter

The Kalman filter (KF) of the previous section is the optimal state estimator for a linear system with predetermined parameters. Unfortunately, it is not directly applicable to fermentation systems due to their nonlinearity and difficulties in determining model parameters. The extended Kalman filter (EKF) was originally developed for state estimation in systems with nonlinear model representation. More importantly, the EKF can also handle situations with unknown, time-varying, or ill-defined model parameters. As the latter characteristics appear frequently in fermentation, EKF theory may have applications to fermentor identification and is reviewed briefly in this section.

3.1 EKF Algorithm

The Kalman filter of the previous section cannot be applied to nonlinear systems because the Gaussian distributions are not preserved in time for such systems. Rigorous treatment of the time evolution of the distribution for nonlinear systems, although feasible, quickly introduces insurmountable algebraic complications. A more practical approach is to linearize the model equations and consider the propagation of only the mean and the variance. This is the approach taken by the EKF, which is a widely accepted data processing algorithm for dynamic models of nonlinear systems. Moreover, in order to minimize the errors brought about by the linearization, the reference trajectory is systematically updated by relinearizing the model equations about the optimal state estimates as they become available:

$$f_x[\hat{x}] \equiv \frac{\partial f[x]}{\partial x}\bigg|_{x=\hat{x}} \tag{19}$$

$$h_x[\hat{x}] \equiv \frac{\partial h[x]}{\partial x}\bigg|_{x=\hat{x}} \tag{20}$$

This yields the following equations for the propagation of the mean and the variance:

$$\dot{\tilde{x}} = f[\tilde{x}, u, t] \qquad \tilde{x}(t_{i-1}) = \hat{x}(t_{i-1}) \tag{21}$$

$$\dot{\tilde{P}} = f_x[\hat{x}]\tilde{P} + \tilde{P}f_x[\hat{x}] + Q \quad \tilde{P}(t_{i-1}) = P(t_{i-1}) \tag{22}$$

The updating equation is accordingly modified as follows:

$$\hat{x}(t_i) = \tilde{x}(t_i) + K(t_i)\{z_i - h[\tilde{x}(t_i)]\} \tag{23}$$

where the optimal gain matrix and the update of the covariance matrix are calculated by

$$K(t_i) = \tilde{P}(t_i)h_x^T[\hat{x}_i]\{h_x[\hat{x}_i]\tilde{P}(t_i)h_x^T[\hat{x}_i] + R\}^{-1} \tag{24}$$

$$P(t_i)^{-1} = \tilde{P}(t_i)^{-1} + h_x^T[\hat{x}_i]R^{-1}h_x[\hat{x}_i] \tag{25}$$

This completes the EKF algorithm. By analogy to the Kalman filter, the same statistical interpretation can be offered for the EKF. The defining equations similarly express a statistically balanced compromise between the current state and incoming measurements.

The EKF algorithm has found applications in the estimation of both the state variables and model parameters. This is done by augmenting the state vector to also include those parameters which have to be identified. Thus, the parameters are treated as additional state variables. In so doing, the dynamics of the parameter change must be modelled, and several possibilities exist. For reliably stationary parameters one obviously uses

$$\dot{\theta} = 0 \tag{26}$$

For parameters varying slowly with time, the following dynamic equation has been employed:

$$\dot{\theta} = \xi(t) \tag{27}$$

where $\xi(t)$ is a random noise sequence. This dynamic model makes the evolution of the conditional density function more susceptible to incoming measurements. In this case, the recursive filter updating is a statistical compromise between the new information in the form of incoming measurements and the best model prediction before the update. Use of a high intensity for the process noise, ξ, has the effect of diffusing with time the covariance matrix of the predicted parameter vector in the course predicted by Eq. (22). This, in turn, tends to make the updates rely more on the new measurements. The estimator is thus alerted to changes in the process dynamics and it adapts accordingly. When it is nearly certain that the model structure is accurately formulated and contains θ as time-invariant parameters, the diffusional intensity of the noise ξ may be reduced.

Although straightforward, the implementation of EKF is actually not a simple matter. As seen in Fig. 1, the filter is designed to compromise the disagreement, if any, between the model prediction and the incoming measurements based on their relative reliability as presented by the inverse of their variances. However, the question of "how unreliable the model representation is" raises some difficulties with its real implementation. This issue is related to "filter tuning" and is particularly important in fermentation processes. Suppose, for example, a metabolic variable in the model is not well known and it is parameterized and allowed to vary with time. It is not easy, in such a case, to know *a priori* how rapidly this parameter will change as fermentation progresses, or how inaccurate the model description will be while this parameter is still being identified. This makes it difficult to predetermine the state noise covariance matrix, Q. Furthermore, there are no straightforward guidelines to optimal tuning of the filter. Filter tuning is specific to individual processes and usually requires preliminary computer simulation. Sometimes it is possible to add an adaptive tuning mechanism which adjusts the magnitude of the state noise variance. During the fermentation, for example, if the model equations predict the system dynamics fairly well, then the variance of the residuals (i.e., the difference between the measurements and the estimates) gives an estimate of the variance of the measurement noise. As the newly calculated residual exceeds the average noise level, the discrepancy is attributed to the change in the system dynamics instead of just being neglected. In proportion to the discrepancy, the variances of the state variables and parameters

are increased to adapt to the changes in fermentation dynamics. This auto-tuning with adaptive noise estimation was proposed by JAZWINSKI (1970).

As a final note, it should be mentioned that the inclusion of the parameters as additional state variables will render a linear problem nonlinear. Furthermore, the computational load in the EKF implementation of the augmented model increases significantly. It is therefore advisable to carry out a sensitivity analysis beforehand in order to select those parameters that do vary during the process and that also have a decisive effect on culture identification.

3.2 EKF Applications to Bioreactor Identification

Case Study 1

An interesting application of the EKF to a hypothetical enzymatic process of cellulose hydrolysis is reported by CAMINAL et al. (1987). The main objective of this work was to identify the bioreactor state in terms of concentrations and extent of hydrolysis. This was accomplished from measurements of reactant and product concentration, but in the face of uncertainties in the rate model parameters and slow enzyme deactivation.

The process considered consists of two sequential enzymatic steps for the overall hydrolysis of cellulose to glucose:

$$\text{cellulose} \xrightarrow{\text{cellulase}} \text{cellobiose}$$

$$\text{cellobiose} \xrightarrow{\beta\text{-glucosidase}} \text{glucose}$$

Both steps are carried out in the same well-mixed reactor from which samples are withdrawn periodically and analyzed off-line for the concentrations of the three key substances. Enzymatic assays are employed which yield in this particular case direct measurements of all state variables. However, due to interfering noise and assay inadequacies, the above measurements are corrupted and do not constitute an accurate representation of the real concentrations. Hence, identification theory was applied to extract the best concentration estimates from these measurements.

Following the mathematical formulation of Eq. (2), the measurements are related to their actual values as follows:

$$C_{A,m} = C_A + v_1 \tag{28}$$

$$C_{B,m} = C_B + v_2 \tag{29}$$

$$C_{C,m} = C_C + v_3 \tag{30}$$

where C_A, C_B, and C_C are the actual levels of cellulose, cellobiose, and glucose, respectively. The measured concentrations denoted by the subscript m differ from the corresponding true values by the random noise sequences, v_1, v_2, and v_3, respectively.

Assuming Michaelis–Menten kinetics with competitive inhibition for the rates of enzymatic reactions, the dynamics of the state is described by the following material balances:

$$-\frac{dC_A}{dt} = \frac{r_m C_A}{K_m(1 + C_B/K_i) + C_A} + \zeta_1 \tag{31}$$

$$\frac{dC_B}{dt} = -\frac{dC_A}{dt} + \frac{dC_C}{dt} + \zeta_2 \tag{32}$$

$$\frac{dC_C}{dt} = \frac{r_m' C_B}{K_m'(1 + C_C/K_i') + C_B} + \zeta_3 \tag{33}$$

with r_m and r_m' representing the activity of the cellulase and β-glucosidase enzymes, K_m and K_m' the corresponding Michaelis constants, and K_i and K_i' the inhibition constants. Random noise sequences ζ_1, ζ_2, and ζ_3 are added linearly to material balances to account for unknown effects, pertubations, and uncertainties unaccounted for in material balances. Eqs. (31) to (33) are equivalent to Eq. (1) for the state dynamics of this example.

The next step in the EKF implementation is the augmentation of the state vector to include uncertain model parameters. In particular, the enzymatic activities r_m' and r_m in Eqs. (31) to (33) are expected to decay at a considerable rate due to thermal deactivation at the working temperature of 50 °C. Consequently, the model structure with constant values for r_m' and r_m is not adequate for a reliable description of the system dynamics. Furthermore, good estimates

of the Michaelis and inhibition constants are not available, especially during the initial stages of the process. CAMINAL et al. (1987) subjected all the model parameters to identification. Following the approach described in Sect. 3.1, the parameters are modelled as nonstationary stochastic processes

$$\dot{\theta} = \xi \tag{34}$$

where θ^T is defined as $[r_m, r_m', K_m, K_m', (K_m/K_i), (K_m'/K_i')]^T$ and ξ^T as $[\zeta_4, \zeta_5, \zeta_6, \zeta_7, \zeta_8, \zeta_9]$. Note that by treating the parameters as additional state variables the algorithm of Eqs. (19) to (25) can be used to obtain the best estimates of the extended state vector, except for the assignment of the tuning factors.

Intuitively, r_m' and r_m appear to show more rapid variation than other kinetic constants, since these two parameters reflect the enzyme thermal deactivation which is certainly anticipated. Thus, the noise processes for these two parameters ζ_4 and ζ_5 require higher diffusional intensity of their distribution than in the case of other parameters. In this work, the tuning matrix is determined by simulation as

$$\zeta^T = [10^{-8}, 0, 10^{-8}, 10^{-5}, 10^{-5}, 10^{-8}, 10^{-8},$$
$$10^{-8}, 10^{-7}] \tag{35}$$

The estimates of the concentrations of each sugar are shown in Fig. 2. The decrease of the estimated β-glucosidase enzymatic activity r_m' becomes significant after the initial period of 400 minutes as shown in Fig. 3. Inclusion of this enzymatic activity in the extended state vector enhanced the ability of the identification algorithm to track changes in the reaction kinetics. This adaptively updated dynamic model served for the unbiased elimination of noises contained in the raw measurements.

Case Study 2

Although the concentration of biomass and of some key components of the abiotic phase are frequently used to describe the state of fermentors, these variables are by no means all-inclusive in their representation of the biological process. Certain parameters are occasionally reported with the above variables, since they

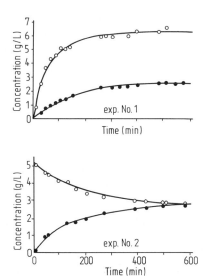

Fig. 2. Case Study 1 of EKF: Raw measurements of cellobiose (○) and glucose (●) concentrations and their estimates by EKF (curves). Experiment 1 (top): initial concentrations of cellulose and cellobiose are 9.95 g/L and 0.0 g/L, respectively. Experiment 2 (bottom): 0.0 g/L and 5.54 g/L, respectively. Figures from CAMINAL et al. (1987).

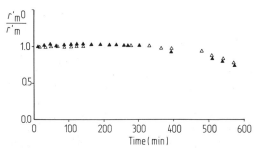

Fig. 3. Case Study 1 of EKF: Estimation of parameter r'_m, the enzymatic activity of β-glucosidase, by EKF. The EKF algorithm identifies the decrease in the enzymatic activity during the reaction. (△) experiment 1 and (▲) experiment 2. Figures from CAMINAL et al. (1987).

are believed to convey additional information about the culture. Such parameters, termed culture parameters (STEPHANOPOULOS and SAN, 1984), include the specific rate of growth, specific metabolic rate, and instantaneous overall fermentation yields.

In another application of EKF to fermentor identification, SAN and STEPHANOPOULOS (1984a) attempted the on-line identification of batch, fed-batch, and continuous fermentation of baker's yeast on glucose-limiting nutrients. This is the first time that rigorous identification procedures were applied to bioreactors, and three novel contributions of this application should be noted. First, an expanded fermentor state was used, including the usual fermentor variables as well as the culture parameters mentioned above. Second, no models were employed for the dependence of the latter on the concentration of biomass and components of the abiotic phase. Instead, culture parameters were identified by employing an adaptive estimation algorithm. Third, several of the measurements employed were only indirectly related to the fermentor state. Specifically, ammonia addition for pH control was measured, and oxygen uptake and carbon dioxide evolution rates were monitored by off-gas analysis. The measurements were combined with stoichiometric balances which allowed the calculation of some key variables and the casting of the problem into the form of Eqs. (1) and (2). In the same context, proton balances have been employed in other applications in order to relate ammonia addition and pH control to the production of acidic and basic products (SAN and STEPHANOPOULOS, 1984b).

In this example, the state vector comprised biomass, substrate, and product concentrations as well as the specific growth rate, μ, and the overall substrate and product yields. The reactor was equipped for the on-line measurement of pH, temperature, and the amount of ammonia added for pH control. Off-gas analysis also produced on-line measurements of the oxygen uptake and carbon dioxide evolution rate, *OUR* and *CER*, respectively.

In order to relate the above measurements to the state variables, stoichiometric relationships were invoked based on the following representation of the fermentation process (WANG et al., 1977):

$$aC_6H_{12}O_6 + bO_2 + cNH_3 \rightarrow$$
$$C_\alpha H_\beta O_\gamma N_\delta + dH_2O + eCO_2 + fC_2H_6O + \text{m.e.p.}$$

This relation depicts the partition of glucose and nitrogen source into yeast biomass, ethanol product, and other miscellaneous extracellular products (m.e.p.). Two approximations were introduced. First, the assumption that the formation of m.e.p. in the fermentation reaction was negligible. Second, that the elemental composition of yeast biomass was constant. The validity of the second approximation was experimentally tested for cells at different phases of fermentation, which were found to be adequately described by the formula $CH_{1.666}O_{0.511}N_{0.168}$.

For known stoichiometry, the above reaction permits the determination of the total biomass growth rate and the total metabolic rates (product formation, substrate consumption) from the measurement of OUR alone. In order to allow for changing stoichiometric coefficients due to variations in the carbon partition, it was proposed to continuously determine such coefficients from the on-line measurement of gas-exchange rates. Including the nitrogen uptake rate measurement, NUR, the four elemental balances and the three on-line measurements of OUR, CER, and NUR are related through the following equation to the unknown stoichiometric coefficients:

$$
\begin{bmatrix}
6 & 0 & 0 & 0 & -1 & -2 \\
12 & 0 & 3 & -2 & 0 & -6 \\
6 & 2 & 0 & -1 & -2 & -1 \\
0 & 0 & 1 & 0 & 0 & 0 \\
0 & 0 & 0 & 0 & 1 & 0 \\
0 & 1 & 0 & 0 & 0 & 0
\end{bmatrix}
\begin{bmatrix}
a \\ b \\ c \\ d \\ e \\ f
\end{bmatrix}
=
$$

$$
=
\begin{bmatrix}
1.000 \\
1.666 \\
0.511 \\
0.168 \\
0.168 CER/NUR \\
0.168 OUR/NUR
\end{bmatrix}
\tag{36}
$$

The above equation allowed the on-line determination of the coefficients a–f, and then the determination of the biomass production rate ($R = OUR/b$), and the substrate-to-biomass ($Y_b = 1/a$) and substrate-to-product ($Y_p = f/a$) yields. Following the formalism of Eq. (2), these variables, i.e., R, Y_b, and Y_p, can be regarded as the new measurements related to the state through the following equations (subscript m denotes the actual measurements of the corresponding quantities):

$$R_m = \mu b + \xi_1(t) \tag{37}$$

$$Y_{b,m} = Y_b + \xi_2(t) \tag{38}$$

$$Y_{p,m} = Y_p + \xi_3(t) \tag{39}$$

The added noise rationalizes the signal interference in the raw measurements and uncertainties in the mapping of the new measurements by Eq. (36).

For the system dynamics, the following material balances were used:

$$\dot{b} = \mu b - Db \tag{40}$$

$$\dot{s} = D(s_f - s) - \frac{1}{Y_b}\mu b \tag{41}$$

$$\dot{p} = \frac{1}{Y_p}\mu b - Dp \tag{42}$$

where b, s, and p are the biomass, glucose, and ethanol in the reactor. Written for a growth associated product, the formalism can be extended in a similar way to non-growth associated products as well. In Eqs. (40), (41), and (42), D is the dilution rate, equal to zero or $\dot{V}/V$ for batch and fed-batch reactors, respectively.

In this form, Eqs. (40) to (42) are little more than a definition of the parameters, μ, Y_b, and Y_p. In order to complete the formulation of the state dynamics, equations for the time rate of change of μ, Y_b, and Y_p are needed. No obvious balances of the type written for b, s, and p are available. One has the option of formulating detailed metabolic models for the culture parameters or relying entirely on the incoming measurements for their identification. In choosing between these two possibilities, it should be noted that the Kalman filter generates optimal estimates by suitable weighting between the accuracy of the incoming measurements and the reliability of the system dynamic equation. When the equations portray the actual system dynamics with adequacy, they play a vital role in the system identification scheme. On the other hand, if models are used which are of limited reliability or involve a large number of parameters that need to be estimated, the identification result may not be any better than that obtained by simply subjecting the parameters themselves to adaptive identification. The choice depends basically on the availability of a reliable model structure. If one is available, its inclusion leads to a more robust filter performance at the expense of increased complication with the implementation algorithm. If there are doubts about the validity of the model, its use is inconsequent and the estimation algorithm relies entirely on the measurements.

In this case study, it was assumed that no model was available and the culture parameters were treated as independently identifiable quantities and allowed to vary with time according to the dynamics

$$\dot{\mu} = \eta_1(t) \tag{43}$$

$$\dot{Y}_b = \eta_2(t) \tag{44}$$

$$\dot{Y}_p = \eta_3(t) \tag{45}$$

The above correspond to Eq. (27) of the previous section for the augmented state vector. With the state balances (40), (41), and (42) and the measurement equations (37) to (39), they can be used in the filter implementation according to Eqs. (19) to (25) of the previous section.

The operation of the implemented filter can be summarized as follows: The available raw measurements are mapped to a more relevant form, the culture parameters, by using the fermentation equation (36). Filtering by Eqs. (43) to (45) follows to eliminate noise. Then they are fed to the deterministic material balances, Eqs. (40) to (42), to compute the changes of the biomass, the substrate level, and the product level. This approach is well justified when the parameters do not vary rapidly under the operating conditions. However, if they change at rates comparable to the state variables and the incoming measurements have significant noise levels, the filter implementation has serious tuning difficulties for proper tracking capability. This is primarily because the model does not contain information about the behavior of these parameters except that they may change during the course of fermentation.

The performance of the implemented filter with a laboratory fed-batch fermentor is shown in Figs. 4 to 9. In these figures, esti-

mates by simple moving-average and intermittent off-line measurements are also shown for comparison. The estimates for biomass, glucose level, and ethanol level show smooth trends and appear to agree well with the off-line measurements. As predicted from the lack of explicit dynamic structures for the substrate and ethanol yields, the estimates of these parameters are mostly dependent on the incoming measurements. Thus, the estimates evidently reflect the noise of the measurements.

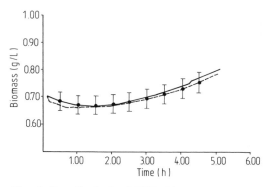

Fig. 4. Case Study 2 of EKF: Biomass concentrations in a fed-batch fermentation of baker's yeast: (●) off-line measurements; (···) estimates by EKF; (—) estimates by moving average. Note that EKF estimates agree well with off-line measurements as fermentation progresses. Figures from SAN and STEPHANOPOULOS (1984b).

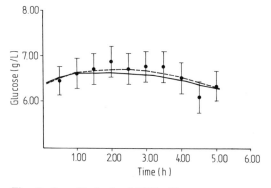

Fig. 5. Case Study 2 of EKF: Glucose concentrations in a fed-batch fermentation of baker's yeast: (●) off-line measurements; (···) estimates by EKF; (—) estimates by moving average. Figures from SAN and STEPHANOPOULOS (1984b).

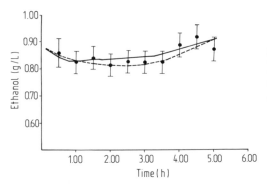

Fig. 6. Case Study 2 of EKF: Ethanol concentrations in a fed-batch fermentation of baker's yeast: (●) off-line measurements; (· · ·) estimates by EKF; (—) estimates by moving average. Figures from SAN and STEPHANOPOULOS (1984b).

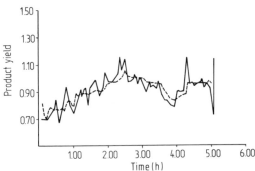

Fig. 8. Case Study 2 of EKF: Product yield as a function of time in a fed-batch fermentation of baker's yeast: (· · ·) estimates by EKF; (—) estimates by moving average. Note that EKF gives smooth estimates. Figures from SAN and STEPHANOPOULOS (1984b).

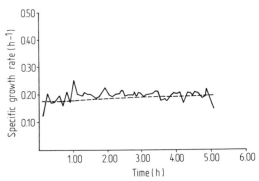

Fig. 7. Case Study 2 of EKF: Specific growth rate as a function of time in a fed-batch fermentation of baker's yeast: (· · ·) estimates by EKF; (—) estimates by moving average. Note that EKF gives noise-free estimates. Figures from SAN and STEPHANOPOULOS (1984b).

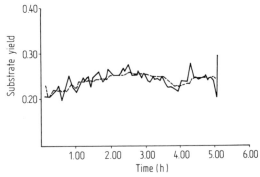

Fig. 9. Case Study 2 of EKF: Substrate yield as a function of time in a fed-batch fermentation of baker's yeast: (· · ·) estimates by EKF; (—) estimates by moving average. Figures from SAN and STEPHANOPOULOS (1984b).

However, the estimate of the specific growth rate shows substantially noise-free behavior, whereas simple averaging manifests unsteady fluctuation. Large non-recurring measurement pertubations are rejected by the filter.

As mentioned earlier, the state vector augmentation results in a nonlinear estimation problem. Therefore, despite the flexibility and straightforward implementation of EKF, the recursive identification of Eq. (3) may not give the same solution as the off-line batch identification process due to approximations introduced by the nonlinearity. This sometimes makes it difficult to establish satisfactory convergence. This point is elaborated further in Sect. 5.

4 Parameter Estimator on State Observer: Sequential State/Parameter Estimation (SSPE)

It should be clear by now that a significant part of the overall fermentor identification is the estimation of unknown or uncertain model parameters. In EKF, the system identification is accomplished by treating the parameters as additional state variables. In this section, another state/parameter identification method is presented. It is based on the observation that, in an adequately formulated model, the model parameters are distinguished from the state variables in that they change markedly slower than the state variables, and sometimes they are, in fact, time invariant.

In order to illustrate this point, consider a model representation in which the parameters are truly time invariant. If these parameter values are accurately known, then the state vector estimation made by the Kalman Filter (or by Extended Kalman Filter for a nonlinear model) and its future projection using the model equation will result in small prediction errors. Conversely, this property can be used for identification of the parameters in the model equation if they are *a priori* unknown. The best parameter values are the ones which result in the smallest prediction errors. This is the underlying concept of the system identification method dealt with in the following section. Algorithms for the implementation of the concept of recursive parameter estimation are presented in Sect. 4.2.

4.1 Parameter Estimation by Least Square Error Method

This method determines the unknown parameters θ by seeking to maximize the conditional distribution density $f_{x(t_i), Z(t_i)|\theta}$ of the state and all the measurements to the present time $(Z(t_i) = [z(t_i),\ z(t_{i-1}), \ldots, z(t_0)])$ conditioned upon the values of the unknown parameters. By Bayes' rule,

$$f_{x(t_i), Z(t_i)|\theta} = f_{x(t_i)|Z(t_i),\theta} \prod_{j=1}^{i} f_{z(t_j)|Z(t_{j-1}),\theta}$$

If one assumes a Gaussian distribution, it is more convenient to maximize the likelihood function L defined as

$$L = \ln f_{x(t_i), Z(t_i)|\theta} =$$
$$= \ln [f_{x(t_i)|Z(t_i),\theta} \prod_{j=1}^{i} f_{z(t_j)|Z(t_{j-1}),\theta}] \quad (46)$$

For a linear system, the likelihood function will be given by the following quadratic sum:

$$L = -\ln \left[(2\pi)^{\frac{n}{2}} (\det Q_x)^{\frac{1}{2}} \right] -$$
$$- \sum_{j}^{i} \ln \left[(2\pi)^{\frac{n}{2}} (\det Q_z(t_j; \theta))^{\frac{1}{2}} \right] - \quad (47)$$
$$- \tfrac{1}{2} [x(t_i) - \hat{x}(t_i; \theta)]^{\mathrm{T}} Q_x^{-1} [x(t_i) - \hat{x}(t_i; \theta)] -$$
$$- \tfrac{1}{2} \sum_{j}^{i} [z(t_j) - H\tilde{x}(t_j; \theta)]^{\mathrm{T}} \cdot$$
$$\cdot Q_z(t_i; \theta)^{-1} [z(t_j) - H\tilde{x}(t_j; \theta)]$$

At time t_i, the optimal state/parameter estimates are obtained by solving the following two simultaneous equations.

$$\left. \frac{\partial L}{\partial \theta} \right|_{\substack{x=\hat{x} \\ \theta=\hat{\theta}}} = 0^{\mathrm{T}} \quad (48)$$

$$\left. \frac{\partial l}{\partial x} \right|_{\substack{x=\hat{x} \\ \theta=\hat{\theta}}} = 0^{\mathrm{T}} \quad (49)$$

It should be noted that Eq. (48) expresses the fact that the optimal estimates of the model parameters are those which would make all the observed measurements up to time t_i most likely to happen. Furthermore, the meaning of Eq. (49) is that the optimal estimate of the state vector is the one that locates the peak of its distribution. Thus, the simultaneous solution of Eq. (48) and Eq. (49) actually becomes a sequential solution. First, the parameter vector is determined by Eq. (48) as that which minimizes the sum of the prediction errors. Second, the state estimation, which is the solution of Eq. (49), is obtained by filtering the measurements with the aid of the updated model by the new parameter values. The second step is identical to the Kalman filter discussed in Sect. 2.2. Therefore, the focus of this approach to system identification is on the development of a tractable recursive least square prediction error algorithm for parameter identification.

4.2 Recursive Least Square Error Algorithm

The development of the previous section is conceptually straightforward and not substantially different from typical static (batch) parameter estimation by least squares. However, it leads to a sizable implementation problem so that a simplifying recursive version of Eq. (48) is sought in this section.

First it is assumed that the covariance matrix Q_z does not depend on the parameter vector. This allows one to discard the normalizing factors from the density functions in Eq. (48) which can now be written equivalently as

$$J = -\frac{\partial}{\partial\theta}\frac{1}{2}\sum_j^i [z(t_j) - H\tilde{x}(t_j;\theta)]^{\mathrm{T}}\varLambda^{-1}(t_j) \cdot$$

$$\cdot [z(t_j) - H\tilde{x}(t_j;\theta)] = 0^{\mathrm{T}} \qquad (50)$$

$z(t_j)$ is the measurement at time t_j and $\tilde{x}(t_j)$ is the prediction of the state at time t_j from the state at time t_{j-1} using the model equation. $\varLambda$ is the covariance matrix of the prediction error (denoted as Q_z in Eq. (47)), whose dependency on the parameter vector is neglected. The parameter vector that satisfies Eq. (48) and solves this nonlinear least square error problem can be found by the Gauss–Newton iterative searching method, as

$$\hat{\theta}_N = \hat{\theta}_{N-1} + S[\hat{\theta}_{N-1}]^{-1}G[\hat{\theta}_{N-1}] \qquad (51)$$

where N is the iteration index and G and S are the first (Gradient) and second (Hessian) derivatives of the likelihood function with respect to θ. By iterating on the parameter update, the parameter values which maximize the likelihood function (or minimize the sum of the least square prediction error) can be located.

At this point it is important to pause and reflect upon the enormity of the problem suggested by Eqs. (50) and (51). At some sampling time t_j, one needs to first formulate the sum of quadratic terms of Eq. (50). This requires the state predictions $\tilde{x}(t_k)$, $0 \le k < j$, i.e., at all previous sampling points. These predictions are obtained by the model equations and depend on the parameter values used in the current iteration. After forming the derivative suggested by Eq. (50) and the Gradient and Hessian matrices of the likelihood function, one needs to iterate Eq. (51) until satisfactory convergence is achieved. Furthermore, as the parameters are updated at every iteration, the computation of the Gradient and Hessian matrices becomes heavily loaded. This is appreciated by looking at the change of the state prediction $\tilde{x}(t_k)$ resulting from changes in the parameter vector. This prediction depends on the parameters when it is projected from the state estimate of the previous point $\hat{x}(t_{k-1})$. The state estimate, in turn, depends on the parameter vector which is obtained as seen in Eqs. (9), (10), (16), and (17). The entire process is repeated at the next sampling point to yield the parameter values that minimize the sum of the prediction error up to that time. If one considers the very large number of sampling points, the strong dependence of Gradient and Hessian matrices on the parameters, and the recursive nature of the state prediction equations, it becomes clear that the rigorous approach for parameter estimation is time-consuming and very intensive, making it inappropriate for on-line implementation.

One approximation that greatly simplifies the search for the optimal set of parameters is to replace the iterative index in Eq. (51) with the recursive time index. This means that instead of iterating batch with a static set of measurements, one now does a dynamic iteration, moving from measurement to measurement in time and updating the parameters by an equation analogous to Eq. (51). The Gauss–Newton search for the parameter vector is executed only once at each sampling time starting from the previous sampling time. This approach is successful with slowly varying parameters. Compared to the rigorous approach of Eqs. (50) and (51), it will converge more slowly on the correct parameter values. However, it may be rendered more adaptive by some mechanisms which will be discussed later in this section. In general, it works well with systems that exhibit smooth variations in time such as fermentors.

The new parameter updating equation that is obtained by replacing the iterative index with the recursive time index is

$$\hat{\theta}(t_i) = \hat{\theta}(t_{i-1}) + S[t_i, \hat{\theta}(t_{i-1})]^{-1} G[t_i, \hat{\theta}(t_{i-1})] \quad (52)$$

G and S in Eq. (52) are still the Gradient and Hessian matrices of the likelihood function. They can also be approximated by expressions that drastically simplify the calculations due to their recursive nature. Referring to LJUNG and SÖDERSTRÖM (1985) for the details involved in such approximations, the update equations for the matrices G and S are given by

$$G[t_i, \hat{\theta}(t_{i-1})] \approx$$
$$\approx \psi(t_i) \hat{\Lambda}^{-1}(t_i) [z(t_i) - H\tilde{x}(t_i)] \quad (53)$$

$$S[t_i, \hat{\theta}(t_{i-1})] \approx$$
$$\approx S[t_{i-1}, \hat{\theta}(t_{i-2})] + \psi(t_i) \hat{\Lambda}^{-1}(t_i) \psi^{\mathrm{T}}(t_i) \quad (54)$$

Matrix $\Lambda(t_i)$, which is the covariance matrix of the prediction error, is approximately updated by the recursive formula

$$(i)\hat{\Lambda}(t_i) = (i-1)\hat{\Lambda}(t_{i-1}) + [z(t_i) - H\tilde{x}(t_i)][z(t_i) - H\tilde{x}(t_i)]^{\mathrm{T}} \quad (55)$$

Note that Eq. (52) indicates that when a discrepancy exists between the prediction and the actual measurement (the prediction error), the parameter vector readjusts itself in an optimally balanced manner. The readjustment is proportional to the prediction error $(z - H\tilde{x})$, and the gain matrix $(S^{-1}\psi\hat{\Lambda}^{-1})$, which is calculated on a statistical basis.

Probably the most important quantity in this recursive parameter estimation algorithm is the sensitivity matrix, also known as ψ-matrix. It appears in Eqs. (53) and (54) in determining the Gradient and Hessian matrices of the likelihood function. It expresses the sensitivity of the state prediction to changes in the

parameter values and is defined as

$$\psi^{\mathrm{T}}(t_i) \equiv \frac{\mathrm{d}}{\mathrm{d}\theta} [H\tilde{x}(t_i)] \quad (56)$$

In order to determine the ψ-matrix, a state prediction equation is first formed by projecting the state estimate at the previous time instant with the transition matrix $\Phi(t_i, t_{i-1})$ derived from the model equation

$$\tilde{x}(t_i) = \Phi(t_i, t_{i-1})\hat{x}(t_{i-1}) \quad (57)$$

The Kalman filter equation is used for the state estimate at time t_{i-1} as in Eq. (16):

$$\hat{x}(t_{i-1}) = [I - K(t_{i-1})H]\tilde{x}(t_{i-1}) + K(t_{i-1})z(t_{i-1}) \quad (58)$$

Combining Eqs. (57) and (58) yields an equation for the propagation of the state prediction with time. Upon introduction into Eq. (56) and differentiation, the following equation is obtained for the sensitivity matrix (here, for the sake of simplicity, the matrix H is taken as an identity matrix):

$$\psi(t_i)|_{\hat{\theta}(t_{i-1})} = \Phi[I - K]\psi(t_{i-1})|_{\hat{\theta}(t_{i-1})} + \frac{\partial\Phi[I-K]}{\partial\theta}\tilde{x}(t_{i-1})\Big|_{\hat{\theta}(t_{i-1})} + \frac{\partial K}{\partial\theta}z(t_{i-1})\Big|_{\hat{\theta}(t_{i-1})} \quad (59)$$

The subscript, $|_{\hat{\theta}(t_{i-1})}$, indicates that the corresponding quantities are evaluated at the parameter vector estimate available at time t_{i-1}. Eq. (59) is a rigorous equation for the time evolution of the ψ-matrix. However, the computation of $\psi(t_{i-1})|_{\hat{\theta}(t_{i-1})}$ causes considerable complications. It is generally accepted to use $\psi(t_{i-1})|_{\hat{\theta}(t_{i-2})}$ instead of $\psi(t_{i-1})|_{\hat{\theta}(t_{i-1})}$ in Eq. (59), leading to an approximate recursive update of the ψ-matrix which, however, can be implemented on-line.

The algorithm of Eqs. (54) through (59) can be summarized as follows. First, the ψ-matrix is determined before the measurement at time t_i is made. Using Eq. (57) for the state prediction and the current measurement $z(t_i)$, the prediction error $(z(t_i) - H\tilde{x}(t_i))$ and its covariance matrix $\Lambda(t_i)$ are calculated. Subsequently, the Gradient and Hessian matrices are calculated and the parameter vector estimate is updated by Eq. (52). This leads to an update of the system dynamic model. Finally, the estimation of the state vector is made with the measurements and the updated model by Kalman filter (or extended Kalman in case of nonlinear model representation). This sequential estimation of parameter and state vectors is shown in Fig. 10.

weighting sequence

$$-\frac{\partial}{\partial\theta}\frac{1}{2}\sum_{j=1}^{t}[z(t_j)-H\tilde{x}(t_j)]^{\mathrm{T}}\cdot$$

$$\cdot\{\alpha_{t,j}\Lambda(t_j)\}^{-1}[z(t_j)-H\tilde{x}(t_j)]=0^{\mathrm{T}} \qquad (60)$$

Use of a weighting sequence $\alpha_{t,j}$, which increases with decreasing time index j, causes the

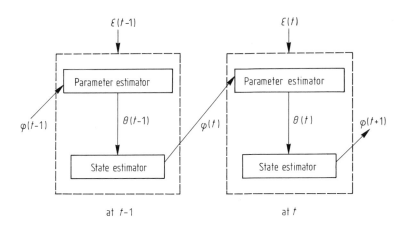

Fig. 10. Operation of a sequential state/parameter estimator.

As mentioned earlier, the above algorithm is applicable to systems with adequate models for which the parameters are not expected to vary significantly with time. If system dynamics exhibit significant variation with time so that the model parameter vector needs to be updated accordingly, the above algorithm may not show satisfactory tracking capability. This is caused by the continuous increase with time of the Hessian matrix, Eq. (54), resulting in a diminishing contribution of incoming measurements to the update of the parameters. Put differently, excessive confidence is gradually being built on current parameter estimates, which makes the algorithm place less weight on more recent observations.

In order to capture important information contained in current measurements and also to improve the ability of the algorithm to handle time-varying parameters, the algorithm is modified to prevent the overconfidence founded on current parameter estimates. This is accomplished by modifying the prediction error covariance matrix through the use of a

covariance of old data to become larger. This discounts the reliability of old data and depends more on recently collected findings for the determination of the parameter estimates. Different tracking schemes can be accommodated by selecting different weighting sequences, such as Fixed Windows and Forgetting Factor (GOODWIN and SIN, 1984). The result is often a more flexible and robust parameter estimation algorithm.

4.3 SSPE Application to Bioreactor Identification

Case Study 1

Use of the SSPE algorithm for regulating glucose concentration in yeast fermentation has been reported (PARK and RAMIREZ, 1989). In a fed-batch reactor, changes in the glucose level can be represented by the following material balances

$$X(t_{i+1}) = (1 + \tau\mu) X(t_i) + \omega_1(t_i) \tag{61}$$

$$S(t_{i+1}) =$$

$$= \frac{\tau\mu}{Y} X(t_i) + S(t_i) + \tau m q(t_i) + \omega_2(t_i) \tag{62}$$

$$V(t_{i+1}) = V(t_i) + \tau q(t_i) + \omega_3(t_i) \tag{63}$$

where X, S, and V are the total biomass, glucose, and culture volume in the fermentor, respectively. They form the state vector $x(t_i) = [X(t_i), S(t_i), V(t_i)]$. The flow rate of the medium into the reactor and its glucose concentration are represented, respectively, by $q(t_i)$ and m. The sampling interval, denoted as τ, was constant. These equations contain two model parameters, the specific growth rate μ and the substrate-to-biomass yield Y. Since the true values of these parameters are unknown, the noise sequence, $\omega(t_i)$, is added to compensate for model inaccuracy when estimates of the parameters are used in Eqs. (61) to (63).

On-line measurements of biomass and glucose concentrations were made by optical methods based on the biomass turbidity and a colorimetric glucose assay. For measurement of the total biomass and the glucose in the fermentor, their measured concentrations were multiplied by the measured culture volume. Then, allowing for noise interference in the measurements, the measurement equation becomes

$$z_m(t_i) = x(t_i) + \xi(t_i) \tag{64}$$

where $z_m(t_i) = [X_m(t_i), S_m(t_i), V_m(t_i)]$, the subscript, m, indicating the measured quantities. Thus, in this example, the measurement matrix H is the identity matrix in Eq. (58).

Before implementing Eqs. (53) to (59) to obtain recursive updates of the parameters μ and Y, the following transformation was introduced:

$$\theta_1 = 1 + \tau\mu \tag{65}$$

$$\theta_2 = -\frac{\tau\mu}{Y} \tag{66}$$

Direct identification of μ and Y results in a nonlinear filter algorithm. The transformed parameters, θ_1 and θ_2, make linear entries into the model equations (61) to (63) and thus simplify the algorithm. Once the estimates of θ_1 and θ_2 are obtained, μ and Y can be easily calculated from Eqs. (65) and (66).

The first step of SSPE implementation is to obtain the recursive update equation for the sensitivity matrix, ψ. Using the system equation Eqs. (61) to (63) with the measurement equation (64), the Kalman filter state predictor of Eqs. (57) and (58) can be obtained as a function of the two unknown parameters, θ_1 and θ_2. First, the system equation of Eqs. (61) to (63) is equivalently expressed as

$$x(t_{i+1}) = \Phi(\theta)x(t_i) + Bq(t_i) + \omega(t_i) \tag{67}$$

where Φ and B are the transition matrix and control-to-state vector represented as

$$\Phi(\theta) = \begin{bmatrix} \theta_1 & 0 & 0 \\ \theta_2 & 1 & 0 \\ 0 & 0 & 1 \end{bmatrix} \quad B = \begin{bmatrix} 0 \\ \tau m \\ \tau \end{bmatrix} \tag{68}$$

In a similar manner to Eqs. (57) and (58), the Kalman filter state prediction is expressed as follows:

$$\tilde{x}(t_{i+1}) = \Phi(\theta)[I - K(\theta)]\tilde{x}(t_i) +$$

$$+ \Phi(\theta)K(\theta)z(t_i) + Bq(t_i) \tag{69}$$

where the Kalman gain matrix $K(\theta)$ is obtained using the following two equations (equivalent to Eqs. (17) and (15), respectively):

$$K(\theta) = \Phi(\theta)[P(t_i, \theta) + R]^{-1} \tag{70}$$

$$P^{-1}(t_i, \theta) = [\Phi(\theta)P(t_{i-1}, \theta)\Phi^T(\theta) + Q]^{-1} + R^{-1} \tag{71}$$

By taking the first derivative of Eq. (69) with respect to θ, the recursive equation for the ψ-matrix of Eq. (59) is obtained as

$$\psi(t_{i+1}) = \Phi[I - K]\psi(t_i) +$$

$$+ \frac{\partial}{\partial\theta}[\Phi(I - K)\tilde{x}(t_i) + Kz(t_i)] \tag{72}$$

In calculating the last term in Eq. (72) at time t_i, the state prediction and the measurement vector are inserted as their values are available at this time. A numerical derivative method is used to obtain the matrix value of the term. Otherwise, a lengthy derivation is required to obtain an explicit expression for the last term in Eq. (72) due to the complicated dependence of the Kalman gain matrix on the parameter vector. The second step is the update of the parameter values. With the ψ-matrix obtained using Eq. (72), the current estimates of Λ, S, and G are produced by employing Eqs. (56), (54), and (55), respectively. Finally, the parameter vector update was made by using Eq. (53). The third step is the state vector estimation by Kalman filter Eq. (58) using the renewed model equations that incorporate the updated parameter values $\hat{\theta}_1$ and $\hat{\theta}_2$. Starting from an initial guess of the parameter values, the above procedure is repeated at each sampling moment.

In Sect. 4 it was noted that the underlying basis for the SSPE algorithm is the distinctively lower rate of parameter variation compared to that of the state variables. Since μ and Y can be strongly dependent upon culture conditions such as the glucose level, dissolved oxygen level, temperature, and pH, the parametrization of these two metabolic variables may not be appropriate for SSPE application. However, if the culture conditions are regulated at constant levels, it is reasonable to expect that these parameters do not show considerable variation. PARK and RAMIREZ (1989) used the SSPE algorithm coupled with regulation of culture conditions, including a model-based optimal glucose-level regulator.

Successful regulation of glucose concentration requires accurate estimates of the state variables and, furthermore, reliable prediction of bioreactor future behavior. The importance of a reliable model becomes even more pronounced at low glucose concentrations due to rapid culture dynamics that may lead to glucose depletion. The objective of robust glucose regulation at low levels was met by designing a control law which appropriately utilized the outputs of the system identification algorithm. In short, the control scheme consists of a feedforward and a Proportional-Integral (PI) feedback control. The feedforward control is determined by using model predictions of the state variables. Thus, it depends heavily on the

accuracy of the model equations to describe the actual fermentation dynamics. As such, the performance of the feedforward control is directly coupled to that of the system identification algorithm. In addition, a PI feedback control is supplemented as a backup in the event the feedforward control does not perform well. This approach of control law design offers the opportunity of a precise control performance with reliable on-line system identification. Also, if the system identification fails, disastrous consequences are prevented by shifting to the PI-backup control. It should be noted that the performance of the PI control is not as satisfactory as that of the optimally performing feedforward control.

Typical results of the performance of the combined system identification (SSPE) and the described control law are shown in Figs. 11 to 14. The glucose concentration is regulated at a setpoint of 0.2 g/L in a 20 L fermentor. In this

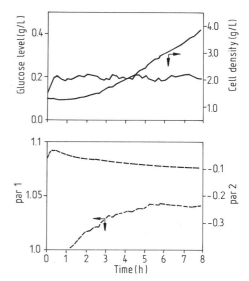

Fig. 11. Case Study 1 of SSPE: Glucose and biomass concentration changes in a glucose controlled fed-batch yeast fermentation. The glucose concentration is regulated at set-point 0.2 g/L by a control law which incorporates the SSPE algorithm. Figures from PARK and RAMIREZ (1989) (above).

Fig. 12. Case Study 1 of SSPE: Adaptation of parameter estimates of θ_1 (par1) and θ_2 (par2) by the SSPE algorithm. Note the rapid adaptation followed by stabilization of the estimates. Figures from PARK and RAMIREZ (1989) (below).

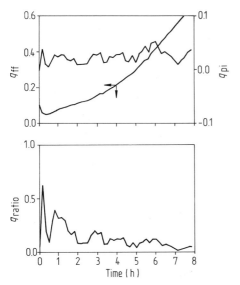

Fig. 13. Case Study 1 of SSPE: Feedforward (q_{ff}) and feedback (q_{PI}) control actions during a glucose-regulated fed-batch fermentation. Feedforward control is mainly determined by the SSPE system identification algorithm. Figures from PARK and RAMIrez (1989) (above).

Fig. 14. Case Study 1 of SSPE: Profile of the ratio ($q_{ratio} = |q_{PI}/q_{ff}|$), the regulation is done primarily by feedback (q_{PI}) and feedforward (q_{ff}) control action. Note that the contribution of q_{PI} decays as time progress, indicating that the SSPE algorithm performs the identification task satisfactorily. Figures from PARK and RAMIrez (1989) (below).

experiment, the Kalman filter gain matrix for state estimation was set relatively high so that the system identification algorithm is more sensitive to incoming measurements. An active update of the system dynamic model is indicated by the extensive change in the estimated parameter values. For the initial parameter values, those which were calculated independently at nearly the same culture conditions were used. However, significant variations during fermentor operation were observed. Based on the updated model, the feedforward control action regulates the culture glucose level precisely at the designated set point. As shown by the ratio of the two control actions, $|q_{PI}/q_{ff}|$, the regulation is done primarily by the feedforward control and the PI-backup control is not activated for most of the time.

When the function of the on-line model update is repressed, the control performance is not as satisfactory. If the Kalman gain matrix is chosen to be of small magnitude, the system identification algorithm becomes conservative and relies more on the initially known system dynamics. As shown in Fig. 15 and Fig. 16, the deviation of the glucose level from the set point becomes significant, indicating that the fermentor model on which the control action is based does not reflect the actual fermentor dynamics. Comparison of this result with the result in Figs. 11 through 14 exemplifies the importance of an active system identification for satisfactory fermentation regulation.

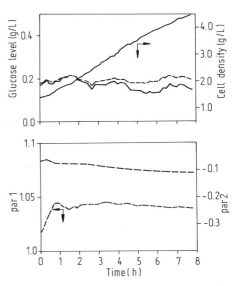

Fig. 15. Case Study 1 of SSPE: Performance of glucose level control when SSPE activity is repressed. Comparison with Fig. 11 demonstrates the importance of active system identification for satisfactory fermentor control. Figures from PARK and RAMIrez (1989) (above).

Fig. 16. Case Study 1 of SSPE: Profile of parameter estimates of θ_1 (par1) and θ_2 (par2) when SSPE activity is repressed. Note the distinctively slower adaptation of estimates compared with those in Fig. 12. Figures from PARK and RAMIrez (1989) (below).

Case Study 2

An algorithm for the detection of bioreactor contamination was reported by CHATTAWAY and STEPHANOPOULOS (1989). The behavior of contaminated cultures is distinguishably different from that of uncontaminated ones. By operating on-line a system identification algorithm, this type of erroneous fermentor performance and its dynamics can be identified. Combined with an appropriate decision criterion, this algorithm can then be used for the timely detection of contaminating species in a fermentor.

Consider the following two growth equations for the desired species and the contaminant in a batch culture:

$$x_{m,t+1} = (1 + \tau \mu_m) x_{m,t} + \zeta_{m,t} \tag{73}$$

$$x_{c,t+1} = (1 + \tau \mu_c) x_{c,t} + \zeta_{c,t} \tag{74}$$

Subscripts m and c designate the desired species and the contaminant, respectively. The specific growth rate of each species is denoted as μ and the sampling interval as τ. Since the specific growth rates are parametrized, the noise sequences ζ are added to compensate for model inaccuracy. Eqs. (73) and (74) form the state model equations.

Using the stoichiometric balances of Sect. 3.2 (Case Study 2), the data from off-gas analysis can be transformed to growth rate measurements. Furthermore, through nitrogen balance and ammonia addition measurements, on-line biomass concentrations were obtained. Accounting for the usual noise, these measurements are cast as follows in the measurement equations for the identification formalism:

$$z_{1,t} = x_{m,t} + x_{c,t} + \xi_{1,t} \tag{75}$$

$$z_{2,t} = \mu_m x_{m,t} + \mu_c x_{c,t} + \xi_{2,t} \tag{76}$$

Eqs. (75) and (76) relate the measurements to the state variables x_m and x_c. It should be noted that in the absence of a distinguishing characteristic for either of the species, neither the biomass nor the growth rate measurement can be partitioned between the desired strain and the contaminating species. In this context, the system of Eqs. (73) through (76) is non-observable, which means that the two measurements of growth rate and biomass do not contain sufficient information to estimate the individual biomass concentrations and to

identify the corresponding specific growth rates.

The non-observability problem can be solved by recalling that the algorithm has a two-fold objective, namely, accurate tracking for as long as a pure fermentation is maintained and providing an alarm signal at the earliest possible moment after the onset of contamination. One can then assign a known and constant value, b, to the specific growth rate of the contaminant and subject x_m and x_c to identification, a definitely legitimate task. By choosing a large value for b, in fact larger than the maximum possible specific growth rate of the desired strain, the growth of the latter is distinguished mathematically from that of the contaminant and, to a certain extent, physically also. If the reactor is not contaminated, system identification reveals that $x_{c,t}$ stay close to zero, for the very rapid growth kinetics of the contaminant is incompatible with the observed fermentor dynamics as expressed by the measurements of the biomass and growth rate. Under such conditions (i.e., $x_{c,t} \approx 0$), the system is observable and accurate estimates of μ_m and x_m are obtained. Upon the appearance of contaminants growing at a high rate, a non-zero biomass must be invoked to explain the faster changes in biomass and total growth rate. This produces an alarm signal when the state and parameter estimates beyond this point deviate from the true values, unless the selected value of b happens to be close to the true specific growth rate of the contaminant. For more sensitive detection, a small value of b should be used.

The SSPE algorithm was used for the identification of the specific growth rate μ_m and the estimation of the biomass concentration. To this end, a recursive equation for the update of the ψ-matrix is required. Combining Eqs. (57) and (58) yields the following equation for the propagation of the state prediction:

$$\tilde{x}(t_{i+1}) = \Phi[I - K(t_i)H]\tilde{x}(t_i) + \Phi K(t_i)z(t_i) \tag{77}$$

The equation for the propagation of the measurement prediction is obtained by substituting $\tilde{x}(t_{i+1})$ by $H^{-1}\tilde{z}(t_{i+1})$ and $\tilde{x}(t_i)$ by $H^{-1}\tilde{z}(t_i)$. Upon substitution and rearrangement, one obtains the following recursive prediction equation for the measurement:

$$\tilde{z}(t_{i+1}) = H\Phi H^{-1}z(t_i) + \\ + H\Phi(H^{-1}-K)(\tilde{z}(t_i)-z(t_i)) \qquad (78)$$

with matrices H and Φ defined as follows for the system of Eqs. (73) to (76):

$$H = \begin{bmatrix} 1 & 1 \\ \mu_m & b \end{bmatrix} \quad \Phi = \begin{bmatrix} 1+\tau\mu_m & 0 \\ 0 & 1+\tau b \end{bmatrix} \qquad (79)$$

It can be seen that the dependence of $H\Phi H^{-1}$ on the parameter μ_m is linear. How-

ever, the dependence of the matrix, $H\Phi$ $(H^{-1}-K)$, is more complicated. In order to reduce the complexity involved, this matrix is replaced by a constant diagonal matrix whose diagonal elements are treated as the tuning factors. From Eqs. (73) to (76) the prediction of the measurement $\tilde{z}_{2,t+1}$ is shown to be

$$\tilde{z}_{2,t+1} = (z_{2,t}-bz_{1,t})\tau\hat{\mu}_m + \\ + z_{2,t}(1+b\tau)-\beta(z_{2,t}-\tilde{z}_{2,t}) \qquad (80)$$

From the above, the equation for the recursive update of the ψ-matrix is given as

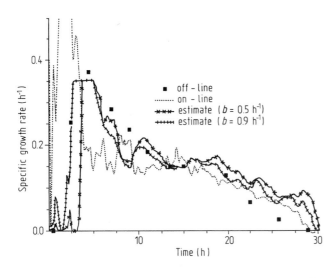

Fig. 17. Case Study 2 of SSPE: Adaptation of parameter estimates of specific growth rate in non-contaminated yeast culture by the SSPE algorithm with different settings of b-value. (■) off-line measurements; ($\cdots$) on-line measurement using Eq. (36). Figures from CHATTAWAY and STEPHANOPOULOS (1989).

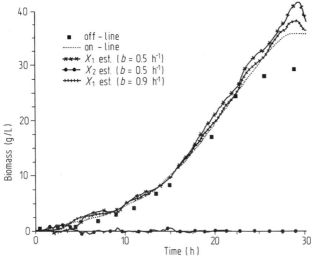

Fig. 18. Case Study 2 of SSPE: Estimation of biomass concentration in non-contaminated yeast culture by the SSPE algorithm with different settings of the b-value. (■) off-line measurements; ($\cdots$) on-line measurement using Eq. (36). Note that the estimates agree well with the off-line measurement and the estimates of the contaminating species remain at zero. Figures from CHATTAWAY and STEPHANOPOULOS (1989).

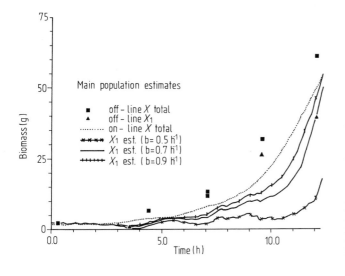

Fig. 19. Case Study 2 of SSPE: Estimation of biomass concentration of yeast in *Escherichia coli* contaminated yeast culture by the SSPE algorithm with different settings of the *b*-value. (■) off-line measurements of total biomass; (···) on-line measurements of total biomass using Eq. (36). (▲) off-line measurements of yeast biomass. Figures from CHATTAWAY and STEPHANOPOULOS (1989).

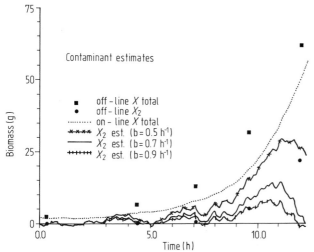

Fig. 20. Case Study 2 of SSPE: Estimation of biomass concentration of *Escherichia coli* in *E. coli* contaminated yeast culture by the SSPE algorithm with different settings of the *b*-value. (■) off-line measurements of total biomass; (···) on-line measurements of total biomass using Eq. (36); (●) off-line measurements of *E. coli* biomass. Note that the algorithm detects the contamination more sensitively with a smaller value of *b* selected. Figures from CHATTAWAY and STEPHANOPOULOS (1989).

$$\psi_{t+1} \equiv \frac{\mathrm{d}}{\mathrm{d}\theta} \tilde{z}_{2,t+1} = (z_{2,t} - bz_{1,t})\tau +$$

$$+ \beta \frac{\mathrm{d}}{\mathrm{d}\theta} \tilde{z}_{2,t} = (z_{2,t} - bz_{1,t})\tau + \beta\psi_t \qquad (81)$$

where β is the corresponding element in the replacing diagonal matrix. Then, by applying Eqs. (52) through (55) the parameter μ_m is updated. With this new estimate of μ_m, the state variables, $x_{1,t}$ and $x_{2,t}$, are estimated by the Kalman filter.

Figs. 17 to 20 show the experimental test of the developed algorithm. As a model system, yeast was used for the main species and *Escherichia coli* as the contaminant. In Figs. 17 and 18, the on-line estimation results of a non-contaminated culture are shown. The biomass estimates are in good agreement with the off-line measurements. Also, the estimated contaminant biomass is close to zero, as expected. In Figs. 19 and 20, the estimation result of a culture that is intentionally contaminated is shown. It is clearly seen that the estimates of the biomass level of the contaminant monotonically increase. This demonstrates that the algorithm can detect the presence of contaminants soon after they first appear in the fermentor. As mentioned before, the sensitivity of detection depends on the value of b selected. Smaller values of b result in a lower threshold of detection at the cost of increasing the probability of false positive occurrences.

5 Concluding Remarks

The scope, underlying theory, and applications of on-line bioreactor estimation were discussed in this chapter. Implementation of the two most widely accepted algorithms, EKF and SSPE, was illustrated by several examples. As noted, the difference between the two algorithms is in the mathematical formalism. Both methods offer systematic means of incorporating model equations and measurements in an integrated identification procedure. Furthermore, these methods are equally capable of filtering out the interfering measurement noise and of obtaining optimal estimates of the state variables and parameters on-line.

In EKF, estimation of both the state variables and model parameters is achieved only by slight additional computational complexity over the linear state filter. However, unlike the SSPE, this parallel augmentation does not take sufficient care of the portion of the Kalman gain matrix related to the update of the parameter vector (LJUNG and SÖDERSTRÖM, 1985). This can cause divergence of estimates, especially when the optimal filter gain matrix depends strongly on the parameters.

System identification was shown to be useful in four case studies. These were also intended to illustrate the application of the algorithms. In the approaches presented, the model equations play a critical role in providing the basis for the reconstruction of relevant bioreactor events. If a reliable model and credible parameters are not available, the approach is not likely to produce satisfactory results. In view of these limitations, a different approach to system identification has recently been taken which is based on pattern-recognition theories and uses historical performance data to establish standard process patterns in a reduced dimensional space (GUTERMANN and STEPHANOPOULOS, 1989). Deviations from this standard path are portrayed in a generalized map that also indicates the identified causes for the observed deviations. Such a map can then be used as guide in generating control law. In the examples examined (GUTERMANN and STEPHANOPOULOS, 1989), the above could be accomplished equally well with on-line data alone, thus allowing the opportunity for direct on-line implementation with fermentation processes.

As a final note, the system identification theory itself is a widely and intensively investigated field. Its applications vary from blackbox to highly structured systems. Among these, the identification algorithms discussed in this chapter, EKF and SSPE, are suitable for a broad spectrum of applications in bioreactor operations.

Acknowledgement

The writing of this chapter was supported by the National Science Foundation through the MIT Biotechnology Process Engineering Center Grant No. CDR-8803014 and Grant No. EET-8711725.

6 References

CAMINAL, G. et al. (1987), Application of extended Kalman filter to identification of enzymatic deactivation, *Biotechnol. Bioeng.* **29**, 366.

CHATTAWAY, T., STEPHANOPOULOS, G. N. (1989), An adaptive state estimator for detecting contaminants in bioreactors, *Biotechnol. Bioeng.* **34**, 647.

GOODWIN, G. C., SIN, K. S. (1984), *Adaptive Filtering Prediction and Control.* Englewood Cliffs, N. J.: Prentice-Hall.

GUTERMAN, H., STEPHANOPOULOS, G. N. (1989), Application of principal component analysis to pattern recognition in fermentation processes, submitted to *Biotechnology and Bioengineering*.

ISERMANN, R. (1984), Process fault detection based on modeling and estimation methods – a survey, *6th Int. Fed. Automatic Control Symp. Identification and System Parameter Estimation,* Washington D. C.

JAZWINSKI, A. H. (1970), *Stochastic Processes and Filtering Theory*. New York: Academic Press.

LJUNG, L., SÖDERSTRÖM, T. (1985), *Theory and Practice of Recursive Identification*. Cambridge, MA: MIT Press.

PARK, S., RAMIREZ, W. F. (1989), Optimal regulatory control of bioreactor nutrient concentration incorporating system identification, accepted by *Chemical Engineering Science*.

SAN, K., STEPHANOPOULOS, G. N. (1984a), Studies on on-line bioreactor identification. II. Numerical and experimental results, *Biotechnol. Bioeng.* **26**, 1189.

SAN, K., STEPHANOPOULOS, G. N. (1984b), Studies on on-line bioreactor identification. IV. Utilization of pH measurement for product estimation, *Biotechnol. Bioeng.* **26**, 1209.

STEPHANOPOULOS, G. N., SAN, K. (1984), Studies on on-line bioreactor identification. I. Theory, *Biotechnol. Bioeng.* **26**, 1176.

WANG, H. Y., COONEY, C. L., WANG, D. I. C. (1977), Computer-aided baker's yeast fermentation, *Biotechnol. Bioeng.* **19**, 69.

YDSTIE, B. E., et al. (1986), Adaptive control session, pp. 421–512, in: *Chemical Process Control - CPCIII, Proc. Third Int. Conf.* Asilomar, CA, (MORARI, M., McAVOY, T., Eds.).

8 Optimization of Sampling

AXEL MUNACK

Braunschweig, Federal Republic of Germany

It may be argued that a suitable choice of sampling period may not be a crucial task for biotechnical systems. The processes show a very slow dynamic behavior, which allows measurements to be made quasicontinuously. However, this fact should not induce an immediate jump to conclusions. It should be kept in mind that data from bioreactor measurements show quite different dynamic behavior; the response of the gas holdup and the oxygen concentration in the liquid phase to changes in the aeration rate is rather rapid and cannot be controlled precisely by taking samples once a minute. This shows that the non-biological control loops for a bioreactor may show fast behavior and must be sampled with sufficient frequency.

On the other hand, one must consider that most of the biological data can only be measured with some lag time and with relatively low sampling rates. This is true even when auto-analyzers are used and is even more so for manual analyses. Thus, the situation arises that, although the biological processes are very slow, sampling may be so infrequent that a good reconstruction of the biological state of the system is not possible.

Since manual analyses are costly and auto-analyzers also require large amounts of material and manpower, one must plan experiments and the type and number of samples carefully. This chapter provides some basic considerations for a suitable choice of sampling time, some advanced calculations for optimization of sampling, and applications to parameter identification problems for models of biotechnical processes.

1 General Remarks Concerning a Suitable Choice of Sampling Time

For a discussion of the various aspects, which must be considered for a reasonable choice of the sampling interval, one should first understand some basic principles of sampled-data control systems. Fig. 1 shows a simple configuration of such a control system. The controlled variable $y(t)$ is compared to the reference input $w(t)$ to form the error $e(t)$. From this continuous signal only a sampled sequence is supplied to a digital computer that performs the necessary calculations to control the plant. These computations result in a discrete time output signal $u(kT)$ that is fed into the digital/analog converter, which holds the output steady during the next sampling interval, resulting in $u(t) = u(kT)$, $kT \leq t < (k+1)T$.

Thus, the process is driven by a staircase function which better approximates a continuous signal, the smaller the sampling interval T is. The purpose of the control loop is twofold: on the one hand, the controlled variable should follow changes in the reference input almost precisely and with good dynamics (tracking problem), and, on the other hand, the influence of various disturbances $n(t)$ on the controlled variable should be kept very small (disturbance rejection).

A rigorous means to study the behavior of discrete-time systems is provided by the z-transform. This is a transformation of signals in the time domain onto complex functions. It permits one to treat systems in the transformed domain in a manner similar to that calculus which is provided by the Laplace transform for continuous-time systems. The transformed signal representation for a discrete-time signal $x(kT)$ is given by

$$\mathfrak{Z}(x(k)) = x(z) = \sum_{k=0}^{\infty} x(kT) \cdot z^{-k} \tag{1}$$

where $z = e^{Ts}$.

It is impossible and at the same time beyond the scope of this chapter, to provide a general background of the theory of these systems. An excellent textbook on this topic was written by ACKERMANN (1985).

In the specific field of application treated here, however, at least one fundamental question should be answered: is it possible to reconstruct the continuous signal $x(t)$ from the samples $x(kT)$? Equivalently, the question may be posed: under what conditions does the inverse z-transform give an unambiguous result? The answer is provided by the sampling theorem (SHANNON, 1949), which states that a reconstruction is possible, provided the Fou-

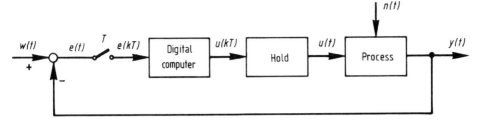

Fig. 1. Configuration of a sampled-data control system.

rier transform $X(j\omega)$ of $x(t)$ holds $X(j\omega)=0$ for $|\omega|\geq\omega_m$, and the sampling period is chosen such that $T\leq\pi/\omega_m$, resulting in a sampling frequency $\omega_s\geq2\omega_m$. This gives a first hint as to how the sampling time may be chosen. However, a different rule of thumb for the choice of the sampling interval is provided by considering not only the signals but also the controlled system. The concept important in this context is that of controllability, which refers to the following property: A state $x(t_0)$ of a linear system is *controllable* if there exists a finite time instant $t_l > t_0$ and an input $u(t_k)$, $0\leq k < l$, *such that the state $x(t_0)$ is transferred to the zero state $x(t_l)=0$.*

The concept of state-space representation of dynamical systems has not yet been introduced. Roughly speaking, the state of a system, denoted by the state-space vector x, may be interpreted as the actual charge of all mass, energy, or momentum reservoirs of the system (or a transformation of it). For example, the biological growth process in Sect. 2 is fully described by the state $x=(x,s)^T$, x and s denoting cell and substrate concentration, respectively (however, this is a nonlinear system).

For a completely controllable linear discrete-time system it can be shown that n steps are always sufficient to transfer the system to the origin, n being the dimension of the state vector. However, the corresponding input signal $u(k)$ will often show large amplitudes in order to achieve this goal, which does not match the real situation of bounded inputs (see Fig. 1) where u is the output of a digital/analog converter which is always bounded. Of course, further bounds are implied by elements of the loop, e.g., input flow pumps, valves, etc. This means that the controllability region of systems with bounded inputs has to be con-

sidered. This is the set of all initial states $x(t_0)$ which may be transferred to the zero state within a certain number of time steps by a bounded input sequence $u(k)$. ACKERMANN (1985) gave several examples of computed controllability regions in which the area of the region is taken as a measure, depending on the step number. These calculations, normalized to that sampling period τ where controllability is entirely lost, demonstrate that a great increase in the controllability region occurs for $T=\tau/2$, $\tau/3$ and $\tau/4$. Afterwards, the region grows further (in the limit – as T approaches zero – to the controllability region of the corresponding continuous-time system). The amount of additional increase, however, is not very large.

Fig. 2. Determination of the sampling period.

Therefore, as a rule of thumb, the sampling frequency should be chosen four times larger than the frequency at which controllability is lost, which corresponds to the frequency $\omega_s = 2\omega_m$ of the sampling theorem. The resulting rule for design of sampled-data control systems is summarized as follows (Fig. 2):

● For a continuous process with eigenvalues s_i, one may draw the smallest possible circle in the *s*-plane around the origin $s = 0$ such that all the poles s_i are encircled. The radius of this circle is r; then $\omega_s = 8r$ is chosen, which means that $T = \pi/4r$.

Further recommendations of ACKERMANN refer to anti-aliasing filters or neglected dynamics. These are relevant to mechanical systems but need not be considered for application to biotechnical processes.

2 Criteria to Optimize Measurements for Model Identification

The control of systems by discrete-time controllers was treated in the foregoing section. This section is concerned with a different aspect of sampling. The open loop aspect of taking measurements from a process for system identification will now be considered. Further information on the estimation of model parameters is provided in Chapter 4 of this book.

The task of model identification may be specified by the following definition given by ZADEH (1962): "Identification is the determination, on the basis of input and output, of a system (=model) within a specified class of systems (=models), to which the system (=process) under test is equivalent." This classical definition states one fact very clearly which the user quite often is not aware of – namely, the problem of specification of a suitable class of models. This must be treated very thoroughly, while keeping in mind the purpose of the model, e.g. diagnosis, monitoring, prediction of future behavior, process control, or process optimization and scale-up (EYKHOFF, 1988).

To start with, a rather large class of models will be discussed that may be used for dealing with all of these goals (although, in some cases, particularly for process control, the resulting model may be too complicated).

The class of models $\mathcal{M}$ contains continuous-time systems with a nonlinear dynamic state-space description and a nonlinear static output relation. The *structure* of the system is fixed by *a-priori* knowledge, which means that only system *parameters* are treated as unknown. In order to keep the presentation relatively simple, lumped-parameter systems are treated; extension to the distributed-parameter case is not crucial, cf. the case study in Sect. 3.

$$\mathcal{M}(P): \dot{x} = f(x, t, u, P), \ x(0) = x_0(P) \tag{2a}$$

$$y(t_i, u, P) = g(x(t, u, P), t_i, P) \tag{2b}$$

Here, $x \in \mathbb{R}^n$ denotes the system state, $u \in \mathbb{R}^q$ the input, and $P \in P_{ad} \subset \mathbb{R}^p$ the vector of unknown system parameters. P_{ad} designates the set of admissible parameters which may be fixed by physical, chemical, or biological considerations. It should be as small as possible in order to permit a unique solution. According to the real situation, the outputs $y \in \mathbb{R}^m$ are discrete-time measurements, and the case of unknown parameters in the output operator g is explicitly included.

To select the best candidate in the class of models, the error between the output $\hat{y}$ of the model under consideration and the measurement y^M taken at the system is computed,

$$e_i = e(t_i, u, \hat{P}) = \hat{y}(t_i, u, \hat{P}) - y^M(t_i, u) \tag{3}$$

and all errors are combined to form the identification functional:

$$J_1(u, \hat{P}) = \sum_{i=1}^{N} e_i^T Q_i e_i \tag{4}$$

where Q_i are positive semidefinite, symmetric matrices. Now the modelling problem may be formulated as a parameter optimization problem: find model parameters $\hat{P}^*$, for which the following holds:

$$J_1(u, \hat{P}^*) \leq J_1(u, \hat{P}) \ \forall \hat{P} \in P_{ad}; \ \hat{P}^* \in P_{ad} \tag{5}$$

Even if the identified model parameters are equal to the system parameters, the identification functional will not be exactly zero. This is due to measurement noise, which simultaneously causes the vector $\hat{P}^*$ to be a random variable: even if a series of identical experiments would be possible, the measured outputs would be different, due to the measurement noise. These different signals then lead to different identified parameters. A very important task of optimization of the measurements is therefore posed by the requirements that bias-free estimates be computed (the mean $E\{\hat{P}*\}$ of infinitely many estimates equals the true parameter vector P), and that the variance of the estimates be as small as possible. The second condition may be formulated mathematically by an optimization problem:

$$\Lambda(E\{(\hat{P}*-P)(\hat{P}*-P)^T\})=\Lambda(V)\rightarrow\min \qquad (6)$$

where $\Lambda:\mathbb{R}^{p\times p}\rightarrow\mathbb{R}$ is a functional used to weight the covariance V.

To solve this problem, questions regarding optimization of the information content of measurements require discussion. This leads to an optimized choice of the weighting matrices Q_i and a discussion of the experimental conditions when taking the measurements. For distributed-parameter systems the question of the best position of the sensors may also be solved.

Some assumptions have to be made concerning the noise that disturbs the measurements. Zero-mean Gaussian white noise is considered with a diagonal covariance matrix $C(t_i)$, which means that the 'true' system outputs y are additively disturbed to give the measurements:

$$y^M(t_i)=y(t_i)+\varepsilon(t_i) \qquad (7a)$$

where

$$E\{\varepsilon(t_i)\}=0, \quad i=1,\ldots,N \quad \text{and} \qquad (7b)$$

$$E\{\varepsilon(t_i)\varepsilon^T(t_j)\}=\delta_{ij}\cdot C(t_i)$$
$$i,j=1,\ldots,N \qquad (7c)$$

After performing a linearized consideration of the influence of small parameter variations on the system's output trajectories along the nominal trajectory (with nominal parameters P), inserting this linearization into the identification functional Eq. (4), and taking the expectation of the functional – where Eqs. (7b/c) are

used – one obtains:

$$E\{J_1(u,P+\delta\hat{P})\}\approx$$
$$\delta\hat{P}^T\left[\sum_{i=1}^N Y_P^T(t_i,u,P)Q(t_i)Y_P(t_i,u,P)\right]\delta\hat{P}+$$
$$+\sum_{i=1}^N \text{tr}(C(t_i)Q(t_i)) \qquad (8)$$

Here,

$$Y_P(t_i,u,P)=\frac{\partial g}{\partial x}\bigg|_{x^o,u,P}X_P+\frac{\partial g}{\partial P}\bigg|_{x^o,u,P} \qquad (9)$$

denotes the output sensitivities with respect to parameter variations, evaluated along the nominal output trajectories. The state sensitivities X_P are computed by solving the linear system

$$\dot{X}_P(t,u,P)=\frac{\partial f}{\partial x}\bigg|_{x^o,u,P}X_P+\frac{\partial f}{\partial P}\bigg|_{x^o,u,P} \qquad (10a)$$

where the derivatives are taken along the nominal trajectories $x^0(t,U,P)$, with the initial condition

$$X_P(0)=\frac{dx_0}{dP} \qquad (10b)$$

If one chooses the weighting matrix $Q(t_i)$ as the inverse of the covariance matrix C, then the matrix formed by evaluating the sum in brackets of Eq. (8) is just the Fisher information matrix F of the estimation problem (LJUNG, 1987). This matrix is the inverse of the parameter estimation error covariance matrix of the best linear unbiased estimator (BLUE). Therefore, it gives an upper bound for the precision of the estimate obtainable by a certain experiment. Eq. (8) gives:

$$E\{J_1(u,P+\delta\hat{P})\}\approx\delta\hat{P}^T F(u,P)\delta\hat{P}+Nm \qquad (11)$$

This demonstrates again that a zero value of the identification functional cannot be expected. On the other hand, a prerequisite for a minimum of the identification functional is clear: the information matrix must be positive definite. If this very basic condition is fulfilled, then the corresponding experiment is said to be *informative* (GOODWIN, 1987). For a non-informative experiment, $\det(F) = 0$ or $\lambda_{\min}(F) = 0$, $\lambda_{\min}$ being the smallest eigenvalue of F, could be used as an indicator. Moreover, not only these qualitative but also quantitative statements may be obtained. To provide the largest possible distance to the singular non-informative case, $\det(F)$ (D-criterion) or $\lambda_{\min}(F)$ (E-criterion) can be maximized. Other criteria known from the literature refer to a maximization of the trace of F (simplified A-criterion) or to a minimization of $\det(S)$ (D-criterion), $\lambda_{\max}(S)$ (E-criterion) or $\mathrm{tr}(S)$ (A-criterion), where $S = F^{-1}$. (In the case of the BLUE, $V = S$ holds.) The A-criterion may be interpreted as a minimization of the identification errors in an arithmetic mean, and the D-criterion optimizes the geometric mean. The E-criterion, however, minimizes the largest error. Whereas the application of the D- or E-criterion is possible with respect to the covariance matrix S or the information matrix F (with the same results), the A-criterion gives different results. The simplest criterion $\mathrm{tr}(F) \rightarrow \max$ should not be used, since it may lead to non-informative experiments (GOODWIN, 1987). Thus, the functional of the optimization problem, Eq. (6), is seen as a D- or E-criterion.

The quantitative evaluation of the information content of measurements from an experiment offers the opportunity to optimize the experimental conditions, which means finding input functions $u(t)$ that lead to the most informative experiments. In the dynamical case, this problem is in principle an infinite-dimensional optimization problem, since functions in the time domain must be optimized. In order to reduce the numerical burden, the input functions may be discretized, such that only values at a finite number of node points must be determined. Between the node points, the function may be considered as linear in time (if such a permanently changing feed-rate is possible with the installed process equipment), or as a constant. Both possibilities provide simple means for transforming the problem of computation of an optimal input function in the time domain into a parameter optimization problem. However, since usually various restrictions must be observed (e.g., input flows are always non-negative and bounded), these may be easily incorporated in such a simple representation of functions. This would not be

the case when using other types of approximations of functions, such as polynomial approximations.

Some further remarks on the measurement noise must be made. Here this noise reflects not only the classical noise generated by sensors but also takes into account the precision of the measurements (though a Gaussian distribution with a zero mean may be somewhat crude to describe this effect). Nevertheless, in some cases one is not able to model these errors by a constant noise level; on the contrary, a relative error would be more adequate. This leads to a consideration of semi-relative sensitivities:

$$(\tilde{Y}_P)_{k,j} = \frac{(Y_P)_{k,j}}{(y^0)_k} \tag{12}$$

which may also be incorporated into Eqs. (8) and (11) by setting

$$(Q(t_i))_{j,j} = (\tilde{C}^{-1})_{j,j} = \frac{(C^{-1})_{j,j}}{(y^0)_j^2} \tag{13}$$

For small values of the nominal output $(y^0)_j$ this relative weighting may lead to overly optimistic results, since the error approaches zero as the trajectory itself goes to zero. Therefore, a modification should be used instead of Eq. (13):

$$(\tilde{C}^{-1})_{j,j} = \frac{(C^{-1})_{j,j}}{(\max((y^0)_j, (y)_{j,\min}))^2} \tag{14}$$

A further modification may be carried out, which takes into account the fact that in general the desired accuracy of the estimated parameters is not an absolute value but a relative one. This is taken into consideration by (fully) relative sensitivities, when Eq. (12) is modified to give

$$(\tilde{Y}_P)_{k,j} = (P)_j \frac{(Y_P)_{k,j}}{(y^0)_k} \tag{15}$$

Below, an application of the above formulated theoretical considerations to a simple biotechnological growth process is considered, following the results presented by MUNACK (1989). The process is described by two ordinary differential equations,

$$\dot{x}(t) = \mu(s) \cdot x(t) - \mu_D \cdot x(t) \qquad (16)$$

$$\dot{s}(t) = -\frac{1}{Y_{x,s}} \cdot \mu(s) \cdot x(t) \qquad (17)$$

where x denotes the cell concentration and s the substrate concentration. Both states form the state vector $y = (x, s)^T$, which – at the instant of measurement – directly gives the undisturbed output vector of the system. μ_D is a decay rate, and $Y_{x,s}$ is the yield coefficient. For $\mu(s)$ a Michaelis–Menten-type nonlinear relation is assumed to hold, which gives

$$\mu(s) = \mu_m \cdot \frac{s}{K_s + s} \qquad (18)$$

where μ_m is the maximum specific growth rate, and K_s is the Michaelis limitation constant. The trajectories of this simple batch process are shown in Fig. 3. The parameters used for simulation refer to HOLMBERG (1982) and are compiled in Tab. 1. The problem to be solved is the determination of the four system parameters, given measurements of $x(t)$ and $s(t)$. This identification problem was theoretically shown to be solvable in the noise-free case by

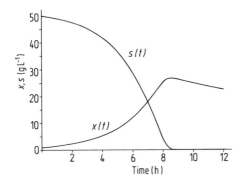

Fig. 3. Trajectories of the batch cultivation.

Tab. 1. Parameters Used for the Batch Cultivation of Fig. 3

$\mu_m = 0.5 \text{ h}^{-1}$	
$\mu_D = 0.05 \text{ h}^{-1}$	$x(0) = 1 \text{ g} \cdot \text{L}^{-1}$
$Y_{x,s} = 0.6$	$s(0) = 50 \text{ g} \cdot \text{L}^{-1}$
$K_s = 3 \text{ g} \cdot \text{L}^{-1}$	

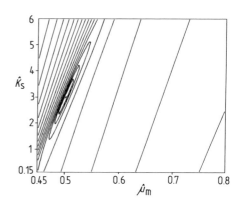

Fig. 4. Plot of the undisturbed identification functional for a batch experiment.

POHJANPALO (1978). In practical situations, however, the states $x(t)$ and $s(t)$ are not measurable without noise and other errors. In the following, only the identification of those two parameters is treated, which are most difficult to identify. This means that μ_D and $Y_{x,s}$ are assumed to be known. The restriction on the case of two parameters enables us to draw a plot of the identification functional $J_1(\mu_m, K_s)$ if there are no disturbances (Fig. 4). This plot exhibits the shape of a flat valley along a certain direction in the (μ_m, K_s) plane. This means that by choosing (μ_m, K_s) parameter values along this direction, the corresponding functional values will be very small. In case of measurement noise or other errors, each experiment will then result in a different parameter estimate which lies along the flat valley. Therefore, the system is practically non-identifiable, a fact which was reported first by HOLMBERG (1982). All estimates result in good agreement of measured calculated data. The identified parameters, however, are biologically meaningless.

Now the above described procedure for optimization of experimental conditions is applied to the process. A batch process offers only one usable degree of freedom to optimize the experiment in order to permit better estimates of the parameters. This is the initial substrate concentration $s(t=0)$. The main effect on the performance of the estimates of parameters, however, is provided by turning the process to fed-batch operation and taking the flow rate as an input variable thus minimizing $J_1(q(t), P, s(t=0))$. This means that the second state equation (Eq. (17)) is modified to give

$$\dot{s}(t) = -\frac{1}{Y_{x,s}} \cdot \mu(s) \cdot x(t) + \frac{S_R}{V} \cdot q(t) \qquad (19)$$

where S_R is the substrate concentration in the feed reservoir, V the reactor volume, and $q(t)$ the feed flow rate.

For simplicity, it is assumed that S_R/V is very high, such that $q(t)$ is very small and does not result in any dilution of the reactor content. A modified E-criterion is used here:

$$\Lambda(F) = \frac{\lambda_{max}(F)}{\lambda_{min}(F)} \rightarrow min \qquad (20)$$

This does not change the results in principle, compared with the original E-criterion; however, the fact should be kept in mind for the interpretation of the numerical results.

The complete results for batch conditions ($T = 8$ h) and various fed-batch durations (12–24 h) are shown in Fig. 5. The shape of the curves, particularly those for 12 h and 18 h, respectively, may be due to the fact that the optimization algorithm did not reach the global optimum but stopped in a local one. The conclusions that may be drawn from the results are:

- There is a best initial substrate concentration which should be computed.
- Compared with the best possible batch, a fed-batch which lasts only four hours longer may give a functional which is one tenth of the batch; this means that the precision of the parameter estimates may be increased by a factor of more than 3.
- The shape of the functional looks very satisfactory. For a feed profile like that of Fig. 6, which is a crude approximation of an optimized feed (MUNACK, 1989), the functional plot is drawn in Fig. 7. The flat valley could be changed almost to a cone.

These results demonstrate very clearly that some of the difficulties encountered in the

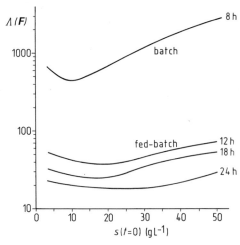

Fig. 5. Plot of $\Lambda(F)$ for batch and various fed-batch conditions.

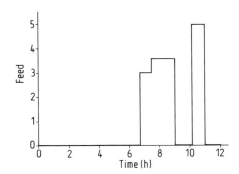

Fig. 6. Approximated optimized feed profile.

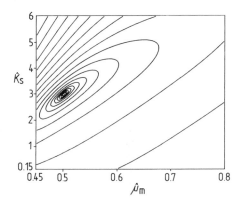

Fig. 7. Plot of the identification functional for optimized fed-batch conditions ($T = 12$ h, feed profile cf. Fig. 6).

identification of parameters may be overcome by a thorough review of the setting of the problem. The treatment here was entirely based on nonlinear, deterministic, time-domain considerations, and the noise characteristics were kept as simple as possible. For a detailed study in frequency domain, we refer to ZARROP (1979).

3 Application to Cultivation in a Tower Loop Reactor

The application of the methods of Sect. 2 to distributed-parameter processes has already been mentioned. Those are processes which show distinct profiles of their state variables along the spatial coordinate(s). For most biotechnological applications, one may think of concentration profiles which occur in reactors that are not of the stirred-tank type. However, temperature profiles may also have to be considered, when induction by temperature shift is carried out in tubular reactors. Distributed-parameter processes are naturally modelled by partial differential equations (PDEs). Most authors agree in preserving the distributed nature of the process in the calculations as long as possible. A 'rapid lumping', which means a direct approximation of the spatial partial derivatives in the PDE by difference formulas and a further treatment of the resulting system of ordinary differential equations, may sometimes be worse than the immediate formulation of a compartmented model. This is due to the nonlinear terms which usually occur in the description of biotechnical processes. Therefore, we are led to the alternatives either to treat partial differential equations as long as possible or to describe the process directly by formulating balances for compartments and their mutual exchange of mass and energy.

In the following, the first strategy is applied to a microbial growth process in a tower loop reactor (bubble column). For a more detailed treatment of this type of reactor, we refer to SCHÜGERL (1985). A schematic diagram of the

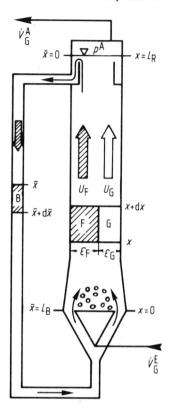

Fig. 8. Schematic diagram of the tower loop reactor.

used double-reactor system (column and loop) is shown in Fig. 8. Since only the column is gassed from the bottom, gas phase (G) balances are formulated only for this part, while liquid phase balances appear in the column (F) and in the bypass (B). For modelling the growth process, only mass balances are taken into account, since isothermal conditions are assured by control, and momentum is negligible. The task of the controller is to achieve maximum cell production at minimal cost, which means lowest possible substrate (S) and oxygen (O) consumption.

A general model, based only on the nutrients O_2 and S, the products X (cell concentration) and CO_2, and the spatially varying gas velocity u_G leads to a nonlinear system of six parabolic partial differential equations and five plug flow equations (LUTTMANN, 1980; LUTTMANN et al., 1985). For control purposes within a limited area of operating conditions, several simplifications may be introduced in order to reduce this large number of system equations. These are:

- Biomass and substrate are well mixed in the liquid phase,
- the substrate concentration is nowhere growth limiting (extended culture),
- the profile of the oxygen mole fraction in the gas phase of the reactor is described by a quadratic function,
- the respiratory quotient equals 1; therefore the velocity of the gas phase is reciprocal to the pressure profile in the reactor,
- residence time in the loop is small compared to that in the column; conditions in the loop are assumed to be quasi-stationary.

With these assumptions, which were proven to be justified for the pilot plant used, the model is drastically reduced to a quasi-linear PDE of parabolic type, describing the dissolved oxygen concentration c_{OF} in the liquid phase of the column, Eq. (21), an algebraic equation for the oxygen concentration c_{OG} in the gas phase of the column, Eq. (22), an implicit algebraic equation for the dissolved oxygen concentration c_{OB} in the loop, Eq. (23), and an ordinary differential equation for the biomass concentration c_X, Eq. (24).

Normalized dissolved oxygen concentration in the liquid phase of the reactor

$$\text{PDE:} \quad \frac{\partial c_{OF}}{\partial t} = \frac{1}{Bo_F} \cdot \frac{\partial^2 c_{OF}}{\partial z^2} - v_F \cdot \frac{\partial c_{OF}}{\partial z} +$$

$$+ k_L a^E \cdot \Psi(K_{St}) \cdot (p \cdot c_{OG} - c_{OF}) -$$

$$- q_{O/Xm} \cdot \frac{c_{OF}}{k_0 + c_{OF}} \cdot c_X$$

IC: $c_{OF}(0) = p(0)$

BC: $\left. \frac{\partial c_{OF}}{\partial z} \right|_{z=0} = Bo_F \cdot v_F \cdot (c_{OF}(0) - c_{OB}(1))$

$\left. \frac{\partial c_{OF}}{\partial z} \right|_{z=1} = 0$

where

$$\Psi(K_{St}) = \begin{cases} e^{-K_{St} \cdot \frac{z}{a}} & 0 \leq z < a \\ e^{-K_{St} - a \cdot K_{St} \cdot (z-a)} & a \leq z \leq 1 \end{cases}$$

$$(a = 0.1) \quad (21)$$

Normalized oxygen concentration in the gas phase of the reactor

$$c_{OG} = 1 - \frac{U_{O_2}}{2 + Bo_G v_G^E} \cdot (2 + Bo_G v_G^E \cdot (2z - z^2))$$

where

$$U_{O_2} = \frac{\overline{OTR}}{q_{O_2}^E} = \frac{1}{q_{O_2}^E} \cdot \left[\frac{\mu_t}{Y_{X/O}} \cdot c_X + \right.$$

$$+ v_F \cdot (c_{OF}(1) - c_{OB}(1)) +$$

$$\left. + q_{O/Xm} \cdot c_X \cdot \int_0^1 \frac{c_{OF}}{k_0 + c_{OF}} \, dz \right] \quad (22)$$

Normalized dissolved oxygen concentration in the loop

$$c_{OB}(\bar{z}) - c_{OB}(0) + k_0 \cdot \ln \frac{c_{OB}(\bar{z})}{c_{OB}(0)} = -q_{O/Xm} \frac{\bar{z}}{v_B} c_X$$

BC: $c_{OB}(0) = c_{OF}(1)$ $\quad (23)$

Normalized concentration of cells

$$\frac{dc_X}{d\tau} =$$

$$= Y_{X/O} \cdot \left[\frac{v_B}{v_B + v_F} \cdot q_{O/Xm} \cdot c_X \cdot \int_0^1 \frac{c_{OF}}{k_0 + c_{OF}} \, dz + \right.$$

$$\left. + \frac{v_B \cdot v_F}{v_B + v_F} \cdot (c_{OB}(0) - c_{OB}(1)) - \frac{\mu_t}{Y_{X/O}} \cdot c_X \right]$$

IC: $c_X(0) = 1$ $\quad (24)$

The system equations are coupled via the states and the boundary conditions. This means that c_X enters into the dissolved oxygen equations, and, on the other hand, c_{OF} and c_{OB} enter into the cell equation. c_{OF} and c_{OG} are also coupled, and the physical connection of reactor column and loop is modelled via coupling terms in the boundary conditions.

Four parameters have turned out to be unknown and/or temporally varying. These are two fluid-dynamic parameters, $k_L a^E$ and K_{St}, both describing oxygen transfer from the gas phase into the liquid phase along the reactor column, and two biological parameters, the metabolic quotient $q_{O/Xm}$ and the yield coefficient $Y_{X/O}$. Measurements can be taken of dissolved oxygen concentration c_{OF} at distinct points in the reactor column, cell concentration and outlet gas mole fractions, allowing computation of the overall oxygen transfer rate $\overline{OTR}$. This computed value is treated as a further measurement. Thus, the identification functional (output least squares error criterion) is formulated as follows, where ˆ denotes model outputs with estimated parameter set $\hat{P} = [\hat{k}_L a^E, \hat{K}_{St}, \hat{q}_{O/Xm}, \hat{Y}_{X/O}]^T$:

$$J_1(\hat{P}) = \int_0^T \left\{ W(t) \left[\sum_{i=1}^N (\hat{c}_{OF}(x_i,t) - c_{OF}^M(x_i,t))^2 + \right. \right.$$
$$\left. \left. + \sigma \cdot (\widehat{\overline{OTR}}(t) - \overline{OTR}^M(t))^2 + \right. \right.$$
$$\left. \left. + \kappa \cdot (\hat{c}_X(t)) - c_X^M(t))^2 \right] \right\} dt \qquad (25)$$

The numerical treatment of this problem of parameter identification, in particular the formulation of very efficient algorithms for its on-line solution, has been discussed in detail by MUNACK (1986). The algorithms refer to theoretical work on identification of parameters in PDEs by CHAVENT (1974), in which the

adjoint state equations were used to compute the gradient of the functional with respect to the unknown parameters.

Details concerning these problems will not be discussed here. Instead, emphasis will be placed on problems concerning the optimization of sampling of measurement data. The questions to be solved in this context are as follows:

● How many sampling points for measuring the dissolved oxygen concentration along the reactor column should be used?
● Which are the best positions along the reactor column to install these sensors?
● Which type of measurement is most valuable for identification of the unknown parameters? Which data set is essential for identification?

Again, the answer is provided by the evaluation of the Fisher information matrix. For a distributed parameter process, this method has been discussed in the literature by QURESHI et al. (1980). A great variety of related methods has been compiled in the survey paper of KUBRUSLY and MALEBRANCHE (1983).

The D-criterion was used to evaluate the information contents of the different possible measurement data sets. The results are summarized below. Fig. 9 shows the computed determinants as a function of the position of the c_{OF} sensor(s) along the reactor column. Curve (1) gives information which may be obtained by using a single c_{OF} sensor as the only measurement device. It can be seen that this sensor is best placed at the bottom of the reactor (the normalized position being $x/L_R = 0$). This result does not change when further measurements are additionally used, namely, the curves $(1 + X)$ for additional cell concentration measurements, $(1 + O)$ for additional $\overline{OTR}$ measurements, and $(1 + X + O)$ for use of all three measurement data. However, as can be seen from Fig. 9, finding an adequate type of measurement is much more critical than an optimal allocation of the c_{OF} sensor. While det (F) varies about 2.5 decades by sensor allocation, it may be increased by more than 8 decades when measuring c_X and $\overline{OTR}$ in addition to c_{OF}. It can also be stated that $\overline{OTR}$ is the

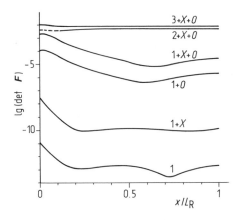

Fig. 9. Information contents of various sensor combinations as a function of the c_{OF} sensor position.

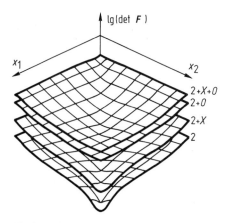

Fig. 10. Complete solution of the two c_{OF} sensor allocation problem.

most valuable measurement of all. While c_X results in an increase by 3 decades, $\overline{OTR}$ gives an increase of 7 decades, and further addition of c_X to $c_{OF}(0)$ and $\overline{OTR}$ adds only 1.5 decades.

Fig. 9 also indicates incomplete results for the allocation problem of two and three c_{OF} sensors. The first sensor is fixed at the bottom and – if there are three sensors – the second is fixed at the top of the reactor. Complete results may also be drawn (cf. Fig. 10), but there the situation is difficult to visualize. Fig. 9 demonstrates that using more than two dissolved oxygen sensors adds little information to the process of identification. Of course, further computations may be carried out for increasing numbers of c_{OF} sensors. These show that even taking eight measurements of the dissolved oxygen concentration does not give as much information as a single c_{OF} sensor near the bottom of the reactor combined with the $\overline{OTR}$ measurement.

When the number of oxygen sensors is increased, the optimal locations are not spread uniformly over the whole spatial area of the reactor. Using more and more sensors partly leads to the same positions as already occupied by the other sensors. This is due to an assumption made for the calculations stating that the noise of different measurements is not correlated. One would have to be careful with the installation of the sensors in order to fulfill this condition in practice.

Until now, off-line identification of the parameters has been optimized, which means that data of the complete fed-batch (extended culture) have been used. The situation may change if on-line identification is needed, e.g., for adaptive optimization of the conditions of cultivation. This means that all parameters are treated as time-varying, and only actual values are identified by the use of data from the most recent measurements. This type of adaptive control was studied for the described process in the tower loop reactor by MUNACK and THOMA (1981). Here one must guarantee that in each identification (or adaptation) interval an estimate of the unknown parameters is possible, which means that instead of the criterion

$$\max_{x_i} \ [\det(F)] \qquad (26)$$

a worst case study must now be performed, leading to a criterion

$$\max_{x_i} \left\{ \min_{j=2,\ldots,T/\tau} \ [\det(F_j)] \right\} \qquad (27)$$

where τ is the length of the adaptation interval. The results of this optimization problem are summarized in Fig. 11. They show a similar behavior when enough measurements are carried out. The cases of pure c_{OF} or $c_{OF}+c_X$ measurements, however, are totally insufficient for identification. This again emphasizes

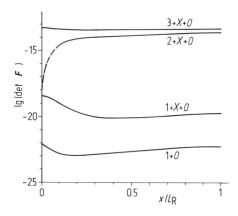

Fig. 11. Information contents of various sensor combinations as a function of the c_{OF} sensor position in case of cyclic identification.

the high information content of $\overline{OTR}$ measurements in the example treated here.

The case study carried out in this chapter demonstrates the benefits gained by a suitable choice of equipment for measurements installed at a bioreactor. It turned out that a 'lumped' measurement – in this case $\overline{OTR}$ – may be of highest value for identification of distributed-parameter systems. This fact would have been underestimated in most cases, even if experts had been asked. In critical situations, particularly for on-line identifications, an analysis of the information provided by different measurements should, therefore, be carried out. However, the computational effort required to gain the results reported here is quite high. There is a definite lack of standardized software to perform the calculations. Further research is needed.

4 Extension of the Results for Process Control and Optimization

Since Chapter 16 of this book is concerned with control of bioreactor systems, only some extensions of the material presented above to control problems will be addressed here. These further considerations are motivated by the fact that the identified parameters may be used in the off-line design of control systems or directly as a part of an explicit adaptive control scheme.

In process or control system design, an objective function is usually defined which measures the performance of the system. Using estimated parameters instead of the true system parameters, it is very important to imagine the effect of parameter errors on the performance of the system. This influence can be evaluated by sensitivity analyses. For a system described by Eqs. (2a/b), let the control performance be defined as

$$J_C = \int_0^T h(x, \hat{P}, u, t)\,\mathrm{d}t \qquad (28)$$

Then the partial derivatives

$$\left.\frac{\partial J_c}{\partial \hat{P}_i}\right|_{\hat{P}}, \quad i = 1, \ldots, p \qquad (29)$$

denote the influence of changes in the control performance due to small changes of the system parameters around the estimated parameter vector $\hat{P}$. It is obvious that parameters that influence the performance very strongly should be identified with higher precision than parameters which result in smaller deviations. This may be achieved by weighting the information matrix before carrying out the optimization (TAKAMATSU et al., 1971), giving

$$\tilde{F}(u, \hat{P}) =$$

$$= \mathrm{diag}_i\left(\left|\frac{\partial J_c}{\partial P_i}\right|^{-1}\right) F(u, \hat{P})\,\mathrm{diag}_i\left(\left|\frac{\partial J_c}{\partial P_i}\right|^{-1}\right) \qquad (30)$$

A minimax strategy for this purpose was formulated in the same paper. For relatively large errors of the parameter estimates, an evaluation of robustness may be more adequate than considerations of sensitivity by Eq. (30).

For enhancement of the control performance, several extensions of the cyclic techniques described in the last part of the case study of Sect. 3 are possible (OLFO controller, receding horizon controller) (MUNACK and THOMA, 1981, and NELLIGAN and CALAM, 1983).

This brings the discussion about optimization of sampling to an end. Sampling has been treated in this chapter under various aspects referring to linear discrete-time systems and, furthermore, to nonlinear models, where the sampling problems have been considered from a completely different point of view. Evaluation of the information contents of measurements was shown to provide a powerful tool to solve many of the problems arising from identification of dynamic systems. A main purpose of this chapter has been to demonstrate the benefits of these analyses in order to save instrumentation costs, to achieve a high precision in control and parameter estimation, and to reduce time for the analyst during the tedious search for errors in ill-posed identification problems.

5 References

ACKERMANN, J. (1985), *Sampled-Data Control Systems,* Berlin: Springer.

CHAVENT, G. (1974), Identification of functional parameters in partial differential equations, in: *Identification of Parameters in Distributed Systems,* pp. 31–48, New York: ASME.

EYKHOFF, P. (1988), A bird's eye view on parameter estimation and system identification, *Automatisierungstechnik* **36**, 413–420, 472–479.

GOODWIN, G. C. (1987), Identification: Experiment design, in: *Systems and Control Encyclopedia,* Vol. 4, pp. 2257–2264, Oxford: Pergamon.

HOLMBERG, A. (1982), On the practical identifiability of microbial growth models incorporating Michaelis-Menten type non-linearities, *Math. Biosci.* **62**, 23–43.

KUBRUSLY, C. S., MALEBRANCHE, H. (1983), A survey on optimal sensors and controllers location in DPS, *3rd IFAC Symp. Control of Distributed Parameter Systems,* Toulouse, pp. 59–73, Oxford–New York: Pergamon.

LJUNG, L. (1987), *System Identification: Theory for the User,* Englewood Cliffs: Prentice-Hall.

LUTTMANN, R. (1980), Modellbildung und Simulation von SCP-Prozessen in Blasensäulenschlaufenfermentern, *Dissertation,* Universität Hannover, FRG.

LUTTMANN, R., MUNACK, A., THOMA, M. (1985), Mathematical modelling, parameter identification and adaptive control of single cell protein processes in tower loop bioreactors, *Adv. Biochem. Eng. Biotechnol.* **32**, 95–206, Berlin–Heidelberg–New York: Springer.

MUNACK, A. (1986), On parameter identification for complex biotechnical systems, *Proc. 1st IFAC Symp. Modelling and Control of Biotechnological Processes,* Noordwijkerhout, pp. 159–165, Oxford, New York: Pergamon.

MUNACK, A. (1989), Optimal feeding strategy for identification of Monod-type models by fedbatch experiments, in: *Computer Applications in Fermentation Technology,* pp. 195–204, London–New York: Elsevier.

MUNACK, A., THOMA, M. (1981), On modelling, identification, and adaptive control of a class of distributed-parameter systems. *Int. J. Policy Inform.* **5**, 39–76.

NELLIGAN, I., CALAM, C. T. (1983), Optimal control of penicillin production, using a mini-computer, *Biotechnol. Lett.* **5**, 561–566.

POHJANPALO, H. (1978), System identifiability based on the power series expansion of the solution, *Math. Biosci.* **41**, 21–33.

QURESHI, Z. H., NG, T. S., GOODWIN, G. C. (1980), Optimum experiment design for identification of distributed parameter systems, *Int. J. Control* **31**, 21–29.

SCHÜGERL, K. (1985), *Bioreaktionstechnik,* Vol. 1, Frankfurt: Salle & Sauerländer.

SHANNON, C. E. (1949), Communication in the presence of noise, *Proc. IRE* **37**, 10–21.

TAKAMATSU, T., HASHIMOTO, I., SHIOYA, S. (1971), Determination of experimental condition taking account of the effect of parameter accuracy on system design, *J. Chem. Eng. Jpn.* **4**, 87–91.

ZADEH, L. A. (1962), From circuit theory to system theory, *Proc. IRE* **50**, 856–865.

ZARROP, M. B. (1979), *Optimal Experiment Design for Dynamic System Identification,* Berlin–Heidelberg–New York: Springer.

III. Modelling of Bioreactor Systems

9 Cell Models

KARL-HEINZ BELLGARDT

Hannover, Federal Republic of Germany

List of Symbols

C	concentration $(g \cdot L^{-1})$ or $(mol \cdot L^{-1})$
c	vector of intrinsic concentrations $(g \cdot g^{-1})$ or $(mol \cdot g^{-1})$
C	vector of concentrations $(g \cdot L^{-1})$ or $(mol \cdot L^{-1})$
D	dilution rate (h^{-1})
e	vector of model and measurement errors
E	enzyme
F	flow rate $(L \cdot h^{-1})$
H	enthalpy $(J \cdot mol^{-1})$
J	performance index, metabolic coordinator
k	vector of constants
K	general constant
K_I	inhibition constant $(g \cdot L^{-1})$
K_B	saturation constant of Blackman kinetics $(L \cdot g^{-1} \cdot h^{-1})$
K_C	saturation constant of Contois kinetics $(g \cdot L^{-1})$
K_M	half saturation constant of growth kinetics $(g \cdot L^{-1})$
K_T	saturation constant of Teissier kinetics $(g \cdot L^{-1})$
m	rate of endogenous or maintenance metabolism $(g \cdot g^{-1} \cdot h^{-1})$ or $(mol \cdot g^{-1} \cdot h^{-1})$
p	vector of model parameters
P	vector of mass exchange gas–liquid $(g \cdot L^{-1} \cdot h^{-1})$
q	vector of specific reaction rates $(g \cdot g^{-1} \cdot h^{-1})$ or $(mol \cdot g^{-1} \cdot h^{-1})$
Q	vector of reaction rates $(g \cdot L^{-1} \cdot h^{-1})$ or $(mol \cdot L^{-1} \cdot h^{-1})$
r	vector of intrinsic specific reaction rates $(g \cdot g^{-1} \cdot h^{-1})$ or $(mol \cdot g^{-1} \cdot h^{-1})$
R	gas constant $(J \cdot mol^{-1} \cdot K^{-1})$
T	temperature (K) or $(°C)$
u	cybernetic variable
v	control variable of cybernetic models
V	volume (L)
W	weighting matrix
X	concentration vector of the biotic phase $(g \cdot L^{-1})$
Y	stoichiometric matrix
Y_{ij}	yield coefficient for substance i from substance j
z	vector of growth-independent specific reaction rates $(g \cdot g^{-1} \cdot h^{-1})$ or $(mol \cdot g^{-1} \cdot h^{-1})$
Z	concentration vector of the abiotic phase $(g \cdot L^{-1})$
α	fraction of viable cells
θ	vector of operating parameters
τ	normalized kinetics
μ	specific growth rate (h^{-1})
1	vector containing all ones

Subscripts

C	carbon dioxide
D	variable associated with dormant cells
E	variable of endogenous metabolism, enzyme, or ethanol
G	growth-associated variable
i	element
I	variable of the inflow of the reactor, or inhibitor
lim	limiting value
L	liquid phase of the reactor
max	maximum value
min	minimum value
M	variable of the model, or metabolite
O	oxygen, or variable of the outflow of the reactor
P	product
S	substrate
tot	total or overall value
X	cell mass, or variable of the biotic phase
Z	variable of the abiotic phase

Superscripts

P	power of initial substrate concentration
R	power of substrate concentration, order of reaction
T	transpose of a matrix or vector

1 Introduction

The history of mathematical models of biotechnological processes began with the famous equations of Blackman (1905), Monod (1942), and Teissier (1942), which related the concentration of the limiting substrate to the growth rate of the microorganisms. Meanwhile, a great number of models was developed for a wide variety of fermentations using different microorganisms. Most of them still contain kinetics of the Monod type, which shows the fundamental nature of that equation.

Within the framework of this chapter one cannot provide a complete survey of all the important or interesting types of models. Therefore, emphasis will be put on introducing the ideas and elementary structure of cell models for biotechnological processes. This frame will be filled in by giving a few examples for models. Practical application of models which are treated in other parts of this series will not be dealt with. Only a few fields will be mentioned where models were proven to be useful. Since mathematical models give a functional relation between the process variables, they are ideally suited for many tasks in process design, e.g.:

- Optimization of plant structure and operating parameters,
- calculation of optimal time profiles for substrate feeding or other variables,
- planning of experiments in order to obtain maximum information with a minimum of time and expense,
- design of control systems,
- estimation of variables or parameters that cannot be measured, and indirect measuring methods.

Many of these tasks impose very high requirements on the accuracy of the model, which are not easy to meet with simple models. The mathematical tools and basic methods for formulation of the models are available due to the work of Roels et al. (1978) and Roels (1982), Frederickson et al. (1970), and Ramkrishna et al. (1967). But there is still no general rule on how to develop a mathematical model for the biological reactions and the underlying mechanisms in a fermentation process, that is a cell model guaranteed to be correct in its structure. In any case, extensive simplifications must be made to obtain a workable model, although the main features of the kinetics of growth and product formation, and of the type of fermentation must be retained in the model. The degree of complexity and the variables chosen for the model also depend directly on the intended use. A good practice in modelling of biotechnological processes is to keep the cell model as simple as possible. Of course, the more complex the model and the more variables and parameters it contains, the better it can be fitted to experimental data. But this must not lead one to believe that such complex models are more accurate. Experimental verification and parameter identification for complex models are very difficult and troublesome, and if one does not evaluate the results very carefully, the model is probably fitted to the errors of the measurement.

In the following, the kinetics and types of biotechnological processes will be briefly discussed with respect to their relation to mathematical modelling. The connection of biological cell models and reactor models, which are the subject of other chapters of this book, will be defined later, and the modelling concepts of the system cell will be introduced for unstructured and structured models. There are other concepts of modelling which cannot be covered in this chapter. Among these are elemental balances, energetics and electron balances, models including age distributions, cell cycle models, and models of multi-species cultivations.

2 Principles of Model Building for Biotechnological Processes

2.1 Kinetics and Types of Fermentations

A time course of a batch fermentation is shown in principle in Fig. 1. With respect to cell mass, the process can be divided into an initial lag phase, a growth phase, a stationary phase, and a declining or death phase. The initial lag phase is due to regulatory phenomena of the microorganisms as an adaption from the preculture to the conditions in the fermentor. Afterwards, exponential growth can usually be observed. In both phases substrate is in excess. In a later stage of the process, substrate limitation, product inhibition, or other phenomena lead to the stationary phase with a zero growth rate. This can be followed by a declining phase in which the cell mass decreases due to lysis or endogenous metabolism.

The behavior is more complicated when considering the product concentration, which was classified by GADEN (1959) into three types. In type I fermentations, growth and product formation are fully associated, the product arises from an essential growth proc-

ess. Therefore, the product increases proportionally to the growth rate, and there is no separate production phase. A certain coupling can also be found in fermentations of type II, overproduction of primary metabolites, but the interrelation follows more complicated kinetics with separate growth and production phases. Type III represents secondary metabolite fermentations, in which there is no simple connection between production of the secondary metabolite and the primary metabolism. Most antibiotic processes can be grouped into this type. Actually the classification into types II and III is a bit hazy, and the mechanisms have not always been fully clarified.

2.2 Types of Biological Models

The function of a biological model is to describe the metabolic reaction rates and their stoichiometry on the basis of present and past fermentor conditions. By reason of metabolic complexity, all models must be a rough simplification of reality. The main difficulty for modelling is to identify the most important factors that influence the growth process, and to find a suitable model structure for the intracellular processes. According to this view of the cells, biological models or growth models can be divided into unstructured and structured types.

Unstructured models (see Sect. 3) are the simplest. They take the cell mass as a uniform quantity without internal dynamics whose reaction rate depends only upon the conditions in the liquid phase of the reactor. Therefore, the models only contain kinetics of growth, substrate uptake, and product formation. This is a good approximation if the response time to environmental changes of the cell is either negligibly small or very long compared with the duration of the fermentation process. The relatively unspecific modelling of the biotic phase in unstructured biological models limits their descriptive potency. Only the average growth behavior of the culture is described in the sense of an averaging over the population of the cells, and a time averaging over the cell division cycle. In many cases this simplification is reasonable because it is impossible to have exact knowledge of the heterogeneous

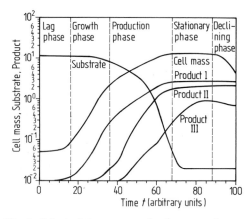

Fig. 1. Principal time course of cell mass, substrate, and product concentration for different types of fermentation.

composition of the biomass and the state of the intracellular systems. But the simplification can also cause errors in the static and dynamic behavior of the model. Particularly, time averaging results in sizeable errors in the presence of metabolic regulation or synchronous growth of the culture, in which metabolic variation during the division cycle becomes globally visible. Model errors from population averaging can appear if the cells cannot adapt at once to new growth conditions.

For inclusion of some of the above effects, and if the cellular relaxation time is of the same magnitude as the duration of the process, the internal state of the cells must also be considered. This leads to so-called structured models. This type of model is presented in Sect. 4. The models can be structured on the basis of biomass components such as concentrations of metabolites, enzymes, or RNA, or by population-related variables, describing different morphological types of cells or cell aging. Models with a structure on the population level are also called segregated biological models.

2.3 The Biological System Cell in the Modelling View

Biotechnological processes generally have the following structural elements: the liquid phase and gas phase, which together form the abiotic phase, and the biotic phase, which consists of the cell population. The properties of these phases are characterized by time-dependent macroscopic variables such as concentrations, or physical variables such as temperature. All the reactions catalyzed by microorganisms take place in the liquid phase. The relevant components are:

- Cell mass, C_X, autocatalytically synthesized from the provided substrates,
- substrates, C_S, as energy and nutrient suppliers,
- products of metabolism, C_P, inside or outside of the cells,
- dissolved gases, mainly oxygen, C_O, and carbon dioxide, C_C, which are connected to the gas phase by mass exchange.

The physical and biochemical processes in bioreactors are described by balances for mass, impulse, and energy, and by the related conservation laws (ROELS et al., 1978). For a unified modelling approach the equations are written in vector notation. Let C be the column vector of concentrations in the liquid phase of the reactor, then in the above example C is defined by:

$$C = (C_S, C_P, C_O, C_C, C_X)^T$$

Generally the concentration vector can be divided into concentrations of the biotic phase, X, and of the abiotic phase, Z:

$$C = \begin{bmatrix} Z \\ X \end{bmatrix} \tag{1}$$

C has to include N different vectors X_i to X_N in multi-species cultivations when not only one, but N cell types are present. In structured models the microorganisms are modelled as a complex system with further sub-components, which, for instance, account for several intracellular concentrations. Then, X is the state vector of the biotic phase, and the total cell mass C_X is the sum of all its components:

$$C_X = I^T \cdot X \tag{2}$$

In unstructured models the microorganisms are viewed as a homogeneous component, and X reduces to a scalar variable:

$$C_x = X \tag{3}$$

The mass balances of the liquid phase for stirred tank reactors are:

$$\frac{dC(t)}{dt} = \frac{F_I(t)}{V_L(t)} \cdot (C_I - C(t)) + P(t) + Q(t) \tag{4}$$

$$\frac{dV_L(t)}{dt} = F_I(t) - F_O(t) \tag{5}$$

where F_I and F_O are the inflow and outflow rate of the reactor, V_L is the liquid phase volume, C_I is the concentration in the inflow, Q is the vector of reaction rates in the liquid phase, and P is the mass exchange vector with the gas phase. With the above model equations different kinds of processes can be described:

Batch fermentations with
$F_I(t) = F_O(t) = 0$, $V_L = $ const,
fed-batch fermentations with $F_I(t) \neq 0$,
$F_O(t) = 0$, $V_L \neq $ const,
semicontinuous fermentations with
$F_I(t) \neq F_O(t) \neq 0$, $V_L \neq $ const,
continuous fermentations with
$D \cdot V_L = F_I(t) = F_O(t) \neq 0$, $V_L = $ const.

After one has chosen the vector of concentrations, C, and established the corresponding balance equations, the next step of modelling is to describe the biological reaction, Q, by the cell model. The aim of this chapter is to introduce the principal ideas for models of the biological system. The transport step, P, will not be considered, and it will be assumed that no other reactions than those catalyzed by the microorganisms take place.

A general characteristic of microbially catalyzed reactions is that the total volumetric reaction rate is proportional to the amount of active biomass. The biological models are advantageously formulated using specific reaction rates, defined by:

$$q(C(t), p(t)) = \frac{Q(C(t), p(t))}{C_X(t)} \tag{6}$$

Besides the concentrations, other time-varying operating parameters can also influence the microbial reactions. This is considered by the

time-dependent vector $p(t)$, but usually p is constant.

Throughout this chapter, the models will be given for specific reaction rates. For a more convenient notation, a distinction will be made between elementary, independent specific reaction rates, r, for which kinetic expressions have to be specified, and the entire reaction vector, q, which can be calculated from r by stoichiometry. Due to conservation laws and stoichiometric laws, the relation between the globally visible net reaction rates in Eq. (6) and the intrinsic specific reaction rates, $r(C(t), p(t))$, within the microorganisms, can be written as:

$$q(C(t), p(t)) = Y(t) \cdot r(C(t), p(t)) \tag{7}$$

where Y is the matrix of stoichiometric or yield coefficients, which may vary with time but is usually taken as constant. In this nomenclature $C(t)$ and $p(t)$ are the input variables of the biological model, and the net reaction rates $q(t)$ are the output variables. In Fig. 2 the corresponding structure of a model of a biotechnological process is shown. It consists of the liquid phase model, a gas phase model, and a cell model. Taking the biological model as a local microkinetic model, it should be independent of the type of reactor and mode of operation, and also independent of the reactor models. In the cell model, the microorganisms are represented by kinetic equations for the reaction vector r. Unstructured cell models are algebraic ones, while structured models also include further differential equations for the biological state vector X.

2.4 Identification of Parameters

Model building is always combined with theoretical studies and practical experiments. The establishment of complex models is a very iterative process (MOSER, 1981). Since the problems in parameter identification and model verification increase rapidly with the complexity of the model, one should begin with maximally simplified assumptions and withdraw them step by step in case that the model quality is not sufficient. Modelling includes not only the selection of the correct model struc-

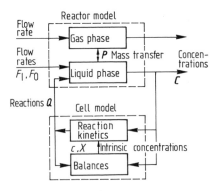

Fig. 2. Structure and elements of mathematical models for biotechnological processes.

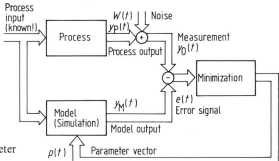

Fig. 3. Block diagram of the procedure of parameter identification by the output error method.

ture, but also the quantitative determination of the model parameters. Unfortunately, their values are often not known or only inexactly known in advance and must be determined by fitting the model to experimental data. For simple models, regression methods such as Lineweaver–Burk plots can be used, but for complex models more general methods of parameter identification have to be employed.

The task of parameter identification is to estimate unknown model parameters by comparing process (experiment) and model according to a given performance criterion. The identification procedure should be able to eliminate disturbances of the measured variables. Offline methods for parametric models are particularly interesting for model building. Usually the very general method of minimizing the variance of the output error (least square estimation) is used (FASOL and JÖRGL, 1980). A schematic diagram of the method is shown in Fig. 3. The parameter identification serves to minimize the performance criterion:

$$J(e,p) = \sum_{i=1}^{N} e^{\mathrm{T}}(t_i) \cdot W \cdot e(t_i) \xrightarrow{p} \min \qquad (8)$$

which depends on the parameter vector p and the error signal:

$$e(t_i) = y_O(t_i) - y_M(t_i) \qquad (9)$$

Here, N is the number of measured values, t_i are the measuring times, y_M the model predictions, y_O the measured values, and W is a weighting matrix. The usual regression method is derived from Eq. (8) as a special case. For complex non-linear systems, numerical simula-

tion techniques in conjunction with numerical optimization methods have to be used.

Parameter identification by output error methods does not evaluate the structure of the model and the accuracy of the estimated parameters. Thus, a good fit may be achieved even with wrong parameters, especially for non-linear systems, high model order, and only a few measured values. The limited accuracy of measurements not only hinders parameter identification, but also discriminates between different models. This should always be kept in mind during model building.

3 Unstructured Models on the Population Level

In unstructured models, the biological reaction depends directly and solely on macroscopic variables that describe the conditions in the fermentor. The only biological state variable is the cell mass concentration, C_X. Nevertheless, many phenomena in biotechnological processes can be covered by this type of model. Besides the cell mass, only those other variables have to be considered in the model that show great variation during the fermentation and which have significant influence on microbial behavior. In the following, the focus will be on models for substrate uptake kinetics, including multiple limitations, inhibition effects, and product formation. Due to the extensive omissions and simplifications of cell metabolism, these kinetics should be taken as formal kinetics which give only an integral description of the process and not a detailed picture of the underlying mechanisms.

3.1 Simple Growth and Substrate Uptake Kinetics

During growth, new cell mass is formed autocatalytically from substrate with the specific growth rate μ:

$$\mu(C_S) = r_X(C_S) \tag{10}$$

which is a function of the concentration of the limiting substrate. Microbial reactions usually show saturation at high substrate concentrations, that is, the reaction rate approaches a maximum value. On the other hand, the reaction rate equals zero if no substrate is available. Therefore, Eq. (10) can be rewritten as:

$$\mu(C_S) = \mu_{max} \cdot \tau(C_S) \tag{11}$$

where τ is a normalized growth kinetics. Several basic types of kinetics for the specific

growth rate have been proposed, as summarized in Tab. 1. In many cases the ratio of cell mass formed to substrate consumed, the yield coefficient, Y_{XS}, is constant as a good approximation. For a single substrate and cell mass as the only product, C is given by:

$$C = (C_X, C_S)^T$$

In this simple model there is only one intrinsic reaction, $r = \mu(C_S)$, and therefore Eq. (7) becomes:

$$q(C_S) = \begin{bmatrix} q_X \\ q_S \end{bmatrix} = \begin{bmatrix} 1 \\ -Y_{XS}^{-1} \end{bmatrix} \cdot \mu(C_S) \tag{12}$$

Since substrate uptake is proportional to the growth rate, the kinetics of Tab. 1 can also be taken as normalized substrate uptake kinetics. Due to the saturation curve, all substances which influence the growth but are present in sufficiently high concentrations fortunately need not be considered in the model.

Although the equations in Tab. 1 should be taken as formal kinetics for the globally observed behavior of the culture, which need not have a close relation to microkinetics of the biological reaction, one can give an interpretation of the special form of certain kinetics. The Monod equation is analogous to the Michaelis–Menten enzyme kinetics. Its applicability to biotechnological processes can be referred to growth rate limiting, carrier-mediated transport systems for the substrate. The equations of MOSER and VAVILIN are similar to the Monod kinetics except that the reaction for the substrate is not of first order. The Vavilin equation, in which C_{S0} is the initial substrate concentration, has found application in processes with toxic substrates. The Blackman equation can be interpreted such that at high substrate concentrations not the substrate uptake, but another metabolic reaction, is rate limiting. The Contois kinetics considers an effect of cell concentration on the growth due either to inhibition by the cells themselves or to diffusional limitation for substrate by a limitation constant which is proportional to the cell concentration. The equations of POWEL (1967) and MASON and MILLES (1976) account for additional diffusion-driven flux of substrate into the cell. A plot of the $\mu(C_S)$ charac-

Tab. 1. Growth Kinetics for a Single Substrate

Name	Year	Normalized Kinetics τ
BLACKMAN	1905	$\min(1, K_B \cdot C_S)$
MONOD	1942	$\dfrac{C_S}{K_M + C_S}$
TEISSIER	1942	$1 - e^{-K_T \cdot C_S}$
MOSER	1958	$\dfrac{C_S^R}{K_M^R + C_S^R}$
CONTOIS	1959	$\dfrac{C_S}{K_C \cdot C_X + C_S}$
POWEL[a]	1967	$\dfrac{C_S - K_1 \cdot \tau(C_S)}{K_M + C_S - K_1 \cdot \tau(C_S)}$
MASON and MILLES	1976	$\dfrac{C_S}{K_M + C_S} + K_D \cdot C_S$
VAVILIN	1982	$\dfrac{C_S^R}{K_M^{R-P} \cdot C_{S0}^P + C_S^R}$

[a] this kinetics is given in implicit form

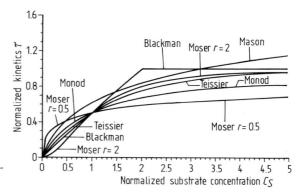

Fig. 4. Normalized $\mu(C_S)$ characteristics of substrate uptake kinetics τ versus substrate concentration C_S, normalized to the half-saturation constant.

teristics of the kinetics is given in Fig. 4. In practice, because of measurement errors, it is difficult to discriminate between the different kinetics, especially when only using batch culture data. Therefore, the Monod kinetics is generally a good choice for the model.

3.2 Substrate-Independent Growth Kinetics

In certain cases the application of substrate-independent kinetics given in Tab. 2 can make sense. The origin of the kinetics can be found in modelling of natural populations. Also in some fermentation processes, the growth rate does not mainly depend on a limiting substrate. For example, in cultivations of mammalian cells on micro-carriers the cells need free surfaces to colonize and grow. Growth stops if no unoccupied surface is available.

Tab. 2. Substrate-Independent Growth Kinetics

Name	Year	Normalized Kinetics τ
Logistic law	1938	$1 - \dfrac{C_X}{C_{X,\max}}$
CUI and LAWSON	1982	$\dfrac{1 - \dfrac{C_X}{C_{X,\max}}}{1 - \dfrac{C_X}{C_{X,\lim}}}$
FRAME and HU	1988	$1 - e^{-K \cdot \frac{C_{X,\max} - C_X}{C_X}}$

Such kinetics can be interpreted as an inhibition by the cell concentration or a space limitation. For this type of fermentation, the kinetics of FRAME and HU (1988) can give a better description than the logistic law because it exhibits a sharper transition to the stationary phase. Another application of these kinetics is for processes in which information on the limiting substrate is not available, e.g., the limitation of an unidentified component in complex media.

3.3 Substrate and Product Inhibition

Besides substrate limitation, inhibition by substrates or products is quite often found in biotechnological processes. Both have been examined by many authors. A short review is given by HAN and LEVENSPIEL (1988). Tab. 3 gives a list of normalized inhibition kinetics. Most of the applied kinetics are extensions of the Monod equation and have been derived from enzyme inhibition kinetics (DIXON and WEBB, 1967). These kinetics cannot predict zero growth for a finite inhibitor concentration. Therefore, other empirical equations have been proposed for the description of this behavior. The last five equations in Tab. 3 predict a zero growth rate at $C_I = K_I$. Product, substrate, and cell inhibition can be obtained by choosing the variable C_I as C_P, C_S, or C_X, respectively. The competitive type only applies for substrate inhibition. For non-toxic substrates there is also no evidence of non-compe-

Tab. 3. Inhibition Kinetics for a Single Inhibitor

Name	Year	Normalized Kinetics τ_I
HALDANE (competitive type)	1965[a] (1930)	$\dfrac{C_S}{K_M + C_S + \dfrac{C_S^2}{K_I}}$
WEBB (competitive type)	1963[a]	$\dfrac{C_S \cdot \left(1 + \dfrac{C_S}{K_{I1}}\right)}{K_M + C_S + \dfrac{C_S^2}{K_I}}$
IERUSALIMSKY (non-competitive type)	1965[a]	$\dfrac{C_S}{K_M + C_S} \cdot \dfrac{1}{1 + \dfrac{C_I}{K_I}}$
YANO et al. (competitive type)	1969[a]	$\dfrac{C_S}{K_M + C_S \cdot \left(1 + \sum_N \left[\dfrac{C_S}{K_I}\right]^N\right)}$
EDWARDS	1970	$\dfrac{C_S}{K_M + C_S} \cdot e^{-C_I \cdot K_I^{-1}}$
YANO and KOYA (generally non-competitive type)	1973	$\dfrac{C_S}{K_M + C_S} \cdot \dfrac{1}{1 + \left[\dfrac{C_I}{K_I}\right]^N}$
WAYMAN and TSENG	1976[a]	$\dfrac{C_S}{K_M + C_S} + K_I \cdot \min(C_{SI} - C_{S0})$
Teissier type		$e^{-C_I \cdot K_I^{-1}} - e^{-K_T \cdot C_S}$
GHOSE and TYAGI DAGLEY and HINSHELWOOD	1952[a]	$\left[1 - \dfrac{C_I}{K_I}\right] \cdot \dfrac{C_S}{K_M + C_S}$
CHEN et al.	1976	$\left[1 - \dfrac{C_S}{K_I}\right] \cdot \dfrac{C_S}{K_M + C_S - C_2 \cdot C_S^2}$
BAZUA and WILKE	1977[a]	$\left[1 - \dfrac{C_I}{K_I}\right]^{1/2} \cdot \dfrac{C_S}{K_M + C_S}$
LEVENSPIEL	1980[a]	$\left[1 - \dfrac{C_I}{K_I}\right]^N \cdot \dfrac{C_S}{K_M + C_S}$
HAN and LEVENSPIEL	1987	$\left[1 - \dfrac{C_I}{K_I}\right]^M \cdot \dfrac{C_S}{K_M \left[1 - \dfrac{C_I}{K_I}\right]^M + C_S}$

[a] cited in HAN and LEVENSPIEL (1987)

titive substrate inhibition. The given equations can be generalized for multiple inhibitions by combining the normalized kinetics to a product with several factors.

3.4 Multiple Limitations

The above kinetics are valid for a single limiting compound. Several attempts have been made to develop models for multiple limitations. Most of them are extensions of the elementary kinetics in Tabs. 1 and 3. There are several possibilities for the interrelation of growth rate and substrate concentration, depending on the substrate itself and on the microorganisms. All substrates can be essential, meaning there is no growth if only one of the substrates is lacking; or one of the substrates is sufficient for growth and others are used up in parallel or in sequence. In any case, compared with simple kinetics, the model has to consider much more information on the structure of metabolism. In the following, the extension of single substrate kinetics to multiple limitations is shown for the mentioned cases. The limits of this approach and how these restrictions can be overcome will be discussed as well.

3.4.1 Essential Substrates

All essential substrates must be present in the medium to allow for growth of the microorganisms. In the interacting model all substrates together determine the growth rate. For N substrates, this can be expressed by a product of N normalized kinetics given in Tabs. 1 and 3:

$$\mu(C_{S1}, C_{S2}, \ldots C_{SN}) = \tag{13}$$
$$= \mu_{max} \cdot \tau_1(C_{S1}) \cdot \tau_2(C_{S2}) \ldots \cdot \tau_N(C_{SN})$$

A slightly different view of multiple limitations is given in the non-interacting model. Here it is assumed that only the substrate with the greatest limitation determines the growth rate. This leads to a selection of the minimum growth rate allowed among all substrates:

$$\mu(C_{S1}, C_{S2}, \ldots C_{SN}) = \tag{14}$$
$$= \min \{\mu_{S1}(C_{S1}), \mu_{S2}(C_{S2}), \ldots, \mu_{SN}(C_{SN})\}$$

where all terms μ_{Si} have the form of Eq. (11). It should be noted that Eq. (14) is equivalent to a multiple Blackman kinetics if one of the terms is set to a constant value, e.g., $\mu_{S1} = \mu_{max}$. In practice there are also only slight differences between the interacting and non-interacting models. As an example, both are compared in Fig. 5, which shows the simulation of a glycerol- and oxygen-limited culture with both of the following kinetics (from SINCLAIR and RYDER, 1984):

Interacting model,

$$\mu(C_S, C_O) =$$
$$= \mu_{max} \cdot \frac{C_S}{K_M + C_S} \cdot \frac{C_O}{K_O \cdot C_X + C_O} \tag{15a}$$

Non-interacting model,

$$\mu(C_S, C_O) =$$
$$= \mu_{max} \cdot \min \left[\frac{C_S}{K_M + C_S}, \frac{C_O}{K_O \cdot C_X + C_O} \right] \tag{15b}$$

In vector notation the entire cell model takes the form:

$$C = (C_X, C_S, C_O)^T \tag{16a}$$

$$r = (\mu(C_S, C_O), \mu_E)^T \tag{16b}$$

$$q = (q_X, q_S, q_O)^T = Y \cdot r \tag{16c}$$

$$Y = \begin{bmatrix} 1 & -1 \\ -Y_{XS}^{-1} & 0 \\ -Y_{XO}^{-1} & 0 \end{bmatrix} \tag{16d}$$

and the liquid phase model is given by Eq. (4), specialized for chemostat processes. Constant yield coefficients for oxygen and substrate were used in the simulation in conjunction with a constant rate of endogenous metabolism, μ_E (see Sect. 3.5). The differences between the models are not significant when compared with errors of measurement.

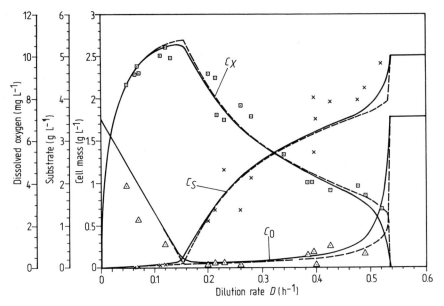

Fig. 5. Simulations of a glycerol–oxygen limited chemostat using the interactive (——) and non-interactive (- - -) model of SINCLAIR and RYDER (1975). Symbols are measured data, taken from the reference, for the concentrations of cell mass C_X (□), glycerol C_S (×), and dissolved oxygen C_O (△).

3.4.2 Growth Enhancing and Alternative Substrates

Several substances can be catabolized in parallel with other substrates and so increase the growth rate. For modelling these kinds of phenomena the entire kinetics can be derived from the sum of elementary kinetics, Eq. (11),

$$\mu(C_{S1}, C_{S2}, \dots C_{SN}) = \qquad (17)$$
$$= \mu_1(C_{S1}) + \mu_{S2}(C_{S2}) + \dots + \mu_{SN}(C_{SN})$$

In reality, most of the kinetics are not purely additive. The biological meaning is that the bottle-neck for growth is not associated with the substrate uptake steps. Usually the cells can achieve the maximum growth rate with one preferred substrate, e.g., glucose. If this is available in excess, no further increase due to the presence of other sugars can be observed. This is only possible if the preferred substrate is partially limiting. In addition, the uptake of those substrates that result in less efficient

growth is subject to catabolite repression by the preferred substrate. Even if several substrates are available, they are taken up sequentially. To account for these effects, Eq. (17) can be extended by additional inhibition terms. For a Monod-type kinetics this yields for two substrates where the first is preferred:

$$\mu(C_{S1}, C_{S2}) = \mu_1(C_{S1}) + \mu_2(C_{S1}, C_{S2}) =$$
$$= \mu_{\text{max},1} \cdot \frac{C_{S1}}{K_{M1} + C_{S1}} +$$
$$\qquad\qquad\qquad (18\,\text{a})$$
$$+ \mu_{\text{max},2} \cdot \frac{C_{S2}}{K_{M2} + C_{S2} + \dfrac{C_{S1}}{K_{I1}}}$$

$$r = (\mu_1, \mu_2)^T \qquad (18\,\text{b})$$

$$q = (q_{S1}, q_{S2}, \mu)^T = Y \cdot r \qquad (18\,\text{c})$$

$$Y = \begin{bmatrix} -Y_{XS1}^{-1} & 0 \\ 0 & -Y_{XS2}^{-1} \\ 1 & 1 \end{bmatrix} \qquad (18\,\text{d})$$

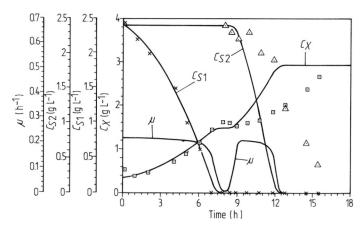

Fig. 6. Simulation with an unstructured model of diauxic growth of *Klebsiella terrigena* on glucose and maltose. Experimental data (symbols) from HOPF in BELLGARDT et al. (1988), concentration of glucose C_{S1} (×), maltose C_{S2} (△), cell mass C_X (□), and specific growth rate μ.

As an example, the simulation of diauxic growth of *Klebsiella terrigena* on glucose and maltose is plotted in Fig. 6. The sequential uptake of the substrates can be described in principle, but the model cannot be fitted to the diauxic lag phase. In this model the delay of growth on the second substrate is due only to an unrealistically low inhibition constant, K_{I1}. For comparison, a simulation of the same process with a structured model is shown in Figs. 12 and 13.

3.4.3 Combinations of Essential and Alternative Substrates

The above kinetics can be combined to a more general form, as proposed by TSAO and HANSON (1975). They grouped the substrates into m classes of essential substrates, and in each such class there were several alternative or enhancing substrates. The entire kinetics is then given by:

$$\mu(C_{S1}, C_{S2}, .. C_{SN}) = (\mu_{S11} + \mu_{S12} .. + \mu_{S1K})$$
$$\cdot (\mu_{S21} + \mu_{S22} .. + \mu_{S2L})$$
$$\cdot$$
$$\cdot$$
$$\cdot$$
$$\cdot (\mu_{Si1} + \mu_{Si2} .. + \mu_{SiI})$$
$$\cdot$$
$$\cdot (\mu_{Sm1} + \mu_{Sm2} .. + \mu_{SmN})$$
$$\tag{19}$$

where I, K, L, and N are the numbers of essential substrates in the ith class.

3.4.4 Comments on Multi-Substrate Kinetics

While the multi-substrate kinetics Eqs. (13) and (14) for essential substrates seem to give a good description of observed phenomena, the case for alternative substrates is not so clear. To the author's knowledge, experimental data to support Eqs. (17) and (19) are very rare. The reason might be a too-simple approximation of the growth kinetics due to the neglect of the regulatory response of the microorganisms. As already mentioned, a sequential uptake is often found with the presence of excess substrate. As long as the preferred substrate can support a sufficient growth rate, it suppresses and inhibits the uptake system of other substrates. After it has become limiting, the enzymes for the substrate with the next lower preference are induced or derepressed. The result is a remarkable lag phase for the adaption to the next substrate, which can only be described by structured models. To make the situation more complicated, the same substrates which are used up sequentially at high concentrations are also used up in parallel if the concentrations are low enough. Furthermore, the uptake of several alternative substrates can be affected by one or more essential substrates such as oxygen. A well-known example is the growth of yeast in batch and continuous culture (see Sect. 4.5). This means that many regulatory phenomena have to be considered for a model of multi-substrate kinetics that are very closely associated with the pathways of the microorganism. It can be concluded that Eqs. (17) and (19) on the one hand are too simple and on the other hand too general to cover the difficult adaptation of the cells. A promising idea of dealing with these phenomena is the cybernetic modelling approach reported in Sect. 4.4.

3.5 Deviations from the Constant Yield Case

At low growth rates one often can observe a decrease in cell yield. This can be explained by an additional growth-independent substrate consumption for maintaining the cell structure, or by lysis processes. HERBERT (1959) took this effect into account by discriminating between observed net growth rate, μ, and true growth rate, μ_G. The difference is the specific rate of endogenous metabolism, μ_E:

$$\mu = \mu_G - \mu_E \tag{20}$$

In the usual vector notation the model is:

$$r = (\mu_G, \mu_E)^T \tag{21a}$$

$$q = (q_X, q_S)^T \tag{21b}$$

$$Y = \begin{bmatrix} 1 & -1 \\ -Y_{XSG}^{-1} & 0 \end{bmatrix} \tag{21c}$$

and by comparing Eq. (21) with Eq. (12) the observed growth dependent yield becomes:

$$Y_{XS}(\mu) = Y_{XSG} \frac{\mu}{\mu + \mu_E} \tag{22}$$

Here, μ_G may be expressed as any of the previously given kinetics. During strong growth limitation the observed growth rate, μ, can become less than zero. In this case the rate μ_E also can be viewed as a specific rate of cell lysis

or the degradation rate of intracellular storage material.

PIRT (1965) and IERUSALIMSKY (1967) referred the phenomena of the decreasing yield to an additional need for substrate consumption for maintenance of the cell structure, expressed by the specific rate m_S. Thus, if the uptake rate of substrate used only for growth is q_{SG}, then the total substrate uptake q_S is:

$$q_S = q_{SG} - m_S \tag{23}$$

(note that production is defined as positive rate) and the model in vector notation

$$r = (\mu, m_S)^T \tag{24}$$

$$q = (q_X, q_S)^T \tag{25}$$

$$Y = \begin{bmatrix} 1 & 0 \\ -Y_{XS}^{-1} & -1 \end{bmatrix} \tag{26}$$

Now the specific growth rate is given by

$$\mu = -q_{SG} \cdot Y_{XSG} \tag{27}$$

and the yield becomes, comparing Eq. (12),

$$Y_{XS}(\mu) = Y_{XSG} \cdot \frac{\mu}{\mu + m_S \cdot Y_{XSG}} \tag{28}$$

By comparing Eqs. (22) and (28) one can see that both are equivalent in the regions of normal growth, if

$$m_S = \frac{\mu_E}{Y_{XSG}}$$

The main drawback of Eq. (23) is that in dynamic simulations the substrate concentration can become less than zero because of the zero-order maintenance reaction. This situation cannot be reached with the model Eq. (20), where with strong limitation the growth rate becomes less than zero.

Besides the mathematical differences between the above models, there are for each some discrepancies with experimental data at very low dilution rates. Under such conditions the observed cell yield can be much higher than that predicted by the simple models, Eqs. (22) and (28), where it tends towards zero when μ approaches zero. To improve the model in this aspect, PIRT (1987) divided the entire population, C_X, into fractions of viable cells, C_{XG}, and of active, dormant cells without substrate turnover, C_{XD}, respectively,

$$C_{XD}(\mu) = (1 - \alpha(\mu)) \cdot C_X \qquad (29)$$

$$C_{XG}(\mu) = \alpha(\mu) \cdot C_X \qquad (30)$$

where $\alpha(\mu)$ is the fraction of viable cells as a function of the specific growth rate. The true specific growth rate μ_G is increased compared with the observed growth rate μ:

$$\mu_G = \frac{\mu}{\alpha(\mu)} \qquad (31)$$

When describing the substrate demand of the growing fraction by the maintenance model Eq. (25), the observed growth yield with respect to the total cell mass is:

$$Y_{XS}(\mu) = Y_{XSG} \frac{\mu}{\mu + m_S \cdot \alpha(\mu) \cdot Y_{XSG}} \qquad (32)$$

This equation can correlate the experimental data for cell yield, fraction of viable cells, and dilution rate (see Fig. 7).

PIRT's approach can be slightly extended to give a mechanistic description for the dependence of the relative fraction of viable cells, $\alpha(\mu)$, on the growth rate during steady states of a chemostat culture, by introducing a specific formation rate of dormant cells from viable cells, K_D. From the steady state balances in a chemostat:

$$0 = \frac{dC_{XG}(t)}{dt} = (\mu_G - K_D - D) \cdot C_{XG} \qquad (33)$$

$$0 = \frac{dC_{XD}(t)}{dt} = K_D \cdot C_{XG} - D \cdot C_{XD} \qquad (34)$$

with $D \equiv \mu$, the ratio of viable cells to dormant cells is obtained from Eq. (34) as:

$$\frac{C_{XG}(\mu)}{C_{XD}(\mu)} = \frac{\mu}{K_D} \qquad (35)$$

On the other hand, dividing Eq. (39) by Eq. (29):

$$\frac{C_{XG}(\mu)}{C_{XD}(\mu)} = \frac{\alpha(\mu)}{1 - \alpha(\mu)} \qquad (36)$$

and combining this with Eq. (35) gives:

$$\alpha(\mu) = \frac{\mu}{K_D + \mu} \qquad (37)$$

By substituting $\alpha(\mu)$ into Eq. (32) one obtains the observed yield during steady states of a chemostat with the presence of dormant cells as a function of the specific growth rate as:

$$Y_{XS}(\mu) = Y_{XSG} \cdot \frac{K_D + \mu}{K_D + \mu + m_S \cdot Y_{XSG}} \qquad (38)$$

which predicts a non-zero observed yield if μ tends to zero. Fig. 7 shows a fit of Eqs. (37) and (38) to experimental data given by PIRT

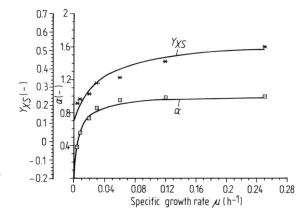

Fig. 7. Plots of the growth yield Y_{XS} from glycerol (Eq. 38) and of the viable cell fraction α (Eq. 37) versus the dilution rate D for a glycerol-limited chemostat culture of *Klebsiella aerogenes* (data from PIRT, 1987).

(1987). The identified parameter values are $K_D = 0.0064$ h^{-1}, $Y_{XSG} = 0.55$, and $m_S = 0.045$ h^{-1}.

3.6 Influence of Other Variables on Growth

In biotechnological processes growth and product formation depend, besides on substrates, on many other variables and operating parameters. A convenient approach is to consider this in the kinetic and stoichiometric parameters, while maintaining the form of the kinetics. In the following, $p(\theta(t))$ is the vector of variable parameters, and $\theta(t)$ is the vector of operating variables that influence the growth. Then the kinetic model may be written as:

$$r = r[C(t), p(\theta(t))] \tag{39}$$

$$Y = Y[p(\theta(t))]$$

A mechanistic model for temperature effects and a general black box model are presented next.

3.6.1 Mechanistic Description of Temperature Effects

The influence of temperature on microbial activity can be explained by the superposition of activation and deactivation effects. As a result of both, an optimum temperature for growth of microorganisms usually exists. Below the optimum temperature, the temperature dependency of the maximum reaction rate in Eq. (39) due to activation can be expressed by the Arrhenius relationship:

$$r_{max}(T) = K_1 c_{EA} e^{-\frac{\Delta H_1}{RT}} \tag{40}$$

where K_1 is a constant, c_{EA} is the relative concentration of the active form of a rate-limiting enzyme, ΔH_1 is the activation enthalpy of its reaction. According to ESENER et al. (1980) the negative influence of temperature on the growth above the optimum temperature can be explained by an inactivation effect on the in-

volved enzymes. For a fast inactivation reaction, there is an equilibrium between the active, c_{EA}, and inactive, c_{EI}, form of the enzyme given by:

$$c_{EI} = K_2 c_{EA} e^{-\frac{\Delta H_2}{RT}} \tag{41}$$

where K_2 is a constant and ΔH_2 is the enthalpy change of the inactivation reaction. Since the sum of the relative enzyme fractions c_{EI} and c_{EA} has to equal one, Eqs. (41) and (42) can be combined into a complete model of the temperature dependency of microbial reactions:

$$r_{max}(T) = \frac{K_1 e^{-\frac{\Delta H_1}{RT}}}{1 + K_2 e^{-\frac{\Delta H_2}{RT}}} \tag{42}$$

In Fig. 8 experimental data from ESENER et al. (1980) for the maximum specific growth rate of *Klebsiella pneumoniae* are compared with the above model. The re-identified parameters are: $K_1 = 6.74 \cdot 10^{14}$ h^{-1}, $K_2 = 1.6 \cdot 10^{48}$, $\Delta H_1 = 86.8$ kJ·mol^{-1}, and $\Delta H_2 = 287.6$ kJ·mol^{-1}.

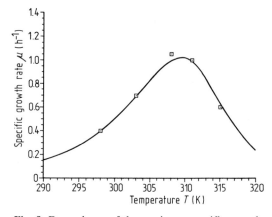

Fig. 8. Dependency of the maximum specific growth rate of *Klebsiella pneumoniae* on the temperature (data from ESENER et al., 1980).

3.6.2 Black-Box Models for Unknown Mechanisms

Often there is no obvious or simple mechanism for the explanation of functional relations between growth activity and certain parameters. In such cases a black-box approach is the easiest way to build a mathematical model directly by fitting it to experimental data. A very flexible choice for the unknown functional relations is a sum of powers of the parameters. For $\theta = (\theta_1, \theta_2, \ldots, \theta_n, \ldots, \theta_N)$ and $p = (p_1, p_2, \ldots, p_1, \ldots, p_L)$, the general function can be

$$p_1(\theta) =$$
$$\sum_{i=0}^{I_1} \sum_{j=0}^{J_1} \cdots \sum_{k=0}^{K_1} C_{ijkl} \cdot \theta_1^i \cdot \theta_2^j \cdots \theta_N^k \qquad (43)$$

with arbitrary summation bounds I_1, J_1 to K_1 and model constants C_{ijkl}, by which the model is fitted to experimental data.

REUSS et al. (1986) developed such a model for biological parameters of the gluconic acid production by *Aspergillus niger*. The dependency of the maximum specific production rate, $q_{P,max}$, on pH was approximated by:

$$q_{P,max}(pH) = q_{P,max}(pH = 5.5) \cdot$$
$$\cdot (-4,07 + 1.84\,pH - 0.167\,pH^2)$$

CHU and CONSTANTINIDES (1988) correlated kinetic parameters to temperature and pH for a cephalosporin C process, thus $\theta = (T, pH)^T$.

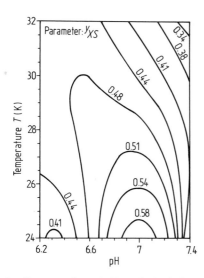

Fig. 9. Contour plot of Y_{XS} of *Cephalosporium acremonium* in the *T*-pH plane (based on CHU and CONSTANTINIDES, 1988).

As an example, the contour plot for cell yield from glucose is shown in Fig. 9. The corresponding correlation function is:

$$Y_{XS}(\theta) = 0.495$$
$$-0.015 \cdot \Delta T - 0.194 \cdot \Delta pH^2$$
$$-0.057 \cdot \Delta T \cdot \Delta pH$$
$$+0.006 \cdot \Delta T^2 \cdot \Delta pH \; +0.039 \cdot \Delta T \cdot \Delta pH^2$$
$$+0.134 \cdot \Delta T \cdot \Delta pH^3 - 0.144 \cdot \Delta pH^3$$
$$-0.006 \cdot \Delta T \cdot \Delta pH^2 - 0.018 \cdot \Delta T^2 \cdot \Delta pH^3$$

with $\Delta T = T - 301\,K$, $\Delta pH = pH - 6.8$

3.7 Primary Metabolite Formation

The production of primary metabolites is directly coupled with the central metabolic processes. Therefore, it can be described by simple equations similar to those for substrate uptake, Eq. (23), known as the Luedeking-Piret equation. The general form for a multi-product process of N products is in vector notation (LAM and OLLIS, 1981):

$$q_P = Y_{PX} \cdot \mu + m_{PX} \qquad (44)$$

where Y_P is the vector of yield coefficients for the growth-associated production and m_P is the vector of the specific production rate due to maintenance metabolism. Furthermore, Eq. (23) for substrate uptake has to be extended to take into account the substrate turnover for product formation:

$$q_S = q_{SG} - m_S - \sum_{i=1}^{N} \frac{q_{Pi}}{Y_{PSi}} \qquad (45)$$

After substitution of Eqs. (44) and (24) and rearrangement, Eq. (45) regains the original Luedeking–Piret form:

$$q_S = \frac{\mu}{Y_{S,tot}} - m_{S,tot} \qquad (46)$$

where $Y_{S,tot}$ is the total yield of substrate consumption for growth and growth-associated product formation:

$$Y_{S,tot} = \left[\frac{1}{Y_{XSG}} + \sum_{i=1}^{N} \frac{Y_{PXi}}{Y_{PSi}} \right]^{-1} \qquad (47)$$

and $m_{S,\text{tot}}$ is the total maintenance coefficient for all maintenance-associated processes, respectively:

$$m_{S,\text{tot}} = \left[m_{SG} + \sum_{i=1}^{N} \frac{m_{PXi}}{Y_{PSi}} \right] \qquad (48)$$

Usually, when evaluating the model parameters, the values for $X_{S,\text{tot}}$ and $m_{S,\text{tot}}$ will be determined, and not the values Y_{XSG} and m_S, which may be taken as "theoretical" parameters. The total yield already accounts for all products, whether they are considered in Eq. (45) or not. The product formation then depends solely on the parameters Y_{PX} and m_{PX}.

4 Structured Models on the Cellular Level

4.1 General Formulation

In contrast to unstructured models, structured models provide information about the physiological state of the microorganisms, e.g., changes in their composition and regulatory adaption to environmental changes. A structured model should only include variables for the most relevant intracellular processes to keep the model simple and the number of parameters as small as possible. Otherwise the difficulties for the experimental model verification and parameter identification become insuperable. Also the behavior of the model becomes more difficult to understand.

Structured models of low complexity are obtained by lumping biological systems of similar function and dynamics together into a few pools. The average behavior of the ith pool is modelled by a representative variable $X_i(t)$ (see Eq. 1). This kind of model introduced by WILLIAMS (1967) is called a compartment model. In a compartment model the total biomass $C_X(t)$ is the sum of the components of the biotic phase, as defined by Eq. (2). In structured models the intrinsic concentrations of the cell are of more importance than the volume-related concentrations $X(t)$, because the

former control the intracellular reactions. When writing the balance equations for intrinsic variables, one has to take special care of the consistency of the model. The derivation presented here is based on that of ROELS (1982).

The intrinsic concentrations are defined as:

$$c(t) = \frac{X(t)}{C_X(t)} \qquad (49)$$

From Eqs. (7) and (49) it follows for the sum of all components of $c(t)$:

$$I^{\text{T}} \cdot c(t) = 1 \qquad (50)$$

and therefore,

$$\frac{d(I^{\text{T}} \cdot c(t))}{dt} = 0 \qquad (51)$$

Without loss of generality, the vector of reaction rates,

$$Q(t) = [Q_Z(t), Q_X(t)]^{\text{T}} \qquad (52)$$

and the matrix of yield coefficients,

$$Y = [Y_Z, Y_X]^{\text{T}} \qquad (53)$$

are split into two parts, the first (index Z) is related to the abiotic phase and the second (index X) to the biotic phase. In the following, only the elements related to the biotic phase, $Q_X(t)$ and Y_X, shall be considered. Usually there is no exchange of the biotic phase with the gas phase, $P_X(t) \equiv 0$. By substituting Eqs. (7) and (49) into Eq. (4), the balance equation for the cellular components is obtained as:

$$\frac{d(C_X(t) \cdot c(t))}{dt} = Y_X \cdot r(t) \cdot C_X(t) + \qquad (54)$$

$$+ \frac{F_I(t)}{V_L} \cdot (X_I(t) - c(t) \cdot C_X(t))$$

It has to be assumed that biomass entering the reactor has the same state as the biomass already growing in the fermentor because the intrinsic concentrations are not additive:

$$X_I(t) = c(t) \cdot C_{XI}(t) \qquad (55)$$

Therefore, the model is only valid for cell recycle from the same fermentor or without inflow of biomass. By partial differentiation of Eq. (54) one obtains:

$$C_X(t) \cdot \frac{d\boldsymbol{c}(t)}{dt} + \boldsymbol{c}(t) \cdot \frac{dC_X(t)}{dt} = \qquad (56)$$

$$= Y_X \cdot \boldsymbol{r}(t) \cdot C_X(t) +$$

$$+ \frac{F_I(t)}{V_L} \cdot \boldsymbol{c}(t) \cdot (C_{XI}(t) - C_X(t))$$

Multiplying $\boldsymbol{I}^T$ by each summand and substituting in Eq. (49) gives:

$$C_X(t) \cdot \frac{d\boldsymbol{I}^T \boldsymbol{c}(t)}{dt} + \boldsymbol{I}^T \cdot \boldsymbol{c}(t) \cdot \frac{dC_X(t)}{dt} = \qquad (57\,\mathrm{a})$$

$$= \boldsymbol{I}^T \cdot (Y_X \cdot \boldsymbol{r}(t)) \cdot C_X(t) +$$

$$+ \frac{F_I(t)}{V_L} \cdot \boldsymbol{I}^T \cdot \boldsymbol{c}(t) \cdot (C_{XI}(t) - C_X(t))$$

or after simplification by using Eqs. (50) and (51):

$$\frac{dC_X(t)}{dt} = \boldsymbol{I}^T \cdot (Y_X \cdot \boldsymbol{r}(t)) \cdot C_X(t) + \qquad (57\,\mathrm{b})$$

$$+ \frac{F_I(t)}{V_L} \cdot (C_{XI}(t) - C_X(t))$$

where

$$\mu(t) = \boldsymbol{I}^T \cdot (Y_X \cdot \boldsymbol{r}(t)) \qquad (58)$$

is the specific growth rate. After considering Eqs. (57 b) and (58), in Eq. (56) the balance equation for the intrinsic concentrations becomes:

$$\frac{d\boldsymbol{c}(t)}{dt} = Y_X \cdot \boldsymbol{r}(t) - \mu \cdot \boldsymbol{c}(t) \qquad (59)$$

The second summand in this equation is due to dilution of intracellular material by cell growth.

4.2 Examples of Structured Models

In the following, two structured models will be presented, one for diauxic growth on two carbon sources and another for the production of an antibiotic.

4.2.1 Comparison of an Implicitly Structured Model with a Two-Compartment Model

Unstructured models cannot provide a satisfactory simulation of processes with lag phase. An example that was already presented in Sect. 3.4.2 is the switch from the first substrate to the second in diauxic growth processes. To improve the model behavior while avoiding structured models, an explicitly time-dependent formal kinetic model similar to Eq. (18) can be applied:

$$\mu(C_{S1}(t), C_{S2}(t)) =$$

$$= \mu_1(C_{S1}(t)) + \mu_2(C_{S2}(t), t)$$

where the second summand is time-dependent,

$$\mu(C_{S2}(t), t) = \begin{cases} \mu_{max}(1 - e^{-Kt})\,\tau(C_{S2}(t)) & t \ge t_1 \\ 0 & t < t_1 \end{cases}$$
$$(60)$$

Here t_1 is the starting time for growth on the second substrate, C_{S2}. The first two factors can be interpreted as a time-varying maximum growth rate:

$$\mu'_{max}(t) = \mu_{max} \cdot (1 - e^{-Kt}) \qquad (61)$$

When applying such kinetics to the simulation of diauxic growth, the problem of determining the switching time, t_1, appears. The kinetics also fail if the normal progress of the lag phase is disturbed, e.g., by adding other substrates, because the model contains no explicit state variable. These drawbacks can be overcome when extending Eq. (61) to an explicitly structured model. Obviously the differential equation

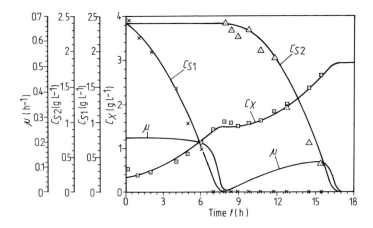

Fig. 10. Simulation with a formal kinetic model of diauxic growth of *Klebsiella terrigena* on glucose and maltose (legend see Fig. 6).

$$\frac{d\mu'_{max}(t)}{dt} = K \cdot (\mu_{max} - \mu'_{max}(t))$$

gives the same time dependence of $\mu'_{max}(t)$ for $t > t_1$ and $\mu'_{max}(t_1) = 0$.

A structured model with two compartments has the same order of complexity as the above differential equation, since due to Eqs. (50) and (51) it is sufficient to describe both compartments by one differential equation. For the model the structure shown in Fig. 11 shall

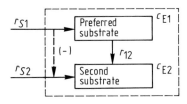

Fig. 11. Block diagram of the two-compartment model of diauxic growth.

be assumed. Compartment 1 represents a constitutive pathway for catabolism of substrate 1, while the pathway for substrate 2 is repressed by substrate 1 – indicated by the dotted line – and induced by substrate 2. The uptake of substrate 2 is controlled by the intrinsic concentration of key enzyme $c_{E2}(t)$. Therefore, using Monod kinetics for substrate uptake, the synthesis rates for the two compartments become:

$$r_{S1}(t) = r_{max,1} \cdot \frac{C_{S1}(t)}{C_{S1}(t) + K_{S1}} \qquad (62)$$

$$r_{S2}(t) = r_{max,2} \cdot \frac{C_{S2}(t)}{C_{S2}(t) + K_{S2}} \cdot$$

$$\cdot \frac{K_{I1}}{C_{S1}(t) + K_{I1}} \cdot c_{E2}(t)$$

Furthermore, constitutive formation of a key enzyme proportional to the concentration of c_{E1} is introduced:

$$r_{12}(t) = K_2 \cdot c_{E1}(t) \qquad (63)$$

In the nomenclature of Sect. 4.1 the elements of the model become:

$$r = (r_{S1}, r_{S2}, r_{12})^T \qquad (64)$$

$$q = (q_{S1}, q_{S2}, \mu_1, \mu_2)^T = Y \cdot r \qquad (65)$$

$$Y = \begin{bmatrix} -Y_{XS1}^{-1} & 0 & 0 \\ 0 & -Y_{XS1}^{-1} & 0 \\ 1 & 0 & -1 \\ 0 & 1 & 1 \end{bmatrix} \qquad (66)$$

where the vector of concentrations is:

$$C = (C_{S1}, C_{S2}, C_{E1}, C_{E2})^T$$

Upon substituting in Eqs. (64) to (66), one obtains from Eq. (59):

$$\frac{dc_{E2}(t)}{dt} = r_{S2}(t) + K_2 \cdot (1 - c_{E2}(t)) - \quad (67)$$

$$- \mu(t) \cdot c_{E2}(t)$$

and from Eq. (58)

$$\mu(t) = r_{\max,1} \cdot \frac{C_{S1}(t)}{C_{S1}(t) + K_{S1}} +$$

$$+ r_{\max,2} \cdot \frac{C_{S2}(t)}{C_{S2}(t) + K_{S2}} \cdot \quad (68)$$

$$\cdot \frac{K_{I1}(t)}{C_{S1}(t) + K_{I1}} \cdot c_{E2}(t)$$

Despite the additional factor c_{E2}, this is analogous to the unstructured model Eq. (18). By comparing the structured model with the formal kinetic model, one can see that on one hand, the complexity is only very moderately increased, while the number of parameters is actually the same. Besides kinetic parameters, the structured model contains the parameters K_{I1} and K_2, and the formal kinetic model K and t_1. On the other hand, the value of the inhibition constant, K_{I1}, is not very critical, and the structured model can have much wider applicability.

Simulations with the formal kinetic model Eq. (60) and the above structured model are plotted in Figs. 10 and 12. Both can describe the diauxic lag phase well and are quantitatively equivalent for this undisturbed process. After exhaustion of the glucose, the growth rate decreases quickly and then rises slowly during adaption to the second substrate. In Fig. 12b the time course of the intrinsic concentrations is additionally shown.

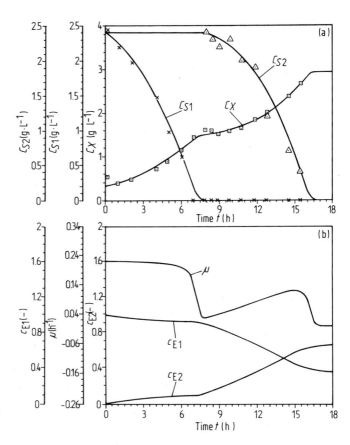

Fig. 12. (a) Simulation with a two-compartment model of diauxic growth of *Klebsiella terrigena* on glucose and maltose (legend see Fig. 6). (b) Intrinsic concentrations c_{E1} and c_{E2} and specific growth rate μ.

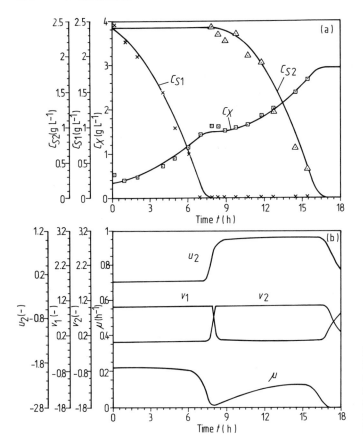

Fig. 13. (a) Simulation with a cybernetic model of diauxic growth of *Klebsiella terrigena* on glucose and maltose (legend see Fig. 6). (b) Control variables v_1 and v_2, cybernetic variable u_2, and specific growth rate μ.

Starting from the given initial conditions, c_{E2} is approaching its constitutive level during the first growth phase. Afterwards, its transient controls the diauxic lag phase.

4.2.2 Compartment Model for Antibiotic Production

The production of secondary metabolites such as vitamins, pigments, or antibiotics is closely related to regulatory effects of the metabolism as a response to impaired growth conditions. For modelling such effects one can use structured models. Besides growth, they can also describe the regulatory state of the cells and the pathways of metabolism leading to products. This section presents a simple mathematical model for the production of the antibiotic myxothiazol (MXF 16), which is formed at low growth rates under oxygen limitation. Since the role of secondary metabolism and its regulation as a part of the entire metabolism is, qualitatively, still not sufficiently clear – because of the complexity of the problem – the model must be based on numerous simplifications and reasonable assumptions.

The structure of the cell model shown in Fig. 14 contains three compartments for the main pathways. The solid lines describe mass flows and the dashed lines the effect of regulation. In catabolism, the substrate is decomposed into the collective metabolite, c_M, from which the further cell components are synthesized. Anabolism is for the production of macromolecules and cellular material such as enzymes. Hence, output of this block must also include enzyme 1. This catalyzes the substrate decomposition process, and thereby controls

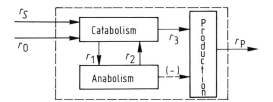

Fig. 14. Block diagram of the three-compartment model of antibiotic production.

the synthesis of the metabolite from the substrate.

A metabolic path, r_3, which is activated at growth limitation, is indicated through the block containing enzyme 2. At the transition to growth limitation the synthesis of enzyme 2 is activated by decreasing concentrations of c_{E1}. Enzyme 2 is responsible for product synthesis so that an increased product concentration will be obtained during growth limitation. Enzyme 1 can be decomposed to metabolite by the rate r_2 to provide material for the formation of enzyme 2. As shown for other antibiotics, it is assumed in the model that product formation is directly affected by the dissolved oxygen concentration. The balance equations of the model for the intrinsic concentrations are:

$$\frac{dc_{E1}(t)}{dt} = K_1 \cdot c_M(t) - K_4 \cdot c_{E1}(t) - \mu(t) \cdot c_{E1}(t)$$

$$\frac{dc_{E2}(t)}{dt} = K_2 \cdot c_M(t) \frac{K_I}{K_I + c_{E1}(t)} - \mu(t) \cdot c_{E2}(t)$$

$$\frac{dc_M(t)}{dt} = K_3 \cdot \frac{C_S(t)}{K'_S + C_S(t)} \cdot \frac{C_O(t)}{K'_O + C_O(t)} \cdot c_{E1}(t) -$$

$$- \left[K_1 + K_2 \cdot \frac{K_I}{K_I + c_{E1}(t)} \right] \cdot c_M(t) +$$

$$+ K_4 \cdot c_{E1}(t) - \mu(t) \cdot c_M(t)$$

The joint effect of substrate, dissolved oxygen, and cell state on the growth rate and product formation is modelled via an extended Monod kinetics for the formation of c_M. For the sake of simplicity the yields for intracellular reactions are assumed to equal one, and the amount of product is neglected in the calcula-

tion of total biomass. With the concentration vector

$$C = (X \qquad , \qquad Z)^T \tag{69}$$
$$= (C_M, C_{E1}, C_{E2}, C_S, C_O, C_P)^T$$

and the specific reactions

$$q = (q_S, q_O, q_P, q_M, q_{E1}, q_{E2})^T \tag{70}$$

the vector of cellular reactions and the matrix of yield coefficients become:

$$r = \begin{bmatrix} r_M \\ r_1 \\ r_2 \\ r_3 \\ r_P \end{bmatrix} = \begin{bmatrix} K_3 \cdot \dfrac{C_S(t)}{K'_S + C_S(t)} \cdot \dfrac{C_O(t)}{K'_O + C_O(t)} \cdot c_{E1}(t) \\ K_1 \cdot c_M \\ K_4 \cdot c_{E1} \\ K_2 \cdot c_M(t) \dfrac{K_I}{K_I + c_{E1}(t)} \\ K_P \cdot \dfrac{C_O(t)}{K'_O + C_O(t)} \cdot c_{E2}(t) \end{bmatrix} \tag{71}$$

$$Y = \begin{bmatrix} -Y_{XS}^{-1} & 0 & 0 & 0 & 0 \\ -Y_{XO}^{-1} & 0 & 0 & 0 & 0 \\ 0 & 0 & 0 & 0 & 1 \\ 1 & -1 & 1 & -1 & 0 \\ 0 & 1 & -1 & 0 & 0 \\ 0 & 0 & 0 & 1 & -1 \end{bmatrix} \tag{72}$$

Fig. 15 shows the simulation for a cultivation of the gliding bacteria *Myxococcus fulvus* on peptone (data from LEHMANN et al., 1979). After an exponential growth phase, the cells are maintained under strong oxygen transfer limitation in order to induce the formation of antibiotics, beginning with $t = 13$ h. The extent of the limitation is gradually changed in the further course of the experiment. This also affects the formation rate of the antibiotics. The course of the three cell states can be seen in Fig. 15b. In the exponential phase the levels of metabolite and enzyme are constant, whereas in the second part of the process a significant response to the limitation appears, which induces product synthesis.

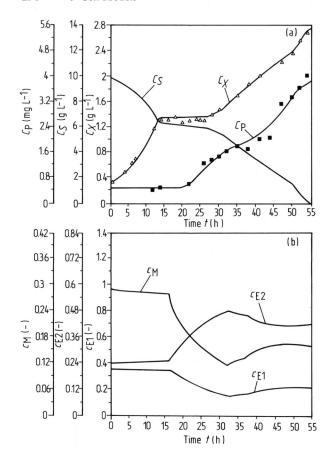

Fig. 15. (a) Simulation with a three-compartment model of antibiotic production of *Myxococcus fulvus*. Experimental data (symbols) from LEHMANN et al. (1979), concentrations of limiting substrate C_S, cell mass C_X ($\triangle$), and product C_P ($\blacksquare$). (b) Intrinsic concentrations c_{E1}, c_{E2}, c_M.

4.3 Cybernetic Models

In Sect. 3.4 it was shown that unstructured models have many drawbacks for multi-substrate kinetics. Sequential and parallel uptake of more than one substrate is accompanied by a complex metabolic regulation, which cannot be included satisfactorily into simple kinetics. But even compartment models will often not give a simple and usable description in this situation, because they have to contain many variables, complicated mechanisms, and kinetics for the metabolic regulation.

The cybernetic modelling approach, introduced by RAMKRISHNA (1982) and revised recently by TURNER et al. (1988), tries to overcome these problems by suggesting another direction for model building. In this case no mechanistic interrelation is proposed for the utilization of different substrates. Each substrate consumption follows its own single-substrate kinetics. This extensive simplification is compensated by assigning an optimal control motive to the metabolic activity of the microorganisms. Control is accomplished by adjusting the level of key enzymes for each substrate uptake system according to an optimality criterion. In this view the microorganisms are always able, by allocating their internal resources, to choose those substrates which best fit their requirements. In the cybernetic approach it is not questioned how the microorganisms can coordinate their metabolic reactions to obtain optimum control. The concept relies on the assumption that evolution has already selected the mechanisms permitting optimal control. Cybernetic models can *predict* sequential and parallel substrate consumption without any assumption about coupled kinetics by extrapolation from experiments on single substrates.

For formulation of the model, the microbial growth on each substrate is represented by a reaction:

$$C_X + C_{Si} \xrightarrow{E_i} (1 + Y_{XSi}) \cdot C_X \qquad (73)$$

where E_i is the representative pool of key enzymes for the catabolism of substrate i. The enzymes have to be formed by the optimal strategy given later. The specific growth rate on substrate i, $\mu_i(C_{Si})$, is a product of three factors, the intrinsic concentration of the key enzyme, c_{Ei}, a kinetic expression τ_{Gi}, chosen from Tab. 1, and an activity controlling variable v_{Gi}:

$$\mu_i(C_{Si}(t), t) = \qquad (74)$$
$$= \mu_{\max,i} \cdot \tau_{Gi}(C_{Si}(t)) \cdot c_{Ei}(t) \cdot v_{Gi}(t)$$

The total growth rate is:

$$\mu(C_{S1}(t), \ldots, C_{SN}(t), t) = \sum_{i=1}^{N} \mu_i(C_{Si}(t), t) \qquad (75)$$

The partition of the control action into one part for long-term response by induction and repression, $c_{E1}(t)$, and another part for immediate response by inhibition and activation, $v_{Gi}(t)$, is necessary, because both play an important part in metabolism. The addition of a preferred substrate during growth on a second substrate is an example. In this case the cells will switch at once to the preferred substrate and stop the catabolism of the other by inhibition, although all its related enzymes are present at high levels. This immediate inhibition will be followed by slower degradation of the enzymes for the less preferred substrate.

In analogy to Eq. (74), the substrate consumption for maintenance can be written as

$$m_i(C_{Si}(t), t) = \qquad (76)$$
$$= m_{\max,i} \cdot \tau_{Mi}(C_{Si}(t)) \cdot c_{Ei}(t) \cdot v_{Mi}(t)$$

and the total maintenance need is

$$m(C_{S1}(t), \ldots, C_{SN}(t), t) = \sum_{i=1}^{N} m_i(C_{Si}(t), t) \qquad (77)$$

The formation rate of the key enzymes is controlled by the cybernetic variables u_i and by substrate kinetics:

$$r_{Ei}(C_{Si}(t), t) =$$
$$= r_{E,\max,i} \cdot \tau_{Ei}(C_{Si}(t)) \cdot u_i(t) + r_{Ei,0} \qquad (78)$$

A constitutive formation rate $r_{Ei,0}$ appears in Eq. (78) besides the inducible formation rate. This is due to mathematical requirements, but is also in agreement with experimental studies showing that there is not complete repression of enzymes. Therefore, the adaption process starts with a minimum speed that is clearly greater than zero. Now the balance equations of the model for a chemostat become:

$$\frac{dC_X(t)}{dt} = \left[\mu(t) - \frac{F(t)}{V_L} \right] \cdot C_X(t) \qquad (79)$$

$$\frac{dC_{Si}(t)}{dt} = -\left[\frac{\mu_i(t)}{Y_{XSi}} + m_i(t) \right] \cdot C_X(t) +$$
$$\qquad (80)$$
$$+ \frac{F(t)}{V_L} \cdot (C_{Sil}(t) - C_{Si}(t))$$

$$\frac{dc_{Ei}(t)}{dt} = r_{Ei}(t) - K_{DEi} \cdot c_{Ei}(t)$$
$$\qquad (81)$$
$$- \mu(t) \cdot c_{Ei}(t)$$

where K_{DEi} is the degradation rate constant for key enzymes. To complete the model, the control variables have to be specified. In analogy to economics, the cybernetic variables u_i are determined through Herstein's matching law, which maximizes the total profit, here μ, by allocating fractions of a fixed resource to alternative pathways i, according to the expected maximum relative profit for the ith pathway, here μ_i:

$$u_i(t) = \frac{\mu_{\max,i} \cdot \tau_{Gi}(C_{Si}(t)) \cdot c_{Ei}(t)}{\sum\limits_{i=1}^{N} \mu_{\max,i} \cdot \tau_{Gi}(C_{Si}(t)) \cdot c_{Ei}(t)} \qquad (82)$$

with

$$\text{(repressed)} \quad 0 \le u_i(t) \le 1 \quad \text{(induced)} \qquad (83)$$

and

$$\sum_{i=1}^{N} u_i(t) = 1 \qquad (84)$$

The control variables v_{Gi} and v_{Mi} for regulation of the key enzymes by inhibition and activation are allocated according to a heuristic strategy, such that pathways yielding a higher growth rate are preferred:

$$v_{Gi}(t) = \frac{\mu_{\max,i} \cdot \tau_{Gi}(C_{Si}(t)) \cdot c_{Ei}(t)}{\max_{j} \{ \mu_{\max,j} \cdot \tau_{Gj}(C_{Sj}(t)) \cdot c_{Ej}(t) \}} \qquad (85)$$

$$v_{Mi}(t) = 1 - \frac{\mu(t)}{\sum_{i=1}^{N} \mu_{\max,j} \cdot u_j(t)} \qquad (86)$$

with

$$0 \le v_{Gi}, v_{Mi}(t) \le 1 \qquad (87)$$
(inhibited)   (activated)

Obviously, the above optimization strategy is a local one. It only considers the actual growth conditions without trying to optimize the response over a certain interval in the future.

An example of the simulation of diauxic growth with the above cybernetic model is given in Fig. 13. Monod kinetics were chosen for the substrate uptake steps. The quality of the model fit is comparable to the simulation with the two-compartment model. But it should be noted that here there is no need to specify any kinetics for metabolic regulation. Also for more complex processes the formalism of cybernetic models remains the same. This constitutes a remarkable simplification of model building.

4.4 The Metabolic Regulator Approach

Another way of modelling with many similarities to the cybernetic models is the metabolic regulator approach proposed by BELL-GARDT et al. (1988). Again the microorganism is modelled as an optimal strategist trying to optimally use the different pathways. As an extension, not only the uptake of different substrates, but also the formation of primary metabolites, is taken into account in this optimization strategy. Metabolic control of the formation of such products depending on growth conditions can often be found. Examples are the production of acetone–butanol and butanol-diol, where besides these main products other solvents, ethanol or acids can be formed in varying ratios. Another well known example is growth of baker's yeast. The yeast can direct its metabolism to any mixture of fermentative growth with ethanol formation or oxidative growth with high cell yield, depending on the available substrate and oxygen.

If products are to be considered in the optimization, a more accurate modelling of the biological reactions is necessary to describe the requirements of resources for the product synthesis and the profit gained from it. This has to be done in terms of balances for material, energy (ATP), and electrons (NADH) for the main pathways of substrate uptake, product formation, and synthesis of cell material. The advantage of a more detailed biochemical model for the reactions is also that it leads in a straightforward way to a correctly structured model with fewer independent parameters. It is assumed that all reactions associated with growth are in the quasi-steady-state with respect to fermentor conditions and intracellular enzyme concentrations. Dynamics are only associated with the latter two. Then the reaction model can be written as a system of algebraic equations:

$$Y \cdot r(t) = z(t) \qquad (88)$$

where Y is a $(N \times M)$-matrix of stoichiometric coefficients, $r(t)$ is a $(N \times 1)$-vector of intracellular specific reaction rates, and $z(t)$ is a $(M \times 1)$-vector of growth-independent specific reaction rates, such as maintenance terms or rates of constitutive reactions. In Eq. (88) the number of modelled reactions, N, will usually be larger than the number of stoichiometric equations, M; otherwise growth does not depend on extracellular conditions, and the microorganisms have no freedom to regulate the relative activity of their pathways. Since the system of Eq. (88) is underdetermined, additional conditions have to be introduced to obtain a predictive model. These are derived in two ways.

The first is to consider inherently rate-limiting steps of the metabolism, such as kinetics of substrate uptake systems or a maximum biosynthetic capacity for macromolecules and cell material. Rate-limiting steps are of special importance for the modelling of microbial adaption due to regulation of key enzymes by induction and repression. Besides this slow regulation, there is a fast regulation by enzyme inhibition. This means the turnover of a regulated pathway may be lower than its maximum turnover, which is determined by the concentration of the key enzyme. Therefore, additional conditions for the actual reaction rates r have to be formulated as inequalities for its elements:

$$r_{\min,i}(t) \leq r_i(t) \leq r_{\max,i}(t) \tag{89}$$

where $r_{\min}$ and $r_{\max}$ are the possible rates of the inherently rate-limiting steps. When the equal sign is valid, the reaction r_i is not inhibited. Expressions for the minimum and maximum rates are given later. Usually $r_{\min}(t)$ equals zero. Only if there is a reaction against the direction defined as mathematically positive, may $r_{\min}(t)$ contain non-zero negative elements.

The second way to derive more conditions and finally to solve Eq. (88) uniquely is to introduce an optimality criterion, the metabolic coordinator J, for the immediate metabolic regulation by inhibition and activation, which has been left unspecified until now:

$$J(r(t)) \xrightarrow[\text{Eqs. (88) and (89)}]{r(t)} \text{optimum} \tag{90}$$

The meaning of Eq. (90) is that the metabolic coordinator optimally determines the actual rates of all pathways under the constraints given by the structure of the metabolism, Eq. (88), and the inherently rate-limiting steps, Eq. (89). $N - M$ reactions r_i will always equal the corresponding elements of $r_{\min}$ or $r_{\max}$. A reaction $r_i(t)$ of $r(t)$ is said to be rate-limiting if one of the following conditions holds:

$$\begin{aligned} r_i(t) &\equiv r_{\min,i}(t) \\ r_i(t) &\equiv r_{\max,i}(t) \end{aligned} \tag{91}$$

The optimization strategy is a local one, since J only depends on the actual reaction rates. As in the cybernetic models, the optimization strategy followed by the microorganisms is assumed to be the maximization of the specific growth rate:

$$J(r(t)) \equiv \mu(t) \tag{92}$$

The metabolic coordinator is functionally equivalent to the control variables $v(t)$ in Eq. (74).

To complete the modelling, the inherently rate-limiting steps in Eq. (89) have to be specified. If $r_i(t)$ is a substrate uptake step, then $r_{\max,i}(t)$ can be chosen as any of the kinetics in Tab. 1 or 3. Also in this approach, there are no coupled kinetics independent of the complexity of the metabolism.

If $r_i(t)$ is a reaction controlled by a key enzyme which is subject to induction or repression, then the dynamics of its regulation must be given. In the cybernetic models this is done by the balance equations (71) together with the cybernetic variables, u_i. In this approach, a strategy is used which is local to the regulated pathway. The metabolism is said to be in the adapted state if none of the reactions $r_i(t)$ which are controlled by key enzymes is rate-limiting. This means that the concentration of the key enzyme has to be increased by further induction, as long as the coordinator J fully uses the capacity of the pathway up to the actual maximum rate. If the actual rate allocated to the pathway is not rate limiting, then the key enzyme is inhibited. In this case the concentration of the enzyme should be reduced by stronger repression to economize resources. These conditions can be met by a system of the form:

$$\frac{dc_{Ej}(t)}{dt} = (K_{1j} - \mu(t)) \cdot c_{Ej}(t) + \\ + K_{2j} \cdot (r_j(t) + r_{0j}) \tag{93}$$

$$r_{\max,j}(t) = K_{3j} \cdot c_{Ej}(t) \tag{94}$$

where c_{Ej} is the intrinsic concentration of the key enzyme for the pathway j, and r_{0j} is a low constitutive level of enzyme synthesis. This system can be viewed as a tracking controller, a metabolic regulator for the concentration of the key enzyme, which is regulated according to actual requirements $r_j(t)$. Metabolism is viewed as a system of feed-back control loops,

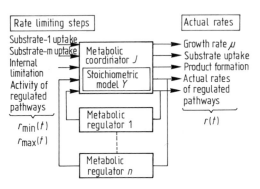

Fig. 16. Block diagram of the metabolic regulator model: metabolism as a system of control loops.

As an example, the modelling approach is applied to the growth simulation of baker's yeast. Yeast has a wide choice of metabolic pathways to respond to environmental conditions. Depending on the substrate and oxygen supply, great changes in the stoichiometry of growth can be observed, accompanied by dynamic metabolic regulation. The model considers the glucose uptake rate r_S, the oxygen uptake rate r_O, the ethanol reaction rate (uptake r_{EC}, or production r_{EP}), and the carbon dioxide production rate r_C. Intracellular reactions included in the model are: the rate of glycolysis r_S, of acetate production r_{AC}, the representative rate for the citric acid cycle r_{TC}, the rate of gluconeogenesis r_G, and the specific rate of biosynthesis $\mu = r_X$. The following stoichiometric model can be derived from the stoichiometry of the metabolic pathways, together with the balances for ATP and NADH:

$$\begin{bmatrix} -1 & 0 & 0 & 0 & K_{EG} & 0 & 1 & K_{B1} & 0 \\ -2 & 0 & -1 & 0 & -1-K_{Ad} & 1 & 2 & 0 & 0 \\ 0 & 0 & -1 & 1 & 0 & 0 & 0 & K_{B3} & 0 \\ -2 & 2 & 1 & 4 & -1-2K_{EG} & 1 & 2 & -K_{B2} & 0 \\ -3 & 2P/O & 0 & 1 & -2K_{EG} & 0 & 2 & -Y_{ATP}^{-1} & 0 \\ 0 & 0 & -1 & -2 & -1-K_{Ad} & 0 & 0 & 0 & 1 \end{bmatrix} \cdot \begin{bmatrix} r_G \\ r_O \\ r_{AC} \\ r_{TC} \\ r_{EP} \\ r_{EC} \\ r_S \\ r_X \\ r_C \end{bmatrix} = \begin{bmatrix} r_S \\ 0 \\ 0 \\ 0 \\ m_{ATP} \\ 0 \end{bmatrix} \quad (97)$$

as shown in Fig. 16, each consisting of a metabolic controller, Eqs. (93) and (94), for exactly one reaction. Additional conditions for the constants can be derived to obtain the behavior of the metabolic regulator described below. If the pathway is not rate-limiting, Eq. (93) has to be stable for all $\mu(t)$ to keep $r_{max,j}(t)$ close to $r_j(t)$, and therefore:

$$K_1 < 0 \quad (95)$$

If the pathway is not rate-limiting, then $r_{max,j}(t) \equiv r_j(t)$, and Eqs. (93) and (94) become an autonomous system, which must be unstable to increase the concentration of the key enzyme; thus it follows that

$$K_1 - \mu(t) - K_2 \cdot K_3 > 0 \quad (96)$$

Although this model can cover most of the complex growth phenomena of yeast, it contains only eight independent stoichiometric parameters, all of which have a biological meaning. In Eq. (97), the maintenance requirements of the microorganisms are modelled analogously to Eq. (27) as a fraction m_{ATP} of the total energy consumption r_{ATP} due to growth processes:

$$r_{ATP} = \frac{r_X}{Y_{ATP}} + m_{ATP} \quad (98)$$

With this general formulation, the maintenance requirements need not be explicitly considered in the substrate uptake or product formation rates.

The inherently rate-limiting steps in the model are the glucose uptake rate, $q_S(C_S) =$

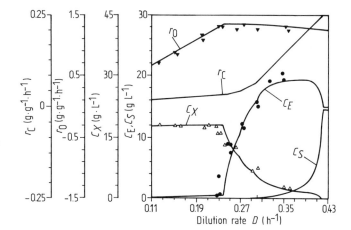

Fig. 17. Simulation with a metabolic regulator model of chemostat growth of *Saccharomyces cerevisiae* on glucose and ethanol. Experimental data (symbols) from RIEGER et al. (1983), concentrations of glucose C_S, ethanol C_E (●), cell mass C_X (△), specific oxygen uptake rate r_O (▼), and specific rate of carbon dioxide production r_C.

$q_{S,max} \cdot \tau(C_S)$, for which a Monod kinetics can be assumed, and the uptake steps for ethanol r_E, and oxygen r_O. For the sake of simplicity, these latter terms are introduced as first-order kinetics. The intracellular rate-limiting steps are limited oxidative capacity, $r_{AC,max}$, and the dynamic regulation of respiration, $r_{O,max}(t)$ and of gluconeogenesis, $r_{G,max}(t)$. For each of the latter two reactions, metabolic regulators according to Eqs. (93) and (94) are introduced into the model. With these assumptions, Eq. (89) assumes the special form:

$$
\begin{aligned}
0 &\leq r_G \leq r_{G,max}(t) \\
0 &\leq r_O \leq \min(K_O \cdot C_O, r_{O,max}(t)) \\
0 &\leq r_{AC} \leq r_{AC,max} \\
0 &\leq r_{TC} \leq \infty \\
0 &\leq r_{EP} \leq \infty \\
0 &\leq r_{EC} \leq K_{E1} \cdot C_E \\
0 &\leq r_S \leq \infty \\
-\infty &\leq r_X \leq \infty \\
0 &\leq r_C \leq \infty
\end{aligned}
\tag{99}
$$

For reactions that cannot become rate-limiting, the maximum boundary was set at infinity. With the above model, growth of yeast in chemostat culture on a mixed substrate of 1.5% glucose and 1.5% ethanol was simulated. The results are given in Fig. 17. At low growth rates, both substrates are used up completely. For $D > 0.23$ h^{-1} some of the ethanol is left in the medium because of the limited capacity of oxidative metabolism, $r_{AC,max}$. In this

range of dilution rates, glucose is the preferred substrate due to more efficient energy production. The increase in ethanol is accompanied by a decrease in cell concentration. For $D > 0.3$ h^{-1} ethanol is even produced. All these phenomena are predicted by the metabolic coordinator J that maximizes the growth rate with respect to the actual inherently rate-limiting steps.

Fig. 18 shows simulation results for an aerobic batch cultivation of the yeast in glucose medium. In the first growth phase, up to 5.5 h, the yeast is growing on the preferred substrate, glucose. Since the oxidative capacity $r_{AC,max}$ is limiting, the cells cannot achieve the maximum growth rate without producing ethanol, although this makes growth inefficient with respect to the yield of cell mass. As long as there is enough glucose, the uptake of ethanol is inhibited by the metabolic coordinator, and the ethanol uptake system is repressed. In the transient phase when the metabolism is switching from glucose to ethanol growth, both regulatory systems for r_G and r_O become rate-limiting for a short period and, therefore, their activities are induced. But growth is limited mainly by low oxygen concentrations during the second growth phase. At high growth rates the energy requirements for maintenance are met by substrate catabolism. In the starvation period, beginning from $t = 18$ h, the cell mass concentration decreases. Here the model predicts a degradation of cell material to produce energy. In this view, the energy mainte-

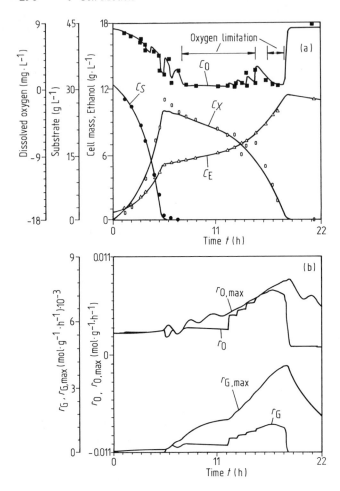

Fig. 18. (a) Simulation with metabolic regulator model of diauxic batch growth of *Saccharomyces cerevisiae* on glucose and ethanol. Experimental data (symbols) from KUHLMANN in BELLGARDT et al. (1983); concentrations of glucose C_S (●), ethanol C_E (△), cell mass C_X (○), and dissolved oxygen C_O (□). (b) input and output variables of the metabolic regulators, for respiration r_O and $r_{O,max}$, and for gluconeogenesis r_G and $r_{G,max}$.

nance concept Eq. (97) is a mechanistic union of HERBERT's (Eq. 20)) and PIRT's models (Eq. (23)).

5 Conclusion

This chapter has been a rush through a few aspects of the modelling of the system cell in a biotechnological process. Unstructured, formal kinetic models with their extremely simplified view of the living cell can often provide an adequate description of growth dynamics, even under non-stationary operation of temperature or other process variables. Unstruc-

tured models only fail when intracellular dynamics must be considered. Then the concepts of structured models can provide a better approximation to cellular metabolism.

Different ways of building structured models were presented. For many processes, the descriptive power of all kinds of structured models may be quite similar, as shown by some examples. Structured models must still be based on oversimplification of reality, and one should be careful when assigning model variables to certain unique substances in the cell. A good point of view for model building is to take the model for what it is: a vague picture of the input–output behavior of a cell population.

Compartment models as a special type of structured models have been successfully developed for many processes. But it seems not to be useful to include more than four compartments. Otherwise the number of parameters and of possible interactions between compartments increases rapidly. In contrast to compartment models, cybernetic models and metabolic regulator models emphasize the dynamics of metabolic regulation. But at the same time they completely neglect the mechanisms of metabolic regulation and substitute a very general optimization strategy. These kinds of models are suprisingly successful in predicting growth on multi-substrates as well as changes in the product spectrum of primary metabolites.

6 References

BELLGARDT, K. H., KUHLMANN, W., MEYER, H. D. (1983), Deterministic growth model of *Saccharomyces cerevisiae,* parameter identification and simulation, in: *Proc. 1st IFAC Workshop on Modelling and Control of Biotechnological Processes, Helsinki,* 1982, New York: Pergamon Press.

BELLGARDT, K. H., HOPF, N., LUTTMANN, R., DECKWER, W. D. (1988), A new approach for structured growth models, in: *Preprints of Fourth Int. Conf. on Computer Applications in Fermentation Technology: Modelling and Control of Biotechnological Processes,* Cambridge, UK, Chichester (UK): Ellis Horwood Ltd. Publishers.

CHEN, B. J., LIM, H. C., TSAO, G. T. (1976), A model for bacterial growth on methanol, *Biotech. Bioeng.* 18, 1629.

CHU, W. B., CONSTANTINIDES, A. (1988), Modeling, optimization and computer control of the cephalosporin C fermentation process, *Biotech. Bioeng.* 32, 277.

CONTOIS, D. E. (1959), Kinetics of bacterial growth: Relationship between population density and specific growth rate of continuous culture, *J. Gen. Microbiol.* 21, 40.

CUI, Q., LAWSON, G. J. (1982), Study on models of single populations: An expansion of the logistic and exponential equations, *J. Theor. Biol.* 98, 645.

DIXON, M., WEBB, E. C. (1967), *Enzymes,* 2nd. Ed., London: Longman.

EDWARDS, V. H. (1970), The influence of high substrate concentrations on microbial kinetics, *Biotech. Bioeng.* 12, 679.

ESENER, A. A., ROELS, J. A., KOSSEN, N. W. F. (1980), The influence of temperature on the maximum specific growth rate of *Klebsiella pneumoniae, Biotech. Bioeng.* 23, 1401.

FASOL, K. H., JÖRGL, M. P. (1980), Principles of model building and identification, *Automatica* 16, p. 505, London: Pergamon Press.

FRAME, K. K., HU, W. S. (1988), A model for density-dependent growth of anchorage-dependent mammalian cells, *Biotech. Bioeng.* 32, 1061.

FREDRICKSON, A. G., MAGEE, R. D., TSUCHIYA, H. M. (1979), Mathematical models for fermentation processes, *Adv. Appl. Microbiol.* 13, 419.

GADEN, E. L. (1959), Fermentation process kinetics, *J. Biochem. Microbiol. Technol. Eng.* 1, 413.

HAN, K., LEVENSPIEL, O. (1988), Extended Monod-equation for substrate, product and cell inhibition, *Biotech. Bioeng.* 32, 430.

HERBERT, D. (1959), in: *Recent Progress in Microbiology* (TENVALL, D., Ed.), Stockholm: Almquist & Wiksell.

IERUSALIMSKY, N. D. (1967), Bottle-necks in metabolism as growth rate controlling factors, in: *Microbial Physiology and Continuous Culture, 3rd International Symposium* (POWELL, E. O., Ed.), p. 23, London: H. M. S. O.

LAM, J. C., OLLIS, D. (1981), Kinetics of multiproduct fermentations, *Biotech. Bioeng.* 23, 1517.

LEHMANN, J., BERTHE-CORTI, L., STEVEN, W., NOTHNAGEL, J., PIEHL, G. W., GERTH, K. (1979), Verfahren zur Vermehrung von *Myxococcus fulvus* DSM 1368, *Ger. Patent Appl.* No. 2924868.

MASON, T. J., MILLES, N. F. (1976), Growth kinetics of yeast grown on glucose of hexadecan, *Biotech. Bioeng.* 18, 1337.

MONOD, J. (1942), *Recherches sur la croissance des cultures bacteriennes,* Paris: Herrmann et Cie.

MOSER, A. (1958), *The dynamics of bacterial populations maintained in the chemostat,* Publication 614, Washington, DC: The Carnegie Institution.

MOSER, A. (1981), *Bioprozeßtechnik,* Wien–New York: Springer Verlag.

PIRT, S. J. (1965), The maintenance energy of bacteria in growing cultures, *Proc. R. Soc. Ser. B* 163, 224.

PIRT, S. J. (1987), The energetics of microbes at slow growth rates: Maintenance energy and dormant organisms, *J. Ferment. Technol.* 65(2), 173.

POWEL, E. O. (1967), *Proc. Microbial Physiology and Continuous Culture, Third Int. Symposium* (POWEL, E. O., Ed.), p. 34, London: H. M. S. O.

RAMKRISHNA, D. (1982), A cybernetic perspective of microbial growth, in: *Foundations of Biochemical Engineering Kinetics and Thermodynamics in Biological Systems* (PAPOUTSAKIS, E., STEPHANOPOULOS, G. N., BLANCH, H. W., Eds.), Washington, DC: American Chemical Society.

RAMKRISHNA, D., FREDRICKSON, A. G., TSUCHIYA, H. M. (1967), Dynamics of microbial population: Models considering inhibitors and variable cell composition, *Biotech. Bioeng.* **9**, 129.

REUSS, M., FRÖHLICH, S., KRAMER, B., MESSERSCHMIDT, K., POMMERENNING, G. (1986), Coupling of microbial kinetics and oxygen transfer for analysis and optimization of gluconic acid production with *Aspergillus niger, Bioprocess Eng.* **1**, 79.

RIEGER, M., KÄPPELI, O., FIECHTER, A. (1983), The role of a limited respiration of *Saccharomyces cerevisiae* in the incomplete oxidation of glucose, *J. Gen. Microbiol.* **129**(3), 653.

ROELS, J. A. (1982), *Energetics and Kinetics in Biotechnology,* New York: Elsevier.

ROELS, J. A., KOSSEN, N. W. F. (1978), On the modelling of microbial metabolism, *Prog. Ind. Microbiol.* **14**, 95.

SINCLAIR, C. G., RYDER, D. N. (1975), Models for continuous culture of microorganisms under both oxygen and carbon limiting conditions, *Biotech. Bioeng.* **17**, 375.

TSAO, G. T., HANSON, T. P. (1975), Extended Monod equation for batch cultures with multiple exponential phases, *Biotechnol. Bioeng.* **17**, 1591.

TURNER, B. G., RAMKRISHNA, D., JANSEN, N. B. (1988), Cybernetic modeling of bacterial cultures at low growth rates: Mixed substrate systems, *Biotech. Bioeng.* **32**, 46.

VAVILIN, V. A. (1982), The theory and design of aerobic biological treatment, *Biotech. Bioeng.* **24**, 1721.

WILLIAMS, F. M. (1967), A model of cell growth dynamics, *J. Theor. Biol.* **15**, 190.

YANO, T., KOYA, S. (1973), Dynamic behavior of the chemostat subject to product inhibition, *J. Gen. Appl. Microbiol.* **19**, 97.

10 Stirred Tank Models

Matthias Reuss

Stuttgart, Federal Republic of Germany

Rakesh Bajpai

Columbia, Missouri 65203, U.S.A.

List of Symbols

a_p	specific surface area liquid–biomass
A	decaying amplitude, Eq. (5)
A_m	minimum acceptable value of the decaying amplitude
Bi	Biot number
c	concentration
C_D	drag coefficient
C_G	concentration of oxygen in the gas phase
C_L	dissolved oxygen concentration
d_i	impeller diameter
d_p	diameter of (hypothetical) diffusion element
D	molecular diffusion coefficient
Da	Damkoehler number
D_T	diameter of tank/vessel
E	discretized circulation time distribution
f	circulation parameter, Eq. (54)
f_j	interphase friction factor
F	flow rate of feed solution
F_G	momentum exchange beween gas and liquid due to interphase drag
Fr	Froude number
g	acceleration of gravity
H	height of fluid in the tank; Henry's law constant
k	kinetic energy
$k_L a$	volumetric gas–liquid mass transfer coefficient
k_p	mass transfer coefficient liquid biomass
K	morphological constant in Eq. (86) representing the strength of mycelial filaments
K_A	amplitude decay rate constant, Eq. (5)
K_M	Michaelis–Menten constant
K_{O_2}	Michaelis–Menten constant for oxygen uptake
K_S	Monod's saturation constant for substrate
L	length of circulation loop
m_{tracer}	mass of tracer liquid introduced into the system at time zero
n	speed of agitation, rpm
n_p	number of cycles

N	number of tanks in series; number of classes for discrete circulation time distribution
N_B	aeration number
Ne	power number
P	power input; pressure; product concentration
P_G	gassed power input
P/V	power input per unit liquid volume
q	fraction of impeller discharge that crosses over from one impeller to the other
Q, Q_p	pumping capacity of the impeller
Q_{CO_2}	volumetric carbon dioxide uptake rate
Q_{O_2}	volumetric oxygen uptake rate
r_{O_2}	specific oxygen uptake rate
R	bubble radius
Re	Reynold's number
RQ^{sp}	respiration quotient at which system operation is desired
S	substrate concentration
S_Φ	source and sink term, Eq. (51)
S_{crit}	critical substrate concentration above which by-product formation takes place
t	time
u	velocity of liquid
U_j	slip velocity
v	velocity
v	characteristic velocity of circulation
v_R	radial velocity
v_{rw}	radial flow at wall
v_s	superficial gas velocity
V	volume of fluid in the tank
V_G	gas flow rate
w_G, w_L	velocity in azimuthal direction
X	biomass concentration (dry weight)
Y	impulse response signal; oxygen fraction in gas
$Y_{CO_2/P}$	carbon dioxide production associated with byproduct formation
$Y_{P/S}$	product yield on substrate
Y_{X/O_2}	biomass yield on oxygen
$Y_{X/S}$	biomass yield on substrate
z	axial coordinate, Eq. (50)
α	geometrical parameter, Eq. (81)
α_G	gas hold-up
β	fractional exchange between two adjacent impeller regions
β	overall mass transfer coefficient

ε	energy dissipation rate
ε_p	volume fraction of particle
η	effectiveness factor
μ	specific growth rate of cells
μ_{eff}	effective viscosity
μ_l	parameters of log-normal distribution
μ_t	turbulent viscosity
ν	specific product formation rate, dynamic viscosity
θ	mean residence time; mean of the circulation time distribution
θ_{mix}	mixing time
ω	backmixing parameter, Eq. (54)
σ^2	variance of the circulation time distribution
σ_θ^2	normalized variance of the circulation time distribution
Φ	Thiele modulus
ψ	general modulus
Φ_p	pumping capacity

exclusive and, therefore, do not allow an exact replication of environmental similarity at any two different scales (AIBA et al., 1973; KOSSEN, 1985; KOSSEN et al., 1985; OLDSHUE, 1983; SWEERE et al., 1987). Under such circumstances, the behavior of microorganisms in different fermentors remains uncertain. As a matter of fact, the use of volumetric properties as scale-up criteria demands a guarantee of uniformity of properties throughout the system, something that may be impossible even in a small reactor.

Tab. 1. Common Criteria for Scale-up of Stirred Bioreactors

Volumetric oxygen transfer coefficient	$k_L a$
Volumetric power input	P/V
Volumetric gas flow	$\dot{V}_G/V$
Impeller tip speed	$n d_i$
Agitation speed	n
Terminal mixing time	θ_∞

1 Introduction

During scale-up, an attempt is made to recreate a physiological and hydrodynamic environment in a large reactor as similar as possible to that established in bench-scale and/or pilot plant vessels. Engineering solutions to this problem include maintaining geometric similarities whenever possible, and also criteria such as constant power per unit volume, volumetric mass transfer coefficient, circulation time, shear rate, tip speed, etc. (Tab. 1). The logic behind the different criteria is that preserving each of these singly represents maintaining the constancy of the corresponding characteristic of the extracellular environment at different scales. And, if this environmental property is the one that most critically influences the desired microbial productivity, a successful scale-up might result. Indeed, a number of microbial systems are in accord with these techniques and permit a production-scale operation reasonably in agreement with that established in the laboratory. However, it is easy to see that these criteria are mutually

Complex interactions between transport phenomena and reaction kinetics characterize bioreactors and determine their performance. While the reaction kinetics is not scale-dependent, the transport processes are. As a result, problems of scale-up may be ascribed to the critical activities in the domain of transport processes. Identification of these activities may be carried out by comparing the different time constants in a process as shown in Tab. 2 (KOSSEN et al., 1985; ROELS, 1983; SWEERE et al., 1987). Such an analysis leads one to conclude that in several bioprocesses the distribution of mass and/or energy is often the critical activity that affects the extracellular environment of the cells.

A quantitative description of these phenomena should consequently rest upon the two interwoven aspects of structured modelling. The first aspect concerns the complex interaction of the substructural elements of the cells, including the mathematical formulation of the reaction rates, and the key regulation of these networks in response to changes in the environment. The second aspect has to do with the structure of the abiotic phases of the bioreac-

Tab. 2. Time Constants for Different Processes Taking Place in a Bioreactor
(OOSTERHUIS, 1984)

Processes		t (in s)[a]
Transport Processes		
a) Gas–liquid mass transfer	$t_{0T} = \dfrac{1}{k_L a}$	5.5–11.2
b) Circulation time	$t_c = \dfrac{V/n_{imp}}{1.5\, n d_i^3}$	12.3
c) Gas residence time	$t_G = (1 - \alpha_G)\dfrac{V}{\dot{V}_G}$	20.6
d) Heat transfer	$t_{HT} = \dfrac{V \rho C_p}{h A}$	330–650
Conversion Processes		
a) Oxygen consumption	$t_{oc} = \dfrac{C_L}{r_{O_2}^{max}}$	0.7–16
b) Substrate utilization	$t_{sc} = \dfrac{S_0}{r_s^{max}}$	$5.5 \cdot 10^4$
c) Biomass growth	$t_G = \dfrac{1}{\mu_{max}}$	$1.2 \cdot 10^4$
d) Heat production	$t_{HP} = \dfrac{\rho C_p \Delta T_{cooling}}{r_{HM} + r_{HS}}$	350

[a] Typical time values for gluconic acid fermentation in a production-scale
reactor

tor in order to analyze the quality of mixing and other transport phenomena such as mass transfer between the phases causing gradients in the concentrations of various substrates and products. Because of the interaction between the biotic and abiotic phases of the system via flow of energy and material, the structures of the cellular kinetics and the structures for the physical transport of momentum, mass, and heat should not be considered alone. Both of these, individually and jointly, affect the final outcome of the process.

Despite this strong interdependence, this chapter will concentrate on structured modelling of the abiotic phases. The discussion will focus on problems of the mechanically agitated bioreactor. An attempt is made to critically discuss the state of the art in describing the distribution of mass and energy, thereby confronting the classical approach based on recirculation time distributions, compartment models, and modern fluid dynamic models based on the numerical solution of turbulent flow equations. With the aid of two simple examples it will be demonstrated how the interactions of mass and energy distributions influence the process outcome at different scales of operation as well as suggest modifications in the operation and design of bioreactors. Aerobic growth of baker's yeast and oxygen transfer in viscous broths will be used as examples.

Let us follow the discussion with some qualitative considerations of distributions in stirred bioreactors in the light of scale-up problems. It is now well known (GÜNKEL and WEBER, 1975; NAGATA, 1975; OKAMOTO et al., 1981) that energy introduced into such a system with the help of an agitator is dissipated mainly in the vicinity of the agitator, i.e., a non-uniform distribution of energy takes place. For small reactors, the region of intense energy distribu-

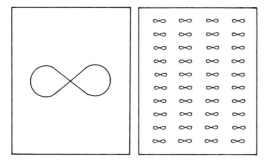

Fig. 1. Hypothetical concepts for scale-up of stirred bioreactors.

tion is significant compared to the rest of the reactor volume, and a uniform energy distribution may be considered possible for practical purposes. However, as the reactor volume increases, a uniform distribution of energy becomes more and more difficult. It can either be achieved by using bigger impellers, as shown in Fig. 1a, which may be a physical impossibility beyond a certain stage, or by using a number of impellers, each dissipating energy into its own microcosmos, as shown in Fig. 1b. This would also be a design impossibility. It seems, therefore, reasonable to assume that some non-uniformity in energy distribution exists at all scales of operation. If the reaction kinetics in question shows an interaction with the energy distribution, the different reactors would then have different performances.

A similar situation exists in a number of bioprocesses in which nutrients are continuously introduced into the broth. For specific nutrients such as oxygen and sometimes other nutrients such as the carbon source, the time constants for their distribution (mixing) may be of the same magnitude as those of their consumption in any reasonably-sized reactor beyond the bench-scale. The time constants for mixing can be reduced by introduction of more power into the system, yet the power levels required may become very high. These may also have a drastic influence upon the microbial activity. These considerations made BRYANT (1977) pose the question: "Should the object be to design the fermentor so that it behaves as a backmix reactor in which every piece ... experiences an identical environment at any time? Alternatively, should the possible

spatial variations be recognized and accepted in the design?" Particularly with respect to bioprocesses wherein the reactants need to be fed in order to achieve a defined extracellular environment, this question and others, such as how many ports of entry should be mounted and where in order to minimize the effects of spatial variations, become all the more important. Since it has been shown by a number of research workers (BRYANT, 1977; KHANG and LEVENSPIEL, 1976; NAGATA, 1975; UHL and GRAY, 1966) that recirculation exists in these mechanically agitated fermentors, will it be enough to have one impeller (Fig. 1a) generating very high circulation rates, or to have multiple impellers (Fig. 2), each with its own circulation patterns that may not interact with those of others? Since very often the major source of energy introduction are the impellers that also create the circulations, their design influences not only the distribution of mass but also that of energy into the system. This coupling of

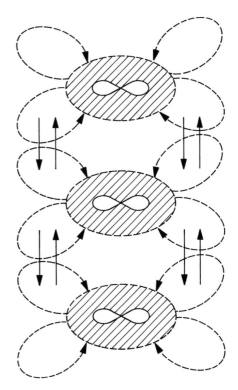

Fig. 2. Distribution of mass and energy in multiple-impeller reactors.

mass and energy distributions may have serious implications for a number of biological systems. It may be added, however, that whereas the discrepancies of mass distribution affect the reaction rate directly by influencing the concentration levels, the problems of energy distributions will also play a role either through the gas–liquid/liquid–solid mass transfer or through the diffusion in morphological forms, or by affecting cellular metabolic capabilities. In any case, a thorough understanding of the problems of mixing and of energy distribution and their interactions with the kinetic processes is required.

2 Characterization of Mass and Energy Distributions in Stirred Tanks

The quantitative characterization of mixing has been a subject of active research in chemical engineering, and a number of publications and review papers have appeared in the past. For continuous flow reactors, the technique of residence-time distributions is generally employed. Based upon the observed residence-time distribution, different models of mixing are proposed, and these are then used to evaluate the effect of mixing upon the performance of the reactor (LEVENSPIEL, 1972; NAUMANN, 1981; SMITH, 1981; WEN and FAN, 1975). Since most of the microbial reactors are operated in a batch or fed-batch (semicontinuous) mode, the well-established and widely used residence-time distribution techniques will not be discussed here.

Mixing in batch reactors is typically characterized with the help of a terminal mixing time, defined as the time required to achieve a given degree of homogeneity of concentrations. This parameter has often been used to design control loops for operating variables such as temperature, pH, etc., and a number of correlations exist relating mixing time to the operating conditions. However, this does not allow a quantitative evaluation of the spatial variation of property values in the vessel and, therefore, cannot be used to account for the effect of non-homogeneous distributions of substrate concentrations upon the microbial metabolism in the fermentors. Moreover, this concept appears to suggest that the goal should be to achieve a uniformity of concentration throughout the vessel, which may not be economical as the scale of operation increases. If a region of intense turbulence exists in the vessel, such as the impeller zone in agitated reactors or the gas distribution zone in an air-lift reactor, the contents of the vessel may be circulated through it at a high enough frequency so that the concentration of the target reactant(s) is kept within bounds along the circulation path. This approach involving circulation time and its distribution manifests itself in recycle models. However, the number of papers dealing with such models for batch and semicontinuous operation is limited. Earlier important contributions were made by HOLMES et al. (1964), KHANG and LEVENSPIEL (1976), MANN and CROSBY (1973), MANN et al. (1974) and VONCKEN et al. (1964). The experimental methods used for the measurement of circulation time and distribution can be categorized either as stimulus–response techniques or as flow-follower techniques.

2.1 Stimulus–Response Methods Based upon Tanks-in-Series Models

In these methods pulse injection of a tracer is made near stirrer tips and the response is measured nearby. Thus, injection and measurement are made in a well-defined region of the tank in a small, well-mixed volume. Using, for instance, ionic tracers and conductivity cells located in the form of a loop around the impeller (Fig. 3), the experimental observations may be interpreted with the aid of various recycle models. As a simple representation of the recirculation flow, KHANG and LEVENSPIEL (1976) suggested a tank-in-series model illustrated in Fig. 4. The impulse response for such a recycle system takes the form:

$$y(t) = 1 - B \exp\left(-\frac{2Nt}{\theta}\right) +$$

$$+ 2 \sum_{k=1}^{M} \exp\left[-N\left(1 - \cos\frac{2\pi k}{N}\right)\frac{t}{\theta}\right] \cdot$$

$$\cdot \cos\left(\frac{tN}{\theta}\sin\frac{2\pi k}{N} + \frac{2\pi k}{N}\right) \qquad (1)$$

where $B = 0$; N odd
$\qquad B = 1$; N even
$\qquad M = (N-1)/2$; N odd
$\qquad M = N/2 - 1$; N even

which can be reduced for large times and $2\pi/N \ll 1$ to the approximation

$$y(t) \simeq 1 + 2 \exp\left(-\frac{2\pi^2}{N\theta}t\right) \cdot$$

$$\cdot \cos\left(\frac{2\pi}{\theta}t + \frac{2\pi}{N}\right) \qquad (2a)$$

This is an equation of a wave with decreasing amplitude

$$A \simeq 2 \exp\left(-\frac{2\pi^2}{N\theta}t\right) \qquad (2b)$$

In the context of flow-follower measurements discussed later in this section, it is also worthwhile to refer to the results of the central limit theorem of statistics. For an arbitrary one-passage residence-time distribution (RTD), and $[f(t)]$ with mean θ and variance σ^2, it can be shown from this theorem (KHANG and LEVENSPIEL, 1976) that the tracer distribution after

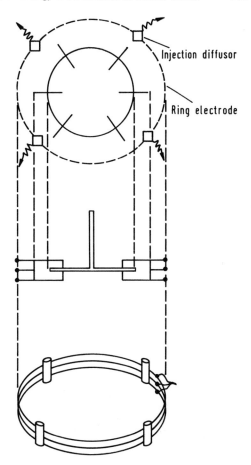

Fig. 3. Injection diffusors and conductance coil for impulse tracer experiments.

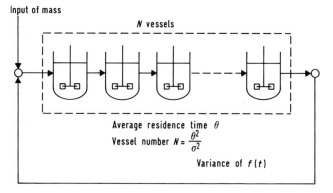

Fig. 4. Recycle model for a batch stirred reactor.

n_p cycles approaches a log-normal distribution as $n_p \to \infty$, or

$$f_{n_p}(t) = \frac{1}{\sqrt{(2\pi n_p \sigma^2)}} \exp\left(-\frac{(t - n_p \theta)^2}{2\pi n_p \sigma^2}\right) \qquad (3)$$

for large n_p

Since the variance of the tank-in-series model is given by

$$\sigma_\theta^2 = \frac{\sigma^2}{\theta^2} = \frac{1}{N} \qquad (3a)$$

Eq. (2) may be written for an arbitrary single-pass distribution in the form

$$y(t) \simeq 1 + 2 \exp\left(-\frac{2\pi^2 \sigma_\theta^2}{\theta} t\right) \cdot$$

$$\cdot \cos\left(\frac{2\pi}{\theta} t + 2\pi\sigma_\theta^2\right) \qquad (4)$$

with the decaying amplitude

$$A \simeq 2 \exp\left(-\frac{2\pi^2 \sigma_\theta^2}{\theta} t\right) = 2 \exp(-K_A t) \qquad (5)$$

Though Eqs. (4) and (5) have been arrived at as a special case (tank-in-series model), according to the central limit theorem the same solution holds for any other single-passage RTD, and these equations are therefore generally applicable (KHANG and LEVENSPIEL, 1976).

From a typical impulse response of a single impeller which is symmetrically placed in the tank, (Fig. 5) the amplitude decay rate constant K_A in Eq. (5) can be obtained by measuring the peak and valley values A from the signal amplitudes and the times of their appearance. Taking logarithms of both sides of Eq. (5), one gets

$$\ln A = \ln 2 - K_A t, \text{ where } K_A = 2\pi^2 \frac{\sigma_\theta^2}{\theta} \qquad (6)$$

Thus the slope of the $\ln A$ vs. t plot gives the K_A value. From measurements in two tanks with several different impellers, KHANG and LEVENSPIEL (1976) suggested a dimensionless equation for turbine impellers:

$$\frac{n}{K_A}\left(\frac{d_i}{D_T}\right)^{2.3} \simeq 0.5$$

for $Re = \frac{n d_i^2}{v} > 2 \cdot 10^3$ \qquad (7)

The advantage of using this concept is that the decay rate constant K_A can be easily related to the terminal mixing time, which is widely used for characterization of the global mixing quality in stirred tanks. Defining a terminal mixing time θ_{mix} as the time which is necessary to reduce a tracer pulse to a required uniformity (equivalent to a final amplitude of the signal response A_m), we may write Eq. (6) as follows:

$$\theta_{mix} = \frac{1}{K_A} \ln\left(\frac{2}{A_m}\right) \qquad (8)$$

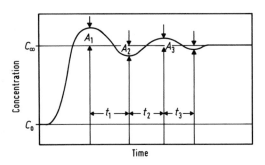

Fig. 5. An example of the impulse response in a reactor with symmetrical impeller position.

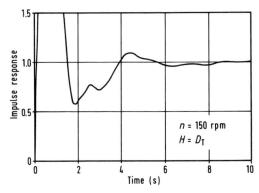

Fig. 6. Measured impulse response for the asymmetrical impeller position shown in system 3 of Fig. 7.

When trying to apply this simple concept to the characterization of the circulation flow in a stirred bioreactor, we are first confronted with the problem that in such vessels the position of the impeller(s) is usually not symmetrical. For an aerated stirred bioreactor, the impeller has to be placed near the bottom. If the liquid height is still equal to the tank diameter, this may result in a geometrical configuration illustrated in system 3 of Fig. 7. Applying the

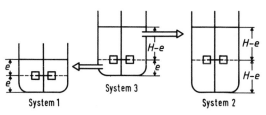

Fig. 7. Replacement of an asymmetrical impeller position by superposition of two symmetrical systems.

method suggested by KHANG and LEVENSPIEL (1976), BERKE (1980) measured tracer responses in such configurations. A typical example for the results obtained is illustrated in Fig. 6. Obviously, the asymmetrical impeller position drastically influences the response signal. Thus, it is no longer possible to treat this signal in the way suggested by KHANG and LEVENSPIEL (1976). The difficulties may be

partly overcome by considering that the system now consists of two different flow regions. The overall system response may thus be predicted from a superposition of the recirculation time distributions in the two different regions similar to the way suggested by MANN and CROSBY (1973) in their theoretical considerations.

Fig. 7 illustrates how the asymmetrical flow regions of system 3 may be replaced by two symmetrical systems. The overall pulse response can then be calculated from

$$y_{\text{system}\,3}(t) = \frac{y_{\text{system}\,1}(t) + y_{\text{system}\,2}(t)}{2} \qquad (9)$$

using Eq. (4) and (5) for the calculation of tracer response in the two symmetrical subsystems 1 and 2.

An example of the comparisons between measured and predicted results is shown in Fig. 8. Though the agreement between the measured and predicted results appears to be satisfactory, it must be emphasized that a number of uncertainties remain when applying this strategy. First of all, it requires identification of four parameters from a single signal response curve. Secondly, an additional exchange of fluid between the individual recirculation flows cannot be entirely excluded. Such an exchange, however, would require further parameters to be estimated from the measured signal, which would finally make such a procedure rather intractable.

2.2 Flow-Follower Techniques

Some of the problems mentioned before may be overcome by making use of flow-follower methods. The advantages of these techniques have been well summarized by BRYANT (1977), and his paper should be consulted for further guidance. The analysis of the frequency pattern for passage of flow followers through an active region — impeller zone in the case of a stirred vessel — provides us with the following useful information:
a) mean circulation time θ,
b) standard deviation of circulation times σ_θ^2, and
c) distribution of circulation times.

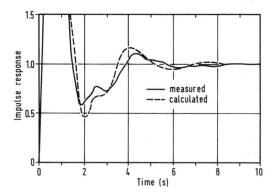

Fig. 8. Example of a comparison between measured and calculated impulse response.

2.2.1 Measurements and Data Analysis

Of the different types of flow followers suggested in the literature, two appear to have been extensively used due to the ease with which they can be coupled to the continuous recording of events and further processing of data. These are the radio-flow followers and the magneto-flow followers. In the radio-flow follower method a small radio-transmitter placed in a hollow plastic sphere is used (BRYANT, 1977; BRYANT and SADEGHZADEH, 1979; MIDDLETON, 1979; OOSTERHUIS, 1984). The density of the sphere is adjusted to that of the fluid. When placed in a stirred vessel, the follower moves with the circulation fluid. Every time this particle passes an aerial which takes the form of a loop around the impeller, a signal from the radio-transmitter is sent to a receiver.

An alternative flow follower, which consists of a small magnet within a hollow plastic sphere, has been used by MUKATAKA et al. (1976, 1980, 1981a, b) for measurements in glass vessels up to 30 liters. Fig. 9 shows a modified layout of this method applied to circulation-time distribution measurements in pilot plants up to a volume of 3000 liters (BOELCKE, 1983; REUSS, 1983; REUSS and BRAMMER, 1985). When the magnetic sphere passes through a conductive coil located around the impeller, it induces an electromotive force whose spikes can be analyzed in a way similar to the signals of a radio-transmitter. A process computer then calculates the time difference between the passages. These differences are classified in 20 different time classes with a fixed time interval Δt.

From t_i (mean circulation time for class i, $i = 1 \ldots N$) and n_i (the number of stored entries in this time class), the probability density f_i of ith class can be calculated from

$$f_i \Delta t = \frac{n_i}{\sum\limits_{j=1}^{N} n_j} \tag{10}$$

Here $\sum\limits_{j=1}^{N} n_j$ is the total number of signals and is at least 1000.

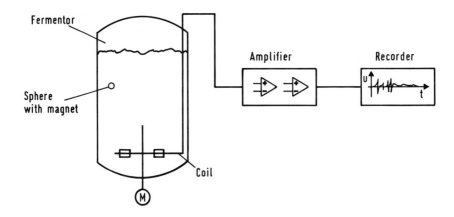

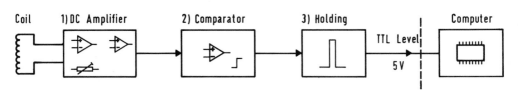

Fig. 9. Layout of the equipment for the magneto-flow follower technique.

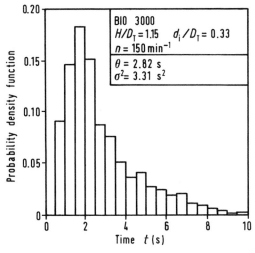

Fig. 10. Example of a histogram for circulation times.

Fig. 10 shows a typical histogram for the circulation times obtained at the 3000 liter scale for a single impeller. The density function may be characterized by the mean circulation time

$$\theta = \sum_{i=1}^{N} \hat{t}_i f_i \Delta t \tag{11}$$

and the variance

$$\sigma^2 = \bar{t}^2 - \theta^2 \tag{12}$$

where

$$\bar{t}^2 = \sum_{i=1}^{N} \hat{t}_i^2 f_i \Delta t \tag{13}$$

The χ^2 test verifies that the log-normal distribution is the best fit of the experimental observations.

Thus

$$f(t) = \frac{1}{\sqrt{2\pi\sigma_1^2}} \frac{1}{t} \exp\left(-\frac{(\ln t - \mu_1)^2}{2\sigma_1^2}\right) \tag{14}$$

where

$$\theta = \exp\left(\mu_1 + \frac{\sigma_1^2}{2}\right) \tag{15}$$

and

$$\sigma^2 = \theta^2 (\exp\sigma_1^2 - 1) \tag{16}$$

or

$$\sigma_1^2 = \ln\left(\frac{\sigma^2}{\theta^2} + 1\right) \tag{17}$$

and

$$\mu_1 = \ln\theta - \ln\left(\frac{\dfrac{\sigma^2}{\theta^2} + 1}{2}\right) \tag{18}$$

This result, however, needs further interpretation in light of the discussions in the previous section concerning the problem of asymmetrical impeller positions. As a matter of fact, a single log-normal distribution cannot account for the circulation-time distribution of an asymmetrically positioned impeller, because the fluid flow consists of two different recirculation streams. Fig. 11, which is a log-normal distribution plot, demonstrates these deviations. Thus, to be correct, one needs to consider the system as divided into two different flow regions, and the circulation time distribution should be calculated from the superposition of the RTD of the individual regions. In the present case this would lead to the following equation for the entire distribution

$$f_\Sigma(t) = q_1 f_1(t) + (1 - q_1) f_2(t) \tag{19}$$

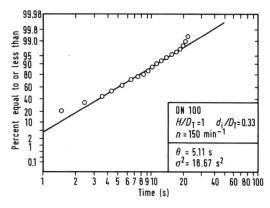

Fig. 11. Log-normal plot of circulation time data.

where q_1 is the fraction of the flow that passes through region 1 having a log-normal distribution f_1 (MANN and CROSBY, 1973). This more accurate procedure, however, would require the adjustment of five parameters, including q_1. For reasons of simplicity, the effect of geometric asymmetry upon the distribution may be neglected. In this case, the measured data are approximated by a single log-normal distribution, which can be used in connection with microbial kinetics.

From systematic measurements of distributions in two different vessels (100 and 3000 liters) at different ratios of liquid height/tank diameter and impeller diameter/tank diameter, respectively, the mean circulation time could be correlated with the geometrical and operational parameters in the following manner (BOELCKE, 1983; REUSS and BRAMMER, 1985; REUSS, 1988):

$$n\theta = 0.76 \left(\frac{H}{D_T}\right)^{0.6} \left(\frac{D_T}{d_i}\right)^{2.7} \tag{20}$$

2.2.2 Estimation of the Mean Circulation Time from Simple Flow Models

A number of researchers have tried to estimate the mean circulation time and/or the terminal mixing time from simple models for the fluid flow in the tank by predicting a characteristic length of the circulation path and using an appropriate expression for an averaged circulation velocity (JOSHI et al., 1982; PANDIT and JOSHI, 1983; MCMANAMEY, 1980). Mean circulation time is then calculated from

$$\theta = \frac{L}{\bar{v}} \tag{21}$$

The characteristic velocity for circulation, v, is usually assumed to be represented by the radial flow velocity at the vessel wall. This velocity is given by (VAN DER MOLEN and VAN MAANEN, 1978):

$$v_{rw} = 0.85 \pi n d_i \left(\frac{d_i}{D_T}\right)^{7/6} \tag{22}$$

Thus, the characteristic flow for circulation of the fluid includes the entrainment flow. The application of magneto-flow follower techniques with a coil located close to the impeller blades results in the detection of the discharge flow from the impeller. Under these conditions it is more reasonable to estimate the corresponding mean circulation velocity from the continuity equation, which takes the form

$$Q_P = C_1 n d_i^3 = C_2 \bar{v} D_T^2 \tag{23}$$

with pumping capacity of the impeller Q_p. Here, it is assumed that the fluid flow through the impeller region is equal to the corresponding circulation through the cross-section of the tank, characterized by the fluid velocity $\bar{v}$. From Eq. (23) we predict

$$\bar{v} = C_1' n d_i \left(\frac{d_i}{D_T}\right)^2 \tag{24}$$

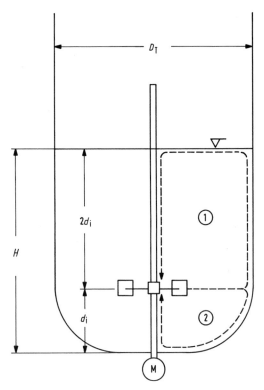

Fig. 12. Circulation paths for a single impeller (asymmetrical position).

The circulation paths are predicted in a similar manner as suggested by McManamey (1980). The lengths of the two circulation paths illustrated in Fig. 12 are given by

$$L_1 = D_T + 2H - 2.5\,d_i \qquad (25)$$

and

$$L_2 = D_T + 1.5\,d_i \qquad (26)$$

This results in an averaged value for the path length

$$\bar{L} = \frac{L_1 + L_2}{2} = D_T + H - 0.5\,d_i \qquad (27)$$

It must be emphasized that McManamey (1980), in trying to estimate the terminal mixing time, followed a concept in which the longest path (L_1) was chosen along with the fluid velocity at the vessel wall (Eq. (22)).

Making use of Eqs. (24) and (27), the mean circulation time (Eq. (21)) can be estimated to be

$$\theta = \frac{\left[1 + \dfrac{H}{D_T} - 0.5\,\dfrac{d_i}{D_T}\right] D_T^3}{n\,d_i^3} \qquad (28)$$

As shown in Fig. 13, the term in brackets in Eq. (28) can be approximated by a simple power function, resulting in the following correlation:

$$\left[1 + \frac{H}{D_T} - 0.5\,\frac{d_i}{D_T}\right] =$$

$$= 1.64 \left(\frac{H}{D_T}\right)^{0.6} \left(\frac{D_T}{d_i}\right)^{0.1} \qquad (29)$$

This correlation holds in a region $0.2 \leq d_i/D_T \leq 0.5$ and $1 \leq H/D_T \leq 2$. The final equation for the mean circulation time θ is, therefore, given by

$$n\,\theta = C_1' \left(\frac{H}{D_T}\right)^{0.6} \left(\frac{D_T}{d_i}\right)^{3.1} \qquad (30)$$

A comparison between this equation and the empirical correlation from the experimental observations, Eq. (20), shows an excellent

agreement in the exponent for (H/D_T). The theoretical value of the exponent for (D_T/d_i) is, however, higher than that observed in the experiments. This deviation is caused by the uncertainties in the estimation of the characteristic circulation velocity.

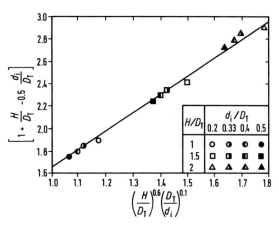

Fig. 13. Graphical presentation of Eq. (29).

2.2.3 Influence of Aeration

Only a few published papers deal with the mixing characteristics in aerated stirred tanks. The reported results for the terminal mixing time under aerated conditions are somewhat controversial. Blakebrough and Sambamurthy (1966) as well as Einsele and Finn (1980) observed higher mixing times with increasing aeration rates. In contrast, Paca et al. (1976) predicted shorter mixing times. Joshi et al. (1982), measuring terminal mixing times in various liquids, reported a slight influence of aeration rate:

$$n\,\theta_{\text{mix}} \simeq \dot{V}_G^{1/12} \qquad (31)$$

Bryant and Sadeghzadeh (1979) were the first to thoroughly investigate the mechanisms of the mixing process in aerated liquids with the aid of the radio flow-follower technique. This method provides the necessary information on how the mean circulation time and the variance of the distribution are affected by

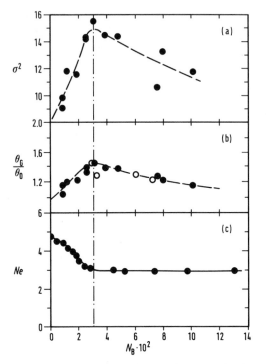

Fig. 14. Influence of aeration on the mean circulation time and variance of the distribution compared with the power characteristics of the aerated impeller (agitation speed is constant, $Fr = n^2 d_i/g = 0.23$; index 0: non-aerated impeller).

aeration. The experimental observations indicated longer mean circulation times with increasing air flow rate. At the same time, mixing in the circulation paths is more intensive (higher variance of the distribution σ^2), indicating that the gas bubbles contribute to the mixing process. Since the terminal mixing time is influenced by both parameters, Eq. (8) together with Eqs. (6) and (3a), the overall efficiency of mixing may be unaffected under certain conditions.

Mean circulation times and variances σ^2 of the distributions obtained from magneto-flow follower measurements as described above are presented in Fig. 14 (RIESMEIER, 1984). In this figure, the ratio of mean circulation times under aerated and unaerated conditions is plotted against the aeration number ($N_B = \dot{V}_G/n d_i^3$) at a constant agitation speed n or the Froude number $Fr = n^2 d_i/g$. In agreement with the ob-

servations made by BRYANT and SADEGHA-ZEDEH (1979), the mean circulation times increase with increasing aeration rates $\dot{V}_G$. Beyond a critical aeration rate, however, the circulation times start to decrease again. It is interesting to compare this behavior of the circulation times with the power characteristics of the impeller which are also presented in this figure. The increase in circulation times corresponds to the decrease in the power number Ne due to the aeration. The observed maximum in the mixing characteristics corresponds, on the other hand, to the critical aeration number at which the power number asymptotically approaches its final value. Since the impeller starts to be flooded at this critical aeration rate, we may interpret the decreasing circulation times beyond this flooding point as a consequence of a mixing process mainly caused by the aeration. The behavior of the system above this point is similar to that of a bubble column which shows improved mixing with increasing gas flow rates. Fig. 15 shows that the decrease in circulation times with increasing air flow rates can be satisfactorily accounted for by the decrease of the power number Ne. The slope of the regression line is -0.35, which is close to the exponent in the equation presented by MUKATAKA et al. (1981b) and the influence of the power number upon the terminal mixing times in nonaerated liquids reported by MERSMANN et al. (1975).

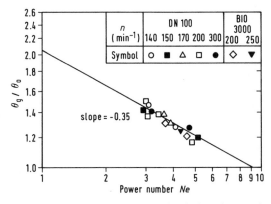

Fig. 15. Correlation between circulation time and power number for the aerated, non-flooded impeller.

As can be seen from Fig. 14, the variance of the distribution σ^2 also increases with increasing air flow rates below the flooding point. When calculating the terminal mixing time by making use of Eqs. (3a), (6), and (8)

$$\theta_{\text{mix}} = \frac{1}{K_A} \ln\left(\frac{2}{A_{\text{m}}}\right) =$$

$$= \frac{\theta}{2\pi\sigma_\theta^2} \ln\left(\frac{2}{A_{\text{m}}}\right) = \frac{\theta^3}{2\pi\sigma^2} \ln\left(\frac{2}{A_{\text{m}}}\right) \quad (32)$$

one observes that this value is only weakly influenced by the aeration rate.

2.3 Tracer Responses from Circulation-Time Distribution

The flow-follower methods result in a direct measurement of the characteristic circulation patterns in a given system. It is obvious from the analysis presented in Sect. 2.1 that this information can be translated into a tracer response obtained by stimulus–response techniques. Eq. (1) through (5) can be used to calculate the response of a non-reactive tracer once the parameters θ and σ^2 are established. For reactive systems, however, the governing equations will be integro-differential in nature. For a set of coupled variables, as are commonly observed in biological systems, the solutions are difficult to obtain. Flow models, on the other hand, result in simpler governing equations. In this section a flow model and some results will be presented.

In a mechanically stirred reactor where the fluid continuously recirculates due to the action of the impeller, a very high degree of turbulence exists in the vicinity of the impeller, and up to 70% of the energy distribution may occur in this region (CUTTER, 1966; GÜNKEL and WEBER, 1975; MOECKEL, 1980; PLACEK et al., 1986). As a result, the fluid elements passing through it are intensely mixed down to the molecular level. This region may be considered a *micromixer* through which the entire fluid in the reactor passes at a frequency dictated by the circulation time distribution. Away from the impeller, the turbulent intensity rapidly decreases, and the interactions between the different fluid elements may range from complete micromixing to complete segregation. The residence time distribution in this zone, called a *macromixer,* is the circulation time distribution in the reactor. This results in a two-environment model (Fig. 16) similar to that proposed by MANNING and coworkers (1965). The different cases of complete, partial, or zero segregation may be simulated by using a tanks-in-series approach or a Monte-Carlo simulation approach. For non-reactive tracers and for first-order reactions, all degrees of segregation give the same results.

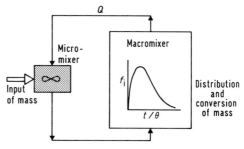

Fig. 16. Two-environment model for mixing in a stirred bioreactor.

BAJPAI and REUSS (1982a) have used a Monte-Carlo simulation method in which the physical system is divided into a number of discrete elements. In each of these elements the reaction process is simulated for a short period of time, at the end of which the system-specific interactions are simulated. As a result of the interactions, the state and the number of elements may change. The simulation process is then repeated again for the next time period. For complete segregation, the process may be schematically represented as in Fig. 17.

The volume of the macromixer is divided into N discrete elements, each having a unique age defined as the time spent in the macromixer. If time is divided into segments of Δt, the volume of a freshly entered element (age zero) will be $Q\Delta t$, where Q is the recirculation rate in the system. At the same time, a recirculating stream of volume $Q\Delta t$ leaves the macromixer. All the existing elements in the macromixer contribute to the existing stream according to the circulation time distribution.

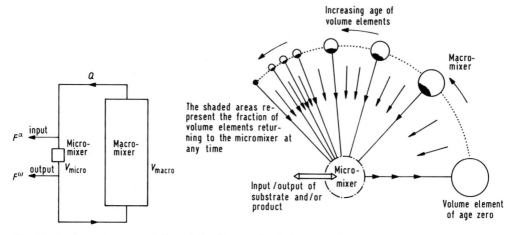

Fig. 17. A schematic representation of the discrete simulation procedure.

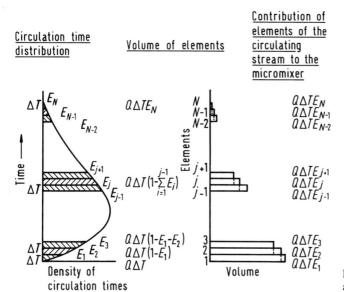

Fig. 18. Circulation time distribution and flow through the macromixer.

These contributions are denoted by dark areas in Fig. 17. As a result, volumes of the elements decrease. The contribution of the Nth element is exactly equal to its volume; in other words, this element leaves the macromixer. At this moment, the ages of all the elements are advanced, i.e., the jth element is termed the $(j+1)$th element. The newly entered element of volume $Q\Delta t$ becomes the first element. The composition of the exit stream is determined by the weighted average of the contributions of each element in the macromixer as in the following equation:

$$C_{\text{exit}} = \frac{\sum\limits_{i=1}^{N} E_i C_i Q \Delta t}{\sum\limits_{i=1}^{N} E_i Q \Delta t} \tag{33}$$

where E is the fraction of the circulation times between $(i-1)\Delta t$ and $i\Delta t$. E_i is related to the circulation time distribution $f(t)$

$$E_i = \int_{(i-1)\Delta t}^{i\Delta t} f(t)\,dt \tag{34}$$

The volume elements in the macromixer are assumed to be completely segregated, and the progress in each of these with age can be calculated with the help of known initial conditions from the time each entered the macromixer. Since each element has a unique age, reactions in each proceed also to a unique extent, and each thus makes a different contribution to the recycle stream. The circulation time distribution and the volumes of the different elements in it are shown in Fig. 18. For further details the original paper by BAJPAI and REUSS (1982a) is recommended. For intermediate degrees of segregation, methods based upon coalescence and redispersion can be used (CURL, 1963; BAJPAI et al., 1977).

The volume of the micromixer is assumed to be negligible compared to that of the macromixer. Hence, the micromixer acts as an ideal mixer of the recirculating stream with any incoming stream. This situation is shown in Fig. 19 for tracer inputs into the reactor. A material balance around the micromixer results in the following expression for the concentration of tracer in the new element:

$$C_{ent} = C_{exit} + \frac{m_{tracer}}{Q\Delta t}\,\delta(t) \tag{35}$$

Typical simulations of tracer responses in the impeller region for log-normal circulation time distributions (corresponding to symmetrically placed impellers) using the discrete simulation procedure are shown in Fig. 20. The predicted oscillating concentrations of tracer in the impeller region are similar to those experimentally observed by others (KHANG and LEVEN-SPIEL, 1976; MIDDLETON, 1979). Similar results were predicted by BRYANT (1977) using a convolution procedure.

Mixing time, defined as the time required for the oscillations to die down to a suitably low level, can be seen from these figures to depend not only upon the mean circulation time, θ, but also upon the variance, σ^2. This contradicts the popular suggestion that the mixing time is a fixed multiple of the mean circulation

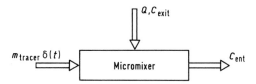

Fig. 19. Material balance around the micromixer in the case of impulse tracer input.

time unless the variance is insensitive to the measurements. The results of MIDDLETON (1979) show this not to be the case.

The different degrees of segregation in the macromixer must also be considered, if reactions are also simultaneously taking place. Specific examples of such cases will be presented later in this chapter.

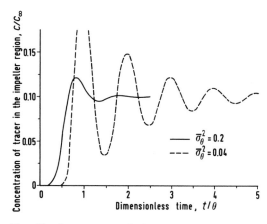

Fig. 20. Computer prediction of the dynamics of dispersion of inert tracer throughout reactor volume.

2.4 Circulation-Time Distributions for Two-Impeller Systems

2.4.1 Measurements Using Flow-Follower Techniques

The circulation flow in a multiple impeller system may be envisaged as a superposition of the circulation flow of the individual impellers and an exchange flow between the impeller zones. The problem of the quantitative analy-

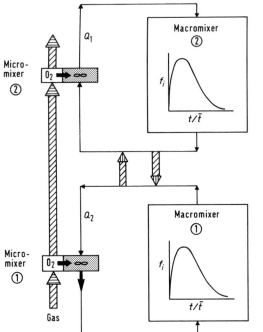

Fig. 21. Two-environment model for mixing in a stirred bioreactor with two impellers.

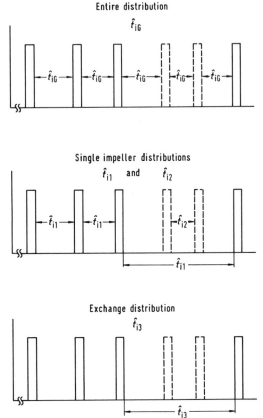

Fig. 22. Scheme for analyzing flow-follower signals in the two-impeller system.

sis of flow-follower measurements, in which each impeller is encircled by its own aerial or coil, can be solved in different ways. The strategy for analysis depends to a great extent upon the manner in which the vessel volume is structured for the purpose of studying the impact of the circulation flow upon the outcome of the microbial reaction.

MUKATAKA et al. (1981a) applied the magneto-flow-follower technique with the coils located outside a 22 liter stirred reactor to a two-impeller system and studied the circulation time distributions in the upper and lower impeller region as well as the exchange flow. The micro-macro mixer model presented in Sect. 2.3 can be extended to the two-impeller system as shown in Fig. 21. In this scheme, each impeller is considered as a micromixer of negligible volume, with the rest of the tank divided into two (equal to the number of impellers) macromixers. The circulation streams from the macromixers are assumed to exchange mass with each other. The exchange flow between

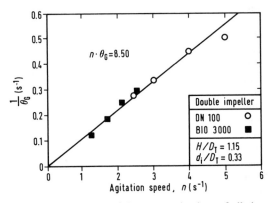

Fig. 23. Correlation of the averaged values of all circulation times in the two-impeller system (geometrical properties, Fig. 24).

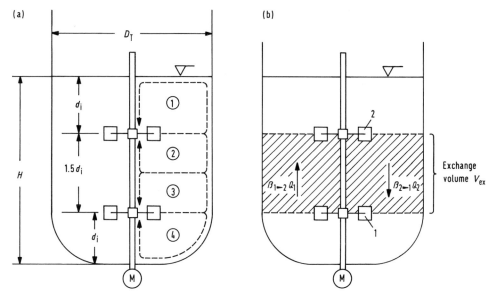

Fig. 24. Impeller configuration and exchange flow in the two-impeller system.

the two macromixers is defined by exchange coefficients in the following manner:

$$\beta_{2\to1} = \frac{q_{2\to1}}{Q_2} \text{ and } \beta_{1\to2} = \frac{q_{1\to2}}{Q_1} \tag{36}$$

Here $q_{2\to1}$ and $q_{1\to2}$ are the exchange flow rates which (because of continuity) must be equal, and Q_1 and Q_2 are the circulation flow rates through macromixers 1 and 2, respectively. The magneto-flow-follower technique has been applied to this system by using two coils, each connected to its own electronic equipment and an A/D channel at the computer interface (BOELCKE, 1983). Thus, the computer detects a sequence of signals which can afterwards be analyzed in different ways, as illustrated in Fig. 22. The measured time differences for the circulation time distributions of the two impellers, as well as the time differences for the exchange between the impellers, are again represented by distribution functions. From the measured distributions the mean circulation times, variances, and exchange times can be calculated. As illustrated in Fig. 23, the overall mean circulation times, which are average values of all circulation times obtained in both circulation regions, can be correlated for the two scales of operation using geometrical properties as illustrated in Fig. 24 with

$$n\theta_\Sigma = 8.5 \tag{37}$$

This value is slightly higher than half of the value expected for a single-impeller system at the same height/diameter ratio. Taking $H/D_T = 1.15$ and $D_T/d_i = 3$ and making use of Eq. (20) results in

$$n\theta_{(single\ impeller)} = 15.4 \tag{38}$$

or

$$\frac{n\theta_\Sigma}{n\theta_{(single\ impeller)}} = 0.55 \tag{39}$$

The higher value in the two-impeller system may again be interpreted by comparing the length of the mean circulation paths. The length for the two-impeller system illustrated in Fig. 24 is given by

$$\bar{L}_2 = \frac{L_1 + L_2 + L_3 + L_4}{4} =$$

$$= \frac{2(D_T + 1.5\,d_i) + 2(D_T + d_i)}{4} \tag{40}$$

Thus, the ratio of the lengths of mean circulation paths in the double- and single-impeller system is

$$\frac{\bar{L}_1}{\bar{L}_2} = \frac{17}{24} = 0.7 \tag{41}$$

a value slightly higher than the measured value of 0.55.

In order to estimate the exchange coefficients β from the measured circulation time distributions, it is assumed that only those fluid elements contribute to the exchange flow which belong to the exchange volume illustrated in Fig. 24b. For the geometrical configuration shown in Fig. 22a, this volume is given by

$$V_{\text{ex}} = \frac{6}{7} V_i \quad i = 1, 2 \tag{42}$$

The exchange coefficients are then calculated from

$$\beta_{2\to1} = \frac{q_{2\to1}}{Q_2} = \frac{3}{7} \frac{V_2}{\theta_{2\to1}} \frac{V_2}{\theta_2} \tag{43}$$

and

$$\beta_{1\to2} = \frac{q_{1\to2}}{Q_1} = \frac{3}{7} \frac{V_1}{\theta_{1\to2}} \frac{V_1}{\theta_1} \tag{44}$$

where $\theta_{2\to1}$ and $\theta_{1\to2}$ are the mean exchange times obtained from their exchange distributions, and θ_1 and θ_2 are the mean circulation times in the lower and upper macromixer, respectively. From the data represented in Tab. 3 it can be seen that

$$\beta_{2\to1} > \beta_{1\to2}$$

but the two flow rates $q_{1\to2}$ and $q_{2\to1}$ are approximately equal in the range of the scatter of the experimental data. The differences in exchange coefficients obtained in this manner can be related to the lower circulation times in the lower impeller zone (θ_1 in Tab. 3). The differences are probably due to the bottom construction of the vessel and to the fact that there are no baffles in the lower circulation loop of impeller 1. In order to see the influence of the geometrical properties, an even

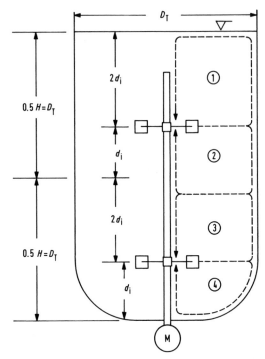

Fig. 25. Impeller configuration for condition: liquid height $H = 2 \times$ tank diameter D_T.

more asymmetrical impeller location was studied (Fig. 25). The results of the measurements presented in Tab. 3 illustrate that the exchange flow rates $q_{1\to2}$ and $q_{2\to1}$ again show reasonable agreement. In contrast to the results with the impeller location of Fig. 24a and b, the exchange coefficients $\beta_{1\to2}$ now have similar values. A better insight into the behavior of the mean circulation times in the upper and lower impeller regions can be found by investigating the length of the corresponding mean circulation paths. The ratio of the mean path lengths in the lower and upper impeller regions for the geometrical configuration illustrated in Fig. 25 is given by

$$\frac{\bar{L}_2}{\bar{L}_1} = \frac{1 + 3\dfrac{d_i}{D_T}}{1 + 2\dfrac{d_i}{D_T}} \tag{45}$$

With the value of $d_i/D_T = 0.333$, a factor of 1.2 is calculated. Multiplying this value by the

Tab. 3. Results of Magneto-Flow Follower Measurements in the Two-Impeller System

Agitation Speed n (min^{-1})	θ_1 (s)	θ_2 (s)	$\theta_{1\rightarrow2}$ (s)	$\theta_{2\rightarrow1}$ (s)	$\beta_{1\rightarrow2}$	$\beta_{2\rightarrow1}$	$q_{1\rightarrow2}$ (s^{-1})	$q_{2\rightarrow1}$ (s^{-1})
Reactor Volume 100 L (Impeller Position Fig. 24a)								
150	2.81	3.90	3.71	3.36	0.32	0.50	3.07	3.46
180	2.39	3.37	3.21	2.18	0.32	0.66	3.62	5.29
240	1.86	2.24	2.74	1.86	0.29	0.52	4.21	6.27
300	1.66	2.16	1.82	1.38	0.39	0.67	6.34	8.34
Reactor Volume 3000 L (Impeller Position Fig. 24a)								
75	5.25	6.66	7.32	6.01	0.31	0.47	33.72	40.30
100	4.07	5.03	6.71	5.74	0.26	0.38	36.48	43.14
125	2.98	4.10	4.49	3.85	0.28	0.46	53.65	64.06
150	2.65	3.45	3.62	3.23	0.31	0.46	66.80	76.13
Reactor Volume 100 L (Impeller Position Fig. 25)								
150	3.44	5.70	8.28	7.36	0.25	0.33	2.82	3.14
200	2.77	4.44	5.86	5.47	0.28	0.35	3.92	4.29
250	2.32	3.43	5.25	5.70	0.27	0.26	4.52	4.12
300	1.92	3.27	4.12	4.17	0.28	0.34	5.66	5.66

factor of 1.3, predicted for the effects of the difference in the lower and upper circulation (Tab. 3), results in an overall ratio of

$$\frac{\theta_2}{\theta_1} = 1.6 \qquad (46)$$

This value is in excellent agreement with the experimentally observed differences between the mean circulation times in the upper and lower impeller regions presented in Tab. 3. From the simple calculation presented above, it may be concluded that reasonable estimates for the mean circulation times can be predicted for different impeller configurations by calculating the mean lengths of the individual circulation paths. This feature should lead to reasonable estimates for the mean circulation times.

2.4.2 Measurements Using Tracer Responses

The concept of using circulation time and its distribution to analyze mixing phenomena appears to be very sound and helpful in explaining many observed phenomena related to the

scale of operation. However, sometimes it may not be possible to employ flow-follower techniques to establish the parameters of circulation, particularly in large-scale industrial processes involving multiple impellers. Specifically, the requirements of sterility and presence of filaments in broth may rule out the introduction of non-sterilizable flow followers and the aerials or coils in the production reactor. In such cases, tracer measurement methods can still be used as exemplified by the work of JANSEN et al. (1978), who measured mixing characteristics with working volumes up to 120 m^3. During operation of the bioreactor, calculated quantities of radioactive tracer (technetium isotope) were introduced at the top, and mixing was followed by measuring the tracer concentration at the bottom. Under these circumstances, a question arises as to whether the circulation and exchange parameters can be reliably estimated from such data.

For a two-impeller system, simulations of tracer responses were conducted using the previously described micro-macro mixer model. For tracer injection in the upper impeller region, typical tracer concentration profiles in the two turbulent zones are shown in Fig. 26. Using discretized tracer concentrations predicted at the lower impeller, the parameters θ,

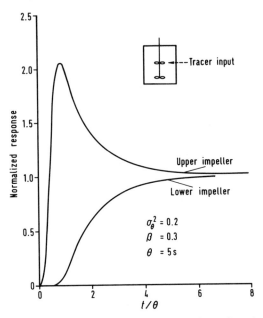

Fig. 26. Computer predictions of the dynamics of dispersion of inert tracer throughout the reactor in the two-impeller system.

acceptable estimate. Such estimation procedures would be very useful in cases of interference with mixing in industrial situations.

2.5 Multi-Compartment Models for Coupling Oxygen Transfer, Mixing and Kinetics

Thus far we have examined models which are based on recirculation time distributions. Another class of models can be designed from aggregations of various compartment structures. The behavior of these models is then described through simulations of the coupled material balance equations. A few examples of these models are introduced below. For a more detailed description of the various models, including their applications, the reader is referred to the original papers.

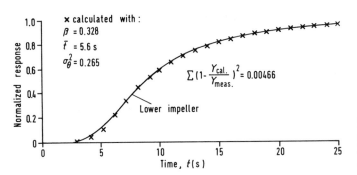

Fig. 27. Results of parameter estimation from a simulated signal response in the lower impeller region.

σ^2, and β were then back-calculated using a modified simplex optimization algorithm (Nelder-Mead method). The results of such an exercise are shown in Fig. 27 where the (so-called) experimental data are plotted as discrete points, and the continuous curve belongs to the predicted profile with optimized parameters. That the crucial parameters θ and β are estimated to within 10% of the actual values shows that the parameters of the two-compartment model are sensitive enough to allow an

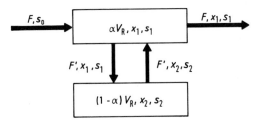

Fig. 28. Two-region mixing model (SINCLAIR and BROWN, 1970).

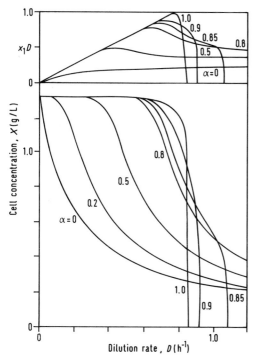

Fig. 29. Exit cell concentration and productivity as a function of the dilution rate at different interchange rates for the two-region model of SINCLAIR and BROWN (1970).

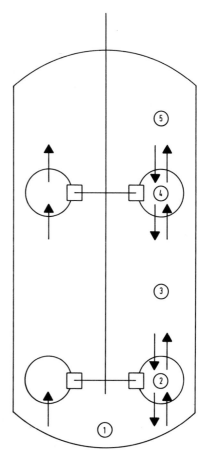

Fig. 30. Five-compartment model (OOSTERHUIS and KOSSEN, 1984).

The two-compartment model of SINCLAIR and BROWN (1970)

This model has been proposed for a continuous stirred tank reactor (CSTR) to explain the influence of the intensity of mixing. The model consists of two well-mixed compartments which interact via exchange flow (Fig. 28). The solution of the material balance equations for biomass and substrate (Fig. 29) illustrates the strong effect of the intensity of mixing on the exit cell concentration in a CSTR.

The five-compartment model of OOSTERHUIS and KOSSEN (1984)

This model is schematically illustrated in Fig. 30. The model consists of two impeller compartments and three compartments which characterize the flow conditions in the rest of the tank. Considering the oxygen kinetics described by Monod's equation, the material balance equations for oxygen in the gas and liquid phases of the ith compartment are given by

$$\dot{V}_G(C_G^{i-1} - C_G^i) - k_L a^i V_{L+G}^i (C_G^i/H' - C_L^i) = 0$$

$$\phi_P(C_L^{i-1} + C_L^{i+1} - 2C_L^i) +$$
$$+ k_L a^i V_{L+G}^i (C_G^i/H' - C_L^i) -$$
$$- r_{O_2}^{max} \frac{C_L^i V_L^i}{K_{O_2} + C_L^i} = 0 \qquad (47)$$

OOSTERHUIS (1984) has proposed the following concept for estimation of the volumetric mass transfer coefficient: for the impeller compartments the $k_L a$ value is calculated from the equation for coalescing systems (VAN'T RIET, 1979):

$$k_L a = 0.0323 \left(\frac{P_G}{V_L}\right)^{0.4} (v_S)^{0.5} \tag{48}$$

$k_L a$ values for the rest of the compartments are estimated from an equation for bubble columns (HEIJNEN and VAN'T RIET, 1982):

$$k_L a = 0.3 (v_S)^{0.7} \tag{49}$$

Furthermore, it is assumed that the liquid flow rate between the compartments is equal to the pumping capacity of the impellers. Typical results from the model are presented in Fig. 31 showing a comparison between calculated and measured data for dissolved oxygen in percentage of saturation.

The multi-turbine model of BADER (1987a, b)

The model, schematically shown in Fig. 32, separates the reactor into a series of mixing zones. Mixing of the liquid phase is represented by the liquid flow between the compartments and is related to the pumping capacity

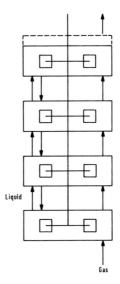

Fig. 32. Mixing cell model for a multi-turbine fermentor (BADER, 1987b).

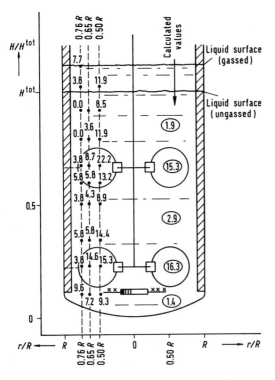

Fig. 31. Comparison between computed and measured dissolved oxygen profiles (dissolved oxygen tension in % saturation) (OOSTERHUIS and KOSSEN, 1984).

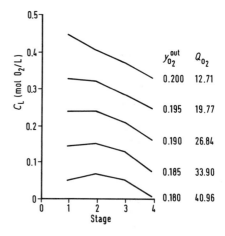

Fig. 33. Model prediction of dissolved oxygen profiles at different uptake rates (BADER, 1987b).

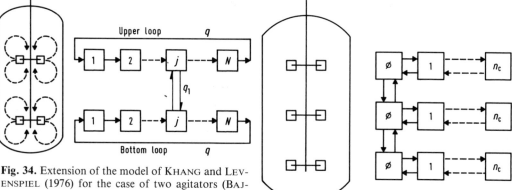

Fig. 34. Extension of the model of KHANG and LEVENSPIEL (1976) for the case of two agitators (BAJPAI and SOHN, 1987).

Fig. 35. Recycle-backmix circulation model (SINGH et al., 1986, 1987, 1988).

of the impellers. Back-mixing of the gas phase is neglected in this model. Fig. 33 illustrates a typical example from simulations of the dissolved oxygen profiles as a function of agitation speed.

The stage model of BAJPAI and SOHN (1987)

In this model the concept of circulation time distribution has been extended to multiple impellers. As schematically illustrated for a two-impeller system in Fig. 34, a single circulation loop has been considered for each impeller. Additionally, the middle tank is assumed to be involved in inter-impeller exchange. The model is capable of predicting the observed tracer responses in multiple impeller systems as long as the measurements are made at an impeller other than the one in which tracer is injected.

The multi-compartment model of
SINGH et al. (1986, 1987, 1988)

As schematically illustrated in Fig. 35, the model structure consists of a series of well-mixed compartments with recycle and back-flow. The first important model parameter is the liquid circulation flow rate, which is alternatively estimated from the pumping capacity or predicted from measured pH transients (SINGH et al., 1986). The second parameter is the number of stagnant compartments for each impeller stage, which is assumed to be a function of the rheological properties of the broth and the level of turbulence. Dissolved

oxygen profiles are predicted from the numerical solution of the coupled system of oxygen balances for the compartments. As suggested by OOSTERHUIS and KOSSEN (1984), the $k_L a$ value for the impeller compartments is estimated from Eq. (48) and the $k_L a$ value for the stagnant compartments from Eq. (49). The volume of the impeller compartments is also predicted in the manner of OOSTERHUIS and KOSSEN (1984). The approach is based upon the radial velocity profile data for a standard turbine impeller presented by COSTES and COUDEREC (1982), which can be fitted to the equation

$$\frac{v_R}{\pi N d_i} = 0.7 \exp\left(-12.16 \frac{z}{d_i}\right) \tag{50}$$

where z is the coordinate in the axial direction. The shear rate is then given by

$$\frac{d v_R}{d z} = -26.74 \, N \exp\left(-12.16 \frac{z}{d_i}\right) \tag{51}$$

For low viscosity systems the shear rate is close to zero at a distance of $z/d_i = 0.5$. This result is then used to predict the size of the impeller region from a torus geometry schematically illustrated in Fig. 36. Next, it is assumed that all the compartments are of equal volume. Thus,

$$n_C = \frac{V_L}{n_i V_i} - 1 \tag{52}$$

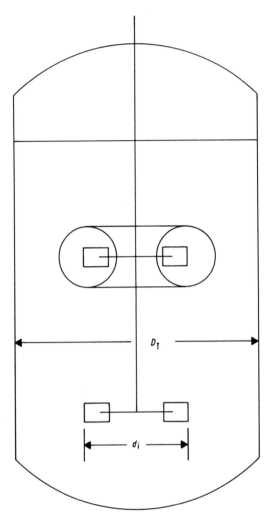

Fig. 36. Torus zone around the impeller.

In the case of highly viscous non-Newtonian fluids, the criterion is introduced that beyond a critical shear rate (2.5 s^{-1}) the pseudoplastic flow behavior can be approximated by a Newtonian viscosity. The size of the impeller region is then predicted from calculating the axial distance z that corresponds to this critical shear rate.

A typical result from model simulations is illustrated in Fig. 37. Oxygen profiles are shown for the non-Newtonian *Penicillium chrysogenum* broth in a 40 m³ fermentor.

The multiple multi-phase compartment model of RAGOT and REUSS (1990)

There are two serious limitations to the application of the compartment models discussed so far. First of all, these models do not account for the influence of backmixing of the gas phase. Secondly, in most of the models the number of compartments is due to the model structure related to the intensity of backmixing or turbulence intensity. When attempting to simulate the dynamic behavior of a microorganism moving through different regions of a bioreactor, serious problems may arise if the compartment structure results in discontinuities in the extracellular concentrations. These discontinuities may cause system-specific dynamic responses of the organisms quite different from those in a structure with smoother changes. Thus, if a structured model for the dynamic behavior of a microorganism is used, which is based upon the intrinsic intracellular enzyme kinetics, the number of compartments will have a profound effect on the dynamics of the entire system (reactor and process). An important criterion for the design of an appropriate model is, therefore, to tackle the problem of the number of compartments in a similar way as, for example, choosing the step change in a numerical integration procedure.

The following approach introduces a new concept that incorporates the mixing of both phases, gas and liquid, as well as mass transfer and kinetics. The model is based upon an appropriate aggregation of well-mixed multi-phase compartments. Fig. 38 schematically illustrates a single multi-phase compartment which consists of gas-, liquid-, and biophase. For simplicity the following discussion is restricted to two-phase compartments. As a consequence, the biophase is taken into account as an unstructured sink for material such as oxygen in the liquid phase. The ratio of gas and liquid fraction in a single compartment is determined by the specific gas holdup:

$$\alpha_G = \frac{V_G}{V_G + V_L} \tag{53}$$

Aggregation is performed by connecting the gas and liquid fractions of the compartments through circulation streams and backflow. The

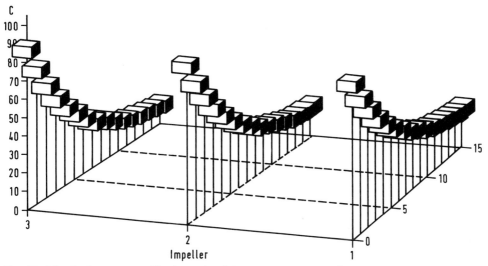

Fig. 37. Dissolved oxygen profiles in a 40 m³ fermentor (1.5 kW/m³ power input) (SINGH et al., 1987).

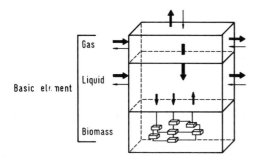

Fig. 38. Basic element for the three-phase multi-compartment model of RAGOT and REUSS (1990).

model structure can thus account for different mixing intensities of the gas and liquid phase in the tank, thereby using the same number of compartments for the two phases. Fig. 39 summarizes a few examples of structures for stirred tank reactors with different numbers of impellers. An important feature of the model is the fact that the number of compartments per impeller is now preset and does not depend upon the mixing intensity.

If we consider the case of oxygen consumption by suspended microorganisms in the liquid phase, the material balance equations for

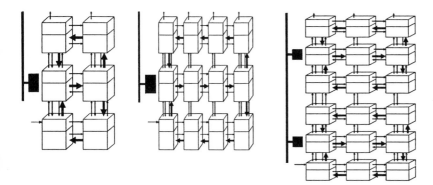

Fig. 39. Aggregation of two-phase basic elements to multi-compartment models for a stirred tank with single and double impeller.

oxygen in the liquid and gas phases are given by

$$
V_{L_{i,j}} \frac{dC_{L_{i,j}}}{dt} =
$$

$$
\dots \underbrace{-(1-\omega)fC_{L_{i,j+1}} + fC_{L_{i,j-1}}}_{\text{liquid circulation}} +
$$

$$
+ \underbrace{k_L a_j V_{L_{i,j}} \left(\frac{y_{G_{i,j}} P_i}{H'} - C_{L_{i,j}} \right)}_{\text{mass transfer}} +
$$

$$
+ \underbrace{r_{O_{2,ij}}(C_{X_{i,j}}, C_{S_{i,j}} \dots) V_{L_{i,j}}}_{\text{reaction}} \quad (54)
$$

$$
V_{G_{i,j}} \frac{dy_{G_{i,j}}}{dt} =
$$

$$
\dots \underbrace{+ \alpha \dot{V}_{G_{i,j-1}} y_{G_{i,j-1}} + \beta \dot{V}_{G_{i+1,j}} y_{G_{i+1,j}}}_{\text{gas circulation}} \dots -
$$

$$
\underbrace{- k_L a_j V_{L_{i,j}} \frac{RT}{P_i} \left(\frac{y_{G_{i,j}} P_i}{H'} - C_{L_{i,j}} \right)}_{\text{mass transfer}} \quad (55)
$$

where f and ω are the circulation and back-mixing parameters in the liquid phase, whereas

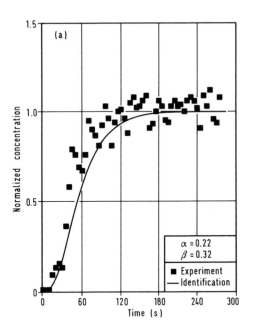

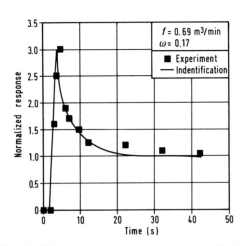

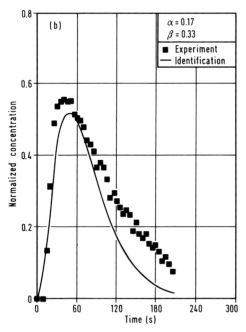

Fig. 40. Estimation of the mixing parameters for the liquid phase. Comparison between model predictions and experimental observations for the dynamic response to a heat pulse injection in a 150 liter tank (data from JURY et al., 1988).

Fig. 41. Estimation of mixing parameters for the gas phase from experimental observations of the dynamic response to a step change (methane) at the gas inlet in a 30 liter fermentor: (a) step change, (b) pulse injection.

α and β denote the corresponding properties of the gas phase.

The model can only be applied if, in addition to gas holdup and mass transfer coefficients, reliable data for f, ω, α, and β are available. In order to show the application of the model, the structure shown in Fig. 40 has been applied to the data of JURY et al. (1988) concerning the response to heat pulse injection in the liquid phase in a 100 liter vessel. The two parameters f and ω were identified by comparison of the dynamic response of the model and the measured data. The estimates were made with the help of Nelder and Mead's simplex algorithm. The two parameters characterizing the mixing of the gas phase can be estimated from the measured residence time distributions of the gas phase. Figs. 41a and b show results from such measurements using methane as a tracer. Measurements of the response to step change (a) and pulse injection (b) were performed in a 30 liter vessel with the aid of a flame ionization detector which was placed in the exhaust gas line. The two parameters α and β were again estimated from the solution of the material balance equations and application of Nelder and Mead's simplex algorithm.

Once the parameters of mixing in the gas and liquid phase have been identified, the combined model can be used to calculate the performance of the stirred tank as a bioreactor. As an example, this is performed in Fig. 42, showing the distribution of oxygen in the

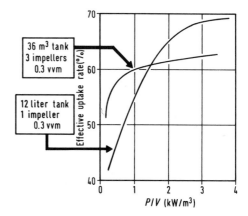

Fig. 43. Oxygen uptake as a function of power input. Comparison between the results for a well-mixed system and the two-phase multi-compartment model.

gas and liquid phases incorporating a Monod type of kinetics for oxygen consumption. It is easy to see that the influence of the two additional parameters, backmixing of the gas phase and hydrostatic pressure, become all the more important with increasing size of operation. One important consequence of these distributions of oxygen in the tank is the fact that classical scale-up rules like $P/V =$ idem may result in wrong conclusions. Fig. 43 shows an example in which the averaged oxygen uptake rate predicted from the distribution in Fig. 42 is plotted against power input and compared with the results from a well-mixed system.

2.6 Turbulence Models

The ultimate goal for modelling the behavior of stirred tank bioreactors is the prediction of the velocity, temperature, and concentration fields in the tank. Though far from such a comprehensive description based upon the solution of the complete set of balance equations for the multi-phase system, there are important first steps regarding numerical computations of multi-dimensional multi-phase flow which show the direction for future research in this field. In the following, only an introduction to this topic is attempted. The interested reader must be referred to the extensive special literature.

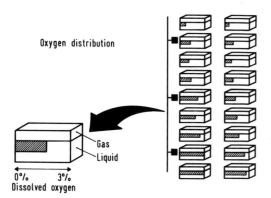

Fig. 42. Computed distribution of oxygen profiles in a 30 m³ tank. Oxygen consumption is predicted with the aid of Monod kinetics.

Predicting the turbulent flow generated by a turbine impeller in a standard baffled vessel is a necessary first step in progressing towards the calculation of the coupled transport phenomena in the reactor. Computation of the turbulent flow in the agitated vessel is most often based on the so-called $k - \varepsilon$ model. Here k stands for the turbulent kinetic energy and ε is its dissipation rate.

2.6.1 Governing Equations for Two-Phase Flow

The model discussed in detail elsewhere (ISSA and GOSMAN, 1981; HARVEY and GREAVES, 1982a, b; ELLUL et al., 1985; LAI and SALCUDEAN, 1987; NALLASAMY, 1987; TRÄGHÅRD, 1988) is merely quoted here. The governing equations will first be presented for the general situation of two-phase flow (gas and liquid). Results for special applications, including the case of single-phase flow, will be demonstrated afterwards.

The equations quoted below are based on the derivation by HARLOW and AMSDEN (1975) and have been summarized by LAI and SALCUDEAN (1987). Considering a cylindrical coordinate system with radial (r), circumferential (θ), and axial (z) distances, the velocity components of gas and liquid are v_G, v_L in the radial, w_G, w_L in the azimuthal, and u_G, u_L in the axial directions. The continuity equation and the momentum equation for the gas- and liquid phase are then given by:

Gas phase

Continuity

$$\frac{\partial(\alpha_G \rho_G u_G)}{\partial z} + \frac{1}{r} \frac{\partial(r \alpha_G \rho_G v_G)}{\partial r} +$$

$$+ \frac{1}{r} \frac{\partial(\alpha_G \rho_G w_G)}{\partial \theta} = 0 \tag{56}$$

Momentum

$$0 = -\alpha_G \frac{\partial P}{\partial z} + F_{G,z} + \alpha_G(\rho_G - \rho_L)g_z \tag{57}$$

$$0 = -\alpha_G \frac{\partial P}{\partial r} + F_{G,r} \tag{58}$$

$$0 = -\alpha_G \frac{1}{r} \frac{\partial P}{\partial \theta} + F_{G,\theta} \tag{59}$$

where g denotes the acceleration of gravity, $F_{G,j}$ is the j-component of the momentum exchange between the phases due to interphase drag, P is pressure, and α_G is the gas hold-up.

Liquid phase

Continuity

$$\frac{\partial(\alpha_L \rho_L u_L)}{\partial z} + \frac{1}{r} \frac{\partial(r \alpha_L \rho_L v_L)}{\partial r} +$$

$$+ \frac{1}{r} \frac{\partial(\alpha_L \rho_L w_L)}{\partial \theta} = 0 \tag{60}$$

Momentum

$$\frac{\partial[\alpha_L(\rho_L u_L \phi - \mu_{eff} \partial\phi/\partial z)]}{\partial z} +$$

$$+ \frac{1}{r} \frac{\partial[r\alpha_L(\rho_L v_L \phi - \mu_{eff} \partial\phi/\partial r)]}{\partial r} +$$

$$+ \frac{1}{r} \frac{\partial[\alpha_L(\rho_L w_L \phi - \mu_{eff} 1/r \partial\phi/\partial \theta]}{\partial \theta} = S_\phi \tag{61}$$

where Φ stands for u_L, v_L, and w_L, and S_ϕ is the source and sink term of the corresponding variable defined in Tab. 4.

The effective viscosity is predicted from the sum of molecular and turbulent transport coefficients:

$$\mu_{eff} = \mu + \mu_t \tag{62}$$

Furthermore, the following equalities must hold:

$$\alpha_G + \alpha_L = 1 \tag{63}$$

and

$$F_{G,j} = -F_{L,j} \quad (j = z, r, \theta) \tag{64}$$

The momentum exchange is assumed to be given by

Tab. 4. Source Terms in Liquid Phase Momentum Equations

ϕ	S_ϕ
u_L	$-\alpha_L \dfrac{\partial P}{\partial z} + F_{L,z}$
v_L	$-\alpha_L \dfrac{\partial P}{\partial r} + F_{L,r} + \alpha_L \left\{ \rho_L \dfrac{w_L^2}{r} - \mu_{\text{eff}} \dfrac{1}{r^2} \left(v_L + 2 \dfrac{\partial w_L}{\partial \theta} \right) \right\}$
w_L	$-\alpha_L \dfrac{1}{r} \dfrac{\partial P}{\partial \theta} + F_{L,\theta} - \alpha_L \left\{ \dfrac{1}{r} \rho_L v_L w_L + \mu_{\text{eff}} \dfrac{1}{r^2} \left(w_L - 2 \dfrac{\partial v_L}{\partial \theta} \right) \right\}$

$$F_{G,j} = f_j (V_{L,j} - V_{G,j}) \tag{65}$$

with the interphase friction factor f_j.

If the gas phase is modelled as spherical bubbles, the interface friction factor can be predicted from

$$f_j = \frac{3}{8} \rho_L \alpha_G C_D \frac{U_j}{R} \tag{66}$$

where R is the bubble radius, U_j the slip velocity, and C_D the drag coefficient:

$$C_D = \frac{48}{Re_j} \tag{67}$$

$$Re_j = 2RU_j \frac{\rho_L}{\mu_L} \tag{68}$$

According to the Prandtl–Kolmogoroff model the turbulent viscosity is related to the kinetic energy k and its dissipation rate ε by

$$\mu_t = C_\mu \rho_L \frac{k^2}{\varepsilon} \tag{69}$$

The quantities k and ε are taken to obey the following transport equations

$$\frac{\partial [\alpha_L (\rho_L u_L \phi - \Gamma_{\phi,\text{eff}} \partial\phi/\partial z)]}{\partial z} +$$

$$+ \frac{1}{r} \frac{\partial [r \alpha_L (\rho_L v_L \phi - \Gamma_{\phi,\text{eff}} \partial\phi/\partial r)]}{\partial r} +$$

$$+ \frac{1}{r} \frac{\partial [\alpha_L (\rho_L \omega_L \phi - \Gamma_{\phi,\text{eff}} 1/r \, \partial\phi/\partial\theta)]}{\partial\theta} =$$

$$= \alpha_L S_\phi \tag{70}$$

where

$$\Gamma_{\phi,\text{eff}} = \frac{\mu_{\text{eff}}}{\sigma_\theta} \tag{71}$$

The source term S_ϕ denotes the generation of turbulent properties and is given by

$$S_k = G - \rho\varepsilon \tag{72}$$

$$S_\varepsilon = C_G \frac{G\varepsilon}{k} - C_L \frac{\rho\varepsilon^2}{k} \tag{73}$$

with

$$\begin{aligned}
G = \mu_t \Bigg\{ & \left(\frac{\partial u_L}{\partial r} + \frac{\partial v_L}{\partial z} \right)^2 + \\
& + \left(\frac{\partial w_L}{\partial z} + \frac{1}{r} \frac{\partial u_L}{\partial \theta} \right)^2 + \\
& + \left[\frac{1}{r} \frac{\partial v_L}{\partial \theta} + r \frac{\partial (w_L/r)}{\partial r} \right]^2 + \\
& + 2 \left[\left(\frac{\partial u_L}{\partial z} \right)^2 + \left(\frac{\partial v_L}{\partial r} \right)^2 + \frac{1}{r^2} \left(r + \frac{\partial w_L}{\partial \theta} \right)^2 \right] \Bigg\}
\end{aligned} \tag{74}$$

The values for the different constants are summarized in Tab. 5.

Numerical solutions of the system of coupled balance equations have been presented

Tab. 5. Values of Constants in the $k - \varepsilon$ Turbulence Model

C_G	C_L	C_μ	σ_k	σ_ε
1.44	1.92	0.09	1.00	1.30

for one- and two-phase flow. HARVEY and GREAVES (1982a, b) predicted turbulent single-phase flow in an agitated vessel by application of the $k-\varepsilon$ model. The computations, however, rely on a rather unrealistic boundary condition for the turbulence parameters in the impeller region. The adopted approach assumes that the gradients of mean velocity components determine the level of turbulence in the vicinity of the impeller. This inadequate oversimplification of the flow situation in the impeller region has been removed in the work of PLACEK et al. (1986). These authors extend the $k-\varepsilon$ model through an incorporation of different turbulent energy scales. Thus, a multiple-scale model is created in which the energy spectrum of turbulence in the vessel is divided into the large-scale vortices produced by the impeller, the intermediate or transfer eddies, and the dissipation eddies. Furthermore, the boundary conditions in the impeller region rely on a discharge flow model suggested by PLACEK and TAVLARIDES (1985), which includes the trailing vortices at the impeller blades observed by VAN'T RIET and SMITH (1975).

ISSA and GOSMAN (1981) tried to predict the three-dimensional turbulent two-phase flow in an agitated and aerated vessel. Again, the most critical point seems to be the boundary condition in the impeller region. The empirically described flow in this region makes the predicted results uncertain. With the aid of a superposition of oxygen transfer from the gas bubbles into the liquid as well as a Monod-like expression for oxygen consumption, TRÄGHÅRD (1988) presented the numerical solution of the equations of motion together with the $k-\varepsilon$ turbulence model for the two-phase flow. As mentioned by the author, the treatment of the gas bubbles as well as the flow conditions in the impeller region must be considered as weaknesses in this model. The most difficult problem and also the most important task for the future is the comprehensive mathematical treatment of the disruption and coalescence of the gas bubbles, including the existence of a size distribution. The other aspect, hitherto not thoroughly investigated, is the analysis of the relative motion of the gas bubbles in the turbulent flow.

It may be concluded that the investigations of the two-phase modelling of turbulence in agitated and aerated vessels is only now beginning. The task is not an easy one, and it seems too early to speculate on coupling the turbulent transport equations with the microbial reaction model which must be structured in order to account for the behavior in a continuously changing environment. Nevertheless, results from models for turbulent two-phase flow, as far as the distribution of kinetic energy and dissipation in the tank is concerned, could be used immediately to improve confidence in the multi-phase compartment models described in the previous section. This strategy would result in a simpler model structure for the abiotic phases with distributed parameters predicted from the two-phase turbulent model. This kind of model reduction is presently a prerequisite to couple more complex metabolic models to the dynamic response of microorganisms to changing environmental conditions.

3 Applications to Microbial Systems

Through the application of regime analysis (SWEERE et al., 1987 and Tab. 2), and also from the concept of relaxation times (ROELS, 1983), it becomes clear that all the processes with time constants for reactions comparable or smaller than those for mixing have the potential for interactions between mass and energy distribution and kinetics. Aerobic processes, in which oxygen must be continuously delivered to the liquid phase via aeration (introduction of mass) and agitation (energy) to meet large oxygen demands of microbial cells, are one category of such systems. A second class of potentially interesting microbial processes are the ones in which manipulations of the extracellular environment are used to direct metabolism in a specific direction (GRIOT et al., 1986). In such cases, the demands of the critical nutrients change during the process. Their supply, therefore, must also be changed appropriately. At smaller scales of operation in which mixing is normally not a problem (for exceptions see HANSFORD and HUMPHREY,

1966; KNOEPFEL, 1972; EINSELE et al., 1978), purely kinetic considerations govern the process. As the scale of operation increases, mixing patterns must also be considered. This whole spectrum of problems has been an area of vigorous research and developmental activity in biochemical engineering.

Applications of models in some simple situations will be discussed in the next sections to demonstrate quantitative investigations of several important operational and scale effects.

3.1 Oxygen Transfer to Molds

Based upon a Sherwood number of 2 for diffusion from bulk to a spherical particle having no relative motion, CALDERBANK (1967) has shown that for unicellular microorganisms as single cells (i.e., not as flocs or films), no external diffusional limitations for the uptake of nutrients should exist. When, however, they do form flocs (clusters of single cells) or films, diffusion of nutrients in these may become limiting under suitable circumstances (low bulk concentrations, large uptake rates, large particle diameter, etc.). ATKINSON (1974) has covered these phenomena in detail. For consideration of these effects, the text of BAILEY and OLLIS (1986) can be also used.

3.1.1 Problem of Energy Distribution

For low viscosity biosuspensions having low cell densities, diffusion may not play an important role. However, in an agitated vessel where fluid continuously recirculates through a turbulent zone, a minimum eddy size exists under all hydrodynamic conditions. Often this eddy size is considerably larger than the size of any single cell, and all the transfers over distances less than the size of the eddy take place via molecular diffusion (BRYANT, 1977; BIRYOKOV, 1983). BOLZERN and BOURNE (1983) have also reported the influence of viscosity upon product distribution of a fast chemical reaction in a hypothetical sphere of the size of a turbulent microscale (Kolmogoroff length). Diffusion in this eddy may be important for

high cell densities in which oxygen consumption rates will be high. For fermentations using molds existing in pellet or filamentous form, diffusional limitations are known to exist even at lower cell densities.

Diffusion of oxygen in pellets has been extensively investigated (YANO et al., 1961; YOSHIDA, 1976; AIBA et al., 1971; KOBAYASHI et al., 1973; ATKINSON, 1974; METZ, 1976; MIURA, 1976; REUSS, 1976; VAN SUIJDAM, 1980). Various asymptotic and numerical solutions to the governing equation

$$\frac{1}{r^2} \frac{d}{dr} \left(r^2 D \frac{dC_L}{dr} \right) - r_{O_2}^{max} \frac{C_L}{K_M + C_L} X = 0 \quad (75)$$

involving molecular diffusion of oxygen and its consumption under pseudo-steady-state conditions have been presented in the literature. This analysis was extended by REUSS (1976) to include the various external transport resistances in which the boundary condition is written as

$$-D \left(\frac{dC_L}{dr} \right)_{r=R} = \tilde{\beta}(C_L^* - C_L|_{r=R}) \quad (76)$$

$\tilde{\beta}$, here, is the overall mass transfer coefficient

$$\frac{1}{\tilde{\beta}} = \frac{\varepsilon_p}{k_L a} + \frac{1}{K_p a_p} \quad (77)$$

Numerical solutions for this case are shown in Fig. 44 in the form of a dimensionless oxygen consumption rate as a function of the Biot number for mass transfer with the Damkoehler number as a parameter. Since the Biot number now includes the gas–liquid as well as the liquid–solid mass transfer resistances, one can calculate an effective oxygen consumption rate for a given set of operating conditions in the bioreactor and a specified pellet diameter. For calculations of different transport coefficients, one should refer to the comprehensive review by HENZLER (1982). It should be noted that Eq. (70) has also been used to discuss the effects of diffusion in eddies (BRYANT, 1977; BIRYOKOV, 1983), where the diffusion radius is governed by the energy input and broth viscosity.

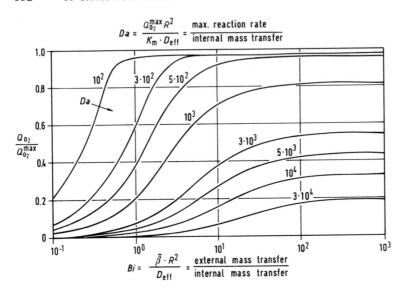

Fig. 44. Effect of external and internal mass transfer limitations on effective oxygen consumption of spherical pellets.

For filamentous morphology, few quantitative data are available dealing with transport resistances. Also the physical nature of the suspension is somewhat unclear (METZ et al., 1979). Several experimental observations, however, show changes in the critical dissolved oxygen concentration for growth as well as for product formation during the course of the fermentation (STEEL and MAXON, 1962, 1966; WANG and FEWKES, 1977; FOX, 1978). Particularly, STEEL and MAXON (1966) as well as WANG and FEWKES (1977) found a strong influence of the impeller to tank-diameter ratio. The latter authors postulated that the mass transfer of oxygen from bulk liquid to the mycelial surface was the controlling factor.

This problem has been quantitatively analyzed by REUSS et al. (1982) with the experimental data for oxygen uptake in *Aspergillus niger* broths. The mycelial mass has been considered to consist of hypothetical spheres of diameter d_p, in which biomass is uniformly distributed. Assuming that the diffusion of oxygen in these hypothetical spheres is the critical transport phenomenon, the known solutions of Eq. (75) could be used. ATKINSON (1974) has provided a pseudo-analytical solution of this problem as

$$Q_{O_2} = \eta\, r_{O_2}^{max} \frac{C_L}{K_M + C_L} X \qquad (78)$$

where η is the effectiveness factor. The effectiveness factor is related to diffusion and reaction parameters as follows:

$$\eta = 1 - \frac{\tanh \phi}{\phi}\left(\frac{\psi}{\tanh \psi} - 1\right) \text{ for } \psi \le 1 \qquad (79)$$

and

$$\eta = \frac{1}{\psi} - \frac{\tanh \phi}{\phi}\left(\frac{1}{\tanh \psi} - 1\right) \text{ for } \psi \ge 1 \qquad (80)$$

where Φ is known as the Thiele modulus, defined as

$$\phi = \frac{\alpha}{3}\frac{d_p}{2}\sqrt{\frac{r_{O_2}^{max} X}{K_M D_{O_2}}} \qquad (81)$$

and ψ as a "general modulus" related to the Thiele modulus and other indices as

$$\psi = \frac{\phi}{\sqrt{2}}\frac{C_L}{K_M + C_L} \cdot$$
$$\cdot \sqrt{\frac{1}{\dfrac{C_L}{K_M} - \ln\left(1 + \dfrac{C_L}{K_M}\right)}} \qquad (82)$$

α is a geometrical parameter whose value for spherical particles was suggested by ATKINSON (1974) as 1.16.

In analogy to turbulent eddies, the sizes of these diffusion elements are governed by a balance of shearing forces and the forces necessary to break mycelial filaments. Through comparisons between suitably designed oxygen uptake measurements and the Eqs. (78) through (82), the sizes of the elements were found to pertain to the inertial subrange of turbulent eddies. In the absence of coalescence of the spherical elements, their sizes are governed by the maximum shearing forces encountered in the vessel, and the following expression for d_p can be obtained:

$$d_p \sim \varepsilon_{max}^{-0.25} \qquad (83)$$

Similar expressions have been reported by METZ (1976) for mycelial pellets and by PARKER et al. (1972) for floc sizes in waste treatment. The maximum energy dissipation rate appearing in Eq. (83) may be replaced by the power input, ε, and the geometric parameters using the following expression

$$\varepsilon_{max} = 0.5\,\bar{\varepsilon} \left(\frac{D_T}{d_i}\right)^3 \qquad (84)$$

obtained from measurements of local energy dissipation in stirred reactors having turbine impellers (LIEPE et al., 1971). Accordingly,

$$d_p = K (0.5)^{-0.25}\, \bar{\varepsilon}^{-0.25} \left(\frac{D_T}{d_i}\right)^{-0.75} \qquad (85)$$

The constant K is related to the strength of mycelial filaments.

BERKE (1980) and RIZZI (1982) measured the oxygen uptake kinetics at different power input levels and impeller to tank diameter ratios. Thiele modulus values calculated from these experiments are plotted in Fig. 45, showing the validity of Eq. (85). The oxygen uptake data obtained with different values of (d_i/D_T) are shown in Fig. 46 as a function of power input. Also plotted are the results of simulations using Eqs. (78) through (85) with $K=8.8\cdot10^{-4}$. The close agreement confirms the proposed influence of geometric parameters upon uptake kinetics through their effect upon local energy dissipation in the impeller region. A large number of experimental data (BERKE, 1980; STEEL and MAXON, 1962,

1966; WANG and FEWKES, 1977; RIZZI, 1982) suggest that for the same energy input smaller impellers are more effective than larger ones in supplying oxygen to filamentous organisms.

3.1.2 Mass and Energy Distribution

STEEL and MAXON (1966), studying the scale-up of the novobiocin fermentation, observed that the pronounced influence of the

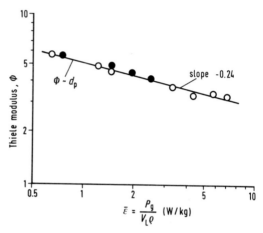

Fig. 45. Thiele modulus as a function of the energy dissipation rate for filamentous morphology (data from batch fermentation with *Aspergillus niger*).

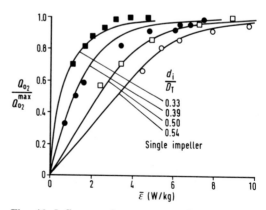

Fig. 46. Influence of mean energy dissipation rate and ratio of impeller-tank diameters on the effective oxygen consumption rate of *Aspergillus niger* growing with filamentous morphology (symbols: measured data; solid lines: predictions).

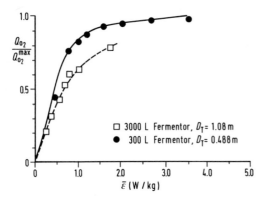

Fig. 47. Influence of scale on the dependency between oxygen uptake rate and mean energy dissipation rate (measurements during growth of *Aspergillus niger*).

impeller to tank-diameter ratio in small fermentors could not be verified in a larger tank having 15 m³ volume when the power/volume ratio was kept constant. Also, the measured specific oxygen uptake rates in the larger vessel were lower than those in the smaller unit for the whole range of impeller diameters. The scale effects can also be clearly seen from Fig. 47, wherein experimental measurements of oxygen uptake rates in *Aspergillus niger* broths in 300 and 3000 liter reactors have been reported (RIZZI, 1982). The experiments were performed by transferring part of the broth from the 3000 liter scale to a 300 liter vessel in order to keep all the morphological and physicochemical conditions identical in both reactors. Obviously, the differences in the uptake rates must be attributed to mixing performances of the reactors.

In the light of the previous discussions in this chapter, a phenomenological picture as shown in Fig. 48 may be proposed for any scale of operations with significant circulation times (REUSS, 1983). In accordance with many experimental observations in highly viscous non-Newtonian systems, it may be assumed that gas–liquid mass transfer takes place only in the impeller region. It is here that the diffusion elements, whose diameters are determined by the maximum energy dissipation rate, Eq. (85), are loaded with oxygen. These loaded elements circulate through the system according to the circulation time distribution until they

return again to the impeller region. During these circulations, progressively decreasing turbulence is encountered as the distance from the impeller increases; hence, some flocculation may take place, as it is a fast process (METZ et al., 1979). Yet the nature of filaments would prevent their coalescence, and for the interior of elements a case of complete segregation may be justified. In the context of the micromixer–macromixer model suggested in Fig. 16, the governing equation for change of dissolved oxygen concentrations in the recirculating fluid elements can be written as

$$\frac{dC_L}{dt} = -\eta r_{O_2}^{max} \frac{C_L}{K_M + C_L} X \tag{86}$$

where t represents time during a circulation at $t = 0$, $C_L = C_L^0$ (the concentration of oxygen in loaded elements entering the macromixer). Due to the assumption of complete micromixing in the micromixer, the concentration of

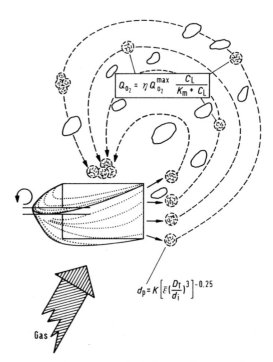

Fig. 48. Phenomenological picture of mass and energy distribution for oxygen supply of highly viscous biosuspensions with oxygen diffusion limitations.

dissolved oxygen in elements entering the macromixer would be uniform and equal to C_L^0.

Here η is the effectiveness factor, accounting for diffusional limitations in diffusion elements of size d_p. η and d_p are related to operating and environmental conditions by Eqs. (78) through (83) and Eq. (85). The constant can be determined by measuring the oxygen uptake kinetics (Q_{O_2} vs. C_L in a small fermenter) through changes in power input and using an optimization procedure to minimize differences in Q_{O_2} observed and Q_{O_2} calculated with Eq. (78).

For a given circulation time distribution, the concentration of oxygen in the exit stream from the macromixer is given by

$$\bar{C}_L = \int_0^\infty C_L(t)f(t)\,dt = \sum_{i=1}^N C_L(t_i)E_i \quad (87)$$

Loading of elements in the micromixer can be described by taking the oxygen balance around it (Fig. 49) as

$$V k_L a(C_L^* - C_L^0) = Q(C_L^0 - \bar{C}_L) = \dot{V}_G(y^\alpha - y) \quad (88)$$

Here Q is the circulation rate ($= V/\theta$), and C_L^* is the solubility of dissolved oxygen. This may be related to y and y^α by using an appropriate assumption concerning gas residence time-distribution in the reactor. If the gas phase is considered well-mixed, then

$$C_L^* = \frac{y}{H}p\big|_{\text{impeller}} \quad (89)$$

Eqs. (86) through (89) can be iteratively solved to predict the influence of the impeller to tank

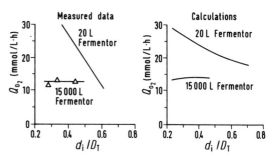

Fig. 50. Comparison of experimental observations (STEEL and MAXON, 1966) and model predictions for oxygen supply in a fermentation broth of *Streptomyces niveus*.

diameter ratio (d_i/D_T) at fixed power input (ε) at any scale. These simulation results show a behavior qualitatively similar to that observed by STEEL and MAXON (1966), Fig. 50. In these simulations, θ values were calculated from a correlation given by MIDDLETON (1979) and $k_L a$ values using a correlation by REUSS (1983). No attempt was made to obtain a quantitative fit due to the diversity of sources of parameter values. The qualitative agreement suggests that the pronounced effects of the scale of operation may be explained by a coupled effect of circulation and diffusion. Although the smaller impeller in the large tank still produces smaller diffusion elements due to a higher maximum energy dissipation rate, these elements stay away from the impeller (loading zone) for a longer time due to reduced pumping capacity. As a result, diffusion and circulation effects are antagonistic in nature and tend to annul each other as the (d_i/D_T) ratio is changed. At the same time, the speed of agitation in the larger vessel is low compared to that in a smaller vessel (constant $\bar{\varepsilon}$), resulting in lower $k_L a$ values. This causes a net reduction in oxygen transfer to cells in the large vessels.

For low-viscosity fermentation broths, the assumption of no mass transfer in the macromixer may not be correct. In this case, Eq. (86) will be modified to include an oxygen supply term involving a gas-liquid mass transfer coefficient in the macromixer. As suggested by OOSTERHUIS (1984) and OOSTERHUIS and KOSSEN (1984), the flow phenomena

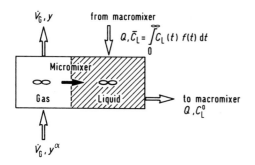

Fig. 49. Oxygen balance around the micromixer.

in the regions away from the impeller resemble those in bubble columns, and one could use gas–liquid mass transfer correlations for bubble columns. A similar proposal was made by ANDREW (1982), SINGH et al. (1986, 1987, 1988) and RAGOT and REUSS (1990).

3.1.3 Multiple-Impeller Systems

Even though multiple impellers are commonly used in laboratory and industrial bioreactors, precious little work dealing with them has been published, at least in the open literature. With regard to mixing and mass transfer, it is clear that when the circulation loops become too large, one should introduce more impellers on the shaft. Some quantitative information as to when exactly it should be done and how can be obtained through the application of the concept here proposed for mass and energy distribution. The information concerning circulation time distributions in each impeller zone, exchanges between the impellers, and power input and oxygen transfer at each impeller will be of crucial importance. Because of little information available concerning the last two, we may assume the im-

pellers to be independent of each other and to have similar gas–liquid mass transfer coefficients.

Considering a simple form of exchange, a micro-macromixing framework for two impellers of Fig. 51 may be used. For the sake of simplicity, the two impellers may be considered identical and to possess the same recirculation flow as well as circulation time distributions. These assumptions may be somewhat drastic in the light of data presented earlier (see Sect. 2.4). As a result of the equal circulation rate, however, the exchange coefficient, β, will also be the same for both impellers. If we used the superscripts "upper" and "lower" to distinguish between concentrations at the two impellers (after exchange) these will be given by

$$C_L^{upper} = (1 - \beta)\bar{C}_L^{upper} + \beta\bar{C}_L^{lower}$$
$$C_L^{lower} = \beta\bar{C}_L^{upper} + (1 - \beta)\bar{C}_L^{lower} \qquad (90)$$

where C_L denotes the exit stream concentrations as per Eq. (87). The balances at the micromixer can be taken in the manner shown by Fig. 51 as

$$\frac{V}{2} k_L a(C_L^{*\ upper} - C_L^{0\ upper}) =$$
$$= Q(C_L^{0\ upper} - \bar{C}_L^{upper}) =$$
$$= \dot{V}_G(y_{in}^{upper} - y_{out}^{upper})$$

$$\frac{V}{2} k_L a(C_L^{*\ lower} - C_L^{0\ lower}) =$$
$$= Q(C_L^{0\ lower} - \bar{C}_L^{lower}) =$$
$$= \dot{V}_G(y_{in}^{lower} - y_{out}^{lower}) \qquad (91)$$

In this case,

$$y_{out}^{lower} = y_{in}^{upper} = H \frac{C_L^{*\ lower}}{p^{lower}} \qquad (92)$$

and

$$y_{out}^{upper} = H \frac{C_L^{*\ upper}}{p^{upper}} \qquad (93)$$

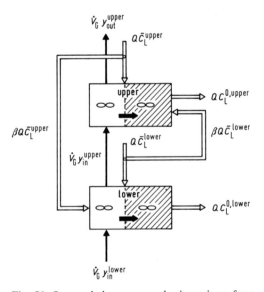

Fig. 51. Oxygen balances around micromixers for a two-impeller system.

Here Q is the circulation rate given by $V/(2\theta)$.

Eq. (91) accounts for possible hydrostatic pressure effects that may be encountered in industrial-scale reactors.

Eqs. (90) through (93) were solved with Eqs. (78) through (82), and the calculated oxygen uptake rates in filamentous broths (complete segregation in the macromixer) are plotted against power input levels using a single impeller in the same configuration in Fig. 52.

As a result, a net reduction in oxygen transfer is predicted in spite of some reduction in the mean circulation time. RIZZI (1982) conducted several experiments in which *Aspergillus niger* broths from a 300 liter fermenter were transferred into two identical 80 liter vessels having one and two impellers, respectively. A typical result comparing the two systems is shown in Fig. 53, which confirms the same trend as the theoretical predictions.

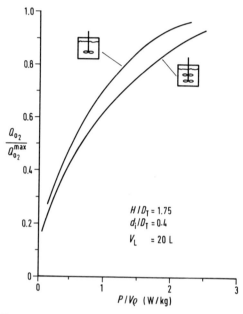

Fig. 52. Comparison of oxygen availability rates between single- and two-impeller systems (model prediction).

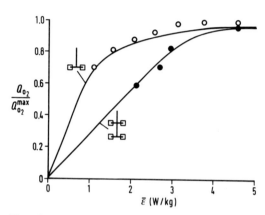

Fig. 53. Comparison of oxygen availability rates between single- and two-impeller systems (experimental) observations during growth of *Aspergillus niger*.

Surprisingly, the two-impeller system is predicted to deliver lower amounts of oxygen than a single impeller for the same $\bar{\varepsilon}$. A more careful consideration of different parameters reveals that the two-impeller system has

a) lower speeds of agitation, n (at constant $\bar{\varepsilon}$),
b) lower values of $k_L a$ (due to higher apparent viscosity in the non-Newtonian system), and
c) a larger diffusion element size, d_p than the single impeller system.

This effect of multiple impellers on oxygen transfer capacity in filamentous broths could be confirmed in all the experiments at this scale, which suggests the validity of the proposed theory. Unfortunately, at this scale of operation, diffusional effects dominate, and circulation does not play any significant role. As the vessel size increases, however, mixing would tend to be more important. Under these circumstances, there may be a critical size above which a two-impeller system would prove better than a single impeller. This, however, requires additional work and remains an unsolved problem. If such limits can be identified for specific kinetic features of a given system, a more rational basis for deciding the number of impellers at different scales of operation could be established.

3.1.4 The Role of Micromixing

While dealing with problems involving mixing and microbial reaction kinetics, mixing at the cellular and even molecular level in certain cases must also be considered. In the previous sections it was assumed that fluid and diffusion elements in their recirculation paths (in the macromixer) are in a state of complete segregation. BAJPAI and REUSS (1982b) have investigated the extreme cases of segregation in the macromixer for the kinetics of uptake of oxyen by filamentous microorganisms. The reaction kinetics for this system are described by Eqs. (78) through (82). The sizes of the diffusion elements depend upon the power input, ε, and the ratio of impeller diameter to tank diameter, d_i/D_T, as in Eq. (85). The schematics of simulations are presented in Fig. 54. Maximum mixedness was simulated using a stirred-tanks-in-series configuration. To be rigorous, this representation corresponds to the case of "sequential mixedness" (WEN and FAN, 1975), because molecular diffusion is not

possible within any two maximally-mixed stirred tanks, whereas in "maximum mixedness" molecules having the same life expectancy from any part of the entire system should be able to mix infinitely fast. However, due to the recycle nature of systems considered here, it is most convenient to use the tanks-in-series configuration to represent the extreme of mixedness. All the tanks were considered to be maximally mixed and had equal volume. For the ith tank in the series, the governing equation for oxygen uptake is

$$\frac{d C_{L_i}}{d t} = \frac{NQ}{V}(C_{L_{i-1}} - C_{L_i}) -$$

$$- \eta r_{O_2}^{\max} \frac{C_{L_i}}{K_M + C_{L_i}} X \quad i = 1, \ldots, N \quad (94)$$

C_{L_o} is the concentration of dissolved oxygen in the micromixer (C_L^0 in Eq. (88)), and C_{L_N} is the concentration in the exit stream of the macromixer (C_L in Eq. (87)). For micromixer balance, Eq. (88) for a single-impeller system holds. For a pseudo-steady state of dissolved

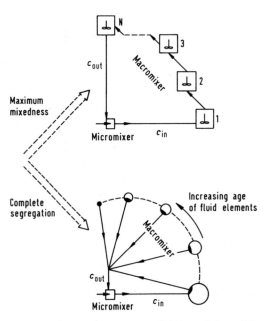

Fig. 54. Two-environment model for mixing with schematic representation of the two extremes of micromixing.

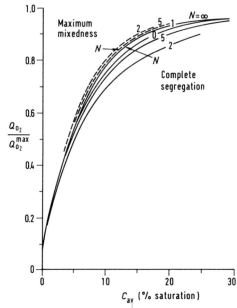

Fig. 55. Predicted oxygen uptake kinetics for different reactor circulation time distributions corresponding to a mean circulation time (0) of 10 s.

oxygen at a given biomass concentration X $(N+1)$, algebraic Eqs. (94) along with the micromixer balance Eq. (88) can be solved and the oxygen uptake rate can be calculated using an overall oxygen balance.

The case of complete segregation may be handled in the manner described in Sect. 2.1.2, and the results are presented in Fig. 55 for the case of a mean circulation time of 10 seconds. Similarly, somewhat more pronounced results were obtained at higher mean circulation times. In this case, the number of vessels-in-series does not appear to play a significant role for maximal mixing due to mass transfer limitations in the diffusion elements. For mycelial broths in which diffusional limitations are present, the assumption of complete segregation results in conservative predictions. Also the effects of the extremes of segregation would be within a few percent of each other for the experimentally useful ranges of mean circulation time. For low-viscosity Newtonian broths, however, the window is wider.

For the interpretation of the effect of micromixing, it is worthwhile to apply the estab-lished principles of chemical reaction engineering to the limiting cases of zero- and first-order kinetics. By averaging the concentrations and the reaction rate values over all the elements of corresponding reactors, one obtains the following relationships for the rates of overall reactions for stirred tank reactors (CSTR, $N=1$) and plug flow reactors (PFR, $N=\infty$):

1. for a first-order reaction, the overall reaction rate is the same for PFR as well as for extreme cases of CSTR.

$$(-r)_{av}|_{CSTR_{no\,segregation}} =$$
$$= (-r)_{av}|_{CSTR_{complete\,segregation}} =$$
$$= (-r)_{av}|_{PFR} = k\,C_{av} \qquad (95)$$

2. for a zero-order reaction, the overall reaction rate for the same average dissolved oxygen concentration shows the order:

$$(-r)_{av}|_{CSTR_{no\,segregation}} \geq (-r)_{av}|_{PFR} \geq$$
$$\geq (-r)_{av}|_{CSTR_{complete\,segregation}} \qquad (96)$$

For zero-order kinetics, such an analysis leads to Fig. 56, where the results for a PFR and a completely segregated CSTR are presented. For a completely mixed CSTR, the solution is trivial, i.e., the maximum rate holds at all $C_{av} > 0$. Comparison of Figs. 55 and 56 suggests that kinetic behavior in the circulation is dominated by a zero-order reaction. This is understandable due to the low K_M value for the oxygen uptake ($=1.28$ µmol/L), although it is corrupted by the first-order diffusive process. For maximum mixedness, both zero- and first-order kinetics contribute.

The results presented in Fig. 55 reveal that reactors with circulation time distributions resembling those of a PFR are better for naturally segregated systems such as highly viscous non-Newtonian fermentation broths. Also, the uncertainties of mixedness or the extent of segregation are far less important in such reactors. This points to the desirability of achieving narrow distributions of draft tubes or of using loop reactors. Such designs (HINES, 1978; BLENKE, 1979, 1985; FAUST and SITTIG, 1980) offer advantages for minimizing dead spaces, increasing reliability of scale-up, and

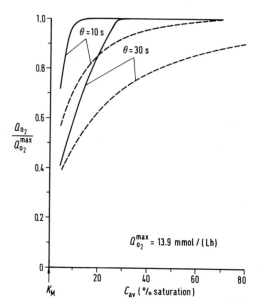

Fig. 56. Predicted kinetics for a zero-order reaction in a continuous stirred tank (----) and plug flow configurations (——) for two different mean residence times.

also providing operational benefits for the viscous non-Newtonian broths.

3.2 Substrate Distribution in Baker's Yeast Fermentation

The second application of the concept of coupling of mixing and microbial reactions presented in this chapter is concerned with problems of uniform distribution of the energy and carbon source in large-scale operations. Thus, referring to Tab. 2, we are now considering a class of processes in which the time constants of substrate consumption ($t_{sc} = S_0/r_S^{max}$) for zero-order reactions are of the same order of magnitude as the circulation times in the bioreactor. The most impressive and interesting example for this problem is the continuous SCP production on methanol as carbon and energy source in the ICI loop reactor at a scale of $2100 \, m^3$ (SENIOR and WINDNASS, 1980; SCOTT, 1983). Here the backmixing of the substrate originally resulted in a decreased biomass yield due to high local methanol concentrations. This problem was finally solved by multiple introduction of the substrate. Many other industrially important processes operated in a fed batch mode may be considered because of their low substrate concentrations in the reactor. Obviously, baker's yeast production belongs to this category of processes. In order to prevent ethanol production and, thus, yield reduction beyond a critical sugar concentration under aerobic conditions (Crabtree effect), sugar must be fed to the batch processes. One of the scale-up problems is then to get a uniform distribution of the concentrated sugar feed, because the relaxation time of the Crabtree effect is known to be extremely short (EINSELE et al., 1978).

3.2.1 Simulations of Fed-Batch Operations Based upon a Fixed Schedule for Sugar Feeding

The effects of mixing dynamics upon the yield and productivity of this process have been theoretically investigated by using the circulation model for flow of the biosuspension

in conjunction with the micro-macro-mixer model as described in Sect. 2.3 (BAJPAI and REUSS, 1982a). For studying the effect of varying glucose concentration during circulation, it is only necessary to consider the microbial reaction in the vicinity of the critical sugar concentration. Thus, a rather simple unstructured empirical model was chosen for the purpose of this study. It involves Monod kinetics for growth

$$\mu = \frac{1}{X} \frac{dX}{dt} = \mu_{max} \frac{S}{K_S + S} \qquad (97)$$

and also a production rate of ethanol above the critical sugar concentration S_{crit}

$$v = \frac{1}{X} \frac{dP}{dt} = 0 \qquad \text{for } S \leq S_{crit}$$

$$= v_{max} \frac{S - S_{crit}}{K_P + (S - S_{crit})} \qquad \text{for } S > S_{crit} \quad (98)$$

Substrate consumption in the individual elements for the macromixer is given by

$$\frac{dS}{dt} = - \frac{1}{Y_{X/S}} \frac{dX}{dt} - \frac{1}{Y_{P/S}} \frac{dP}{dt} \qquad (99)$$

The model parameters were estimated from VON MEYENBURG's data for continuous culture of *Saccharomyces cerevisiae* (VON MEYENBURG, 1969) and presented in the original paper (BAJPAI and REUSS, 1982a). By integrating the balance equations for a short time period Δt in each of the volume elements present in the macromixer, the concentrations of the individual contributions to the recycle stream to the micromixer can be calculated. These concentrations are then averaged in the micromixer according to

$$S_{av} = \sum_{j=1}^{N} S_j E_j$$

$$X_{av} = \sum_{j=1}^{N} X_j E_j$$

$$P_{av} = \sum_{j=1}^{N} P_j E_j \qquad (100)$$

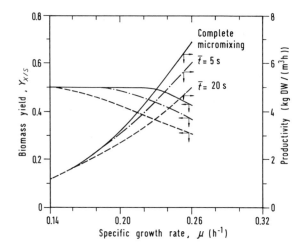

Fig. 57. Simulations of fed-batch production of baker's yeast on glucose. Glucose concentration in the feeding stream = 350 kg/m³.

in which E_j is again the fraction of the circulation times between $(j-1)\Delta t$ and $j\Delta t$ (Eq. (34)). The calculation for the next time interval is then repeated after affecting a balance across the micromixer. Here it is assumed that the concentrated sugar solution is fed to the impeller region. The calculated composition of the newly formed elements, which are the initial conditions for the first element in the macromixer for the next time interval, are then given by:

$$S = \frac{QS_{av} + FS_0}{Q+F}$$

$$X = \frac{QX_{av}}{Q+F}$$

$$P = \frac{QP_{av}}{Q+F} \tag{101}$$

where F and S denote the feeding rate and its concentration. Fig. 57 shows the results of simulations of fed-batch operations. The predicted integral productivities after 15 hours of fed-batch operation against specific growth rate are plotted in this figure. This specific growth rate may be considered as a set point for an open loop control based upon a fixed schedule for sugar feeding (Zulauf process). Complete micromixing was simulated by solving Eqs. (97) through (99) without circulation time distribution and it displays the upper lim-

its for the system. Due to the very high sugar concentration in the feed ($S_0 = 350$ kg/m³), the effect of circulation is also significant and results in a decrease in the critical growth rate at which ethanol appears in the system. These predictions are in qualitative agreement with those observed in industrial practice, where the exponential feeding program for sugar at the industrial scale usually differs from the optimal conditions predicted in the laboratory.

3.2.2 Influence of Mixing on Computer-Controlled Feeding

As far as the experimental verification of these model predictions is concerned, special emphasis is given to the computer-aided feed forward–feedback control of the baker's yeast production (REUSS and BRAMMER, 1985). This example also opens the door into another field of industrial importance for the application of the concept of coupling mixing and microbial reaction, namely the scale-up of control strategies for fermentation processes.

For the control of sugar feeding, the strategy suggested by WANG et al. (1979) was applied in a slightly simplified form. In this strategy the respiration coefficient (RQ) is used as a set point. Feedback control is implemented through a proportional controller which reduces the feeding rate, based upon the addi-

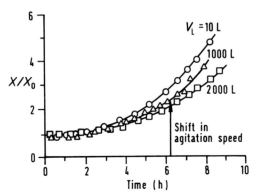

Fig. 58. Growth curves of *Saccharomyces cerevisiae* in computer-controlled fed-batch cultivations at different scales of operation.

tional carbon dioxide production due to ethanol formation at sugar levels beyond the critical concentration. Thus,

$$F = F_0 [1 - K'_P (Q_{CO_2} - R Q^{sp} Q_{O_2})] \tag{102}$$

The feed-forward part of the controller is calculated from

$$F_0 = \frac{\mu (X V)}{Y_{X/S} S_0} \tag{103}$$

where the specific growth rate μ and the biomass concentration X can be estimated on-line by making use of the elemental balances and the continuously measured oxygen uptake and CO_2 evolution rates. Fig. 58 summarizes the experimentally observed growth curves by applying this strategy to three different scales of operation (10, 1000, and 2000 liter working volumes). It must be emphasized that the same set point ($RQ = 1.05$) was chosen for the experiments at the different scales. As expected from the model simulations presented in the previous section for the open-loop feed-forward strategy (Fig. 57), the productivity at the larger scale is reduced because of larger circulation times. The observed increase in growth rate after making a shift in the speed of agitation in the 1000 liter fermentor (Fig. 58) is a further proof of this hypothesis. Since oxygen was always present in excess, increased oxygen transfer at higher speeds of agitation can be

ruled out as the cause of the increase in specific growth rate. Another experimental result, shown in Fig. 59, serves to strengthen these arguments. Here two experiments were conducted in an identical fashion except that one involved feeding the sugar solution into the impeller region, while the sugar in the other was fed from the top at the surface of the liquid. Several repeated experiments confirmed this observation, which is clearly the case in point.

A simplified model version was selected for the special purpose of applying the framework of coupling the intensity of substrate distribution and microbial reaction to the experimental conditions of the computer-controlled feeding, using a constant respiration coefficient.

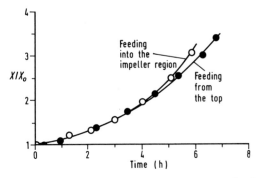

Fig. 59. Effect of the feeding point on growth of *Saccharomyces cerevisiae* in computer-controlled fed-batch cultivations (10 L scale).

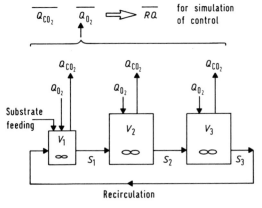

Fig. 60. Simplified model structure for recirculation flow in an agitated batch reactor.

The idea behind this model reduction is the long-term objective of these investigations. One of the problems of interest will be the simulation of the system dynamics for designing control strategies for large-scale operations. This application requires the non-stationary solution of the model, including the gas–liquid mass transfer. Obviously, the basic structure of the model should be simplified for this complex task.

The simplified model is shown in Fig. 60, where the entire vessel is divided into three CSTRs in series. The first CSTR (V_1), which accounts for a small fraction of the volume, is used to model the conditions in the vicinity of the injection ring of the concentrated sugar feed. It is assumed that the volume of this vessel is equal to the product of the cross-section of the reaction times and the width of the impeller blades. A similar approach has been suggested by MARINI et al. (1984) for studying the effects of mixing on the outcome of a polymerization reaction. The two other vessels account for the conditions in the largest part of the reaction volume. The number of two has been estimated from the measured recirculation time distributions (Sect. 2.2.1), according to

$$N = \theta^2/\sigma^2 \qquad (104)$$

Assuming a constant biomass concentration for a short time period of the process, the substrate balance equations in the three vessels can be written in the following form

$$\frac{dS_1}{dt} = \frac{FS_0}{V_1} + \frac{Q}{V_1}S_3 - \frac{Q+F}{V_1}S_1 -$$
$$- \frac{F}{V_{\text{total}}}S_1 - \mu_1\frac{X}{Y_{X/S}} - \nu_1\frac{X}{Y_{P/S}}$$

$$\frac{dS_2}{dt} = \frac{Q+F}{V_2}(S_1 - S_2) - \frac{F}{V_{\text{total}}}S_2 -$$
$$- \mu_2\frac{X}{Y_{X/S}} - \nu_2\frac{X}{Y_{P/S}}$$

$$\frac{dS_3}{dt} = \frac{Q+F}{V_3}(S_2 - S_3) - \frac{F}{V_{\text{total}}}S_3 -$$
$$- \mu_3\frac{X}{Y_{X/S}} - \nu_3\frac{X}{Y_{P/S}} \qquad (105)$$

with the recirculation stream

$$Q = \frac{V_{\text{total}}}{\theta} \qquad (106)$$

μ and ν represent the kinetics for growth, and ethanol formation rates are then calculated from

$$\mu_{\text{total}} = \frac{V_1\mu_1 + V_2\mu_2 + V_3\mu_3}{V_{\text{total}}} \qquad (107a)$$

and

$$\nu_{\text{total}} = \frac{V_1\nu_1 + V_2\nu_2 + V_3\nu_3}{V_{\text{total}}} \qquad (107b)$$

The entire oxygen consumption rate is then predicted with

$$\bar{Q}_{O_2} = \mu_{\text{total}} \frac{X}{Y_{X/O_2}} \qquad (108)$$

and the CO_2 evolution rate with

$$\bar{Q}_{CO_2} = R Q^{\text{sp}} \bar{Q}_{O_2} + \nu_{\text{total}}\frac{X}{Y_{CO_2/P}} \qquad (109)$$

where Y_{X/O_2} and $Y_{CO_2/P}$ are the yield coefficients in kg biomass per mol O_2 and mol CO_2 per kg ethanol, respectively.

For simulation of the entire growth curve, the system is considered in a pseudo-steady state for $\Delta t = 5$ min process time. Thus, the

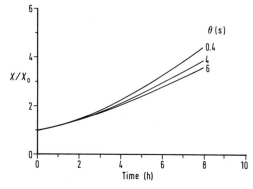

Fig. 61. Simulations of computer-controlled fed-batch cultivations at different mean circulation times θ.

system of Eq. (105) is iteratively solved with $dS/dt = 0$. The three substrate concentrations S_1, S_2, and S_3 are then used to predict the average growth and ethanol production rate, Eqs. (107a), (107b), oxygen consumption and carbon dioxide evolution rates, Eqs. (108), (109), and finally the respiration coefficient $RQ = Q_{CO_2}/Q_{O_2}$. With the aid of a search algorithm, the feeding rate $F_i(t_i)$ is calculated afterwards, which minimizes the deviation between actual RQ and the set point ($RQ^{sp} = 1.05$). Finally from

$$\frac{dX}{dt} = \mu_{total} - \frac{X F_i}{V_{total_i}} \tag{110}$$

and

$$V_{total_{i+1}} = V_{total_i} + F_i \Delta t \tag{111}$$

biomass concentrations and reaction volumes for the next 5 min time interval are calculated. The results of the simulations using the circulation times calculated from Eq. (20), and taking into account the effect of aeration (Fig. 15), are presented in Fig. 61. As observed in the experiments, the increase in circulation time results in a corresponding reduction of growth rate. Due to the reduced recirculation stream, the concentrated sugar solution is not diluted as much as for shorter circulation times, and local spots of high concentrations will appear in the first vessel of the scheme in Fig. 60. The control strategy, therefore, reduces the feeding rate as a result of which sugar concentrations in the larger part of the vessel are below the critical value to match the critical growth rate.

4 Conclusions

Problems related to the geometry and the scale of reactors may arise wherever transport processes are present along with the kinetics of the reaction system. The concept of coupling transport phenomena and reaction kinetics is commonly used to predict the observed reactor performances for situations involving defined hydrodynamics. Stirred bioreactors operated in the batch and fed-batch mode, however, defy such a treatment due to a lack of precise quantitative description of their hydrodynamics. The capability of characterizing fluid movement in stirred bioreactors with the help of circulation time distributions permits such a coupling. These distributions can be measured by using flow-followers or by injection of non-reaction tracers. The latter method appears to be particularly suitable for viscous non-Newtonian fermentation broths. With existing correlations for single-impeller systems, this measurement method can be utilized for precise estimation of mass exchange between impellers in a multi-impeller system. In viscous fermentation broths, distribution of energy introduced into the fluid also plays an important role.

In order to understand the influence of the geometry and scale of operation, the distributions of mass and energy should be coupled with kinetic processes in bioreactors.

5 References

AIBA, S., KOBAYSHI, K. (1971), Comments on oxygen transfer into mycelial pellets, *Biotechnol. Bioeng.* **13**, 583–588.

AIBA, S., HUMPHREY, A. E., MILLIS, N. F. (1973), *Biochemical Engineering*, 2nd Ed., Tokyo: Tokyo University Press.

ANDREW, S. P. S. (1982), Gas-liquid mass transfer in microbiological reactors, *Trans. Inst. Chem. Eng.* **60**, 3–13.

ATKINSON, B. (1974), *Biochemical Reactors*, London: Pion, Ltd.

BADER, F. G. (1987a), Modelling mass transfer and agitator performance in multiturbine fermentors, *Biotechnol. Bioeng.* **30**, 37–51.

BADER, F. G. (1987b), Improvements in multi-turbine mass transfer models, in: *Biotechnology Processes, Scale-up and Mixing* (HO, C. S., OLDSHUE, J. Y., Eds.), pp. 96–106, New York: American Institute of Chemical Engineers.

BAILEY, J. E., OLLIS, D. F. (1986), *Biochemical Engineering Fundamentals*, 2nd Ed., New York: McGraw Hill.

BAJPAI, R. K., REUSS, M. (1982a), Coupling of mixing and microbial kinetics for evaluation of the performance of bioreactors, *Can. J. Chem. Eng.* **60**, 384–392.

BAJPAI, R. K., REUSS, M. (1982b), Considerations of macromixing and micromixing in semi-batch stirred tank bioreactors, in: *Chemical Reaction Engineering* (WEI, J., GEORGAKIS, C., Eds.), pp. 555–565, Washington, D. C.: Boston ACS Symposium Series 196.

BAJPAI, R. K., SOHN, P. U. (1987), Stage models for mixing in stirred bioreactors, in: *Biotechnology Processes: Scale-up and Mixing* (HO, C. S., OLDSHUE, J. Y., Eds.), pp. 13–21, New York: AIChE Publ.

BAJPAI, R. K., RAMKRISHNA, D., PROKOP, A. (1977), Effect of drop interactions on the performance of hydrocarbon fermentors, *Biotechnol. Bioeng.* **19**, 1761–1772.

BERKE, W. (1980), Einfluß der Diffusion im Mycel auf die Sauerstoffkinetik von *Aspergillus niger, Diploma Thesis,* Technische Universität, Berlin.

BIRYOKOV, V. V. (1983), Computer control and optimization of microbial metabolite production, in: *Modelling and Control of Biochemical Processes* (HALME, A., Ed.), pp. 135–144, Helsinki: IFAC-Publication, Pergamon Press.

BLAKEBROUGH, N., SAMBAMURTHY, K. (1966), Mass transfer and mixing rates in fermentation vessels, *Biotechnol. Bioeng.* **8**, 25–42.

BLENKE, H. (1979), Loop reactors, *Adv. Biochem. Eng.* **13**, 121–214.

BLENKE, H. (1985), Biochemical loop reactors, in: *Biotechnology* (REHM, H.-J., REED, G., Eds.), Vol. 2, pp. 465–569, Weinheim–Deerfield Beach/Florida–Basel: VCH.

BOELCKE, C. (1983), Experimentelle Untersuchungen zur Zirkulationsverteilung der Flüssigphase in Rührreaktoren, *Diploma Thesis,* Technische Universität, Berlin.

BOLZERN, O., BOURNE, J. R. (1983), Mixing and fast chemical reaction – IV. Extension of the reaction zone, *Chem. Eng. Sci.* **38**, 999–1003.

BRYANT, J. (1977), The characterization of mixing in fermenters, *Adv. Biochem. Eng.* **5**, 101–123.

BRYANT, J., SADEGHZADEH (1979), Circulation rates in stirred and aerated tanks, in: *Proc. 3rd Eur. Conf. on Mixing,* 4–6th April, *York,* Vol. I, pp. 325–336.

CALDERBANK, P. H. (1967), Mass transfer in fermentation equipment, in: *Biochemical and Biological Engineering Sciences* (BLAKEBROUGH, N., Ed.), Vol. 1, New York: Academic Press.

COSTES, J., COUDEREC, J. P. (1982), Pumping capacity and turbulence intensity in baffled stirred tanks; influence of the size of the pilot unit, *Proc. 4th Eur. Conf. on Mixing,* 28–29th April, *Noordwijkerhout,* Paper B2, pp. 25–34.

CURL, R. L. (1963), Dispersed phase mixing: I. Theory and effects in simple reactors, *AIChE J.* **9**, 175–181.

CUTTER, L. A. (1966), Flow and turbulence in a stirred tank, *AIChE J.* **12**, 35–45.

EINSELE, A., FINN, R. K. (1980), Influence of gas flow rates and gas hold up on blending efficiency in stirred tanks, *Ind. Eng. Chem. Proc. Des. Dev.* **19**, 600–603.

EINSELE, A., RISTROK, D. L., HUMPHREY, A. E. (1978), Mixing times and glucose uptake measured with a fluorometer, *Biotechnol. Bioeng.* **20**, 1487–1492.

ELLUL, I. R., ISSA, R. I., LOONEY, M. K. (1985) Numerical computation of multi-dimensional multi-phase flow, in: *Numerical Methods in Laminar and Turbulent Flow, Proc. 4th Int. Conf., Swansea,* pp. 1218–1231.

FAUST, U., SITTIG, W. (1980), Methanol as carbon source for biomass production in a loop reactor, *Adv. Biotechnol. Eng.* **17**, 63–99.

FOX, R. I. (1978), The applicability of published scale-up criteria to commercial fermentation processes, *First Eur. Congr. Biotechnology, Interlaken,* Preprints Part 1, pp. 1/80–1/83.

GRIOT, M., MOES, J., HEINZLE, E., DUNN, I. J., BOURNE, J. R. (1986), A microbial culture for the measurement of macro and micro mixing phenomena in biological reactors, *Proc. Conf. on Bioreactor Fluid Dynamics, Cambridge,* Paper 16, pp. 203–216.

GÜNKEL, A. A., WEBER, M. E. (1975), Flow phenomena in stirred tanks, *AIChE J.* **21**, 931–949.

HANSFORD, G. S., HUMPHREY, A. E. (1966), The effect of equipment scale and degree of mixing on continuous fermentation yield at low dilution rates, *Biotechnol. Bioeng.* **8**, 85–96.

HARLOW, F. H., AMSDEN, A. A. (1975), Numerical calculation of multiphase fluid flow, *J. Comput. Phys.* **17**, 19–52.

HARVEY, P. S., GREAVES, M. (1982a), Turbulent flow in an agitated vessel. Part I: A predictive model, *Trans. Inst. Chem. Eng.* **60**, 195–200.

HARVEY, P. S., GREAVES, M. (1982b), Turbulent flow in an agitated vessel. Part II: Numerical solution and model predictions, *Trans. Inst. Chem. Eng.* **60**, 201–210.

HEIJNEN, J. J., VAN'T RIET, K. (1982), Mass transfer, mixing, and heat transfer phenomena in low-viscous bubble column reactors, *Proc. IVth Eur. Conf. on Mixing,* 27–29th April, *Noordwijkerhout.*

HENZLER, H.-J. (1982), Verfahrenstechnische Auslegungsgrundlagen für Rührbehälter als Fermenter, *Chem. Ing. Tech.* **54**, 461–476.

HINES, D. A. (1978), The large scale pressure cycle fermenter configuration, *Proc. First Eur. Congr. on Biotechnology,* Survey Lectures, DECHEMA-Monographien, Vol. 82, Nr. 1693–1703, pp. 55–64, Weinheim: Verlag Chemie.

HOLMES, D. B., VONCKEN, R. M., DEKKER, J. A. (1964), Fluid flow in turbine-stirred, baffled tanks – I. Chem. Eng. Sci. 19, 201–208.

ISSA, R. I., GOSMAN, A. D. (1981), The computation of three-dimensional turbulent two-phase flows in mixer vessels, in: Numerical Methods in Laminar and Turbulent Flow, pp. 829–839, Swansea: Prineridge Press.

JANSEN, H., SLOT, S., GÜRTLER, H. (1978), Determination of mixing times in large-scale fermenters using radioactive isotopes, in: Preprints First Eur. Congr. on Biotechnology Part II, pp. 80–83, Poster Papers, Frankfurt: DECHEMA.

JOSHI, J. B., PANDIT, A. B., SHARMA, M. M. (1982), Mechanically agitated gas-liquid reactors, Chem. Eng. Sci. 37, 813–844.

JURY, W., SCHNEIDER, G., MOSER, A. (1988), Modelling approach to industrial bioreactors, Proc. 6th Eur. Conf. on Mixing, 24–26th May, Pavia, pp. 451–456.

KHANG, S. J., LEVENSPIEL, O. (1976), New scale-up and design method for stirred agitated batch mixing vessels, Chem. Eng. Sci. 31, 569–577.

KNOEPFEL, H. P. (1972), Zum Crabtree-Effekt bei Saccharomyces cerevisiae and Candida tropicalis, Dissertation 4906, ETH Zürich.

KOBAYASHI, T., VAN DEDEM, G., MOO-YOUNG, M. (1973), Oxygen transfer into mycelial pellets, Biotechnol. Bioeng. 15, 27–45.

KOSSEN, N. W. F. (1985), Bioreactors: Consolidation and innovations, Review/Preview Lecture presented at the 3rd Eur. Congr. on Biotechnology, München, Vol. IV, pp. 257–282, Weinheim-Deerfield Beach/Florida–Basel: VCH Verlagsgesellschaft.

KOSSEN, N. W. F., OOSTERHUIS, N. M. G. (1985), Modelling and scaling up of bioreactors, in: Biotechnology (REHM, H.-J., REED, G., Eds.), Vol. 2, pp. 572–605, Weinheim–Deerfield Beach/Florida–Basel: VCH Verlagsgesellschaft.

LAI, K. Y. M., SALCUDEAN, M. (1987), Computer analysis of multi-dimensional, turbulent, buoyancy-induced, two-phase flows in gas-agitated-liquid reactors, Comput. Fluids 15, 281–295.

LEVENSPIEL, O. (1972), Chemical Reaction Engineering, London: John Wiley & Sons.

LIEPE, F., MOECKEL, H.-O., WINKLER, H. (1971), Untersuchungen in Rührmaschinen, Chem. Tech. (Leipzig) 23, 231–237.

MANN, U., CROSBY, E. J. (1973), Cycle time distribution in circulating systems, Chem. Eng. Sci. 28, 623–627.

MANN, U., CROSBY, E. J., RUBINOVITCH, M. (1974), Number of cycles distribution in circulating systems, Chem. Eng. Sci. 29, 761–765.

MANNING, F. S., WOLF, D., KEAIRNS, D. L. (1965), Model simulation of stirred tank reactors, AIChE J. 11, 723–727.

MARINI, L., GEORGAKIS, C. (1984), Low-density polyethylene vessel reactors. Part I: Steady state and dynamic modelling, AIChE J. 30, 401–408.

MCMANAMEY, W. J. (1980), A circulation model for batch mixing in agitated, baffled vessels, Trans. Inst. Chem. Eng. 58, 271–276.

MERSMANN, A., EINENKEL, W.-D., KAEPPEL, M. (1975), Auslegung und Maßstabsvergrößerung von Rührapparaten, Chem. Ing. Tech. 47, 953–964.

METZ, B. (1976), From Pulp to Pellet, Ph. D. Thesis, University of Delft.

METZ, B., KOSSEN, N. W. F., VAN SUIJDAM, J. C. (1979), The rheology of mould suspensions, Adv. Biochem. Eng. 11, 103–156.

VON MEYENBURG, K. (1969), Katabolit-Repression und der Sporungszyklus von Saccharomyces cerevisiae, Dissertation 4279, ETH Zürich.

MIDDLETON, J. C. (1979), Measurement of circulation within large mixing vessels, Paper A2, 3rd Eur. Conf. on Mixing, 4–6th April, York, Vol. I, pp. 15–36.

MIURA, Y. (1976), Transfer of oxygen and scale-up in submerged aerobic fermentation, Adv. Biochem. Eng. 4, 3–40.

MOECKEL, H. O. (1980), Die Verteilung der örtlichen Energiedissipation in einem Rührwerk, Chem. Tech. (Leipzig) 32, 127–129.

VAN DER MOLEN, K., VAN MAANEN, H. R. E. (1978), Laser-Doppler measurements of the turbulent flow in stirred vessels to establish scaling rules, Chem. Eng. Sci. 33, 1161–1168.

MUKATAKA, S., KATAOKA, H., TAKAHASHI, J. (1976), Measurement of mean circulation time of a mycelial suspension in a stirred tank, Kagaku Kogaku Ronbunshu 2, 628–630.

MUKATAKA, S., KATAOKA, H., TAKAHASHI, J. (1980), Effects of vessel size and rheological properties of suspensions on the distribution of circulation times in stirred vessels, J. Ferment. Technol. 58, 155–161.

MUKATAKA, S., KATAOKA, H., TAKAHASHI, J. (1981a) Circulation times of various fluids in stirred fermentors, in: Advances in Biotechnology (MOO-YOUNG, M., Ed.), Vol. I, pp. 523–528, Toronto: Pergamon Press.

MUKATAKA, S., KATAOKA, H., TAKAHASHI, J. (1981b), Circulation time and degree of fluid exchange between upper and lower circulation regions in a stirred vessel with a dual impeller, J. Ferment. Technol. 59, 303–307.

NAGATA, S. (1975), Mixing Principles and Applications, New York: John Wiley & Sons.

NALLASAMY, M. (1987), Turbulence models and

their applications to the prediction of the internal flows: A Review, *Comput. Fluids* **15**, 151–194.

NAUMANN, E. B. (1981), Continuous mixing, *Chem. Eng. Commun.* **8**, 53–127.

OLDSHUE, J. Y. (1983), *Fluid Mixing Technology*, New York: McGraw Hill.

OKAMOTO, Y., NISHIKAWA, M., HASHIMOTO, K. (1981), Energy dissipation rate distribution in mixing vessels and its effects on liquid–liquid dispersion and solid–liquid mass transfer, *Int. Chem. Eng.* **21**, 88–94.

OOSTERHUIS, N. M. G. (1984), Scale-up of Bioreactors. *Ph. D. Thesis*, University of Delft.

OOSTERHUIS, N. M. G., KOSSEN, N. W. F. (1984), Dissolved oxygen concentration profiles in a production-scale bioreactor, *Biotechnol. Bioeng.* **26**, 546–550.

PACA, J., ETTLER, P., GREGR, V. (1976), Hydrodynamic behaviour and oxygen transfer rate in a pilot plant fermenter. I. Influence of viscosity, *J. Appl. Chem. Biotechnol.* **26**, 309–317.

PANDIT, A. B., JOSHI, J. B. (1983), Mixing in mechanically agitated gas-liquid contactors, bubble columns and modified bubble columns, *Chem. Eng. Sci.* **38**, 1189–1215.

PARKER, D. S., KAUFMANN, W. J., JENKINS, D. (1972), Floc breakup in turbulent flocculation processes, *J. San. Eng. Div.* **SA1**, 79–99.

PLACEK, J., TAVLARIDES, L. L. (1985), Turbulent flow in stirred tanks. Part I: Turbulent flow in the turbine impeller region, *AIChE J.* **31**, 1113–1120.

PLACEK, I., TAVLARIDES, L. L., SMITH, G. W., FORT, I. (1986), Turbulent flow in stirred tanks. Part II: A two scale model of turbulence, *AIChE J.* **32**, 1771–1786.

RAGOT, F., REUSS, M. (1990), A multi-phase compartment model for stirred bioreactors incorporating mass transfer and mixing, Poster, Paper in: *2nd Int. Symp. on Biochemical Engineering*, 5–7th March, *Stuttgart*.

REUSS, M. (1976), Bioverfahrenstechnik – Sauerstofftransport in Bioreaktoren, *VDI-Fortschr. Verfahrenstech.* **14**, 551–563.

REUSS, M. (1983), Mathematical models for coupled oxygen transfer and microbial kinetics in bioreactors, in: *Modelling and Control of Biotechnical Processes* (HALME, A., Ed.), pp. 33–45, Oxford–New York–Toronto–Sydney–Paris–Frankfurt: Pergamon Press.

REUSS, M. (1988), Influence of mechanical stress on the growth of *Rhizopus nigricans* in stirred bioreactors, *Chem. Eng. Technol.* **11**, 178–187.

REUSS, M., BRAMMER, U. (1985), Influence of substrate distribution on productivities in computer controlled baker's yeast production, in: *Modelling and Control of Biotechnological Processes*

(JOHNSON, A., Ed.), pp. 119–124, Oxford–New York–Toronto–Sydney–Paris–Frankfurt: Pergamon Press.

REUSS, M., BAJPAI, R. K., BERKE, W. (1982), Effective oxygen consumption in fermentation broths with filamentous organisms, *J. Chem. Technol. Biotechnol.* **32**, 81–91.

RIESMEIER, B. (1984), Einfluß der Begasung auf das Rezirkulationsverhalten der Flüssigphase in gerührten Bioreaktoren, *Diploma Thesis*, Technische Universität, Berlin.

VAN'T RIET, K. (1979), Review of measuring methods and results in non-viscous gas–liquid mass transfer in stirred vessels, *Ind. Eng. Chem. Proc. Des. Dev.* **18**, 367–375.

VAN'T RIET, K., SMITH, J. M. (1975), The trailing vortex system produced by Rushton turbine agitators, *Chem. Eng. Sci.* **30**, 1093–1105.

RIZZI, M. (1982), Einfluß des Energieeintrages auf die Sauerstoffaufnahme von *Aspergillus niger* in gerührten Bioreaktoren, *Diploma Thesis*, Technische Universität, Berlin.

ROELS, J. A. (1983), *Energetics and Kinetics in Biotechnology*, Amsterdam: Elsevier Biomedical Press.

SCOTT, R. (1983), Design and evaluation of experiments to provide scale-up information for the ICI single cell protein process, in: *BIOTECH 83*, Northwood, UK: Online Publ. Ltd.

SENIOR, P. J., WINDNASS, G. (1980), The ICI single cell protein process, in: *Biotechnology, a Hidden Past, a Shining Future, 13th Int. TNO Conf., The Hague*.

SINCLAIR, C. G., BROWN, D. E. (1970), Effect of incomplete mixing on the analysis of the static behaviour of continuous cultures, *Biotechnol. Bioeng.* **12**, 1001–1017.

SINGH, V., HENSLER, W., FUCHS, R. (1986), On-line determination of mixing parameters in fermentors using pH transient, in: *Bioreactor Fluid Dynamics*, 15–17th April, *Cambridge*, Paper 18, pp. 231–256.

SINGH, V., FUCHS, R., CONSTANTINIDES, A. (1987), A new method for fermentor scale-up incorporating both mixing and mass transfer effects – I. Theoretical basis, in: *Biotechnology Processes, Scale-up and Mixing* (HO, C. S., OLDSHUE, J. Y., Eds.), New York: American Institute of Chemical Engineers, pp. 200–214.

SINGH, V., FUCHS, R., CONSTANTINIDES, A. (1988), Use of mass transfer and mixing correlation for the modelling of oxygen transfer in stirred tank fermentors, in: *Bioreactor Fluid Dynamics, Proc. 2nd Int. Conf.*, 21–23th Sept., *Cranfield, UK*, pp. 95–115.

SMITH, J. M. (1981), *Chemical Engineering Kinetics*, 3rd Ed., New York: McGraw Hill.

STEEL, R., MAXON, W. D. (1962), Some effects of turbine size on novobiocin fermentations, *Biotechnol. Bioeng.* **4**, 231–240.

STEEL, R., MAXON, W. D. (1966), Dissolved oxygen measurements in pilot- and production scale novobiocin fermentations, *Biotechnol. Bioeng.* **8**, 97–108.

VAN SUIJDAM, J. C. (1980), Mycelial Pellet Suspension, Biotechnological Aspects, *Ph. D. Thesis,* University of Delft.

SWEERE, A. P. J., LUYBEN, K. C. A. M., KOSSEN, N. W. F. (1987), Regime analysis and scale down: Tools to investigate the performance of bioreactors, *Enzyme Microb. Technol.* **9**, 386–398.

TRÄGHÅRD, C. (1988), A hydrodynamic model for the simulation of an aerated agitated fed-batch fermentor, *Proc. 2nd Int. Conf. on Bioreactor Fluid Dynamics,* 21–23th Sept., *Cambridge UK,* pp. 117–134.

UHL, V. W., GRAY, J. B. (1966), *Mixing,* Vols. 1 and 2, New York: Academic Press.

VONCKEN, R. M., HOLMES, D. B., DEN HARTOG, H. W. (1964), Fluid flow in turbine-stirred, baffled tanks – II. Dispersion during circulation, *Chem. Eng. Sci.* **19**, 209–213.

WANG, D. I. C., FEWKES, R. C. J. (1977), Effect of operating and geometric parameters on the behaviour of non-Newtonian, mycelial antibiotic fermentation, *Dev. Ind. Microbiol.* **18**, 39–56.

WANG, H. Y., COONEY, C. L., WANG, D. I. C. (1979), Computer control of baker's yeast production, *Biotechnol. Bioeng.* **21**, 975–995.

WEN, C. Y., FAN, L. T. (1975), *Models for Flow Systems and Chemical Reactors,* New York: Marcel Dekker.

YANO, T., KODAMA, T., YAMADA, K. (1961), Fundamental studies on the aerobic fermentation. Oxygen transfer within a mold pellet, *Agric. Biol. Chem.* **25**, 580–584.

YOSHIDA, T. (1967), Studies on submerged culture of Basidiomycetes. The oxygen transfer within pellets of *Leutinus edodos, J. Ferment. Technol.* **45**, 1119–1129.

11 Tower Reactor Models

JOSÉ C. MERCHUK

Beer Sheva, Israel

List of Symbols

A cross-sectional area (m^2)
a interfacial area per unit volume (m^{-1})
B biomass concentration (kg m^{-3})
B_A dimensionless biomass concentration $(B/Y_x S_f)$
Bo Bodenstein number, Eq. (60)
C_B concentration of component B (mol L^{-1})
C_j liquid phase concentration in stage j
C_0 distribution parameter
C_L constant in the friction factor correlation
C^* saturation concentration of oxygen in the liquid phase (mol L^{-1})
D diffusivity coefficient (m^2 s), Eq. (79)
D tower diameter (m)
D_h hydraulic diameter (m)
D_z liquid phase dispersion coefficient (m s^{-2})
d_s Sauter diameter (m)
E energy dissipation in the liquid (W)
f friction factor ($-$)
f_b constant in Eq. (87)
fr fraction of reversed flow gas in the downcomer ($-$)
g gravitational acceleration (m s^{-2})
H height (m)
He Henry's constant (Pa m^3 mol^{-1})
h_1 effective width, in Eq. (52) (m)
J_G superficial gas velocity (m s^{-1})
J_L superficial liquid velocity (m s^{-1})
J^D drift velocity (m s^{-1}), Eq. (24), Eq. (25)
K friction coefficient; consistency index in power law model (Pa s^n)
K_1 loss coefficient for change in velocity direction ($-$)
K_2 loss coefficient for change in velocity magnitude ($-$)
K_A dimensionless group $(K_M S_f)$
K_B constant in Eq. (45), Eq. (70)
K_d dimensionless group (K_M/S_i)
K_M Michaelis–Menten constant (mol L^{-1})
K_{ST} coalescence factor, Eq. (89)
k_L mass transfer coefficient (m s^{-1})
k_{L0} initial mass transfer coefficient (m s^{-1}), Eq. (88)
k_L' local value of k_L

k_1 first-order kinetic constant (s^{-1})
L ungassed liquid height (m)
L_e equivalent length (m)
l equivalent length of the bottom section (diameters)
l_m mixing length (m)
M productivity ratio, Eq. (110)
m ratio of gas to liquid flow rate, Eq. (111)
m_f molar flow rate (mol s^{-1})
N total number of cells
n flow index in power law model; parameter of gas holdup radial profile, Eq. (17)
n_0 inlet molar flow rate (mol s^{-1})
P pressure (Pa)
Pe Peclet number ($-$), Eq. (59)
Q radial point of inversion of liquid velocity
Q_G gas flow rate (m^3 s^{-1})
Q_L liquid flow rate (m^3 s^{-1})
q constant in Eq. (6)
R radius (m); recycle ratio ($-$); ideal gas constant
R_1 dimensionless stream function, Eq. (11)
R_A dimensionless reaction group, Eq. (109)
Re^* Reynolds number, Eq. (84)
RX respiration rate of microorganisms (mol O$_2$ L^{-1} s^{-1})
r radial distance (m); fraction of entrained gas in Eq. (98)
S substrate concentration (kg m^{-3})
S_A dimensionless concentration of substrate (S/S_s)
s ratio of mean residence times in the perfectly mixed and plug-flow zones, Eq. (65)
Sc Schmidt number, Eq. (83)
Sh Sherwood number, $k_L a D^2/D$
St_G Stanton number for the gas phase, Eq. (102) ($-$)
St_L Stanton number for the liquid phase, Eq. (103) ($-$)
T temperature (K)
t time (s)
t_b batch time (s)
t_c exposure time (s)
U velocity (m s^{-1})
U_0 slip velocity (m s^{-1})
U_∞ bubble terminal velocity (m s^{-1})
u_s velocity of slug (m s^{-1})
V volume (m^3)

Y mole fraction in the gas phase
y_x yield coefficient
y_1 dimensionless oxygen concentration in the gas phase (Y/Y_i)
y_2 dimensionless oxygen concentration in the liquid phase (C/C^*)
z axial coordinate (m)

Greek letters

α pressure change group, Eq. (93), Eq. (104) $(-)$
β flowing volumetric concentration, Eq. (48) $(-)$
β_r volume-to-interfacial area ratio, Eq. (113) $(-)$
γ shear rate (s^{-1}); backflow ratio $(-)$
γ_r Hatta number, Eq. (112)
δ film width (m)
Δ difference
ε energy dissipation per unit mass $(W\ kg^{-1})$
ζ ratio between cross-sectional areas of riser and downcomer $(-)$
η axial distribution of concentrations $(-)$, Eq. (107)
η_r fractional conversion $(-)$
μ specific growth rate (s^{-1})
ν kinematic viscosity $(m^2\ s^{-1})$
ρ density $(kg\ m^{-3})$
σ surface tension $(N\ m^{-1})$; variance $(-)$
τ shear stress $(kg\ m^{-1}\ s^{-2})$
ϕ fractional gas holdup $(-)$
φ dimensionless radius (rR^{-1}) $(-)$
ψ stream function
ψ_0 maximum value of ψ
ω vorticity

Subscripts

a   annulus
B   bottom
b   back
CL   circulation
c   center
ch   charge
cy   cycle
d   downcomer
dch   discharge
dr   draft tube

e   equilibrium
f   friction; feed; flow
G   gas
i   inlet
L   liquid
M   molecular
max   maximum
min   minimum
r   riser
s   slug
t   turbulent
to   total
T   top
w   wall

Superscripts

*   dimensionless
$-$ mean
$'$ fluctuating component
i   inlet

1 Introduction

The mathematical representation of a reactor is recognized as a basic tool in chemical engineering and has two different uses: a) design of a new reactor, which fulfills given requirements of production rate and product quality, or b) simulation of the performance of given equipment, in order to predict its behavior under various operating conditions. Sometimes both aspects interact and complement each other.

In order to describe a bubble tower reactor, three basic categories of information are required: fluid dynamics of the multiphase system, mass and heat transfer characteristics, and kinetics of the chemical reactions in the system. While under certain conditions either the mass transfer rate or the reaction rates may be limiting, making information about the other rate irrelevant, the fluid-dynamics of the tower reactor is always required as a cornerstone in the building of a model for the system.

2 Flow Modelling

2.1 Flow Configuration

The possible combinations of gas and liquid flow in a tower are as follows:

A   Gas flow upwards. No net liquid flow. This is the typical arrangement in batch bubble columns.

B   Cocurrent gas and liquid upflow. This arrangement is the most common in continuous systems, especially in cases where a solid reagent or catalyst must be held in suspension in the liquid.

C   Cocurrent gas and liquid downflow. This form of operation is not very common; its main application is in processes where a very high gas throughput is required leading to irregularities in upflow operation.

D   Countercurrent gas upflow–liquid downflow. This is the situation in the tray of a distillation column. In taller systems, the downward liquid velocity may increase the gas holdup and residence time.

ALR Air lift reactors are in fact a combination of three different sections, each having a different fluid dynamic behaviour. Indeed, one can find type A flow in the gas separator, type B flow in the riser, and type C flow in the downcomer section.

The most common form of operation in bubble columns corresponds to type A. Cocurrent upflow (type B) and countercurrent (type D) do not depart very much from the general flow characteristics of type A, as long as the liquid flow rate is relatively low. This is usual, since the required liquid residence times are normally orders of magnitude smaller for the liquid than for the gas phase due to the density difference. Most of the studies on tower reactor hydrodynamics are based on this type of flow.

Although about a dozen different gas–liquid flow configurations have been recognized (BARNEA and TAITEL, 1986), only two of them are of interest in tower reactors (WALLIS, 1969; WISWANATHAN, 1986):

a) Homogeneous bubbly flow regime, where bubbles are relatively small and uniform in diameter, and turbulence is low.

b) Churn-turbulent regime, where a wide range of bubble sizes coexists within a very turbulent liquid. This flow regime can be reached from the homogeneous bubbly flow by increasing the gas flow rate. Another way of reaching a churn-turbulent flow zone is by starting from slug flow and increasing the liquid turbulence by increasing the flow rate or the diameter of the tower reactor (Fig. 1). This latter flow configuration, the

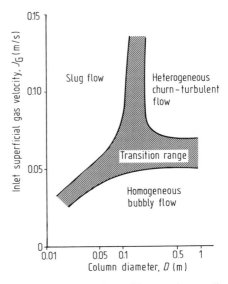

Fig. 1. Flow configurations of interest for gas–liquid tower reactors (from WISWANATHAN, 1986).

slug flow, is important only as a situation to be avoided by all means, since large bubbles bridging the entire tower cross-section offer very poor capacity for mass transfer and liquid mixing.

2.1.1 Flow Regime Transitions

BARNEA and TAITEL (1986) have reviewed the criteria for predicting the transitions from one flow configuration to the other.

2.1.1.1 Homogeneous Bubbly Flow–Slug Flow

This transition can be depicted as a series of states starting with small bubbles generated at low rates, which behave as rigid spheres rising rectilinearly. But as gas flow rate increases, so does the bubble diameter. Above a certain diameter, which is 0.002–0.003 m in air/water systems at atmospheric pressure, spherical caps appear and with them zigzag motion. Considerable turbulence develops, with random collisions between bubbles and consequent coalescence. Larger Taylor-type bubbles may appear, rising much faster than the smaller ones. As they take over those smaller bubbles in their path, they increase in volume until they finally bridge the whole cross-section of the tower (TAYLOR, 1954). However, this occurs, as can be seen in Fig. 1, only for a tower diameter smaller than 0.10–0.15 m. For larger diameters, slug flow as defined above is not observed. Instead, an increase in superficial gas velocity leads directly to the churn-turbulent regime. This configuration offers a smaller gas–liquid interfacial area than the bubble flow, but the mass transfer coefficient k_L is enhanced because of the turbulence, balancing at least partially the loss of interfacial area.

BARNEA and TAITEL (1986) have proposed a series of criteria for characterizing the transition between the flow configurations in two-phase flow. For the case of the transition from bubble to slug flow, it is accepted that a slug will form when a very high gas holdup is attained. GRIFFITH and SNYDER (1964) found this to happen when the gas holdup ϕ is around 0.25 to 0.30. In this situation it is assumed that the bubbles are so close to one another that the frequency of collision and coalescence increases sharply. BARNEA and TAITEL consider the maximum allowable packing of the bubbles. The maximum fraction of rigid spheres packed in a cubic lattice is 0.52. But since the bubbles are not perfect spheres and

are moving on oscillatory paths, they will collide frequently before this fraction is attained. If it is assumed that the minimal distance between bubbles that allows them freedom of movement is half their radius, then a maximal gas fraction of 25% is obtained.

Defining the superficial gas and liquid velocities as:

$$J_G = U_G \phi \tag{1}$$

$$J_L = U_L(1-\phi) \tag{2}$$

The relative velocity of the gas phase is:

$$U_0 = U_G - U_L \tag{3}$$

In the range of bubble size of interest, the relative velocity U_0 depends only very slightly on the bubble diameter. Taking HARMATHY's (1960) expression

$$U_0 = 1.53 \left[\frac{g(\rho_L - \rho_g)\sigma}{\rho_L^2} \right]^{1/4} \tag{4}$$

and $\phi = 0.25$, Eqs. (1) to (4) give

$$J_L = 3.0 J_G - 1.15 \left[\frac{g(\rho_L - \rho_G)\sigma}{\rho_L^2} \right]^{1/4} \tag{5}$$

which becomes the limit between bubbly flow and slug flow on the $(J_L - J_G)$ plane. This limit depends on liquid properties only, since for the situations of interest in tower reactor design ρ_G is usually negligible, and geometric parameters do not appear in Eq. (5).

2.1.1.2 Slug Flow–Dispersed Bubble Flow

The authors then define a boundary between slug flow and dispersed bubble flow, a flow configuration where small bubbles are dispersed into a very turbulent liquid. The condition is:

$$2 \left[\frac{0.4\sigma}{(\rho_L - \rho_G)g} \right]^{1/2} \left(\frac{\rho_L}{\sigma} \right)^{3/5} \left[\frac{2}{D} C_L \left(\frac{D}{v_L} \right)^{-q} \right]^{2/5} \cdot$$

$$\cdot u_s^{2(3-q)/5} = 0.725 + 4.15 \left(\frac{J_G}{u_s} \right)^{0.5} \tag{6}$$

and is based on considering a balance between surface tension and forces due to turbulent fluctuations. The transition from slug flow to dispersed bubble flow would occur when the turbulent breakup process is able to disintegrate the Taylor bubble into bubbles small enough to remain spheric. However, the high liquid velocities required for that process exceed the usual operational range of tower reactors, and, therefore, this boundary is of little relevance for our purposes.

2.1.1.3 Slug Flow–Churn-Turbulent Flow

For the transition between slug flow and churn-turbulent flow, the expressions given by BARNEA and TAITEL (1986) can be written as:

$$J_L = \left(\frac{0.825}{\phi} - 1\right) J_G - 0.22 \sqrt{gD} \qquad (7)$$

In Eq. (7), the diameter of the tower, D, plays an important role. As D increases, the superficial liquid velocity, J_L, required for transforming a slug flow into churn-turbulent flow for a given superficial gas velocity, J_G, decreases. The three boundaries given by Eqs.

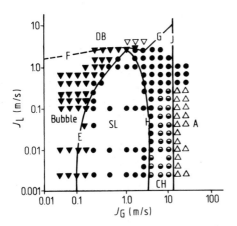

Fig. 2. Flow pattern transitions for a 5.1 cm diameter tube, upward air–water flow. 0.1 MPa, 25 °C; ▼ bubble or dispersed bubble flow; ● slug flow; ○ churn flow. △ annular flow E, Eq. (5); G, Eq. (6); H, Eq. (7) (from BARNEA and TAITEL, 1986).

(5), (6), and (7) can be seen in Fig. 2. This figure is for a 5.1 cm diameter column. As the diameter increases to sizes representative of bench scale or pilot plant size, the slug flow regime disappears and instead becomes a diffuse and uncertain area between bubble flow and churn-turbulent flow. Most of the cases of interest for chemical and biochemical tower reactor design fall either into this area or into the bubble flow region.

2.2 Bubble Column Flow Models

In the case of bubble column operation, where there is little or no net liquid flow, the bubbly flow configuration covers two sub-regions.

At low gas superficial velocity, the bubbles generated in the gas sparger rise without much coalescence. Bubble breakup is also negligible in this region, and the initial bubble diameter is maintained, changing only because of hydrostatic changes or interchange of gas with the liquid phase. Under these conditions, most of the mass transfer takes place near the sparger (ZHAO et al., 1988). As superficial gas velocity increases, bubble size, bubble oscillations, and liquid turbulence increase, coalescence and breakup appear in a certain measure, but the general behavior does not change very much. The rate of the gas holdup increase in this region is rapid and nearly linear. But as the superficial gas velocity increases further, the so-called recirculation sub-regime is attained. Under these conditions, a central ascending and oscillating liquid path develops with recirculation cells at both sides of it, as can be seen in Fig. 3. These patterns can be easily seen in two-dimensional bubble columns. MERCHUK (1986) compared the two-dimensional flow patterns in a bubble column and an airlift reactor.

When operating in the bubble column mode and at low gas velocities, bubbles ascend in straight lines, especially in the lower half of the column. In the top half of the column, they begin to oscillate (Fig. 3A), apparently due to the increase in bubble diameter caused by the decrease of the hydrostatic pressure and by some degree of coalescence. However, as the gas velocity increases, the oscillations of

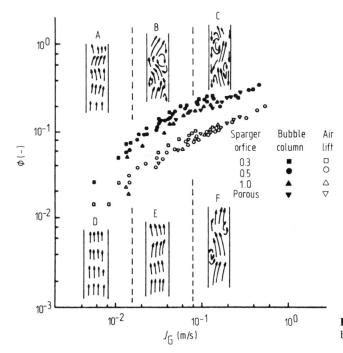

ϕ (—)

J_G (m/s)

Sparger orfice	Bubble column	Air lift
0.3	■	□
0.5	●	○
1.0	▲	△
Porous	▼	▽

Fig. 3. Recirculation patterns in a bubble column (from MERCHUK, 1986).

the bubbles increase in amplitude, and a mean stream is generated characterized by high local velocity and larger bubbles. Recirculation cells with small bubbles that oscillate in accordance with the mainstream line appear at the sides of the column (Fig. 3 B). With a further increase in gas velocity, the frequency of oscillations increases sharply while the amplitude remains constant or increases slightly. This leads to a sharp increase in turbulence. The recirculation flow rate in the side cells increases at the expense of the central path, which, under these conditions, consists of the largest bubbles present in the system (Fig. 3 C). A clear downstream pattern develops, bringing the smallest bubbles to the bottom of the column. At even larger values of the superficial gas velocity, the relative importance of the main central path increases. The amplitude of the oscillations decreases, and most of the gas passes rapidly through the center. The loops at the sides become even more turbulent, and the downstream of recirculating liquid is clearly visible at both sides of the column.

Flow configurations B and C correspond to the "transition regime" between the uniform bubbling regime and the liquid recirculating regime. They exist in bubble columns having a sparger hole diameter of less than about 0.01 m (WALLIS, 1969). One of the characteristics of the transition regime is an increase in the gas holdup. A critical superficial velocity has been defined (SAKATA and MIYAUCHI, 1980; MARUYAMA et al. 1981) as the superficial gas velocity at maximum holdup, and this is supposed to be the point at which incipient regular recirculation begins.

The modelling of the recirculation regime is very important for prediction of the mixing characteristics of bubble columns.

One of the more widely accepted models for the hydrodynamics of bubble columns is the circulation cell model by JOSHI and SHARMA (1979). It is based on the concepts of stream function and of vorticity, and on their relationship as given by LAMB (1932):

$$w = -\frac{1}{r}\left(\frac{\partial^2 \psi}{\partial z^2} + \frac{\partial^2 \psi}{\partial r^2} - \frac{1}{r}\frac{\partial \psi}{\partial r}\right) \tag{8}$$

This model assumes axial symmetry, and the height of each circulation cell is obtained by means of a criterion of minimum vorticity and an energy balance. It was found that the minimum vorticity appeared when the height of the circulation cell was equal to the tower diameter, for any values of the diameter. The number of circulation cells is therefore equal to the total dispersion height divided by the diameter (Fig. 4). The model predicts the following axial and radial components for the veloci-

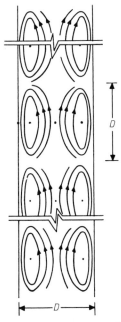

Fig. 4. Multiple circulation cells in a bubble column (JOSHI and SHARMA, 1979).

ty as a function of the tower diameter, r, and the axial position, z:

$$U_z = \frac{\psi_0}{(D/2)^2} \frac{1}{r^*} \frac{dR_1}{dr^*} \sin(\pi z^*) \tag{9}$$

$$U_r = -\frac{2\pi\psi_0}{DH} \frac{R_1}{r^*} \cos(\pi z^*) \tag{10}$$

where U_z and U_r are the axial and radial components of the liquid velocity, respectively, R_1 is the dimensionless stream function:

$$R_1 = \psi/\psi_0 \tag{11}$$

and ψ_0 is the maximum value of the stream function. This function is calculated under the assumption that all the kinetic energy associated with the downward flow of the liquid is dissipated in the turbulence. JOSHI et al. (1986) give relationships for the energy dissipation by the liquid, E, obtained from energy balances, in which the kinetic energy related to the inflowing gas, the energy associated with dissipation at the column wall, and the energy used for bubble breakup are neglected. The integration of Eq. (8) to give Eqs. (9) and (10) is facilitated by the fact that, as shown by LAMB (1932), the vorticity divided by the radial distance from the axis depends on the stream function only:

$$\frac{\omega}{r} = k_2\psi \tag{12}$$

where k_2 is a constant.

This model does not agree with observations on two-dimensional bubble columns (MERCHUK, 1986), especially because of the assumed axisymmetry. Also, the axial component at the wall, Eq. (9), would change direction in passing from one cell to another because of the function sinus. This contradicts the widely accepted fact that the flow at the wall is downwards all along the tower in the recirculation regime. This has been criticized by VAN DER AKKER and RIETEMA (1982). However, JOSHI and SHARMA (1979) have shown that their model allows a satisfactory prediction of the axial velocity and the axial dispersion coefficient. Later on, JOSHI (1980) modified the model in order to take into account interaction between adjacent cells. JOSHI et al. (1986) have recently reviewed the achievements of this model when applied to bubble columns, gas–liquid–solid sparged reactors, and gas–solid fluidized beds. A model of similar characteristics was applied by VAN DER AKKER and RIETEMA (1982) to liquid–liquid systems.

The recirculation model has lately been extended to a three-dimensional version which assumes cylindrical eddies piled up transversely to each other in the column as depicted in Fig. 5 (ZEHNER, 1986a). This arrangement solves the problem of interference between eddies if a downward path along the wall is accepted. ZEHNER derives an equation for the

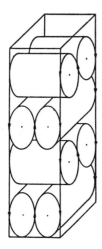

Fig. 5. Cylindrical eddies model (from ZEHNER, 1986a).

mean liquid velocity in the recirculation path in the bubble column based on the simplified model:

$$\bar{U}_L = \sqrt[3]{\frac{1}{f} \frac{\Delta\rho}{\rho_L} gDJ_G} \qquad (13)$$

where f refers to the pressure drop related to the liquid–gas mixture circulation in the tower.

An alternative approach to the description of the fluid dynamics of bubble columns in recirculating turbulent flow is the description of a single internal loop near the walls of the tower. This has been done (HILLS, 1974; UEYAMA and MIYAUCHI, 1979) starting with the equation of motion:

$$-\frac{1}{r}\frac{d}{dr}(\tau) = \frac{d\bar{p}}{dz} + (1-\bar{\varphi})\rho_L g \qquad (14)$$

The shear stress τ is then related to the velocity through the molecular and turbulent viscosity by:

$$\tau = -(\nu_M + \nu_t)\rho_L \left(\frac{dU}{dr}\right) \qquad (15)$$

UEYAMA and MIYAUCHI (1979) demonstrated that for a bubble column with $L/D \gg 1$, the turbulent shear stress is given by the following equation:

$$-\nu_t \frac{dU}{dr} = (1-\varphi)\bar{u}_r'\bar{u}_z' \qquad (16)$$

where $\bar{u}_r'$ and $\bar{u}_z'$ are the fluctuating components of the liquid velocity in the r and z directions.

The profile of the gas holdup along the radius is assumed to be given by:

$$\phi/\bar{\phi} = \left[\frac{n+2}{n}\right][1-(r/R)^n] \qquad (17)$$

UEYAMA and MIYAUCHI (1979) obtained the time-averaged liquid velocity for the general case of net liquid flow through the tower as a function of n, which represents the profile of the gas holdup. Fig. 6 shows the agreement of their model with experimental data. It can be seen that the velocity profile intersects the zero velocity line within the tower cross-section.

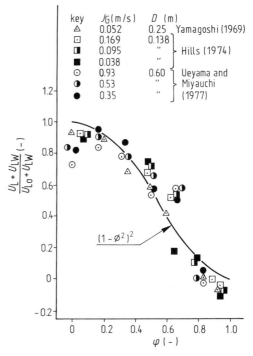

Fig. 6. Distribution of liquid velocities as a function of the dimensionless radius φ as predicted by UEYAMA and MIYAUCHI (1979).

This point of zero velocity has been used by other investigators as the main parameter, instead of the turbulent viscosity used by UEYAMA and MIYAUCHI (WALTER and BLANCH, 1983; YANG et al., 1986). In this way, a clear advantage is obtained since flow inversion can be measured directly, in contrast to turbulent viscosity, which has to be obtained by model fitting.

The solution to the equation of motion, Eq. (14), with the holdup distribution given by Eq. (17), has been obtained by YANG et al. (1986) with the boundary conditions of symmetry in the center of the tower, and zero liquid velocity at a point where $\varphi = r/R = Q$ (flow inversion point). The correct expression is:

$$\frac{U_L}{U_{Lmax}} = 1 + Q^{-2}\varphi^2 +$$
$$+ \left(\frac{2 - Q^{-2}}{Q^n - \dfrac{4}{(n+4)}} \right) (\varphi^{n+2} - Q^n \varphi^2) \qquad (18)$$

This is a very simple expression depending on two parameters: n, which represents the profile of gas holdup, and Q which therefore encompasses the influence of geometry and liquid properties. It is tempting to consider that, since gas holdup depends on the same variables, it should be possible to correlate Q with φ to obtain a single parameter model. However, at the present time, not enough independently obtained data are available.

Eq. (18) is similar to the result given by WALTER and BLANCH (1983). YANG et al. (1986) go further and extend their model to co-current columns by simply adding the superficial liquid velocity to U_L. This is, however, an oversimplification. Net liquid velocity induces changes both of the gas holdup and the flow pattern. It follows that both n and Q, the parameters of the model, are in fact functions of U_L. This is not apparent in the model by YANG et al. (1986), and such a function is neither given nor suggested.

CLARK et al. (1987) presented a model for turbulent circulation in bubble columns that predicts the radial distribution of the axial liquid velocity in tall columns with fully developed turbulent flow. Their analysis is based on the mixing length theory (SCHLICHTING,

1968). They use published expressions for the axial shear stress across the diameter of the tower (LEVY, 1960), for the radial distribution of the gas holdup (BANKOFF, 1960) and for the mixing length (SCHLICHTING, 1968), and finally obtain velocity profiles that fit the experimental data obtained by HILLS (1974).

The model by CLARK et al. (1987) has no analytical solution so that the velocity distribution must be obtained by numerical integration. The solution is a function of the shear at the wall. By integration of the velocity profile over the cross-sectional area, the net liquid flow rate is obtained:

$$Q_L = \int_0^R 2\pi r U(r) dr \qquad (19)$$

Since Q_L is usually known, or measurable, the wall shear corresponding to the system is obtained.

While the recirculation regime is the usual flow state in tower reactors in batch operation or with low liquid velocities, visual observations show clearly that air lift operation leads to much higher liquid velocities. VERLAAN et al. (1988) extended the observations of MERCHUK and STEIN (1981a) and gave a criterion for the transition between bubble column (recirculation regime) and air lift flow. When the liquid circulation velocity, as calculated from JOSHI and SHARMA's model (1979),

$$U_{CL} = 3.0 [D(J_G - \phi U_0)]^{1/3} \qquad (20)$$

is much higher than the liquid superficial velocity:

$$U_{CL} \gg J_L \qquad (21)$$

the recirculation regime predominates and the above described models are applicable. On the other hand, if

$$U_{CL} \ll J_L \qquad (22)$$

the recirculation has a minor role in the fluid dynamics of the system, and other types of models, as described in the next section, should be used.

The criteria proposed by VERLAAN et al. (1986) can be evaluated with the data in Fig. 7.

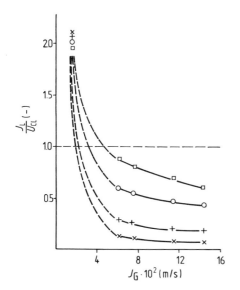

Fig. 7. Superficial liquid velocity and circulation velocity as a function of superficial gas velocity (from VERLAAN et al., 1988).

The different symbols in the figure indicate different degrees of energy losses in the loop. The higher the energy losses, the lower J_L/U_{CL}, indicating that the liquid is preferably recirculated. This agrees with visual observations in two-dimensional systems (MERCHUK, 1986).

2.3 Air Lift Reactor Flow

2.3.1 The Drift Flux Model

One of the models for the representation of two-phase flow that has proven to be extremely useful is the drift flux model as presented by ZUBER and FINDLAY (1965). They derived general expressions useful for prediction of the gas holdup and for interpretation of experimental data, applicable to non-uniform radial distributions of liquid velocity and gas fractions.

The drift velocities are defined as the difference between the velocity of each phase and the volumetric flux density of the mixture, J:

$$J = J_G + J_L \qquad (23)$$

The drift velocities are then:

$$J_G^D = U_G - J \quad \text{and} \qquad (24)$$

$$J_L^D = U_L - J \qquad (25)$$

Assuming that no phase changes occur in the system, and that densities remain constant, ZUBER and FINDLAY (1965) derived a relationship between the weighted mean velocity of the gas phase and the weighted mean value of the flux of the gas phase. The weighted mean value of a variable F is defined by them as:

$$\bar{F} = \frac{\langle \phi F \rangle}{\langle \phi \rangle} \qquad (26)$$

where the brackets indicate the average value of a variable F over the cross-sectional area of the tower:

$$\langle F \rangle = \frac{1}{A} \int_A F \, dA \qquad (27)$$

The above mentioned relationship between the weighted means of velocity and flux, from the definition of drift velocities is:

$$\bar{U}_G = \frac{\langle \phi J \rangle}{\langle \phi \rangle} + \frac{\langle \phi J_G \rangle}{\langle \phi \rangle} \qquad (28)$$

The following equation is obtained by dividing and multiplying the first term of the right-hand side by $\langle J \rangle$:

$$\bar{U}_G = \frac{\langle J_G \rangle}{\langle \phi \rangle} = C_0 \langle J \rangle + \frac{\langle \phi J_G \rangle}{\langle \phi \rangle} \qquad (29)$$

where C_0 is called the distribution parameter,

$$C_0 = \frac{\langle \phi J \rangle}{\langle \phi \rangle \langle J \rangle} \qquad (30)$$

and is the inverse of the "flow parameter" defined by BANKOFF (1960). C_0 depends on the radial distribution of the gas holdup and the superficial liquid velocity, the former being the most influential. If the gas holdup profile is uniform across the section, $C_0 = 1$; if ϕ is higher at the center of the section than at the walls, C_0 is larger than unity, and it is smaller than unity if the reverse is true.

The second term in the right-hand side of Eq. (29) depends on both the distribution of ϕ and on the local drift flux of the gas. For the case when interaction between bubbles is not negligible, ZUBER and HENCH (1962) give the following expression:

$$J_G = 1.53 \left[\frac{\sigma g \Delta \rho}{\rho_L^2} \right]^{1/4} (1 - \phi)^{3/2} \qquad (31)$$

which is valid for the bubbly churn-turbulent regime. After integration,

$$\frac{\langle \phi J_G \rangle}{\langle \phi \rangle} = 1.53 \left[\frac{\sigma g \Delta \rho}{\rho_L^2} \right]^{1/4} = U_\infty \qquad (32)$$

where U_∞ indicates the terminal velocity of a bubble in a quiescent liquid. Eq. (29) now becomes:

$$\frac{\langle J_G \rangle}{\langle \phi \rangle} = \alpha \langle J \rangle + U_\infty \qquad (33)$$

which has been shown to be an adequate expression for correlation of gas holdup measurements in tower reactors with high liquid velocities, as in the case of air lift reactors (MERCHUK and STEIN, 1981a).

2.3.2 Air Lift Reactor Flow Models

Air lift reactors are tower reactors where the liquid, rather than finding its circulation pattern randomly, flows through channels especially designed for this purpose. Those channels may be an internal concentric draft tube, a straight baffle splitting the vessel into riser and downcomer, or an external loop. The considerable importance of air lift technology in modern biochemical engineering research and practice has established the need for specific models for the corresponding fluid dynamics, especially for liquid velocity and mixing.

Two main methods have been used for the modelling of the two-phase flow in air lift reactors: energy balances and momentum balances.

CHAKRAVARTY et al. (1974) used the energy balance approach in order to obtain a link between superficial gas velocity, holdup, and liquid velocity, which allows an evaluation of the latter from easily measurable variables. In their analysis, the energy input is the potential energy of the gas bubbles once inside the reactor, thus neglecting the kinetic energy associated with gas injection. This energy input is equal to the sum of the energy dissipation due to the downward movement of entrained bubbles in the downcomer, the energy converted into kinetic energy of the liquid, the energy losses due to the change of direction and magnitude of liquid velocity at the bottom of the reactor, and the energy losses due to friction of the gas–liquid mixture on the walls. The authors present the final expression of the balance as follows:

$$a \bar{U}_L^3 + b \bar{U}_L^2 + c U_L + d = 0 \qquad (34)$$

where

$$a = (K + ML) + \frac{\phi_r}{\phi_d} \qquad (35)$$

$$b = \frac{J_G}{\phi_r} - U_\infty \frac{A_b \phi_d}{A_r \phi_r} \qquad (36)$$

$$c = 2 g L \frac{(\phi_r - \phi_d)}{1 - \phi_r} \qquad (37)$$

$$d = \frac{2 g L}{1 - \phi_r} \left\{ 2 U_\infty \frac{A_d \phi_d}{A_r} - J_G (1 - \phi_r) \right\} \qquad (38)$$

In these expressions, the superficial gas velocity J_G is based on the gas flow rate from the sparger over the riser cross-section. The coefficient K depends on coefficients for changes of direction, K_1, and magnitude of velocity, K_2,

$$K = K_1 \left[1 + \left(\frac{A_r}{A_d} \right)^2 \right] + K_2 \qquad (39)$$

and M encompasses the friction factors for the two-phase flow in the riser, f_r, and downcomer, f_d:

$$M = 4 \left[\frac{f_r}{D_r} + \frac{f_d}{D_d} \left(\frac{A_r}{A_d} \right)^2 \right] \qquad (40)$$

LEE et al. (1986) used a slightly different analysis to predict the liquid velocity in air lift reactors. They obtained the following expression from the energy balance:

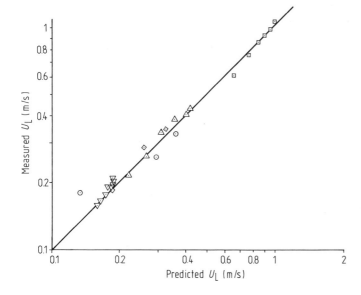

Measured U_L (m/s) — Predicted U_L (m/s)

Fig. 8. Prediction of liquid velocity U_L by the energy balance method (from LEE et al., 1986).

$$Q_G P \ln[1 + \rho_L g L / P] = A_r \rho_L g L \phi_r U_\infty$$
$$+ \tfrac{1}{2} \rho_L [\bar{U}_{Lr}^3 f_r A_r (1 - \phi_r) + \bar{U}_{Ld} f_d A_d (1 - \phi_d)]$$
$$+ A_d \rho_L g L \phi_d U_{Ld}, \tag{41}$$

where L is the ungassed liquid height. From the conservation of the liquid flow rates in the riser and downcomer, the following expression is obtained:

$$U_{Lr} A_r (1 - \phi_r) = U_{Ld} A_d (1 - \phi_d) \tag{42}$$

LEE et al. (1986) calculated U_L for a series of published data in both concentric and external loop air lift reactors and their own results in split vessels. This is shown in Fig. 8. It should be considered, however, that most of the friction coefficients are adapted to fit the data, rather than being obtained from independent sources.

JONES (1985) managed to express the results of his energy balance, based on previous works of NIKLIN (1962) and FREEDMAN and DAVIDSON (1969), in an expression free of empirical constants. His results, however, fit the experimental data only qualitatively. The fit is satisfactory only for very small diameters. An improvement of this method was suggested by CLARK and JONES (1987), taking into account the radial distribution of gas holdup through the drift flux model, Eq. (29). But the values

of the distribution coefficient C_0 needed for satisfactory fitting of the experimental data for the larger diameters are far from the range usual in this type of flow.

CHISTI et al. (1988) extended a model originally proposed by BELLO (1981) and based on an energy balance over the air lift loop as follows:

$$E_i = E_r + E_d + E_B + E_T + E_f \tag{43}$$

where the energy input due to isothermal gas expansion E_i is set equal to the sum of the energies dissipated due to wakes behind the bubbles in the riser, (E_r), the energy loss due to stagnant gas in the downcomer (E_d), the energy losses due to liquid turnaround at the bottom and top of the loop (E_B and E_T), and the energy loss due to friction in the riser and the downcomer.

CHISTI et al. (1988) manage to write E_r as a function of E_i, thus eliminating the latter from the overall energy balance. This allows them to obtain an expression for the liquid velocity in quadratic form, rather than the cumbersome cubic form of Eqs. (34) and (41). Their explicit expression of the average liquid velocity is:

$$U_{Lr} = \left[\frac{2gH(\phi_r - \phi_d)}{\dfrac{K_T}{(1 - \phi_r)} + K_B \zeta^2 \dfrac{1}{(1 - \phi_d)^2}} \right]^{0.5} \tag{44}$$

This equation has the particularity that the gas flow rate, the main and many times the only manipulable variable in the operation, is not present directly, but exerts its influence through the gas holdup. CHISTI et al. (1988) show that most of the published data on liquid velocity for the different types of air lift reactors can be satisfactorily correlated by Eq. (44) by choosing adequate values for the friction coefficients in each case. Only one coefficient needs to be adjusted, since the authors assume that K_T, the friction coefficient at the top of the loop, is negligible in concentric tube reactors, and that in external loop reactors K_T can be taken as equal to K_B, the friction coefficient for the bottom of the loop.

The authors present an empirical correlation of K_B for their fitting:

$$K_B = 11.4 \left(\frac{A_d}{A_B}\right)^{0.79} \qquad (45)$$

where A_B is the minimal cross-section at the bottom of the airlift reactor.

Although the fitting of the data is acceptable, it seems that the method of adapting K_B must be improved in order to use it for scale-up and design.

CHISTI and MOO-YOUNG (1988) further extended this model in order to predict the liquid circulation in air lift reactors operating with pseudoplastic fluids such as mold suspensions. This is a very important improvement, since many commercially useful fermentations involve such non-Newtonian liquids.

The other technique, used by several researchers to predict liquid velocity, is the momentum balance in the air lift reactor. This method has been used by BLENKE (1979), HSU and DUDUKOVIC (1980), KUBOTA et al. (1978), BELLO (1981), and KOIDE et al. (1984).

MERCHUK and STEIN (1981a) presented a simple model for the prediction of the liquid velocity as a function of the gas input in an air lift reactor. They assumed that the pressure drop at the bottom of their external loop reactor could be expressed as a continuation of the downcomer, using an equivalent length L_e. The frictional pressure drop in the different sections of the loop can be defined as:

$$\Delta P_i = 2 C_{fi} \rho_L J_{Li} (J_{Li} + J_{Gi}) \frac{L}{D_{hi}} \qquad (46)$$

where i can be riser, downcomer, and bottom section. In the case of the latter, L is replaced by L_e, the equivalent length.

The pressure drop balance gives:

$$L \rho_L g (\phi_r - \phi_d) = \Sigma \Delta P_i \qquad (47)$$

which, assuming the same frictional coefficient for all sections, leads to the following relationship between holdup, velocities, and geometric characteristics of the air lift reactor:

$$\phi_r - \phi_d = \frac{2 C_L J_{Ld}^2}{g}$$
$$\left[\left(\frac{1}{D_{hd}} + \frac{l}{L}\right) \left(1 + \frac{\beta R}{1 - \beta}\right) + \frac{1}{D_{hr} \zeta^2 (1 - \beta)} \right] \qquad (48)$$

where l is the equivalent length of the bottom section in terms of the hydraulic diameter, R is the recycle ratio,

$$R = Q_{Gd}/Q_{Gr} \qquad (49)$$

β is the "flowing volumetric concentration" defined by ZUBER and FINDLAY (1965),

$$\beta = \frac{Q_G}{Q_G + Q_L} \qquad (50)$$

applied in this case to the riser, and ζ is the ratio between cross-sectional areas of riser and downcomer,

$$\zeta = \frac{A_r}{A_d} \qquad (51)$$

MERCHUK and STEIN (1981a) considered the special case of their experimental system, where no gas was recirculated, and obtained a justification for their experimental finding that the liquid velocity was proportional to the superficial gas velocity to the power of 0.4.

KUBOTA et al. (1978) have used a similar approach to the analysis of ICI's "deep shaft" reactor. This is one of the largest air lift reactors ever built. It consists of a long, vertical shaft more than 100 meters deep. KUBOTA et al. (1978) presented a model for liquid circulation in this reactor, which has the characteristic that under steady-state operation the air is injected not at the bottom, but in the downcomer at a certain depth, which slightly complicates the analysis. The authors were able to simulate the operation of the reactor and to predict the condition of minimum air supply required to avoid flow reversal.

VERLAAN et al. (1986) used a similar model, combined with ZUBER and FINDLAY's (1965) expression for the gas velocity, Eq. (29), to calculate the frictional coefficients for a wide range of reactor volumes from experimental data reported by several authors. Fig. 9 shows that the calculated liquid velocity agreed with the experimental values reported, mainly for low gas recirculation rates.

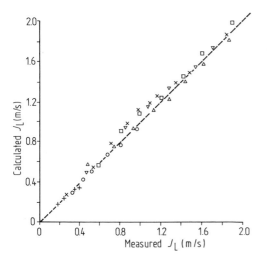

Fig. 9. Calculated and experimental liquid velocities in an airlift loop reactor (from VERLAAN et al., 1986).

KOIDE et al (1984) presented an analysis of the liquid flow in a concentric tube air lift reactor based also on a momentum balance. The main difference compared with the model of MERCHUK and STEIN (1981a) is that they used a convergence–divergence flow model for the bottom and the top of the loop. At the bottom, the effect of flow reversal on the pressure drop is included in an effective width of the gas–liquid flow path under the lower end of the draft tube h_1, which is smaller than the actual gap. The pressure drop at the bottom is calculated from:

$$
\Delta P_{\mathrm B} =
$$

$$
0.45\left[\frac{\rho_{\mathrm L}}{2}\left\{\frac{A_{\mathrm a} U_{\mathrm{La}}}{\pi D_{\mathrm{dr}} h_1 (1-\phi_{\mathrm a})}\right\}^2\right]\left(1-\frac{\pi D_{\mathrm{dr}} h_1}{A_{\mathrm a}}\right)+
$$

$$
+\frac{\rho_1}{2}\left(\frac{U_{\mathrm{Ldr}}}{1-\phi_{\mathrm{dr}}}\right)^2\left\{\frac{A_{\mathrm{dr}}(1-\phi_{\mathrm{dr}})}{\pi D_{\mathrm{dr}} h_1(1-\phi_{\mathrm a})}-1\right\}^2 \tag{52}
$$

and h_1 is given by an empirical equation based on the experimental results of the authors for several geometrical arrangements.

Later on, KOIDE et al. (1988) proposed to add to this expression the term:

$$
4510\left(\frac{\mu_{\mathrm L} U_{\mathrm{Gdr}}}{D_{\mathrm{dr}}}\right) \tag{53}
$$

in order to obtain a better fit of their experimental results, but without further justification. The constant 4510 was obtained by the direct search method using their data.

The pressure drop at the top of the liquid loop is considered again in terms of a convergence–divergence model, as

$$
\Delta P_{\mathrm T} =
$$

$$
0.685\rho_{\mathrm L}\left(\frac{U_{\mathrm{Ldr}}}{1-\phi_{\mathrm{dr}}}\right)^2\left\{1-\frac{A_{\mathrm{dr}}(1-\phi_{\mathrm{dr}})}{A_{\mathrm{to}}(1-\bar\phi)}\right\}^2+
$$

$$
+0.945\rho_{\mathrm L}\left(\frac{U_{\mathrm{La}}}{1-\phi_{\mathrm a}}\right)^2\left\{1-\frac{A_{\mathrm a}(1-\phi_{\mathrm a})}{A_{\mathrm{to}}(1-\bar\phi)}\right\}^2 \tag{54}
$$

where the average gas holdup is calculated as

$$
\bar\phi = (A_{\mathrm{dr}}\phi_{\mathrm{dr}}+A_{\mathrm a}\phi_{\mathrm a})/(A_{\mathrm{dr}}+A_{\mathrm a}) \tag{55}
$$

KOIDE et al. (1988) later proposed to use for the pressure drop at the top section an expression similar to that used at the bottom, based on a convergence–divergence model, and they gave an empirical expression for the effective height of the liquid above the draft tube.

Eq. (54) predicts pressure drops at the top, $\Delta P_{\mathrm T}$, of the order of those predicted at the bottom, $\Delta P_{\mathrm B}$, predicted by Eq. (52). In this sense, KOIDE et al. (1984, 1988) assign much more importance to $\Delta P_{\mathrm T}$ than has been the case before.

2.4 The Axial Dispersion Model

A complete model of a tower reactor would require velocity vectors at each point of the system, for both gas and liquid phases. This information, as seen in the discussion above, is not always available with a high degree of certainty because of the complicated nature of the fluid dynamics in two-phase flow. Furthermore, the solution of this type of model demands a high degree of sophistication and computational effort. For this reason, it is customary to use models for the residence-time distribution and mixing which allow the concentration of all the mixing characteristics within a few parameters.

The axial dispersion model has the advantage of having a single parameter and is widely accepted for the representation of tower reactors.

This model is based on a visualization of the mixing process in the tower reactor as a random, diffusion-like eddy movement superimposed on a plug flow. The axial dispersion coefficient D_z is the only parameter in the formulation,

$$\frac{\partial C}{\partial t} = D_z \frac{\partial^2 C}{\partial z^2} + \bar{U} \frac{\partial C}{\partial z} \tag{56}$$

where C is the concentration of a tracer. The boundary conditions depend on the specific type of tower reactor.

For a batch bubble column operation, the net average velocity is nil, and Eq. (55) is reduced to:

$$\frac{\partial C}{\partial t} = D_z \frac{\partial^2 C}{\partial z^2} \tag{57}$$

This is a closed system as far as the liquid is concerned. Using the appropriate boundary and initial conditions, OHKI and INOUE (1970) gave the following solution:

$$\frac{C}{C_e} =$$

$$1 + 2 \sum_{n=1}^{\infty} \left[\left(\cos \frac{n\pi}{H} \cdot z \right) \exp \left(- \frac{n^2 \pi^2 D_z t}{H^2} \right) \right] \tag{58}$$

from which the dispersion coefficient, D_z, can be obtained if the response of the system to a pulse of tracer is measured. The results are usually presented as the Peclet number:

$$Pe = U_L H / D_z \tag{59}$$

or as the Bodenstein number:

$$Bo = U_L D / D_z \tag{60}$$

In the case of the net liquid flow in the column, Eq. (56) must be considered. This case was solved by VAN DER LAAN (1958) following the original treatment by WEHNER and WILHELM (1956). The solution can be expressed as a relationship between the variance of the temporal change of the concentration of a tracer after a pulse injection and the Peclet number:

$$\tau^2 = \frac{2}{Pe} + \frac{2}{Pe^2} [1 - \exp(-Pe)] \tag{61}$$

In the case of airlift reactors, the closed system boundary conditions which lead to Eq. (59) are no longer valid.

The case of a loop reactor with open boundary was dealt with by VONCKEN et al. (1964). The mathematical procedure was modified by MURAKAMI et al. (1982), and a simplified approximation was offered. But this solution refers to a system which has a constant value of

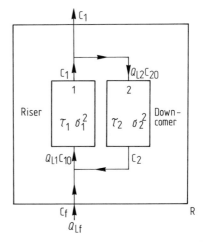

Fig. 10. Model for a two-section loop reactor (WARNECKE et al., 1985).

the dispersion coefficient over the whole volume. This is not true for airlift reactors because of the different hydrodynamic characteristics explained in Sect. 2.3.2.

WARNECKE et al. (1985) presented a model showing the importance of considering each section as a separate unit, as well as the interaction between the sections (Fig. 10). The model is based on two sections R_1 and R_2 (riser and downcomer) with liquid flow rates Q_{L1} and Q_{L2}, with a recycle ratio:

$$r = Q_{L2}/Q_{L1} \tag{62}$$

and with mean residence times and variances τ_1, τ_2, σ_1^2 and σ_2^2. Thus, this model covers the case of net liquid flow through the reactor (continuous operation). In this case, the overall variance is:

$$\sigma^2 = \frac{\sigma_1^2 + r\sigma_2^2}{1-r} + r\left(\frac{\tau_1 + \tau_2}{1-r}\right)^2 \tag{63}$$

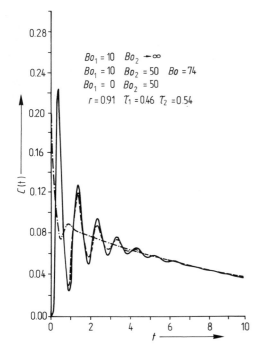

Fig. 11. Response of WARNECKE's (1985) model to a pulse for several values of the Bodenstein number in each one of the sections.

and the mean time is:

$$\tau = \frac{\tau_1 + \tau_2}{1-r} \tag{64}$$

WARNECKE et al. (1985) showed that this model can reproduce the behavior of an airlift reactor, assuming that R_1 and R_2 are represented by reactors ranging from plug flow to perfect mixing (Fig. 11).

A model completely opposed to the above was presented by MERCHUK and YUNGER (1990) for an air lift reactor. Instead of a riser and a downcomer directly interconnected, and each with a different dispersion coefficient, as assumed by WARNECKE et al. (1985), they assumed that both riser and downcomer behave as plug-flow sections, and all the mixing occurs solely in the gas separator, presented as a perfectly mixed stage. The response of the system to a pulse input is given by:

$$\frac{C}{C_e} = \sum_{n=0}^{k} \frac{1}{n!} \left(\frac{t}{\tau_2} - ns\right)^n \exp(ns - t/\tau_2) \tag{65}$$

where k is the integer value of t/τ_1, and τ_1, τ_2 are the mean residence times of the perfectly mixed and plug-flow regimes, respectively.

This model was presented in order to stress the importance of considering the influence of each section of the reactor on overall mixing. Indeed, the model simulates quite adequately the behavior of an airlift reactor, the only parameter being the ratio of the residence times in the gas separator and plug flow regions. Typical curves and relationships between the ratio of the residence times and the Bodenstein number can be seen in Figs. 12 and 13, which indicate how much care must be taken in interpretation of experimental data, since completely different models can appear satisfactory under given operational conditions. However, only a model describing the real fluid dynamics of the system is adequate for scale-up.

In all of these models, the dispersion coefficient is a phenomenological constant, which must be found experimentally. One can, however, obtain D_z from a mathematical model of the flow in the system. One of the simplest methods is the circulation cell model by JOSHI and SHARMA (1979) that can be used to calcu-

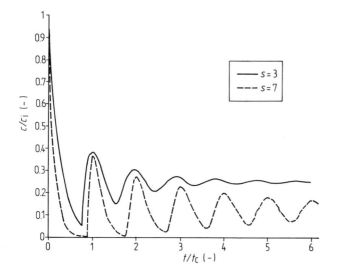

Fig. 12. Response of the simple two-zone model to a pulse input (MERCHUK and YUNGER, 1990).

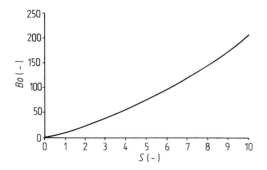

Fig. 13. Correlation of the ratio of residence times in the plug flow region and the perfectly mixed region with the Bodenstein number (MERCHUK and YUNGER, 1990).

late the liquid circulation velocity in bubble column reactors:

$$U_{CL} = 3.1 [H(U_g - \varepsilon U_0)]^{1/3} \tag{66}$$

Eq. (66) can be used in the relation proposed by JOSHI and SHARMA (1979) for calculating the dispersion coefficient in the liquid phase:

$$D_z = 0.33 \, D \, U_{CL} \tag{67}$$

ZEHNER (1986b) derived the following theoretical expression from his model of cylindri-

cal eddies:

$$D_z = \frac{D}{2} \sqrt[3]{\frac{1}{f} \frac{\Delta\rho}{\rho_L} g D J_G} \tag{68}$$

Both Eqs. (67) and (68) are shown by the respective authors to fit available data on mixing in bubble column tower reactors.

MATSUMOTO et al. (1988) derived a theoretical expression for the dispersion coefficient in the liquid phase of a bubble column, including the case of suspended solids.

This model is based on the mixing length theory, expressing the axial dispersion coefficient as:

$$D_z = |U_z'| \, l_m \tag{69}$$

where U_z' is the averaged fluctuation velocity of the slurry and l_m is the mixing length for mass transfer. Assuming spherical bubbles of uniform diameter in a cylindrical tower, an overall energy balance and appropriate simplifications allow the derivation of the following expression:

$$\frac{D_z}{\sqrt{g D^3}} =$$

$$K_B \left[\left(1 - \frac{\phi}{1-\phi} \frac{J_L}{J_G}\right) \left(\frac{\rho_m - \rho_g}{\rho_m}\right) \right]^{1/2} \left(\frac{\phi^{1/2}}{1-\phi}\right) \tag{70}$$

where K_B is a coefficient that must be determined experimentally and that originates in the definition of the mixing length for mass transfer:

$$l_m = K_B \sqrt{2} D (1-\phi)^{-1} \tag{71}$$

Here ρ_m is the apparent density of the slurry phase, which equals the liquid density when solids are not present. The authors found satisfactory agreement between their model and experimental results with both two-phase and three-phase systems. D_z depends on the superficial gas velocity via the gas holdup.

Another theoretical development relevant to three-phase tower reactors is due to WACHI et al. (1987). They extended the model originally presented by MIYAUCHI et al. (1981) for a bubble column to allow for net liquid flow. Starting from a momentum balance in an annular element, they integrate to obtain the superficial liquid velocity:

$$J_L = \int_0^R 2\pi r (1-\phi) \frac{U_L}{\pi R^2} dr \tag{72}$$

where the local velocity U_L is given by:

$$U_L = \frac{\tau_w R}{2 v \rho_L} (1-\varphi^2) +$$

$$+ g \frac{R^2 \phi}{8v} (1-\varphi^2)^2 + U_{Lw} \tag{73}$$

The liquid velocity at the wall U_{Lw} is:

$$U_{Lw} = \frac{J_L}{1-\phi} - \frac{\tau_w R}{12 \rho_L v} \left(\frac{3-4\phi}{1-\phi} \right) -$$

$$- g \frac{R^2 \phi}{48 v} \left(\frac{2-3\phi}{1-\phi} \right) \tag{74}$$

which is an explicit equation, since τ_w depends on the liquid velocity at the wall, as given by UEYAMA and MIYAUCHI (1976):

$$\tau_w = \frac{\rho_L}{(11.63)^2} U_{Lw} |U_{Lw}| \tag{75}$$

Their final equation for the axial dispersion coefficient is:

$$D_z = \frac{(1-\phi)\varphi^2}{8v} q_r \left\{ \frac{11}{96} \left(\frac{\tau_w \varphi}{\rho_L v} \right)^2 + \right.$$

$$+ \frac{73}{17\,280} \left(\frac{g \phi \varphi^2}{v} \right)^2 + \left(U_{wL} - \frac{J_L}{1-\phi} \right) +$$

$$+ \frac{1}{160} \frac{g \phi \tau_w \varphi^3}{\rho_L v} + \frac{g \phi \varphi^2}{8v} \left(U_{Lw} - \frac{J_L}{1-\phi} \right) +$$

$$\left. + \frac{2 \tau_w R}{3 \rho_L v} \left(U_{Lw} - \frac{J_L}{1-\phi} \right) \right\} + q_z \frac{v}{1-\phi} \tag{76}$$

where q_r and q_z are correction factors which take account of the local derivations of the holdup and of the turbulent kinematic viscosity, v.

Fig. 14 shows quite satisfactory agreement of experimental data and theory, where the influence of both gas and liquid superficial velocities is presented. The values of the correction factors q_r and q_z are shown in Fig. 14.

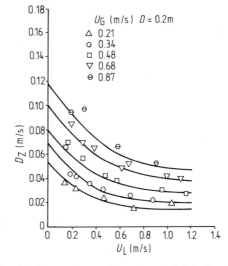

Fig. 14. Comparison of experimental data for the axial dispersion coefficient, D_z, with the predictions of the model by WACHI et al. (1987).

2.5 Summary of Flow Models

The two-phase flow in bubble columns is a very complex phenomenon, and exact description is still beyond our knowledge. But the use of sensible assumptions has permitted the formulation of several mathematical models which allow acceptable approximations for both the liquid velocity and the mixing characteristics of the system. In particular, the axial dispersion model is one of the most accepted models for the representation of mixing, and models have been published for the different types of tower reactors.

3 Mass Transfer Modelling

The determination of mass transfer coefficients in laboratory or industrial equipment can be based either on steady-state or on unsteady-state mass transfer experiments. For gas–liquid systems, these experiments can be either absorption or desorption studies.

Steady-state experiments are associated either with the measurement of small differences in the concentration of continuous phases or, in the case of gas absorption in batch systems, with the provision of a sink for the absorbed gas by means of a chemical reaction. Such chemical reactions imply the presence of reactive solutes in the liquid that may greatly influence the physical characteristics of the system. Therefore, most of the experimental results are largely restricted to the system being studied. This is due to the sensitivity of gas dispersion characteristics, especially of interfacial areas and gas holdup, to variations in density, viscosity, ionic strength, surface tension, etc. Since the hydrodynamics of the system are so heavily dependent on gas holdup, the entire reactor performance is affected by changes in the physico-chemical properties of the liquid phase.

Unsteady-state methods for determining mass transfer coefficients, $k_L a$, are complicated by the possibility of distortion of the measured values by the probe response dynamics. Probe response dynamics become an influential factor in $k_L a$ calculations when the value for $k_L a$ is approximately equal to or greater than the inverse of the value for the probe response time. These factors complicate the modelling.

Whatever the method used, the basis of the technique is the comparison of a measured variable with the value (or values) predicted by a mathematical model of the process. It has already been pointed out that the choice of the model is very important and a poor assumption of gas or liquid phase flow characteristics may lead to errors and deviations from the true values (KEITEL and ONKEN, 1981; MERCHUK and SIEGEL, 1988).

All published data on mass transfer in tower reactors are based on fitting very simple fluid dynamic models to experimental values of concentration, measured either in steady or unsteady state operation. These models assume either plug flow or complete mixing of the two phases present.

The most common assumption is complete mixing (CM) in the liquid phase and plug flow (PF) in the gas phase. This approach is justified by considering the relaxation times of the processes involved (ROELS, 1983).

The ratio defined by DECKWER (1986),

$$\phi = \frac{\text{liquid phase mixing time}}{\text{mass transfer time}} =$$

$$= \frac{L^2/\phi_L D_{zL}}{1/k_L a} \tag{77}$$

may be taken as a sufficient indicator. A low value of this ratio implies that mixing is much faster than mass transfer, and therefore the concentration is homogeneous throughout the reactor. In such a case, the CM assumption seems to be justified, and therefore the values of the mass transfer coefficient obtained from fitting the experimental data to the model are reliable and valid for scale-up. But problems may arise when applying these data to the modelling of an absorption with chemical reaction, where the mass transfer rate is increased because of the modification of the profiles of the absorbed component near the gas–liquid interphase (DANCKWERTS, 1970). The theory of absorption with chemical reaction allows the prediction of a newly increased coefficient, on the basis of the $k_L a$ without chemical reaction, the kinetics and the diffusion constants. But the problem that is usually not addressed is that under the chemical reaction regime the first assumption of perfect mixing in the bulk of the liquid phase may no longer be valid, since $k_L a$ in Eq. (75) is increased and therefore the mass transfer coefficient based on this fluid-dynamic model may also no longer be valid.

Had the experimental mass transfer coefficient $k_L a$ been obtained by the fitting of a realistic fluid-dynamic model, $k_L a$ would be valid under any condition irrespective of the kinetic regime. But, as stated before, the state of the art is still far from this achievement.

The key for further improvements in this direction lies in recognizing that mass transfer rates depend strongly on the hydrodynamics, and in the quite common case of sparingly soluble gases, specifically on the dynamics of the liquid phase near the gas–liquid interface. It follows that a connection must be found between the models presented above for the flow in a tower reactor and the mass transfer rate in it.

3.1 Prediction of Mass Transfer Rates

The factors potentially affecting the volumetric mass transfer coefficient are the interfacial area, related to the gas phase holdup, and the dynamics of the liquid near the interphase (DANCKWERTS, 1970). It is clear that holdup and bubble size have a strong effect on the mass transfer rate. It has also been shown by KALISCHEWSKY and SCHÜGERL (1978) that the mass transfer coefficient itself can vary by an order of magnitude when the viscosity of the liquid is increased. This suggests that the fluid dynamics of both the liquid and gas phases should be the basis for the modelling of mass transfer in tower reactors.

Several attempts have been made to describe the mass transfer coefficient in terms of fluid dynamic characteristics (LEVICH, 1962; BANERJEE et al. 1970; LAMONT and SCOTT, 1970; FORTESCUE and PEARSON, 1967; HENSTOCK and HANRATTY, 1979). Recently, KAWASE et al. (1987) extended the previous models to the treatment of non-Newtonian liquids obeying the power law model:

$$\tau = -K\gamma^n \tag{78}$$

The model is based on the penetration theory, which predicts the following simple dependence of the mass transfer coefficient k_L on the diffusivity and the contact time of an element of liquid with the gas:

$$k_L = \frac{2}{\sqrt{\pi}} \sqrt{\frac{D}{t_c}} \tag{79}$$

The contact time is obtained by use of Kolgomoroff's length scale, which results in a function of the energy dissipation in the tower reactor, ε, finally giving:

$$k_L = \frac{2}{\sqrt{\pi}} \sqrt{D} \left\{ \frac{\varepsilon}{K/P} \right\}^{\frac{1}{2(1+n)}} \tag{80}$$

where n is the index in the power law model.

The interfacial area per unit volume, a, was expressed by use of CALDERBANK's (1958) equation for bubble size and an equation given by KAWASE and MOO-YOUNG (1986) for the gas holdup. The volumetric mass transfer coefficient is then given in a dimensionless form as:

$$Sh = 12 C_4 \frac{1}{\sqrt{\pi}} \sqrt{1.07} \, n^{1/3} Sc^{1/2}$$

$$Re^{(2+n)/2(1+n)} Fr^{(11n-4)/39(1+n)} Bo^{3/5} \tag{81}$$

where

$$Sh = \frac{k_L a D^2}{D}, \quad \text{Sherwood number} \tag{82}$$

$$Sc = \frac{K/\rho D^{1-n}}{J_G^{1-n} D}, \quad \text{Schmidt number} \tag{83}$$

$$Re^* = \frac{D^n J_G^{2-n}}{K/\rho}, \quad \text{Reynolds number} \tag{84}$$

$$Fr = \frac{J_G^2}{g D}, \quad \text{Froude number} \tag{85}$$

$$Bo = \frac{g D^2 \rho}{\sigma}, \quad \text{Bodenstein number} \tag{86}$$

and the constant C_4 is given tentatively as:

$$C_4 = 0.0645 \, n^{2/3}$$

KAWASE et al. (1987) showed a reasonable agreement of experimental data with Eq. (81), for both non-Newtonian and Newtonian $(n-1)$ fluids. They found the experimental results presented by various researchers to fall within $\pm 50\%$ of the predicted values.

While the above model is certainly encouraging, it should be understood that it is not the ideal connection between the field of flow and the mass transfer rate. The model by KAWASE et al. (1987), as with others previously published, is based on the specific energy dissipation in the reactor for the fluid-dynamic evaluation. However, the specific energy dissipation is a macroscopic parameter that lumps all the volume into a single value. It is obvious that different geometric arrangements will give different distributions of flow characteristics for the same energy input. The model is not able to discriminate between them.

3.2 Structured Models for Mass Transfer in Tower Reactors

Several attempts have been made to establish models of a bubble column reactor which take account of the fluid dynamic structure of the reactor. Such structured models consider that different areas whithin the reactor may have distinctive mass transfer characteristics due to the flow in the region.

DECKWER et al. (1978) conducted a very comprehensive study of carbon dioxide transfer in two bubble columns by measuring the profiles of the absorbed gas along the col-

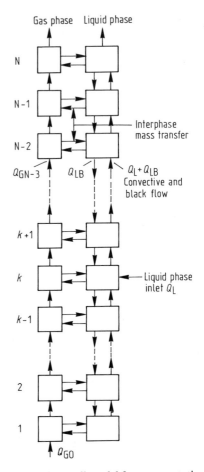

Fig. 16. Back flow cell model for representation of a tower reactor (DECKWER et al., 1980).

umns. They found that a constant value of the mass transfer coefficient, while giving a satisfactory interpretation of the overall mass transfer rate in the tower, could not predict the concentration profile. In order to achieve this it was assumed that $k_L a$ was highest near the gas sparger and then diminished along the tower. The following profile was used:

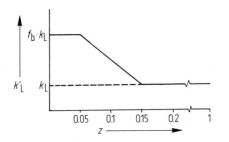

Fig. 15. Proposed profile for the variation of the mass transfer coefficient, k_L, along the axis of the column (DECKWER et al., 1978).

$$k'_L = f_b k_L \qquad\qquad 0 \leq z/H < 0.05$$

$$k'_L = k_L (1 + 10(f_b - 1)(0.15 - z/H)) \qquad (87)$$

$$0.05 \leq z/H \leq 0.15$$

$$k'_L = k_L \qquad\qquad 0.15 \leq z/H \leq 1$$

The profile can be seen in Fig. 15. This approach permitted a satisfactory fitting of the experimental carbon dioxide profiles. DECKWER et al. (1978) recommended the value of $f_b = 2$.

The spatial variation of $k_L a$ has also been considered by other authors, and has been proven experimentally by ZHAO et al. (1988). A model of a bubble column based on back flow cells, which can be seen in Fig. 16, was proposed by DECKWER et al. (1980). In order to explain the absorbed CO_2 profiles satisfactorily, the authors postulated the following variation of the mass transfer coefficient as a function of cell number:

$$k_L = k_{LB}[1 + a\exp - b(j - k^2)] \tag{88}$$

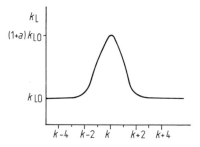

Fig. 17. Dependence of the mass transfer coefficient, k_L, on cell number in the back flow cell model (DECKWER et al., 1980).

where j is the cell number and k denotes the cell where the gas is injected into the system and where most of the energy is dissipated (Fig. 17).

LUTTMAN et al. (1983b) found similarly that a constant value of the mass transfer coefficient for oxygen absorption could not explain the dissolved oxygen concentration profiles during yeast cultivation. They explained this phenomenon as due to a strong change in the specific interfacial area caused by increased coalescence of bubbles near the gas sparger. The mass transfer coefficient was given by:

$$k_L a(z, t) =$$
$$\begin{cases} k_L a^i(t) \exp(-K_{ST}(t)z/\alpha H) & 0 \le z \le \alpha H \\ k_L a^i(t) \exp(-K_{ST}(t)) & \alpha H \le z \le H \end{cases}$$
$$\tag{89}$$

where K_{ST} is defined as the coalescence factor, $k_L a$ is the maximum value of the mass transfer coefficient and is valid only at the gas inlet. From the sparger up it decreases until $z = \alpha H$, where it reaches the constant value $k_L a^i(t) \exp(-K_{ST})$. Fig. 18 shows how consideration of the axial variations of the mass transfer rate improves the prediction of oxygen concentration profiles.

Still another aspect of this problem was considered by SCHUMPE and DECKWER

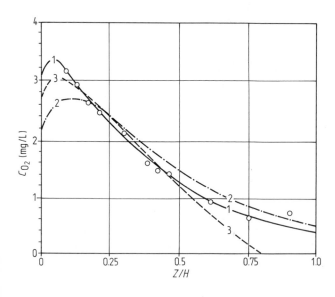

Fig. 18. Comparison of the measured profile of dissolved oxygen with the prediction of different models: (1) coalescence function and Monod kinetics with regard to oxygen axial changes; (2) space-independent $k_L a$ and Monod kinetics with respect to oxygen; (3) coalescence function and space-independent growth rate (LUTTMAN et al., 1983b).

(1980). In their critical analysis of the chemical method for the determination of interfacial areas, they consider the distribution of bubble size and its effect on the overall mass transfer rate. The same idea had been presented earlier by MERCHUK and RONCO (1966), who studied the effect of the physical mass transfer rate on the enhancement factor, and concluded that the existence of areas of different kinetic regimes may originate from local variations of the physical mass transfer coefficient, $k_L a$.

In airlift reactors, the different flow characteristics in each part of the reactor (riser, downcomer, and gas separator) indicate that a structured model should be used for the system. Since the dependence of the mass transfer coefficient on the fluid dynamics is well established, it is obvious that the minimum desirable analysis of such a system would involve a different $k_L a$ for each region. This argument has not yet been tested experimentally, and all of the mass transfer coefficients published are in fact based on the assumption of complete mixing in the liquid phase.

MERCHUK and SIEGEL (1988) proposed a method for the simultaneous determination of the mass transfer rates in the three regions of an airlift reactor by measuring the transient concentration at three points of the system after a pulse input. No data have yet been reported on the matter, but it is a promising method for providing data on $k_L a$ that can be used for modelling and scale-up of airlift reactors.

concentrations, temperatures, pH, etc. Recently, it has been suggested (MERCHUK, 1990) that, for biochemical reactions, the shear stress has an effect on the kinetics itself, therefore adding a new dimension to the problem, since the distribution of shear over the volume of the reactor would become important.

4.1 Bubble Column Modelling

Process modelling in bubble column reactors has been reviewed very clearly by DECKWER (1986). The classification proposed for the models is based on the assumptions of the flow of the gas and the liquid phases. Fig. 19 shows the categories considered by DECKWER.

		Gas phase			
		CM	AD	PM	0
Liquid phase	CM	X	X	X	X
	AD		X	X	X
	PF			X	X
	0	X	X	X	X

Fig. 19. Classification of tower reactor models based on the flow characteristics of each phase (DECKWER, 1986): CM, completely mixed; AD, axially dispersed; PF, plug flow; 0, irrelevant. X indicates relevance of the model.

4 Process Models in Tower Reactors

The modelling of a chemical or biochemical process in a tower reactor implies the integration of flow, mass transfer, and kinetics and is, therefore, based on the respective models for each of the three aspects. The interaction between mass transfer and kinetics is also taken into account, while the interaction between flow characteristics and kinetics is expressed by taking into account the local values of the

In Fig. 19 CM indicates complete mixing, PF plug flow, AD axial dispersed, and 0 indicates that, due to the reaction regime, the concentrations of the reactants in the phase do not influence the overall reaction rate, and therefore the fluid dynamics of the phase are irrelevant.

DECKWER's classification covers all the possibilities of bubble column modelling using either plug flow, complete mixing, or axial dispersed fluid dynamic models for gas and liquid phases. Because of their relevancy, only the cases marked with an X in Fig. 19 are presented by the author. For instance, the assumption that the Bodenstein number for the gas phase is substantially larger than the Bodenstein number for the liquid phase,

$$Bo_L = \frac{H U_L}{\phi_L D_{zL}} \ll Bo_G = \frac{H U_G}{\phi_g D_{zL}} \qquad (90)$$

excludes the combinations in which the mixing of the gas phase is better than that of the liquid phase; thus, only the cases marked with an X in Fig. 19 have been analyzed by DECKWER (1986).

The general procedure presented is as follows. First, a balance is written for the main reactant, A, in the gas phase. For a completely mixed liquid phase, this balance can be integrated directly to give the axial profile of A in the gas. This allows the calculation of the gas absorbed per unit volume or the specific absorption rate. This is a function of the concentration of A in the bulk of the liquid. A balance for A over the liquid film allows the solution of the problem, depending on the kinetic regime. For this DECKWER (1986) used the solutions given by CHAUDARI and RAMACHANDRAN (1980) based on the film model. In the case of spatial variation of concentrations in the liquid phase, the procedure must be modified to take this into account.

Axial dispersion models for bubble columns will not be discussed in this chapter. Interested readers are referred to DECKWER's excellent work.

An alternative way of presenting the mixing in a reactor is the use of series of completely mixed cells. It is known that a series of completely mixed cells approaches plug flow when the number of cells is very large, assuming unidirectional connections between successive cells. A more complex behavior can be modelled on the basis of this scheme, as in the back flow cell model proposed by DECKWER et al. (1980), which can be seen in Fig. 16. An equal number of cells is assumed for gas and liquid, which are interconnected at each level via interfacial mass transfer. This does not imply equal mixing characteristics in both phases, because while the gas-side cells are connected in such a way that convective mass transfer is only possible upwards, the liquid-side cells also allow back flow. A mass balance for the absorbing component A around the cell j in the gas phase gives:

$$\left[1 + \frac{\alpha}{(j-1)} \right] Y_{(j-1)} U_{G(j-1)}/U_{GB} -$$

$$- [1 + \alpha_j] Y_j U_{Gj}/U_{GB} -$$

$$- St_{Gj} \left[(1 + \alpha_j) Y_j - C_j \frac{He}{P_B} \right] = 0 \qquad (91)$$

where the Stanton number for the gas phase at cell j is:

$$St_{Gj} = (k_{Lj}/N)(6\bar{\phi}\varphi_j/d_s)(H/U_{GB}) \cdot$$
$$\cdot (RT/He) \qquad (92)$$

and α is the coefficient for the linear variation of hydrostatic pressure along the column:

$$\alpha = -\rho g (1 - \bar{\phi}) H/(N P_B) \qquad (93)$$

The function φ_j describes the spatial distribution of the gas holdup in the cell j, and N is the total number of cells, identical for both gas and liquid phase.

In the liquid phase, the balance for the case of cocurrent flow gives the following:

$$(1 + \gamma) C_{j-1} - (1 + 2\gamma + St_{Lj}) C_j + \gamma C_{j+1} +$$
$$+ St'_{Lj} Y_j = 0 \qquad (94)$$

where γ is the back flow ratio, C_j is the concentration of the considered component in the liquid phase, stage j, and the two Stanton numbers are given by:

$$St_{Lj} = (k_{Lj}/N)(6\bar{\phi}\varphi_j/N)(H/U_L)/d_s \qquad (95)$$

$$St'_{Lj} = St_{Lj} P_B (1 + \alpha_j)/He \qquad (96)$$

Similarly, equations were derived for countercurrent gas–liquid flow. As explained in the mass transfer section, a variation of the mass transfer coefficient was assumed by DECKWER et al. (1980) in order to account for an increased mass transfer rate near the gas sparger. The authors reported satisfactory fitting of data on absorption of CO_2, where both the total mass transfer rate and the concentration profiles were monitored.

Sometimes, the chemical process is of such a nature that very simple hydrodynamic models can combine successfully with the kinetics. This is the case with the model presented by

TANG et al. (1987) for phenol degradation in a draft tube gas–liquid–solid fluidized bed. A modified draft-tube airlift reactor was used in the experimental work, but the mathematical model was based on a CM approximation for the liquid phase, and in this sense it can be considered a bubble column. The kinetics of the diffusion and reaction of both phenol and oxygen in the biofilm and in the activated carbon particles are taken into account, and the resulting model seems to explain the transient behavior of the system when a sudden change in phenol concentration is applied.

Thus, it can be seen that the model used for the hydrodynamics of the tower reactor does not always correspond to the experimental apparatus used. Recycles may be taken into account in the liquid phase of a bubble column, whereas total mixing may be assumed for an airlift reactor. The choice will depend on the characteristic times of the process (mass transfer, kinetics) and on the sophistication required to accomplish the aims of the modelling.

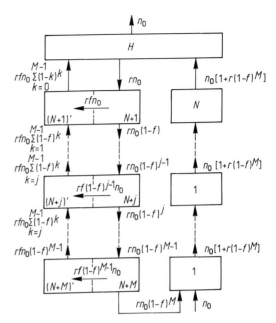

Fig. 20. Gas phase circulation in an airlift tower reactor, considering upflow of coalesced bubbles in the downcomer. Model of Ho et al. (1977).

4.2 Air Lift Models

The published simulations of processes in air lift reactors refer mainly to biological processes, especially single-cell protein production. Ho et al. (1977) presented a model for oxygen transfer in an air lift fermenter, where the absorbed oxygen is consumed at a constant rate by growing microorganisms. The model is structured in the sense that the flow structure of the reactor is considered. Indeed, the particular characteristics of the three different parts of the reactor, riser, gas separator, and downcomer, are taken into account. This is a model based on the concept of perfectly mixed compartments in series, similarly to others presented before. Starting from the gas sparger, N stages are counted for the riser, one for the gas separator, and M for the downcomer. At each level, oxygen transfer is considered between the gas and the liquid compartment. While the liquid in both the riser and downcomer is represented by a series of stages connected unidirectionally, the flow patterns of the gas in the downcomer are more complicated (Fig. 20). In addition to the down-flow of the entrained

gas, the model allows for an up-flow of gas, which represents the larger, coalesced bubbles that rise countercurrently, a phenomenon which is observed in air lift reactors with relatively low liquid velocity in the downcomer. It is assumed that a fraction, fr, of the volumetric gas flow reverses direction due to coalescence, constant for all stages. The mass balances in the riser are in the stage i:

$$m_{i-1} Y_{i-1} - n_i Y_i - $$
$$- k_L a_i [(P_i/He) Y_i - C_i] (1 - \phi_i) V_i = 0 \qquad (97)$$

for the gas phase, and

$$Q C_{i-1} - Q C_i + $$
$$+ k_L a_i [(P_i/He) Y_i - C_i] (1 - \phi_i) V_i - $$
$$- R X (1 - \phi_i) V_i = 0 \qquad (98)$$

for the liquid phase.

In the downcomer, Ho et al. (1977) represented the gas by two balances, one for the down-flowing stream and the other for the up-flowing stream. In fact, as can be seen in Fig.

20, they considered two separate gas phases: an entrained gas phase flowing downward,

$$n_{j-1} Y_{j-1} - n_j Y_j - n_j^T Y_j -$$
$$- k_L a_j [(P_j/He) Y_j - C_j](1 - \phi_j) V_i = 0 \qquad (99)$$

and a gas phase flowing upward

$$n'_{j+1} Y'_{j+1} + n_j^T Y_j - n'_j Y'_j -$$
$$k_L a'_j [(P_j/He) Y'_j - C_j](1 - \phi_j) V_i = 0 \qquad (100)$$

The mass balance for the liquid phase is similar to Eq. (98), except that two mass transfer terms appear, one for each gas phase.

The molar flow between phases can be obtained assuming that a fraction r from the inlet gas flow rate is entrained into the downcomer, and remembering that a fraction fr will reverse direction due to coalescence:

$$n_j^T = n_0 r fr (1 - fr)^{j-1} \qquad (101)$$

Fig. 21 shows the effect of tower height on the dissolved oxygen profiles for several gas flow rates, as predicted by the model for an oxygen consumption rate of 0.5 mg $L^{-1} s^{-2}$ O_2. It can be seen that an increase in tower height would be beneficial at higher gas flow rates. The predicted dissolved oxygen profiles are more homogeneous at lower tower heights.

Perfectly mixed cells in series have also been used to model chemical processes in air lift tower reactors. A rather elaborate model was presented by WACHI and MORIKAWA (1986) for the chlorination of ethylene in a boiling bubble column reactor. The fluid dynamic description of the reactor is similar to the model by DECKWER et al. (1980) (Fig. 16), but with the addition of a loop connecting the first and the last liquid phase cells (Fig. 22). This follows the model proposed by MIYAUCHI and VERMEULEN (1963). In addition, heat transfer between gas and liquid phases is considered

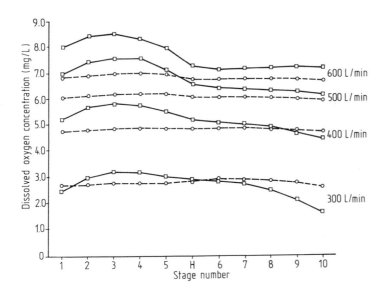

Fig. 21. Profiles of oxygen concentration predicted by the model of HO et al. (1977) for several gas flow rates and two different heights, and an oxygen consumption rate of 0.5 mg $L^{-1} s^{-1} O_2$.

The disengagement top section is considered a single perfectly mixed stage for both phases. The mass transfer coefficient was assumed to vary due to changes in holdup in the downcomer, but was regarded as constant all along the riser due to the balancing effects of gas absorption and expansion because of pressure changes.

along with the variation in gas holdup due to the absorption of gaseous species, vaporization of solvent, and axial variations in both temperature and pressure.

The process consists of the addition of chlorine to ethylene to give 1,2-dichloroethane, and simultaneous absorption and reaction of both gases is considered via both mass and

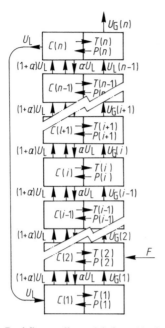

Fig. 22. Backflow cell model for chlorination of ethylene in a boiling reactor (from WACHI and MORIKAWA, 1986).

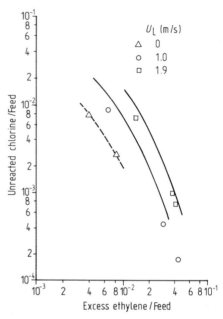

Fig. 23. Predicted and experimental conversions of chlorine and ethylene in a boiling bubble column reactor (from WACHI and MORIKAWA, 1980).

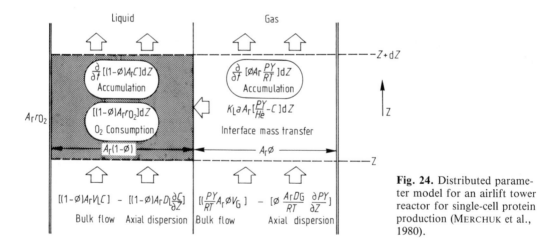

Fig. 24. Distributed parameter model for an airlift tower reactor for single-cell protein production (MERCHUK et al., 1980).

heat energy balances. The model was able to represent the conversions of ethylene and chlorine satisfactorily, as shown in Fig. 23.

A distributed parameter gas model has been used (MERCHUK et al., 1980; MERCHUK and STEIN, 1981 b) for the modelling of single-cell protein production in an airlift reactor. Continuous operation was considered. The oxygen consumption was represented by a Monod-type expression for biomass production and substrate and oxygen consumption. Fig. 24 shows a mass balance in a differential of the height dz. Although equations were presented for axially dispersed flow of both phases, calculations were done assuming plug flow.

The differential equations representing the riser are as follows for the oxygen concentration in the gas phase:

$$\frac{dy_1}{d\eta} = -St_G \frac{(1-y_1 Y^i)^2}{(1-Y^i)} \left[\left(\frac{1+\alpha-\alpha\eta}{1+\alpha} \right) y_1 - y_2 \right] \tag{102}$$

and for the oxygen dissolved in the liquid phase:

$$\frac{dy_2}{d\eta} = St_L \left[\left(\frac{1+\alpha-\alpha\eta}{1+\alpha} \right) y_1 - y_2 \right]$$
$$- (1-\phi) R_A D_A \left(\frac{B_A S_A}{S_A + K_A} \right) \tag{103}$$

where y_1 and y_2 are the dimensionless concentrations of oxygen in the gas and liquid phases, η is the dimensionless axial coordinate, Y^i is the inlet oxygen fraction in the gas, and $(1+\alpha)$ is the ratio of the hydrostatic pressure at the bottom and top of the tower.

$$\alpha = g\rho_L(1-\phi)H/2P_T \tag{104}$$

The Stanton numbers for the gas and liquid phases are defined as follows:

$$St_G = (Hk_L a/J_G)(RT/He) \tag{105}$$

$$St_L = Lk_L a/J_L \tag{106}$$

In Eq. (103), B_A and S_A are the dimensionless concentrations of biomass and substrate. The axial distribution of these concentrations is given by:

$$\eta = \frac{1}{(1-\phi)R_A} \cdot$$
$$\cdot \left[\frac{K_A}{S_A^0 + B_A^0} \ln \left(\frac{S_A^0 B_A}{B_A^0 S_A^0 - B_A B_A^0 + (B_A^0)^2} \right) + \right.$$
$$\left. + \ln \left(\frac{B_A}{B_A^0} \right) \right] \tag{107}$$

and

$$S_A = S_A^0 - B_A + B_A^0 \tag{108}$$

The reaction group R_A is given by:

$$R_A = H\mu/J_L \tag{109}$$

The gas separator was represented by a well-mixed stage, both in the gas and in the liquid phase. The gas holdup was related to the superficial gas velocity via a momentum balance as given in Eq. (47). Simulations were carried out by integration of the system for the special case of no gas entrainment in the downcomer. This would be the most critical operation mode because of the danger of anoxia in the downcomer.

Fig. 25 shows the profiles of oxygen concentration along the airlift reactor. At $\eta = 0.5$, the

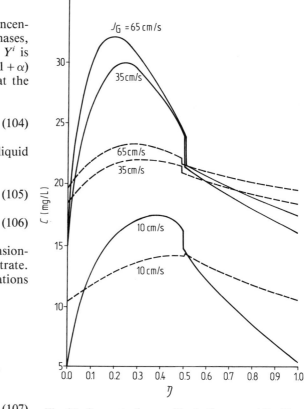

Fig. 25. Concentration profiles in the gas and liquid phases of an airlift reactor during single-cell protein production in continuous operation for two tower heights: —— $H = 12$ m; ---- $H = 4$ m (MERCHUK and STEIN, 1981 b).

jump in concentration indicates the gas separation zone. The concentrations are calculated with Monod type kinetics for several superficial gas velocities and two heights, 4 and 12 meters. It can be seen that the drop of dissolved oxygen concentration in the downcomer is dangerously large for the smallest flow rate, but for higher J_0 it remains at high levels over the entire length of the reactor. This model predicts a stronger dependence of the profiles of dissolved oxygen on reactor height than the model based by Ho et al. (1977) on interconnected cells.

The model of MERCHUK and STEIN (1981b) was improved by including both axial dispersion terms in the calculations and by considering the transient terms of the differential equations, thus allowing for the modelling of a batch process (LUTTMAN et al., 1982, 1983a). The model was used for parameter identification and simulation of single-cell protein production.

5 Operation Policies

Tower reactors are operated in industry in both batch and continuous mode. In between these two methods of operation, there is the possibility of operating under variable volume conditions. LUND and SEAGRAVE (1971) showed that this type of operation in a homogeneous reactor may increase the yield of the desired products. HALLAILE and MERCHUK (1986) showed that in addition other advantages are expected from the cyclic operation of gas–liquid and gas–liquid–solid reactors, especially in catalytic reactions where the catalyst may be reused without treatment.

The cyclic operation of the reactor is shown in Fig. 26. The vessel is first filled up to a volume V_{max} at a constant feed flow rate Q_f. The reactor is then operated batchwise during a period of time $t_2 - t_1$ and emptied at a constant flow rate Q, but only down to a certain volume V_{min} whereupon the cycle starts again. After a number of cycles, the system reaches a pseudo-steady-state characterized by the fact that if the flow rates and times remain the

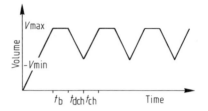

Fig. 26. Cyclic operation of a reactor.

same, the composition of all the components will repeat itself in each cycle.

The model was integrated under the assumption of perfect mixing in all phases, which may be a reasonable assumption since the main interest in this type of operation lies in slow chemical reactions in which the concentrations in the bulk of the liquid play an important role. For a chemical reaction of the first order in the absorbed gas, some of the solutions presented can be seen in Fig. 25. The ratio of the productivities of cyclic operation to batch operation, M, is defined as:

$$M = [(Q_{Lf} C_{Bf} t_{ch} - Q_L \int C_{BL} dt_{ach})/t_{cy}]/$$

$$\left(\frac{C_{Bf} V_{max} \eta}{t_b}\right) \tag{110}$$

where C_{Bf} is the concentration of the liquid phase reactant B in the feed.

The ratio of gas to liquid feed rate is given by:

$$m = Q_L P/He Q_G \tag{111}$$

The results are presented as graphs of M vs. the fractional conversion η_r for several values of the Stanton number St and m (Fig. 27). In general, the ratio decreases with η_r, indicating that batch operation becomes more attractive as the degree of conversion increases. This effect is stronger as the residual volume V_{min} becomes larger, $V_{min} = 0.5$. For $V_{max} = 0.1$, the productivity ratio M becomes much less sensitive to η_r. In case of high St and low m, the pattern is reversed and M increases slightly with η_r (Fig. 27A). In Fig. 27C–D it is shown that at high values of m the cyclic operation of the reactor is the best choice over a wide range of η_r and St. At low values of m, (Fig. 27A–B), batch operation may be preferred, especially at high St.

Fig. 27 was calculated for a slow chemical reaction. In general, the approach is focused on the case of non-zero concentration of the

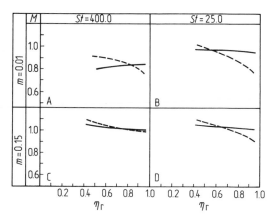

Fig. 27. Ratio of cyclic-operated to batch-operated reactor, M, as a function of the fractional conversion η_r, for two values of the Stanton number St and two values of m. —— $V_{min}/V_{max} = 0.1$; ---- $V_{min}/V_{max} = 0.5$.

reacting gas in the liquid phase, which is true only for a slow reaction regime. Mathematical expressions for the enhancement factor for the different operating conditions were obtained as a function of the Hatta number

$$\gamma_r = \sqrt{D k_1 C_{BL}}/k_L \qquad (112)$$

and the volume-to-interfacial area ratio, β_r,

$$\beta_r = \phi/a\delta \qquad (113)$$

Several factors affect the processes in a rather complicated way, which makes it difficult to predict the effect of changes in the parameters of the system and which stresses the importance of model simulations.

HALLAILE and MERCHUK (1986) presented the required equations for the calculations of the productivity in a gas–liquid process under various operational conditions that allow the selection of the optimal operation mode.

6 Closing Remarks

An overview of the models published for the simulation of tower reactors has been presented in the preceding pages. Most of the efforts in mathematical modelling have been devoted to the hydrodynamics as a primary phenomenon. The degree of sophistication reached in this area has not yet been applied in the field of mass transfer. Models for chemical and biochemical processes usually use a very simple description of the fluid dynamics in the reactor. This is sometimes justified by the requirements of the system. However, it becomes obvious that much is still to be done, especially in the matching and connecting of heat and mass transport to momentum transport.

7 References

BANKOFF, S. G. (1960), A variable density single fluid model for two-phase flow with particular reference to steam–water flow, *ASME J. Heat Trans.* **82**, 265–272.

BANERJEE, S., SCOTT, D. S, RHODES, E. (1970), Studies on cocurrent gas liquid flow in helically coiled tubes. II. Theory and experiments with and without chemical reaction, *Can. J. Chem. Eng.* **48**, 542–551.

BARNEA, D., TAITEL, Y. (1986), Flow pattern transition in two-phase gas–liquid flow, in: *Encyclopedia of Fluid Mechanics* (CHEREMISINOFF, N. P., Ed.), Chap. 16, pp. 103–491, Houston: Gulf Publ. Co.

BELLO, R. A. (1981), *Ph. D. Thesis,* University of Waterloo, Ontario, Canada.

BLENKE, H. (1979), Loop reactors, *Adv. Biochem. Eng.* **13**, 121–215.

CALDERBANK, P. H. (1958), Physical rate processes in industrial fermentation, *Trans. Inst. Chem. Eng.* **36**, 443–449.

CHAKRAVARTY, M., SINGH, H. D., BARUAH, J. N., IYENGAR, M. S. (1974), Liquid velocity in a gas-lift column, *Indian Chem. Eng.* **16**, 17–22.

CHAUDARI, R. V. RAMACHANDRAN, P. A. (1980), Three-phase slurry reactors, *AIChE J.* **26**, 177–201.

CHISTI, M. Y., MOO-YOUNG, M. (1988), Prediction of liquid circulation velocity in airlift reactors with biological media, *J. Chem. Tech. Biotechnol.* **42**, 211–219.

CHISTI, M. Y., HALARD, B., MOO-YOUNG, M. (1988), Liquid circulation in air-lift reactors, *Chem. Eng. Sci.* **43**, 451–457.

CLARK, M., JONES, A. G. (1987), On the prediction of liquid circulation in a draft-tube column, *Chem. Eng. Sci.* **42**, 378–378.

CLARK, N. N., ATKINSON, C. M., FLEMMER, R. L. C. (1987), Turbulent circulation in bubble columns, *AIChE J.* **33**, 515–518.

DANCKWERTS, P. V. (1970), *Gas–Liquid Reactions,* New York: McGraw Hill.

DECKWER, W. D. (1986), Design and simulation of bubble column reactors, in: *Chemical Reactor Design and Technology* (DE LASA, H. I., Ed.), pp. 411–461, *NATO ASI Series,* Dordrecht–Boston–Lancester: Martinus Nijhoff.

DECKWER, W. D., ADLER, I., ZAIDI, A. (1978), A comprehensive study on CO_2 interface mass transfer in vertical cocurrent and countercurrent gas–liquid flow, *Can. J. Chem. Eng.* **56**, 43–55.

DECKWER, W. D., HALLENSLEBEN, J., POPOVIC, M. (1980), Exclusion of gas sparger influence on mass transfer in bubble columns, *Can. J. Eng.* **58**, 190–197.

FREEDMAN, W., DAVIDSON, J. F. (1969), Holdup and liquid circulation in bubble columns, *Trans. Inst. Chem. Eng.* **47**, T251–T262.

FORTESCUE, G. E., PEARSON, F. R. A. (1967), *Chem. Eng. Sci.* **22**, 1163.

GRIFFITH, P., SNYDER, C. A. (1964), *MIT Report* 5003-29 (TID 20947).

HALLAILE, M., MERCHUK, J. C. (1986), Operation policies for a gas–liquid stirred tank reactor, *Chem. Eng. Commun.* **46**, 179–196.

HARMATHY, T. Z. (1960), Velocity of large drops and bubbles in media of infinite or restricted extent, *AIChE J.* **6**, 281–288.

HENSTOCK, W. H., HANRATTY, T. J. (1979), Gas absorption by a liquid layer flowing on the wall of a pipe, *AIChE J.* **25**, 122–131.

HILLS, J. H. (1974), Radial non-uniformity of velocity and voidage in a bubble column, *Trans. Inst. Chem. Eng.* **52**, 1–9.

HO, C. S., ERICKSON, L. E., FAN, L. T. (1977), Modeling and simulation of oxygen transfer in airlift fermentors, *Biotechnol. Bioeng.* **19**, 1503–1522.

HSU, Y. C., DUDUKOVIC, M. P. (1980), *Chem. Eng. Sci.* **35**, 135.

JONES, A. G. (1985), Liquid circulation in a draft-tube bubble column, *Chem. Eng. Sci.* **40**, 449–462.

JOSHI, J. B. (1980), Axial mixing in multiphase contactors – A unified correlation, *Trans. Inst. Chem. Eng.* **58**, 155–165.

JOSHI, J. B., SHARMA, M. M. (1979), A circulation cell model for bubble columns, *Trans. Inst. Chem. Eng.* **57**, 244–251.

JOSHI, J. B., PANDIT, A. B., RAO, R. (1986), in: *Encyclopedia of Fluid Mechanics* (CHEREMISINOFF, N. P., Ed.), Chap. 16, Houston: Gulf Publ. Co.

KALISCHEWSKY, K., SCHÜGERL, K. (1978), Determination of mass transfer coefficients of absorption of oxygen and carbon dioxide in different culture media, *Ger. Chem. Eng.* **1**, 140–149.

KAWASE, Y., MOO-YOUNG, M. (1986), Influence of non-Newtonian flow behavior on mass transfer in bubble columns with and without draft tubes, *Chem. Eng. Commun.* **40**, 67–83.

KAWASE, Y., HALARD, B., MOO-YOUNG, M. (1987), Theoretical prediction of volumetric mass transfer coefficients in bubble columns for Newtonian and non-Newtonian fluids, *Chem. Eng. Sci.* **42**, 1609–1617.

KEITEL, G., ONKEN, U. (1981), Errors in determinations of mass transfer in gas liquid dispersions, *Chem. Eng. Sci.,* **36**, 1927–1981.

KOIDE, K., IWAMOTO, S., TAKASAKA, Y., MATSURA, S., TAKASASHI, E., KIMURA, M., KUBOTA, H. (1984), Liquid circulation, gas holdup and pressure drop in a bubble column with draught tube, *J. Chem. Eng. Jpn.* **17**, 611–618.

KOIDE, K., KIMURA, M., NITTA, H., KAWABATA, H. (1988), Liquid circulation in bubble column with draught tube, *J. Chem. Eng. Jpn.* **21**, 393–399.

KUBOTA, H., HOSONO, Y., FUJIE, K. (1978), Characteristic evaluation of ICI air-lift type deep shaft reactor, *J. Chem. Eng. Jpn.* **11**, 319–325.

LAMB, H. (1932), *Hydrodynamics,* Cambridge (UK): Cambridge University Press.

LAMONT, J. C., SCOTT, D. S. (1970), An eddy cell model of mass transfer into the surface of a turbulent fluid, *AIChE J.* **16**, 513–519.

LEE, C. H., GLASGOW, L. A., ERICKSON, L. E., PATEL, S. A. (1986), Liquid circulation in airlift fermentors, *8th AIChE Annual Meeting,* Miami Beach, November 2–7, paper 8.d.

LEVICH, V. G. (1962), *Physicochemical Hydrodynamics,* Engelwood Cliffs, N. J.: Prentice Hall.

LEVY, S. (1960), Steam slip – Theoretical prediction from momentum model, *Trans. ASME J. Heat Transfer* **82**, 113–124.

LUND, M. M., SEAGRAVE, R. C. (1971), Optimal operation of a variable volume stirred tank reactor, *AIChE J.* **17**, 30–37.

LUTTMAN, R., BUCHHOLZ, H., ZAKREZEWSKI, W., SCHÜGERL, K. (1982), Identification and mass-transfer parameters and process simulation of SCP production in airlift tower reactions with an external loop, *Biotechnol. Bioeng.* **24**, 817–835.

LUTTMAN, R., THOMA, M., BUCHHOLZ, H., SCHÜGERL, K. (1983a), Model development, parameter identification, and simulation of the SCP production process in air lift tower bioreactors with external loop – I. Generalized distributed parameter model, *Comput. Chem. Eng.* **7**, 43–50.

LUTTMAN, R., THOMA, M., BUCHHOLZ, H., SCHÜGERL, K. (1983b), Model development, parameter identification, and simulation of SCP production processes in air lift tower reactors with external loop – II. Extended culture modelling and quasi-steady-state parameter identification, *Comput. Chem. Eng.* **7**, 51–63.

MARUYAMA, T., YOSHIDA, S., MIZUSHINA, T. (1981), The flow transition in a bubble column, *J. Chem. Eng. Jpn.* **14**, 352–357.

MATSUMOTO, T., HIKADA, N., KAMIMURA, H., TSUCHIYA, M., SHIMIZU, T., MOROOKA, S. (1988), Turbulent mixing-length model for axial turbulent diffusion of liquid in three-phase fluidized bed, *J. Chem. Eng. Jpn.* **21**, 256–261.

MERCHUK, J. C. (1986), Gas hold-up and liquid velocity in a two-dimensional air lift reactor, *Chem. Eng. Sci.* **41**, 11–16.

MERCHUK, J. C. (1990), Shear effects in suspended cells, *Adv. Biochem. Eng. Biotechnol.,* submitted.

MERCHUK, J. C., RONCO, J. J. (1966), Transfert de masse avec réaction chimique simultanée, *Genie Chim.* **96**, 71–79.

MERCHUK, J. C., SIEGEL, M. H. (1988), Air-lift reactors in chemical and biological technology, *J. Chem. Tech. Biotechnol.* **41**, 105–120.

MERCHUK, J. C., STEIN, Y. (1981a), Local hold-up and liquid velocity in air-lift reactors, *AIChE J.* **27**, 377–388.

MERCHUK, J. C., STEIN, Y. (1981b), A distribution parameter model for an air-lift fermentor. Effects of pressure, *Biotechnol. Bioeng.* **23**, 1309–1324.

MERCHUK, J. C., YUNGER, R. (1990), *Chem. Eng. Sci.,* in press.

MERCHUK, J. C., STEIN, Y., MATELES, R. (1980), A distributed parameter model for an air-lift fermentor, *Biotechnol. Bioeng.* **22**, 1189–1211.

MIYAUCHI, T., VERMEULEN, T. (1963), Diffusion and back-flow models for two phase axial dispersion, *Ind. Eng. Chem. Fund.* **2**, 304–310.

MIYAUCHI, T., FURUSAKI, S., MOROOKA, S., IKEDA, Y. (1981), Transport phenomena and reaction in fluidized catalyst beds, *Adv. Chem. Eng.* **11**, 275–448.

MURAKAMI, Y., HIROSHE, T., ONO, S., NISHIJIMA, T. (1982), Mixing properties in loop reactor, *J. Chem. Eng. Jpn.* **15**, 121–125.

NIKLIN, D. J. (1962), Two-phase bubble flow, *Chem. Eng. Sci.* **17**, 693–702.

OHKI, Y., INOE, H. (1970), Longitudinal mixing of the liquid phase in bubble columns, *Chem. Eng. Sci.* **25**, 1–16.

ROELS, J. A. (1983), *Energetics and Kinetics in Biotechnology,* Amsterdam: Elsevier Biomedical Press.

SAKATA, M., MIYAUCHI, T. (1980), Liquid recirculation in bubble columns, *Kagaku Kogaku Ronbunshu* **6**, 428.

SCHLICHTING, H. (1968), *Boundary Layer Theory,* New York: McGraw Hill.

SCHUMPE, A., DECKWER, W. D. (1980), Analysis of chemical methods for determination of interfacial areas in gas-in-liquid dispersions with nonuniform bubble sizes, *Chem. Eng. Sci.* **35**, 2221–2233.

TANG, W. T., WISECARVER, K., FAN, L. (1987), Dynamics of a draft tube gas–liquid–solid fluidized bed bioreactor for phenol degradation, *Chem. Eng. Sci.* **42**, 2123–2134.

TAYLOR, G. I. (1954), The dispersion of matter in turbulent flow in a pipe, *Proc. R. Soc. London* **A223**, 446–468.

UEYAMA, K., MIYAUCHI, T. (1976), Recirculation in bubble columns, *Kagaku Kogaku Ronbunshu* **2**, 595–560.

UEYAMA, K., MIYAUCHI, T (1977), Behavior of bubbles and liquid in a bubble column, *Kagaku Kogaku Ronbunshu* **3**, 19–27.

UEYAMA, K., MIYAUCHI, T. (1979), Properties of recirculating turbulent two phase flow in gas bubble columns, *AIChE J.* **25**, 258–266.

VAN DER AKKER, H. E. A., RIETEMA, K. (1979), Detection of multiple generation patterns generated by the application of a positive temperature gradient, *Trans. Inst. Chem. Eng.* **57**, 147–155.

VAN DER AKKER, H. E. A., RIETEMA, K. (1982), Correspondence on the paper by JOSHI-SHARMA, *Trans. Inst. Chem. Eng.* **60**, 255–256.

VAN DER LAAN, E. T (1958), Notes on the diffusion-type model for the longitudinal mixing in flow, *Chem. Eng. Sci.* **7**, 187–191.

VERLAAN, P., TRAMPER, J., VAN'T RIET, K., LUYBEN, K. C. A. M. (1986), A hydrodynamic model for an airlift-loop bioreactor with external loop, *Chem. Eng. J.* **33**, B43–B53.R.

VERLAAN, P., VOS, J. C., VAN'T RIET, K. (1988), From bubble column to airlift-loop reactor: Hydrodynamics and axial dispersion of the transition flow regime, in: *Bioreactor Fluid Dynamics II* (KING, R., Ed.), pp. 259–275. London: Elsevier.

VONCKEN, R. M., HOLMES, D. B., DEN HARTOG, H. W. (1964), Fluid flow in turbine stirred baffled tanks, *Chem. Eng. Sci.* **19**, 209–213.

WACHI, S., MORIKAWA, H. (1986), Chlorination of ethylene in a boiling bubble column reactor, *J. Chem. Eng. Jpn.* **20**, 238–245.

WACHI, S., MORIKAWA, H., UEYAMA, K. (1987), Gas holdup and axial dispersion in gas liquid concurrent bubble column, *J. Chem. Eng. Jpn.* **20**, 309–316.

WALLIS, G. B. (1969), *One-Dimensional Two-Phase Flow.* New York: McGraw Hill.

WALTER, J. F., BLANCH, H. W. (1983), Liquid circulation patterns and their effect on gas holdup and axial mixing in bubble columns, *Chem Eng. Commun.* **25**, 243–262.

WARNECKE, H. J., PRUSS, J., LANGEMANN, H. (1985), On a mathematical model for loop reactors – I. Residence time distributions, moments and eigenvalues, *Chem. Eng. Sci.* **40**, 2321–2326.

WEHNER, J. F., WILHELM, R. H. (1956), Boundary conditions of flow reactor, *Chem. Eng. Sci.* **6**, 89–93.

WISWANATHAN, K. (1986), Flow patterns in bubble columns, in: *Encyclopedia of Fluid Mechanics* (CHEREMISINOFF, N. P., Ed.). Chap. 38, pp. 1180–1215. Houston: Gulf Publ. Co.

YAMAGOSHI, T. (1969), *B. S. Thesis,* Dept. of Chem. Eng., University of Tokyo, Japan.

YANG, Z., RUSTEMEYER, V., BUCHHOLZ, R., ONKEN, U. (1986), Profile of liquid flow in bubble columns, *Chem. Eng. Commun.* **49**, 51–67.

ZEHNER, P. (1986a), Momentum, heat and mass transfer in bubble columns. Part 1. Flow model of the bubble column and liquid velocities, *Int. Chem. Eng.* **26**, 22–29.

ZEHNER, P. (1986b), Momentum, mass and heat transfer in bubble columns. Part 2. Axial blending and heat transfer, *Int. Chem. Eng.* **26**, 29–35.

ZHAO, M., NIRANJAN, K., DAVIDSON, J. F. (1988), Mass transfer in short bubble columns, in: *Bioreactor Fluid Dynamics II* (KING, R., Ed.), pp. 291–308, London: Elsevier.

ZUBER, N., HENCH, J. (1962), *Steady State and Transient Void Fraction of Bubbling Systems and Their Operating Limits,* General Electric Co. Report, 626, Lido, N. J.

ZUBER, M., FINDLAY, J. A. (1965), Average volumetric concentration in two-phase flow systems, *Trans. ASME J. Heat Transfer* **87c**, 453–468.

12 Process Models:
Optimization of Yeast Production
– A Case Study

Karl-Heinz Bellgardt

Hannover, Federal Republic of Germany

Jingqi Yuan

Shanghai, People's Republic of China

List of Symbols

A_R	cross-sectional area of reactor, m^2
B	duration of budding phase, h
$b(i)$	discrete age distribution, number of budding cells in ith class of cycling age
C_{b1}, C_{b2}	parameters in cycling phase equations for budding cells
C_C	concentration of dissolved carbon dioxide, $g\,m^{-3}$
C_{d1}, C_{d2}	parameters in cycling phase equations for daughter cells
C_O	concentration of dissolved oxygen, $g\,m^{-3}$
C_{p1}, C_{p2}	parameters in cycling phase equations for parent cells
C_1, C_2	parameters of the regulation model
CTR	carbon dioxide transfer rate, $g\,m^{-3}\,h^{-1}$
CU	currency unit
D	diffusion constant, $m^2\,s^{-1}$
$d(i)$	discrete age distribution, number of daughter cells in ith class of cycling age
d_S	Sauter diameter, m
E	ethanol concentration, $kg\,m^{-3}$
E_1, E_2	specific enzyme concentrations
ETR	ethanol transfer rate, $kg\,m^{-3}\,h^{-1}$
F	substrate flow rate, $m^3\,h^{-1}$
F_G	aeration rate, $m^3\,min^{-1}$
F_{GO}	gas flow rate at the outlet of the reactor, $m^3\,min^{-1}$
$\boldsymbol{F}$	system vector of the regulation model
FBC	fraction of budded cells, %
FDC	fraction of unbudded daughter cells, %
FPC	fraction of unbudded parent cells, %
$\boldsymbol{f}$	input vector of the regulation model
g	acceleration of gravity, $m\,s^{-2}$
H	Henry constant, Pa^{-1}
HTR	water transfer rate, $kg\,m^{-3}\,h^{-1}$
J	performance index
K_{D1}	specific costs for centrifugation, $CU\,m^{-3}$
K_F	rate of fixed costs, $CU\,h^{-1}$
K_Q	quality index for profit
K_S	half saturation concentration for substrate, $kg\,m^{-3}$
K_{S1}	factor of reduction of sugar concentration by addition of media components

K_{S2}	economic constant
K_1	weight of product in performance index
$k_{CL}a$	volumetric mass transfer coefficient for carbon dioxide, liquid side, h^{-1}
$k_{OL}a$	volumetric mass transfer coefficient for oxygen, liquid side, h^{-1}
$k_{EG}a$	volumetric mass transfer coefficient for ethanol, gas side, h^{-1}
M	molecular weight, $kg\,kmol^{-1}$
m_{ATP}	rate of endogenous ATP metabolism, $mol\,g^{-1}\,h^{-1}$
N	particle number, mol
n	exponent of k_L–D-correlation
n_b	number of cycling age intervals in budding cycling phase
n_d	number of cycling age intervals in unbudded daughter cycling phase
n_p	number of cycling age intervals in unbudded parent cycling phase
OTR	oxygen transfer rate, $g\,m^{-3}\,h^{-1}$
P	profit, $CU\,h^{-1}$
P_{Ae}	costs for aeration, CU
P_D	costs for downstream processing, CU
P_{DC}	costs for centrifugation, CU
P_{DD}	costs for drying, CU
P_E	total economic proceeds, CU
P_F	fixed costs, CU
P_{Mo}	costs for molasses, CU
P_S	costs for medium, CU
P_{Sa}	costs for salts and media components, CU
P_{Sp}	costs for media preparation and cleaning, CU
p	pressure, Pa
$p(i)$	discrete age distribution, number of parent cells in ith class of cycling age
p_{Ae}	specific costs for aeration, $CU\,kg^{-1}$
p_E	specific selling price for yeast, $CU\,kg^{-1}$
p_{Mo}	specific costs for molasses, $CU\,kg^{-1}$
p_{Sa}	specific costs for media components, $CU\,kg^{-1}$
q_S	specific sugar uptake rate, $mol\,g^{-1}\,h^{-1}$
R	gas constant, $J\,mol^{-1}\,K^{-1}$
$R(r)$	$=\begin{cases} r & r>0 \\ 0 & r<0 \end{cases}$ for switching function
R_C	volumetric carbon dioxide production rate, $kg\,m^{-3}\,h^{-1}$

R_E	volumetric ethanol reaction rate, kg m^{-3} h^{-1}
R_O	volumetric oxygen uptake rate, kg m^{-3} h^{-1}
R_S	volumetric substrate uptake rate, kg m^{-3} h^{-1}
R_X	volumetric growth rate of cell mass, kg m^{-3} h^{-1}
RQ	respiratory quotient
r	specific reaction rate, mol g^{-1} h^{-1}
r	vector of specific reaction rates, mol g^{-1} h^{-1}
S	total sugar concentration, kg m^{-3}
T	cell number doubling time, h
T_C	discrete cycling age interval, h
T_F	fermentation period, h
T_I	temperature of the inlet air flow, K
T_L	temperature of the liquid phase, K
T_P	time for preparation and downstream processing, h
T_T	length of the process cycle, h
t	time, h
t_M	dead time of measurements, h
U_d	duration of unbudded daughter phase, h
U_p	duration of unbudded parent phase, h
V_G	volume of gas phase in fermentor, m^3
V_L	volume of liquid phase in fermentor, m^3
V_R	volume of fermentor, m^3
w_{SG}	superficial gas velocity, m s^{-1}
X	dry cell mass concentration, kg m^{-3}
x	mole fraction
X_N	cell number concentration, m^{-3}
Y	stoichiometric matrix
y	state vector of the regulation model
z	vector of growth-independent specific reaction rates, mol g^{-1} h^{-1}
ε_G	mean relative gas hold-up
ρ_W	density of water, kg m^{-3}
μ	specific growth rate, h^{-1}

Subscripts

C	carbon dioxide
E	ethanol
F	value at the end of the fermentation period
G	gas phase of the reactor
I	variable related to the inlet air flow
L	liquid phase of the reactor
M	measured value
max	maximum value
min	minimum value
N	nitrogen
O	oxygen
R	variable related to the substrate inflow
S	substrate
sat	saturation value
W	water
X	cell mass
0	initial value at the beginning of the fermentation period

Superscripts

T	transpose of a matrix or vector
*	saturation concentration

1 Introduction

A biotechnological process for the production of a certain substance consists, in addition to fermentation, of many elementary steps, unit processes, and unit operations. A process model should describe the most important steps of a biotechnological process. A general process structure is shown schematically in Fig. 1.

Production starts with the cleaning and sterilization of reactors and with preparation of substrate, media components, and inoculum from the strain being maintained. Furthermore, process energy and compressed air have to be supplied. During fermentation the microorganisms are propagated in a series of reactors. The batch or fed-batch mode of operation is preferred for most industrial processes. The reactors usually have an increasing volume, beginning with the smallest seed tank at a 10 L scale, up to the final production tank on a 10–100 m^3 scale, where the largest amount of cell mass and product is formed. The fermentation broth of one reactor is used as an inocu-

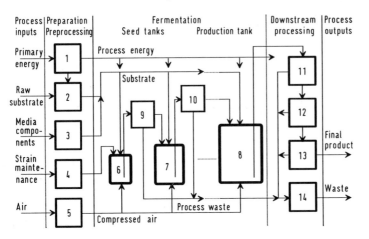

Fig. 1. Schematic diagram of a biotechnological production process: (1) Process energy supply, (2) substrate preparation, (3) preparation of media additives, (4) inoculum preparation, (5) compressed air supply, (6, 7) seed tanks, (8) production tank, (9, 10) storage and intermediate processing, (11) cell separation and disruption, (12) product separation, (13) product processing, (14) waste processing.

lum for the next stage. Between them there can be additional processing steps such as filtration and storage. The final fermentation stage is followed by downstream processing steps, separation of cells and media, disruption of cells for release of intracellular products, purification of the primary raw product, and additional processing of the raw product to obtain the final product for dispatch. The treatment of waste is only indicated by one block in Fig. 1, but it is nevertheless an important part of the biotechnological process. Waste material arises from the cleaning of reactors and from downstream processing. The waste must be sterilized and the load of organic material reduced. Depending on downstream processing it may be necessary to remove solvents or other reactants.

A process model should provide a coherent description of the entire process on the level of plant operation. The degree of complexity of the model or the possibility of simplified modelling of some parts of the plant is determined by the intended application of the model. The model may be used to answer several interesting questions: What will be the output of product per unit time for a given input of raw material and primary energy? What are the costs of production and of waste treatment? What is the optimum mode of operation for the reactors? Under what dynamic control of manipulating variables is the product obtained with high productivity and the desired quality? How can the profit be maximized? Very sim-

plified balances may be sufficient as models for the calculation of the total conversion of the process, e.g., yield of cell mass or product. Greater modelling effort and a more detailed description of the most important unit operations are generally required for a model-based process optimization. One should be aware that the result of such theoretical optimizations may be significantly influenced by the model accuracy, especially for the determination of an optimal dynamic control.

To develop an accurate and complete model is not an easy task for several reasons. Sometimes there are only very simplified models available for parts of the plant, e.g., fermentation. Moreover, many interdependencies between the elementary units of the process are often qualitatively and quantitatively unknown. Fermentation is influenced in a complicated manner by medium composition, substrate quality and preprocessing, and inoculum preparation. The fermentation itself may influence downstream processing by varying the rheology of the fermentation broth and product properties. Model building is further complicated because for batch and fed-batch operation the process is never in a steady state and the model must consider the process dynamics, at least for the fermentation part. Therefore, the modelling of the biological system is obviously an important aspect of process models for biotechnological systems.

To our knowledge no such comprehensive model has been published in the literature.

Also in this chapter we intend only to deal with certain aspects of an ideally complete process model. The baker's yeast process will be taken as an example in order to show the philosophy behind process models. As an additional, often neglected aspect, a description of the quality of the final product will be included in the process model. The model will be used to investigate strategies for optimal control of the production process. Economic optimization requires a global view of interconnected processing steps of the whole process and a detailed look at the fermentation steps.

In the following sections the baker's yeast production process is briefly described. Then biological properties of the yeast requiring consideration in the model are discussed in order to lead to our topic, the process model. Based on this model the optimization goal is defined, and results of the subsequent optimization are finally presented. The optimum control strategy is a compromise between the high productivity and yield of the fermentation part and good product quality after downstream processing.

2 The Baker's Yeast Process

2.1 Development of the Yeast Process

Baker's yeast production in submerged culture is one of the earliest biotechnological processes. Its development began in the 19th century when an increasing demand for yeast in bread production could not be satisfied by the limited availability of yeast from traditional sources: beer and distillery processes. The reason for the shortage was mainly the change to new strains for beer production, ones that were unsuitable for baking. Initially, the baker's yeast process was a copy of alcoholic fermentations, with very low productivity and yield – anaerobic batch cultivation on substrates from grains. In the second half of the 19th century there was great progress in the

technology of yeast production, accompanied by the development of microbiology. Following Pasteur's discovery continuous aeration was applied (in England, about 1886), and the method of fed-batch cultivation was introduced (Germany and Denmark, 1915–1920) to avoid the Crabtree effect. Further progress was achieved by the use of pure cultures and by the substitution for grain material of the cheaper molasses in the 1920s. Since then the process has been improved more in detail than in general layout. More recently attention has turned to automatic control of the process. One problem is to control the substrate supply in a way that keeps the process at high productivity, but avoids loss of yield due to sugar over-feeding and ethanol production. In 1961 the principle of automatic control of molasses feed rate by the ethanol concentration in the exhaust gas was proposed by RUNGELDIER and BRAUN. This control was first realized only in 1971 due to the lack of cheap and reliable sensors. Another method of automatic control of the molasses feed rate uses the respiratory quotient. Compared to control by ethanol concentration, this method requires more exacting measuring devices and very accurate gas-phase balancing to prevent stability problems. For further information on the history and development of the yeast process the reader should refer to the literature (REED and PEPPLER, 1973; SKINNER et al., 1980; REED and NAGODAWITHANA, 1990).

2.2 Biological Properties of the Yeast

Baker's yeast *Saccharomyces cerevisiae* is a single-celled microfungus. Vegetative multiplication by budding is most important for its propagation in technical processes. For growth in aqueous media the cells need carbon and nitrogen sources, vitamins, and fatty acids. As a eukaryotic organism the yeast has a wide choice of metabolic pathways and regulatory systems to react in a suitable manner to the availability of different substrates and oxygen. Only subjects important for our modelling goal will be mentioned here. Surveys of yeast metabolism can be found in MILLS and KREBS

(1986), COONEY (1981), ROSE and HARRISON (1987), and KOCKOVÁ-KRATOCHVÍLOVÁ (1990).

2.2.1 Metabolic Types of Yeast Growth

Oxygen supply is one of the main determining factors of yeast metabolism and is the limiting parameter for the production process. In excess of oxygen, yeast can achieve an *oxidative metabolism*. The carbon source is completely oxidized to carbon dioxide and water for energy production via the Krebs cycle and the respiratory chain. These pathways can yield up to 36 mol of ATP per mol of glucose, resulting in a very high yield for cell mass up to 0.5 g dry cell mass per g of sugar. Oxidative growth is therefore the preferred metabolic type for baker's yeast production.

In the case of absence or lack of oxygen deprivation, yeast changes to *fermentative metabolism*, and ethanol is produced as a final proton acceptor for NADH. The chemical energy comes mainly from glycolysis, with a yield of 2 mol of ATP per mol of sugar. Due to this low efficiency compared to oxidative growth and due to the loss of carbon in the product ethanol, the cell yield is also very low. To obtain the same amount of cell mass as during oxidative growth much greater amounts of carbon source must be metabolized which is accompanied by the production of a large volume of carbon dioxide.

Yeast metabolism in the dough is also fermentative. The fermentative activity – the ability to produce a certain amount of CO_2 in a given time (usually 2 or 3 hours) – is, besides storage properties and resulting flavor of the dough, an important quality index for the yeast product, because the produced carbon dioxide raises and loosens the dough.

2.2.2 Regulatory Effects During Yeast Growth

For the directed coordination of its metabolism linked to the substrate and oxygen supply,

yeast utilizes a complex regulatory system on the epigenetic level by enzyme induction/repression and on the reaction level by enzyme activation/inhibition. This regulation affects the uptake of sugars, glycolysis, respiration, glycerol metabolism, gluconeogenesis, and alcohol dehydrogenases. The most important effects included in the model are described below.

The *Pasteur effect,* the suppression of fermentative activity by respiration, was the earliest regulatory effect discovered in yeast. Its connection with energy metabolism is well accepted, but the biochemical details and their relation to observed growth kinetics and other regulatory systems are still under discussion (LAGUNAS, 1986). Enzymes of the respiratory pathways, which are repressed during fermentative growth, are induced under aerobic conditions. For the modelling of the Pasteur effect it will be assumed that the cells always fully utilize their oxidative pathways up to maximum capacity. This is determined by the oxygen supply or by the actual induction state of key enzymes of the respiratory pathway. Excess substrate is directed to fermentative metabolism.

The *Crabtree effect* is referred to as respiro-fermentative metabolism under an excess of oxygen and sugar. The Crabtree effect was originally considered to be a repression of respiratory enzymes by glucose or glucose catabolites. However, the view of saturation of the respiratory system is increasingly accepted and is also adopted for the model in this chapter. The result of the Crabtree effect is very similar to the inverse Pasteur effect. Both are very important for the control of baker's yeast production, because an over-supply of sugar relative to a limited capacity of oxidative metabolism – either by internal saturation (Crabtree effect) or by external oxygen supply (inverse Pasteur effect) – leads to fermentative growth with low yield of cell mass.

The *glucose effect* is the repression of uptake systems or pathways needed for other substrates at high concentrations of glucose. As a result of this repression glucose is the preferred substrate for initial catabolism, even when other substrates are present at high concentrations. For baker's yeast production the glucose effect with respect to ethanol, as a possible substrate for oxidative growth, is most important, because ethanol can be produced during the fermentation. Ethanol and sugar can be taken up in parallel; then both compete for the oxidative pathways. In the model it is assumed that ethanol can be used by the cells up to a rate determined by the remaining capacity of respiration left by sugar catabolism. The glucose effect with respect to maltose, the second main carbon source in molasses, need not to be considered for production strains that possess a constitutive maltase system. Such strains also have an increased fermentative activity in dough.

2.2.3 The Cell Cycle of Yeast

The cell cycle during yeast multiplication by budding is a dynamic process of a series of ordered steps as shown schematically in Fig. 2: growth of parent cells (G_1 phase), bud formation and bud growth (S, G_2, and M phases) and separation of bud and parent cell (G_1^* phase). The period for the completion of the cycle is different for parent and daughter cells. A newly formed daughter cell needs a longer time in the G_1 phase for the formation of its first bud than for the subsequent buds. Therefore, yeast populations are heterogeneous mixed populations with a particular age distribution of distinguishable daughter cells, parent cells, and budding cells. The genealogical age of a cell can be determined easily by counting the number of bud scars on the cell surface. For exponentially growing cultures these data can be used to calculate the age distribution of a population and the duration of the cell-cycle phases (LORD and WHEALS, 1980; THOMPSON and WHEALS, 1980). The duration of the G_1 phase is positively correlated with cell-number doubling time. It is found, in contrast to the G_1 phase, that the duration of the budded phase ($S + G_2 + M + G_1^*$) is almost constant and independent of growth conditions.

The dynamics of the cell cycle must be considered in the process model, since it is desirable to predict some aspects of the product quality, which depends on the fermentative activity and storage properties. TAKAMATSU et al. (1985) have found that fermentative activity is correlated with the fraction of budding

cells (*FBC*) in the yeast population, as shown in Fig. 3. Since storage properties are also related to *FBC* (see our data in Fig. 3, where the cells were stored at 3–5 °C for 27 days), this single parameter can be used in a process model to describe several quality aspects. A low *FBC* at the end of the final fermentation stage results in good product quality. There are other parameters influencing the quality, such as stability of the yeast cells on storage or osmose-sensitivity for use of yeast in sweet doughs. The extension of the model to these parameters remains a task for the future.

To establish the correct structure for the cell model, including the cell-cycling model and the metabolic model, one should know the interaction between metabolic processes and cell division. Numerous investigations in this field have been summarized in WHEALS (1987). It

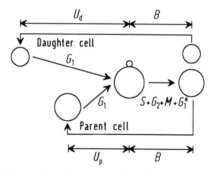

Fig. 2. Graphical representation of the cell cycle of yeast.

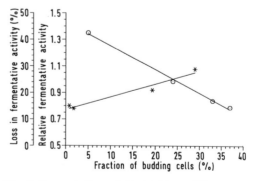

Fig. 3. Fermentative activity (○) (data of TAKAMAT-ZU et al. 1985) and storage properties (∗) (own data) depending on the fraction of budding cells, *FBC*.

was found that most of the proteins, cell-wall components, and RNA are synthesized at a constant rate during the cell cycle. The activity of most enzymes is also not strongly modulated or else their concentrations are much higher than needed for catalytic activity. The rate-limiting step of the cell cycle is obviously the growth process, i.e., the increase of cell mass or volume, which is directly connected with the continuous processes of metabolism. The cells need to reach a certain mass or volume to initiate the budding. The genetic program for cell division can normally be executed faster than the accumulation step. This also explains why the G_1 phase of the first generation is longer than that of parent cells: the daughter cells are smaller in volume.

The following properties facilitate the model building because they support the following simplifying assumptions:

- Parent cells, daughter cells, and budding cells have the same metabolic activity, which is described by a unique specific growth rate determined by substrate and oxygen supply.
- The cell-division cycle is controlled by the rate of increase of cell mass, the specific growth rate.
- There is no feedback from the cell-cycling model to the metabolic model.

2.3 Process Description

In industry, baker's yeast is produced in a multi-stage reactor system. The general structure of a yeast process follows the layout in Fig. 1 with some modifications. Fermentation is typically carried out in a series of up to six reactor stages. The first two are operated in batch mode under sterile conditions, the others in fed-batch mode to avoid loss of yield. The ethanol formed in the batch stages is used as a substrate together with the sugar in stages 3 and 4. The feed rate of the substrate molasses must be adjusted accordingly in order to obtain a full ethanol conversion at the end of stage 4. Ethanol reduces the risk of infections; the accompanying disadvantage, loss of yield, can be kept low by the parallel conversion of molasses and ethanol. After the fourth stage

cells are concentrated by centrifugation and stored in a separate tank for the pitching yeast. This storage gives additional flexibility in scheduling the final production stages, where two (stage 5) or four (stage 6) tanks are operated in parallel with some offset in inoculation times. This procedure reduces the peaks in energy and air consumption and leads to more continuous downstream processing. The reactors in stage 5 and 6 are similar in size, ranging from 60 m³ to more than 200 m³.

The profile of the molasses feed in these stages somewhat resembles the exponential–linear scheme that will be discussed further in the following sections: in the first exponential growth phase under an excess of oxygen a constant specific growth rate of about 0.25 h⁻¹ – just below the critical value for the Crabtree effect – is set by an exponentially increasing substrate flow. In the second phase the growth rate must be limited by an almost constant feed rate according to the maximum oxygen transfer to the reactor in order to avoid oxygen limitation and ethanol production. After all substrate is fed to the production tank, the reactor is usually further aerated for about one or two hours without sugar feeding. Ethanol that could have accumulated is thereby reduced, and the cells are conditioned for good storage properties. The *FBC* remains unchanged during this period. Therefore, a low *FBC* for good quality must have already been attained during the feed phase. In the downstream processing steps following fermentation, centrifugation and/or dehydration by rotating vacuum filters, the final yeast product is obtained.

3 Process Model

This section presents the dynamic process model that will be used to investigate several optimum control strategies in the baker's yeast process and to determine optimal operating parameters. For optimization, in addition to productivity, yield, and profit, the quality of the yeast product will be taken into account. For simplicity, only the final production tank is considered here in detail, because it has major influence on productivity and product quality, and also because it uses a large percentage of the energy resources. A complete model for the entire series of cultivations in the multi-stage process can be built by multiplying the elementary model.

The model consists of a reactor model and a cell model. Additional model equations are introduced to consider upstream and downstream processing steps in a simplified form. This is done by an additional dead time for the whole production cycle and by calculating the costs of the processing steps. To obtain significant results for the description of yeast quality, the cell model must be quite complex, and it will be explained in more detail.

According to the general structure of models for biotechnological processes shown in Chapter 9, the presented model should describe both the reactor system, including gas and liquid phases, and the biological system yeast. This cell model includes a kinetic model for the metabolic pathways, a dynamic model for the main regulatory systems of metabolism, and a model for the cell multiplication cycle which permits prediction of product quality.

3.1 Reactor Model

The reactor model dynamically describes concentrations in the gas and liquid phases of the reactor as governed by initial conditions, manipulating variables, and biological reactions of the yeast cells. For simplicity it is assumed that both phases are ideally mixed. Therefore, the corresponding sub-models include only ordinary differential equations. A third sub-model describes the mass exchange between gas and liquid phases. The validity of the model was verified for an industrial-scale process, and unknown model parameters were identified with data from production runs.

3.1.1 Liquid Phase Model

The liquid phase model for the main components cell mass, X, substrate molasses, S, ethanol, E, dissolved oxygen, C_O, dissolved carbon

dioxide, C_C, and liquid volume, V_L, is derived from mass balances for the fed-batch reactor. The model equations

$$\frac{dX(t)}{dt} = R_X(t) - \frac{F(t)}{V_L(t)} \cdot X(t) \tag{1}$$

$$\frac{dS(t)}{dt} = -R_S(t) + \frac{F(t)}{V_L(t)} \cdot (S_R - S(t)) \tag{2}$$

$$\frac{dE(t)}{dt} = R_E(t) - \frac{F(t)}{V_L(t)} \cdot E(t) + ETR(t) \tag{3}$$

$$\frac{dC_O(t)}{dt} = -R_O(t) + \frac{F(t)}{V_L(t)} \cdot$$
$$\cdot (C_{OR} - C_O(t)) + OTR(t) \tag{4}$$

$$\frac{dC_C(t)}{dt} = R_C(t) + \frac{F(t)}{V_L(t)} \cdot$$
$$\cdot (C_{CR} - C_C(t)) + CTR(t) \tag{5}$$

$$\frac{dV_L(t)}{dt} = F(t) \tag{6}$$

include the accumulation in the liquid phase, the biological reactions, and the mass transport which can be divided into two parts. The first term in every equation is due to the inflow of substrate. The second term in the balances for oxygen, carbon dioxide, and ethanol comes from mass exchange with the gas phase. The molasses flow rate, F, is the main manipulating variable of the reactor. It determines the increase in volume of the liquid phase and the related dilution effect for the process variables. The sugar concentration in the feed, S_R, is an operating parameter. The reaction rates for cell growth, R_X, substrate and oxygen uptake, R_S and R_O, and ethanol and carbon dioxide production, R_E and R_C, are given by the cell model in Sect. 3.2. The mass transfer rates for oxygen, carbon dioxide and ethanol, OTR, CTR, and ETR, are determined by the mass transfer model (Sect. 3.1.3). Water strip-off by the air flow is neglected in the mass balances of the liquid phase. The temperature is assumed to be constant.

3.1.2 Gas Phase Model

The main components of the gas phase are oxygen, carbon dioxide, nitrogen, ethanol, and water. They are advantageously described by their mole fractions, x_O, x_C, x_N, x_E, and x_W, respectively. The model equations derived from molar balances of the gas phase components are:

$$\frac{dx_O(t)}{dt} = \frac{p_I T_L F_G(t)}{p T_I V_G(t)} x_{OI}(t) -$$
$$- \frac{F_{GO}(t)}{V_G(t)} x_O(t) -$$
$$- \frac{R T_L V_L(t)}{M_O p V_G(t)} OTR(t) \tag{7}$$

$$\frac{dx_C(t)}{dt} = \frac{p_I T_L F_G(t)}{p T_I V_G(t)} x_{CI}(t) -$$
$$- \frac{F_{GO}(t)}{V_G(t)} x_C(t) -$$
$$- \frac{R T_L V_L(t)}{M_C p V_G(t)} CTR(t) \tag{8}$$

$$\frac{dx_N(t)}{dt} = \frac{p_I T_L F_G(t)}{p T_I V_G(t)} x_{NI}(t) -$$
$$- \frac{F_{GO}(t)}{V_G(t)} x_N(t) \tag{9}$$

$$\frac{dx_E(t)}{dt} = - \frac{F_{GO}(t)}{V_G(t)} x_E(t) -$$
$$- \frac{R T_L V_L(t)}{M_E p V_G(t)} ETR(t) \tag{10}$$

$$\frac{dx_W(t)}{dt} = \frac{p_I T_L F_G(t)}{p T_I V_G(t)} x_{WI}(t) -$$
$$- \frac{F_{GO}(t)}{V_G(t)} x_W(t) -$$
$$- \frac{R T_L V_L(t)}{M_W p V_G(t)} WTR(t) \tag{11}$$

The positive direction of the mass transfer streams, OTR, CTR, ETR, and HTR, is directed to the liquid phase. It is assumed that no nitrogen is exchanged between gas and liquid phases and no ethanol is present in the air flow at the inlet.

The resulting outlet gas stream, F_{GO}, is usually not measured, but it can be calculated, since the summation over all mole fractions must equal one,

$$x_O + x_C + x_N + x_E + x_W \equiv 1 \tag{12}$$

and therefore:

$$\frac{dx_O(t)}{dt} + \frac{dx_C(t)}{dt} + \frac{dx_N(t)}{dt} + \frac{dx_E(t)}{dt} +$$
$$+ \frac{dx_W(t)}{dt} \equiv 0 \tag{13}$$

After summing up Eqs. (7) to (11) and using the conditions Eqs. (12) and (13) we have the outlet gas stream as

$$F_{GO}(t) = \frac{p_I \cdot T_L \cdot F_G(t)}{p T_I} - \frac{R \cdot T_L \cdot V_L(t)}{p} \cdot$$
$$\cdot \left[\frac{OTR(t)}{M_O} + \frac{CTR(t)}{M_C} + \frac{ETR(t)}{M_E} + \frac{WTR(t)}{M_W} \right] \tag{14}$$

The gas phase model can be further simplified under the assumption that the outlet gas is saturated with water:

$$x_W(t) = x_{W\text{sat}}(t) = 0.027 \text{ at } 303 \text{ K}$$

By using Eq. (11) under the steady-state assumption and Eq. (14), the water strip-off can be calculated as:

$$\frac{WTR(t)}{M_W} = \frac{p \cdot F_G(t)}{T_I \cdot R \cdot V_L(t)} \frac{x_{WI} - x_{W\text{sat}}}{1 - x_{W\text{sat}}} +$$
$$+ \frac{x_{W\text{sat}}}{1 - x_{W\text{sat}}} \left[\frac{OTR(t)}{M_O} + \frac{CTR(t)}{M_C} + \frac{ETR(t)}{M_E} \right] \tag{14a}$$

The above model gives the mole fractions of the dispersed gas phase. If one wants to control the process by means of exhaust gas measurements, the additional time delay in the head space of the reactor, the exhaust gas pipes, and the measurement devices must be considered. This can be done by simple delay models, with the measurement equations:

$$\frac{dx_{OM}(t)}{dt} = k_M(x_O(t-t_M) - x_{OM}(t)) \quad (15a)$$

$$\frac{dx_{CM}(t)}{dt} = k_M(x_C(t-t_M) - x_{CM}(t)) \quad (15b)$$

$$\frac{dx_{EM}(t)}{dt} = k_M(x_E(t-t_M) - x_{EM}(t)) \quad (15c)$$

where k_M is a reciprocal time constant and t_M a dead time. For simplicity, both parameters were assumed to be equal for the three components. Finally, the measured respiratory quotient can be calculated under a steady-state assumption as

$$RQ_M = \frac{x_{CE}(1-x_{OM}) - x_{CM}(1-x_{OE})}{x_{OE}(1-x_{CM}) - x_{OM}(1-x_{CE})} \quad (16)$$

The water and ethanol content of the outlet gas stream was neglected in Eq. (16). The measured respiratory quotient is an approximation of the metabolic respiratory quotient:

$$RQ = \frac{R_C M_O}{M_C R_O} \approx RQ_M$$

and thus can be used to estimate the metabolic type of growth for the yeast:

$RQ > 1$ indicates growth is fermentative
$RQ = 1$ indicates respiratory growth on sugar substrate
$RQ < 1$ can be found in the case of ethanol consumption

Because of these properties it is possible to use the respiratory quotient as a control variable for the substrate flow. A setpoint of $RQ \approx 1.2$ at slightly fermentative growth is a good compromise between performance of the control loop and yield.

3.1.3 Mass Transfer Model

The mass transfer rate between gas phase and liquid phase is proportional to the concentration gradient in the interfacial area and to the volumetric mass transfer coefficient. It can be calculated by the film model. For oxygen and carbon dioxide the major mass transfer resistance is located at the liquid side, and for ethanol at the gas side; therefore:

$$OTR(t) = k_{OL}a(t) \cdot (C_O^*(t) - C_O(t)) \quad (17)$$

$$CTR(t) = k_{CL}a(t) \cdot (C_C^*(t) - C_C(t)) \quad (18)$$

$$ETR(t) = k_{EG}a(t) \cdot (x_C(t) - x_C^*(t)) \quad (19)$$

where * indicates the saturation concentration in the gas–liquid interface. The saturation concentration for the dissolved gases depends on the mole fractions in the gas phase. Following Henry's law, here given for oxygen:

$$x_{OL}^*(t) = H_O p_O(t) = H_O p x_O(t) \quad (20)$$

the mole fraction in the liquid phase at saturation is proportional to the partial pressure in the gas phase. The mole fraction in the liquid phase can be approximated by:

$$x_{OL}^* = \frac{N_O^*}{N_{tot}} \approx \frac{N_O^*}{N_W} = \frac{m_O^* M_W}{m_W M_O} \quad (21)$$

Combining Eq. (20) and Eq. (21) one obtains

$$x_{OL}^* = \frac{m_O^*}{V_L} \cdot \frac{V_L}{m_W} \cdot \frac{M_W}{M_O} = \frac{C_O^*}{\rho_W} \frac{M_W}{M_O} \quad (22)$$

Finally, the saturation concentration for oxygen is:

$$c_O^*(t) = \frac{H_O \rho_W M_O p}{M_W} \cdot x_O(t) \quad (23)$$

In a similar way the saturation concentration of carbon dioxide can be derived:

$$c_C^*(t) = \frac{H_C \rho_W M_O p}{M_W} \cdot x_C(t) \quad (24)$$

For exact calculations, especially for carbon dioxide, one must additionally consider the salt- and ion-content of the fermentation broth. The mass transfer resistance depends on the hydrodynamics in the reactor and the physical properties of the fermentation broth. Of special interest for further modelling is the correlation of the mass transfer coefficient with the molecular diffusion constant,

$$k_L \sim D^n \quad (25)$$

where for small bubbles

$n = 0.5$ (Higby's model)

and for large bubbles

$n = 0.66$ (Danckwerts' model)

With this equation the volumetric mass transfer coefficient for carbon dioxide can be calculated, while that for oxygen is known:

$$k_{CL}a = \left[\frac{D_C}{D_O}\right]^n k_{OL}a \qquad (26)$$

In air-lift fermentors the hydrodynamics are controlled by the aeration rate, which also influences the mass transfer. The following equation from HEIJNEN and VAN'T RIET (1984) for air-lift fermentors can be used to incorporate this effect into the model:

$$k_{OL}a = 1512 \cdot w_{SG}^{0.7} \qquad (27)$$

where

$$w_{SG} = \frac{F_G}{A_R} \qquad (27a)$$

is the superficial gas velocity. The maximum working volume of the reactor is limited by the mean relative gas holdup,

$$\varepsilon_G = \frac{V_G}{V_R}$$

which can be estimated (SCHÜGERL et al., 1977) as

$$\varepsilon_G = 0.91 \left[\frac{w_{SG}}{g^{0.5} \cdot d_S}\right]^{1.19} \qquad (28)$$

The mean Sauter diameter is $d_S = 0.008$ m. The parameters of Eqs. (27) and (28) were re-estimated for a 60 m^3 air-lift reactor.

3.2 Cell Model

The cell model for the yeast *Saccharomyces cerevisiae* determines the metabolic activity, expressed by the specific reaction rates, as a function of reactor concentrations and process history. The cell model should describe quantitatively the growth kinetics – considering the biological properties that were summarized in Sect. 2.1 – and the cell-division cycle. The model includes three parts that are presented in the next sub-sections. The metabolic model for kinetics and stoichiometry of growth and the regulation model for metabolic long-term regulation both represent the continuous reactions of growth. The cell cycle model covers the discontinuous processes of cell multiplication and the age distribution of the population. As discussed in Sect. 2.1, cell metabolism as a whole does not vary greatly during the cycling process, and the increase of cell mass is the rate-limiting step of cell multiplication. Therefore, the metabolic model and the cell cycle model are almost completely separated. The only connection is by the specific growth rate, r_X, that is transferred from the metabolic model to the cell cycling model. The combined cell model and reactor model relates the cell growth and cycling process to commonly controllable manipulating variables, e.g., substrate feeding rate and aeration rate.

3.2.1 Metabolic Model

This study utilizes the model of BELLGARDT, that is presented in detail in Chapter 9, Sect. 4.4. Only the important equations are given here. The stoichiometry of growth is described by a system of equations (Eq. (88) in Chapter 9):

$$Yr = z \qquad (29)$$

where Y is the matrix of stoichiometric coefficients and

$$r = (r_G, r_O, r_{Ac}, r_{TC}, r_{EP}, r_{EC}, r_S, r_X, r_C)^T \qquad (30)$$

is the vector of specific molar reaction rates which includes in this order the rates of glycerol secretion, oxygen uptake, and acetyl-CoA production, the net-rate of the Krebs cycle, the rates for ethanol production and consumption, the rate of glycolysis, the growth rate, and the carbon dioxide production rate. The vector of growth independent reaction rates:

$$z = (q_S, 0, 0, m_{ATP})^T \qquad (31)$$

includes the substrate uptake rate,

$$q_S = \frac{q_{Smax}}{K_S + S}$$

and the specific rate of ATP consumption for endogenous metabolism. Not all elements of r are needed for the calculation of the following volumetric reaction rates in the reactor model:

$$R_X(t) = r_X(t) X(t) = \mu(t) X(t) \qquad (32)$$

$$R_S(t) = M_S q_S(t) X(t) \qquad (33)$$

$$R_E(t) = M_E(r_{EP}(t) - r_{EC}(t)) X(t) \qquad (34)$$

$$R_O(t) = M_O r_O(t) X(t) \qquad (35)$$

$$R_C(t) = M_C r_C(t) X(t) \qquad (36)$$

The growth kinetics are determined by the optimal strategy:

$$r_X(t) \rightarrow \text{maximum} \qquad (37)$$

under the additional restrictions (Eq. (99) in Chapter 9) for rate-limiting steps of metabolism,

$$r_{min} \leq r \leq r_{max} \qquad (38)$$

The metabolic model, Eqs. (29)–(31) and Eqs. (33) and (38), is a static one for the instantaneous adaption of cells to the actual supply of

Tab. 1. Parameters in the Metabolic Model

Parameter	Value	Units
q_{Smax}	0.018	mol g^{-1} h^{-1}
K_S	0.45	g L^{-1}
K_E	0.5	L g^{-1} h^{-1}
K_O	0.4	mol mg^{-1} h^{-1}
K_{B1}	0.003	mol g^{-1}
K_{B2}	0.034	—
K_{B3}	0.01	—
K_{EG}	0.05	—
K_{Ad}	0.024	—
P/O	2	mol$_{ATP}$ atom$_O^{-1}$
Y_{ATP}	11	g mol^{-1}
m_{ATP}	0.001	mol g^{-1} h^{-1}
r_{Acmax}	0.0175	mol g^{-1} h^{-1}

substrate and oxygen. The model accounts for the glucose effect, Crabtree effect, and Pasteur effect. The model parameters for baker's yeast fed-batch processes are given in Tab. 1.

3.2.2 Dynamic Regulation Model

Many growth phenomena of yeast can be described by the static model given in the previous section. It covers the rapid metabolic regulation on the level of enzyme activity by using the optimal strategy, Eq. (37). But the respiratory activity and the pathways for gluconeogenesis are subject to long-term regulation by enzyme induction and repression. The resulting lag phases of growth during phases of regulatory adaptation must be described by corresponding dynamic models.

Extending the presentation in Chapter 9, Sect. 4.4, the model for regulation of respiratory activity is here established in state space representation as:

$$\frac{dy(t)}{dt} = F(\mu) y(t) + f(\mu) r_O(t) \qquad (39)$$

where the elements of the state vector

$$y = (r_{Omax}, E_1, E_2)^T$$

are the maximum specific oxygen uptake rate, r_{Omax}, and two fictitious enzymes E_1 and E_2. The matrix elements of F and f are non-linear functions of model parameters and the specific growth rate $\mu(t)$ (BELLGARDT et al., 1986). The model equation describing the induction of gluconeogenesis for ethanol growth is:

$$\frac{dr_{Smax}(t)}{dt} = [C_1 - \mu(t)] r_{Smax}(t) - \\ - C_2 [r_{Smin} - R(-r_S(t))] \qquad (40)$$

The output variables of the regulation model are possible rate-limiting steps of the metabolism. They enter the metabolic model via the corresponding elements r_{Smax} and r_{Omax} of the vector r_{max} in Eq. (38). For more details refer to Chapter 9.

3.2.3 Cell Cycle Model

The unequal cycling mechanism during growth of baker's yeast as illustrated in Fig. 2 is generally accepted. The daughter cell is smaller than the parent cell and must increase in size before it initiates a new bud. This asymmetric mode of dividing determines the specific cycling age distribution of the cell population. The cell cycling age refers here to the relative position of a cell in its cycling phase with respect to time. In this model the complicated cycling process of yeast is divided into only three cycling phases: unbudded daughter cell cycling phase, unbudded parent cell cycling phase, and budding cell cycling phase. The budding phase is equal for both budding daughter cells and budding parent cells. While the cells are growing, the cycling age of unbudded daughter cells, unbudded parent cells and budding cells changes from 0 to U_d, U_p, and B, respectively. These are the durations of the cycling intervals. Numerous experiments on a variety of strains have demonstrated that U_d, U_p, and B can be linearly correlated with the cell number doubling time, T (LORD and WHEALS, 1980; HARTWELL and UNGER, 1977). These correlations:

$$U_d = C_{d1} \cdot T + C_{d2}$$
$$U_p = C_{p1} \cdot T + C_{p2} \quad (41)$$
$$B = C_{b1} \cdot T + C_{b2}$$

are called cycling phase equations. For given strain and culture conditions, the parameters in Eq. (41) can be determined experimentally. The values used here for the strain *Saccharomyces cerevisiae* H620 growing at a temperature of 305 K are shown in Tab. 2.

Tab. 2. Parameters of the Cycling Phase Equations

Parameter	Value	Unit
C_{d1}	1.2	
C_{d2}	-2	h
C_{p1}	0.44	
C_{p2}	-0.43	h
C_{b1}	0.023	
C_{b2}	1.5	h

To simplify and speed up the simulation of the age distribution model the continuous cycling phases are divided into small intervals with constant width T_C (YUAN et al., 1991). The number of these discrete intervals in each cycling phase is defined as n_d, n_p, and n_b, respectively,

$$n_d = \frac{U_d}{T_C}$$

$$n_p = \frac{U_p}{T_C} \quad (42)$$

$$n_b = \frac{B}{T_C}$$

All cells are assumed to be distributed in these cycling age intervals. The resulting structure of the model is shown in Fig. 4. The variables

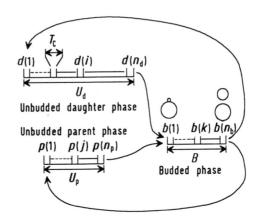

Fig. 4. Structure of the discrete cell cycling model.

$d(i)$, $p(j)$, and $b(k)$ represent the number of cells in the ith, jth and kth cycling age intervals. In each simulation step of length T_C, the cells are shifted from one interval to the next according to the following rules:

$$
\begin{aligned}
d(i+1) &\leftarrow d(i), & i &= 2, 3, \ldots, n_d \\
p(j+1) &\leftarrow p(j), & j &= 2, 3, \ldots, n_p \\
b(k+1) &\leftarrow b(k), & k &= 2, 3, \ldots, n_b \quad (43)\\
d(1) &\leftarrow b(n_b + 1) \\
p(1) &\leftarrow b(n_b + 1) \\
b(1) &\leftarrow d(n_d + 1) + p(n_p + 1)
\end{aligned}
$$

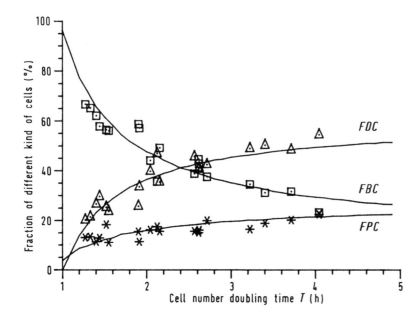

Fig. 5. Comparison of experimental data from LORD and WHEALS (1980) for *FBC* (□), *FDC* (△), and *FPC* (*) of exponentially growing cultures with simulation results (lines) by the cell cycling model: Dependency on the cell number doubling time, *T*.

which define two connected cycles for parent cells, daughter cells, and budding cells. To consider the smoothing effect of random fluctuation of the generation time of single cells in the population, a moving-average filtering was introduced in every simulation step for the cells in all cycling intervals. For exponential growth the cycling process can be simulated by model Eqs. (41) to (43). Fig. 5 shows a comparison of experimental data with this model. In good agreement with the data, the simulated fraction of budding cells increases with shorter doubling times. The fractions of cells in the three cycling phases are calculated by summation over all age intervals. Thus, the fraction of daughter cells is:

$$FDC = \sum_{i=1}^{n_d} \frac{d(i)}{X_N} \tag{44}$$

the fraction of parent cells is:

$$FPC = \sum_{j=1}^{n_p} \frac{p(j)}{X_N} \tag{45}$$

and the fraction of budding cells is:

$$FBC = \sum_{k=1}^{n_b} \frac{b(k)}{X_N} \tag{46}$$

The cell number is calculated as

$$X_N = \sum_{i=1}^{n_d} d(i) + \sum_{j=1}^{n_p} p(j) + \sum_{k=1}^{n_b} b(k) \tag{47}$$

The input variable, *T*, of the cell cycling model must be related to the specific growth rate given by the metabolic model. Experimental data indicate that at relatively low growth rates, e.g., $\mu < 0.3\,h^{-1}$, the mean cell volume of yeast cells undergoes only small changes. In this case the assumption is reasonable that cell number doubling time is equal to cell mass doubling time, and therefore:

$$T(t) = \frac{\ln(2)}{\mu(t)} \tag{48}$$

In batch or fed-batch fermentations doubling time is time-dependent and the model must adapt to it. When *T* increases or decreases, the number of age intervals, n_d, n_p, and n_b, must be increased or decreased correspondingly, due to constant T_C, and the cells must be redistributed over more or fewer cycling intervals. The cells in the U_p and *B* phases may be uniformly redistributed without changing the relative positions in the cycling phase because the volumes of the unbudded parent and budding cells are relatively constant. However, for the unbudded daughter cells the cell volume was observed to vary significantly. This may require a non-uniform redistribution during the transients of growth, a subject not further discussed here.

3.3 Consideration of Upstream and Downstream Processing Steps in the Model

In this study major emphasis was put on the model for the fermentation part because this governs the dynamics of the entire process. Nevertheless, for an economic optimization the other parts of the process cannot be neglected, although detailed modelling is not required. With the optimization goal in mind it is sufficient to estimate the approximate costs for medium preparation, cleaning, inoculum propagation, aeration, operation, and downstream processing. This model is presented next.

The fixed costs for operation, capital investment, and staff are proportional to the process cycle:

$$P_F = K_F T_T \tag{49}$$

where

$$T_T = T_F + T_P \tag{50}$$

is the total duration for completing of the process cycle, T_F is the fermentation period, and T_P the additional time consumption for preparation and downstream processing. The costs for medium:

$$P_S = P_{Mo} + P_{Sa} + P_{Sp} \tag{51}$$

include the price for substrate:

$$P_{Mo} = [V_{L0} S_0 + K_{S1}(V_{LF} - V_{L0})S_R]p_{Mo} \tag{52}$$

the price for salts and other media components:

$$P_{Sa} = (1 - K_{S1})(V_{LF} - V_{L0})p_{Sa} \tag{53}$$

and the costs for media preparation and storage, including the costs of water consumption and waste water treatment:

$$P_{Sp} = K_{S2} V_{LF} \tag{54}$$

where p_{Mo} and p_{Sa} are the specific prices for 1 kg of pure molasses and media components. The aeration rate is assumed constant. Thus, the costs are proportional to the fermentation period:

$$P_{Ae} = F_G T_F p_{Ae} \tag{55}$$

where p_{Ae} is the specific cost for production of 1 m^3 of compressed air.

The downstream processing includes centrifugation and drying. The effort for the first is proportional to the liquid volume to be processed:

$$P_{DC} = K_{D1} V_{LF} \tag{56}$$

and for the latter proportional to the dried cell mass:

$$P_{DD} = V_{LF} X_F p_{DD} \tag{57}$$

with the specific costs, p_{DD}, for drying 1 kg of yeast product. The total balance for downstream processing is:

$$P_D = P_{DC} + P_{DD} \tag{58}$$

Finally, one must determine the proceeds from product sold. The effort of propagating of the pitching yeast in the multi-stage process for the inoculation of the production tank is significant and should be taken into account for the optimization, although only the final production tank is explicitly considered in the model. A good estimate for the related costs is the selling price of the product itself. This must be subtracted from the proceeds. In addition, the quality of the yeast product must be considered. This is related to the fraction of budding cells by an empirical correlation. Thus, the total proceeds can be expressed as:

$$P_E = (X_F - X_0)V_{LF}(1 - K_Q F B C_F)p_E \tag{59}$$

where K_Q is a quality index of the selling price

and p_E the specific maximum selling price for 1 kg of best quality dry yeast.

3.4 Verification of the Process Model

Before the process model can be used in process optimization, it must be compared with experimental data to prove that it is sufficiently accurate. Normally, not all model parameters are known in advance, so the unknown parameters must be identified from suitable experiments. The full range of metabolic types and growth conditions must be covered by these experiments and, therefore, the fermentation schedule may deviate from industrial production. As an example of model verification a single simulation by the process model of a combined batch and fed-batch experiment at laboratory scale is presented in Fig. 6. The batch-part lasts from 0 to 15 h and the fed-batch part from 15 to 25 h. At the beginning, the sugar concentration is high and

ethanol is produced due to the Crabtree effect. The uptake of ethanol is suppressed by the glucose effect until all sugar is used up. This behavior leads to the well-known growth-diauxy of yeast. Under fed-batch operation the Crabtree effect can be avoided by low feeding rates of sugar at the beginning. Only after the last step-increase of the substrate flow is there overfeed and ethanol production. In Fig. 6b the cycling model is compared with experimental data for FBC. Besides the tendency of FBC to increase with higher growth rates, some small oscillations in FBC can be observed. The latter property will be used later for the optimum control strategy. The final value of FBC is about 20%.

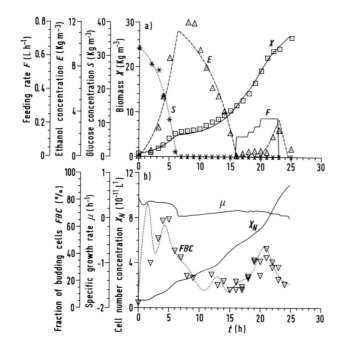

Fig. 6. Combined batch and fed-batch cultivation of baker's yeast in a 20 L reactor: Comparison of experimental data (symbols) and simulation results (lines) by the process model.
a) Variables of the reactor model: Sugar feed rate F and concentrations of cell mass X (□), substrate S (∗), and ethanol E (△).
b) Variables of the cell cycling model: Specific growth rate μ, cell number concentration X_N, and fraction of budding cells FBC (▽).

ered by these experiments and, therefore, the fermentation schedule may deviate from industrial production. As an example of model verification a single simulation by the process model of a combined batch and fed-batch experiment at laboratory scale is presented in Fig. 6. The batch-part lasts from 0 to 15 h and the fed-batch part from 15 to 25 h. At the beginning, the sugar concentration is high and

Besides this simulation a number of other experiments from laboratory scale to production scale under different operating conditions and feeding policies were simulated. The model was able to describe them all equally well with almost identical sets of parameters. This can be taken as a proof of the validity of the model, and the following optimization can be expected to give significant results.

4 Application of the Process Model for Optimization

4.1 Goals for Optimization

All economic indices for a process, such as quality, productivity, and efficiency, are interconnected. An optimization of one also influences the others in a complex way. It is thus better to take the total economic profit as the essential criterion for optimization. Such a procedure is applied to an industrial scale airlift fermentor with a volume $V_R = 60 \text{ m}^3$ by using the process model presented before. The total economic profit per unit time, P, is calculated by using Eqs. (49) to (59) as:

$$P = \frac{P_E - P_S - P_{Ae} - P_D - P_F}{T_T} \tag{60}$$

The specific costs are chosen according to an industrial process. The resulting economic balance is given in Tab. 3.

Tab. 3. Approximate Economic Balance in a Baker's Yeast Production Process (in total consumption)

Costs	Relative Value
Inoculum	2%
Raw materials	80%
Energy	9%
Fixed and operation	9%

Based on the process model the total economic profit of the baker's yeast production given by Eq. (60) can be optimized. Subject to optimization are the following constant operating parameters, substrate concentration in the feed and inoculum, S_R and S_0, aeration rate, F_G, cell mass in the inoculum, X_0, and fermentation period, T_F. For the molasses feed $F(t)$ an optimum control profile is determined. Simulations showed that an hourly change of flow rate is sufficient. Such infrequent changes

also facilitate industrial operation. The corresponding hourly values of the flow rate and the other operating parameters were numerically optimized by means of a multi-variable simplex method. Here, the aeration rate is assumed to be constant to reduce the dimension of the optimization problem. In practice, in the early stages of the process it is adjusted to the demand of the yeast cell. This simplification is not a serious drawback, because time-varying but sufficient aeration rates do not change the course of the fermentation.

The choice for the values of the variables to be optimized is not completely free for technical reasons. Therefore, several constraints on the optimization must be taken into account. The maximum working volume is limited to:

$$V_{L\max} \le (1 - \varepsilon_G) V_R \tag{61}$$

assuming for simplicity that the reactor can be filled completely and that no additional space is required in the case of foaming. Condition Eq. (61) also sets a limit for the fermentation period:

$$T_F \le T_{F\max} \tag{62}$$

where $T_{F\max}$ is defined by:

$$V_{L\max} = V_{L0} + \frac{1}{T_F} \int_0^{T_{F\max}} F(t)\,dt \tag{63}$$

The aeration rate is limited on both sides,

$$F_{G\min} \le F_G \le F_{G\max} \tag{64}$$

$F_{G\min}$ is the minimum air flow rate that will ensure sufficient mixing in the reactor and prevent flooding of the air sparger. The maximum aeration rate, $F_{G\max}$, is given by the capacity of the air compressor. The substrate pump sets a limit on the molasses flow rate:

$$0 \le F(t) \le F_{\max} \tag{65}$$

Finally, the maximum substrate concentration in the feed is given either by the sugar content of the molasses or the maximum viscosity that can be handled:

$$S_R \le K_{S1} S_{R\max} \tag{66}$$

The equation also considers by the factor K_{S1} the dilution effect through adding other media components.

4.2 Principal Strategy for Quality Optimization

Prior to economic optimization with the complete process model, which demands considerable computer time, the principle control strategy for optimizing only the product quality will be investigated. Separate analysis of this parameter can assist in clarifying the results of the economic optimization with its interdependent economic indices.

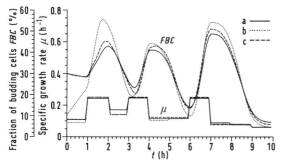

Fig. 7. Optimum control of final FBC by the specific growth rate: Simulation by the cell cycling model for different initial age distributions. Curve a: uniform distribution with $FDC = 40\%$, $FPC = 30\%$, $FBC = 30\%$. Curve b: uniform distribution with $FDC = 50\%$, $FPC = 40\%$, $FBC = 10\%$. Curve c: exponential distribution with $FDC = 40\%$, $FPC = 30\%$, $FBC = 30\%$.

According to the experimental results in Fig. 3, the fermentative activity is inversely proportional to FBC, here taken as a quality index for the product. Unfortunately, good quality is achieved with long doubling times and low growth rates. This means that productivity is very low (see Fig. 5). The results of Fig. 5 are obtained under steady conditions for exponential growth. For unsteady operation and step changes in growth rate the cell-cycling process of the yeast population tends to synchronize. As a result, oscillations in FBC can be observed. Because of these dynamics of the cycling process, maximization of fermentative activity of the final product under high productivity may be realized by suitable dynamic control of the specific growth rate, μ (DAIRAKU et al., 1982). The result of such a quality optimization of a fed-batch baker's yeast cultivation with $T_F = 10$ h is shown in Fig. 7. For this simulation only the cell-cycling model was used. An upper limit of 0.25 h^{-1} was introduced for μ to deal approximately with the Crabtree effect. Under the conditions applied here, the optimal control policy is in principle to force synchronization of population growth by three short periods of increased growth rate. Pulse synchronization leads to strong oscillations in FBC. The cells are harvested in the dynamic minimum of FBC at the end of the fermentation. This minimum is much lower than the stationary value of about 20% for the averaged specific growth rate. In this way productivity and quality can be decoupled to some extent in order to optimize both.

For practical realization of quality control it is important to know the sensitivity of the optimum strategy to changes in the initial age distribution of the population. A high sensitivity would require an analysis of the age distribution at the beginning of each fermentation run and a recalculation of the optimum strategy. In Fig. 7 the optimal μ-control is shown for three different initial age distributions. As the simulation results suggest, the cycling process is to some extent self-stabilizing, and the same control strategy can be applied regardless of the initial distribution. This is a great advantage for practical operation.

One cannot expect that the same control strategy will result from an optimization with the complete process model, because other indices enter the performance criterion. Furthermore, the ideal μ-control may not be realizable because the specific growth rate is only indirectly manipulated by the molasses feed rate. Oxygen limitation sets an upper bound on the growth rate and the mass storage properties of the liquid phase limit the speed of μ-alterations.

4.3 Optimum Operation of Baker's Yeast Production with Consideration of Product Quality

The optimization of the economic profit Eq. (60) with the process model was carried out as explained in Sect. 4.1 under the constraints given in Eqs. (61) to (66). In this section the optimum flow rate profile is presented for an industrial production tank and a laboratory fermentor, while the dependence of the opti-

mum strategy on the operating parameters and economic indices is discussed in the following sections.

With the given economic indices the point of optimal operation is obtained for a fermentation period $T_F = 10$ h, aeration rate $F_G = 90$ m^{-3} min^{-1}, and substrate concentration $S_R = 114$ kg m^{-3}. The simulation result for a 60 m^3 air-lift reactor is given in Fig. 8. The time course of flow rate and fraction of budding cells clearly resembles the pattern of the pure quality optimization, although the

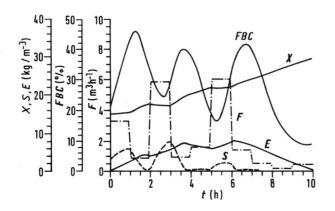

Fig. 8. State and manipulating variables under optimal economic operation: Simulation of a production run in a 60 m^3 airlift-reactor at $V_G = 90$ m^3 min^{-1}, $S_R = 114$ kg m^{-3}.

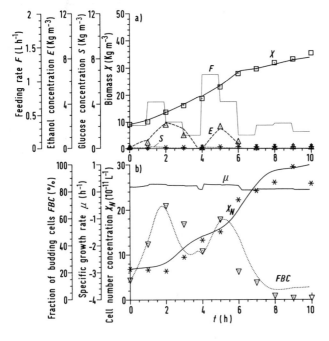

Fig. 9. State and manipulating variables under optimal economic operation: Comparison of model predictions (lines) and experimental data (symbols) for a 20 L reactor.
a) Variables of the reactor model: Sugar feed rate F and concentrations of cell mass X ($\square$), substrate S ($*$) and ethanol E ($\triangle$).
b) Variables of the cell cycling model: Specific growth rate μ, cell number concentration X_N ($*$), and fraction of budding cells FBC ($\triangledown$).

reactor model and the metabolic model put further constraints on the system. As will be analyzed in Sect. 4.5, the actual economic indices in the profit criterion lead to a slight preference of productivity over yield. This gives rise to some discrepancies to usual exponential-constant feeding schedules. Under this control strategy, there is an over-feed of sugar – even with some sugar accumulation – and ethanol accumulation in the first half of the fermentation. This increases the productivity greatly. The loss in yield can be kept low because ethanol is consumed in the second half of the fermentation. Under the conditions shown in Fig. 8 ethanol metabolism is very efficient because sugar is also fed. Therefore, ethanol is mainly linked to the Krebs cycle for energy production and not to ATP-consuming gluconeogenesis.

In Fig. 9 simulation results for a 20 L reactor are compared with experimental data. The optimum profile of the flow rate and the liquid phase concentrations are given in the upper part (a), and the variables of the cycling model in the lower part (b). Again, oscillations of FBC are forced by pulses on the flow rate. Unlike in Figs. 7 and 8, the optimum policy here requires a double peak on the flow rate. This change in the optimum scheme must be ascribed to operating parameters and process dynamics determined by the reactor, which were not considered for optimum μ-control.

The finally reached FBC of 2% is even lower than in the simulation. However, the good agreement of simulation and measurement shows the validity of the presented model.

4.4 Influence of Operating Conditions on Optimum Operation

The optimum mode of operation presented in the previous section represents only one point in the multi-dimensional parameter space. To investigate the behavior of the optimal solution, its dependency on systematic variations of operating parameters can be studied by simulation. Only partial results are presented. In Fig. 10 the profit after Eq. (60) is plotted over the fermentation interval and the aeration rate. The diagram was calculated as follows. For every fixed pair of F_G and T_F optimization was carried out with all other parameters and with the substrate feed rate $F(t)$, and the resulting sub-optimal profit was plotted. This means that the initial conditions, the inlet substrate concentration, and the feeding policy are different for every point in the F_G-T_F plane.

As expected, aeration rate has a strong influence on profit since available oxygen is one of the major limiting factors for productivity. In principle, profit increases with increasing

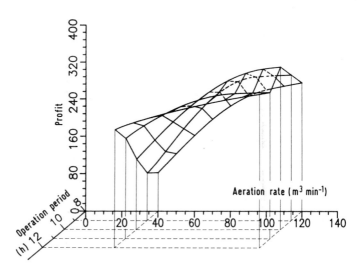

Fig. 10. Optimized economic profit after Eq. (60) under different operation periods and aeration rates.

aeration rate, but only to a certain extent when costs for compressed air or reduced quality become dominant. For fermentation periods of less than ten hours profit diminishes because of lower final yeast production. For longer fermentation periods the sub-optimum is more flat with respect to aeration rate because lower aeration rates can be compensated by a proper feeding strategy, which reduces growth rate. But profit decreases together with productivity.

4.5 Influence of the Performance Criterion on Optimum Operation

In the previous sections optimum control of baker's yeast production was investigated with an economic criterion that included product quality. The optimum policy for substrate feed deviated greatly from usual flow profiles, especially in the case of pure quality optimization, as presented in Fig. 7. The optimum also depends on the relative value of the other economic indices entering the profit margin, e.g., productivity or the costs for substrate and energy. For an investigation of these influences on the feed schedule, the previous results in Fig. 8 are contrasted with the optimum solution for a modified simpler criterion:

$$J = K_1 \frac{m_{X\text{tot}}}{T_F V_R} + \frac{m_{X\text{tot}}}{m_{S\text{tot}}} \tag{67}$$

where the first summand is proportional to the productivity and the second is the yield. The total sugar consumption is:

$$m_{S\text{tot}} = S_0 V_{L0} + S_R (V_{LF} - V_{L0}) \tag{68}$$

and the produced total cell mass is:

$$m_{X\text{tot}} = X_F V_{LF} - X_0 V_{L0} \tag{69}$$

For an exact comparison of all results summarized in Tab. 4, initial cell mass, $X_0 \cdot V_{L0}$, amount of total sugar, $m_{S\text{tot}}$, fermentation period, T_F, and the aeration rate were maintained at the same values as in Fig. 8. Profit is always calculated by Eq. (60).

The simulation results of cases 1 to 3 are given in Fig. 11. If yield is the most important aspect of yeast production (cases 1 and 2), the exponential-linear feed profile is optimal because it avoids oxygen limitation and ethanol production. The exponential increase at the beginning becomes more distinct in case 2 with slightly higher weight on productivity. But here also productivity is low because the growth rate must be maintained at relatively low values to avoid loss of yield. It is interesting that in the last interval from nine to ten hours the flow rate is decreased in order to convert the remaining ethanol.

If emphasis is placed on high productivity (case 3) the optimization results in a different two-phase strategy. In the first phase there is strong over-feeding with ethanol production

Tab. 4. Comparison of Yield, Productivity, and Profit for Different Optimization Criteria

Case	Criterion	Yield	Productivity	Profit	Figure
1	Eq. (67) $K_1 = 0.06$	0.49	4.8	136	11a
2	Eq. (67) $K_1 = 0.08$	0.48	4.9	140	11b
3	Eq. (67) $K_1 = 0.1$	0.42	6.6	195	11c
4	Eq. (60) $K_Q = 0$	0.42	6.5	200	similar to 11c
5	Eq. (60)	0.45	6.4	250	8
6	— Batch operation	<0.32	≈6.5	—	—

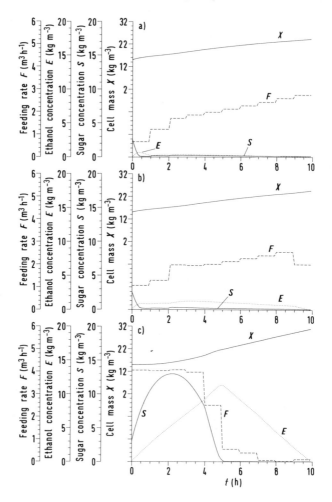

Fig. 11. Influence of the optimization criterion on the optimum control policy for the flow rate: Simulations according to Tab. 4 of the concentrations of cell mass, substrate and ethanol.
a) Maximum yield (case 1)
b) Yield and productivity (case 2)
c) Maximum productivity (case 3).

and sugar accumulation, and the growth conditions correspond to batch operation. In the second phase the feed rate is lowered to small values so that the ethanol can be completely depleted with relatively small loss of total yield. This strategy is very similar to the combined batch/fed-batch scheme of DE LOFFRE, proposed in 1941, which had been applied by some baker's yeast factories until 1961. The result of economic optimization when disregarding product quality (case 4) is quite close to the previous high-productivity case, case 3. This indicates that with the given economic indices operation cost is relatively important compared with the substrate price. Increase of profit in economic optimization is mainly due to improved product quality with yield and

productivity in a similar range as before. Obviously, quality can be improved with slight additional effort and small cut-off on the other economic indices.

For completeness, pure batch fermentation data are given in case 6. This was simulated under the assumptions that the reactor is filled completely at the beginning of the fermentation and that no ethanol or substrate inhibition occurs. Therefore, values for yield and productivity are estimates for the best case, and actual values would be lower. The initial substrate concentration

$$S_0 = \frac{m_{S\,tot}}{V_{LF}}$$

is chosen so that the total amount of converted sugar is equal to that in the other cases. Productivity can be compared with case 3, but the yield is much lower since in the first phase of diauxic growth metabolism is more fermentative than under fed-batch operation. The second diauxic growth phase is also more inefficient because ethanol growth is not supported by additional sugar supply. High ethanol concentration lowers the yield further by a stronger stripping effect of the air flow.

5 References

BAILEY, J. E., OLLIS, D. F. (1983), *Biochemical Engineering Fundamentals,* New York: McGraw-Hill.

BELLGARDT, K. H., MEYER, H. D., KUHLMANN, W., SCHÜGERL, K., THOMA, M. (1986), Application of an extended Kalman-filter for state estimation of a yeast fermentation, *IEE-Proceedings* 133, No. 5, pp. 226–234.

COONEY, C. L. (1981), Growth of microorganisms, in *Biotechnology* (REHM, H.-J., REED, G., Eds.), Vol. 1, p. 76ff, Weinheim: Verlag Chemie.

DAIRAKU, K., IZUMOTO, E., MORIKAWA, H., SHIOYA, S., TAKAMATZU, T. (1982), Optimal quality control of baker's yeast fed-batch culture using population dynamics, *Biotechnol. Bioeng.* 24, 2661–2674.

HARTWELL, L. H., UNGER, M. W. (1977), Unequal division in *Saccharomyces cerevisiae* and its application for control of cell division, *J. Cell. Biol.* 75, 422–435.

HEIJNEN, J. J., VAN'T RIET, K. (1984), Mass transfer, mixing and heat transfer phenomena in low viscosity bubble column reactors, *Chem. Eng. J.* 28, 21–42.

KOCKOVÁ-KRATOCHVÍLOVÁ, A. (1990), *Yeasts and Yeast-like Microorganisms,* Weinheim: VCH.

LAGUNAS, R. (1986), Misconceptions about the energy metabolism of *Saccharomyces cerevisiae, Yeasts* 2, 221–228.

LORD, P. G., WHEALS, A. E. (1980), Asymmetrical division in *Saccharomyces cerevisiae, J. Bacteriol.* 142, 808–818.

REED, G., NAGODAWITHANA, W. T. (1990), *Yeast Technology,* 2nd Ed., New York: Van Nostrand Reinhold.

REED, G., PEPPLER, H. J. (1973), *Yeast Technology,* Westport, Connecticut: The AVI Publishing Company, Inc.

ROSE, A. H., HARRISON, J. S. (Eds.) (1987), *The Yeasts,* Vols 1 and 2. London: Academic Press.

RUNGELDIER, K., BRAUN, E. (1961), Method and Device for Controlling and Growth of Microbial Cultures, *U.S. Patent* 3,002,894, October 3.

SCHÜGERL, K., LÜCKE, J., OELS, U. (1977), Bubble column bioreactors. Tower bioreactors without mechanical agitation, *Adv. Biochem. Eng.* 7, 1–84.

SKINNER, F. A., PASSMORE, S. M., DAVENPORT, R. R. (Eds.) (1980), *Biology and Activities of Yeast,* London: Academic Press.

TAKAMATSU, T., SHIOYA, S., CHIKATANI, H. (1985), Comparison of simple population models in baker's yeast fed-batch culture, *Chem. Eng. Sci.* 40, 499–507.

THOMPSON, P. W., WHEALS, A. E. (1980), Asymmetrical division in glucose limited chemostat culture, *J. Gen. Microbiol.* 121, 401–409.

WHEALS, A. E. (1987), Biology of the cell cycle of yeast, in *The Yeasts* (ROSE, A. H., HARRISON, J. S., Eds.), Vol. 1, pp. 283–390, London: Academic Press.

YUAN, J. Q., BELLGARDT, K. H., JIANG, W. S., DECKWER, W. D. (1991), A dynamic cell cycling model for growth of baker's yeast and its application in profit optimization, *Bioprocess Eng.* 6 (6).

13 Aerobic Wastewater Process Models

GRAHAM F. ANDREWS

Idaho Falls, Idaho 83415, U.S.A.

List of Symbols

Acronyms

ATP	adenosine triphosphate
BOD	biochemical oxygen demand
COD	chemical oxygen demand
CMAS	complete mix activated sludge
MLVSS	mixed liquor volatile suspended solids
PHB	polyhydroxybutyrate, an intracellular storage product
RBC	rotating biological contactor
TOC	total organic carbon
TSS	total suspended solids

Algebraic

a	surface area of biofilm per unit volume of reactor
a'	wetted surface area of biofilm per unit flow rate
B	$(K+S)/KY$
C	nutrient concentration in a biofilm
D	diffusivity of a nutrient in the biofilm
d	floc diameter
E	$1-(DYC_i)_l/(DYC_i)$
k_1, k_2	rate constants in model of Tab. 2
K	Monod half-velocity constant
k_d	biomass decay coefficient
k_m	biomass maintenance coefficient
k_l	mass transfer coefficient in the liquid phase
L	thickness of a biofilm
M	"molecular weight" of a carbon equivalent of biomass
m_A	mass of active biomass in a floc
m_S	mass of stored substrate in a floc
N	nutrient uptake rate per unit area of biofilm
n	number concentration of flocs
q	specific substrate uptake rate
(P/O)	oxidative phosphorylation ratio
R	outer disc radius
RR	recycle ratio
R_1	radius of non-wetted region of a disc
r	radius
r_j	volumetric rate of consumption or production of constituent j
S	substrate concentration in liquid

t	time
U	settling velocity
v	hydraulic loading or superficial velocity
w	rotational speed of an RBC
x	mass fraction of non-volatile material in biomass
X	biofilm density
X_A	active biomass concentration
X_S	stored substrate concentration
Y	biomass yield
Y_{ATP}	biomass produced per mole of ATP used
y	m_S/m_A, composition of a floc particle
y_m	maximum possible value of y
y_∞	value of y under quasi-steady state conditions
y_{AV}	X_S/X_A; average value of y in reactor
z	distance from reactor inlet

Greek

α	angle on an RBC disc
β	available electrons per carbon equivalent
γ	dimensionless support particle radius
δ	liquid film thickness
ε	porosity of biofilm
ε_b	biofilm hold-up in reactor
ε_s	total solids hold-up in reactor
η	effectiveness factor
μ	net biomass specific growth rate
μ_{max}	maximum value of μ
v	general yield value
ρ_k	rate of process k
τ	hydraulic residence time
θ_x	mean cell residence time
θ	defined by Eq. (18)
θ_f	$L(\bar{q}X/DK)^{1/2}$ dimensionless biofilm thickness

Subscripts

c	contact tank
i	concentration at biofilm interface or reactor inlet
j	identifies a constituent of the wastewater or biomass
l	limiting nutrient
o	oxygen
s	stabilization tank

1 Introduction

1.1 What is in a Model?

All models of chemical and biochemical processes consist of four parts: assumptions, rate equations, yield equations, and mass balance equations.

1.1.1 Assumptions

Reality is much too complex to be described exactly by a simple set of mathematical relations. This is true for physical and chemical processes but it is doubly true for biological wastewater treatment due to the extreme complexity of the basic phenomena. Wastewater consists of an unknown and variable mixture of chemical species, some of molecular size, some colloidal and some particulate in size. This mixture is acted on by a variety of different microbial species. They may consume the chemicals in the wastewater simultaneously or sequentially (CHAIN and DeWALLE, 1975), and the mechanisms involved in the consumption of colloids may be different from those used for soluble organics. Furthermore, some of the microorganisms may be predators, preying on other species.

The first step in writing a mathematical model must be to create a more tractable picture out of this complexity. This is done by assumptions, i.e., statements about phenomena we believe to be relatively important and that can safely be "lumped" with other phenomena or left out of the model altogether. The most important assumptions in wastewater treatment modelling concern the structure of wastewater and of biomass.

The simplest assumption that can be made about wastewater is that its concentration can be described by a single parameter such as *BOD*, *COD*, or *TOC*. This "pseudo single solute" assumption is obviously a considerable over-simplification (except for industrial wastewaters that contain a single dominant type of organic chemical). More sophisticated models for municipal sewage discard this as-

sumption and divide the total organic matter into "soluble biodegradable", "soluble non-biodegradable", "colloidal biodegradable", and "colloidal non-biodegradable" fractions (EKAMA et al., 1986).

The simplest, so-called "unstructured" models are based on the assumption that the biomass concentration can also be adequately described by a single parameter (volatile suspended solids, biofilm thickness, etc.). In more advanced models the biomass is given either a biological structure by dividing it into the different types of organisms it contains (aerobic heterotrophs, filamentous bacteria, nitrifying autotrophs, denitrifiers, protozoa etc.) or a chemical structure by dividing it into RNA, stored substrate, biopolymer, inert matter, protoplasm etc. It is sometimes stated that unstructured models require the assumption that the biomass composition is constant, but this is unnecessarily restrictive. Unstructured models describe a condition called balanced growth. This is a quasi-steady state assumption, requiring only that the environment of the biomass (composition and concentration of wastewater, pH, etc.) changes sufficiently slowly so that the biomass can adjust its internal composition to adapt to the changes. This adaption requires both that the internal composition of each microorganism (its chemical structure) remains perfectly acclimated to its environment, and that the ratios of the different types of microorganisms (the biological structure of the biomass) remain optimal for dealing with the prevailing conditions. The time scales over which these two types of acclimation occur are obviously different, and an unstructured model will only apply rigorously if they are short compared to the time scales over which the environment of the biomass varies.

This is important because different treatment systems naturally give rise to balanced or unbalanced growth conditions. In a conventional complete-mix activated sludge (CMAS) system, for example, if we ignore microbial activity in the settler (a common assumption), the conditions to which a floc particle is exposed are always the same. This remains approximately true even if there are fluctuations in the influent wastewater (flow, concentration, or composition) because the mixing tends to damp out these fluctuations. Thus, balanced growth is a valid assumption for this system. It would not be valid for a plug-flow activated sludge system, nor for the contact stabilization system in which biomass acclimated to the low-substrate concentrations in the stabilization tank is suddenly dumped into the high-substrate conditions of the contact tank. Modelling the contact stabilization system is inherently more difficult than the CMAS system because it necessarily requires a structured model, although this has not always been appreciated [see the discussion by OR-HON (1977) of the model proposed by BENE-FIELD and RANDALL, (1986)].

The same distinction should be made in other systems. The environment of a bacterium in a trickling filter with liquid recycle changes slowly as the biofilm grows and the bacterium gets buried deeper in it. Balanced growth is a reasonable assumption. Compare this with a rotating biological contactor where the surface of the biofilm is constantly switching between the wastewater (low dissolved oxygen) and the air (high dissolved oxygen). In a batch culture the lag phase is inherently unbalanced growth (unless the inoculum is perfectly acclimated to the conditions it finds in the culture, in which case the lag disappears) and is consequently not predicted by unstructured models. The exponential phase is inherently balanced growth, and the declining growth phase may involve either type of growth depending on how fast the substrate concentration decreases, which depends on the initial concentrations of biomass and substrate. This is why caution is needed when applying parameter values determined from batch culture to actual treatment systems, or when applying unstructured kinetic models to sequencing batch reactors (ORHON et al., 1986).

1.1.2 Rate Equations

Once the constituents of the wastewater and biomass in the model have been specified, the next step is to define the processes by which these constituents are created and destroyed. For a simple unstructured model there is only one such process, the growth of biomass accompanied by consumption of its substrate

$$CH_mN_nO_p + aO_2 + bNH_3 \rightarrow$$
substrate
$$dCH_rN_sO_t + eCO_2 + fH_2O \qquad (1)$$
biomass

We need to know two things about these processes, their rates and their yields. The rate can be defined in terms of the rate of production or consumption of any of the constituents, and the yields then allow us to predict the production rates of all the others. Since the processes are catalyzed by the presence of active biomass, the most fundamental way of defining rates is as specific rates, that is the rate per unit mass of active biomass. The usual choice is to specify the specific growth rate, μ, but the most common unstructured model (LAURENCE and MCCARTY, 1980) starts with the specific substrate consumption rate q:

$$q = \frac{\bar{q}S}{K+S} \qquad (2)$$

The volumetric reaction rate, that is the rate of reaction per unit volume of reactor, can be found by multiplying the specific rate by the active biomass concentration. There has long been a tendency in the literature to call the volumetric rate of substrate consumption $(-dS/dt)$. This nomenclature should be strongly discouraged. $(-dS/dt)$ means by definition "the rate of decrease of substrate concentration with respect to time", and is equal to the volumetric rate of substrate consumption only in a batch culture. It's use for this purpose in other reactor systems results from intellectual confusion between rate equations and mass balance equations (Sect. 1.1.4). The recent commission on nomenclature (GRAU et al., 1987) recommended r_j for the net volumetric rate of consumption of constituent j.

A modern structured model divides the biomass and wastewater into many constituents giving rise to many different processes, each with its own rate and yield equations. The model proposed by a recent IAWPRC (International Association for Water Pollution Research and Control) task group (HENZE et al., 1987), for example, has thirteen constituents including alkalinity and various forms of nitrogen in the wastewater, and three types of bacteria and adsorbed colloidal matter in the biomass. They are produced and consumed by eight different processes including nitrification, denitrification, and cell decay. Actually there are nine processes, but the ninth, adsorption of colloidal organic matter onto the biomass, is assumed to be very rapid compared to the subsequent enzymatic hydrolysis of the adsorbed material, so there is no need to specify a rate equation for it. This model is neatly summarized in matrix form (Tab. 1) with one column for each constituent and one row for each process. The rate equations for each process are listed (as volumetric rates) down the right side of the matrix.

The form of the rate equations adopted for the individual processes in a structured model is obviously important and, to a certain extent, arbitrary. The choice is usually based on a scrutiny of the literature, experience with simple systems, intuition about rate-determining steps (as in the hydrolysis of adsorbed colloidal material discussed above), and ultimately the testing of the model against experimental data. An additional, less used, test is to analyze mathematically the prediction of the set of rate equations under quasi-steady state, balanced growth conditions. This can be illustrated by the simple structured model proposed by ANDREWS and TIEN (1977). The biomass is divided into "active biomass" and "stored substrate", the latter lumping together intracellular storage products, adsorbed colloidal material, and some part of the extracellular biopolymer.

ues. This exploration of the behavior predicted under quasi-steady state, balanced growth conditions is a purely mathematical exercise that should be applied to all structured models.

1.1.3 Yield Equations

A relationship between the yields found in unstructured models can be found by performing element balances on Eq. (1):

$$\frac{\beta_S}{Y_S} = \frac{4}{Y_o} + \beta_b \tag{4}$$

$Y_S = d$, cell yield

$$= \frac{\text{carbon equivalent of biomass produced}}{\text{carbon equivalent of substrate consumed}}$$

$Y_o = \dfrac{d}{a}$, oxygen yield

$$= \frac{\text{carbon equivalent of biomass}}{\text{mole } O_2 \text{ consumed}}$$

Here $\beta_S = 4 + m - 3n - 2p$ is the number of available electrons per carbon equivalent of substrate, a measure of the oxidation/reduction state of the substrate. β_b is a similar quantity for the organic fraction of the biomass, and this analysis is useful only because β_b has been found to be surprisingly constant among microbial species, growth states, etc. (ANDREWS, 1989). These quantities are related to COD values. β_S is four times the number of moles of O_2 required for complete oxidation

$$\text{substrate} \xrightarrow{\text{uptake}} \text{stored substrate} \xrightarrow{\text{growth}} \text{active} + \text{respiration} \tag{3}$$
$$\text{biomass} \nearrow \text{products}$$
$$\text{maintenance}$$

The rates and yields are shown in matrix form in Tab. 2. It can be shown (PADUKONE and ANDREWS, 1989) that under balanced growth conditions these equations reduce to the basic unstructured model (Eqs. (2) and (9), see Sect. 3.2). While this does not prove that these simple equations are correct (other equations may also reduce to this form), it does show that they are at least consistent with the considerable experimental evidence that exists on balanced growth. It also simplifies parameter evaluation, since the model parameters can be related to the known Monod parameter val-

of a carbon equivalent of substrate, or 1/8 the COD (in grams) of a carbon equivalent.

$$\frac{M}{8(1-x)} \frac{Y_S}{\beta_S} = Y_{COD} \frac{\text{g biomass produced}}{\text{g } COD \text{ consumed}} \tag{5}$$

Applying similar reasoning to Y_o and β_b reduces Eq. (4) to the well-known formula for calculating the oxygen requirements of a process:

COD removed = O_2 consumed
+ COD in biomass produced (in g) (6)

Tab. 1. The IAWPRC[a] Model (HENZE et al., 1987)

j / Process ↓	1 S_I	2 S_S	3 X_I	4 X_S	5 $X_{B,H}$	6 $X_{B,A}$	7 X_P	8 S_O	9 S_{NO}	10 S_{NH}
1 Aerobic growth of heterotrophs		$-\dfrac{1}{Y_H}$			1			$-\dfrac{1-Y_H}{Y_H}$		$-i_{XB}$
2 Anoxic growth of heterotrophs		$-\dfrac{1}{Y_H}$			1				$-\dfrac{1-Y_H}{2.86\,Y_H}$	$-i_{XB}$
3 Aerobic growth of autotrophs						1		$-\dfrac{4.57-Y_A}{Y_A}$	$\dfrac{1}{Y_A}$	$-i_{XB}-\dfrac{1}{Y_A}$
4 "Decay" of heterotrophs				$1-f_P$	-1		f_P			
5 "Decay" of autotrophs				$1-f_P$		-1	f_P			
6 Ammonification of soluble organic nitrogen										1
7 "Hydrolysis" of entrapped organics		1		-1						
8 "Hydrolysis" of entrapped organic nitrogen										

Stoichiometric parameters:
Heterotrophic yield: Y_H
Autotrophic yield: Y_A
Fraction of biomass yielding particulate products: f_P
Mass N/mass COD in biomass: i_{XB}
Mass N/mass COD in products from biomass: i_{XP}

Component units (by column):
1. Soluble inert organic matter (mol COD/L)
2. Readily biodegradable substrate (mol COD/L)
3. Particulate inert organic matter (mol COD/L)
4. Slowly biodegradable substrate (mol COD/L)
5. Active heterotrophic biomass (mol COD/L)
6. Active autotrophic biomass (mol COD/L)
7. Particulate products arising from biomass decay (mol COD/L)
8. Oxygen (negative COD) (mol ($-COD$)/L)
9. Nitrate and nitrite nitrogen (mol N_2/L)
10. $NH_4^+ + NH$ nitrogen (mol N_2/L)

[a] International Association for Water Pollution Research and Control

Tab. 1. (Continued)

11 S_{ND}	12 X_{ND}	13 S_{ALK}	Process Rate, $\rho_j\,(mL^{-1}s^{-1})$
		$\dfrac{-i_{XB}}{14}$	$\hat{\mu}_H \left(\dfrac{S_S}{K_S+S_S}\right)\left(\dfrac{S_O}{K_{O,H}+S_O}\right) X_{B,H}$
		$\dfrac{1-Y_H}{14\cdot 2.86\,Y_H}$	$\hat{\mu}_H \left(\dfrac{S_S}{K_S+S_S}\right)\left(\dfrac{K_{O,H}}{K_{O,H}+S_O}\right)\left(\dfrac{S_{NO}}{K_{NO}+S_{NO}}\right)\eta_g X_{B,H}$
		$\dfrac{-i_{XB}}{14}$	
		$-\dfrac{i_{SB}}{14}-\dfrac{1}{7\,Y_A}$	$\hat{\mu}_A \left(\dfrac{S_{NH}}{K_{NH}+S_{NH}}\right)\left(\dfrac{S_O}{K_{O,A}+S_O}\right) X_{B,A}$
	$i_{XB}-f_P i_{XP}$		$b_H X_{B,H}$
	$i_{XB}-f_P i_{XP}$		$b_A X_{B,A}$
-1		$\dfrac{1}{14}$	$k_a S_{ND} X_{B,H}$
			$k_h \dfrac{X_S/X_{B,H}}{K_X+(X_S/X_{B,H})}\left[\left(\dfrac{S_O}{K_{O,H}+S_O}\right) + \eta_h\left(\dfrac{K_{O,H}}{K_{O,H}+S_O}\right)\left(\dfrac{S_{NO}}{K_{NO}+S_{NO}}\right)\right] X_{B,H}$
1	-1		$\rho_1(X_{ND}/Y_S)$

Soluble biodegradable organic nitrogen (mol N_2/L)

Particulate biodegradable organic nitrogen (mol N_2/L)

Alkalinity (molar units)

Kinetic parameters:

Heterotrophic growth and decay:
   $\hat{\mu}_N, K_S, K_{O,H}, K_{NO}, b_H$

Autotrophic growth and decay:
   $\hat{\mu}_A, K_{NH}, K_{O,A}, b_A$

Correction factor for anoxic growth of heterotrophs: n_g

Ammonification: k_a

Hydrolysis: k_h, K_X

Correction factor for anoxic hydrolysis: n_h

Tab. 2. A Simple Structured Model (ANDREWS and TIEN, 1977)

Process				Specific Rates (per unit active biomass)
Uptake	-1	1		$q = k_1 S (1 - y/y_m)$
Growth		$-1/Y$	1	$\mu = k_2 y$
Maintenance		-1		k_m
Component	Substrate	Stored substrate	Active biomass	
Concentration	S	X_S	X_A	$y = m_S/m_A$ Floc composition parameter

This analysis can be extended to actually calculating yield values by use of the Y_{ATP} hypothesis, which states that producing a gram of biomass requires the cell to expend a fixed amount of energy in the form of ATP. This is superior to the attempt to predict yields based on the free energy of formation of the substrate, because there is no guarantee that microorganisms convert free energy into a form they can use (ATP) with a constant efficiency. In aerobic metabolism the production of ATP by substrate-level phosphorylation is usually small compared to that produced by oxidative phosphorylation. Thus the consumption of 1 mole of O_2 will produce $2(P/O)$ moles of ATP and $2 Y_{ATP}(P/O)$ g of biomass. It follows that:

$$\frac{M}{(1-x)} Y_o = 2 Y_{ATP}(P/O) \tag{7}$$

From Eqs. (4), (5), and (7):

$$Y_{COD} = \frac{Y_{ATP}(P/O)}{8(2 + \beta_b(1-x) Y_{ATP}(P/O)/M)} \tag{8}$$

The important feature of this equation is that all the quantities on the right side are approximately constant. Taking the commonly reported values $Y_{ATP} = 10$ g/mol, $(P/O) = 2.8$, $\beta_b = 4.2$, $x = 0.08$, $M = 23$ (ANDREWS, 1989) gives $Y_{COD} = 0.52$ g cells/g COD. In practice Y_{COD} is indeed found to be approximately constant, but the observed yield is closer to 0.38 for carbohydrates and Krebs cycle intermediates and 0.34 for aromatic and aliphatic acids (SERVIZI and BOGAN, 1964). To account for the difference the yield equation in the basic unstructured model is written as:

$$\mu = (Yq) - k_d \tag{9}$$

Y in this equation is a yield coefficient (based on COD, BOD, or TOC depending on the units of q), a parameter characteristic of the types of biomass and wastewater. The observed cell yield (biomass produced/substrate consumed) in an actual reactor system is always less than Y, and depends not only on the values of Y and k_d but also on the type of reactor and the biomass/substrate ratio it contains. Finding accurate values of Y and k_d from observed yield data in batch cultures is particularly difficult (ANDREWS, 1984).

The biomass decay parameter k_d is a good example of how several phenomena of secondary importance can be lumped together into a single parameter with reasonable results. k_d incorporates all of the following in an approximate way:

Maintenance. Microorganisms, like people, need substrate not only for growth but also for staying alive; repairing spontaneous damage to their macromolecules, producing energy for activated transport of nutrients across cell membranes, etc. In a true "maintenance model" the yield equation is identical to Eq. (9) with k_d replaced by $k_m Y$ where k_m is the maintenance coefficient. However, the rate equation is the Monod equation for the specific growth rate μ:

$$\mu = \frac{\bar{\mu} S}{K + S} \tag{10}$$

This model is identical to our "basic unstructured model", Eqs. (2) and (9), as long as $S \gg K k_d / \bar{\mu}$ which is usually true. The basic unstructured model is preferred because it works better at very low substrate concentrations. The maintenance model will not predict a death phase in batch culture, and it tends to predict negative substrate concentrations because the substrate uptake rate is still positive ($q = k_m$) when $S = 0$.

Endogenous metabolism. When no substrate is available, a microorganism will start to catabolize its own protoplasm in order to obtain the energy required for maintenance. This endogenous metabolism is well represented by the biomass decay term.

Predation. The net effect of protozoa grazing on the bacteria is that less total biomass is produced than would be expected from the yield coefficient value and the amount of substrate consumed. This process is represented in a semi-quantitative way by the decay term.

Cell lysis. The cell decay term in the basic unstructured model can also describe the effects of cell lysis on the biomass. However, it can not describe the other effects of lysis: release of protoplasm into the liquid where it becomes substrate for other organisms and the residue of inert organic matter left in the biomass. These can only be properly described by a structured model that has a term "inert biomass". This is one of the reasons why we should not expect the basic unstructured model to adequately describe systems like aerobic sludge digestion where the cell decay processes are the primary, not the secondary, activity.

In a complex structured model a given constituent may be produced or consumed by several different processes. The matrix presentation again allows all the relevant yield values to be presented in a compact form. Consider for example the process "decay of heterotrophs" in the IAWPRC model (Tab. 1). Each

unit of "active heterotrophic biomass" consumed (column 5; a negative sign indicates consumption) produces f_p units of "particulate products" (column 6), $(1 - f_p)$ units of "slowly degradable substrate" due to cell lysis (column 4), and $(i_{x,B} - f_p i_{x,p})$ units of "particulate biodegradable organic nitrogen" (column 12). The units commonly used are COD equivalents. The total volumetric production rate of each constituent is found by summing up a matrix column

$$r_j = \sum_k \rho_k v_{j,k} \tag{11}$$

1.1.4 Mass Balance Equations

The rate and yield equations are characteristic of the type of biomass and wastewater. They describe how the biomass reacts to changes in its environment (substrate composition and concentration, temperature, etc.) and should be the same whatever type of reactor the biomass is in. This is why parameter values determined in a batch culture, for example, should be applicable to an activated sludge plant if all the assumptions underlying the model are satisfied in both reactor systems. The type of reactor is described in a model by mass balance equations. A complete model consists of the rate and yield equations coupled with mass conservation equations for each constituent they involve.

Writing mass balance equations requires assumptions about the mixing conditions for the liquid and solid phases in the reactor. The equations are quite simple for plug flow (Sect. 2.3), pure diffusion (Sect. 2.1) and completely-mixed tanks (except when combined with structured kinetics models when the residence time distribution of the biomass becomes important; Sect. 3.2). However, real reactors rarely conform to these idealizations. More complex reactor models such as a series of stirred tanks with backflow (TERASHIMA and ISHIKAWA, 1985) give much better descriptions of real, large-scale reactors.

The failure to separate mass balance, yield, and rate equations and to clearly state the assumptions on which each is based is an error that can create confusion. Use of the nomenclature $(-\mathrm{d}S/\mathrm{d}t)$ for uptake rate of substrate by biomass is one example of this. Another was demonstrated by the experimental observation that cell yields in chemostats were always lower than in batch culture. This caused debate (JAMES, 1982) despite the fact that it is an inevitable conclusion of the basic unstructured model (ANDREWS, 1984). Attempts to predict flocculation in activated sludge plants from results on extracellular biopolymer production in batch culture are another example. A microorganism can produce large amounts of polysaccharides late in a batch culture because, in its recent past, it was exposed to high concentrations of substrate. This is not true of an organism in a CMAS system. Mathematically, the mass balance equations are completely different. The correct procedure would be to formulate a hypothesis about when bacteria produce biopolymer, construct a kinetic model in which biomass is structured into cells and biopolymer (an unstructured model can convey no information about the recent history of a microorganism), solve the model for a batch culture, compare it with the data to determine whether the hypothesis is correct and what the parameter values should be, and finally to solve the model with the mass balance equations for a CMAS and see what it predicts. This has yet to be done, although some semi-quantitative results are available (Sect. 3.4).

1.2 Objectives and Advances in Modelling

The basic models available for aerobic wastewater treatment systems can be found in several textbooks (GRADY and LIM, 1980). The objective of this chapter is to review more recent developments and to show the directions in which modelling is moving and should move in the future.

1.2.1 Complexity versus Utility

In any field there is a direct relationship between the ability of a model to faithfully describe real systems and its mathematical complexity. The need for compromise was well described in GLEICK's (1987) book on non-linear dynamics:

"You can make your model more complex and more faithful to reality or you can make it simpler and easier to handle. Only the most naive scientist believes that the perfect model is one that perfectly represents reality. Such a model would have the same drawbacks as a map as large and detailed as the city it represents, a map depicting every park, every street, every building, every tree, every pothole, every inhabitant and every map. Were such a map possible its specificity would defeat its purpose: to generalize and abstract."

In recent years the wide availability of powerful personal computers has vastly increased our ability to deal with difficult mathematics, so the tendency has been to make fewer assumptions and produce more complex models. This has allowed the modelling of the interrelated processes of soluble and particulate *BOD* removal, nitrification, and denitrification. Such models are now available for activated sludge systems (HENZE et al., 1987; DOLD and MARAIS, 1986), oxidation ditches (TERASHIMA and ISHIKAWA, 1985), and biofilms (WANNER and GUJER, 1985; SEIGRIST and GUJER, 1987). Modern models also deal more realistically with liquid mixing in the reactor and are usually of the unsteady-state type to allow for the diurnal variation in the flow and concentration of wastewater (RITTMAN, 1985). This allows for theoretical investigation of different control strategies for dealing with this variation (BARTON and MCKEOWN, 1986). A start has been made on dropping the pseudo-single solute assumption and modelling the removal of individual pollutants (WATKIN and ECKENFELDER, 1984). The ultimate application of computers is the concept of "flexible modelling" in which the program itself "decides" on the appropriate structure for the model from the information fed to it (SCHEFFER et al., 1985).

However, the goal of a single model capable of optimizing the design and control strategies of all wastewater treatment systems has not yet been achieved (JAMES, 1982), and GLEICK's warning of the dangers of complexity should not be forgotten. "The goal of computing is insight not numbers" (HAMMING, 1973), and important insights that are immediately apparent from the algebra of a simple model may be lost in the details of a computer program. A good example is the application of the basic unstructured model, Eqs. (2) and (9), to the mass balances for biomass over a CMAS system, assuming no microbial activity in the settler and no biomass in the settler overflow.

$$\frac{S}{K} = \frac{1 + k_d \theta_x}{\theta_x (\bar{q} \, Y - k_d) - 1} \quad (12)$$

This equation shows that the outlet concentration is independent of the inlet concentration and establishes the mean cell residence time as the most important parameter in the design and operation of activated sludge plants, two non-obvious and very important pieces of information.

Another difficulty is that as the complexity of a model increases so does the number of undetermined parameters, and the problem of model discrimination becomes more severe. A set of experimental data can be described equally well by more than one combination of model structure, rate equations, and parameter values. For example, in the early history of the contact stabilization system it was observed that the MLVSS increased in the contact tank and decreased in the stabilization tank. This could be explained either by the growth/decay hypothesis in which bacteria grew in the contact tank and decayed in the stabilization tank, or by the storage/metabolism hypothesis in which substrate stored in the biomass in the contact tank was metabolized in the stabilization tank. The mathematical description of either hypothesis required a structured kinetic model; active/inert biomass in the first case and active biomass/stored substrate in the second (note that JENKINS and ORHON, 1972, ignored the accumulation of inert matter that accompanies biomass decay). Both types of model could explain the MLVSS data. The only way to choose between them was to collect more detailed data after deciding what specific experimental measurements correspond to the abstract categories "stored substrate" and "inert biomass". Both hypotheses were found to be partly correct and modern models, following BUSBY and ANDREWS (1975), incorporate both types of structure.

Different models serve different purposes, and the rule is to select the simplest model capable of describing the system or phenomenon under study within the precision of the data being collected. Unstructured models, strictly the quasi-steady state reductions of structured models, will work for the CMAS system (and may avoid some numerical instabilities in computer solutions of the complete structured model) but not for contact stabilization. The objective of modelling is also important. A model designed for optimization of nitrogen removal in activated sludge systems (VAN HAANDEL et al., 1982), for example, will necessarily over-simplify some of the factors that are important in modelling reactor/settler interactions (SHEINTUCH, 1987). This is why three models are used in this chapter. The basic unstructured model provides a baseline of expected behavior for a system. The IAWPRC model (Tab. 1) represents the state of the art in modelling the removal of organics and nitrogen by activated sludge systems. The simple structured model (Tab. 2) will be used for illustrative purposes to provide insight into the form of rate equations (Sect. 1.1.2), flocculation (Sect. 3.4), and problems arising from residence time distribution of flocs (Sect. 3.2).

1.2.2 Empirical and Mechanistic Models

The "mechanistic" models described above start from descriptions of the basic physical and biological phenomena in a process. Many models in the literature are not like this. They are purely empirical, mathematical functions that have been found to correlate experimental data (JONES, 1970). Empirical models give little insight into the effects of the system variables or how the system should be optimized or scaled-up. This is best done by a combination of experiment and mechanistic modelling, using experimental data to validate the model, and then using the model to see how the system can be improved and thus what experiments should be done next.

Consider, for example, the common empirical model for the variation of wastewater concentration with depth in a trickling filter:

$$\frac{S}{S_i} = e^{-Az} \tag{13}$$

An identical mechanistic equation can be derived from the basic unstructured model by assuming a thick bacterial film, plug flow of liquid, and an inlet concentration low enough to prevent oxygen limitation and allow the use of first-order kinetics (Eqs. (31) and (34)). However, A is no longer unknown but is related to the system variables by $A = (D\bar{q}X/K)^{1/2}a/v$. This identifies the hydraulic loading, v, as the critical operating parameter and shows that any increase in the interfacial area, a, obtained with improved packing can be matched by a proportional increase in v.

When further experiments show that A is proportional to $v^{-3/4}$ rather than to v^{-1}, a reexamination of the assumptions in the model is required, particularly that of plug flow which has been shown to be incorrect (SUSCHKA, 1987). Since plug flow is a best-case assumption, this is a case where the model has set up an "ideal reactor" that we should try to achieve in practice. Performance improvements are clearly possible if a packing can be found that gives hydraulics closer to plug flow.

Finally, having derived the mechanistic model, we know that it will only apply at low inlet concentrations (which is realistic for municipal sewage and high recycle ratios) and are not surprised when it breaks down with concentrated waste (OLESZKIEWICZ, 1977). At very high concentrations, activity in the biofilm is limited solely by the availability of oxygen, and a mechanistic model would predict $v(S_i - S) = $ a constant, Eq. (37), which again agrees with the experiment.

2 Biofilm Processes

2.1 Biofilm Kinetics

Mass conservation and the limiting nutrient. A microorganism embedded in a biofilm can receive nutrients, including oxygen, only by molecular diffusion from the surface. If the diffusivity is constant throughout the biofilm the mass balance equation for a nutrient in a biofilm of thickness L growing on a flat plate is:

$$D_j \frac{d^2 C_j}{dz^2} = q_j X \tag{14}$$

$$C_j = C_{j,i} \quad \text{at } z = 0$$

$$\frac{dC_j}{dz} = 0 \quad \text{at } z = L$$

There is one such equation for each nutrient, and the specific uptake rate for each nutrient, q_j, is in general a function of all of the nutrient concentrations. Solving this set of equations is a mathematical challenge, and two approaches have generally been adopted. For inherently complex processes like nitrification, a realistic solution can only be obtained numerically

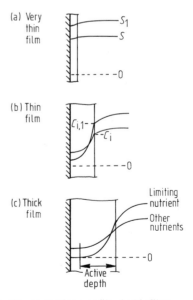

Fig. 1. Concentration profiles in biofilms.

(SEIGRIST and GUJER, 1987). For simpler processes, including the degradation of soluble organics, it is usually assumed that a single nutrient controls the metabolic rate in the biofilm. Only one equation, Eq. (14), need then be solved but an exact algebraic solution is still impossible if q is given by the non-linear Monod equation (2). The usual approach has been to find approximate solutions by taking linear approximations to the Monod equation (ATKINSON and WILLIAMS, 1971; SHIEH et al., 1981; ANDREWS and TRAPASSO, 1985).

The analysis given here, based on the work of ANDREWS (1988a), represents a compromise between the accuracy of the first approach and the simplicity of the second. The first step is to identify the two nutrients that have the most influence on the metabolic rate. To do this the q_j values are eliminated using the yield Eq. (9) and $(\mu + k_d)$ is eliminated between the resulting equations, which can then be integrated to give

$$D_1 Y_1 (C_{1,i} - C_1) = D_2 Y_2 (C_{2,i} - C_2)$$
$$= D_3 Y_3 (C_{3,i} - C_3) = \ldots \qquad (15)$$

It follows that the limiting nutrient, the one whose concentration reaches zero first in the biofilm (Fig. 1), is the one with the lowest value of $(D Y C_i)_j$. The two most probable candidates are organic matter (as *BOD*) and oxygen. Since $Y_{BOD}/Y_o = 1$ by definition, diffusivities in biofilms are proportional to their values in water. Since the diffusivity of O_2 in water is 2.5×10^{-5} cm²/s and the diffusivities of soluble organic compounds in water are close to 1×10^{-5} cm²/s, it follows that oxygen will be limiting if $C_{i,BOD}/C_{i,o} > 2.5$ (approximately). For well-agitated, air-saturated water ($C_{i,o} = 8$ mg/L) this corresponds to $C_{i,BOD} > 20$ mg/L.

However, the limiting nutrient is not necessarily the only one that influences the metabolic rate in the biofilm. Consider the situation when $C_{i,BOD} = 25$ mg/L, $C_{i,o} = 8$ mg/L. Clearly oxygen is limiting, yet at the point inside the film where $C_{BOD} = 7.5$ mg/L, Eq. (15) gives $C_o = 1$ mg/L. Now if the Monod half-velocity constants are $K_{BOD} = 5$ mg/L and $K_o = 0.1$ mg/L (GRADY and LIM, 1980), the reduction in metabolic rate due to oxygen restriction is $q/\bar{q} = (C_o/K_o + C_o) = 0.9$, while the corresponding value for *BOD* restriction is $(C_{BOD}/$

$K_{BOD} + C_{BOD}) = 0.6$. The *BOD* has a larger effect on the rate even though oxygen is the limiting nutrient. To cover this possibility the rate-controlling nutrient is defined as the one whose concentration first reaches the value that restricts the metabolic rate to 90% of its value under liquid-phase conditions. It can be shown that this nutrient has the lowest value of

$$\left[D Y C_i \left(\frac{C_i + K}{C_i + 10K} \right) \right]_j \qquad (16)$$

Note that this is the same as the limiting nutrient if $C_i \gg K$ for all nutrients, but otherwise they may be different.

Eq. (14) for the rate-controlling nutrient (which is given no subscript) can now be written

$$D \frac{d^2 C}{dz^2} = \bar{q} X \frac{C}{K + C} H(C_l) \qquad (17)$$

$$C = C_i \quad \text{at} \quad z = 0$$

$$\frac{dC}{dz} = 0 \quad \text{at} \quad z = L \text{ or when } C = E C_i$$

$$E = 1 - (D Y C_i)_l / (D Y C_i)$$

The two forms of the second boundary condition correspond respectively to a thin film where the limiting nutrient is present throughout the film, and a thick film in which microbial activity is stopped deep inside the film by exhaustion of the limiting nutrient (Fig. 1). The second form arises by setting $C_l = 0$ in Eq. (15). The parameter E gives the relative importance of the limiting and rate-controlling nutrients in fixing the total metabolic rate in the biofilm. It varies between $E = 0$ when one nutrient is both limiting and rate-controlling and $E = 0.9$ (a consequence of the definition of the rate-controlling nutrient) when one nutrient is not limiting but is very rate controlling (this happens if its Monod constant K is very large).

H in Eq. (17) is the Heaviside function which equals 1 if $C_l > 0$ and equals 0 when $C_l = 0$. Mathematically it is the limit of the Monod function as $K \to 0$, and it is what stops metabolism in the film at the point at which the limiting nutrient is exhausted.

Chemical engineers have considerable experience with this problem from their studies of heterogeneous catalysis. These studies show that the key is a correct definition of the dimensionless film thickness (BISCHOFF, 1965). The correct definition for the kinetic equation used here is

$$\theta = \frac{\theta_f C_i}{K + C_i} \left[2 \left(\frac{C_i}{K} (1 - E) - \right. \right.$$

$$\left. \left. - \ln \left(\frac{K + C_i}{K + EC_i} \right) \right) \right]^{-1/2} \tag{18}$$

The results are commonly expressed in terms of an effectiveness factor η:

$$\eta = \frac{\begin{array}{c}\text{actual consumption rate of a}\\\text{nutrient in the film}\end{array}}{\begin{array}{c}\text{consumption rate if entire film}\\\text{were exposed to conditions}\\\text{in the liquid at the interface}\end{array}}$$

$$= \frac{N}{L X \bar{q} \dfrac{C_i}{K + C_i}} \tag{19}$$

N = consumption rate per unit area of film

$$= -D \frac{dC}{dz}\bigg|_{z=0}$$

Eq. (17) is actually solved for two limiting cases:

(a) First-order kinetics:
    $C_i \ll K : \theta = \theta_f (1 - E^2)^{-1/2}$

This is the low-concentration asymptote. The form of θ is found by taking a Taylor series expansion of the logarithm term in Eq. (18). After solving Eq. (17) and substituting the solution into Eq. (19) the result is

thin films: $\tanh\theta_f < 1 - E^2$ $\eta_f = \dfrac{\tanh\theta_f}{\theta_f}$

thick films: $\tanh\theta_f > 1 - E^2$ $\eta_f = \dfrac{1}{\theta_f}$ \hfill (20)

(b) Zero-order kinetics:
    $C_i \gg K : \theta = L (\bar{q} X / 2 D C_i)^{1/2}$

This is the high-concentration asymptote. It was pointed out above that if $C_i \gg K$ for all nutrients then the limiting and rate-controlling nutrients are necessarily the same. Also the logarithm term in Eq. (18) is negligible. The result is

thin films: $\theta < 1$ $\eta_\infty = 1$

\hfill (21)

thick films: $\theta > 1$ $\eta_\infty = \dfrac{1}{\theta}$

Note that this solution corresponds to the limit of Eq. (20) as $E \to 1$, that is, as the concentration of the limiting nutrient approaches zero. This is as expected, because this nutrient necessarily becomes rate-controlling (when $E = 0.9$), and the kinetics for this nutrient are inherently zero-order (the H function). The thick film result says that the effectiveness factor is proportional to $C_i^{1/2}$, a result sometimes referred to as half-order kinetics (JANSEN and HARREMOES, 1985).

These results are all plotted in Fig. 2. The important feature of this graph is not the mathematical detail, but the fact that with θ defined by Eq. (18) all the intermediate solutions fall between the two asymptotic solutions derived above. These two solutions are so close together that the graph can be used as an approximate solution for all values of E, C_i/K, and film thickness. The accuracy of this procedure is usually adequate for wastewater treatment work, since accuracy in this field is always constrained by the need to measure concentrations by "pseudo-single-solute" parameters such as COD. When greater accuracy is needed it can be obtained from simple interpolation formulae between the asymptotic solutions. For example, when the limiting and rate-controlling nutrients are the same ($E = 0$), a comparison with available numerical solutions of Eq. (17) suggests the formula:

$$\eta = \frac{6 K \eta_f + C_i \eta_\infty}{6 K + C_i} \tag{22}$$

The intermediate lines in Fig. 2 were plotted from this formula. Interpolations of this type are clearly needed in the transition region between thin and thick biofilms (i.e., θ close to 1).

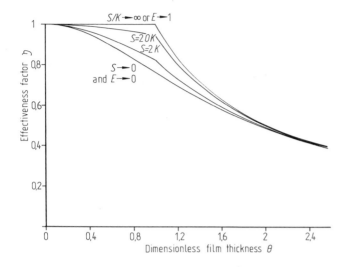

Fig. 2. Effectiveness factor.

The analysis given here is summarized in terms of concentration profiles through the biofilm in Fig. 1. The interfacial concentrations (C_i and $C_{i,1}$) may be less than the bulk-liquid concentrations (S_i and $S_{i,1}$) due to mass-transfer limitations in the liquid phase. In a very thin film (Fig. 1a), essentially all of the microorganisms are exposed to the interfacial concentrations, so $\theta = 1$ by definition (Eq. (19)). As the film gets thicker (Fig. 1b), the rate of consumption of nutrients increases and, since this consumption is fed by diffusion, large concentration gradients appear in the biofilm. For zero-order kinetics ($C_i \gg K$), the drop in concentration does not cause any decrease in metabolic rate so that η still equals one. For other inherent kinetics this biofilm mass transfer resistance does cause a drop in metabolic rate and $\eta < 1$. A truly mass-transfer limited film is shown in Fig. 1c. The film is so thick that the limiting nutrient concentration reaches zero, stopping metabolism. This can happen despite a liquid-phase concentration higher than that of the rate-controlling nutrient, either because it has a lower diffusivity or because more of it is consumed by the film (lower Y). The active depth of the film corresponds to $\theta = 1$. It has been shown both by calculation and by oxygen microprobe measurements that for an aerobic biofilm in contact with air-saturated water this is a film thickness of roughly 150 μm (ATKINSON and FOWLER,

1974). Adding more film beyond this thickness will give no increase in the substrate uptake rate.

2.2 Advances in Biofilm Modelling

Progress is needed and is being made in three areas: finding mathematical simplifications to the analysis given in the previous section in order to facilitate its use in reactor design; producing more complex models to provide insight into important processes such as nitrification and sloughing; and finally procedures for model verification and parameter measurement for particular wastewaters.

2.2.1 Mathematical Simplification

Potentially the most useful simplication is that proposed by KORNEGAY and ANDREWS (1968) who showed experimentally that the substrate uptake rate per unit area of film could be correlated by

$$N = \bar{q} X L \frac{S}{K_A + S} \tag{23}$$

K_A is an apparent Monod constant whose value varies with film thickness. The equation is clearly valid with $K_A = K$ (the inherent Monod constant) for very thin films (Fig. 1a) without liquid-phase mass transfer resistance, because all of the biomass is then exposed to a concentration S of the rate-controlling nutrient. It must also be true as $S \to \infty$, because if S is high enough, all of the film will have adequate nutrients to keep its metabolic rate at its maximum value $\bar{q}$. The question is whether this equation interpolates well between these limiting cases, and if so, how is K_A related to the film thickness, liquid-phase mass transfer coefficients, etc.?

If the rate-controlling and limiting nutrients are the same ($E = 0$: this assumption is inherent in Eq. (23)) and liquid-phase mass transfer resistance can be ignored, these questions can be answered as follows. For given values of film thickness, θ_f, and nutrient concentration, S/K, the effectiveness is calculated from Eq. (22),

would certainly fit within the normal scatter of data from a pilot-scale trickling filter, for example.

The discontinuity in slope seen for $\theta_f = 4$ and $\theta_f = 6$ in Fig. 3 corresponds to $\theta = 1$, the point at which the film switches from being "thick" (i.e., mass transfer limited; Fig. 1c) to "thin". The results for the non-mass transfer limited, thin film sections (high S/K), can be correlated by

$$\frac{K_A}{K} = 1 + 2\theta_f^2 \frac{(3 + \theta_f^2)}{(6 + \theta_f^2)} \qquad (24)$$

When working solely in the thick-film regime (low S/K), the result is

$$\frac{K_A}{K} = \theta_f \qquad (25)$$

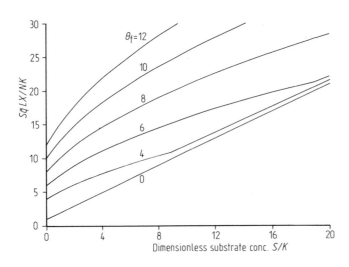

Fig. 3. Model equivalence.

and thus a dimensionless uptake rate $N/\bar{q}LX$ from Eq. (19) is obtained. If Eq. (23) is valid, a plot of the results as $\bar{q}LXS/KN$ versus S/K will give a series of straight lines with intercept K_A/K. This purely theoretical exercise says that the slopes of the lines should all be one, but in practice the quantity $\bar{q}X$ would be an undetermined parameter for fitting experimental data, so some variation in slope is allowed. The results (Fig. 3) show that Eq. (23) works surprisingly well over a wide range of film thickness and substrate concentration. It

2.2.2 Nitrification

Seigrist and Gujer (1987) have given a complete analysis of a biofilm that is nitrifying in the complete absence of organic matter. Four diffusion/consumption Eqs. (14) are solved simultaneously for NO_2^-, HCO_3^-, NH_4^+, and O_2. The interaction between pH, alkalinity, and metabolic rate makes this process extremely complex, but the model agrees well with experimental data.

WANNER and GUJER (1985) considered the more common situation in which nitrification occurs in the presence of low levels of organic substrate. In this model the biofilm is "structured" into heterotrophs and nitrifying autotrophs, and the growth kinetics of each are described by a double Monod type of equation including O_2 and bacterial substrate. The main insight produced by the model is the importance of competition for oxygen between the two types of bacteria. Since the heterotrophs have an inherently higher growth rate, they win the competition and "crowd out" the autotrophs whenever large amounts of organic substitute are present, as in the upper part of a trickling filter. However, in the lower parts of a filter the situation is reversed due to the low level of organics and high level of NH_4^+. Nitrifying bacteria come to dominate the biofilm in this region. The model also predicts a large region between these extremes where the two types of bacteria coexist symbiotically in the biofilm, the autotrophs consuming the NH_4^+ and CO_2 produced by the heterotrophs.

2.2.3 Effect of Colloids

The entire diffusion/consumption analysis of biofilms given above applies only to soluble substrates. Colloidal and particulate organic matter is transported from the bulk liquid to the biofilm surface by different mechanisms including interception, sedimentation, and inertial impaction (BOUWER, 1987). Once at the surface they must stay there because they are too large to diffuse into the biofilm. Application of the IAWPRC hypothesis concerning particulate matter in activated sludge systems (since an individual microorganism cannot know if it is in a biofilm or a floc particle) suggests that the adsorbed particulate matter would slowly hydrolyze, some of the resulting soluble organics diffusing into the film. The particulate matter could also be engulfed as the biofilm grows outward. No mathematical models of these scenarios are available. Experimental data (SARNER and MARKLUND, 1985) suggest that the presence of colloidal organics (starch and sewage colloids) reduces the degradation rate of glucose in biofilms at high

glucose concentrations but increases it slightly at low concentrations.

2.2.4 Steady-State Biofilm Thickness

The analysis in Sect. 2.1 treated the film thickness, L, as a known fixed quantity. In fact it is not. The film thickness that develops in a reactor is the result of a balance between biofilm growth and decay processes including predation, endogenous metabolism and cell lysis deep in the biofilm, and wash-off of cells from the biofilm surface. In aerobic processes the microbial growth rate is high, so that the resulting steady-state film is usually much thicker than the active depth. This allows the use of the thick-film asymptote for the effectiveness factor ($\eta = 1/\theta$) which gives the substrate consumption rate as:

$$N = \left\{ 2D\bar{q}X \left[C_i(1-E) - K\ln\left(\frac{K+C_i}{K+EC_i} \right) \right] \right\}^{1/2}$$

(26)

In systems such as fluidized beds, where the shear stress on the biofilm interface is relatively high, the wash-off of microbes happens continuously. All the decay processes are then lumped together in a single conventional decay term (ANDREWS, 1982) giving the net biofilm growth rate as

$$\frac{dL}{dt} = YN - k_d L$$

(27)

Here Y is a yield coefficient in cm^3 of biofilm produced per gram of substrate consumed. This equation is explored qualitatively in Fig. 4. As a biofilm develops, the substrate uptake rate N initially increases rapidly with film thickness, but then levels off at the value given by Eq. (26) as the film becomes mass-transfer limited. The biofilm develops as long as the growth rate YN is less than the decay rate $k_d L$, until it reaches a steady-state value where the two are equal. Fig. 4 illustrates three important points. First, since growth rates in aerobic processes are high, the steady-state film thickness is usually much larger than the

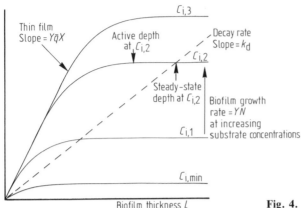

Fig. 4. Steady-state biofilm thickness.

active depth. Second, k_d is a critical parameter in that small changes in its value can produce quite large changes in the steady-state thickness (see line for $C_{i,1}$). The shortage of reliable values for k_d, as a function of shear stress, wastewater type, etc., therefore, restricts the use of this model. Third, there exists a substrate concentration $C_{i,min} = k_d K/(\bar{q} X - k_d)$ below which the "growth" and "decay" lines cross only at the origin, and the model predicts that no film will grow. This implies that for many biofilm reactors (those without mixing of the biofilm phase) there is a definite lower limit to the effluent concentration of organics. This has been extensively studied in a series of papers by RITTMAN (1982). The results are useful in a semi-quantitative way, although the $C_{i,min}$ idea is clearly not strictly correct since thin (monolayer) biofilms are found on rocks in mountain streams, a very high-shear, low-substrate environment. This is probably due to adsorption of substrates at the surface, and to strong microbe–surface interactions (as opposed to the microbe–microbe interactions in a true biofilm).

In low shear reactors including trickling filters and rotating biological contactors (RBCs), the slow, continuous wash-off of microorganisms is less important. The biofilm grows until it can no longer support its own weight and whole sections of it then "slough" off completely. The weakness that produces sloughing occurs in a region not yet considered, the region below the active depth. The theoretical model (Fig. 1 c) was based on no microbial ac-

tivity in this region, but this is obviously only an assumption.

The type of activity depends critically on the identity of the limiting nutrient. If oxygen is limiting, carbon and nitrogen sources are still available, and anaerobic activity can continue. This can produce a variety of gases (N_2, H_2S, CH_4) capable of dislodging the film. Also the anaerobic bacteria will not necessarily produce the extracellular biopolymer that holds a biofilm together. If the organic matter is limiting, then the metabolism in the region of interest must be aerobic and endogenous. The microbes are in the same situation as in aerobic digestion. They will first oxidize any intracellular stored substrate, and then in order to generate energy to stay alive they will start to hydrolyze the extracellular polysaccharide. These changes are reflected in measurements of the variation of biofilm density with thickness, which show a maximum when the thickness equals the active depth (HOEHN and RAY, 1973; note that this casts doubt on the assumption in Sect. 2.1 that diffusivities in a biofilm are constant). However, attempts to build simple "sloughing criteria" related to the active depth into unstructured biofilm models have failed to explain the thick (order of mm) biofilms found in practice (ATKINSON and WILLIAMS, 1971). A consistent sloughing criterion may only be found in a structured model that divides the film into "active biomass" and "extracellular polysaccharide". Such a model may also be able to predict the observed variation in biofilm density.

2.2.5 Parameter Estimation

A device for the experimental measurement of biofilm kinetics, validation of the analysis of Sect. 2.1, and evaluation of parameter values for a particular wastewater, needs to be carefully designed. It must provide a large surface area of biofilm of measurable thickness all exposed to the same conditions, including substrate concentration and shear stress. Providing adequate aeration is another difficulty, and several researchers have chosen to work with denitrifying biofilms.

The concentric-cylinder device (KORNEGAY and ANDREWS, 1968) satisfies all of these requirements except measurement of the biofilm thickness. This device is essentially a Couette viscometer, and rotation of the inner cylinder causes a constant shear over the entire biofilm surface. JANSEN and HARREMOES (1985) used the device to study both aerobic and denitrifying biofilms with methanol, acetic acid, and glucose as carbon sources. They confirmed the 1/2-order kinetic model for thick biofilms. Measuring the biofilm thickness is much easier in the rotating-disc device used by SHIEH and MULCAHY (1985), which is fitted with small elements that can be removed for microscopic examination. As with the concentric-cylinder device, the liquid-phase mass transfer resistance can be eliminated by increasing the rotational speed. However, the rotating disc has the disadvantage that the shear stress on the biofilm surface increases with radius. ANDREWS and TIEN (1981) used a fluidized bed of coal coated with biofilm with liquid recycle to ensure constant substrate conditions in the bed. The average biofilm thickness could be inferred from the way the bed expanded as the biofilm developed, but there is no guarantee that the film thickness was the same throughout the bed.

Compared with models for dispersed-growth systems, biofilm models require one extra parameter, the diffusivity of the nutrient into the biomass. Most of the experiments quoted in this section give values in the range of 50 to 100% of the values in water, although some lower (JANSEN and HARREMOES, 1985) and higher (OMUNA and OMURA, 1982) values have been reported. Assuming a diffusivity of 80% of the water value is acceptable for most purposes since this is not a sensitive parameter; only its square root appears in the equations.

2.3 Biofilm Reactors

The uptake rate expressions derived in the previous sections are applicable to biofilms in any type of reactor. In order to produce a complete mechanistic process model, they must be combined with mass balances for the various nutrients and expressions for the mass transfer rates between the liquid and the biofilm interface. These reactor-specific equations are studied in this section. Assuming plug flow of the liquid phase, the steady-state mass conservation for nutrient j in the reactor is

$$v \frac{dS_j}{dz} = (k_L a)_g (S_j^* - S_j) - \varepsilon_b X \eta q_j(C_i)$$

$$S_j = S_{j,i} \quad \text{at} \quad z = 0 \tag{28}$$

The $(k_L a)_g$ term accounts for possible transfer of the nutrient from the gas phase into the liquid. The final term accounts for nutrient consumption in the biofilm, which must equal the rate at which the nutrient is transferred from the liquid to the biofilm interface:

$$\varepsilon_b X \eta q_j(C_i) = (k_L a)_i (S_j - C_{i,j}) \tag{29}$$

Values of the mass transfer coefficients at the gas–liquid interface, $k_{l,g}$, and the biofilm-liquid interface, $k_{l,i}$, depend mainly on the level of turbulence in the liquid and vary considerably between reactors. They are evaluated from empirical correlations reviewed recently by KISSEL (1986). In doing these calculations it is important not to confuse the nutrient that is limiting in the biofilm with the limiting nutrient for the reactor, defined as the nutrient which will be exhausted first in the liquid phase. The best example of this problem concerns oxygen. As long as the reactor is aerated, the dissolved oxygen concentration can never be zero (although it can be very small), thus, oxygen is not the limiting nutrient for the reactor. It can be, and often is, the limiting nutrient in the biofilm. This problem also extends to nutrients that are not supplied continuously. The yield equation (9) gives $(\mu + k_d) = Y_1 q_1 = Y_2 q_2 = \ldots$. Replacing q_j in Eq. (28) with $(\mu + k_d)/Y_j$, eliminating $(\mu + k_d)$ between the equations for different nutrients and integrating the result gives

$$Y_1(S_{1,i} - S_1) = Y_2(S_{2,i} - S_2) = \ldots \qquad (30)$$

The limiting nutrient for the reactor is, therefore, the one with the lowest value of $(Y S_i)_j$. It would necessarily be the same as the limiting nutrient in the biofilm (which may change through the reactor) only if the diffusivity values were the same for all nutrients.

2.3.1 The Trickling Filter

The biofilm support in a trickling filter consists of plastic packing or rocks whose radius of curvature is very large compared to the biofilm thickness. The biofilm can, therefore, be treated as a flat plate and the biomass holdup $\varepsilon_b = aL$ where $a =$ surface area of packing per unit reactor volume. The conservation equation for organic matter (Eqs. (19) and (28)) is then

$$-v\frac{dS}{dz} = aN \qquad (31)$$

Various solutions are possible depending on the assumptions made about film thickness, limiting nutrients, etc.

Thick film, low BOD. The discussion in Sect. 2.1 suggests that, given efficient oxygen transfer, the organic matter is rate-controlling in the biofilm when $S < 30$ mg BOD/L (approximately). Even at these low concentrations, the biofilm will grow past its active depth of approximately 150 μm. We can, therefore, use the thick film asymptote for N (Eq. (26)) in Eq. (29) which can be written

$$N = k_{L,i}(S - C_i) \qquad (32)$$

In principle, this equation can be solved for the interfacial concentration, C_i, and the uptake rate, N. In practice this cannot be done

algebraically except when C_i is small enough so that the logarithm term can be approximated by two terms of its series expansion

$$\ln\left(\frac{1 + C_i/K}{1 + EC_i/K}\right) = \frac{C_i}{K}(1 - E) -$$
$$- \frac{1}{2}\left(\frac{C_i}{K}\right)^2(1 - E^2) + \ldots \qquad (33)$$

The solution is then

$$N = \frac{FS}{1 + F/k_{L,i}}$$
$$F = \left[\frac{D\bar{q}X(1 - E^2)}{K}\right]^{1/2} \qquad (34)$$

The quantity E allows oxygen to be the limiting nutrient in the biofilm despite the organic matter being rate-controlling. It will decrease through the filter. Fortunately, the quantity $(1 - E^2)^{1/2}$ is close to one even when E is quite large $(0 \le E \le 0.9)$ so this variation has little effect on the solution and is ignored. Substituting N into Eq. (31) and integrating gives the solution discussed previously in Eq. (13). Its main prediction is that the fractional removal depends only on the hydraulic loading and not on the inlet concentration. This has been found to be true up to surprisingly high values of S, although this is due in part to the fact that semi-logarithmic graph paper on which the data are plotted, spreads out the low S data points and, therefore, tends to give them more weight.

Thick film, high BOD. For $S > 30$ mg BOD/L (approximately) Sect. 2.1 suggests that oxygen becomes both limiting and rate-controlling in the biofilm. θ is now defined in terms of dissolved oxygen concentrations and N can be found only from $N = N_o Y_o/Y_S$ (note that with organic matter measured as BOD, $Y_o/Y_S = 1$ by definition). In the oxygen mass balance equation the consumption by the biofilm is balanced by transfer from the gas phase, and the dissolved oxygen concentration changes only slowly through the filter. The left side of Eq. (28) can, therefore, be set to zero, and together with Eq. (29) and the thick film asymptote the result is:

$$N_o = \bar{k}_{L,o} a (S_o^* - C_{i,o}) =$$

$$= \left\{ 2 D_o \bar{q}_o X \left[C_{i,o} - K_o \ln \left(1 + \frac{C_{i,o}}{K_o} \right) \right] \right\}^{1/2} \quad (35)$$

$\bar{k}_{L,o}$ is an overall mass transfer coefficient given by

$$\frac{1}{\bar{k}_{L,o} a} = \frac{1}{(k_{L,o} a)_i} + \frac{1}{(k_{L,o} a)_g} \quad (36)$$

In practice, the interfacial areas of the packing, the biofilm and the gas–liquid interface (a, a_i, a_g, respectively) will be similar but not identical.

As before, Eq. (35) can be solved for the interfacial concentration $C_{i,o}$ and the uptake rate N_o. The important result here is that given sufficient airflow (which can affect the saturation dissolved oxygen concentration S_o^*) all the parameters in Eq. (35) are constants. It follows that N_o, and therefore N, do not vary through the filter, and the solution of Eq. (31) is

$$1 - \frac{S}{S_i} = \frac{z a}{v S_i} \left(\frac{Y_o N_o}{Y_S} \right) \quad (37)$$

In contrast to the low *BOD* equation, the fractional removal is now inversely proportional to organic loading $v S_i / z$.

Eqs. (13) and (37) have both been used extensively in the literature. They are derived rigorously here both to demonstrate the assumptions on which they are based and because trickling filter modelling generally has had an empirical rather than a mechanistic basis.

OLESZKIEWICZ (1977), for example, presented data in the *COD* range of several hundred mg/L, tried various empirical correlations, rejected those with the form of Eq. (13), and found that a plot of S/S_i versus (a/organic loading) gave the best results at high loadings. He also found, with other data, that some plots showed a discontinuity in the slope at a *BOD* of 20 to 30 mg/L. All of this is readily explained by the mechanistic model.

A more complete model (equivalent to the IAWPRC model for the activated sludge process) would fill the gap between Eqs. (13) and (37) and could lead to new insights and process improvements. Its main features would be as follows. The true liquid mixing condition would be represented by a "stirred-tanks in se-

ries" approach, as used in RITTMAN's (1985) model for the effect of fluctuating loads. Each "tank" would be examined iteratively to determine which nutrients were limiting and rate-controlling in the biofilm. The film would be structured first to predict nitrification (Sect. 2.2.2) and second to provide criteria for the effect of sloughing (Sect. 2.2.4). Finally, some consideration would be given to the different mechanisms for removal of soluble, colloidal, and particulate organic material.

2.3.2 Fluidized-Bed Bioreactors

Reactors in which biofilms grow on the surface of solid particles in a liquid-fluidized bed have been applied extensively for *BOD* removal and nitrification. Oxygen is provided either by bubbling air directly through the bed (FAN et al., 1987) or by aerating a liquid recycle stream (the Oxitron system). These reactors have a large potential size advantage over trickling filters because the very small particles used give much larger biofilm surface areas. A bed of 20 mesh sand fluidized to a porosity of 60% has $a = 3800$ m^{-1}, compared to typical values of 62 m^{-1} for crushed-rock trickling-filter packings and $a = 110$ m^{-1} for plastic packings. Growing biomass does not clog a fluidized bed, as it would a packed bed of sand, because a fluidized bed can expand to accommodate the biofilm volume.

A recent analysis (ANDREWS, 1988b) suggests that the potential advantage of fluidized beds is being wasted by inefficient design. It showed that for a given size and density of the support particle there exists a single optimum size and shape of the bed, which is far from the dimensions of the reactors actually being built. This is a good example of how modelling can guide experiments to produce better designs, because it would take decades to evaluate all combinations of particles size, particle density, bed height, and bed diameter purely empirically. The main features of the analysis will be described; the detailed equations can be found in the original paper.

The liquid velocity in a fluidized bed can be adjusted to keep the total solids holdup (ε_s = (biomass volume + support particle volume)/ reactor volume) constant. For spherical supports of radius R coated with a biofilm of thickness L this is related to the biomass hold-up ε_b, by:

$$\frac{\varepsilon_b}{\varepsilon_s} = 1 - \frac{1}{(1+L/R)^3} = 1 - \frac{1}{(1+\theta/\gamma)^3} \qquad (38)$$

γ is a dimensionless particle radius, defined like θ but with L replaced by R. This is a generalization that allows for curvature of the biofilm. For $L \ll R$ it reduces to the "flat plate" equation $\varepsilon_b = aL$ used for the trickling filter (for spheres $a = \varepsilon_s 3/R$).

Now consider what happens to the product $\eta \varepsilon_b$ in the substrate consumption term of the mass balance Eq. (28) as the biofilm grows thicker. For thin films $\eta = 1$ (Fig. 2) while ε_b increases so the consumption rate increases. For very thick films $(\theta \gg \gamma)$ $\varepsilon_b = \varepsilon_s$ is a constant, while η decreases (Fig. 2) so the consumption rate decreases. Somewhere between these extremes there must be an optimum biofilm thickness. This is in contrast to the flat plate case (Fig. 4) where the consumption rate rises to some maximum value. Furthermore, if the particle radius, γ, is too large, η starts to decrease while ε_b is still small and the maximum consumption rate is low. This is a mathematical statement of the obvious fact: to be effective a bioreactor must contain a large amount of biomass, but if the biofilm is too thick some of it becomes inactive due to mass transfer limitations. (Note that Fig. 2 is inexact in this case due to curvature effects; ANDREWS (1988b) gives an exact analysis based on a different definition of the effectiveness factor.)

The values $\gamma = 2$, $\theta = 1$ are a reasonable compromise between these considerations and having unmanageably small particles. Growing this much biofilm would cause the bed to expand approximately 250% above the height of a bed of clean support particles. This is far more expansion than is allowed in practice.

A fluidized bed is more difficult to design than a packed bed, mainly because the support particle and the liquid velocity cannot be set independently. After fixing the particle radius $(\gamma = 2)$ and the particle density, the velocity is fixed by the need to fluidize the particles to the required solids holdup. The height of the bed is then fixed by the contact time needed to satisfy the substrate removal requirement. Choosing a dense particle produces a high liquid velocity and a tall thin bed (and, incidentally, high mass transfer coefficients at the biofilm interface), while a light particle produces a short wide bed. This is illustrated in Fig. 5, the result of a design study for 90% removal of *BOD* from a waste stream of 10 m³/h containing 100 g *BOD*/m³ (ANDREWS, 1988b). Oxygen is provided by saturating a liquid recycle stream with pure oxygen. Repeating the study with air aeration produced very short, wide beds except with very small, dense support particles. This type of calculation can eliminate many obviously unpromising designs

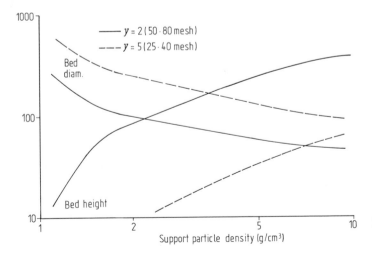

Fig. 5. Bed height and diameter from design study.

and produce "ideal reactors" against which the performance of real reactors can be judged. Note that all of the possible reactors in Fig. 5 have bed volumes an order of magnitude smaller than the present generation of fluidized beds.

There remains the question of how the optimum film thickness ($\theta = 1$) can be achieved in practice. There are two possible approaches. The first is to choose the support particle density so that $\theta = 1$ is the steady-state thickness discussed in Sect. 2.2.4. Higher particle densities give higher shear rates at the biofilm surface and thus higher k_d values. The second is to use mono-sized particles with $\gamma = 2$ (at the average conditions in the bed). Since biofilm is very light, the growth of a biofilm tends to reduce the settling velocity of a particle. Liquid fluidized beds have a natural tendency to stratify with low-settling velocity particles above those with higher settling velocity. This puts particles with thick biofilms at the top of the bed from which they can be removed, washed to remove excess biofilm, and returned to the bed. This is the common approach, but note that it will not work properly if the support particles are not mono-sized, because bed stratification will then occur on the basis of particle size, not biofilm thickness.

2.3.3 Rotating Biological Contactors (RBC)

The liquid in a single RBC stage can be assumed to be completely mixed. The unsteady-state mass balance equation for organic matter is therefore

$$\tau \frac{dS}{dt} = S_i - S - a' \bar{N} \tag{39}$$

The difficulty is that the uptake rate per unit area, N, is not constant over the disc. Some of the film is submerged in the trough, some is just emerging from it with a thin film of liquid attached to it, and some is just returning to it with the liquid film now thoroughly oxygenated after long exposure to the air. The value $\bar{N}$ is an average over these continuously varying conditions:

$$\bar{N} = \frac{1}{\pi(R - R_1)^2} \int_{R_1}^{R} \int_{0}^{2\pi} Nr \, dr \, d\alpha \tag{40}$$

In these equations R_1 is the vertical distance from the disc center to the liquid surface so the region $0 < r < R_1$ is never submerged in the liquid. a' is the wetted surface area per unit of inflow $= 2\pi n(R^2 - R_1^2)/F$, where n is the number of discs. It can be argued that the region $0 < r < R_1$ is actually wetted by the liquid film draining down from above, in which case N should be found by averaging over the whole disc area. This is a matter of opinion, but the latter view can be included without changing the equations by setting $R_1 =$ shaft radius in *all* of the above.

The difficulties in predicting N as a function of r and α are formidable. A set of assumptions that gives a realistic mechanistic model without excessive mathematical complexity has yet to be formulated. Some of the problems and potential solutions are as follows.

The environment at a point on the biofilm/liquid interface changes as the disc rotates. The trough is usually high in substrate and low in dissolved oxygen. As the point emerges from the trough with a liquid film attached to the biofilm, oxygen starts to diffuse from the air through the liquid film while the organic substrate is depleted in the liquid film due to consumption in the biofilm. It is possible that these changes happen so fast that the balanced growth assumption is not valid. Even if this is not the case and unstructured kinetic models can be used, the steady-state diffusion/consumption equations derived in Sect. 2.1 would no longer apply. If radial and tangential diffusion in the biofilm can be neglected, Eq. (14) can be replaced by

$$D_j \frac{d^2 C_j}{dz^2} - X q(C_j) = \frac{\varepsilon \, dC_j}{dt} \tag{41}$$

where $\varepsilon =$ porosity of the biofilm. This equation must be solved twice for each nutrient, once for the time the biofilm is submerged, and once for the time it spends in the air. However, choosing the nutrients for which it must be solved would be difficult. It may be that the organic matter is limiting and/or rate controlling over some parts of the disc while oxygen becomes limiting in others.

The other major problem concerns the hydrodynamics of the liquid film. A common assumption has been to ignore film drainage and centrifugal flow effects and to picture the film as a "slab" of liquid of constant thickness attached to the biofilm and not moving relative to it. There then exists an equation (41) for each nutrient in the liquid film, but with $\varepsilon = 1$, D_j = diffusivity in water and, if microbial activity in the liquid phase can be ignored (another common assumption), $X = 0$. With z = distance from the biofilm interface and δ = liquid film thickness the associated boundary conditions are:

$$t = 0 \quad C_j = \text{concentration in trough}$$

$$z = 0 \quad D_j \frac{dC_j}{dz} = N_j$$

$$z = \delta \quad C_j = S_j^* \quad \text{(for oxygen)}$$
$$\frac{dC_j}{dz} = 0 \quad \text{(for other nutrients)}$$

$$(42)$$

After some time $t = 2(\pi - \cos^{-1} R_1/r)/w$ the liquid film returns to the trough and mixes with it.

Even with this greatly simplified picture of the hydrodynamics, the model requires the simultaneous solution of three equations of the form of Eq. (41) for each nutrient that may become rate-controlling. This has not been attempted. Reactor design is based on purely empirical models, and this has led to disputes about the relative significance of disc rpm, peripheral velocity, trough dissolved oxygen, etc., in scaling-up the reactor (CHESNER and MOLOF, 1977).

The reasoning applied to limiting nutrients in trickling filters in Sect. 2.3.1 applies equally well to RBCs. At low *BOD* concentrations the organic matter is the rate-controlling nutrient in the biofilm and first-order kinetics give reasonable results. HANSFORD et al. (1978) considered this situation. They used all the assumptions listed above, ignored concentration gradients across the liquid film, and worked in terms of a single substrate concentration (a function of tangential but not radial position) in the biofilm. This is valid for first-order kinetics only, because the active depth of the film is then independent of C_i. The results

from this model identify the hydraulic loading per unit area of biofilm ($1/a'$ in Eq. (39)) as the critical operating parameter and predict that, if this is held constant, the fractional removal is independent of inlet substrate concentration loading, flow rate and disc diameter. These are identical to the low *BOD* results for trickling filters. Increases in removal caused by increasing rotational speed, w, were smaller than those observed in practice because the model treats the liquid film thickness, δ, as an independent parameter. In practice δ is an increasing function of w, and this gives an extra improvement in removal. Biofilm thickness was also shown to be a significant parameter, although no attempt was made to predict the steady-state film thickness with a decay or sloughing type of model. GRADY and LIM (1980) offered a similar but more comprehensive model that overcomes some of these objections and includes the effect of rotational speed on mass transfer coefficients in the trough.

At the other extreme of high *BOD* concentrations, oxygen will become the rate-controlling nutrient in the biofilm. As in the trickling filter, the *BOD* removed will then be a constant ($Y_o N_o/Y$) independent of inlet concentration and dependant solely on the oxygen transfer capacity of the system. This has been studied with a complex convection/diffusion model by VAIDYA and PANGARKER, (1987). This includes the liquid film thickness as a function of r, α and w. Several predictions are made, including the interesting result that the total oxygen transfer reaches a maximum when $R_1/R = 0.25$.

3 Suspended Growth Systems

In modelling oxidation ponds, activated sludge systems, etc., the mass transfer resistances discussed in the previous section are usually ignored. This great simplification allows the characteristics of the wastewater and biomass to be described by much more detailed yield and rate equations. Many sets of equations have been proposed, and the IAWPRC model (Tab. 1) represents a consensus of the important phenomena and how they should be described mathematically (see also SHEINTUCH, 1987; SCHEFFER et al., 1985; DOLD and MARAIS, 1986; DOLD et al., 1980; BUSBY and ANDREWS, 1975).

3.1 Mass Transfer Resistance in Flocs

Most of the flocs in an activated sludge system have radii of 50–100 μm. How justified is the assumption that the effectiveness factor equals one? Repeating the calculations from Sect. 2 in spherical coordinates shows that it is approximately valid if the liquid is well-agitated and θ (based on floc radius) $<\sqrt{3}$ (ANDREWS, 1988a). The corresponding floc size depends on the identity and concentration of the limiting nutrient. When oxygen is rate-controlling, $D=2\times10^{-5}$ cm^2/s (80% of the free water value), the liquid is saturated so that $S_o=8$ mg/L, which is much larger than K_o, the floc density $X=70$ mg dry wt per cm^3 (HOEHN and RAY, 1973), and $\bar{q}_o=0.2$ g O$_2$ per g dry wt per hour, hence the limiting floc radius is $R=(6DS_o/\bar{q}_oX)^{1/2}=157$ μm. Mass transfer limitations are, therefore, insignificant. However, taking lower but still acceptable values of D and S_o or higher but still acceptable values of $\bar{q}_o$ and X produces the opposite conclusion. MIKESELL (1984), for example, calculated average floc sizes from the settling velocity data of BISOGNI and LAWRENCE (1971) and then calculated effectiveness factors (with $D=10\%$ of free-water value) as low as 0.05.

The safest conclusion is that intra-floc diffusion limitations are sometimes significant in activated sludge systems. The results of Sect. 2.2.1 are very relevant here. They show that the effect of these diffusional limitations is not to invalidate the form of the Monod function but to increase the apparent value of the half-velocity constant K_A. This is certainly one reason for the wide range of published values for this parameter (SHIEH, 1980).

3.2 Mass Balance Equations for a Stirred Tank

BODE and SEYFRIED (1985) have studied the residence time distributions in activated sludge systems. As in all real reactors they fall between the limits of plug-flow and complete mixing. They are usually modelled as a series of stirred tanks (FIJIE et al., 1988), the number of tanks being somewhere between 1 for complete mixing and infinity for plug flow. This model is chosen for its apparent simplicity. The unsteady-state mass balance for constituent j over a completely-mixed tank is

$$\frac{dS_j}{dt} = \frac{S_{j,i}-S_j}{\tau} - r_j \tag{43}$$

There is one such equation for each constituent in the model (active biomass, colloidal substrate, etc.). τ is the nominal hydraulic residence time and r_j is the volumetric rate of consumption of component j. For an unstructured model, r_j is proportional to the biomass concentration and dependent in some way on the component concentrations S_j. The set of Eqs. (43) can always be solved, although not necessarily algebraically. For a structured model the situation is far more complex. r_j now also depends on the composition of a floc particle, and the flocs in a reactor do not all have the same composition. How can this be accounted for?

Consider what happens to a floc that falls at time $t=0$ into a stirred tank operating at steady state. For the purpose of illustration a simple structured model will be used which divides the floc into mass m_A of active biomass and m_S of stored substrate. If the floc breaks up or bits are washed off, these masses include all of the resulting fragments. The mass conservation equations for the floc are:

active biomass: $\dfrac{dm_A}{dt} = \mu m_A$

$$\tag{44}$$

stored substrate: $\dfrac{dm_S}{dt} = \left(q - \dfrac{\mu}{Y} - k_m\right) m_A$

The rates of substrate uptake, q, and growth, μ, depend on the substrate concentration in the tank, S, which is constant, and the composition of the floc particle, which varies with the time t it spends in the reactor. If the composition is described by the ratio of stored substrate to active biomass $y = m_S/m_A$, its variation is given by:

$$\frac{dy}{dt} = \frac{1}{m_A}\left(\frac{dm_S}{dt} - y\frac{dm_A}{dt}\right) = q - \left(\frac{1}{Y} + y\right)\mu - k_m \tag{45}$$

This may be difficult to solve mathematically (depending on how q and μ vary with y), but what happens physically is familiar. The floc enters with a certain composition, y_i, and there is a lag during which the floc becomes acclimated to its new environment. Mathematically, acclimation means that the floc composition, y, reaches a value that depends on its environment, S, and thereafter, since S is constant, $dy/dt = 0$. This is the quasi-steady state discussed in Sect. 1.1.2.

For the simple, linear rate equations used in the model of Tab. 2, the solution to Eq. (45) is:

$$y = \frac{y_\infty(y_i + y_\infty + B) + (y_\infty + B)(y_i - y_\infty)e^{-k_2 t(B + 2y_\infty)}}{(y_i + y_\infty + B) - (y_i - y_\infty)e^{-k_2 t(B + 2y_\infty)}}$$

$$y_\infty = \frac{B}{2}\left[\left(1 + \frac{4}{k_2 B^2}(k_1 S - k_m)\right)^{1/2} - 1\right] \tag{46}$$

$$B = \frac{K + S}{KY} \quad K = \frac{k_2 y_m}{k_1 Y} = \begin{array}{l}\text{equivalent}\\\text{Monod constant}\end{array}$$

As $t \to \infty$, $y \to y_\infty$, which is the quasi-steady, "balanced growth" solution for the floc composition. In virtually all cases $k_2 B^2 \gg (k_1 S - k_m)$ and the above definition of y_∞ (taking two terms in the series expansion of the square root term) reduces the rate equations from Tab. 2 to the basic unstructured growth model given by Eqs. (2) and (9) (PADUKONE and ANDREWS, 1989). Since there is considerable experimental evidence for this unstructured model under balanced growth conditions, this analysis supports the form of the rate equations in Tab. 2. A similar analysis could usefully be applied to other structured models (note that the solution for y_∞ can be derived directly from Eq. (45) by setting $dy/dt = 0$ and solving for y).

In the simplest possible case the influent of a stirred tank contains a number concentration of n flocs/liter, all of which are identical in size and composition. It will be necessary to assume that if a floc breaks up due to shear or the wash-off cells, then all the fragments leave the reactor at the same time. Unless $y_i = y_\infty$ as in a CMAS system (in which case growth is "balanced" and the structured model is unnecessary), the flocs in the reactor are not all the same. Some will just have entered the reactor and therefore have a size and composition similar to the entering flocs ($m_A = m_{A,i}$, $y = y_i$), while others will have been in the reactor for several days and will be much larger and have a composition $y = y_\infty$. The fraction of the flocs that have been in the reactor for a time between t and $(t + dt)$ is given by the residence time distribution function as $e^{-t/\tau}dt/\tau$. Several important quantities must be defined by integrating over all the flocs in the reactor.

active biomass concentration

$$X_A = \frac{n}{\tau}\int_0^\infty m_A e^{-t/\tau}dt \tag{47}$$

stored substrate concentration

$$X_S = \frac{n}{\tau}\int_0^\infty m_S e^{-t/\tau}dt$$

volumetric substrate consumption rate

$$r_S = \frac{n}{\tau}\int_0^\infty m_A q\, e^{-t/\tau}dt \tag{48}$$

volumetric microbial growth rate

$$r_A = \frac{n}{\tau}\int_0^\infty m_A \mu\, e^{-t/\tau}dt$$

Fortunately, these equations can be considerably simplified for the rate equations given in Tab. 2. Since the composition of a floc is given by $y = m_S/m_A$, Eqs. (48), become:

$$r_S = \frac{n}{\tau}\int_0^\infty k_1 S\left(m_A - \frac{m_S}{y_m}\right)e^{-t/\tau}dt =$$

$$= \frac{n k_1 S}{\tau}\left[\int_0^\infty m_A e^{-t/\tau}dt - \right.$$

$$\left. -\frac{1}{y_m}\int_0^\infty m_S e^{-t/\tau}dt\right] = \tag{49}$$

$$= k_1 S\left(X_A - \frac{X_S}{y_m}\right) = X_A k_1 S\left(1 - \frac{y_{AV}}{y_m}\right)$$

$$r_A = \frac{n}{\tau}\int_0^\infty k_2 m_S e^{-t/\tau}dt = k_2 X_S = X_A k_2 y_{AV}$$

This shows that the volumetric reaction rates r_S and r_A can be found by substituting the average floc composition $y_{AV} = X_S/X_A$ into the rate equations from Tab. 2. It is important to realize that this is valid only because the rate equations in Tab. 2 are *linear* functions of the biomass composition y. Repeating the steps of Eq. (49) for the IAWPRC model (Tab. 1) would not give this result, because the rate of process 7 "hydrolysis of entrapped organics" is a non-linear (Monod-type) function of biomass composition (X_S/X_{BA}). The correct procedure in this case would be to solve Eq. (45) to find how composition varies with the residence time of the floc (the equivalent of Eq. (46)). Substituting this function into the rate equations allows Eq. (48) to be integrated to give the exact result for the volumetric rates.

In summary, there may be no alternative to assuming that the reaction rates in a CSTR (continuous stirred tank reactor) can be based on the average biomass composition in the reactor. The practical situation in which the reactor influent contains a range of floc sizes and compositions, and inert matter is added to the biomass structure (thus requiring two parameters to define "biomass composition") could probably not be analyzed without this assumption. However, it is exact only when the rate equations are linear functions of floc composition. For other models it introduces an unknown amount of error.

3.3 The Contact Stabilization Process

As a further illustration of the use of structured models, the equations derived above will be used to illustrate the storage/metabolism hypothesis (JONES, 1970) for a contact stabilization system treating municipal sewage. In this hypothesis, substrate is stored in the biomass in the contact tank as adsorbed colloidal material or intracellular storage products (and possibly extracellular biopolymer) and is metabolized by the biomass later in the contact tank. Both tanks will be assumed to be completely mixed.

Since there are two tanks, mass balance equations must be written for each tank or for

one tank and the entire system. The former approach is simpler because the equations are identical for each tank; only the input parameters are different. For the model of Tab. 2, three Eqs. (43) are needed for a tank, one for substrate, one for active biomass, and one for stored substrate. Combined with Eqs. (49) the steady-state solution is

$$\frac{S_i}{S} - 1 = k_1 X_A \tau \left(1 - \frac{y_{AV}}{y_m} \right)$$

$$X_A = \frac{X_{A,i}}{1 - k_2 \tau y_{AV}} \tag{50}$$

$$y_{AV} = \frac{y_i + B k_2 \tau y_\infty}{1 + k_2 \tau (B + y_i)}$$

Note that since B is a function of S, considerable computation is needed to obtain results in a useful form. A complete model would consist of three such equations for the contact tank (subscript c), three for the stabilization tank (subscript s), the obvious constraints on biomass composition $y_{i,c} = y_{AV,s}$, $y_{i,s} = y_{AV,c}$ and substrate composition $S_{i,s} = S_c$, and finally the mass balance equations for mixing the recycle stream with the influent wastewater:

$$S_{i,c}(1 + RR) = S_{i,n} + RR S_s$$

$$X_{A,i,c}(1 + RR) = RR X_{A,s} \tag{51}$$

The dimensionless substrate concentration, S/S_i, and biomass composition $y' = y_{AV}/y_m$ are plotted as functions of dimensionless residence time $\tau' = k_2 y_m \tau$ and inlet biomass composition $y_i' = y_i/y_m$ in Fig. 6a–d. The parameters are the dimensionless concentrations of substrate $S_i' = S_i/K$ and active biomass $X_i' = X_{A,i}/KY$ at the tank inlet. The values chosen are typical for a system treating a wastewater with $S_{i,n}$ in the range of 400–500 mg COD/L.

Consider first the *contact tank*. In a conventional activated sludge system it has a residence time of approximately 5 hours which corresponds to $\tau' = 0.5$. Fig. 6a shows that this high residence time gives excellent substrate removal, and also that there would be no point in aerating the return sludge. The substrate removal is not a function of inlet biomass composition, y_i', at these high τ' values. The con-

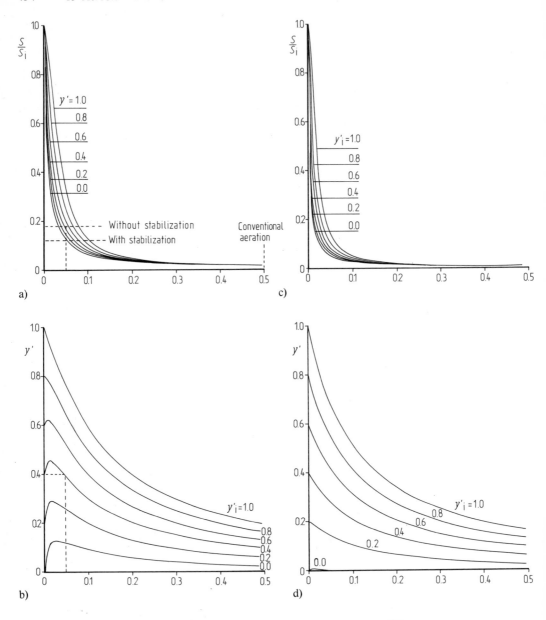

Fig. 6. Contact stabilization model $S'_{i,c} = 33.7$, $X'_{A,i,c} = 160$, $S'_{i,s} = 3.6$, $X'_{A,i,s} = 375$.
(a) Substrate concentration in contact tank, (b) biomass composition in contact tank, (c) substrate concentration in stabilization tank, (d) biomass composition in stabilization tank.

tact tank in a contact stabilization system may only be one tenth this size ($\tau' = 0.05$), and at these low values aerating the return sludge does make a difference. If there is no stabilization tank and negligible metabolic activity in the settler then we require that $y'_{i,c} = y'_c$. Fig. 6b shows that for $\tau' = 0.5$ this happens with $y'_{i,c} = y'_c = 4.0$. Fig. 6a gives the corresponding removal as 82%, while sludge stabilization that reduced $y_{i,c}$ to zero would give 88% removal. In the storage/metabolism hypothesis considered here, this improved removal is due to the ability of stabilized biomass to take up substrate rapidly and store it as adsorbed colloidal material or intracellular storage products.

The substrate concentration and biomass composition at the inlet to the stabilization tank are the same as those at the outlet of the contact tank. Thus if $y'_{i,c} = 0$, $S'_{i,s} = 3.6$ (Fig. 6a) and $y'_{i,s} = 0.12$ (Fig. 6b). The inlet sludge concentration $X'_{A,i,s}$ is fixed by how well the sludge settles, and the value of 375 used in Figs. 6c and 6d is fairly typical. The little substrate present is rapidly consumed in the high biomass/substrate conditions found in the stabilization tank (Fig. 6c). More importantly, it can be seen by interpolating in Fig. 6d that, for $y'_{i,s} = 0.12$, the stabilization tank requires a residence time of approximately $\tau' = 0.3$ (3 h) in order to reduce y' back to zero before the biomass is returned to the contact tank.

Repeating this type of calculation for different values of parameters $X'_{A,i,s}$, $y'_{i,c}$, etc., can suggest experiments to validate the model and procedures for optimizing the design and operation of real systems. A notable feature of the model is the absence of the system mean cell residence time. It can be related to the model parameters by:

$$\theta_x = \frac{\text{total biomass in system}}{\text{wastage rate}} =$$

$$= \frac{\tau_c + \tau_s \dfrac{X_{A,i,c}}{X_{A,c}} \dfrac{(1 + y_{AV,s})}{(1 + y_{AV,c})}}{1 - \dfrac{X_{A,i,c}}{X_{A,c}} \dfrac{X_{A,i,s}}{X_{A,s}}} \qquad (52)$$

θ_x does not appear as a central feature of this model in the way it does for the conventional CMAS process (Eq. (12)). In fact, the use of θ_x

as the main design and control parameter for a contact stabilization system is based solely on extrapolation from conventional systems. While this type of extrapolation can provide useful input into semi-empirical models (ALEXANDER et al., 1980), there are obvious dangers in extrapolating results from the balanced-growth CMAS system to the unbalanced-growth contact stabilization system. The fundamental mechanisms are different, and little basic insight can be gained. Truly mechanistic models of the contact stabilization process, like the simple example given above, may produce more optimal control strategies based on other process parameters. It may, for example, be preferable to keep the recycle flow (and thus τ_s) constant and adjust sludge wastage so as to maximize the sludge concentration from the settler ($X_{A,i,s}$), whatever the resulting θ_x value. The impact of this type of strategy on the sludge settling behavior would obviously have to be considered in a comprehensive model.

3.4 Modelling Sludge Settling

Activated sludge systems are usually conservatively designed from the point of view of substrate removal. Operational difficulties tend to arise not from low dissolved oxygen levels or high *BOD* transients in the effluent, but from difficulties with sludge separation in the secondary settler. The need to incorporate settler behavior into a comprehensive process model has been understood for many years. BUSBY and ANDREWS (1975) presented an excellent model and used it to evaluate various control strategies based on the height of the sludge blanket. However, settler performance was described by a purely empirical correlation between the *TSS* in the effluent and the *MLVSS* in the influent, which could not predict changes in settling behaviour due to biomass composition, sludge age, etc. There are several difficulties in producing a purely mechanistic model of settling.

The settling velocity of a floc particle depends on three things: its size, its density, and its sphericity. DA HONG and GANCZARCZYK (1988) measured the settling velocity of flocs of 100–700 µm and found that

$$U \text{ (mm/s)} = 0.35 + 1.77 \, d \text{ (mm)} \qquad (53)$$

Since flocs fall in the Stokes regimes (MIKESELL, 1984) we would normally expect U to be proportional to d^2. This suggests that the larger flocs were less dense, a result partly supported by the measurements of porosity. (The effect of sphericity is unpredictable; a disc, for example, has a lower sphericity than a sphere but falls considerably faster edgewise.) How can any of these factors be predicted by a model?

Predicting the size of a floc is a similar problem to predicting sloughing and the steady-state thickness of a biofilm (Sect. 2.4). Floc size results from an equilibrium between the growth of biomass in the floc, biomass decay processes, and the wash-off of cells from the floc surface and possible fracture of flocs. The latter processes are caused by shear in the mixing and aeration devices, but a floc's susceptibility to shear depends on what might be called its "cohesiveness" which is a function of its composition.

Flocs are held together by extracellular biopolymers produced by the microorganisms, so the amount of biopolymer produced must be very important in determining cohesiveness of flocs. SHEINTUCH (1987) has reviewed the evidence for activated sludge failure resulting from deflocculation in the settler caused by lack of natural flocculent. The production of biopolymers has been the subject of considerable research and some controversy. Results from batch experiments with pure cultures show that bacteria first accumulate intracellular storage products such as PHB, and only later start to produce extracellular biopolymers (PARSONS and DUGAN, 1971). This is in agreement with the correlation between bioflocculation and PHB content reported by CRABTREE et al. (1966). A high C/N ratio in the media produces less biomass, due to nitrogen limitation, but copious amounts of biopolymer.

This suggests that extracellular biopolymer can best be incorporated into models as a form of stored substrate. In this view, adsorbed colloidal material and PHB are short-term storage products that the biomass can accumulate quickly (PADUKONE and ANDREWS, 1989).

Given sufficient time and the right conditions, mainly excess carbon source, bacteria will then go on to accumulate extracellular polymers as a long-term storage product. These different forms of stored substrate are consumed in the same order. Adsorbed colloids are hydrolyzed and used continuously. When organic matter is no longer available from the liquid, the organisms start to metabolize PHB and similar intracellular storage products. Eventually, over the time scales involved in aerobic digestion, this runs out, and the only energy source available to the microorganisms is the extracellular polymer. The results of PAVONI et al. (1972) show that these polymers are normally resistant to hydrolysis, but that "old" cultures become acclimated to them which results in floc dispersion.

Incorporating this information into a model would strictly require the biomass to be structured into protoplasm and three categories of stored substrate, adsorbed colloids, intracellular storage products, and extracellular biopolymer. This has not yet been attempted (but see SHEINTUCH, 1987), although such a model would be very useful both for activated sludge systems and for aerobic digestion, which is usually analyzed only in terms of simple, first-order kinetics (BHARGAVA and DATAR, 1988). Some qualitative insight can be obtained from models in which all stored substrate is lumped into a single category. For example, the model of Tab. 2 applied to the biomass balance for the conventional CMAS system gives

$$\theta_x = \frac{1}{k_2 y} \qquad (54)$$

The total amount of stored substrate decreases with increasing mean cell residence time, but from the previous discussion we expect that the fraction of the stored substrate that is extracellular polymer would increase as θ_x increases. This would produce an optimum in sludge settling characteristics as a function of θ_x, which is precisely what was found experimentally by BISOGNI and LAWRENCE (1971). For $0.25 < \theta_x < 2$ days they found predominantly dispersed growth with practically no zone settling. This corresponds to our expectation that at low θ_x values the flocs will contain considerable adsorbed material and PHB, but little biopolymer. For $2 < \theta_x < 6$ days the flocs settled well, but for $6 < \theta_x < 30$ days (extended aeration) the particles tended to defloculate and form pin-point floc that settled poorly. Eq. (54) would predict this. Even though all the stored substrate may now be in the form of biopolymer, there is still not enough of it to hold the flocs together.

SHEINTUCH (1987) also showed that a model containing a purely "chemical" biomass structure such as active biomass/stored substrate would never be sufficient to describe all sludge settling problems. Electrostatic effects may also have a role in the process of bioflocculation. Sludge bulking due to the appearance of filamentous bacteria is a purely microbiological phenomenon, and would require the biomass to be structured into "good" zoogleal bacterial species and "bad" filamentous ones. The usual explanation for this problem is that the filamentous organisms have both a lower maximum growth rate and lower Monod half velocity constants for oxygen and organics than the floc formers. This gives them a growth rate advantage with either low dissolved oxygen or low organic substrate concentration, and under these conditions they tend to dominate the biomass. EKAMA and MARAIS (1986) have reviewed recent results and models of this problem and shown how they lead to different strategies for preventing it.

4 References

ALEXANDER, W. V., EKAMA, G. A., MARCACS, G. V. R. (1980), The activated sludge process, Part 2, Application of the general kinetic model to the contact stabilization process, *Water Res.* **14**, 1737–1747.

ANDREWS, G. F. (1982), Fluidized-bed fermentors and analysis, *Biotechnol. Bioeng.* **24**, 2013–2022.

ANDREWS, G. F. (1984), Parameter estimation from batch culture data, *Biotechnol. Bioeng.* **26**, 824.

ANDREWS, G. F. (1988a), Effectiveness factors for bioparticles with Monod kinetics, *Biochem. Eng. J.* **37**, B1–B37.

ANDREWS, G. F. (1988b), Fluidized-bed bioreactors, *Biotechnol. Genet. Eng. Rev.* **6**, 151–178.

ANDREWS, G. F. (1989), Estimating cell and product yields, *Biotechnol. Bioeng.* **33**, 256–266.

ANDREWS, G. F., TIEN, C. (1977), A new approach to bacterial kinetics in wastewater, *J. Environ. Eng. Div. ASCE,* 1057–1074.

ANDREWS, G. F., TIEN C. (1981), Bacterial film growth on adsorbent surfaces, *AIChE J.* **23**, 182–189.

ANDREWS, G. F., TRAPASSO, R. (1985), The optimal design of fluidized bed bioreactors, *J. Water Pollut. Control Fed.* **57**, 143–152.

ATKINSON, B., WILLIAMS, D. A. (1971), The performance characteristics of a trickling filter with hold-up of microbial mass controlled by periodic washing, *Trans. Inst. Chem. Eng.* **49**, 215–224.

ATKINSON, B., FOWLER, H. W. (1974), The significance of microbial films in fermentors, in: *Advances in Biochemical Engineering* (GHOSE, T. K., Ed.) Vol. 3. Berlin–New York: Springer.

BARTON, D. A., MCKEOWN, J. J. (1986), Evaluation of an aerator control strategy utilizing time varying mathematical model simulations, *Water Sci. Technol.* **18**, 189–201.

BENEFIELD, L. D., RANDALL, C. W. (1986), Design procedure for the contact stabilization activated sludge process, *J. Water Pollut. Control Fed.* **48**, 147–152.

BHARGAVA, D. S., DATAR, M. T. (1988), Progress and kinetics of aerobic digestion of secondary sludges, *Water Res.* **22**, 37–47.

BISCHOFF, K. B. (1965), Effectiveness factors for general reaction rate forms, *AIChE J.* **11**, 351–375.

BISOGNI, J. J., LAWRENCE, A. W. (1971), Relationships between biological solids retention time and settling characteristics of activated sludge, *Water Res.* **5**, 753–763.

BODE, H., SEYFRIED, C. F. (1985), Mixing and detention time distribution in activated sludge tanks, *Water Sci. Technol.* **17**, 197–208.

BOUWER, E. J. (1987), Theoretical investigation of particle deposition systems, *Water Res.* **21**, 1488–1498.

BUSBY, J. B., ANDREWS, J. F. (1975), Dynamic modelling and control strategies for the activated sludge process, *J. Water Pollut. Control Fed.* **47**, 1055–1080.

CHAIN, E. S., DEWALLE, F. B. (1975), Sequential substrate removal in activated sludge systems, *Prog. Water Technol.* **7**, 235–241.

CHESNER, W. H., MOLOF, A. H. (1977), Biological rotating disc scale-up design, *Prog. Water Technol.* **9**, 811–819.

CRABTREE, K., BOYLE, W., MCCOY, E., ROHLICH, G. A. (1966), A mechanism of floc formation by *Zooglea ramigera, J. Water Pollut. Control Fed.* **38**, 1968–1980.

DA HONG, J., GANZARCZYK, J. L. (1988), Strobo-scopic determination of settling velocity, size and porosity of activated sludge flocs, *Water Res.* **21**, 257–262.

DOLD, P. L., MARAIS, G. V. R. (1986), Evaluation of the general activated sludge model by the IAWPRC Task Group, *Water Sci. Technol.* **18**, 63–90.

DOLD, P. L., EKAMA, G. A., MARAIS, G. R. (1980), The activated sludge process, Part 1, A general model for the activated sludge process, *Prog. Water Technol.* **12**, 254–268.

EKAMA, G. A., MARAIS, G. R. (1986), The implications of the IAWPRC hydrolysis hypothesis on low F/M bulking, *Water Sci. Technol.* **18**, 11–19.

EKAMA, G. A., DOLD, P. L., MARAIS, G. R. (1986), Procedures for determining influent COD fractions and maximum specific growth rate of heterotrophs in activated sludge systems, *Water Sci. Technol.* **18**, 91–114.

FAN, L. S., KIGIE, K., LONG, T. R., TANG, W. T. (1987), Characteristics of a draft-tube gas–liquid solid fluidized bed reactor for phenol-degradation, *Biotechnol. Bioeng.* **30**, 498–504.

FIJIE, K., TSUBONE, T., MORIYA, H., KUBOLA, H. (1988), A simplified kinetic model to simulate soluble organic substance removal in an activated sludge aeration tank, *Water Res.* **22**, 29–36.

GLEICK, J. (1987). *Chaos,* p. 278. London: Penguin Books.

GRADY, L. C. P., LIM, H. (1980), *Biological Wastewater Treatment.* New York: Marcel Dekker.

GRAU, P., SUTTON, P. M., HENZE, M., ELMACH, S., GRADY, C. P., GUYER, W., KOLLER, J. (1987), Notation for use in the description of wastewater treatment processes, *Water Res.* **21**, 135–139.

HAMMING, R. W. (1973), *Numerical Methods for Scientists and Engineers,* 2nd Ed. New York: McGraw Hill.

HANSFORD, G. S., ANDREWS, J. F., GRIEVES, C. G., CARR, A. D. (1978), A steady-state model for the rotating biological disc reactor, *Water Res.* **12**, 493–504.

HENZE, M., GRADY, C. P. L., GRUYER, W., MARAIS, G. V. R., MATSUO, T. (1987), A general model for single sludge wastewater treatment systems, *Water Res.* **21**, 505–515.

HOEHN, R., RAY, A. D. (1973), Effects of thickness on microbial film, *J. Pollut. Water Control Fed.* **45**, 2302–2315.

JAMES, A. (1982), Some perspectives on the modelling of the biological treatment of wastewater, *Water Sci. Technol.* **14**, 227–240.

JANSEN, J., HARREMOES, P. (1985), Removal of soluble substrates in fixed films, *Water Sci. Technol.* **17**, 1–13.

JENKINS, D., ORHON, D. (1972), The mechanism and design of the contact stabilization activated sludge process, *Adv. Water Pollut. Res.* **6**, 353–364.

JONES, P. H. (1970), A mathematical model for the contact stabilization modification of the activated sludge system, *Adv. Water Pollut. Res.* **5**, Paper II/5.

KISSEL, J. C. (1986), Modelling mass transfer in biological wastewater treatment processes, *Water Sci. Technol.* **18**, 35–45.

KORNEGAY, B. H., ANDREWS, J. F. (1968), Kinetics of fixed-film biological reactors, *J. Water Pollut. Control Fed.* **40**, 460–467.

LAURENCE, A. W., MCCARTY, P. L. (1980), Unified basis for biological treatment and design, *J. Sanit. Eng. Div. ASCE* **96**, 757–778.

MIKESELL, R. D. (1984), A new activated sludge theory: Steady-state, *J. Environ. Eng. Div. ASCE* **110**, 141–151.

OLESZKIEWICZ, J. (1977), Theory and design of high-rate plastic media trickling filters, *Prog. Water Technol.* **9**, 777–785.

OMUNA, M., OMURA, T. (1982), Mass transfer characteristics with microbial systems, *Water Sci. Technol.* **14**, 443–461.

ORHON, D. (1977), Discussion of design procedure for contact stabilization activated sludge process, *J. Water Pollut. Control Fed.* **49**, 865–869.

ORHON, D., CIMSIT, Y., TUNAY, O. (1986), Substrate removal mechanisms for sequencing batch reactors, *Water Sci. Technol.* **18**, 21–33.

PADUKONE, N., ANDREWS, G. F. (1989), A single mathematical model for the activated sludge process and its variants, *Water Res.* **23**, 1535–1541.

PARSONS, A. B., DUGAN, P. R. (1971), Production of extracellular polysaccharide matrix by *Zooglea ramegora, Appl. Microbiol.* **21**, 657–661.

PAVONI, J. L., TENNEY, M. W., ECHLELBERGER, W. F. (1972), Bacterial exocellular polymers and biological flocculation, *J. Water Pollut. Control Fed.* **44**, 414–431.

RITTMAN, B. E. (1982), Comparative performance of biofilm reactor types, *Biotechnol. Bioeng.* **24**, 1341–1370.

RITTMAN, B. E. (1985), The effect of load fluctuations on the effluent concentration produced by fixed-film reactors, *Water Sci. Technol.* **17**, 1–14.

SARNER, E., MARKLUND, S. (1985), Influence of particulate organics on the removal of dissolved organics in fixed-film biological reactors, *Water Sci. Technol.* **17**, 15–26.

SCHEFFER, M. S., HIRAOKA, M., TSUMURA, K. (1985), Flexible modelling of the activated sludge system: theoretical and practical aspects, *Water Sci. Technol.* **17**, 247–258.

SEIGRIST, H., GUJER, W. (1987), Demonstration of mass transfer and pH effects in a nitrifying biofilm, *Water Res.* **21**, 1481–1487.

SERVIZI, J. A., BOGAN R. H. (1964), Thermodynamics aspects of biological oxidation and synthesis, *J. Water Pollut. Control Fed.* **36**, 607.

SHIEH, W. K. (1980), The effect of internal mass transfer resistances on the interpretation of substrate removal data in the suspended growth system, *Water Res.* **14**, 695–699.

SHIEH, W. K., MULCAHY, L. T. (1985), Experimental determination of intrinsic kinetic coefficients for biological wastewater treatment systems, *Water Sci. Technol.* **18**, 1–10.

SHIEH, W. K., MULCAHY, L. T., LaMOTTA, E. J. (1981), Effectiveness factors for spherical particles coated with biofilm, *Trans. Inst. Chem. Eng.* **59**, 121–126.

SHEINTUCH, M. (1987), Steady-state modelling of reactor-settler interactions, *Water Res.* **21**, 1463–1472.

SUSCHKA, J. (1987), Hydraulic performance of percolating biological filters and consideration of O_2 transfer, *Water Res.* **21**, 865–873.

TERASHIMA, Y., ISHIKAWA, M. (1985), The kinetic analysis of BOD and nitrogen removal in an oxidation ditch, *Water Sci. Technol.* **17**, 291–302.

VAIDYA, R. N., PANNGARKAR, V. G. (1987), Convective diffusion model for mass transfer in a RBC, *Water Res.* **21**, 1499–1503.

VAN HAANDEL, A. C., DOLD, P. L., MARAIS, G. R. (1982), Optimization of nitrogen removal in the single sludge activated sludge process, *Water Sci. Technol.* **14**, 443–452.

WANNER, O., GUJER, J. (1985), Competition in biofilms, *Water Sci. Technol.* **17**, 27–44.

WATKIN, A. T., ECKENFELDER, W. W. (1984), Development of pollutant specific models for toxic organic compounds in the activated sludge process, *Water Sci. Technol.* **17**, 279–289.

14 Anaerobic Waste Water Process Models

Dirk Schürbüscher
Christian Wandrey

Jülich, Federal Republic of Germany

List of Symbols

Latin letters

a		model parameter
A^-	mol L^{-1}	anion concentration
Ac^-	mol L^{-1}	acetate ion concentration
b		model parameter
c		model parameter
c	mol L^{-1}	concentration
C^+	mol L^{-1}	cation concentration
CO_2	mol L^{-1}	carbon dioxide concentration
CO_3^{2-}	mol L^{-1}	carbonate concentration
COD	g L^{-1}	chemical oxygen demand
D	h^{-1}	dilution rate
f		state function
F	L h^{-1}	fluid flow rate
G	L h^{-1}	gas flow rate
G	kJ mol^{-1}	free reaction enthalpy
h		measuring function
H	mol L^{-1} Pa^{-1}	Henry constant
H^+	mol L^{-1}	hydrogen ion concentration
HAc	mol L^{-1}	concentration of undissociated acetic acid
HCO_3^-	mol L^{-1}	bicarbonate concentration
H_2CO_3	mol L^{-1}	concentration of carbonic acid
I	mol L^{-1}	inhibitor concentration
J		criterion, square error sum
k		discrete time
$\mathbf{K}$		amplification matrix
K_{AC}	mol L^{-1}	dissociation constant of acetic acid
K_C	mol L^{-1}	dissociation constant of carbonic acid
K_I	mol L^{-1}	substrate inhibition constant
$K_L a$	h^{-1}	mass transfer coefficient
K_S	mol L^{-1}	Monod constant
K_W	mol^2 L^{-2}	ionic product of water
K_∞		steady state parameter
m	mol g^{-1} h^{-1}	maintenance coefficient
M	g mol^{-1}	molecular weight
n		order of reaction
Na^+	mol L^{-1}	sodium ion concentration
OH^-	mol L^{-1}	hydroxide concentration
p	N m^{-2}	pressure
$\mathbf{P}$		covariance matrix
Q		weighting factor
Q	mol L^{-1} h^{-1}	mole flow
$\mathbf{Q}$		covariance matrix
r	mol g^{-1} h^{-1}	specific substrate consumption rate
R	J mol^{-1} K^{-1}	general gas constant
s		variance
S		covariance matrix
S	mol L^{-1}	substrate concentration
t	s	time

T	K	absolute temperature
T_0	s	cycle time
u		input variable
u		adjustment of controller input
U		controller input
V	L	volume
V_m	L mol^{-1}	mole volume of ideal gases
x		state vector
x		stoichiometric coefficient
X		system matrix
X	g L^{-1}	biomass concentration
y		measured variable
y		measured value deviation
y		stoichiometric coefficient
Y		measured value
$Y_{X/S}$	g mol^{-1}	yield
z		stoichiometric coefficient
Z^+	mol L^{-1}	net concentration of strong ions

Greek letters

β	h^{-1}	specific maintenance metabolism rate
γ		correction vector
η		thermodynamic biological efficiency
θ		parameter vector
λ		forgetting factor
μ	h^{-1}	specific growth rate
μ^*	h^{-1}	true specific growth rate
σ		mass fraction of carbon in a molecule
τ	h	residence time

Subscripts

C	carbon
cal	calculated
crit	critical
D	differential part
des	desired value
dis	dissolved
exp	experimental
F	flow
G	gas
in	input
k	discrete time
max	maximum
meas	measured
min	minimum
M	mass flow meter
opt	optimal
R	reactor
stat	steady state
S	substrate
theor	theoretical

tot	total
x	linearized variable
X	produced biomass
∞	steady state

Superscripts

*	saturation value
^	estimated value

Abbreviations

ATP	adenosine triphosphate
BOD	biochemical oxygen demand
COD	chemical oxygen demand
DOC	dissolved organic carbon
TIC	total inorganic carbon
TOC	total organic carbon

1 Introduction

Anaerobic waste water purification processes are increasingly used on an industrial scale. With these processes there is no need for an energy-intensive addition of oxygen, less sewage sludge will arise, and the biogas resulting from the degradation of organic compounds can be used to cover the plant's energy requirements. Furthermore, due to the corresponding accumulation of biomass it is possible to achieve volume-related degradation rates impossible with aerobic operation owing to the lack of oxygen feed. This introduction on a commercial scale also means an increase in interest in modelling these processes as well as special measuring and control techniques. Modelling proves very difficult for anaerobic waste water purification processes, since as a rule mixed cultures are present and their interaction is fairly complex. Moreover, the degradation processes in anaerobic degradation proceed sequentially due to the various microorganisms present, in contrast to aerobic waste water purification where they proceed in parallel; i.e., a modelling error in the first degradation stage will be encountered in the final stage in an intensified form. The application of sophisticated methods for process control is al-

ways easier if mathematical models are available for the systems to be optimized. This is complicated by the fact that technical systems for anaerobic waste water purification often operate as fixed-bed or fluidized-bed reactors, and thus the classic stirred tank balances cannot be applied. The model approaches presented in the following are therefore concerned exclusively with stirred tank processes on a laboratory scale.

2 Anaerobic Waste Water Purification

2.1 Advantages of Anaerobic Waste Water Purification

There are basic differences between the aerobic and anaerobic process sequences in eliminating high-energy organic substances from waste water: aerobically, the organic impurities are largely reduced simultaneously to carbon dioxide and water during respiration by the mixed bacterial culture by means of "biological combustion". Anaerobic degradation is similarly carried out by a mixed culture. However, in this case the waste water constituents are largely fermented one after the other by the various bacteria into methane and carbon dioxide by a type of "biological pyrolysis".

The necessity of an interlinkage of the sequential degradation steps in the anaerobic process by various microorganisms means that the various steps must proceed at the same speed in order to avoid disturbances. This means that anaerobic processes are much more sensitive to disturbing influences than aerobic

processes. The different products of anaerobic and aerobic degradation also present alternative possibilities of obtaining energy for the microorganisms. For example, in the case of acetic acid, the most important intermediate in the anaerobic food chain (WANDREY and AIVASIDIS, 1983) the energy balances are: aerobic (corresponding to the reactions of the citrate cycle and the respiration chain) (Eq. 1):

$$CH_3COO^- + H^+ + 2O_2 \rightarrow 2CO_2 + 2H_2O \quad (1)$$
$$\Delta G = -870 \text{ kJ mol}^{-1}$$

with ΔG standard reaction enthalpy,
and anaerobic (Eq. 2):

$$CH_3COO^- + H^+ \rightarrow CO_2 + CH_4 \quad (2)$$
$$\Delta G = -31 \text{ kJ mol}^{-1}$$

Due to the great energy content of the products, only 4% of the energy obtained aerobically is released anaerobically. The ratio of the energy storable in ATP (adenosine triphosphate), the bacterial accumulator, between the aerobic and anaerobic reaction is similar to that of the free energy; however, it depends on the thermodynamic efficiency of the individual organism (Fig. 1).

The synthesis of new cell substance depends on the energy released for the cell. Thus, during continuous anaerobic waste water purification up to 50% of the organic carbon obtained is incorporated into the cell mass, compared to

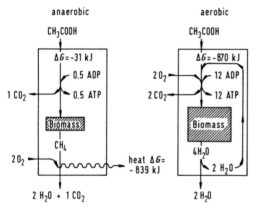

Fig. 1. Energy balance of the anaerobic and aerobic degradation of acetic acid (WANDREY and AIVASIDIS, 1983).

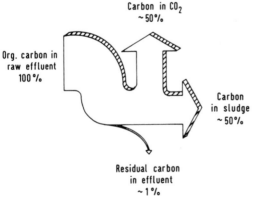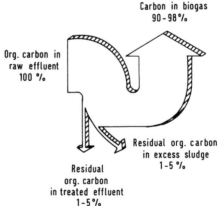

Fig. 2. Comparison of the aerobic (left) and anaerobic (right) degradation of carbon (WANDREY and AIVASIDIS, 1983).

only approximately 3% during aerobic purification (Fig. 2).

Since the elementary compositions of aerobic and anaerobic microorganisms do not differ significantly, the high growth rate of the aerobic microorganisms is accompanied by a correspondingly high additional demand for nitrogen and phosphorus.

The following advantages for anaerobic waste water purification result from the characteristics of the two processes described above: little excess sludge, no aeration necessary (energy savings and simpler reactors), energy obtained by utilizing the biogas (direct combustion, electricity generation, or input into the natural gas grid after methane enrichment), and heavy metal precipitation in the reactor by conversion into insoluble sulfides (landfill sites).

The disadvantages are the slow growth of the microorganisms (low performance in the stirred tank reactor, time-consuming start-up procedure) and the sequential product degradation (unstable system).

Anaerobic waste water purification plants are becoming increasingly interesting due to their simplicity and low energy requirements. TEMPER et al. (1986) have provided a survey of the industrial applications of this process.

2.2 Microorganisms and Metabolic Pathways

The anaerobic conversion of polymeric organic substrates into methane and carbon dioxide is a complex process in which various microbial populations play a part because of their different substrate and product specificities (food chains) (see NYNS, 1986). Nine different steps can be differentiated (HARPER and POHLAND, 1987).

1. Organic polymers are hydrolyzed by extracellular enzymes of facultative or obligate anaerobic bacteria, e.g., *Clostridium* (degradation of compounds containing cellulose and starch) and *Bacillus* (degradation of proteins and fats), to monomeric constituents (amino acids, fatty acids, and sugar).

2. Conversion of the organic monomers by, e.g., *Acetobacterium* and *Pseudomonas* to hydrogen, bicarbonate, short-chain organic acids, alcohols, and methylamine.

3. Oxidation of the reduced organic substrates to bicarbonate and acetate by obligate hydrogen-producing anaerobes, for example, *Desulfovibrio* and *Desulfobacterium*.

4. Formation of acetate from carbon dioxide and hydrogen according to Eq. (3) by homoacetate fermenters, e.g., by *Acetogenium kivui* and *Acetobacterium wieringae*:

$$2\,HCO_3^- + 4\,H_2 \rightarrow$$
$$\rightarrow CH_3COO^- + 4\,H_2O \qquad (3)$$

5. Oxidation of the reduced organic substances to bicarbonate and acetate by sulfate- (*Desulfovibrio*) and nitrate-reducers (*Veillonella alcalescens*).

6. Oxidation of acetate to bicarbonate by sulfate- or nitrate-reducers.

7. Oxidation of hydrogen by sulfate- or nitrate-reducers with the reduction of sulfate to hydrogen sulfide or of nitrate to nitrite, nitrogen, or ammonium.

8. Acetoclastic splitting of acetate into carbon dioxide and methane according to Eq. (4), for example, by *Methanosarcina* and *Methanothrix*.

$$CH_3COOH \rightarrow CH_4 + CO_2 \qquad (4)$$

9. Methanogenesis of carbon dioxide and hydrogen by methane bacteria according to Eq. (5), for example, by *Methanobacterium* and *Methanosarcina*:

$$CO_2 + 4\,H_2 \rightarrow CH_4 + 2\,H_2O \qquad (5)$$

Fig. 3 shows a greatly simplified model of the food chain of complex organic macromolecules up to biogas, in which only three degradation steps are differentiated: acid formation from complex organic molecules (see above 1.), acetic acid formation with the simultaneous formation of hydrogen and carbon dioxide (see above 2., 4.), and methane formation (see above 8., 9.).

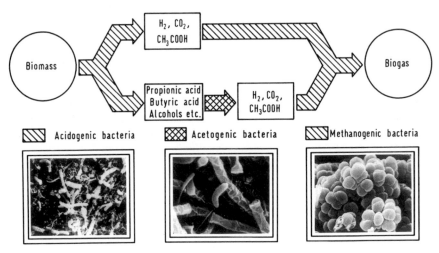

Fig. 3. Scheme for the anaerobic formation of methane from complex organic matter (WANDREY and AIVASIDIS, 1986).

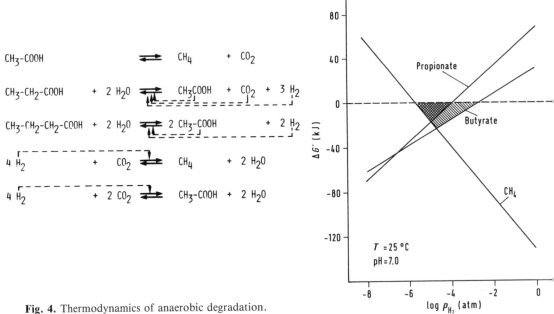

Fig. 4. Thermodynamics of anaerobic degradation.

The first two of these steps must be in dynamic equilibrium with the methanogenesis in order to proceed with maximum speed. The hydrogen concentration plays a special role. On the one hand, it must be as high as possible in order to facilitate methane formation from hydrogen and carbon dioxide, on the other hand, it must be as low as possible for the degradation of propionic acid formed from the preceding links in the food chain. The reactor could become sour from the accumulation of propionic acid. A narrow thermodynamic window must therefore be maintained with respect to the hydrogen concentration in order to permit methane formation but not inhibit propionic acid degradation. This bottleneck can best be prevented by the use of a microorganism population particularly active with respect to hydrogen utilization (Fig. 4).

Measurements of the hydrogen concentration can therefore provide important information about the interaction of the various types of microorganisms, and may also be used for process control (ARCHER et al., 1986; PHELPS et al., 1985; ROBINSON and TIEDJE, 1984).

$$CO_2 + 4 H_2 \rightarrow CH_4 + 2 H_2O \qquad (6)$$
$$\Delta G°' = -138.8 \text{ kJ mol}^{-1}$$

with $\Delta G°'$, the standard reaction enthalpy under physiological conditions.

Methane formation can also be regarded as a precursor of respiration in a primitive reduction atmosphere. The hydrogen does not originate from the dehydration of a substrate, as is the case with cell respiration, but rather from the atmosphere. It does not make use of free oxygen as an acceptor but rather the oxygen bound in the carbon dioxide.

Tab. 1. Thermodynamic Values for the Anaerobic Degradation of Organic Compounds in Vapor Condensates (FREYER, 1988)

Reaction	$\Delta G°'$ (kJ/Reac.)	$\Delta G°'$ (kJ/mol CH$_4$)
$CH_3COO^- + H^+ \rightarrow CO_2 + CH_4$	-31	-31
$4 CH_3OH \rightarrow 3 CH_4 + CO_2 + 2 H_2O$	-310	-103
$CH_3OH + H_2 \rightarrow CH_4 + H_2O$	-112	-112
$4 H_2 + CO_2 \rightarrow CH_4 + 2 H_2O$	-139	-139

Since methane bacteria are responsible for the final stage in the anaerobic degradation of hydrogen, the microorganisms belonging to the kingdom Archaebacteria are of decisive significance for the entire process. The Archaebacteria form an independent group, which is phylogenetically differentiated from the so-called classical bacteria (Eubacteriales) as well as from the higher microorganisms (eukaryotes). Archaebacteria live under extreme conditions not unlike those that prevailed during the earlier developmental stages of the earth's surface. They are found wherever organic substances are decomposed in the absence of oxygen: in swamps, sediments of bodies of water, the digestive tracts of animals and, of course, in anaerobic waste water purification plants. Isolates have also been obtained from geothermal springs and hydrothermal craters. The chemoautotrophic, strictly anaerobic methane bacteria represent the group with the greatest variety of forms, which obtain the energy for CO_2 evolution from the reduction of carbon dioxide to methane (Eq. 6):

The Methanosarcinae, approx. 1–2 μm in diameter, have the greatest metabolic versatility of the methane bacteria. Most varieties can utilize methane/carbon dioxide, methanol, methylamine, or acetate as methanogenic substrates. They have the property of forming agglomerates. Tab. 1 shows the energy gains obtainable with anaerobic substrate degradation by methane bacteria. $\Delta \hat{G}°'$ represents the standard reaction enthalpy under physiological conditions (THAUER et al., 1977). This explains the differences from the values given in Fig. 1 for standard reaction conditions (25 °C, 10^5 Pa).

If one assumes that about 37 kJ mol^{-1} of free energy is required for the formation of ATP, then – depending on the thermodynamic efficiency – 2 to 3 mol of ATP can be generated per mole of methane formed with the above substrates, with the exception of acetate. At most, half a mole of ATP is synthesized per mole of acetate. Since ATP is necessary for anabolism, this explains the slow growth and low cell yields of microorganisms on acetate. Various metabolic pathways exist for methanol, depending on the hydrogen partial pressure. If sufficient hydrogen is present, then methanol is reduced directly to methane. In the absence of hydrogen, the necessary reduction equivalents are obtained by oxidizing part of the methanol.

In order to achieve high bioreactor efficiency, and in particular to accelerate the growth behavior, the biomass concentration in the reactor must be increased in comparison to that of the stirred tank. Various possibilities present themselves for this purpose: pelletization, filtration, adsorption, covalent bonding, and colonization (AIVASIDIS and WANDREY, 1985; HEIJNEN, 1984). Immobilization achieves a high biomass concentration in the reactor and a decoupling of the residence times of the substrate and bacteria (WANDREY and AIVASIDIS, 1983). In this respect fixed-bed reactors have the disadvantage that in the course of time the biomass grows out of the carriers and into the free space, thus resulting in channel formation and a poorer substrate supply to the microorganisms. This type of reactor must therefore be regenerated at intervals of a few months. The excess biomass is sheared off and flushed out either by a short-term increase in the power of the circulating pump or by blowing in inert gas. As an alternative reactor design, fluidized-bed reactors do not have this disadvantage.

3 Measuring Techniques

For measuring physical parameters, such as temperature and pressure, use can be made of the sensors generally applied in biotechnology. For this reason they will not be discussed in detail here. However, the specific boundary conditions of anaerobic waste water purification need to be considered for a large number of variables. The most important conditions are non-sterile operation and the unfavorable environment. A brief survey of the various measuring methods will be presented in the following sections, and particular attention will be paid to their suitability for on-line application in process control.

3.1 Biological Phase

3.1.1 Biomass Concentration

Apart from its activity, the concentration of the biomass is the most important parameter for reactor performance, and at the same time the most difficult to determine. In order to determine the biomass, the solids contained in a liquid sample are centrifuged or filtered off. Those solids remaining after drying contain, among other constituents, the biomass. After weighing, the biomass can be selectively dissolved by the addition of acid or alkali, or else by incineration at 550 °C. After weighing the residue, the difference in the two measurements can be assumed to be the biomass.

The biomass can be measured indirectly, e. g., by measuring the total nitrogen or the total organic carbon of the filtered solids.

3.1.2 Biomass Activity

The fluorescence of the coenzyme F_{420}, present in all methanogenic bacteria, can be used to measure the activity (SCHNECKENBURGER et al., 1984). This measurement can rarely be used for technical media because of the inherent turbidity.

3.1.3 Identification of Methanogenic Bacteria

Methanogenic bacteria can be identified both qualitatively and quantitatively by using antibodies (MACARIO and CONWAY DE MACARIO, 1985).

3.2 Liquid Phase

Great significance is attached to the measurement of summation parameters for the organic pollution of waste water.

3.2.1 pH Value

Apart from the temperature, the pH value represents the most important variable in anaerobic waste water purification. This is due not only to the specific preferences of the individual microorganisms but also to its function as an indirect value for measuring acid production or decomposition. It represents the central value for various control designs, such as pH-auxostatic and pH-chemostatic control, which will be discussed in detail later. For this reason the pH measurement should if possible be designed redundantly. Appropriate measuring amplifiers permitting two electrodes to be connected are commercially available. Although only one actually controls the system, the measured difference between the two electrodes is nevertheless displayed. In addition, the pH value should be determined at least daily by sampling and external measurement. Care must be exercised so that a certain quantity of the first portion of the liquid is discarded before the actual sample is taken. Furthermore, it must be remembered that due to the evolution of carbon dioxide gas the pH value measured in the laboratory may be up to 1 pH unit higher than the pH value in the reactor.

The pH electrodes suffer particularly under the unfavourable environment in anaerobic waste water purification, largely due to various sulfur compounds. The (not sterilizable) jellied electrodes now available require little maintenance and maintain long-term stability with residence times of several months (MONZAMBE et al., 1988).

More "refined" information about the pH value is obtained by titration methods, in which the total concentration of volatile fatty acids is determined by the buffer capacity. The alkalinity is defined as the quantity of acid necessary for a titration to pH 4.5. Under stress or shock conditions the acid production rate may exceed the consumption rate without the pH value falling significantly, although the buffer capacity is already exhausted (POWELL and ARCHER, 1989).

3.2.2 Redox Potential

Whereas the proton activity in the liquid can be measured by the pH electrode, the redox potential represents a measure of the "electron activity" (BÜHLER, 1985), and can thus be utilized for precise measurements of the oxygen partial pressure, especially in anaerobic systems. The measured data as such without any reference values are, in contrast to the temperature or pH value, not particularly informative and cannot be used in physical models. It is generally true that a strongly negative value of the redox potential (-300 to -400 mV) indicates a good anaerobic environment. Nevertheless, attempts can be made to draw up a correlation with other measured values by means of a long-term observation of the redox potential, or the signal may be used as a summation parameter for qualitative statements about the actual process state.

3.2.3 Concentrations of Volatile Substances

The most important measuring instrument for determining the concentrations of volatile components is the gas chromatograph. It permits determination of the most important components such as methanol, acetic acid, butyric acid, etc., within the relatively short interval of 10 minutes. This brief time also makes the instrument interesting for quasi-online application. However, problems occur with sample preparation for long-term use. Moreover, gas chromatographs are very expensive, which does not encourage their use, particularly in waste water purification, for a single process.

3.2.4 Concentrations of Dissolved Gases

The measurement of gas concentrations in outflowing biogas will be discussed in detail later. However, since there is always a more or less large space above the liquid phase of the reactor hindering the timely recognition of changes in gas composition and thus of the reactor state, it is also meaningful to measure dissolved gases in the liquid. This can be done by taking gas specimens via suitable membrane systems in conjunction with mass spectrometers (JOUANNEAU et al., 1980; SCOTT et al., 1983; WHITMORE et al., 1987).

3.2.5 Carbon Determination

Three types of situation must be distinguished in the quantitative determination of carbon compounds in aqueous solution, for each of which several analytical methods are available (EHRENBERGER, 1979):

Total Inorganic Carbon (*TIC*): inorganic carbon comprises the dissolved carbon in the form of carbonate, bicarbonate, and carbon dioxide.

Dissolved Organic Carbon (*DOC*): after removing the TIC, the dissolved organic carbon can be detected by oxidizing the dissolved carbon compounds and determining the resulting carbon dioxide. The carbon contained in particles (e.g., microorganisms) is not included.

Total Organic Carbon (*TOC*): after oxidation at a very high temperature in a catalyst furnace, the carbon contained in the particles can also be determined.

Titrimetric measurements or analysis by infrared absorption are suitable methods for determining the carbon dioxide. The overall analysis requires a certain amount of sophisticated equipment but it can be implemented quasi-on-line.

3.2.6 Biochemical Oxygen Demand (*BOD*)

The biochemical oxygen demand is the parameter most commonly used for quantifying biologically degradable water pollution, and as such it had to be standardized. For this purpose the concept of *BOD*$_5$ was defined as the oxygen demand of a *Pseudomonas* culture over five days. The limitation of the time interval necessarily results in an information mix from waste water pollution and inhibition of the biological system, so this has to be decoupled by means of a toxicity test. It is possible to calculate the inhibiting effects by determining *BOD*$_5$ with different dilution stages of the waste water.

Due to the long measuring time, this analytical method is not suitable for on-line process control but only for determining degradability. On the other hand, further measuring procedures can be introduced and calibrated on the basis of this definition. It is relatively difficult to correlate the *BOD*$_5$ value with other measured values such as the *TOC*, and it depends in particular on the substrate used.

A short-time toximeter with an analysis duration of 15 minutes is appropriate for on-line use. The oxygen consumption of a microbial population in a test reactor is kept constant by controlling the feed ratio of one waste water and one dilution water pump. This inlet ratio is then the measure of the waste water pollution (KÖHNE et al., 1986).

BOD probes make use of immobilized microorganisms as receptors and an oxygen electrode as the transducer (SCHÜGERL et al., 1987).

The use of data obtained with aerobic microorganisms for anaerobic populations is problematic in view of the completely different environment and the different substrate spectrum. Due to the superior reproducibility of the measured data, the *BOD* determination is frequently replaced by a determination of the Chemical Oxygen Demand (*COD*).

3.2.7 Chemical Oxygen Demand (*COD*)

The Chemical Oxygen Demand specifies the volume-related oxygen quantity required to completely oxidize the constituents of the waste water. The *COD* value is suitable as a summary measure for determining the pollution of the waste water or the quality of the purification procedure. It is the most important parameter in the waste water legislation. For Germany the method of determination is mandated in the DIN 38409 (OLIVIER and SCHENDEL, 1983). It is possible to reduce the analysis time to two hours, but nevertheless this analytical method can only be taken into consideration for slow stirred-tank processes for quasi-continuous process control (RENARD et al., 1988). Oxidation is carried out in a defined sulfuric acid–potassium dichromate solution ($K_2Cr_2O_7$). Potassium dichromate is reduced by the organic constituents of the waste water to Cr^{3+}. The excess potassium dichromate quantity is inversely proportional to the *COD* determination based on photometry.

On-line measuring instruments have recently become commercially available, permitting *COD* determination within 3–5 minutes. In these instruments hydrogen peroxide or ozone are used for oxidation.

3.2.8 Degradability Tests

A more reliable basis for the process development of anaerobic waste water purification plants can be obtained from anaerobic degradability tests with the microbial population of the industrial process. In this procedure the gas production is measured directly or indirectly by the internal pressure in a closed container (SHELTON and TIEDJE, 1984; DOLFING and BLOEMEN, 1985; CONCANNON et al., 1988; SCHOTT, 1988). The methane fraction in the gas can be measured after sampling. The final pressure level and the type of curve provide information about the total pollution and toxicity of the waste water. Work on an on-line implementation of such degradability tests for the monitoring of waste water purification plants has been carried out by SCHÜRBÜSCHER (1989).

3.3 Gas Phase

Apart from the pH value, the gas measurement can be regarded as the simplest and most important measurement. This includes measurement of the gas quantity and flow, and also of the gas composition.

3.3.1 Quantity and Volumetric Flow Rate

Several instruments are commercially available making use of different procedures for measurements of gas quantities in the larger

measuring ranges (>20 L h^{-1}). For this reason attention will only be drawn here to a few principles covering the lower measuring range. Wet gas meters are frequently used and are characterized by a high precision of about $\pm 5\%$ to $\pm 1\%$ over a wide range of 1:20 (STROHRMANN, 1980) as well as by insensitivity to fluctuations of the gas composition and corrosiveness of the gas constituents. However, these meters measure the gas quantity, not the gas flow.

Such instruments are also available with a slotted disk so that pulses can be counted photoelectrically. In order to calculate a volume flow, the pulsed signals must then be filtered and mathematically differentiated. ERDMAN and DELWICHE (1985) provide instructions on how to construct such a slotted disk, including evaluation electronics.

In a method developed by BEAUBIEN et al. (1988) a glass flask is filled to a certain pressure with biogas which is then discharged. The gas volume flow can be determined automatically over a range of 1–100 mL min^{-1} by the number of pulses, the dead volume of the flask, and the gas space in the fermenter.

The gas is also frequently collected in a flask from which a liquid has been displaced. Acidification of the liquid prevents the carbon dioxide from dissolving in it. Conversely, in the case of batch experiments, the gas can be fed into a flask filled with alkali solution (Mariotte flask). The carbon dioxide is dissolved in the liquid phase, whereas the methane gas is collected in the head space of the flask and pushes the liquid into a measuring cylinder through an outlet located at the bottom.

Quasi-continuous methods based on the displacement principle have also evolved in which the filling level of the flask can be interrogated by a limit detector. After automatic deaeration of the flask, the displaced liquid can flow back again according to the principle of communicating vessels. The cycles can be counted and can serve as a measure of the gas quantity (MOLETTA and ALBAGNAC, 1982; GLAUSER et al., 1984).

3.3.2 Mass Flow

In the low gas flow range (< 20 L h^{-1}), the measuring principle of the thermal mass flow meter is one of the (few) possibilities for obtaining a direct on-line signal for the gas flow. The gas flows through a capillary tube equipped with two resistance thermometers and a heating coil. The heating spiral is located in the center between the two measuring sensors. The gas is warmed at the heating spiral as it flows through the capillary. The resulting temperature difference between the two sensors is proportional to the gas flow and causes a change in sensor resistance. The current necessary for balance via a bridge circuit is used as the measuring signal. Since the temperature difference is dependent on the heat capacity of the respective gas, the sensor must be calibrated for the gas or gas mixture to be measured. The instruments are therefore adjusted to a particular gas mixture by the manufacturer (e. g., 50% CO_2, 50% CH_4). The gas is dried before entering the sensor in order to avoid measuring distortions and corrosion. Unfortunately, in spite of these measures such sensors are affected over the course of time by the organic sulfur compounds contained in the biogas, so that in the final analysis this solution is not satisfactory.

3.3.3 Composition

The most important compounds, methane and carbon dioxide, are measured by the infrared absorption of the gas constituents. The wavelength of the absorption bands characterize the gas type, whereas the extent of absorption is a measure of the concentrations of the components. The gas flows through a measuring cuvette for each gas component to be measured. The measuring cuvettes are irradiated one after the other with light of wavelengths at which the component to be measured has its absorption maximum and no absorption, respectively. In this way one obtains in succession a concentration-independent and a concentration-dependent signal. The difference is a stable measure of the concentration.

The methane concentration in the gas is particularly important for compiling mathemati-

cal models, since carbon dioxide is very soluble in water. Especially with non-steady-state processes, this leads to a delayed transition into biogas.

Apart from methane, hydrogen is the most important gas component. Measuring instruments developed for medical applications are used for the measurements.

3.4 Level Measurements

Due to the low demands made on the bioreactor (no sterility but, on the other hand, high corrosiveness), reactors of glass or plastic are frequently used for studies on a laboratory scale. In this case, the filling level can be easily controlled by measuring the liquid level (volume regulation).

In this connection, conductometric measuring principles can be applied. One sensor consists of two wires, one of which must be in continuous contact with the liquid. The other is guided vertically through the reactor lid in such a way that its tip marks the desired filling level. When the contact is closed, the discharge pump is switched on by an electronic device. These instruments have the disadvantage that they react to foam. Furthermore, the switching pumping process causes a noise signal in the gas flow so that it first has to be filtered before further mathematical processing (left part of Fig. 5).

Another possibility would be to use an analog filling level sensor providing a signal proportional to the filling level. Such sensors make use of, for example, the principle of the Wheatstone bridge. The sensor consists of a steel probe dipped into the liquid and thus offers an infinite number of resistances. A signal proportional to the depth of submergence is generated. The performance of the discharge pump can then be continuously controlled by a suitable PID controller (right part of Fig. 5).

3.5 Substrate Metering

An important parameter for continuous biotechnical processes, in particular for studies on a laboratory scale, is the residence time or the flow rate. Since the waste water is sometimes

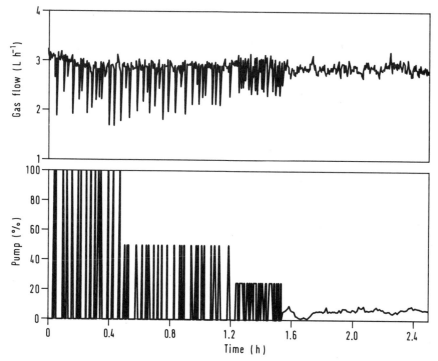

Fig. 5. Elimination of noise in the gas measurement.

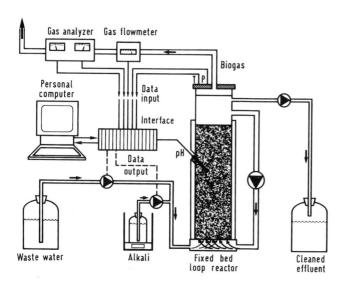

Fig. 6. Scheme of the experimental setup.

corrosive and polluted with solids, not all feed systems can be taken into consideration. Flexible tube pumps are particularly suitable for this purpose.

A metering system permits a very accurate determination of the substrate flow in the course of which a flexible tube pump, gear pump, or membrane pump transports the substrate from an intermediate store on a balance into the reactor. The balance and the microprocessor-controlled pump are connected to each other so that it is possible to exchange data (VETTER and CHRISTEL, 1988). The metering system uses the weight decrease of the substrate on the balance as the controller input.

3.6 Example of an Experimental Assembly

Fig. 6 gives an example of an assembly for a laboratory facility. The variables pH value, temperature, pressure, gas flow, and composition are measured on a standard basis. The substrate is continuously metered with a flexible tube pump, and the filling level is controlled with a level sensor by a discharge pump. The contents of the column-shaped reactor is stirred by a circulating pump that continuously pumps liquid from the top of the reactor to the bottom. The substrate pump and the reactor temperature can be controlled by a connected personal computer.

The pH value is checked off-line daily. The COD value is determined by a cuvette test, and the individual concentrations of volatile components by a gas chromatograph.

4 Mathematical Models

Since anaerobic processes display very long time constants, particular importance is attached to mathematical modelling and the implementation of studies by means of simulations. First, the possibilities provided by the basic mass and energy balances will be present-

ed. A model will then be given that assumes a constant pH value, followed by a model also suitable for simulating changes in the pH value. All models are derived for continuous reactors with ideal stirred tank behavior for biomass and substrate.

4.1 Theoretical Mass and Energy Balances

The composition of the biogas can be calculated from the mass balance by the stoichiometry for the anaerobic degradation of a complex substrate $C_nH_xO_y$ (Eq. 7).

$$C_nH_xO_y + \left[n - \frac{x}{4} - \frac{y}{2}\right] H_2O \rightarrow$$

$$\rightarrow \left[\frac{n}{2} + \frac{x}{8} - \frac{y}{4}\right] CH_4 +$$

$$+ \left[\frac{n}{2} - \frac{x}{8} + \frac{y}{4}\right] CO_2 \qquad (7)$$

By analogy, the chemical oxygen demand for the same substrate results from Eq. (8):

$$C_nH_xO_y + \left(2n + \frac{x}{4} - y\right) O_2 \rightarrow nCO_2 +$$

$$+ \frac{x}{2} H_2O \qquad (8)$$

A chemical oxygen demand of $2n + \frac{x}{4} - y$ mol O_2/mol $C_nH_xO_y$ thus results.

In the case of a generalized application of the balances with the introduction of the degree of reduction, including not only the chemical elements but also the electrons available, the electric charge and the energy are also taken into consideration (SCHÜGERL, 1985). The degree of reduction of a given organic substance $C_nH_xO_yN_z$ is calculated as:

$$\gamma = (4 + x - 2y - 3z)/n \qquad (8a)$$

The mass fraction of carbon in the molecule results from:

$$\sigma = \frac{M_C}{M} \qquad (8b)$$

The maximum theoretical product yield from a substrate can thus be calculated as follows:

$$Y_{P/S}^{\text{theor}} = \frac{\sigma_S \gamma_S}{\sigma_P \gamma_P} \qquad (8c)$$

The following value for the thermodynamic biological efficiency results for the production of methane from acetic acid (ROELS, 1980):

$$\eta = \frac{Y_{P/S}^{\text{exp}}}{Y_{P/S}^{\text{theor}}} = 0.70 \qquad (8d)$$

Extensive discussions of the theoretical effectiveness of anaerobic processes and theoretical energy gains resulting from methane combustion can be found in ANDREWS (1989), HAVLIK et al. (1984), ROELS (1983), ONER et al. (1984), and SOBOTKA et al. (1983).

4.2 Substrate Consumption and Growth Kinetics

The biomass in the stirred tank reactor increases due to growth, and is flushed out during continuous operation (Eq. 9).

$$\frac{dX}{dt} = (\mu(S) - D)X \qquad (9)$$

with D dilution rate
S substrate concentration
t time
X biomass concentration
μ specific growth rate.

The substrate is fed into the reactor convectively. The residual substrate leaves the reactor at the same flow rate. Part of the substrate is converted by the microorganisms (Eq. 10).

$$\frac{dS}{dt} = (S_{\text{in}} - S)D - r(S)X \qquad (10)$$

with $r(S)$ specific substrate consumption rate and
S_{in} input substrate concentration.

Gas is produced during degradation of the substrate (Eq. 11):

$$G = 2 V_m V_R r(S) X \qquad (11)$$

with G biogas flow
V_m mole volume of ideal gases
V_R reactor volume.

The parameters can be experimentally determined by adjusting steady states for various flow rates in succession and determining the substrate and biomass concentrations after sampling in the laboratory. Furthermore, the gas production can also be measured. The following can then be calculated as steady-state values (Eq. 12)

$$\mu = D$$

$$r(S) = \frac{(S_{\text{in}} - S)}{X} D$$

$$Y_{X/S} = \frac{X}{S_{\text{in}} - S} \qquad (12)$$

with $Y_{X/S}$ as yield.

The specific substrate consumption rate can accordingly be represented by the measurable biomass growth and the yield of biomass based on substrate consumption (Eq. 13).

$$r(S) = \frac{\mu(S)}{Y_{X/S}} \qquad (13)$$

In the following we shall make use of the data measured by TRETTER (1987) for a culture of *Methanosarcina barkeri* during the degradation of acetic acid in 5 L stirred tank reactors (pH 6.4, temperature 38 °C, see Tab. 2).

Let the substrate consumption rate be coupled to the substrate concentration S by the Monod kinetics (Eq. 14).

$$r(S) = r_{\text{max}} \frac{S}{K_S + S} \qquad (14)$$

with K_S Monod constant and
r_{max} maximum specific substrate consumption rate.

The classical form of evaluating such kinetics is represented by methods that convert the non-linear kinetics into a graphically easily interpretable straight line by means of linear transformations (EISENTHAL and CORNISH-BOWDEN, 1974). The best known approach is

Tab. 2. Experimental Values (TRETTER, 1987)

S_{in} (mol L^{-1})	D (h^{-1})	X (g L^{-1})	S (mol L^{-1})	$r(S)$ (mol g^{-1} h^{-1})	$Y_{X/S}$ (g mol^{-1})	G (L h^{-1})
0.3	0.0102	0.421	0.0048	0.0072	1.426	0.611
0.3	0.0121	0.430	0.0067	0.0083	1.466	0.755
0.3	0.0140	0.462	0.0162	0.0086	1.628	0.811
0.3	0.0158	0.453	0.0272	0.0095	1.661	0.916
0.3	0.0182	0.432	0.0378	0.0111	1.648	1.060
0.3	0.0201	0.463	0.0499	0.0108	1.851	1.177
0.6	0.0102	0.922	0.0106	0.0065	1.564	1.321
0.6	0.0121	0.880	0.0149	0.0080	1.504	1.506
0.6	0.0140	0.888	0.0188	0.0092	1.528	1.771
0.6	0.0159	0.888	0.0323	0.0102	1.564	1.910
0.6	0.0183	0.890	0.0474	0.0114	1.611	2.110
0.6	0.0204	0.879	0.0717	0.0122	1.664	2.343
0.6	0.0220	0.851	0.0897	0.0132	1.668	2.560
0.9	0.0102	1.263	0.0106	0.0072	1.420	1.988
0.9	0.0124	1.425	0.0164	0.0077	1.613	2.266
0.9	0.0139	1.413	0.0214	0.0086	1.608	2.626
0.9	0.0159	1.396	0.0334	0.0099	1.611	2.968
0.9	0.0178	1.407	0.0423	0.0109	1.640	3.207
0.9	0.0198	1.426	0.0811	0.0114	1.742	3.376
0.9	0.0221	1.411	0.1430	0.0119	1.864	3.602

the double reciprocal plot after Lineweaver–Burke (Eq. 15):

$$\frac{1}{r(S)} = \frac{K_S}{r_{max}} \frac{1}{S} + \frac{1}{r_{max}} \qquad (15)$$

However, due to distortions these transformations do not lead to statistically perfect statements (CORNISH-BOWDEN and EISENTHAL, 1974). The Lineweaver–Burke method is in-

deed the poorest of the linear transformations. For this reason, methods of non-linear optimization are applied almost exclusively today; in the following, mainly the Rosenbrock method (HOFFMANN and HOFMANN, 1971). Fig. 7 shows the substrate consumption kinetics calculated according to both methods with the following values: $r_{max} = 0.0125$ mol g^{-1} h^{-1} and $K_S = 0.0067$ mol L^{-1}.

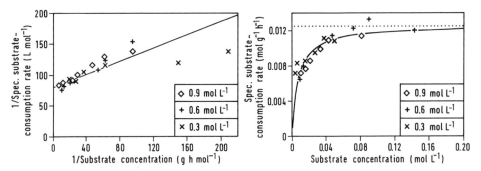

Fig. 7. Kinetics of substrate consumption.

It can be seen that with linear plotting the small substrate concentrations have an inappropriately large influence on the result. Up to this point is was possible to evaluate the data without any statement about the nature of the coupling between microbial growth and substrate consumption.

with μ^*_{max} true specific growth rate and
β specific maintenance metabolism rate.

The measurable growth represents the growth converted into biomass, the concept of true growth is a fictive growth including the maintenance metabolism (BERGTER, 1983).

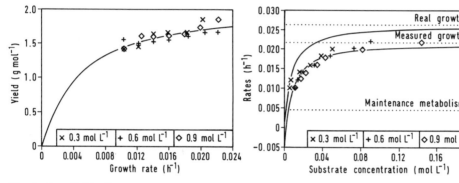

Fig. 8. Growth kinetics.

If the maintenance metabolism is taken into consideration, the substrate consumption rate results from the growth of the biomass and its maintenance metabolism (Eq. 16):

$$r(S) = \frac{\mu(S)}{Y_{X/S}} = \frac{\mu(S)}{Y_{X/S\,max}} + m \qquad (16)$$

with m as the maintenance metabolism coefficient.

Non-linear optimization of Eq. (17)

$$Y_{X/S} = Y_{X/S\,max} \frac{\mu(S)}{\mu(S) + Y_{X/S\,max}\,m} \qquad (17)$$

results in the parameters $Y_{X/S\,max} = 2.104$ g mol^{-1} and $m = 0.0022$ mol g^{-1} h^{-1} (Fig. 8).

The growth rate can then be calculated according to:

$$\mu = r_{max} Y_{X/S\,max} \frac{S}{S+K_S} - m Y_{X/S\,max} \qquad (18)$$

or more briefly according to (Fig. 8, left)

$$\mu(S) = \mu^*_{max} \frac{S}{S+K_S} - \beta \qquad (19)$$

The following values result for the parameters $\mu^*_{max} = r_{max} Y_{X/S\,max} = 0.0263$ h^{-1}, $K_S = 0.0067$ mol L^{-1}, $\beta = m Y_{X/S\,max} = 0.0046$ h^{-1}.

A survey of the derived equations and the parameters is given in Tab. 3.

4.3 Steady State

The concentrations in the non-trivial steady state for substrate and biomass result from the biomass balance (Eq. 20) and the substrate balance (Eq. 21). The substrate concentration can be calculated directly from Eq. (9). The biomass concentration follows after substituting the result of Eq. (10):

$$S = \frac{D + m Y_{X/S\,max}}{r_{max} Y_{X/S\,max} - (D + m Y_{X/S\,max})} K_S \qquad (20)$$

$$X = \frac{D Y_{X/S\,max}}{D + m Y_{X/S\,max}} (S_{in} - S) \qquad (21)$$

It can be seen in Fig. 9 that due to the maintenance metabolism the biomass concentration at $D = 0$ also tends towards zero.

Tab. 3. Mathematical Model (all values at 38 °C)

Gas Phase	
$G_{CH_4} = V_m V_R r(S) X$	$V_R = 5$ L
	$V_m = 25.5$ L mol^{-1}

Liquid Phase	
$\dfrac{dS}{dt} = (S_{in} - S)D - r(S)X$	$r_{max} = 0.0125$ mol g^{-1} h^{-1}
$r(S) = r_{max} \dfrac{S}{K_S + S}$	

Biological Phase	
$\dfrac{dX}{dt} = (\mu(S) - D)X$	$\mu^*_{max} = 0.0263$ h^{-1}
	$K_S = 0.0067$ mol L^{-1}
$\mu(S) = \mu^*_{max} \dfrac{S}{S + K_S} - \beta$	$\beta = 0.0046$ h^{-1}

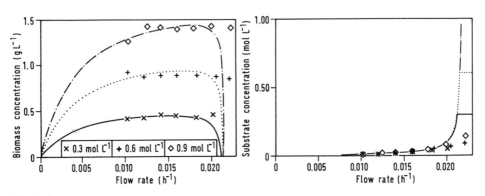

Fig. 9. Steady states.

The minimum substrate concentration that just covers the requirements for maintenance metabolism can be calculated from $\mu = 0$ (Eq. 22).

$$S_{min} = \frac{m}{r_{max} - m} K_S = 0.0014 \text{ mol L}^{-1} \qquad (22)$$

Curves typical of a continuous process can be seen in the plots versus the flow rate. If the substrate concentration in the inlet is increased, an increase in biomass concentration results, but not an increase in substrate concentration.

However, Fig. 9 also shows a definite discrepancy between the model and measured values, particularly at high flow rates. This can be attributed to growth on the wall, which occurs especially if sessile microorganisms are used. This has a particularly distorting effect if microorganisms with very long generation times are involved. As a rule, experiments with small flow rates are not implemented due to the long time constants. For example, a flow rate of

0.005 h^{-1} would correspond to a residence time of 200 h. In order to adjust 99% of a stationary operating point, five residence times would be required, i. e., in this case a period of 42 days. However, in the case of continuous laboratory operation we know from previous experience that experimental problems frequently occur within such a period. On the basis of these boundary conditions, the restricted range of validity of these models should always be kept in mind.

The critical flow rate up to which the process can be operated may be calculated from Eq. (18) to Eq. (23):

$$D_{crit} = \mu(S_{in}) =$$
$$= \left(r_{max} \frac{S_{in}}{K_S + S_{in}} - m\right) Y_{X/S\,max} \qquad (23)$$

The maximum critical flow rate results from:

$$S_{in} \to \infty: \quad D_{crit,max} = (r_{max} - m) Y_{X/S\,max}$$
$$= 0.0217 \text{ h}^{-1} \qquad (24)$$

4.4 Gas Production

Fig. 10 shows the measured gas production plotted against the gas production calculated from the substrate consumption rate (Eq. 25):

$$G_{cal} = 2 V_m V_R r(S) X \qquad (25)$$

with $V_m = 25.5$ mol L^{-1} (at 38 °C).

Only approx. 88% of the theoretical gas production can be measured. This can be explained by the high solubility of carbon dioxide in the liquid. That is to say, part of the carbon dioxide does not leave the reactor with the biogas but is flushed out with the waste water (Eq. 26):

$$0.88 = \frac{G_{exp}}{G_{cal}} = \frac{Q_{exp}}{Q_{cal}} =$$
$$= \frac{Q_{exp}}{Q_{G,CH_4} + Q_{G,CO_2} + Q_{F,CO_2}} =$$
$$= \frac{Q_{exp}}{Q_{exp} + Q_{F,CO_2}} \qquad (26)$$

with Q_{cal} theoretical total carbon flow
      Q_{exp} measured carbon flow in the gas

Q_{F,CO_2} theoretical carbon flow in the liquid
Q_{G,CH_4} measured methane flow
Q_{G,CO_2} measured carbon dioxide flow.

From this it follows that $Q_{F,CO_2} = 0.136\,Q_{exp}$ and $Q_{cal} = 1.136\,Q_{exp}$. The carbon dioxide fraction in the gas is accordingly given by Eq. (27):

$$Q_{G,CO_2} = Q_{cal} - Q_{G,CH_4} - Q_{F,CO_2} =$$
$$= (1.136 - 0.5 \times 1.136 - 0.136) Q_{exp} =$$
$$= 0.432\,Q_{exp} \qquad (27)$$

The carbon dioxide fraction of 43% corresponds to the fraction measured by continuous gas analysis by infrared gas absorption. It may therefore be assumed that the methane production can be calculated as follows:

$$G_{CH_4} = V_m V_R r_{max} \frac{S}{K_S + S} X \qquad (28)$$

From Eq. (11), (20), and (21) it thus follows for the methane flow in the stationary state:

$$G_{CH_4} =$$
$$V_m F_{in} \left(S_{in} - K_S \frac{F_{in} + \beta V_R}{\mu_{max}^* V_R - F_{in} - \beta V_R}\right) \qquad (29)$$

with F_{in} as the fluid flow into the reactor.

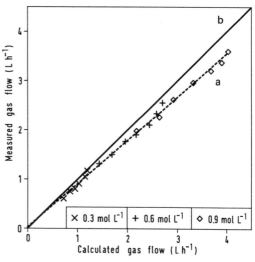

Fig. 10. Measured (curve a) and calculated (curve b) gas production.

4.5 Application of the Steady-State Model to Non-Steady-State Experiments

Three experiments were carried out by TRETTER (1987) according to the program shown in Tab. 4.

Fig. 11 shows the experimental runs and the simulations made with the model determined during stationary experiments. It can be clearly seen that these results are not satisfactory. In experiment I, substrate concentrations are reached which were not set during the steady-state experiments to determine the kinetics. This first inhibits the microorganisms, but

Tab. 4. Non-Steady-State Experiments

	I		II		III	
	S_{in} (mol L^{-1})	D (h^{-1})	S_{in} (mol L^{-1})	D (h^{-1})	S_{in} (mol L^{-1})	D (h^{-1})
from	0.3	0.02	0.6	0.0125	0.9	0.02
to	0.6	0.02	0.6	0.0200	0.3	0.02

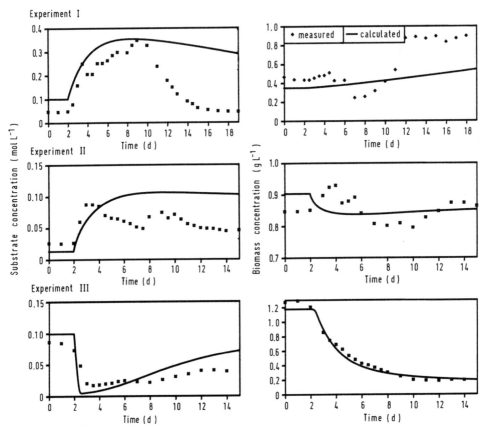

Fig. 11. Non-steady-state courses of experiment and simulation.

after an adaptation phase lasting a few days they begin to grow again. It would not be possible to explain this behavior even by introducing inhibition kinetics, and at least one further differential equation is required. Whereas no definite conclusion can be drawn from experiment II due to the widely scattered measurements, the experiment and prediction are in good agreement in experiment III.

These experiments indicate that the range of validity of the models in question must always be taken into consideration, and models determined from steady-state experiments can only be transferred to non-steady-state conditions to a limited extent. Models of greater complexity would therefore be necessary to describe these processes. However, this would rapidly lead to unacceptably great effort, particularly if only slight knowledge of the inner process sequences is available. Models with time-variant parameters would present an alternative. They could be continuously adapted on-line to the process sequence by means of the measured values. Useful information could be obtained in this way with a minimum of *a priori* information (ISERMANN, 1986; see also Sect. 5).

4.6 Extended Model with pH Calculation

Probably the best known and most frequently quoted model for anaerobic waste water purification was developed by GRAEF and ANDREWS (1973). It also permits the pH value to be calculated. It similarly describes acetic acid degradation as the most important step, since it is limiting in waste water purification. For reasons of simplicity, MATHER (1986) structured the model according to the phases involved:

> biological phase (solid),
> liquid phase,
> transition from liquid to gas phase, and
> gas phase.

The calculations carried out in this chapter refer to a vapor condensate from the cellulose industry with a *COD* value of 41 g L^{-1}.

4.6.1 Biological Phase

The biomass balance is used again in the form introduced above:

$$\frac{dX}{dt} = (\mu(S) - D) X \tag{30}$$

In developing the model for pH calculation, inhibition due to low pH values must be taken into consideration. One possibility consists of multiplying the Monod term used in Eq. (18) by a further pH-dependent factor that describes the pH dependence of the growth rate, e.g., in the form of a parabola with its maximum at the optimal pH value (BASTIN and WANDREY, 1981). GRAEF and ANDREWS (1973) proposed that the pH value should be included by considering as the substrate not the total concentration of the acetic acid, but only the undissociated fraction. They proposed the use of Haldane's approach to the kinetics, which includes substrate inhibition (Eq. 31). Maintenance metabolism is not taken into consideration.

$$\mu(HAc) = \mu_{max} \frac{HAc}{HAc + K_S + \dfrac{K_I}{HAc^2}} \tag{31}$$

with HAc as concentration of the undissociated acetic acid and

K_I substrate inhibition constant.

This approach seems to be generally accepted in the literature even if the determined kinetic parameters are difficult to compare with each other (WIESMANN, 1988). YANG and OKOS (1987) derive the parameters from batch experiments, and DINOPOULOU et al. (1988) derive them from continuous experiments. WITTY and MÄRKL (1985) show that an improved parameter correlation can be achieved by a corresponding conversion of the kinetic parameters relative to the total acetic acid (which was, however, recorded for various pH values).

Eq. (32) follows, since the production of carbon dioxide and methane proceeds in each case equimolarly to the degradation of acetic acid.

$$Q_{X,CO_2} = Q_{X,CH_4} = \frac{\mu(HAc)}{Y_{X/S}} X \tag{32}$$

4.6.2 Liquid Phase

The substrate balance has the familiar form, but the substrate consumption rate is now a function of the undissociated acetic acid (Eq. 33):

$$\frac{dS}{dt} = (S_{in} - S)D - r(HAc)\,X$$

$$r(HAc) = \frac{\mu(HAc)}{Y_{X/S}} \tag{33}$$

A further differential equation permits the simulation of titration with sodium hydroxide solution. It is assumed that the sodium hydroxide solution is always completely dissociated:

$$\frac{dNa^+}{dt} = (Na_{in}^+ - Na^+)D \tag{34}$$

In order to calculate the pH value, the net concentration of strong ions

$$Z^+ = C^+ - A^-$$

is introduced into the charge balance:

$$Z^+ + Na^+ + H^+ =$$
$$= OH^- + Ac^- + HCO_3^- + 2\,CO_3^{2-} \tag{35}$$

$$S \quad = Ac^- + HAc$$

$$Ac^- = S\,\frac{K_{Ac}}{H^+ + K_{Ac}} \tag{36}$$

with A^- anion concentration
Ac^- acetate ion concentration
C^+ cation concentration
CO_3^{2-} carbonate concentration
H^+ hydrogen ion concentration
HCO_3^- bicarbonate concentration
K_{Ac} dissociation constant of acetic acid
    $= 1.74 \times 10^{-5}$ mol L^{-1} (at 25 °C)
    $= 1.70 \times 10^{-5}$ mol L^{-1} (at 38 °C)
Na^+ sodium ion concentration
OH^- hydroxide ion concentration
Z^+ net concentration of strong ions.

Z^+ is a waste-water-specific variable whose concentration in the reactor changes correspondingly to the stirred tank equation if a different waste water is used (Eq. 37).

$$\frac{dZ^+}{dt} = (Z_{in}^+ - Z^+)D \tag{37}$$

In calculating the pH value, the carbon dioxide produced by the microorganisms during degradation of the acetic acid and thus dissolved in the waste water must be taken into consideration. Carbon dioxide is present in various dissolved forms in the liquid:

$$CO_{2tot} = CO_{2dis} + H_2CO_3 + HCO_3^- + CO_3^{2-} \tag{38}$$

with CO_{2tot} total concentration of all forms of CO_2 in the liquid phase
CO_{2dis} concentration of carbon dioxide dissolved in the liquid
H_2CO_3 concentration of carbonic acid.

Due to the low stability of carbonic acid, its concentration is negligible in comparison with the dissolved CO_2. This is taken into consideration in defining the dissociation equilibrium between bicarbonate and the total concentration:

$$HCO_3^- = \frac{K_{C1}\,CO_{2dis}}{H^+} \tag{39}$$

with K_{C1} dissociation constant of bicarbonate $(5.01 \times 10^7$ mol L^{-1} at 38 °C).

The following is true for the dissociation constant of bicarbonate to carbonate:

$$K_{C2} = \frac{CO_3^{2-}\,H^+}{HCO_3^-} =$$
$$= 10^{-10.21} \text{ mol L}^{-1} =$$
$$= 6.17 \times 10^{-11} \text{ mol L}^{-1} \text{ (at 25 °C)}$$
$$= 10^{-10.22} \text{ mol L}^{-1} =$$
$$= 6.03 \times 10^{-11} \text{ mol L}^{-1} \text{ (at 38 °C)} \tag{40}$$

Eq. (41) thus follows at 38 °C:

$$\frac{CO_3^-}{HCO_3^-} = \frac{K_{C2}}{H^+} = 10^{pH - pK_{C2}} =$$
$$= 6 \times 10^{-5} \quad \text{(at pH = 6)}$$
$$= 6 \times 10^{-4} \quad \text{(at pH = 7)} \tag{41}$$

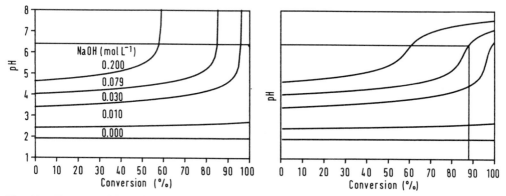

Fig. 12. pH course without (left) and with (right) the influence of carbon dioxide solubility.

The carbonate concentration can thus also be neglected; Eq. (42):

$$CO_{2tot} = CO_{2dis} + HCO_3^-$$ (42)

After combining Eqs. (39) and (42) the following is obtained for the dissolved carbon dioxide:

$$CO_{2dis} = \frac{CO_{2tot}}{\left(1 + \dfrac{K_{C1}}{H^+}\right)}$$ (43)

The saturation concentration for the dissolved carbon dioxide resulting from contact with the gas phase can be calculated by the Henry–Dalton law (Eq. 44):

$$CO_{2dis}^* = H p_{CO_2}$$ (44)

with H Henry constant =
3.01×10^{-7} mol L^{-1} Pa^{-1}
(at 25 °C)
2.11×10^{-7} mol L^{-1} Pa^{-1}
(at 38 °C)

p_{CO_2} CO$_2$ partial pressure in the gas phase.

The concentration of Z^+ in the waste water can be calculated from the ion balance with the pH value of the substrate present (here pH = 1.9). At this pH value the OH$^-$ concentration can be neglected due to the ionic product of water at 25 °C; $K_W = H^+ \cdot OH^- = 10^{-14}$ (see Eq. 35):

$$Z^+ = Ac^- - H^+$$ (45)

The following calculations assume an acetic acid concentration of $S_{in} = 0.36$ mol L^{-1}. It thus follows that: $Z^+ = -0.012$ mol L^{-1}. These negatively charged ions are mainly sulfate and sulfite. Fig. 12 shows the pH values arising in the reactor with certain specific conversion rates. Since a pH value of 6.4 is regarded as optimum according to experience, the waste water is pre-titrated to pH = 4 (left figure) in order to achieve this value at degradation rates of more than 80% (right figure).

The quantity of sodium hydroxide solution required is calculated as:

$$Na^+ = Ac^- - H^+ - Z^+$$ (46)

with $Ac^- = 0.067$ mol L^{-1} and $H^+ = 10^{-4}$ mol L^{-1} from which it follows that $Na^+ = 0.079$ mol L^{-1}.

Fig. 13 shows a titration curve calculated with the derived concentration in comparison with the actual measurement. At first glance the agreement seems very good, but particularly in the pH range between 6 and 7, which is of interest here, there are great deviations. Thus, for example, a slight deviation in the sodium hydroxide solution can result in a calculated pH value of between 7 and 13. In comparison, the real system is much more strongly buffered. Simulations undertaken with this model can therefore only provide qualitative statements.

With pH values adjusted in the reactor between 6 and 7, the hydrogen ion and hydroxide ion concentrations can be neglected.

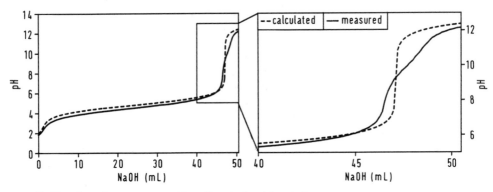

Fig. 13. Titration of waste water with sodium hydroxide solution.

After the dissociation equilibria for acetic and carbonic acid have been substituted in the ion balance (Eq. 35), the following results as a simplified balance:

$$Z^+ + Na^+ = S\frac{K_{Ac}}{H^+ + K_{Ac}} + \frac{K_{C1}C_{dis}}{H^+} \qquad (47)$$

Finally, in order to calculate the pH value, it follows that:

$$H^+ = -\frac{a}{2} + \sqrt{\left(\frac{a}{2}\right)^2 - b}$$

$$a = \frac{K_{Ac}(Z^+ + Na^+ - S) - K_{C1}C_{dis}}{Z^+ + Na^+}$$

$$b = -\frac{K_{C1}C_{dis}K_{Ac}}{Z^+ + Na^+} \qquad (48)$$

The pH value that would have to be achieved if the substrate pump were to be switched off and the total acetic acid converted can also be calculated from this equation. According to experience, a pH value of 7.9 is then obtained for this waste water and a carbon dioxide content in the gas phase of 20%. However, after substitution a calculated pH value of only 6.85 is obtained. This is due, on the one hand, to the difference in the pH curves, but on the other to the inaccuracy of the equation with the Henry coefficient, which depends on so many factors (e.g., pressure, temperature, composition of the medium, salt content, ad-

sorption effects) that it is very difficult to calculate even with model solutions (SCHUMPE, 1985). Further information on calculations of pH values can be found in BLIEFERT (1978), PONS et al. (1990), and WELLINGER (1985).

The total balance for the carbon dioxide in the liquid is now as follows:

$$\frac{dCO_{2tot}}{dt} = Q_{X,CO_2} - CO_{2tot}D - Q_{G,CO_2} \qquad (49)$$

with Q_{G,CO_2} CO$_2$ converted into the gas phase
and
Q_{X,CO_2} CO$_2$ produced by the microorganisms.

4.6.3 Mass Transport

CO$_2$ transport between the liquid and gas phase is described by the mass transfer law:

$$Q_{G,CO_2} = K_L a(CO_{2dis} - CO_{2dis}^*) \qquad (50)$$

with $K_L a$ as the mass transfer coefficient.

The following values are given in the literature for the $K_L a$ value:

4.2 h^{-1} (GRAEF and ANDREWS, 1973)
1500 h^{-1} (MATHER, 1986)

GRAEF and ANDREWS (1973) assumed that, due to good mixing, transport resistances within the gas and liquid volumes can be neglected.

This means that, e.g., a homogeneous gas phase is assumed and that both the gas bubbles within the liquid as well as the gas in the reactor head are of the same composition. MATHER (1986) also differentiated between these two gas phases and additionally defined surface a as dependent on the instantaneous gas production. The model from Eq. (50) will be used for the following considerations.

4.6.4 Gas Phase

The change of carbon dioxide concentration in the gas phase results, assuming a constant gas volume, from the difference between the stream of carbon dioxide flowing out of the liquid into the gas phase and that leaving the reactor together with the biogas (Eq. 51):

$$V_G \frac{dCO_{2G}}{dt} = V_R \left(Q_{G,CO_2} - \frac{p_{CO_2}}{p} Q_G \right) \quad (51)$$

with CO_2 concentration of CO_2 in the gas
  p total pressure in the reactor
  Q_G total gas flow
  V_G gas volume
  V_R liquid volume.

The total gas flow is the sum of the sub-flows, whereby only the most important compounds, namely methane and carbon dioxide, are to be taken into consideration:

$$Q_G = Q_{G,CO_2} + Q_{X,CH_4} \quad (52)$$

The CO_2 concentration in the gas can be calculated from the ideal gas equation:

$$CO_{2G} = \frac{p_{CO_2}}{RT} \quad (53)$$

with R general gas constant
    (8.31 J mol^{-1} K^{-1}) and
  T absolute temperature.

The following is thus the partial pressure of CO_2 in the gas phase from Eq. (51):

$$\frac{dp_{CO_2}}{dt} = \frac{RT}{p} \frac{V_R}{V_G} (p Q_{GCO_2} - p_{CO_2} Q_G) \quad (54)$$

The biogas flow can then be calculated according to Eq. (55):

$$G = G_{CO_2} + G_{CH_4}$$

$$G_{CO_2} = V_m V_R \frac{p_{CO_2}}{p} Q_G$$

$$G_{CH_4} = V_m V_R Q_{X,CH_4} \quad (55)$$

Constant reactor pressure is assumed. A survey of the formulas derived here is given in Tab. 5.

4.7 Further Models

In the models above, the degradation of acetic acid to methane was represented by the entire summary formula (Eq. 33), which is resolved into a number of steps inside the cell, each contributing towards the total degradation. BHADRA et al. (1984) proposed various models and compared their predictions for acetate degradation with the measured values (acetic acid, dissolved CO_2, HCO_3^-, CO_3^{2-}).

HILL and BARTH (1977) extended the model to a mixed culture consisting of acid and methane producers, and applied it to real waste water from stock farming with high organic and ammonium pollution. DROSTE and KENNEDY (1988) applied a model for acid and methane production to a solid film reactor. BHATIA et al. (1985) have been concerned with modelling the inhibition of the methanogenesis of acetic acid by propionic and butyric acid in flocculated biomass.

A number of contributions report studies of the start-up and transition behavior of anaerobic reactors with the aid of mathematical models for the essential degradation steps of acid hydrolysis and methane formation (BRYERS, 1985; KLEINSTREUER and POWEIGHA, 1982; TORRE and STEPHANOPOULOS, 1986).

The influence of hydrogen on acid degradation, particularly the aspect of hydrogen diffusion to the biofilm, has been dealt with in experiments and models by DENAC et al. (1988a) and OZTURK et al. (1989).

In verifying complex mathematical models with the aid of experimental sequences, attention is rarely paid to the fact that too many model parameters make it difficult to identify the parameters, and thus local secondary optima may be produced. A stable solution there-

Tab. 5. Mathematical Model (all values at 38 °C)

Gas Phase

$$\frac{dp_{CO_2}}{dt} = \frac{RT}{p}\frac{V_R}{V_G}(pQ_{GCO_2} - p_{CO_2}Q_G)$$

$$G = V_m V_R \left(\frac{p_{CO_2}}{p}Q_G + Q_{X,CH_4}\right)$$

$$G_{CH_4} = V_m V_R Q_{X,CH_4}$$

$$Q_G = Q_{GCO_2} + Q_{X,CH_4}$$

$V_R = 5$ L

$V_m = 25.5$ L mol^{-1}

$p = 10^5$ Pa

$R = 8.3143$ J mol^{-1} K^{-1}

Mass Transfer

$$Q_{GCO_2} = K_L a(CO_{2dis} - CO^*_{2dis})$$

$$CO^*_{2gel} = Hp_{CO_2}$$

$K_L a$

$H = 2.11 \times 10^{-7}$ mol L^{-1} Pa^{-1}

Liquid Phase

$$\frac{dS}{dt} = (S_{in} - S)D - \frac{\mu(HAc)}{Y_{X/S}}X$$

$$\frac{dZ^+}{dt} = (Z^+_{in} - Z^+)D \qquad \frac{dNa^+}{dt} = (Na^+_{in} - Na^+)D$$

$$\frac{dCO_{2tot}}{dt} = Q_{X,CO_2} - CO_{2tot}D - Q_{GCO_2}$$

$$CO_{2dis} = \frac{CO_{2tot}}{\left(1 + \dfrac{K_{C1}}{H^+}\right)}$$

$$H^+ = -\frac{a}{2} + \sqrt{\left(\frac{a}{2}\right)^2 - b}$$

$$a = \frac{K_{Ac}(Z^+ + Na^+ - S) - K_{C1}CO_{2dis}}{Z^+ + Na^+} \qquad b = -\frac{K_{C1}CO_{2dis}K_{Ac}}{Z^+ + Na^+}$$

$K_{Ac} = 1.70 \times 10^{-5}$ mol L^{-1}

$K_{C1} = 5.01 \times 10^{-7}$

$S_{in} = 0.36$ mol L^{-1}

$Z^+ = -0.012$ mol L^{-1}

$Na^+ = 0.065$ mol L^{-1}

Biological Phase

$$\frac{dX}{dt} = (\mu(HAc) - D)X$$

$$\mu(HAc) = \mu_{max}\frac{HAc}{HAc + K_S + \dfrac{K_I}{HAc^2}}$$

$$Q_{X,CO_2} = Q_{X,CH_4} = \frac{\mu(HAc)}{Y_{X/S}}X$$

$Y_{X/S} = 2.104$ g mol^{-1}

μ_{max}, K_S, K_I

---

fore requires a selection of the really important parameters to be adapted and exclusion of others that can be assumed to be constant (BASTIN et al., 1983; BOLLE et al., 1986; HAVLIK et al., 1984).

5 Observation of Non-Measurable Variables

5.1 Kalman Filter

A greater amount of information can be achieved by adding mathematical models to the measured values. In this way, for example, it is possible to calculate state variables that represent the instantaneous system state and whose time course is defined by the differential equations, but which cannot be directly measured.

The principle shown in Fig. 14 is common to all the principles, such as the Luenberger observer or the Kalman filter.

The real system (Eq. 56) is simulated on the computer by a process and sensor model (Eq. 57). Alterations of the system input as well as the resulting changes in the measuring variable are communicated to the observer. The difference between the true measured variable and the precalculated measured variable serves to adapt the model to the true measured curve via the matrix K:

$$\frac{dx}{dt} = f(x, u) \tag{56}$$

$$y = h(x)$$

$$\frac{d\hat{x}}{dt} = f(\hat{x}, u) + K(y - h(\hat{x})) \tag{57}$$

$$\hat{y} = h(\hat{x})$$

with: f state function
  h measuring function
  K amplification matrix
  u input variable
  x state vector
  y measured variable.

The Luenberger observer is in the first instance only applicable to linear systems. For non-linear systems it would have to be linearized once by a defined operating point or else continuously along the trajectory. A non-linear observer proposed by ZEITZ (1977) can be generally solved for second-order systems. It cannot be applied to higher order systems due to non-linear growth and substrate kinetics and the associated considerable analytical mathematical efforts. It is possible to use non-linear systems with the Kalman filter, although a numerical differential equation solver (e.g., Runge-Kutta) is then required.

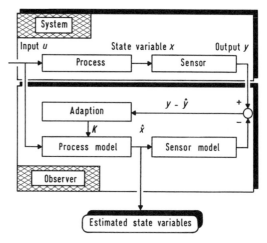

Fig. 14. Principle of Kalman filtering.

The procedures also differ in their adaptability to the system dynamics. Whereas the Luenberger observer can only be adapted to the measuring noise and the differing sensitivities of the state variables through its eigenvalue, the Kalman filter offers the possibility of filtering out noisy measured values as well as adjusting for sensitivity for each state variable, however, at the price of additional differential equations (in the case of an nth order system and additional $n(n+1)/2$ differential equations). STEPHANOPOULOS and SAN (1984) gave a very useful introduction to the theory of the Kalman filter and its applications to biotechnology.

The theory of the Kalman filter is based on the assumption that the systems to be observed are not deterministic but rather stochastic. The Kalman filter then calculates its optimum eigenvalues itself on the basis of the given values by means of the noise behavior of the system. The following results for the amplification matrix:

$$K = P h_X^T(\hat{x}) S^{-1} \tag{58}$$

The covariance matrix P enters noise into the deterministic differential equation system and thus represents a measure of the inaccuracy of the mathematical model. The standard deviations of the individual state variables can be calculated by extracting the root from the elements of the principal diagonal so that they can, for example, be included in the graphic representation (GILLES and SCHULER, 1981). A further differential equation system results in order to calculate the covariance matrix:

$$\frac{dP}{dt} = f_X(\hat{x}) P + P f_X^T(\hat{x}) + Q - K h_X(\hat{x}) P \tag{59}$$

with Q as the covariance matrix.

Since it is assumed that no couplings are present between the noise terms of the individual state values, the covariance matrix Q is only occupied on the main diagonal:

$$Q = \begin{bmatrix} q_{11} & 0 \\ 0 & q_{22} \end{bmatrix} \qquad S = s \tag{60}$$

This means that only one triangular matrix of P must be calculated. The matrix S is simplified to $S = s$ if only one measured value is available. The values represent the variances, i.e., the squares of the standard deviations. These noise terms must be given by the user. Their orders of magnitude can be worked out interactively on the basis of measured values already included by the computer. In doing so, the parameters are assumed to be greater if large changes over time are permitted, and smaller if the value is kept more rigid.

In the concrete case this means that if the measured variable of the methane gas flow is very noisy then s must be assumed to be greater than if it can be calculated exactly. Since s enters into the amplification matrix or, in this case, into the amplification vector as the reciprocal value in the case of greater noise

(large s), this leads to a lower weighting of the error of observation.

A variance is similarly allocated to the state variables through the matrix Q. Its values have a directly proportional effect on the increase of the matrix P over time and thus once again on the amplification vector K.

5.2 On-Line Calculation of the Biomass Concentration

In the case of anaerobic waste water purification, for example, the Kalman filter can be used to attempt to calculate the biomass concentration and its growth rate from the change in the biogas flow:

$$\frac{d\hat{X}}{dt} = (\hat{\mu} - D)\hat{X} \tag{61}$$

$$\frac{d\hat{\mu}}{dt} = 0$$

$$G_{CH_4} = V_m V_R \frac{\hat{\mu} \hat{X}}{Y_{X/S}}$$

The growth rate is thus introduced as a time-variant constant by a further differential equation. In this case, the value for the biomass concentration within Q is assumed to be correspondingly smaller than that for the growth rate.

5.3 On-Line Calculation of the Substrate Concentration

The substrate concentration cannot be calculated with the aid of an observer without a statement about the basic kinetics or further measurements, since otherwise there would be no coupling between the biomass and the substrate balance. However, if the substrate concentration in the inlet is known, then it can be calculated by a balancing of the biogas. It must be remembered that this method of substrate calculation is not an observation in the classical sense, since not all available information is used for the calculation. Since the difference between the calculated and measured methane flow is not included, a rapid convergence of the calculated substrate concentrations cannot be achieved. After substitution, Eq. (64) follows from the differential equation for the substrate concentration (Eq. 62) and the measuring equation (Eq. 63):

$$\frac{dS}{dt} = (S_{in} - S)D - \frac{\mu(S)X}{Y_{X/S}} \quad \text{or}$$

$$\frac{dS}{dt} = (S_{in} - S)D - r(S)X \quad (62)$$

$$G_{CH_4} = V_m V_R \frac{\mu(S)X}{Y_{X/S}} \quad \text{or}$$

$$G_{CH_4} = V_m V_R r(S)X \quad (63)$$

$$\frac{dS}{dt} = (S_{in} - S)D - \frac{G_{CH_4}(t)}{V_m V_R} \quad (64)$$

After the integration of

$$\int_{S(t=0)}^{S(t)} \frac{dS}{(S_{in} - S)D - \dfrac{G_{CH_4}}{V_m V_R}} = \int_0^t dt \quad (65)$$

with $\overline{G}_{CH_4} = \dfrac{1}{t}\int_0^t G_{CH_4}(t)\,dt$,

the relations follow, assuming that G_{CH_4} remains constant for sufficiently short times:

$$D \neq 0: \quad S(t) = S(t=0)e^{-Dt} +$$
$$+ \left(S_{in} - \frac{G_{CH_4}}{D V_m V_R}\right)(1 - e^{-Dt})$$

$$D = 0: \quad S(t) = S(t=0) - \frac{G_{CH_4}}{V_m V_R}t \quad (66)$$

or, in the time-discrete notation:

$$D \neq 0: \quad \hat{S}_k = \hat{S}_{k-1}e^{-DT_0} +$$
$$+ \left(S_{in} - \frac{G_{CH_4}}{D V_m V_R}\right)(1 - e^{-DT_0})$$

$$D = 0: \quad \hat{S}_k = \hat{S}_{k-1} - \frac{G_{CH_4}}{V_m V_R}T_0 \quad (67)$$

with T_0 as the cycle time.

It can be seen that the influence of the error of estimation for the initial concentration decreases exponentially with time, i.e., convergence is ensured in any case. For short cycle times, the differential equation can also be directly converted into a difference equation by rectangular integration:

$$\hat{S}(k) = \hat{S}(k-1) +$$
$$+ \left[(S_{in} - \hat{S}(k-1))D - \frac{G_{CH_4}}{V_m V_R}\right]T_0 \quad (68)$$

In the experiment, the measured methane flow can be corrected by factors:

$$\hat{G}_{CH_4} = k_1 k_2 k_3 G_{CH_4} \quad (69)$$

In order to calculate the factor k_1, the gas meter is read at irregular intervals. The value should if possible be 1, and represents a criterion for the quality of the long-term accuracy of the mass flow meter:

$$k_1 = \frac{\Delta V_G}{\sum_i G_{M_i} T_i} \quad (70)$$

with G_{M_i} gas flow measured by the mass flow meter during time interval i

V_G gas volume measured through the gas meter.

The correction factor k_2 takes the composition of the waste water into consideration. Since the balances for acetic acid were set up as reference components, the biogas fraction attributable to conversion of the acetic acid is estimated by means of this factor.

$$k_2 = \frac{COD_{Ac}}{COD_{tot}} = 0.72 \quad (71)$$

The substrate balance is fitted to the laboratory analyses by the factor k_3. The acetic acid concentration $\hat{S}(t_1)$ estimated at time t_1 is verified by means of the concentration $S_{meas}(t_1)$ determined by a laboratory analysis. If deviations occur here it leads to adaptation of the factor k_3 by means of the amplification factor K:

$$k_3(t_2) = k_3(t_1) + K(\hat{S}(t_1) - S_{meas}(t_1)) \quad (72)$$

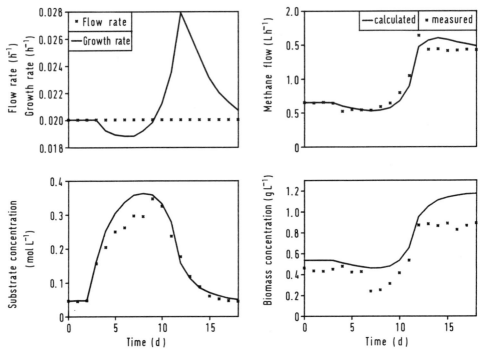

Fig. 15. Observation of state variables (experiment I).

RENARD et al. (1988) employ a comparable procedure; however, their model is based on *COD* analyses taken at strict two-hour intervals.

5.4 Results

The fact that it is worthwhile to apply the Kalman filter even with very few measured points has been shown in the evaluation of the non-steady-state experiments already presented in Fig. 11. The value for the gas flow averaged over one day served as the data base for the evaluation in Figs. 15, 17, and 18. The gas composition was not measured in this experiment. The measured methane flow was therefore assumed to be 57% of the measured total gas flow (Eq. 27).

Since the experiments were started from the stationary state, the actual flow rate $\hat{u} = D$ can be selected as the initial value for estimating the growth rate. The initial value for estimat-

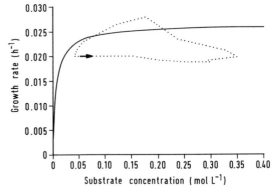

Fig. 16. Comparison of non-stationary with steady-state growth rates (experiment I).

ing the biomass concentration was calculated from the actual biogas flow:

$$\hat{X} = \frac{0.57\,G}{Y_{G/x}\hat{\mu}} \tag{73}$$

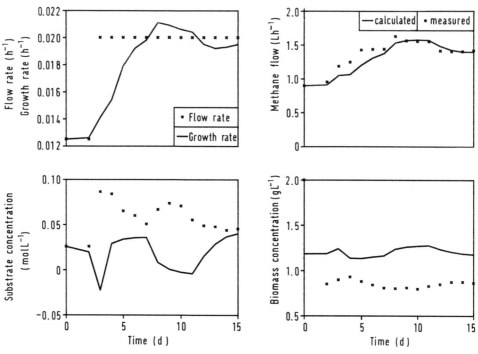

Fig. 17. Observation of state variables (experiment II).

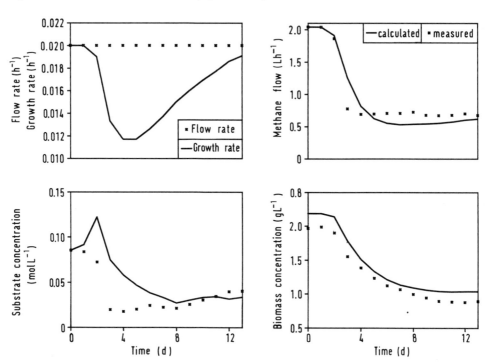

Fig. 18. Observation of state variables (experiment III).

In experiment I an excellent quantitative agreement between measurement and calculation was obtained; therefore, this can be regarded as a realistic calculation of the growth rate (Fig. 15).

Comparison of the growth rates calculated here with those determined for the same system by steady-state experiments shows that they are in the same range (Tab. 2). The exactly closed hysteresis occurring during the course of the non-steady-state experiment is in very good agreement with the theoretical consideration that a higher substrate input concentration only brings about a higher biomass concentration in the reactor, but not a higher substrate concentration (Fig. 16).

In experiment II, correlation with the measured values is very poor, as can be seen in Fig. 11, so other influences must be presumed here (Fig. 17).

The estimate of biomass concentration in experiment III is in very good agreement with the measurements, whereas the calculated substrate concentration converges only very slowly towards the measured concentrations (Fig. 18).

On the whole, however, it is apparent that application of the Kalman filter, also off-line, represents both a valuable supplement to and a replacement for the recorded measurements.

6 Process Control in Anaerobic Waste Water Purification

Process control in anaerobic waste water purification has two essential tasks to fulfill: stable operation of the reactor and continued high performance. Due to the small number of measured variables available, great significance is attached to their interpretation and the feedback of information on the conditions of the microbiological culture (ARCHER, 1983).

6.1 pH Regulation

Regulation of the pH to a physiologically desirable value is the simplest form of process control in anaerobic waste water purification. In doing so, use is made of the pH shift occurring during the degradation of organic compounds. A basic differentiation is made between the following possibilities (GRAEF and ANDREWS, 1973):

– recycling of cells
– recycling of biogas after separating the carbon dioxide
– variation of the flow rate: pH-auxostatic regulation
– addition of sodium hydroxide solution: pH chemostatic regulation.

During pH-auxostatic operation the inflow of acid waste water is controlled so that a certain pH value is maintained in the reactor.

Due to the degradation of acid by the microorganisms, a higher pH value is established in the reactor than is found in the inlet. A pre-neutralization may be meaningful in order to achieve the desired degradation rates. With this method the inflow is controlled, but a buffer vessel must be installed in order to reduce the fluctuations in the inlet. The substrate consumption permits monitoring of the degradation rate. CZAKO et al. (1987) proposed an improvement of the method by a vacuum desorption of the carbon dioxide.

In the case of pH-chemostatic operation, the substrate inflow is not controlled, but the pH value in the reactor is adjusted via a correction medium (sodium hydroxide solution). However, there is a danger that the biomass may be flushed out if the flow rate is too great. The degradation rate can be monitored here by the consumption of correction medium (DENAC et al., 1988b).

In neither strategy is there a direct observation of the degradation rate, since the pH value is an indirectly measured variable. The degradation rate must therefore be additionally monitored. Only then can a decision be made about an adaptation of the effective parameters.

As described above, in pH-auxostatic operation, the pH value in the reactor is controlled by the addition of acid waste water through the substrate pump. Switching controllers are used to turn the pump on and off. If the flow rate is to be used as a measured value for more complex mathematical evaluations, then it is best to use continuous controllers permitting continuous operation of the pump. In this way, the control signals can also be used as indirectly measured variables without further filtering. The system dynamics are altered in the course of time due to the growth of the biomass and the necessary periodic regeneration of the fixed-bed reactor, and also to activity fluctuations and variations in temperature or waste water. If a conventional PID controller were used, its parameters would have to be continuously adapted to the operating state. The use of a parameter-adapted controller would therefore be appropriate. These controllers make use of a mathematical model of the process, continuously adapting its parameters to the actual course of the process, and on the basis of a predefined control criterion can correspond to the instantaneous system dynamics.

HOLT and KÜMMEL, 1979; ISERMANN, 1987b; JACOBS et al., 1980) (Fig. 19).

For this purpose a black-box model describing the process must be found whose parameters are then continuously adapted to the input and output behavior of the process in a parameter identification carried out in parallel with the process. This type of experimental modelling (identification) replaces the theoretical formulation of mechanistic models wherever they would involve unacceptably great efforts in order to achieve a certain accuracy. This is true more of chemical engineering processes than of mechanical and electrical processes, and accordingly should be of great value, particularly in biotechnology (HENSEL et al., 1986).

In the present case it is not possible to describe the control section by a physical model due to the complex pH behavior. A SISO (single-input single-output) model can be used as a black-box model with the substrate flow as the input and the pH value as the output variable. Since digital computers read the data in fixed cycle times, i.e., discretely, time-discrete difference equations must be introduced for the computer-internal representation of continuous differential equations. A linear process without dead time as the simplest form of system behavior can be described by an ordinary difference equation of the nth order. If the parameters for the difference equation system of the nth order are combined, then the deviation values generally found in automatic control engineering are:

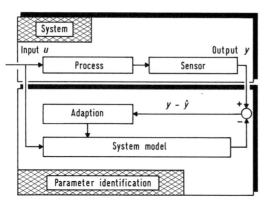

Fig. 19. Principle of parameter identification for model-based controllers.

namics. Such controllers have been successfully used for titration processes in the chemical industry as an alternative to conventional controllers and, due to their excellent control behavior, they contribute towards savings in operating materials (SHINSKEY, 1974; BUCH-

$$y(k) + a_1 y(k-1) + \ldots + a_n y(k-n) =$$
$$= b_1 u(k-1) + \ldots + b_n u(k-n)$$

$$y(k) = Y(k) - Y_\infty$$

$$u(k) = U(k) - U_\infty \qquad (74)$$

with a, b parameters
$\quad k \quad$ discrete time
$\quad n \quad$ system order
$\quad u \quad$ adjustment of controller input
$\quad U \quad$ controller input (here: substrate flow)
$\quad y \quad$ measured value deviation
$\quad Y \quad$ measured value (here: pH value).

The cycle time is no longer explicitly given, but is included in the parameters. This must be remembered if a different cycle time is to be used.

Since the equivalent for the controller output, i.e., the substrate flow to be adjusted in the stationary state, is not known *a priori*, it is similarly estimated, in this case in the form of the implicit equivalent estimate (ISERMANN, 1987a). Eq. (75) results after substitution:

$$Y(k) = -a_1 Y(k-1) - \ldots - a_n Y(k-n) + \\ + b_1 U(k-1) + b_n U(k-n) + K_\infty$$

$$K_\infty = (1 + a_1 + \ldots + a_n) Y_\infty - \\ - (b_1 + \ldots + b_n) U_\infty \qquad (75)$$

The system order n specifies how many "old" measured values are required in order to make a reliable statement about the pH value one cycle time later. The a parameters characterize the system behavior without any influence by the control variable, and the b parameters characterize the influence of the control variable on the system.

The system order can be determined experimentally. For this purpose, the pH value is stimulated by the substrate flow in such a way that the information content of the measured data course is sufficient for a system analysis. For example, pseudo-noise binary signals generated by a shift register may be used (ISERMANN, 1974).

Second-order models are thus used for stirred tank reactors and first-order models for fixed-bed reactors (Eq. 76).

$$\text{pH}(k) = -a \, \text{pH}(k-1) + b F_{\text{in}}(k-1) + K_\infty \quad (76)$$

The parameter vector θ can be calculated in parallel with the process by a sliding, time-weighted regression using a recursive method (HSIA, 1977) (Eq. 77):

$$\theta(N+1) = \theta(N) + \gamma(N+1) P(N) x(N+1) \cdot \\ \cdot [Y(N+1) - x^T(N+1)\theta(N)]$$

$$P(N+1) = \frac{1}{\lambda} [P(N) - \gamma(N+1)P(N)x(N+1) \cdot \\ \cdot x^T(N+1)P(N)]$$

$$\gamma(N+1) = 1/[1 + x^T(N+1)P(N)x(N+1)] \quad (77)$$

with the measuring vector x and the parameter vector θ (Eq. 78):

$$\theta(N) = [a_1(N), \ldots, a_n(N), b_1(N), \\ \ldots, b_n(N), K_\infty]^T$$

$$x(N+1) = [-Y(N), \ldots, -Y(N+1-n), U(N), \\ \ldots, U(N+1-n), 1]^T \qquad (78)$$

The forgetting factor λ serves to fit the parameter identification to the noise behavior of the process $(0 < \lambda < 1)$. Although the parameters are more rapidly adapted with a small λ, nevertheless they are then much noisier. A value close to 1 should first be selected, and it can then be adapted to the noise behavior of the process at any time. Upon starting up, the matrix P is occupied on the principal diagonal by very large positive values in order to ensure a rapid transient behavior.

The parameter-adaptive controller can now be derived by a reversal of the section equation (Eq. 76). By adjusting the substrate flow it will thus be possible to set the pH value for the next cycle time (Eq. 79).

$$F_{\text{in}}(k) = \frac{\text{pH}_{\text{des}} + a \, \text{pH}(k) - K_\infty}{b} \qquad (79)$$

This controller will attempt to adjust the desired pH value within the next time cycle and will thus result in strong controller output variations. In order to limit these variations, the controller must satisfy a criterion J, a combination of control performance and energy requirements. The two fractions can be weighted by a factor Q (Eq. 80).

$$J = (\text{pH}(k+1) - \text{pH}_{\text{des}})^2 + \\ + Q^2 (F_{\text{in}}(k) - F_{\text{in}}(k-1))^2 \qquad (80)$$

By minimizing this control area, the following is obtained for the controller:

$$F_{\text{in}}(k) = \\ \frac{F_{\text{in}}(k-1) Q^2 + b(\text{pH}_{\text{des}} + a \, \text{pH}(k) - K_\infty)}{b^2 + Q^2} \qquad (81)$$

The probable steady-state value for the substrate flow can thus be calculated (Eq. 82).

$$F_{\text{in},\infty} = \frac{(1+a)\,\text{pH}_{\text{des}} - K_\infty}{b} \tag{82}$$

Fig. 22 shows controlled pH curves with an automatic temperature optimization in a fixed-bed reactor.

6.2 Regulation of Gas Production

BASTIN et al. (1983) similarly made use of a first-order model and a parameter-adaptive controller to regulate the gas production of an anaerobic facility serving primarily for gas production. By analogy to Eq. (76), the equation for biogas is as follows:

$$G(k) = -a\,G(k-1) + b\,F(k-1) + K_\infty \tag{83}$$

However, the quality criterion to be minimized is defined slightly differently. Not the difference between the new and old controller output $F_{\text{in}}(k) - F_{\text{in}}(k-1)$ is taken as the controller output difference but the difference between the new controller output and that belonging to the desired steady-state gas flow (Eq. 84).

$$J = (G(k+1) - G_{\text{des}})^2 + Q^2 (F_{\text{in}}(k) - F_{\text{in},\infty})^2 \tag{84}$$

The probable steady-state controller output can, on the other hand, be pre-calculated from the difference equation by analogy to Eq. (81).

$$F_{\text{in},\infty} = \frac{(1+a)\,G_{\text{des}}K_\infty}{b} \tag{85}$$

The following thus results for the controller:

$$F_{\text{in}}(k) = \frac{F_{\text{in}}(k-1)Q^2 + b(G_{\text{des}} + a\,G(k) - K_\infty)}{b^2 + Q^2} \tag{86}$$

6.3 Substrate Regulation

If the reactor is primarily used for waste water purification, then regulation of the substrate concentration in the outlet to a certain permissible value is a possible control goal. It must be assumed that the capacity of the facil-

ity is sufficient, and that the waste water purification facility can be uncoupled from the waste water source for brief periods by a buffer tank connected in series. Due to the non-linearities, model-supported adaptive controllers are also appropriate here. Starting from the model for the methane stage:

$$\frac{dX}{dt} = (\mu(t) - D)X$$

$$\frac{dS}{dt} = (S_{\text{in}} - S)D - \frac{\mu(t)}{Y_{X/S}}X$$

$$G = Y_{G/X}\mu X \tag{87}$$

It is assumed that the substrate concentrations in the reactor inlet and outlet can be regularly measured, that the biogas flow is measured as an auxiliary measured variable and that the process is controlled by the flow rate. Let the growth rate and the yield coefficients be unknown. It follows from Eq. (87) after substitution:

$$\frac{dS}{dt} = (S_{\text{in}} - S)D(t) - \frac{Y_{X/G}}{Y_{X/S}}G \tag{88}$$

After replacing the differential quotient by a weighted difference quotient Eq. (89) is obtained:

$$C_1(t)(S_{\text{des}} - S(t)) = (S_{\text{in}} - S)D(t) - \hat{K}G$$

$$\hat{K} = \frac{Y_{X/G}}{Y_{X/S}} = Y_{S/G} \tag{89}$$

from which it follows for $D(t)$:

$$D(t) = \frac{C_1(t)[S_{\text{des}} - S(t)] + \hat{K}G(t)}{S_{\text{in}}(t) - S(t)}$$

$$0 < D(t) < D_{\text{max}} \tag{90}$$

$\hat{K}$ is continuously adapted by a further differential equation:

$$\frac{d\hat{K}}{dt} = C_2 G(t)[S_{\text{des}} - S(t)] \tag{91}$$

After approximating the differential quotient with an Euler's differential quotient the following results for $\hat{K}$:

$$\hat{K}(k+1) = \hat{K}(k) + T C_2 G(k+1)[S_{des} - S(t)] \quad (92)$$

If the process is started from the stationary state, then $\hat{K}$ can be estimated from Eq. (89) as follows:

$$\hat{K}(0) = \frac{D(0)(S_{in} - S(0))}{G(0)} \quad (93)$$

Experiments with this control have been implemented by RENARD et al. (1988). They checked the *COD* value every two hours. The value C_2 was substituted as a constant, whereas the variable C_1 was substituted in various experiments sometimes as a constant and sometimes as proportional to the gas flow $C_1(t) = C_1 G(t)$. $\hat{K}$ was little changed during on-line fitting.

COSTELLO et al. (1989) therefore kept this parameter constant during their simulations. They demonstrated in simulations that in the case of measuring errors in $G(t)$ and $S_{in}(t)$ a permanent control deviation from Eq. (90) results for the controller; therefore, they proposed a control law extended by an integral part (Eq. 94).

$$\frac{dS}{dt} = C_1(t)[S_{des} - S(t)] + \\ + C_2(t) \int (S_{des} - S(t)) dt \quad (94)$$

The following thus results for the controller:

$$D(t) = \frac{C_1(t)[S_{des} - S(t)] + C_2(t) \int (S_{des} - S(t)) dt + \hat{K} G(t)}{S_{in}(t) - S(t)} \quad (95)$$

DOCHAIN and BASTIN (1985) had already derived such an integral approach for a two-stage reactor (methanation stage with an acidification stage connected in series).

6.4 Hydrogen Regulation

Since hydrogen plays an important part in the interactions between the individual micro-

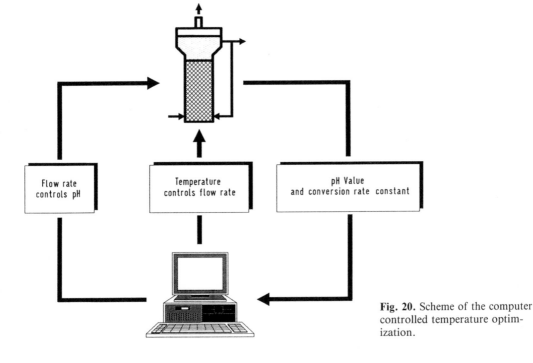

Fig. 20. Scheme of the computer controlled temperature optimization.

bial populations, and the rise in its concentration is an important indication of an overloading of the fermenter, regulation of the hydrogen concentration may be a possible operating strategy. WHITMORE et al. (1987) measured the dissolved hydrogen and methane concentrations with a membrane and a mass spectrometer. It was demonstrated in experiments that acidification of the reactor can be prevented by regulating the dissolved hydrogen concentration to 1 μmol by means of the flow rate.

6.5 Optimization of Operating Parameters

Greatly differing data are given in the literature for the optimum temperature and pH ranges in anaerobic waste water purification. Furthermore, data are rarely given for the method used to determine the optimum values. For this reason, experiments were undertaken to determine such parameters by computer. These methodological developments are naturally of general significance for biotechnology. However, they can be tested particularly effectively with the anaerobic waste water purification process, since this process is not operated continuously or under sterile conditions, and the pH value can be easily measured on-line as an indirect measured variable for the residual substrate concentration and the biogas flow. MÄRKL et al. (1983) proposed a computer-automated pH optimization which was tested by MATHER (1986) in simulations. EBERHARDT (1989), FREYER (1988), and SCHÜRBÜSCHER (1989) developed a computer-controlled temperature optimization that functions according to the diagram shown in Fig. 20.

The temperature is varied by the computer according to the diagram shown in Fig. 21. A change in temperature causes an altered degradation rate and would cause a change in pH if the parameter-adaptive pH controller discussed in Sect. 6.1 did not keep this value constant. An increase in the flow rate is thus obtained as a valuable signal for reactor performance. The algorithm also recognizes so-called characteristic temperatures in the case of a performance change in which the rise in performance is particularly steep. Fig. 22 shows two experiments carried out with a fixed-bed reactor with different temperature increase gradients.

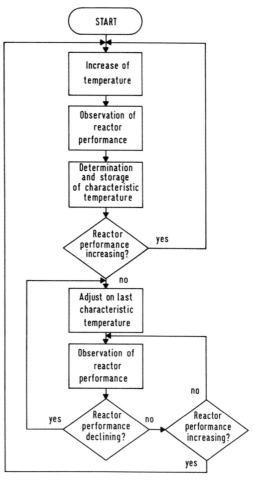

Fig. 21. Flow diagram for temperature optimization.

In spite of different experimental sequences, the experiments converge towards an optimum temperature of 42 °C with a performance increase between 25 and 30%. The time constants of the system must be taken into consideration with the configuration of the optimization sequence. The influence of temperature on cell reproduction is not accounted for with-

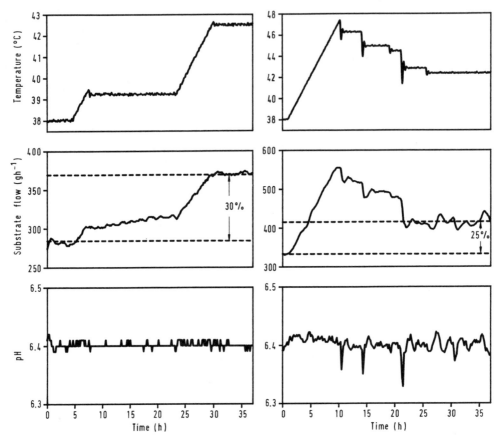

Fig. 22. Experimental results of temperature optimization.

in the selected experimental period in the short-time range. However, it was possible to confirm the results by long-time experiments at the higher temperatures. This method can be used analogously, for example, in order to adapt microorganisms to toxic concentrations.

7 Summary

Interest in anaerobic waste water purification is currently increasing due to a number of advantages in comparison to the classic aerobic method:

- no energy-intensive oxygen transfer
- less excess sludge
- production of combustible biogas.

This also leads to the desire to model and control these processes. However, the models can only be as good as the measuring technology used to verify the results.

It is very difficult to measure the biomass concentration due to the turbidity of the media, their pollution with solids, or the fixation of microorganisms.

In the waste water itself, the pH value and also the summation parameters *COD* and, by more sophisticated measuring procedures, the *BOD* or *TOC* can in any case be used as measured variables. Biologically relevant statements can only be obtained by degradability tests.

It is particularly simple to measure the biogas flow. The biogas composition with respect to carbon dioxide and methane and also hydrogen provides valuable indications of the performance of the reactor.

Models for the methane stage of waste water purification in the stirred tank have been presented. Models including mass transport permit, within limits, the calculation of the pH value.

Observers, in particular Kalman filters, can be employed in anaerobic waste water purification to calculate the biomass concentration and the growth rate in the stirred tank reactor.

The simplest controllers make use of the pH value and thus the pH shift in the individual waste water purification stages as the control variable. In order to regulate the gas production as a function of the substrate concentration, more complex model-supported controllers are used.

8 References

AIVASIDIS, A., WANDREY, C. (1985), Biomasseabtrennung in der anaeroben Abwasserreinigung, gwf-wasser/abwasser **126**, 56–66.

ARCHER, D. B. (1983), The microbiological basis of process control in methanogenic fermentation of soluble wastes; Enzyme Microb. Technol., May, 162–170.

ARCHER, D. B., HILTON, M. G., ADAMS, P., WIECKO, H. (1986), Hydrogen as a process control index in a pilot scale anaerobic digester, Biotechnol. Lett. **8**, 197–202.

BASTIN, K., WANDREY, C. (1981), Process dynamic aspects in continuous anaerobic fermentation, Ann. N. Y. Acad. Sci., 135–145.

BASTIN, G., DOCHAIN, D., HAEST, M., INSTALLE, M., OPDENACKER, P. (1983), Identification and adaptive control of a biomethanization process, in Modelling and Data Analysis in Biotechnology and Medical Engineering, Amsterdam: North-Holland Publishing Company.

BEAUBIEN, A., JOLICOEUR, C., ALARY, J. F. (1988), Automated high sensitivity gas metering system for biological processes, Biotechnol. Bioeng. **32**, 105–109.

BERGTER, F. (1983), Wachstum von Mikroorganismen, p. 70, Weinheim: Verlag Chemie.

BHADRA, A., MUKHOPADHYAY, S. N., GHOSE, T. K. (1984), A kinetic model for methanogenesis of acetic acid in a multireactor system, Biotechnol. Bioeng. **26**, 257–264.

BHATIA, D., VIETH, W. R., VENKATASUBRAMANIAN, K. (1985), Steady-state and transient behavior in microbial methanification, Biotechnol. Bioeng. **27**, 1192–1207.

BLIEFERT, C. (1978), pH-Wert-Berechnungen, Weinheim: Verlag Chemie.

BOLLE, W. L., VAN BREUGEL, J., VAN EYBERGEN, G. C., KOSSEN, N. W. F., VAN GILS, W. (1986), Kinetics of anaerobic purification of industrial wastewater, Biotechnol. Bioeng. **28**, 542–548.

BRYERS, J.-D. (1985), Structured modeling of the anaerobic digestion of biomass particulates, Biotechnol. Bioeng. **27**, 638–649.

BUCHHOLT, F., KÜMMEL, M. (1979), Self-tuning control of a pH-neutralization process, Automatica **15**, 665–671.

BÜHLER, H. (1985), Messen in der Biotechnologie, p. 83, Heidelberg: Hüthig Verlag.

CONCANNON, F., REYNOLDS, P. J., HENNIGAN, A., COLLERAN, E. (1988), Development of a computerized continuous assay for specific methanogenic activity measurement, Poster-papers of the Fifth International Symposium of Anaerobic Digestion, Bologna, Italy, pp. 177–181, Bologna: Monduzzi Editore.

CORNISH-BOWDEN, A., EISENTHAL, R. (1974), Statistical considerations in the estimation of enzyme kinetic parameters by the direct linear plot and other methods, Biochem. J. **139**, 721–730.

COSTELLO, D. J., LEE, P. L., GREENFIELD, P. F. (1989), Control of anaerobic digesters using generic model control, Bioprocess Eng. **4**, 119–122.

CZAKO, L., MIHALTZ, P., MORVAI, L., HOLLO, J. (1987), Development of new neutralization procedure of highly acidic wastewaters during anaerobic treatment, in Proc. 4th Eur. Congr. Biotechnology, Vol. 1, pp. 69–71, Amsterdam: Elsevier Science Publishers B. V.

DENAC, M., MIGUEL, A., DUNN, I. J. (1988a), Modeling dynamic experiments on the anaerobic degradation of molasses wastewater, Biotechnol. Bioeng. **31**, 1–10.

DENAC, M., GRIFFIN, K., LEE, P. L., GRENFIELD, P. F. (1988b), Selection of controlled variables for a high rate anaerobic reactor, Environ. Technol. Lett. **9**, 1029–1040.

DINOPOULOU, G., RUDD, T., LESTER, J. N. (1988), Anaerobic acidogenesis of a complex wastewater, Biotechnol. Bioeng. **31**, 958–978.

DOCHAIN, D., BASTIN, G. (1985), Stable adaptive

controllers for waste treatment by anaerobic digestion, *Environ. Technol. Lett.* **6**, 584–593.

DOLFING, J., BLOEMEN, W. (1985), Activity measurements as a tool to characterize the microbial composition of methanogenic environments, *J. Microbiol. Methods* **4**, 1–12.

DROSTE, R. L., KENNEDY, K. J. (1988), Dynamic anaerobic fixed-film reactor model, *J. Environ. Eng.* **114**, 606–620.

EBERHARDT, R. (1989), *Rechnergestützte Regelung biotechnischer Prozesse,* Dissertation in preparation, Institut für Biotechnologie, Forschungszentrum Jülich KFA.

EHRENBERGER, F. (1979), Zur Bestimmung von Sauerstoffbedarfs- und Kohlenstoff-Kennzahlen (TOD, TOC, DOC usw.) in der Wasserqualitätsbestimmung, *GIT Fachz. Lab.* **23**, 738–747.

EISENTHAL, R., CORNISH-BOWDEN, A. (1974), The direct linear plot, *Biochem. J.* **139**, 715–720.

ERDMAN, M. D., DELWICHE, S. R. (1985), Low-cost digital counting interface for fermentation gas measurement, *Biotechnol. Bioeng.* **27**, 569–571.

FREYER, S. (1988), *Rechnergestützte Optimierung eines anaeroben Abwasserreinigungsprozesses,* Diplomarbeit, Institut für Biotechnologie, Forschungszentrum Jülich KFA.

GILLES, E. D., SCHULER, H. (1981), Zur frühzeitigen Erkennung gefährlicher Reaktionszustände in chemischen Reaktoren, *Chem. Ing. Tech.* **53**, 673–682.

GLAUSER, M., JENNI, B., ARAGNO, M. (1984), An inexpensive, automatic gas meter for laboratory-scale methane digesters and other gas-evolving systems, *J. Microbiol. Methods* **2**, 159–164.

GRAEF, S. P., ANDREWS, J. F. (1973), Mathematical modeling and control of anaerobic digestion, *Water,* 101–131.

HARPER, S. R., POHLAND, F. G. (1987), Enhancement of anaerobic treatment efficiency through process modification, *J. WPCF* **59**, 152–161.

HAVLIK, I., VOTRUBA, J., SOBOTKA, M., VOLESKY, B. (1984), Parametric sensitivity in modeling of anaerobic digestion, *Biotechnol. Lett.* **6**, 607–610.

HEIJNEN, J. J. (1984), Technik der anaeroben Abwasserreinigung, *Chem. Ing. Tech.* **56**, 526–532.

HENSEL, H., ISERMANN, R., SCHMIDT-MENDE, P. (1986), Experimentelle Identifikation und rechnergestützter Regler-Entwurf bei technischen Prozessen, *Chem. Ing. Tech.* **58**, 875–887.

HILL, D. T., BARTH, C. L. (1977), A dynamic model for simulation of animal waste digestion, *J. WPCF,* October, 2129–2143.

HOFFMANN, H., HOFMANN, U. (1971), Einführung in die Optimierung, p. 106, Weinheim: Verlag Chemie.

HSIA, T. C. (1977), System Identification, pp. 69 ff, 83, Lexington: Lexington Books.

ISERMANN, R. (1974), Prozeßidentifikation, pp. 45–49, Berlin: Springer-Verlag.

ISERMANN, R. (1986), Experimentelle Identifikation und rechnergestützter Regler-Entwurf bei technischen Prozessen, *Chem. Ing. Tech.* **58**, 875–887.

ISERMANN, R. (1987a), *Digitale Regelsysteme,* Berlin: Springer-Verlag.

ISERMANN, R. (1987b), Stand und Entwicklungstendenzen bei adaptiven Regelungen, *Automatisierungstechnik* **35**, 133–143.

JACOBS, O. L. R., HEWKIN, M. A., WHILE, C. (1980), Online computer control of pH in an industrial process, *IEE Proc.* **127**, 161–168.

JOUANNEAU, Y., KELLEY, B. C., BERLIER, Y., LESPINAT, P. A., VIGNAIS, P. M. (1980), Continuous monitoring, by mass spectrometry, of H_2 production and recycling in *Rhodopseudomonas capsulata, J. Bacteriol.* **143**(2), 628–636.

KLEINSTREUER, C., POWEIGHA, T. (1982), Dynamic Simulator for anaerobic digestion processes, *Biotechnol. Bioeng.* **24**, 1941–1951.

KÖHNE, M., SIEPMANN, F. W., TE HEESEN, D. (1986), Der BSB_5 und der kontinuierliche Kurzzeit-BSB (BSB-M3) im Vergleich, *Korrespondenz Abwasser* **33**, 787–793.

MACARIO, A. J. L., CONWAY DE MACARIO, E. (1985), Antibodies for methanogenic biotechnology, *Trends Biotechnol.* **3**, 204–208.

MÄRKL, H., MATHER, M., WITTY, W. (1983), Meß- und Regeltechnik bei der anaeroben Abwasserreinigung sowie bei Biogasprozessen, *Münchener Beiträge zur Abwasser-, Fischerei- und Flußbiologie,* Vol. 36, München: R. Oldenbourg Verlag.

MATHER, M. (1986), Mathematische Modellierung der Methangärung, *Fortschr. Ber. VDI,* Reihe 14, No. 28. Düsseldorf: VDI-Verlag.

MOLETTA, R., ALBAGNAC, G. (1982), A gas meter for low rates of gas flow: Application to the methane fermentation, *Biotechnol. Lett.* **4**, 319–322.

MONZAMBE, K. M., NAVEAU, H. P., NYNS, E.-J., BOGAERT, N., BÜHLER, H. (1988), Problematics and stability of on-line pH measurements in anaerobic environments: The jellied combined electrode, *Biotechnol. Bioeng.* **31**, 659–665.

NYNS, E. J. (1986), Biomethanation processes, in *Biotechnology* (REHM, H.-J, REED, G., Eds.), Vol. 8, pp. 207–268, Weinheim: VCH.

OLIVIER, M., SCHENDEL, F.-A. (1983), Gesetzgebung (Stand und Trend); analytische Erfassung der Abwasserinhaltsstoffe, *Papers Verfahrenstechnik der Biologischen Abwasserreinigung,* Krefeld: VDI-Gesellschaft Verfahrenstechnik und Chemieingenieurwesen.

ONER, M. D., ERICKSON, L. E., YANG, S. S. (1984), Estimation of yield maintenance, and product formation kinetic parameters in anaerobic fermentations, *Biotechnol. Bioeng.* **26**, 1436–1444.

OZTURK, S. S., PALSSON, B. O., THIELE, J. H. (1989), Control of interspecies electron transfer flow during anaerobic digestion: Dynamic diffusion reaction models for hydrogen gas transfer in microbial flocs, *Biotechnol. Bioeng.* **33**, 745–757.

PHELPS, T. J., CONRAD, R., ZEIKUS, J. G. (1985), Sulfate-dependent interspecies H_2 transfer between *Methanosarcina barkeri* and *Desulfovibrio vulgaris* during coculture metabolism of acetate or methanol. *Appl. Environ. Microbiol.* **50**(3), 589–594.

PONS, M. N., GARRIDO-SANCHEZ, L., DANTIGNY, P., ENGASSER, J. M. (1990), pH Modelling in fermentation broths, *Bioprocess Eng.* **5**, 1–6.

POWELL, G. E., ARCHER, D. B. (1989), On-line titration method for monitoring buffer capacity and total volatile fatty acid levels in anaerobic digesters, *Biotechnol. Bioeng.* **33**, 570–577.

RENARD, P., DOCHAIN, D., BASTIN, G., NAVEAU, H., NYNS, E.-J. (1988), Adaptive control of anaerobic digestion processes – a pilot-scale application, *Biotechnol. Bioeng.* **31**, 287–294.

ROBINSON, J. A., TIEDJE, J. M. (1984), Competition between sulfate-reducing and methanogenic bacteria for H_2 under resting and growing conditions, *Arch. Microbiol.* **137**, 26–32.

ROELS, J. A. (1980), Application of macroscopic principles to microbial metabolism, *Biotechnol. Bioeng.* **22**, 2457–2514.

ROELS, J. A. (1983), *Energetics and Kinetics in Biotechnology,* Amsterdam: Elsevier Biomedical Press.

SCHNECKENBURGER, H., REUTER, B. W., SCHOBERTH, S. M. (1984), Time-resolved fluorescence microscopy for measuring specific coenzymes in methanogenic bacteria, *Anal. Chim. Acta* **163**, 249–255.

SCHOTT (1988), Schott-Anaerob-Testeinheit. *Produkt-Information Nr. 6191d,* Mainz: Schott Glaswerke.

SCHÜGERL, K. (1985), *Bioreaktionstechnik I,* Frankfurt am Main: Sauerländer.

SCHÜGERL, K., LÜBBERT, A., SCHEPER, T. (1987), On-line-Prozeßanalyse in Bioreaktoren, *Chem. Ing. Tech.* **59**, 701–714.

SCHUMPE, A. (1985), Gas solubilities in biomedia, in *Biotechnology* (REHM, H.-J., REED, G., Eds.), Vol. 2, p. 159, Weinheim: VCH.

SCHÜRBÜSCHER, D. (1989), Rechnergestützte Optimierung und Prozeßführung kontinuierlicher biotechnischer Prozesse untersucht am Beispiel

der anaeroben Abwasserreinigung, *Berichte der Kernforschungsanlage Jülich* Nr. **2325**.

SCOTT, R. I., WILLIAMS, T. N., LLOYD, D. (1983), Oxygen sensitivity of methanogenesis in rumen and anaerobic digester populations using mass spectrometry, *Biotechnol. Lett.* **5**(6), 375–380.

SHELTON, D. R., TIEDJE, J. M. (1984), General method for determining anaerobic biodegradation potential, *Appl. Environ. Microbiol.,* 850–857.

SHINSKEY, F. G. (1974), Adaptive pH controller monitors nonlinear processes, *Control Eng.,* Feb., 57–59.

SOBOTKA, M., VOTRUBA, J., HAVLIK, I., MINKEVICH, J. G. (1983), The mass-ernergy balance of anaerobic methane production, *Folia Microbiol.* **28**, 195–204

STEPHANOPOULOS, G., SAN, K.-Y. (1984), Studies on on-line bioreactor identification, *Biotechnol. Bioeng.* **26**, 1176–1218.

STROHRMANN, G. (1980), *Einführung in die Meßtechnik im Chemiebetrieb,* p. 257, München – Wien: R. Oldenbourg Verlag.

TEMPER, U., PFEIFFER, W., BISCHOFSBERGER, W. (1986), Stand und Entwicklungspotentiale der anaeroben Abwasserreinigung unter besonderer Berücksichtigung der Verhältnisse in der Bundesrepublik Deutschland, *Bericht aus Wassergütewirtschaft und Gesundheitsingenieurwesen,* Technische Universität München.

THAUER, R. K., JUNGERMANN, K., DECKER, K. (1977), Energy conservation in chemotrophic anaerobic bacteria, *Bacteriol. Rev.* **41**, 100.

TORRE, A., STEPHANOPOULOS, G. (1986), Mixed culture model of anaerobic digestion: Application to the evaluation of startup procedures, *Biotechnol. Bioeng.* **28**, 1106–1118.

TRETTER, M. (1987), *Untersuchungen zur Reaktionstechnik des anaeroben mikrobiellen Essigsäureabbaus und zur Wachstumsentkopplung methanogener Mikroorganismen am Beispiel von Methanosarcina barkeri,* Diplomarbeit, Institut für Biotechnologie, Forschungszentrum Jülich KFA.

VETTER, G., CHRISTEL, W. (1988), Durchflußkontrolle kleiner Dosierpumpen bei stetiger und pulsierender Strömung, *Chem. Ing. Tech.* **60**, 672–685.

WANDREY, C., AIVASIDIS, A. (1983), Zur Reaktionstechnik der anaeroben Fermentation, *Chem. Ing. Tech.* **55**, 516–524.

WELLINGER, A. (1985), Process parameters affecting methane production in mesophilic farm digesters, *Process Biochem.,* October, 131–137.

WHITMORE, T. N., LLOYD, D., JONES, G., WILLIAMS, T. N. (1987), Hydrogen-dependent control of the continuous anaerobic digestion process, *Appl. Microbiol. Biotechnol.* **26**, 383–388.

WIESMANN, U. (1988), Kinetik und Reaktionstechnik der anaeroben Abwasserreinigung, *Chem. Ing. Tech.* **60**, 464–474.

WITTY, W., MÄRKL, H. (1985), Reaktionstechnische Aspekte der Methangärung am Beispiel der Vergärung von Penicillinmycel, *Chem. Ing. Tech.* MS 1400/85.

YANG, S. T., OKOS, M. R. (1987), Kinetic study and mathematical modeling of methanogenesis of acetate using pure culture of methanogens. *Biotechnol. Bioeng.* **30**, 661–667.

ZEITZ, M. (1977), Nichtlineare Beobachter für chemische Reaktoren, *VDI-Fortschrittsberichte,* Reihe 8, No. 27.

15 Bioprocess Kinetics and Modelling of Recombinant Fermentation

DEWEY D. Y. RYU

JEONG-YOON KIM

SUN BOK LEE

Davis, California 95616, U.S.A.

List of Symbols

a, b, c parameters or constants defined by Eq. (8) or Eq. (51)

d, e, f parameters defined by Eq. (53)

A, B parameters of the Leudeking–Piret product formation model, or parameters defined by Eq. (23)

A', B' parameters defined by Eq. (33)

D dilution rate (h^{-1})

E energy of activation

$\hat{G}_p$ intracellular plasmid concentration (mg DNA per mL cell volume)

G_p plasmid concentration (mg DNA per g cell)

ΔH enthalpy of deactivation

k_0 constant

k_m^0 rate constant used in Eq. (1) for mRNA synthesis

k_p^0 rate constant used in Eq. (2) for protein synthesis

k_{-m} decay constant for mRNA used in Eq. (1)

k_{-p} decay constant for product used in Eq. (2)

$\hat{m}$ intracellular mRNA concentration

$\hat{p}$ intracellular protein concentration

p concentration of product per unit volume

q_p specific production rate (mg protein per mg cell per h)

r_p rate of plasmid synthesis

R gas law constant

S substrate concentration (g/L)

ΔS entropy of deactivation

t culture time (h)

T temperature (K)

X cell concentration (g/L)

X_T total cell concentration ($=X^+ + X^-$)

α ratio of specific growth rate of plasmid-harboring cells to that of plasmid-free cells ($=\mu^+/\mu^-$)

β proportionality constant defined by Eq. (61)

γ dimensionless parameter related to Γ

ε gene expression efficiency ($=\eta\zeta$)

η transcription efficiency

μ specific growth rate

θ relative plasmid loss rate, or ratio of the specific rate of generation of

plasmid-free cells to the specific growth rate of the plasmid-harboring cells ($=\Theta/\mu^+$)

ρ_b cell density

σ, ξ parameter defined by Eq. (38) and Tab. 2

τ dimensionless time parameter as defined by Eq. (38) and Tab. 2

ζ translation efficiency

Δ difference in specific growth rate between plasmid-free and plasmid-harboring cells (h^{-1}) ($=\mu^- - \mu^+$)

Φ fraction of plasmid-harboring cells

Γ parameter representing ($D - \mu^{app}$)

Θ specific rate of generation of plasmid-free cells or plasmid loss rate ($=\mu^+\theta$)

Ψ fraction of plasmid-harboring cells among the colony-forming cells

Ω ratio of plasmid-free cells to plasmid-harboring cells ($=X^-/X^+$)

Subscripts and Superscripts

1 first stage

2 second stage

T total cell concentration

+ phenotypes of plasmid-harboring cells

− phenotypes of plasmid-free cells

app apparent value

1 Introduction

Recent developments and advances in the areas of genetic engineering and biotechnology have shown that there are unlimited possibilities for their potential applications to various gene products by means of: (1) transferring genes to bacterial host organisms from other gene sources, (2) improving gene expression efficiencies, (3) amplifying gene copy numbers, (4) altering metabolic pathways, (5) improving fermentation yield, and (6) improving purification efficiency.

Within less than two decades immediately following the first demonstration of the recombinant DNA (rDNA) technique by COHEN et al. (1974), there have been exciting developments in the areas of pharmaceutical and medical applications. Some of these gene products are already available on the market, and others are well on the way. The important front-runners among these new gene products include: insulin, growth hormones, interferons, interleukin-2, tissue plasminogen activator, vaccines, diagnostics, amino acids, and enzymes (WEBBER, 1985; OFFICE OF TECHNOLOGY ASSESSMENT, 1982, 1984). Significant progress has also been made in the areas of more traditional "engineering biotechnology" or bioprocess engineering, which include fermentation technology, enzyme technology, and bioseparation technology.

Successful commercial development of gene products requires not only recombinant DNA technology and bioprocess engineering but also many other strategic planning tasks related to clinical testing, handling of regulatory affairs, marketing, and legal and business aspects involved in the commercial development of new products and processes. In recent years we have gained considerable experience in these areas, and they should not be underestimated; this chapter, however, deals with only the scientific and technological aspects of bioprocess kinetics and modelling of recombinant fermentation.

Recombinant DNA technology enables us to identify, clone, and transform the desired genes, and to have the host cell express them very efficiently. On the other hand, bioprocess technology, if and when properly developed and optimized, enables us to produce gene products economically by making good use of recombinant fermentation, purification, and bioseparation technologies. Although many underestimated the importance of bioprocess technology during the early period of biotechnological development, it is now widely recognized, and many of us, especially in the biochemical engineering community, are now focusing on the problems, namely scale-up problems related to recombinant fermentation and bioseparation of gene products. Fig. 1 highlights the important strategic planning task components required for the commercial development of gene products.

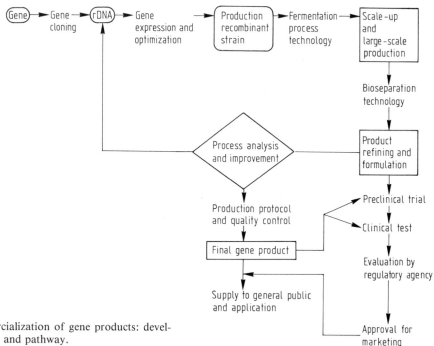

Fig. 1. Commercialization of gene products: development strategy and pathway.

It is also appropriate to consider the "systems engineering" approach to the biotechnology process optimization task, since the rDNA technique used in the "front end" of the research and the bioprocess technology used in the "rear end" of the process development are highly interactive, and a well-coordinated and concerted research and development endeavor could bring about a very efficient commercial development of the gene products. The selection of host organism, gene expression system, and excretion system, and their product separation processes, are good examples.

Recently, as more and more of the second- and third-generation gene products were well on their way to large-scale production and commercialization, the importance of biotechnology process engineering and the scarcity of a related knowledge base came to be recognized. Once a fundamental biochemical engineering study is established for many host–vector systems of practical importance, our in-depth knowledge base can be significantly advanced, and a rational strategy for improving and/or optimizing recombinant fermentation processes can be developed for practical application.

2 Bioprocess Kinetics

Once the recombinant fermentation process is optimized, we must then deal with many more genetic parameters of recombinants. Thus, a deeper understanding of the recombinant fermentation process and quantitative analyses of bioprocess kinetics are necessary for optimization.

2.1 Bioprocess Modelling and Simulation

Our bioprocess model is organized according to five levels of complexity: (1) the molecular level model, (2) the single-cell model, (3) the population model, (4) the bioreactor model, and (5) the bioplant model (BAILEY et al.,

1983; MOSER, 1985) (see Fig. 2). The first two levels are microscopic process kinetics, and the next two levels are macroscopic process kinetics; a combination of all the levels gives the overall bioprocess kinetics. We have very little information about the first two levels of complexity in bioprocess systems.

2.1.1 Molecular Model

Recent advances in molecular biology have enhanced our understanding of the regulatory mechanisms of microbial metabolism and have enabled us to analyze them more accurately than ever before. Analysis of the molecular model provides insight into metabolic regulations taking place at the molecular level. The molecular level models have been built up based on genetic control and regulatory mechanisms. For instance, the control mechanisms of the *lac* promoter–operator and the λdv plasmid replication, which are now well understood, have been studied in detail using molecular mechanism models (LEE and BAILEY, 1984a, b, c, d).

Our ultimate aim is to develop the strategy that gives the best gene expression efficiency. The *trp* or λP_L promoters with the autorepressor control system have been found to yield high gene expression efficiency. Some model simulation studies of the effect of rDNA design on the transcription efficiency indicate that the best strategy is cloning the repressor gene with the target gene on the plasmid and putting the repressor formation under the autorepression control system.

2.1.2 Single-Cell Model

It is convenient to use the specific growth rate as a lumped parameter representing the overall metabolic activities of the whole cell system, although other representations can also be used. Some attempts have been made to develop a cellular model that relates the specific growth rate to the plasmid DNA and other biosynthetic activities of recombinant organisms (LEE and BAILEY, 1984e). Model simulation of λdv replication shows that the plasmid DNA content decreases with an increase in

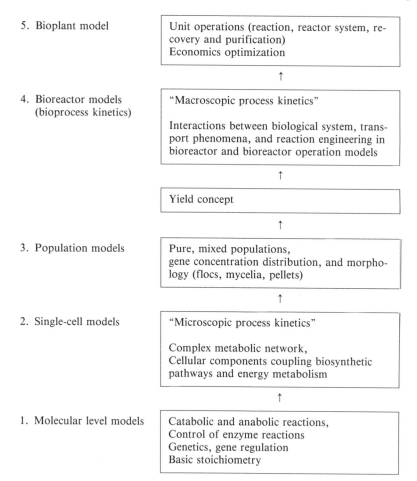

5. Bioplant model — Unit operations (reaction, reactor system, recovery and purification) / Economics optimization

↑

4. Bioreactor models (bioprocess kinetics) — "Macroscopic process kinetics" / Interactions between biological system, transport phenomena, and reaction engineering in bioreactor and bioreactor operation models

↑

Yield concept

↑

3. Population models — Pure, mixed populations, gene concentration distribution, and morphology (flocs, mycelia, pellets)

↑

2. Single-cell models — "Microscopic process kinetics" / Complex metabolic network, Cellular components coupling biosynthetic pathways and energy metabolism

↑

1. Molecular level models — Catabolic and anabolic reactions, Control of enzyme reactions, Genetics, gene regulation, Basic stoichiometry

Fig. 2. An overview of research problem areas relevant to bioprocess modelling (BAILEY et al., 1983; MOSER, 1985).

the specific growth rate in accordance with the experimental results obtained with plasmids R1, pBR322, and ColE1 (ENGBERG and NORDSTROM, 1975; STUEBER and BUJARD, 1982; SIEGEL and RYU, 1985) (see Fig. 3).

2.1.3 Population Model and Bioreactor Model

The recombinant fermentation bioreactor system contains both plasmid-harboring and plasmid-free cells, and mixed cell population dynamics must be considered in the design and analysis of recombinant fermentation processes. From the model simulation there appears to be an optimal gene concentration that corresponds to maximum productivity (AIBA et al., 1982; LEE et al., 1985). The optimal gene concentration may change according to transcription and translation efficiency, which depend on the promoter strength, ribosome binding site, and the specific growth rate. Although the rate of gene product formation is related to gene concentration, too high a gene concentration can put a severe metabolic strain on the primary metabolism of the host cells.

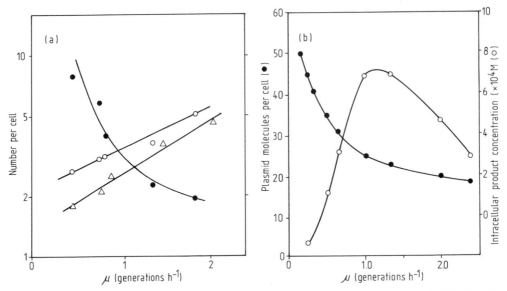

Fig. 3. Effect of the growth rate μ on the plasmid copy number. (a) Experimental data for R1 plasmid; ($\bullet$) plasmid, ($\circ$, $\triangle$) chromosome (ENGBERG and NORDSTROM, 1975). (b) Simulation results of λdv model (BAILEY et al., 1983; LEE and BAILEY, 1984c).

2.2 Kinetics of Gene Product Formation

Many recombinant fermentation processes have been developed to make gene products, and the kinetic expressions can be formulated for transcription and translation. These expressions include genetic and microbiological process design parameters and can be used to evaluate the relationships among these parameters through modelling and simulation.

The equations for the rate of biosynthesis, the rate of degradation, and the rate of dilution by cell growth for the mRNA and the product protein are:

$$\frac{d\hat{m}}{dt} = k_m^0 \eta \hat{G}_p - k_{-m}\hat{m} - \mu\hat{m} \tag{1}$$

$$\frac{d\hat{p}}{dt} = k_p^0 \zeta \hat{m} - k_{-p}\hat{p} - \mu\hat{p} \tag{2}$$

We may obtain a quasi-steady state solution for $\hat{m}$ from Eq. (1),

$$\hat{m} = \frac{k_m^0 \eta \hat{G}_p}{k_{-m}+\mu} \tag{3}$$

and solving Eqs. (1) and (2),

$$\frac{d\hat{p}}{dt} = \hat{G}_p \eta \zeta \left(\frac{k_m^0 k_p^0}{k_{-m}+\mu}\right) - k_{-p}\hat{p} - \mu\hat{p} \tag{4}$$

Now the intracellular parameters can be replaced by the easily measurable bulk parameters:

$$p = \hat{p}X^+/\rho_b \tag{5}$$

$$\frac{dp}{dt} = \frac{1}{\rho_b}\left(X^+\frac{d\hat{p}}{dt} + \hat{p}\frac{dX^+}{dt}\right)$$

$$= \frac{1}{\rho_b}\varepsilon \hat{G}_p f(\mu)X^+ - k_{-p}p \tag{6}$$

where

$$\varepsilon = \eta\zeta \tag{7}$$

$$f(\mu) = \frac{k_m^0 K_p^0}{k_{-m}+\mu} \cong k_0(\mu^+ + b) \tag{8}$$

Assuming that k_{-p} is negligible,

$$\frac{dp}{dt} = \frac{1}{\rho_b} \varepsilon \hat{G}_p f(\mu) X^+ = q_p X^+ \tag{9}$$

$$q_p = \frac{1}{\rho_b} \varepsilon \hat{G}_p f(\mu)$$

$$= \frac{1}{\rho_b} k_0 \varepsilon \hat{G}_p (\mu^+ + b)$$

$$q_p = k_0 \varepsilon G_p (\mu^+ + b) \tag{10}$$

Now the product formation kinetics may be expressed as:

$$\frac{dp}{dt} = k_0 \varepsilon G_p \frac{dX^+}{dt} + k_0 \varepsilon G_p b X^+ \tag{11}$$

$$= A \frac{dX^+}{dt} + B X^+ \tag{12}$$

$$A = k_0 \varepsilon G_p \text{ and } B = A b \tag{13}$$

Eq. (12) is in the form of a Leudeking–Piret equation, and the coefficients A and B show biological significance far more meaningfully than do the empirical constants proposed by LEUDEKING and PIRET (1959).

3 Determination of Genetic Parameters

In recombinant fermentation systems we have a heterogeneous population. Productive and non-productive populations coexist. Those cells harboring plasmids (X^+, plasmid-harboring cells) are productive, and those without plasmids (X^-, plasmid-free cells) are non-productive, since biosynthesis and its regulation depend on the genetic information contained in the recombinant plasmid and the host–vector interaction. Thus, basic understanding of those genetic parameters and their effects on the dynamics of a heterogeneous recombinant population is essential to the economic processing of recombinant fermentation systems.

Determination of key genetic parameters and assessment of their effects on productivity of genetically engineered recombinant organisms are also very important to the design, control, and optimization of large-scale recombinant fermentation processes. These key genetic parameters include: the relative plasmid loss rate ($\theta = \Theta/\mu^+$), which is the ratio of the specific rate of generation of plasmid-free cells to the specific growth rate of plasmid-harboring cells; the specific plasmid loss rate, Θ, which is equal to $\mu^+ \theta$; the gene expression efficiency (ε), a lumped parameter representing the combined efficiency of promoter strength, transcription, and translation; and the growth ratio ($\alpha = \mu^+/\mu^-$), the ratio of specific growth rate of plasmid-harboring cells, μ^+, to that of plasmid-free cells, μ^-.

The productivity of recombinant organisms depends not only on these genetic parameters and other genetic characteristics of recombinants but also on the environmental and operating conditions in a bioreactor system, such as medium, temperature, and dilution rate. An overview of the relationships between the important kinetic parameters affecting the productivity of recombinants is shown in Tab. 1. Thus, it is important to assess the effect of operating conditions on these genetic parameters and, in turn, the effect of the genetic parameters on the productivity of recombinant organism-

Tab. 1. An Overview of the Relationships between the Important Kinetic and Genetic Parameters that Affect the Productivity of Recombinant Organisms

---

Specific productivity of recombinants (q_p)
Gene product concentration (P)

$\uparrow$

Cellular level: heterogeneous population, plasmid-harboring cell (X^+), plasmid-free cell (X), $\Phi = X^+/(X^+ + X^-)$, specific growth rates (μ^+ and μ^-), Growth ratio ($\alpha = \mu^+/\mu^-$)
Gene level: Plasmid concentration (G_p), Specific plasmid loss rate (Θ), Gene expression efficiency (ε), Gene expression kinetics, etc.

$\uparrow$

Environmental and operational parameters: temperature (T), dilution rate (D_1 and D_2), substrate concentration (S_i), etc.

---

isms when recombinant fermentation processes are to be designed and optimized.

From the point of view of bioprocess engineering, it will be extremely useful to: (1) develop a methodology by which those genetic parameters of recombinants can be accurately determined, (2) assess how the parameters are affected by the operating conditions such as temperature and dilution rate, (3) assess the effect of the genetic parameters on the productivity of recombinants in terms of gene expression efficiency, and (4) provide a rational strategy for design, scale-up, control, and optimization of large-scale recombinant fermentation processes.

3.1 Development of the Method – Kinetic Model

A number of investigators have attempted to analyze the population dynamics of recombinant fermentation systems with only limited success (AIBA et al., 1982; LEE et al., 1988; LAUFFENBURGER, 1985). From the point of view of bioprocess engineering, it is very important to determine those kinetic parameters (such as the ones shown in Tab. 1 including θ and α) that are related to the plasmid DNA concentration (G_p) and plasmid-harboring cell fraction (Φ), and also to evaluate their effects on productivity and performance of recombinant fermentation systems. However, due to the difficulties involved in the measurement of these important kinetic parameters, there exist no such data bases. Thus, the method of determining genetic parameters of key importance to the productivity of recombinant fermentation systems will be developed from both a theoretical analysis and an experimental evaluation using a continuous culture system.

A kinetic model for a two-stage continuous culture system and the model for heterogeneous cell population dynamics of a recombinant fermentation system have been derived and reported by LEE et al. (1988). A genetically structured kinetic model which can predict the genetic parameters for a given "host cell/vector" system is derived here from the theoretical analysis of a recombinant fermentation system.

For a chemostat culture system with a heterogeneous recombinant cell population, one can obtain the following mass balance equations:

$$\frac{dX^+}{dt} = -DX^+ + \mu^+ (1-\theta)X^+ \tag{14}$$

$$\frac{dX^-}{dt} = -DX^- + \mu^+ \theta X^+ + \mu^- X^- \tag{15}$$

By introducing Ω as the ratio of plasmid-free to plasmid-harboring cell concentrations at a given time,

$$\Omega = X^-/X^+ \tag{16}$$

the change in Ω with fermentation time can be obtained:

$$\frac{d\Omega}{dt} = \frac{1}{X^+} \left(\frac{dX^-}{dt} - \frac{X^-}{X^+} \frac{dX^+}{dt} \right) \tag{17}$$

Substitution of Eqs. (14) and (15) into (17) gives:

$$\frac{d\Omega}{dt} = \{(\mu^- - \mu^+) + \mu^+ \theta\}\Omega + \mu^+ \theta \tag{18}$$

For simplification, the following variables are introduced:

$$\Delta = \mu^- - \mu^+ \text{ and } \Theta = \mu^+ \theta \tag{19}$$

where Δ is the specific growth rate difference and Θ the specific plasmid loss rate. By substituting Δ and Θ, Eq. (18) is simplified to:

$$\frac{d\Omega}{dt} = (\Delta + \Theta)\Omega + \Theta \tag{20}$$

In general, Δ and Θ are complex functions of genetic characteristics, cell physiology, and environmental conditions. However, under the apparent steady-state conditions (i.e., when the variations in the total cell concentration and limiting substrate concentration with time are negligible) the values of Δ and Θ may be assumed to be nearly constant. Using the initial conditions

$$\Omega = \Omega_0 \text{ at } t=0 \tag{21}$$

the solution for Eq. (20) can be obtained:

$$\Omega = \left\{\Omega_0 + \frac{\Theta}{\Delta + \Theta}\right\} \exp\{(\Delta + \Theta)t\} - \frac{\Theta}{\Delta + \Theta} \tag{22}$$

Rearrangement of Eq. (22) gives

$$\ln(\Omega + A) = Bt + \ln(\Omega_0 + A) \tag{23}$$

where A and B represent

$$A = \frac{\Theta}{\Delta + \Theta} \text{ and } B = \Delta + \Theta \tag{24}$$

Once A and B are determined experimentally, the genetic parameters, Δ and Θ, can be calculated from the following relationships:

$$\Theta = AB \tag{25}$$

$$\Delta = B - AB \tag{26}$$

The method described so far needs one numerical value of specific growth rate, either μ^+ or μ^-, to estimate θ (see Eq. (19)). Since the cell population is continuously changing in this heterogeneous population, and the limiting substrate concentration varies with the dilution rate, it is not easy to measure μ^+ or μ^- accurately. Thus, it is possible to estimate the value of θ only when μ^+ or μ^- are almost the same.

An alternative approach is to find one more equation containing two variables among the three unknowns, provided that one of these parameters is easily measurable. This can be done by incorporating a new parameter, α, the growth ratio ($\alpha = \mu^+/\mu^-$), and measuring it from a new correlation.

When α is introduced, Eq. (18) becomes:

$$\frac{d\Omega}{dt} = \mu^- \{(1 - \alpha + \alpha\theta)\Omega + \alpha\theta\} \tag{27}$$

The material balance for the total cell population can be obtained by combining Eqs. (14) and (15):

$$\frac{dX_T}{dt} = -DX_T + \mu^{app} X_T \tag{28}$$

where

$$X_T = X^+ + X^- \text{ and } \mu^{app} = \frac{\mu^+ X^+ + \mu^- X^-}{(X^+ + X)} \tag{29}$$

At apparent steady state with respect to the total cell population, Eq. (28) gives:

$$D = \mu^{app} \tag{30}$$

The expression for the apparent specific growth rate, μ^{app}, can be further modified to contain Ω and α:

$$\mu^{app} = \mu^- \left\{\frac{(\alpha + \Omega)}{(1 + \Omega)}\right\} \tag{31}$$

From Eqs. (27), (30), and (31), we find that:

$$\frac{d\Omega}{dt} = D\left\{\frac{(1 + \Omega)}{(\alpha + \Omega)}\right\}\{(1 - \alpha + \alpha\theta)\Omega + \alpha\theta\} \tag{32}$$

Eq. (32) can be solved by using the method of separation of variables with an appropriate initial condition, provided that α and θ are practically constant during a certain period of cultivation under well-defined experimental conditions. The solution of Eq. (32) with the initial condition of $\Omega = \Omega_0$ at $t = 0$ (after the culture system reaches an apparent steady state) can be expressed as:

$$\ln(\Omega + A') = B't' + \ln(\Omega_0 + A') \tag{33}$$

where

$$A' = \frac{\alpha\theta}{1 - \alpha + \alpha\theta} \tag{34}$$

$$B' = \frac{1 - \alpha + \alpha\theta}{\alpha - \alpha\theta} \tag{35}$$

$$t' = Dt - \ln\left(\frac{1 + \Omega}{1 + \Omega_0}\right) \tag{36}$$

From Eqs. (33) through (36), the values of α and θ can be estimated, once the values of A' and B' are determined experimentally.

$$\alpha = \frac{(1 + A'B')}{(1 + B')} \text{ and } \theta = \frac{A'B'}{(1 + A'B')} \tag{37}$$

Tab. 2. Comparison of Parameters Found in Eq. (38) with the Corresponding Parameters Found in Eqs. (23) and (33)

Eq. (38)	Eq. (23)	Corresponding Parameters	Eq. (33)
σ	$B \; (=\Delta+\Theta)$		$B'\left(=\dfrac{1-\alpha+\alpha\theta}{\alpha-\alpha\theta}\right)$
ξ	$A\left(=\dfrac{\Theta}{\Delta+\Theta}\right)$		$A'\left(=\dfrac{\alpha\theta}{1-\alpha+\alpha\theta}\right)$
τ	t		$t'\left\{=Dt-\ln\left(\dfrac{1+\Omega}{1+\Omega_0}\right)\right\}$

Growth parameter: $\Delta = B - AB$		$B'\left(=\dfrac{1-\alpha+\alpha\theta)}{\alpha-\alpha\theta}\right)$
Instability parameter: $\Theta = AB$		$\theta = \dfrac{A'B'}{1+A'B'}$

Eq. (23) $\ln(\Omega+A) = Bt + \ln(\Omega_0+A)$
Eq. (33) $\ln(\Omega+A') = B't' + \ln(\Omega_0+A')$
Eq. (38) $\ln(\Omega+\xi) = \sigma\tau + \ln(\Omega_0+\xi)$ (as a generalized equation)

A comparison of Eqs. (23) and (33) reveals that the form of these equations is identical and can be given in the following generalized form:

$$\ln(\Omega+\xi) = \sigma\tau + \ln(\Omega_0+\xi) \tag{38}$$

where $\xi = A$ or A', $\sigma = B$ or B', and $\tau = t$ or t'.

In Tab. 2, parameters found in Eqs. (23) and (33) and their counterpart parameters σ, ξ, and τ in Eq. (38) are summarized for convenient comparison. From the comparison of parameters obtained in this theoretical analysis it is found that:

$$A = A' \text{ and } B = \{\mu^+(1-\theta)\}\,B' \tag{39}$$

When $\Omega \gg \xi$, which is a condition that usually prevails during the prolonged cultivation time, the generalized equation, Eq. (38), becomes:

$$\ln\Omega = \sigma\tau + \ln(\xi+\Omega_0) \tag{40}$$

Thus, the $\ln\Omega$ vs. τ plot should yield a straight line, and the values of σ and ξ can be determined from the slope and intercept, respectively. We can then estimate the values of Δ and Θ

Tab. 3. The Kinetic and Genetic Parameters that are Important for Recombinant Fermentation as Determined According to the Method Described in Sect. 3.1

Dilution Rate (h^{-1})	Kinetic and Genetic Parameters of Recombinants				
	σ	ξ	Δ	α	θ
0.2	0.029	0.59	0.012	0.94	0.086
0.5	0.017	0.37	0.011	0.98	0.013
0.7	0.015	0.63	0.0056	0.99	0.014

from Eqs. (23) ~ (26), or the values of α and θ from Eqs. (26) ~ (37) by using the functional relationships among these parameters as shown in Tab. 3.

It is also possible to simplify Eq. (38) when $\Omega \ll \xi$. This condition usually prevails at the initial period of cultivation, when the plasmid-harboring cell population is very large or predominant. If we take a Taylor series expansion of Eq. (38), we obtain

$$\frac{1}{\Omega} = \left\{\frac{1}{\sigma(\xi+\Omega_0/2)}\left(1-\frac{\Omega_0}{\Omega}\right)\frac{1}{\tau}\right\} - \frac{1}{2\xi} \tag{41}$$

The plot of $\dfrac{1}{\Omega}$ vs. $\left(1-\dfrac{\Omega_0}{\Omega}\right)\dfrac{1}{\tau}$ should yield a straight-line relationship, and the values of σ and ξ can be determined from the slope and intercept of the plot. Here we find the following relationships:

$$\text{slope} = \frac{1}{\sigma(\xi+\Omega_0/2)} \quad \text{and}$$

$$\text{intercept} = -\frac{1}{2\xi} \qquad (42)$$

The σ value can be obtained from the slope of the $\ln\Omega$ vs. τ plot during the time when $\Omega\gg\xi$, and the ξ value is obtained from the intercept of the Ω vs. τ plot or the $(1/\Omega)$ vs. $(1-\Omega/\Omega_0)(1/\tau)$ plot during the time when $\Omega\ll\xi$. Since the ξ value is measured in another experiment, the seed culture with a high value of Ω can be inoculated, and the measurement of the Ω value at a high range of Ω can be performed without a long wait. When $\Omega_0\gg\xi$, Eq. (38) can be simplified to give

$$\ln\Omega = \sigma\tau + \ln\Omega_0 \qquad (43)$$

when the Ω_0 value is very large and X^+ is negligible compared to X^-, and Eq. (43) reduces to:

$$\ln X^+ = -\sigma\tau + \ln X_0^+ \qquad (44)$$

Then the σ value can also be determined from the slope of the $\ln X^+$ vs. τ plot when $\Omega_0\gg\xi$.

The methods described above can be very powerful tools, convenient for use in the accurate estimation of these genetic parameters.

3.2 Application of the Method

Based on the theoretical analysis illustrated in the previous section, we have conducted some experiments in order to ascertain the usefulness and validity of the methods of determining the genetic and kinetic parameters of recombinants that are critically important to the design, scale-up, and optimization of recombinant fermentation processes.

The most accurate method described above can be illustrated by using Figs. 4 and 5 and Eqs. (38), (40), and (41). The values of the kinetic and genetic parameters (σ, ξ, Δ, α, and Θ) are obtained for the different periods of cultivation time that correspond to the high and low values of Ω. When $\Omega\gg\xi$, which corresponds to a condition that prevails during the prolonged cultivation time, the value of σ can be determined from the slope of Fig. 4 using the relationship shown in Eq. (40). When

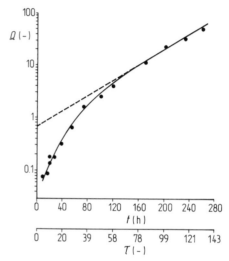

Fig. 4. The ratio of plasmid-free to plasmid-haboring cells $\Omega(=X^-/X^+)$ plotted against real time (t) and dimensionless time (τ). Experimental data at $T=35\,^\circ\mathrm{C}$ and $D=0.5\,\mathrm{h}^{-1}$.

$\Omega\ll\xi$, which corresponds to a condition that prevails during the early period of cultivation time and when a plasmid-harboring cell population is predominant or very large, the value of ξ can be determined from the slope of Fig. 5 using the relationship shown in Eqs. (41) and (42). The values of Δ, Θ, and α can then be estimated from the values of σ and ξ determined experimentally.

The results of an estimate of the kinetic parameters for the model recombinant used (*Escherichia coli* K12 ΔH1 $\Delta trp/$ pPLc23*trp*A1) under the varying conditions of dilution rate are presented in Tab. 3. The values of growth ratio (α) and relative plasmid loss rate (Θ) range from 0.96 to 0.99 h^{-1} and from 0.086 to 0.014, respectively.

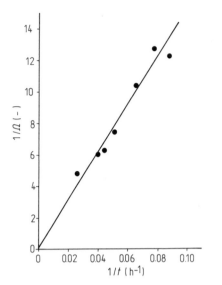

Fig. 5. The plot of $1/\Omega$ vs. $1/t$. The experimental data in the early time period are plotted. When $\Omega_0 = 0$ in the early time period, Eq. (41) becomes $1/\Omega = (1/\sigma\xi)(1/\tau) - 1/2\xi$.

Tab. 4. Microbiological Process Design Parameters

1. Gene Dosage (1, 2, 3, 4, 5,6)
   - Plasmid copy number
   - Regulation of replication
2. Transcription Efficiency (2, 7, 8)
   - Promoter strength
   - Regulation of transcription
3. Translation Efficiency (9, 10, 11, 12, 13, 14)
   - Nucleotide sequence of ribosome binding site (Shine–Dalgarno sequence)
   - Distance between Shine–Dalgarno sequence and ATG or GTG initiation codon
   - Secondary structure of mRNA
   - Codon usage
4. Stability of recombinant DNA (15, 16, 17, 18)
   - Stability of cloned gene
   - Stability of plasmid
5. Stability of mRNA (19, 20)
   - Structure of mRNA
   - Nuclease
6. Stability of protein (21, 22, 23, 24)
   - Structure of protein or gene product
   - Protease
7. Host cell
   - Host/vector interaction
   - Metabolic activity of host cell
8. Others (25, 26, 27, 28, 29)
   - Protein secretion efficiency
   - Proper termination between genes

1 AIBA et al. (1982), 2 THOMPSON (1982), 3 UHLIN et al. (1979), 4 YARROTON et al. (1984), 5 SEO and BAILEY (1986), 6 YAMAKAWA et al. (1989), 7 DE BOER et al. (1983), 8 SHIRAKAWA et al. (1984), 9 GOLD et al. (1981), 10 SCHOTTEL et al. (1984), 11 GROSJEAN and FIERS (1982), 12 LANG et al. (1989), 13 SPANJAARD et al. (1989), 14 SHEPARD et al. (1982), 15 BAILEY et al. (1983), 16 STUEBER and BUJARD (1982), 17 SIEGEL and RYU (1985), 18 LEE et al. (1985), 19 NILSSON et al. (1984), 20 PURVIS et al. (1987), 21 FISH and LILLY (1984), 22 TALMADGE and GILBERT (1982), 23 SHEN (1984), 24 PARSELL and SAUER (1989), 25 GUZZO et al. (1990), 26 BREITLING et al. (1989), 27 PALVA et al. (1983), 28 GRAY et al. (1984), 29 SINGH et al. (1984)

4 Optimization of Recombinant Fermentation (Selected Examples)

Both the microbiological process parameters and the fermentation process parameters affect productivity and must be carefully studied and evaluated when a recombinant fermentation process is to be scaled up and optimized. The selection list of important microbiological process design parameters in optimizing recombinant fermentation processes is presented in Tab. 4. The important fermentation process parameters include the medium design and optimization, the bioreactor design and operation, and the state and control variables involved in the recombinant fermentation processes. However, very little is known about scaling up and optimizing these processes, and a great deal of work needs to be done in this area in order to commercialize many more new gene products in the future. In this section we will discuss how we can improve the recombinant fermentation processes through optimization of dilution rate and temperature control in a two-stage continuous culture model system.

4.1 Optimization of Dilution Rate in a Two-Stage Continuous Culture System

A two-stage continuous culture system has been used to study recombinant fermentation processes (SIEGEL and RYU, 1985; LEE et al., 1988; PARK and RYU, 1990). The system shown in Fig. 6 is composed of the growth stage and the production stage. The separation of the growth and production stages is a design strategy to decrease plasmid instability and in-

crease productivity. Fig. 7 clearly shows that a two-stage continuous culture system has advantages over a single-stage continuous culture system. A brief discussion of how the recombinant fermentation processes can be optimized by controlling the dilution rate in a two-stage continuous culture system is presented below.

The control of the dilution rate in a continuous culture system means the control of the apparent specific growth rate of a heterogeneous population of plasmid-free and plasmid-harboring cells.

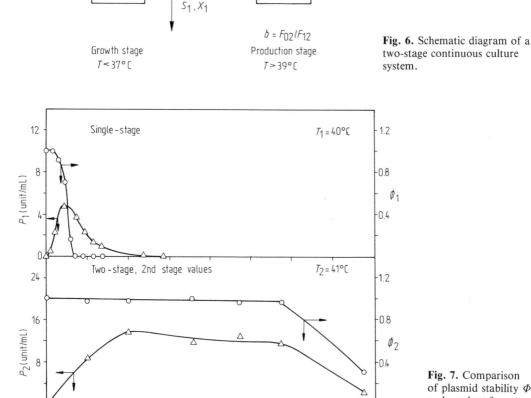

Fig. 6. Schematic diagram of a two-stage continuous culture system.

Fig. 7. Comparison of plasmid stability Φ and product formation P (SIEGEL and RYU, 1985).

For the growth stage:

$$\frac{dX_{1T}}{dt} = (\mu_1^{app} - D_1)X_{1T} \tag{45}$$

where $\mu_1^{app} = \dfrac{(\mu_1^+ X_1^+ + \mu_1^- X_1^-)}{X_{1T}}$

and at the apparent steady state, $\dfrac{dX_{1T}}{dt} = 0$:

$$\mu_1^{app} = D_1 \tag{46}$$

For the production stage:

$$\frac{dX_{2T}}{dt} = D_{12}X_{12} + (\mu_2^{app} - D_2)X_{2T} \tag{47}$$

where $\mu_2^{app} = \dfrac{(\mu_2^+ X_2^+ + \mu_2^- X_2^-)}{X_{2T}}$ and $D_{12} = \dfrac{F_{12}}{V_2}$,

and at the apparent steady state, $\dfrac{dX_{2T}}{dt} = 0$:

$$\mu_2^{app} = D_2 - D_{12}\left(\frac{X_{1T}}{X_{2T}}\right) \tag{48}$$

All the important parameters such as plasmid content, plasmid stability, and gene expression efficiencies are subject to the specific growth rate of host cells. It is therefore important to investigate the optimum dilution rate by studying the effect of the specific growth rate on productivity.

4.1.1 Effect of Specific Growth Rate on Plasmid Content

The plasmid content of the cells, expressed as mg plasmid per g of cells, was measured as a function of the apparent growth rates for the repressed stage (SIEGEL and RYU, 1985). Using a dilution rate of $D_1 = 0.9 \text{ h}^{-1}$, the plasmid contents of the cells for the expressed stage or production stage were measured (PARK et al., 1990). These data are shown in Fig. 8. The plasmid concentration under the repressed condition decreases with the increasing appar-

ent growth rate, while the plasmid concentration under the expressed condition exhibits a maximum.

The steady-state plasmid concentrations for both stages may be described from a plasmid balance as described below.

For the growth stage:

$$D_1 G_{p1} = r_{p1}/\rho_b \tag{49}$$

where r_{p1} is the rate of plasmid synthesis within the cell when there is no gene expression or under the repressed condition.

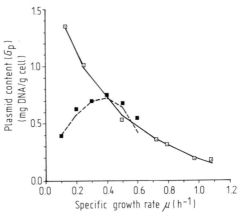

Fig. 8. Effect of specific growth rate on plasmid content. Symbols ⊡ and ■ are experimental data from the repressed stage and the expressed stage, respectively. Lines — and - - - are from Eqs. (52) and (53), respectively (SIEGEL and RYU, 1985; PARK et al., 1990).

For the expressed stage:

$$D_2 G_{p2} - D_{12} G_{p1} = r_{p2}/\rho_b \tag{50}$$

where r_{p2} is the rate of plasmid synthesis under the expressed condition in the production stage. SATYAGAL and AGRAWAL (1989), in their survey of experimental plasmid concentration studies, showed that a possible kinetic expression for the plasmid replication rate may be written as:

$$r_p/\rho_b = \frac{a G_p \mu}{(b + G_p)(c + \mu)} \tag{51}$$

This expression was also found compatible with a rate expression derived from a molecular mechanism for plasmid replication. Using this rate expression for the repressed state and $D_1 = \mu_1$, the plasmid concentration becomes:

$$G_{p1} = \frac{a}{(c + \mu_1)} - b \tag{52}$$

The equation describes a hyperbolic decrease of plasmid content with growth rate. The parameters for the experimental system (PARK, 1988) were evaluated from a least squares fit as $a = 0.98$ mg/g cells/h, $b = 0.52$ mg/g cells, and $c = 0.40$ h. The plasmid concentration data for the second stage for the condition $D_1 = 0.9$, or in terms of the plasmid concentration of the stream coming from the first stage, $G_{p1} = 0.2$ mg plasmid/g cell, can be estimated from Eq. (50) with a rate expression of the form:

$$r_{p2}/\rho_b = \frac{d\,G_{p2}\mu_2}{(e + \mu_2)} - f \tag{53}$$

The rate expression may be considered a simplification of Eq. (51) for the case when $b \gg G_p$ or for low plasmid concentrations. The values of the parameters evaluated from a least squares fit are $d = 1.02$ h, $e = 0.05$ h, and $f = 0.29$ mg/g cells/h.

4.1.2 Effect of Specific Growth Rate on the Gene Expression Rate

Based on the molecular mechanism of transcription and translation, LEE et al. have derived a mathematical model for gene product formation with which the efficiency of gene expression can be evaluated (LEE and BAILEY, 1984 a–e; LEE et al., 1988; RYU et al., 1985). Eq. (54) represents the specific production rate of the gene product (q_p) as a function of the gene concentration (G_p), the efficiency of gene expression (ε), the specific growth rate of plasmid-harboring cells (μ^+), the overall rate constant for gene product biosynthesis (k_0), and another lumped specific rate parameter implicit in contributions of nongrowth-asso-

ciated parameters toward the product biosynthesis (b):

$$q_p/G_p = k_0\varepsilon\,(\mu^+ + b) \tag{54}$$

A two-stage continuous culture system was employed to evaluate these kinetic parameters accurately (SIEGEL and RYU, 1985; LEE et al., 1988). Under the apparent steady state conditions of a two-stage continuous culture system, q_p and μ^+ in Eq. (54) are given:

$$q_p \cong (D_2 p)/X_2^+ \tag{55}$$

$$\mu_2^+ \cong \mu_2^{app} = D_2 - D_{12}\left(\frac{X_{1T}}{X_{2T}}\right) \tag{56}$$

where D_2 and μ_2^{app} represent the dilution rate and apparent specific growth rate in the second stage, and X_{1T} and X_{2T} are the cell concentrations in the first and second stage (Fig. 6). The productivity equation is rearranged in Eq. (57):

$$q_p/G_p = k_0\varepsilon\,(\mu_2^{app} + b) \tag{57}$$

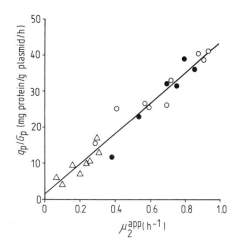

Fig. 9. Relationship between q_p, G_p, and μ_2^{app} (LEE et al., 1988; SEO and BAILEY, 1986).

Based on experimental values the parameters q_p/G_p are correlated with μ_2^{app} as shown in Fig. 9. From a correlation of the type shown in Fig. 9, the values of $k_0\varepsilon$ and b were determined as 41.8 mg protein gene product per mg DNA and 0.036 h^{-1}, respectively. The regression coefficient for this correlation was 0.973, and the plasmid used in the experiment was pPLc23*trp*A1 (LEE et al., 1988; REMAUT et al., 1983). Since the value of b is about 0.036 h^{-1}, one may safely conclude that the specific production rate of the given gene product was strongly growth-associated. The value of $k_0\varepsilon$ estimated above is equivalent to 5.7×10^3 molecules of protein gene product per molecule of DNA.

Under optimal conditions the average value of k_0 for *E. coli* K-12 strain is estimated to be 17.5 mg protein per mg DNA independent of the growth rate (LEE et al., 1988). Assuming that the host *E. coli* K-12 cell used in this experiment is approximately equal to the *E. coli* K-12 strain cited above in protein biosynthetic activity, the value of ε can be estimated to be 2.4. This suggests that the efficiency assessed in gene expression of TrpA protein is approximately twenty times higher than that for the total protein encoded in the host cell chromosomal DNA, since the *trp*A gene which was under the control of the λP_L promoter is about one-eighth the size of pPLc23*trp*A1.

4.1.3 Effect of Specific Growth Rate on Plasmid Stability

The gene-product yield is also a direct function of the plasmid-harboring cell fraction in both stages. Even though the total cell concentration may remain constant, because of plasmid instability the fraction of plasmid harboring cells will continue to decrease with time, and the rate of decrease depends on many factors. The stability of the plasmid-harboring cells is usually characterized by two kinetic parameters related to growth rate and genetic characteristics, namely, the growth ratio α and the plasmid loss rate θ.

A mathematical analysis for the two-stage continuous culture system has been carried out in detail by LEE et al. (1988). According to their study, the fraction of a plasmid-harboring cell population at infinite time, $\Phi_2(\infty)$, becomes:

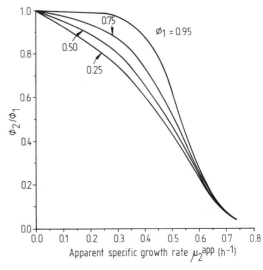

Fig. 10. Effect of μ_2^{app} on $\Phi_2(\infty)/\Phi_1$ at various conditions of Φ_1 ($\alpha_2 = 0.5$, $\theta_2 = 0.05$) (PARK et al., 1990).

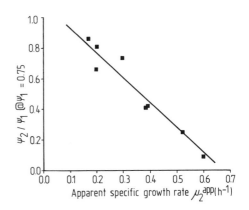

Fig. 11. Effect of μ_2^{app} on plasmid stability ($\Psi_1 = 0.75$, $T_2 = 40\,°C$, $D_2 = 0.7\,h^{-1}$) (PARK et al., 1990).

$$\Phi_2(\infty) = 2\gamma\Phi_1/[(1 - \alpha_2 + \alpha_2\theta_2 + \gamma) + \sqrt{(1 - \alpha_2 + \alpha_2\theta_2 + \gamma)^2 - 4(1 - \alpha_2)\gamma\Phi_1}] \quad (58)$$

where

$$\alpha_2 = \frac{\mu_2^+}{\mu_2^-}, \quad \theta_2 = \frac{\Theta_2}{\mu_2^+}, \quad \text{and } \gamma = \frac{D_2 - \mu_2^{app}}{\mu_2^-}$$

The ratio $\Phi_2(\infty)/\Phi_1$ was plotted against μ_2^{app} in Fig. 10. Here, Φ_1 was varied in the range of 0.25 to 0.95 and α_2 and θ_2 were fixed at 0.5 and 0.05, respectively. The figure shows that the early decrease in $\Phi_2(\infty)/\Phi_1$ values at low μ_2^{app} is steeper as Φ_1 values are lower. The experimental results showing the effect of μ_2^{app} on plasmid stability are given in Fig. 11. Here, Ψ, the fraction of plasmid-harboring cells among the colony-forming cells, is used instead of Φ because some X^+ cells lose their colony-forming ability before X^+ cells lose the ability to produce the cloned-gene product or completely lose their viability (DIPASQUANTANIO et al., 1987; SIEGEL, 1985), and it is very difficult to measure the X^+ concentration accurately (PARK and RYU, 1990). Ψ_2/Ψ_1 values at a constant value of 0.75 are comparable to those shown in Fig. 10.

4.2 Optimization of Temperature Control in a Two-Stage Continuous Culture System

Induction time and inducer concentration for gene expression were shown to be important in optimizing the recombinant fermentation processes (BOTTERMANN et al., 1985). When the λP_L promoter is controlled by the temperature-sensitive repressor (cI$_{857}$), the gene expression is regulated by changing the temperature. Usually a temperature lower than 37 °C is employed for cell growth, and a slightly higher temperature for the expression of the cloned gene product. A two-stage continuous culture system in combination with the temperature-controlled gene switching system has been used to optimize the productivity of the foreign protein (LEE et al., 1988; PARK and RYU, 1990; PARK et al., 1990). This section will deal with the optimization of temperature control for the production of the cloned gene product using kinetic models with parameters evaluated from the experimental data.

4.2.1 Effect of Temperature on the Gene Expression Rate

Transcription starting from the λP_L promoter is induced by inactivation of the temperature-sensitive repressor molecules at temperatures above 38 °C. ACKER et al. (1982) showed that the probability of repression of the λP_R promoter can be quantitatively related to the change of repressor concentration. Likewise, we assume that the transcription efficiency of the λP_L promoter will vary from 0 to 1.0, depending on the degree of repressor inactivation at a constant specific growth rate. It is also assumed that the translation efficiency is constant when the specific growth rate is maintained constant; hence, the gene expression rate is limited at the level of transcription. Using the experimental data of PARK (1988), an adequate kinetic model is established as follows:

$$q_p/G_p = \eta\zeta\{k_p k_m/(k_{-m}+\mu_2^+)\} = \eta(T)K(T) \tag{59}$$

$$\eta(T)=\exp\left(\frac{A}{R\left(\frac{1}{T_{max}}-\frac{1}{T}\right)}\right) \tag{60}$$

$$K(T)=\beta T\exp\frac{\left(\frac{-E}{RT}\right)}{\left(1+\exp\frac{\Delta S}{R}\exp\frac{-\Delta H}{RT}\right)} \tag{61}$$

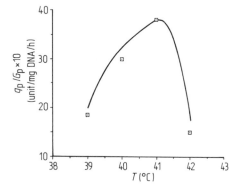

Fig. 12. Specific gene expression rate as a function of temperature. Symbols are experimental data (PARK, 1988).

where

$T_{max} = 314$ K
$A/R = 43.3 \times 10^3$ K
$\beta = 1.148 \times 10^{30}$ units protein/mg plasmid/ h/K
$E/R = 20.0 \times 10^3$ K
$\Delta S/R = 430$
$\Delta H/R = 135 \times 10^3$ K

The comparison of experimental data and calculated values from the kinetic equation are shown in Fig. 12.

4.2.2 Effect of Temperature on the Plasmid-Harboring Cell Fraction

It is convenient to calculate the plasmid-harboring cell concentration from the total cell concentration and the plasmid-harboring cell fraction:

$$X_2^+ = X_{2T} \Phi_2 \tag{62}$$

The product concentration can be expressed as:

$$p_m = p/X_{2T} \tag{63}$$

The decrease of the plasmid-harboring cell fraction with the cell generation time due to segregational instability has been characterized in terms of a relative segregation rate parameter θ and the growth rate ratio α. In terms of these parameters the transient behavior of the plasmid-harboring cell fraction for the growth and production stages is given by:

$$\frac{d\Phi_1}{dt} = (1 - \alpha_1) \Phi_1^2 \mu_1^- - (1 - \alpha_1 + \alpha_1 \theta_1) \Phi_1 \mu_1^- \tag{64}$$

$$\frac{d\Phi_2}{dt} = (1 - \alpha_2) \Phi_2^2 \mu_2^- - (1 - \alpha_2 + \alpha_2 \theta_2) \cdot$$
$$\cdot \Phi_2 \mu_2^- + \Gamma(\Phi_1 - \Phi_2) \tag{65}$$

It was shown that the fraction of plasmid-harboring cells in the repressed stage is not temperature dependent, while in the expressed state the plasmid-harboring cell fraction decreases with increase in temperature (PARK,

1988). LEE et al. (1985) have shown that in a heterogeneous population of plasmid-harboring and plasmid-free cells the growth rate ratio is adversely affected by an increase of plasmid content and product formation:

$$\alpha = \mu^+ / \mu^- = g(G_p) h(p) =$$

$$= \left(1 - \frac{G_p}{G_{pmax}}\right)^n \left(1 - \frac{p}{p_{max}}\right)^m \tag{66}$$

The experimental data for plasmid-harboring cell fractions at different temperatures were adequately modelled by Eqs. (64) and (65) using constant values of $\theta_1 = 0.001$ and $\theta_2 = 0.01$ and growth ratios calculated from Eq. (66) with $n = 0.5$, $m = 0.7$, $G_{pmax} = 2$, and $p_{max} = 100$ (LEE et al., 1988; PARK, 1988). Steady-state

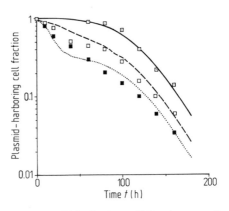

Fig. 13. Plasmid-harboring cell fraction as a function of temperature and time. Symbols are experimental data ($\boxdot$ 35 °C, $\square$ 39 °C, $\blacksquare$ 41 °C) (PARK, 1988).

product concentrations at various temperatures were taken from PARK (1988). The comparison of experimental data and values calculated from the model equations is shown in Fig. 13.

5 Conclusion

Bioprocess kinetics and modelling discussed here show an example of how to develop and optimize a rational strategy for recombinant fermentation processes. Basic understanding of the relationship between the physiology of host cell growth and overproduction of protein is of primary importance to the improvement of recombinant fermentation processes. Based upon physiological and molecular knowledge, kinetic studies and modelling can be established by assessing genetic and environmental parameters affecting the productivity of a cloned gene product. Application of the bioprocess kinetics and modelling approaches presented in this chapter to practical recombinant fermentation process design and development leads to significant progress in scaling-up recombinant fermentation processes and commercializing important gene products.

6 References

ACKER, G. K., JOHNSON, A. D., SHEA, M. A. (1982), Quantitative model for gene regulation by λ phage repressor, *Proc. Natl. Acad. Sci. USA* **79**, 1129–1133.

AIBA, S., TSUNEKAWA, H., IMANAKA, T. (1982), New approach to tryptophan production by *Escherichia coli*: Genetic manipulation of composite plasmids *in vitro*, *Appl. Environ. Microbiol.* **43**, 289–297.

BAILEY, J. E., HJORTSO, M., LEE, S. B., SRIENC, F. (1983), Kinetics of product formation and plasmid segregation in recombinant microbial populations, *Ann. NY Acad. Sci.* **413**, 71–87.

BOTTERMANN, J. H., DE BUYSER, D. R., SPRIET, J. A., ZABEAU, M., VANSTEENKISTE, G. C. (1985), Fermentation and recovery of the EcoRI restriction enzyme with a genetically modified *Escherichia coli* strain, *Biotechnol. Bioeng.* **27**, 1320–1327.

BREITLING, R., GERLACH, D., HARTMANN, M., BEHNKE, D. (1989), Secretory expression in *Escherichia coli* and *Bacillus subtilis* of human interferon alpha genes directed by staphylokinase signals, *Mol. Gen. Genet.* **217**, 384–391.

COHEN, S. N., CHANG, A. C. Y., BOYER, H. W., HELLING, R. B. (1974), Construction of biologi-cally functional bacterial plasmids *in vitro*, *Proc. Natl. Acad. Sci. USA* **70**, 3240–3244.

DE BOER, H. A., COMSTOCK, L. J., VASSER, M. (1983), The *tac* promoter: A functional hybrid derived from the *trp* and *lac* promoters, *Proc. Natl. Acad. Sci. USA* **80**, 21–25.

DI PASQUANTANIO, V. M., BETENBAUGH, M. J., DHURJATI, P. (1987), Improvement of product yields by temperature-shifting of *Escherichia coli* cultures containing plasmid pOU140, *Biotechnol. Bioeng.* **29**, 513–519.

ENGBERG, B., NORDSTROM, K. (1975), Replication of R-factor R1 in *Escherichia coli* K-12 at different growth rates, *J. Bacteriol.* **123**, 179–186.

FISH, N. M., LILLY, M. D. (1984), The Interactions between fermentation and protein recovery. *Bio/Technology* **2**, 623–627.

GOLD, L., PRIBNOW, D., SCHNEIDER, T., SHINEDLING, S., SINGER, B. S., STORMO, G. (1981), Translational initiation in prokaryotes, *Annu. Rev. Microbiol.* **35**, 365–403.

GRAY, G. L., MCKEOWN, K. A., JONES, A. J. S., SEEBURG, P. H., HEYNEKER, H. L. (1984), *Pseudomonas aeruginosa* secretes and correctly processes human growth hormone, *Bio/Technology* **2**, 161–165.

GROSJEAN, H., FIERS, W. (1982), Preferential codon usage in prokaryotic genes: The optimal codon-anticodon interaction energy and the selective codon usage in efficient expressed genes, *Gene* **18**, 199–209.

GUZZO, J., MURGIER, M., FILLOUX, A., LAZDUNSKI, A. (1990), Cloning of the *Pseudomonas aeruginosa* alkaline protease gene and secretion of the protease into the medium by *Escherichia coli*, *J. Bacteriol.* **172**, 942–948.

LANG, V., GUALERZI, C., MCCARTHY, J. E. (1989), Ribosomal affinity and translational initiation in *Escherichia coli*. *In vitro* investigations using translational initiation regions of differing efficiencies from the *atp* operon, *J. Mol. Biol.* **210**, 659–663.

LAUFFENBURGER, D. A. (1985), Stability of colicin plasmids in continuous culture: Mathematical model and analysis, *Biotechnol. Prog.* **1**, 53–59.

LEE, S. B., BAILEY, J. E. (1984a), Genetically structured models for *lac* promoter-operator function in the *Escherichia coli* chromosome and in multicopy plasmids: *lac* operator function, *Biotechnol. Bioeng.* **26**, 1372–1382.

LEE, S. B., BAILEY, J. E. (1984b), Genetically structured models for *lac* promoter-operator function in the chromosome and in multicopy plasmids: *lac* promoter function, *Biotechnol. Bioeng.* **26**, 1383–1389.

LEE, S. B., BAILEY, J. E. (1984c), A mathematical

model for λdv plasmid replication: Analysis of wild-type plasmid, *Plasmid* **11**, 151–165.

LEE, S. B., BAILEY, J. E. (1984d), A mathematical model for λdv plasmid replication: Analysis of copy number mutants, *Plasmid* **11**, 166–177.

LEE, S. B., BAILEY, J. E. (1984e) Analysis of growth rate effects on productivity of recombinant *Escherichia coli* populations using molecular mechanism models, *Biotechnol Bioeng.* **26**, 66–73.

LEE, S. B., SERESSIOTIS, A., BAILEY, J. E. (1985), A kinetic model for product formation in unstable recombinant populations, *Biotechnol. Bioeng.* **27**, 1699–1709.

LEE, S. B., RYU, D. D. Y., SIEGEL, R., PARK, S. H. (1988), Performance of recombinant fermentation and evaluation of gene expression efficiency for gene product in two-stage continuous culture system, *Biotechnol. Bioeng.* **31**, 805–820.

LEUDEKING, R., PIRET, E. L. (1959), A kinetic study of the lactic acid fermentation. Batch process at controlled pH, *J. Biochem. Microbiol. Technol. Eng.* **1**, 393–412.

MOSER, A. (1985), General strategy in bioprocessing, in: *Biotechnology* (REHM, H.-J., REED, G., Eds.), Vol. 2, pp. 173–197, Weinheim–Deerfield Beach/Florida–Basel: VCH.

NILSSON, G., BELASCO, J. G., COHEN, S. N., VON GABAIN, A. (1984), Growth-rate dependent regulation of mRNA stability in *Escherichia coli, Nature* **312**, 75–77.

OFFICE OF TECHNOLOGY ASSESSMENT (1982), *3. Genetic Technology: A New Frontier,* Washington, D. C.

OFFICE OF TECHNOLOGY ASSESSMENT (1984), *4. Commericial Biotechnology: An International Analysis,* Washington, D. C.

PALVA, I., LEHTOVAARA, P., KAARIAINEN, L., SIBAKOV, M., CANTELL, K., SCHEIN, C. H., KASHIWAGI, K., WEISSMANN, C. (1983), Secretion of interferon by *Bacillus subtilis, Gene* **22**, 229–235.

PARK, S. H. (1988), *Ph. D. Thesis,* University of California at Davis.

PARK, S. H., RYU, D. D. Y. (1990), Effect of operating parameters on specific production rate of a cloned-gene product and performance of recombinant fermentation process, *Biotechnol. Bioeng.* **35**, 287–295.

PARK, S. H., RYU, D. D. Y., KIM, J. Y. (1990), *Biotechnol. Bioeng.,* accepted.

PARSELL, D. A., SAUER, R. T. (1989), The structural stability of a protein is an important determinant of its proteolytic susceptibility in *Escherichia coli, J. Biol. Chem.* **264**, 7590–7595.

PURVIS, I. J., BETTANY, A. J., LOUGHLIN, L., BROWN, A. J. (1987), The effects of alterations within the 3' untranslated region of the pyruvate kinase messenger RNA upon its stability and translation in *Saccharomyces cerevisiae, Nucleic Acids Res.* **15**, 7951–7962.

REMAUT, E., STANSSENS, P., FIERS, W. (1983), Inducible high level synthesis of mature human fibroblast interferon in *Escherichia coli, Nucleic Acids Res.* **11**, 4677–4688.

RYU, D. D. Y., SIEGEL, R., LEE, S.B. (1985), Kinetics of gene expression in recombinant *E. coli* K-12 ΔH1 Δ*trp*/pPLc23*trpA1,* Paper presented at the Annual Meeting of Am. Inst. Chem. Eng.

SATYAGAL, V., AGRAWAL, P. (1989), A generalized model of plasmid replication, *Biotechnol. Bioeng.* **33**, 1135–1144.

SCHOTTEL, J. L., SNINSKY, J. J., COHEN, S. N. (1984), Effects of alterations in the translation control region on bacterial gene expression: Use of *cat* gene constructs transcribed from the *lac* promoter as a model system, *Gene* **28**, 177–193.

SHEN, S.-H. (1984), Multiple joined genes prevent product degradation in *Escherichia coli, Proc. Natl. Acad. Sci. USA* **81**, 4627–4631.

SHEPARD, H. M., YELVERTON, E., GOEDDEL, D. V. (1982), Increased synthesis in *E. coli* of fibroblast and leukocyte interferons through alterations in ribosome binding sites, *DNA* **1**, 125–131.

SHIRAKAWA, M., TSURIMOTO, T., MATSUBARA, K. (1984), Plasmid vectors designed for high-efficiency expression controlled by the portable *recA* promoter-operator of *Escherichia coli, Gene* **28**, 127–132.

SEO, J.-H., BAILEY, J. E. (1986), Continuous cultivation of recombinant *Escherichia coli:* existence of an optimum dilution rate for maximum plasmid and gene product concentration, *Biotechnol. Bioeng.* **28**, 1590–1594.

SINGH, A., LUGOVOY, J. M., KOHR, W. J., PERRY, L. J. (1984), Synthesis, secretion and processing of α-factor-interferon fusion proteins in yeast, *Nucleic Acids Res.* **12**, 8927–8938.

SIEGEL, R. (1985), *Ph. D. Thesis,* University of California at Davis.

SIEGEL, R., RYU, D. D. Y. (1985), Kinetic study of instability of recombinant plasmid pPLc23*trpA1* in *E. coli* using two-stage continuous culture system, *Biotechnol. Bioeng.* **27**, 28–33.

SPANJAARD, R. A., VAN DIJK, M. C., TURION, A. J., VAN DUIN, J. (1989), Expression of the rat interferon-alpha 1 gene in *Escherichia coli* controlled by the secondary structure of the translation-initiation region, *Gene* **80**, 345–351.

STUEBER, D., BUJARD, H. (1982), Transcription from efficient promoters can interfere with plasmid replication and diminish expression of plasmid specified genes, *EMBO J.* **1**, 1399–1404.

TALMADGE, K., GILBERT, W. (1982), Cellular location affects protein stability in *Escherichia coli*, *Proc. Natl. Acad. Sci. USA* **79**, 1830–1833.

THOMPSON, R. (1982), Plasmid and phage M13 cloning vectors, in: *Genetic Engineering* (WILLIAMSON, R., Ed.) pp. 1–52, New York: Academic Press.

UHLIN, B. E., MOLIN, S., GUSTAFSSON, P., NORDSTROM, K. (1979), Plasmids with temperature-dependent copy number for amplification of cloned genes and their products, *Gene* **6**, 91–106.

WEBBER, D. (1985), Consolidation begins for biotechnology firms, *Chem. Eng. News,* 25–60, November 18.

YAMAKAWA, M., SUGISAKI, K., MORIMOTO, M., TANAKA, M., YAMAMOTO, M. (1989), Effects of gene dosage on the expression of human growth hormone cDNA in *Escherichia coli, Biochim. Biophys. Acta* **1009**, 156–160.

YARRANTON, G. T., WRIGHT, E., ROBINSON, M. K., HUMPREYS, G. O. (1984), Dual-origin plasmid vectors whose origin of replication is controlled by the coliphage lambda promoter P_L, *Gene* **28**, 293–300.

IV. Control and Automation

16 Control of Bioreactor Systems

Henry C. Lim
Kyu-Sung Lee

Irvine, California 92717, U.S.A.

List of Symbols

CER carbon dioxide evolution rate
D dilution rate
DO dissolved oxygen concentration
Dx cellular productivity
F feed flow rate
F_c coolant flow rate
K Kalman gain matrix
K_c proportional gain
K_p process gain
K_u ultimate gain
k specific aging rate for product, kinetic constant
k_d death rate coefficient
m manipulated variable
OUR oxygen uptake rate
P error covariance matrix
P_u ultimate period
p, P product concentration; performance index
Q system error covariance matrix
Q starch addition rate, Eq. (95)
q vector of model parameters, Eq. (157)
R measurement error covariance matrix
R_a aeration rate
RQ respiration quotient
s substrate concentration
T temperature
T_d time delay
v volume
v measurement noise vector
w system noise vector
x state vector
x cell mass concentration
Y yield coefficient
y vector of output data

Greek letters

α smoothing constant
α_D step-size dilution rate change
α_T step-size temperature change
γ intermediate parameter-updating variable, Eq. (161)
ε error signal
η, θ defined in Eq. (62)
λ forgetting factor
μ specific growth rate

σ specific substrate uptake rate
π specific product formation
τ_I integral time
τ_D derivative time
τ_p process time constant

Subscripts

0 initial
i inlet
d desired
F feed, final
f final
m measured
max maximum
min minimum
opt optimum
s singular
u ultimate
x partial derivation with respect to x

Superscripts

$^\circ$ time derivative
$^\wedge$ estimate
$^\sim$ error between true value and estimate
T transpose
* saturation value
– smoothed value

1 Introduction

The last decade saw significant cost reductions and improvements in the speed and reliability of computer hardware and the increased availability of software. These factors have resulted in a significant increase in the use of computers for control and optimization of fermentation processes. The growing economic pressure to improve the yield, productivity, and quality control of bioreactors, and the advent of powerful personal computers and workstations have resulted in real-time computer applications to fermentation processes. These computers provide, besides traditional data logging, inexpensive and reliable means to integrate computer control into fermentation processes and to carry out more sophisticated tasks such as on-line identification and adaptive control and optimization of fermentation processes.

Until recently the fermentation industry has lagged behind other process industries in implementing control and optimization technology. There are many reasons for the delay. Fermentation processes are much more complex than other industrial processes, involving a large number of complex and dynamic biochemical reactions and transport phenomena, many of which are not well understood. Therefore, it is very difficult to develop a realistic model involving a modest number of key variables. In addition, on-line measurement of these key physiological and biochemical parameters is very difficult if not impossible. Consequently, instruments able to measure more than a few simple parameters are not available. Finally, the organisms have intracellular regulatory mechanisms through which they perform internal regulation. Without proper understanding of these mechanisms we can only manipulate the extracellular environment, hoping to affect intracellular mechanisms in order to optimize bioreactor performance.

The primary objectives of control systems are to provide quality assurance and economic incentives. Slow dynamics associated with microbial bioreactors can result in off-specification products over a significant period for continuous operation or after a long period for batch operations. Unforeseen disturbances in a continuous bioreactor can result in a failure (washout) of the bioreactor, requiring new start-ups. Failures or off-specification products can have catastrophic economic consequences. Therefore, there is a strong economic incentive for proper control and optimization of bioreactors.

The purpose of control is to manipulate the control variables to: (1) maintain the desired outputs at a constant desired value (a regulation or a setpoint problem) by suppressing the influence of external disturbances and/or forcing the outputs to follow a desired profile (a servo or profile problem), (2) stabilize unstable or potentially unstable processes such as continuous cultures (a stabilization problem), and (3) optimize the performance as defined by measures such as yield, productivity, or profit

(an optimization problem). These objectives are to be achieved under various constraints such as safety, environmental regulations, limited resources, and operational constraints.

Problems associated with the design of a control system include:

(a) What are the objectives?
(b) What should be measured?
(c) What should be manipulated?
(d) Which control variables should be paired with which measured output? Should some disturbances be measured and used to manipulate control variables?
(e) How should controllers (typically PID controller parameters) be tuned?
(f) How does one decide the best operating conditions (constant values or variable profiles for the setpoint)?

We best illustrate these points by considering a specific problem, that of controlling the temperature (T) and the dissolved oxygen concentration (DO) of a fermentor at some desired values (T_d, DO_d) by manipulating the coolant flow rate (F_c) and the aeration rate (R_a) in the presence of external disturbances such as changes in the inlet coolant temperature (T_i), the oxygen concentration of the inlet gas (O_i), or the microbial growth rate, which can alter the oxygen demand and the temperature. Here the problem statement itself provides the answers to the questions concerning the objective and manipulation, items (a) and (c) above. Obviously T and DO must be measured, which answers the question concerning measurement (b). The question of proper pairing (d) is a simple matter. Intuition tells us that the temperature (T) should be paired with the coolant feed rate (F_c) and the DO measurement with the aeration rate (R_a) so that the temperature is controlled by the coolant flow rate and the dissolved oxygen concentration by the aeration rate. In certain cases the answer is not obvious and one must resort to a systematic method such as a relative gain array method to obtain a proper pairing (BRISTOL, 1966). The next question is how to tune the controllers, item (e), which is dealt with in many references (COUGHANOWR and KOPPEL, 1965; DOUGLAS, 1972; SHINSKEY, 1979; COHEN and

COON, 1953). The remaining question, item (f), is how to determine the optimal operating conditions; here the desired setpoint values (T_d, DO_d), are covered in Sect. 3.

In this chapter we cover very briefly the basic ideas of control. We then pay more attention to advanced control techniques and the optimization of batch, fedbatch, and continuous bioreactors.

2 Low-Level Control

Many types of control are used for bioreactors in the fermentation industry: manual, automatic, and computer controls. While more recent fermentation plants rely more on automatic and computer controls, there are still many old fermentation plants that rely on manual control. *Manual control* is the simplest type of control. Instead of pneumatic or electrical signals, human operators manipulate control elements such as a coolant valve, an aeration valve, and a feed pump. Although it is a very poor regulatory technique, manual control is still important in start-up and shutdown procedures and for backing up the classical automatic and modern computer controllers. *Classical automatic controls* include an analog, an on-off, a sequence, and a feedback control, all of which rely upon signals other than a human operator. *Computer controls* use a computer in a loop, which handles the measurement signals, generating a control algorithm and activating control elements to implement the control action. Various types of computer controls are used: a direct digital control (DDC), a supervisory control, a setpoint control (SPC), and a modern control.

A classical automatic control system in the form of *feedback control* is shown in Fig. 1. Since classical control techniques are covered extensively in standard references (COUGHANOWR and KOPPEL, 1965; DOUGLAS, 1972; SHINSKEY, 1979), the treatment given here for the sake of completeness is at most cursory. Almost all feedback control systems use negative feedback. The controller generates an error signal, ε, by subtracting the measured process output (controlled variable) from a de-

sired value (setpoint), and calculates the control signal by applying a certain algorithm to the error signal. This control signal then manipulates a process input (control element) to reduce the error. The controllers are named after the particular algorithm they use on the error signal to generate the manipulated variable, m. The most common controller is called a PID controller and uses proportional (P), integral (I), and derivative (D) actions. It is satisfactory for about 80% of control applications.

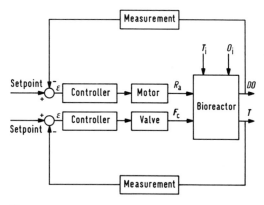

Fig. 1. Feedback control.

The manipulated variable, m, generated by a PID controller, is given by the following algorithm:

$$m = K_c \left(\varepsilon + \frac{1}{\tau_I} \int_0^t \varepsilon \, d\tau + \tau_D \frac{d\varepsilon}{dt} \right) \quad (1)$$

where K_c, τ_I, and τ_D are the proportional, integral, and derivative constants, respectively. Simpler controllers are P, PD, and PI controllers that lack the derivative and integral actions, the integral action, and the derivative action, respectively. The resulting manipulated variables, m, are obtained from the above by setting to zero $1/\tau_I$ and τ_p, τ_p, and $1/\tau_I$, respectively. A special case of a proportional-only control (P) is *on-off* control. Here the controller gain, K_c, is infinite, but the control action is restricted physically within the limits of the control element. On-off controllers are usually used with on-off control elements (e.g., solenoid valves, fixed-speed pumps, constant-load heaters, etc.). This type of control is satisfactory when the changes in the controlled variables are small and vary slowly with time. In many fermentation processes the temperature and pH loops have these characteristics, and thus on-off control is adequate. Although most recent applications of PID controllers utilize analog electronics or digital computers, there are many pneumatic controllers in operation. This is due to two factors. In the past, pneumatic instrumentation was the mainstay of the process industry. Due to their ruggedness and reliability, many existing installations still have pneumatic controllers. Second, pneumatic valves have many desirable characteristics, and they are still specified for some new installations.

2.1 Sequence Control

Since batch and fedbatch processes vary with time, the controller system must deal with time- and event-based process conditions and transition phenomena. In batch and fedbatch fermentations there are many sequential operations that need to be followed in each run. In a batch fermentation the bioreactor must first be cleaned, filled with the fermentation medium, sterilized, brought to proper fermentation conditions, inoculated with the inoculum, allowed to ferment for a specified time or to an end event, and emptied. The starting of pumps, the opening of valves, the completion of charging the bioreactor with medium and inoculum, and completion of fermentation are all events in time, requiring different control actions. Sometimes the control system includes automatic start-up and shut-down procedures.

Sequential control includes all control functions (discrete and continuous/regulatory) triggered by time or events. In sterilization, for example, the fermentor may be brought to a desired temperature, maintained there for a specified period, and then cooled to a desired fermentation temperature. Clearly the steam valve must initially be opened and a temperature controller must operate to reach the desired temperature. When the latter is reached, or when the desired event takes place, the timer is then triggered to maintain the tempera-

ture for the desired period. When the desired time has lapsed, the timer triggers the next process. Cooling the fermentor to the desired temperature is an event that can trigger the opening of a valve for inoculation.

With the advent of inexpensive and compact microprocessors, mechanical devices such as cam controllers and timers formerly used to control sequential operations have been replaced with a modern version of *sequence control* that involves programmable logic systems and computers. Using hardware or software, various basic discrete control functions (direct on/off control, interlock control, sequential control) are developed to control discrete devices such as on-off valves, pumps, or agitators, based on the status (ON-OFF) of equipment or values of process variables.

A discrete device can have only two states, such as ON/OFF, OPEN/CLOSE, and ENABLED/DISABLED. The triggers can be time-based or event-based (temperature, pH, pressure, level, etc.). Sequencing requires reaching an end condition (trigger) before the system can proceed to the next step. In sequential control one trigger or a combination of triggers may decide a step transition.

Improved sequencing can result in significant economic gains. Since the sequencing operations may be done discretely, computerization is an excellent method for their attainment. The time base for the sequence is more accurate and does not need to be calibrated with a computer. The time base and the logic can also be readily changed by computer.

2.2 Single-Loop Control of Environmental Variables

Examples of some common single-loop controls, such as flow control, pressure control, temperature control, and dissolved oxygen control, are given below.

2.2.1 Flow Control

Control loops for flow of gas or liquid are characterized by fast response times (seconds) with essentially no delays. Most disturbances result from high-frequency noise, though generally not of large magnitude, due to stream turbulence, valve changes, and pump vibrations. Proportional-Integral (PI) flow controllers are generally used. The presence of recurring high-frequency noise rules out the use of derivative action.

2.2.2 Gas Pressure Control

It is relatively easy to control gas pressure, because the gas-pressure process is self-regulating unless the gas is in equilibrium with a liquid. PI controllers are normally used with a small amount of integral control action. Due to the short response time compared with other process operation times, derivative action is normally not used.

2.2.3 Temperature Control

Due to the variety of heat transfer equipment, and because processes differ significantly in time scales, general guidelines for temperature control loops cannot be provided. The presence of time delays and/or multiple thermal capacitances (high-order process with or without dead time) leads to an upper limit on controller gain. PID controllers are commonly used to provide more rapid responses than those obtainable with PI controllers.

2.2.4 Dissolved Oxygen Control

Dissolved oxygen can be regulated by aeration rate, agitator speed, or a combination of the two. In industrial-scale fermentors agitator speed may be fixed, in which case the only means of controlling the dissolved oxygen concentration is by varying the aeration rate. A PI controller may be used for this purpose. In research-scale bioreactors dissolved oxygen is often regulated by the cascaded control system that is discussed in Sect. 2.4.2.

2.3 Controller Selection and Tuning Methods

After determination of types of control and location of controllers, it is necessary to specify the modes of control and the parameter values of controller settings (K_c, τ_I, and τ_D) necessary to obtain a satisfactory response. Given below are the general characteristics of various controller modes and some practicable tuning methods used in industry.

2.3.1 Controller Selection

No fixed criteria exist to determine which modes to use in a specific application. Frequently a PID controller is selected simply because it is most likely to accomplish satisfactory control. But it is beneficial to note through a typical process response to load changes the following general characteristics (COUGHANOWR and KOPPEL, 1965):

(1) Proportional control results in a response with a maximum offset, a high maximum deviation, and a moderate period of oscillation that ceases gradually.
(2) Proportional-integral control results in no offset, but at the expense of a higher maximum deviation, a longer period of oscillation, and a longer time required for oscillations to cease than in the case of proportional control.
(3) Proportional-derivative control generally results in the shortest time to steady state with the least oscillation and the smallest maximum deviation, but at the expense of an offset which, though smaller than that for the proportional control, is still significant.
(4) Proportional-integral-derivative control may result in a combination of the advantages of proportional-integral and proportional-derivative controls. The integral action and the derivative action serve to eliminate the offset and lower the maximum deviation, respectively. The derivative action also eliminates some oscillations that occur in PI control.

It is important to emphasize, however, that the above observations do not always apply and that exceptions occur. Nevertheless, the basic characteristics attributed to each mode of control generally persist even in a complex system, and therefore selection of a particular mode of controller should be guided by the above observations.

2.3.2 Tuning Methods

Controller tuning is the adjustment of controller settings to obtain a satisfactory performance (response) of the control system. Although the manufacturers of fermentor instrumentation packages usually supply their controller's settings, on-site tuning is normally required to meet individual requirements.

Acceptable overshoot, settling time, or steady-state oscillation depend on the overall process dynamics, user requirements, and the objectives of the process. For example, in temperature control it may be desirable to operate the process at or near the optimum temperature for cell growth and/or metabolite production. This is characteristically near the maximum temperature for microorganisms. In this situation an overshoot of even a few degrees above the setpoint can be disastrous. Also, a significant deviation of pH from the setpoint can lead to a disastrous result. On the other hand, the requirement for dissolved oxygen control may merely be to maintain the *DO* level at or above some particular value. In these circumstances, overshoot may or may not be critical.

Since controller tuning is often done by trial and error, it can be tedious and time-consuming. Experience with similar control loops may be helpful in making the initial guess. If process model or frequency-response data are available, a systematic design method can be used to calculate controller settings (COHEN and COON, 1953), but on-site tuning may still be required to fine-tune the controller, since the available process information is very seldom accurate or complete. The most widely used controller tuning methods are the *ultimate gain method* and the *process reaction curve method*. The ultimate gain method uses a closed-loop test, whereas the process reac-

tion curve method uses an open-loop test to obtain process dynamic information.

2.3.2.1 Ultimate Gain Method

First proposed by ZIEGLER and NICHOLS (1942), this has been called the *loop tuning* or the *continuous cycling method*. The controller is operated in a closed-loop with the system to be controlled. The first step is to determine experimentally the ultimate gain by using the proportional-only controller and increasing K_c by small increments until continuous cycling of the system variables first occurs. This value of K_c is called the *ultimate gain* and is denoted by K_u. The period of the resulting sustained oscillation is called the *ultimate period*, P_u. The controller settings are then calculated from K_u and P_u using the Ziegler–Nichols (Z–N) tuning rules (the original settings in Tab. 1), which were empirically developed to provide a quarter-decay ratio. These tuning methods were commonly used in industry and served as a basis for comparing other control schemes. Although the original Z–N settings

Tab. 1. Controller Settings Based on the Ultimate Gain Method
(ZIEGLER and NICHOLS, 1942)

Controller	K_c	τ_I	τ_D
Original			
P	$0.5\,K_u$		
PI	$0.45\,K_u$	$P_u/1.2$	
PID	$0.6\,K_u$	$P_u/2$	$P_u/8$
Modified[a]			
PID (Some overshoot)	$0.33\,K_u$	$P_u/2$	$P_u/3$
PID (Some overshoot)	$0.2\,K_u$	$P_u/3$	$P_u/2$

[a] PERRY and GREEN (1984)

give a significant safety margin, for some control loops the degree of oscillation associated with the quarter-decay ratio and the corresponding large overshoot for setpoint changes may be undesirable. Thus, more conservative settings are often required, such as the modified Z–N settings in Tab. 1.

While this method is fairly rapid and simple, it does have several disadvantages, such as large time consumption, lost productivity or poor product quality due to unstable operation or a hazardous situation, and the necessity to repeat the entire tuning procedure if a change is made in some portion of the loop.

2.3.2.2 Process Reaction Curve Method

ZIEGLER and NICHOLS (1942) also proposed an open-loop tuning technique to obtain the characteristic parameters of the process, the process reaction curve method. A small step change of magnitude M in the manipulated variable is introduced to the opened control loop, and the response in the measured variable is recorded with respect to time. This step response, called the *process reaction curve*, is characterized by two parameters: S, the slope of the tangent through the inflection point, and T_d, the time at which the tangent intersects the time axis. The Ziegler and Nichols tuning rules for the process reaction curve method are given in Tab. 2, where $S^* = S/M$ denotes the

normalized slope. These tuning relations were developed empirically to give closed-loop responses with a quarter-decay ratio.

COHEN and COON (1953) observed that the step response of most processing units, consisting of the process, the final control element, and the measuring element, had a sigmoidal shape, which can be adequately approximated by the response of a first-order transfer function with dead time:

$$\frac{\overline{y_m}(s)}{\overline{c}(s)} = G_f(s)G_p(s)G_m(s) = \frac{K_p e^{-T_d s}}{\tau_p s + 1} \tag{2}$$

where $\overline{y_m}$ is the measured value of the controlled variable and $\overline{c}$ is the controller output, both expressed as deviation variables. Model parameters K_p, τ_p, and T_d can be readily estimated from:

$$K_P = B/M \tag{3}$$

and

$$\tau_P = B/S \tag{4}$$

Tab. 3 summarizes the theoretical values of controller settings to give responses having quarter-decay ratio, minimum offset, and minimum area under the load-response curve, and favorable properties.

The process reaction curve (PRC) method offers certain advantages over the ultimate gain method in that only a single experimental test is required instead of a trial and error procedure. Also, the controller settings are easily calculated. But, it also has certain disadvantages, such as the difficulty in determining accurately the slope at the inflection point, the

Tab. 2. Controller Settings Based on the Process Reaction Curve Method (ZIEGLER and NICHOLS, 1942)

Controller	K_c	τ_I	τ_D
P	$1/T_d S^*$		
PI	$0.9/T_d S^*$	$3.33\,T_d$	
PID	$1.2/T_d S^*$	$2\,T_d$	$T_d/2$

Tab. 3. Controller Settings after COHEN and COON (1953)

Controller	K_c	τ_I	τ_D
P	$\dfrac{1}{K}\dfrac{\tau}{T_d}\left(1 + \dfrac{T_d}{3\tau}\right)$		
PI	$\dfrac{1}{K}\dfrac{\tau}{T_d}\left(0.9 + \dfrac{T_d}{12\tau}\right)$	$T_d\dfrac{30 + 3\,T_d/\tau}{9 + 20\,T_d/\tau}$	
PID	$\dfrac{1}{K}\dfrac{\tau}{T_d}\left(\dfrac{4}{3} + \dfrac{T_d}{12\tau}\right)$	$T_d\dfrac{32 + 6\,T_d/\tau}{13 + 8\,T_d/\tau}$	$T_d\dfrac{4}{12 + 2\,T_d/\tau}$

tendency for responses to be oscillatory since the settings were developed to provide a quarter-decay ratio, and the possibility of test results being significantly distorted since the experimental test is performed under open-loop conditions.

COURT (1988) used the process reaction curve method to obtain PID controller settings for the control of dissolved oxygen level in a batch *Escherichia coli* culture by agitation speed. With one set of PID settings it was difficult to maintain the *DO* level at various setpoints. RADJAI et al. (1984) also used the process reaction curve method to investigate redox potential control by manipulating the agitation speed in batch amino acid production by an auxotrophic mutant of *Corynebacterium glutamicum*. Both the process gain (K_p) and the process time constant (τ_p) were found to be functions of fermentation time. These authors also applied the ultimate-gain method to determine K_u and P_u. However, the time-varying nature of K_p and τ_p made this approach tedious.

Closed-loop process reaction curve methods have been proposed to avoid the major disadvantages of the two widely used methods (YUWANA and SEBORG, 1982; JUTAN and RODRIGUEZ, 1984). A closed-loop process reaction curve was obtained by making a small step change in the setpoint during proportional-only control. First-order Pade approximation (YUWANA and SEBORG, 1982) and higher order approximation (JUTAN and RODRIGUEZ, 1984) for the time delay term in Eq. (2) were used. The model parameters in Eq. (2) were then calculated from the closed-loop response. A disadvantage of these closed-loop process reaction curve methods is the more complex model parameter calculation than in the standard process reaction method.

2.3.2.3 Automatic Tuning

The process reaction curve technique is probably the easiest and most convenient way to estimate the controller setting in linear systems. This is not true, however, for nonlinear control schemes such as the control of *DO* level. As an alternative to the Ziegler–Nichols continuous cycling (ultimate-gain) method dis-

cussed previously, ASTROM and HAGGLUND (1988) described an automatic tuning (autotuning) method that used a relay with a dead zone to generate the process oscillation. The system is forced by a relay controller, which causes the system to oscillate with a small amplitude. The amplitude of the process output oscillation can be reduced by adjusting the relay amplitude. The ultimate period, P_u, is obtained by simply measuring the period of the process oscillation. The ultimate gain, K_u, is given by:

$$K_u = \frac{4d}{\pi a} \tag{5}$$

where d is the operator-specified relay amplitude and a is the measured amplitude of the process oscillation. The controller settings are then found using the Ziegler–Nichols rules in Tab. 1.

In their application RADJAI et al. (1984) either increased or decreased the steady-state value of agitation speed by the amount of d as the error in redox potential changed sign from positive to negative. The measured variable thus oscillated around the setpoint with an amplitude a. The value of d was also adjusted as K and τ changed. They investigated the PI controller performance with and without autotuner after the air supply to the bioreactor was cut off for 10 seconds. With the autotuner, the controlled variable was dampened back to its original value in 6 minutes. The system led to instability without the autotuner.

Self-tuning controllers are available that introduce appropriate inputs to the loop and measure the response; from these measurements the controller settings are then calculated and set automatically. This type of controller is very desirable in highly nonlinear processes such as batch or fedbatch fermentation, for which the controller constants may have to be changed with time. The details of self-tuning controllers are given in Sect. 2.4.3.

2.4 Advanced Process Control

In most situations simple single-loop feedback control is adequate. There are situations in which simple single-feedback control is inadequate, however, and more advanced control is required. More advanced control can improve the quality of control over that achievable by simple feedback control. Advanced controls frequently used in industry are discussed here.

2.4.1 Feedforward Control

In this scheme (Fig. 2), unlike in normal feedback control, one does not wait until a disturbance actually affects the output. One instead measures the disturbances (T_i, O_i) and applies corrective control actions (F_c, R_a) in anticipation of the expected effect. This is a better control for eliminating the effect of disturbance. However, to implement a feedforward control scheme, the disturbances must be

measurable, and the effect of the disturbances must be known *a priori*. In addition, feedforward control scheme is rarely used alone, but is combined with feedback control, because without the feedback control there is no way to correct errors caused by imperfect knowledge in predicting the effect of disturbances on the output. A combination of feedback and feedforward control is shown in Fig. 3. This permits the major effect of the disturbance to be corrected by the feedforward controller leaving the fine trimming to the feedback loop.

2.4.2 Multiloop Controls

In practice, most processes involve more than one input (manipulated variables) and more than one output (measured variables). Fig. 1 depicts a two-input–two-output system. The outputs (T, DO) as affected by the disturbances (T_i, O_i) are measured and fed back to a comparator where they are compared to the desired setpoints. The differences (ε) between the desired and measured variables are used to generate controller signals that in turn manipulate the control variables (F_c, R_a) to force the outputs to match the desired value. In this situation coolant flow rate, F_c, affects primarily temperature, T, and the aeration rate, R_a, affects primarily dissolved oxygen. Thus, there is very little interaction, and single-loop techniques (COHEN and COON, 1953; ZIEGLER and NICHOLS, 1942) can be applied to the separate loops. In other situations any one input can significantly affect more than one output; therefore, the interaction is significant and the usual single-loop techniques cannot be used. Sometimes even the problem of proper pairing of the output to input variables cannot be resolved intuitively as we have done in Fig. 1. A relative gain array method (BRISTOL, 1966) and a singular value analysis (SMITH et al., 1981; MOORE, 1986) are used to determine the best pairing. Proper pairing and tuning for multiple-input–multiple-output systems are found in some references (SEBORG et al., 1989).

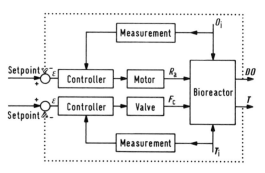

Fig. 2. Feedforward control.

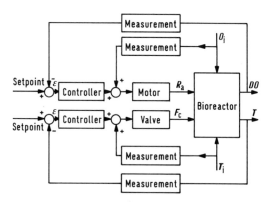

Fig. 3. Feedforward–feedback control.

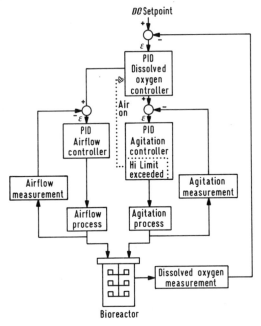

Fig. 4. Cascade control.

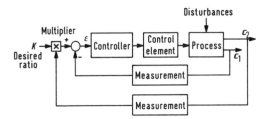

Fig. 5. Ratio control.

2.4.2.1 Cascade Control

This is an improved control scheme in which more than one (two) measurement is used to complete more than one (two) control loop and using only one manipulated variable. The basic idea is to measure an intermediate variable and initiate corrective action, not waiting until the effects of disturbance appear in the outputs. Conceptually, this scheme tries to achieve the same objective as the feedforward scheme, except that implementation is different using a feedback control scheme. This is illustrated for a *DO* control scheme (Fig. 4) that

has one inner loop (a slave) involving the measurement of agitation speed (R_a) and an outer loop (a master) involving the dissolved oxygen measurement (*DO*). The output of the *DO* loop controller serves as the setpoint for the agitation loop. Such a cascade control system is used to regulate the *DO* by manipulating aeration rate. In actuality, when the agitation speed reaches the upper limit, control may be switched to the aeration loop.

2.4.2.2 Ratio Control

The purpose of ratio control is to maintain a desired ratio of two or more variables instead of controlling the actual variables themselves. This is a special type of feedforward control. This type of control configuration is widely used in blending operations and in feeding reactants to a reactor in fixed proportion. Fig. 5 shows a block diagram for ratio control. There are two process variables: c_1 which is controlled, and c_2 which is left uncontrolled by the manipulated variable. The process variables, c_1 and c_2, should not depend on each other, and only c_1 is affected by the manipulated variable.

As shown in Fig. 5 there is a feedback loop for variable c_1, and variable c_2 is constantly changing the setpoint of the c_1 loop so that c_1 is changing to maintain constancy of the c_1/c_2 ratio. It should be noted that the dynamic response of c_2 (and of the ratio) is affected by the form and parameters of the controller.

2.4.3 Adaptive Control

When the characteristics of processes change with time, the operating conditions may need to be changed. These include controller parameters and the setpoints themselves. Adaptive control systems adjust the controller parameters automatically to compensate for variations in the process characteristics. Living microorganisms may go through different life cycles. This in conjunction with their internal control mechanisms leads to time-variant characteristics. For example, in a batch fermentor the cell concentration increases with time, and therefore the controller parameters that were "best" for *DO* control at low cell concentrations, for example, may need to be altered at high cell concentrations. As shown in Fig. 6, a typical adaptive control system has two major elements, one to identify the changes (process identification) from the input and output data,

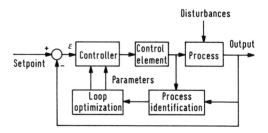

Fig. 6. Adaptive control.

and the other to perform loop optimization (optimizer). The details and examples concerning the process identification and optimizer for a continuous bioreactor are found in Sect. 5.1.

2.4.3.1 Self-Tuning Control

There are two general categories of adaptive control. The first covers situations in which process changes can be either measured or anticipated so that it is possible to adjust the controller settings systematically, based upon the measured or anticipated process change (programmed adaptation). This is a form of feedforward control. The second category refers to situations in which process changes cannot be measured or predicted, so adaptive control is implemented in a feedback manner on-line. Such controllers are implemented through computer control and are called self-tuning controllers (ASTROM and WITTENMARK, 1988).

For example, in Fig. 6 the process identification scheme carries out parameter estimations in a process model using the input–output data acquired on-line. These parameter values are then supplied to the loop optimization scheme, which in turn makes control calculations based on the new parameter values to determine the best controller settings. The con-

troller settings in the feedback control are automatically adjusted by the loop optimizer. This type of controller is called self-tuning or self-adaptive. In most real-time parameter-estimation schemes an external forcing function that excites the process most effectively is introduced to obtain accurate process model parameters.

2.4.4 Computer Control

For a fermentation plant the primary objective of computer control is to produce products as economically as possible. The computer is generally used to provide quality control, save operator time, furnish automatic documentation, and decrease per-loop control costs. Computer control has found its way not only into plants, where the economic advantages are more obvious, but also into pilot plants and research laboratories. In the latter cases the computer provides fast and efficient data acquisition, the ability to monitor and control experimental conditions, and flexibility in the operation of the system.

One situation in which the capabilities of the computer can truly be realized is the implementation of advanced control and optimization strategies. It was not until the 1970s that much of the modern control theory developed in the 1950s and 1960s could be applied practically. This change was sparked primarily by tremendous economic and technological advances in computer hardware, especially the advent of microcomputers. Computer control of fermentation has been an extremely active area, as evidenced by many reviews in the literature (ARMIGER and HUMPHREY, 1979; BULL, 1983; DORBY and JOST, 1977; HAMPEL, 1979; HATCH, 1982; JEFFERIS, 1975; ROLF and LIM, 1982; WEIGAND, 1978; ZABRISKIE, 1979).

Setting the control configuration of the low-level loops is one of the major decisions to be made in computer control. The low-level control loops are those feeding back directly upon one of the basic process measurements, for example, when the manipulated variable of acid or base addition is used to control pH. The decision is whether to use a classical hardware controller with the computer providing

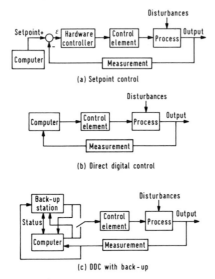

(a) Setpoint control

(b) Direct digital control

(c) DDC with back-up

Fig. 7. Computer control configuration.

the setpoint (SPC; setpoint control), or to replace the hardware controllers with DDC (direct digital control), in which case the computer implements low-level control algorithms as well (see Fig. 7a and b). The setpoints can also be provided by interaction of the human operator with the computer, as in supervisory control.

2.4.4.1 Supervisory Control

In supervisory control, the setpoints are supplied by computer–operator interaction, predefined values or profiles, or economic and scheduling considerations. As an example, consider the bioreactor temperature-control loop. In many situations this loop may be handled by an on/off control in which the controller has three states: cooling, off, and heating. With this type of double-sided on/off control, oscillations can occur; to eliminate these oscillations, the supervisory computer program can decide at any particular time whether to control with cooling or heating. It would be convenient to design the low-level controller so that a positive-valued setpoint would effect control by heating and a negative-valued setpoint would control by cooling. When it would be desirable to have the bio-

reactor temperature change according to a predefined profile resulting from an optimization routine, the supervisory program could then determine the magnitude and sign of the setpoint to force the temperature profile to closely match the given optimum.

2.4.4.2 Setpoint Control (SPC)

With SPC the computer is involved only in providing the setpoints for the low-level loops. Therefore if computer failure occurs, the low-level loops remain operational. Less reliance on the computer for these low-level tasks is desirable, since it frees computer time for other, more complex tasks. One disadvantage of SPC is that types of control techniques for hardware controllers are limited. This is not critical, however, since most low-level loops can be handled by simple controller techniques such as PID and on/off.

2.4.4.3 Direct Digital Control (DDC)

With DDC the computer contains the controller as a program or subroutine. Often one subroutine will handle several control loops. This is sometimes called table-driven DDC because the inputs, outputs, and controller parameters for each loop are stored in a software table. Several guides are available for designing and programming DDC algorithms (BRISTOL, 1977; GOFF, 1966; WEBB, 1980; WILLIAMS et al., 1973). A discrete version of the classical PID control, known as the position form, is:

$$m[n\Delta T] = K_c \left\{ \varepsilon[n\Delta T] + \frac{\Delta T}{\tau_I} \sum_{i=0}^{n} \varepsilon[(n-i)\Delta T] + \right.$$
$$\left. + \frac{\tau_D}{\Delta T} (\varepsilon[n\Delta T] - \varepsilon[(n-1)\Delta T]) \right\} \qquad (6)$$

where ΔT is the sample time (controller time increment), $m[n\Delta T]$ and $\varepsilon[n\Delta T]$ are the current manipulated variable and error, respectively, and $\varepsilon[(n-i)\Delta T]$ are previous errors. This is known as the position algorithm, since it calculates the position of the manipulated variable (control element). A more convenient and therefore more frequently used form is the so-called velocity algorithm, which calculates the incremental change in the manipulated variable,

$$m[n \Delta T] = K_c \left\{ \varepsilon[n \Delta T] - \varepsilon[(n-1) \Delta T] + \right.$$

$$+ \frac{\Delta T}{\tau_1} \varepsilon[n T] + \frac{\tau_D}{\Delta T} (\varepsilon[n \Delta T] -$$

$$\left. -2\varepsilon[(n-1) \Delta T] + \varepsilon[(n-2) \Delta T]) \right\} \qquad (7)$$

The velocity algorithm has the advantage that it can be modified to prevent integral overshoot and subsequent oscillation. Additionally, it is convenient when the control element is of the incremental or integrating type.

Tab. 4. Advantages of Direct Digital Control

Flexibility and versatility
  Control techniques are not limited
  Control algorithms are modified readily
  Controller parameters are adjusted readily
  One control routine for several loops

Reduced hardware requirements
  No need to design, construct, calibrate, or service
  analog controllers

Improved response in some cases
  Better regulation
  Smoother start-up and shut-down
  Bumpless transfer in switching with manual systems

Automatic documentation of control actions, output responses and manipulated variables

---

The advantages offered by DDC are listed in Tab. 4. Due to these advantages, there is a general trend to utilize DDC for fermentation processes. If DDC is used, however, those low-level control loops that are critical to the operation of the process must have a back-up in case of computer failure (see Fig. 7c). Another disadvantage of using DDC for low-level control loops is that the computer must always be available and cannot be diverted to another function while the DDC programs are run-

ning. The evolution of microcomputers and hierarchical computer systems has been a tremendous help in this area. Low-level control can be carried out by microcomputers while higher computers do other tasks more efficiently. Also, in multi-computer systems, the burden of computer failure is reduced because failure of one computer will not cripple the entire process or plant.

3 Optimization and High-Level Control

From the standpoint of process control, the desired setpoint can be either fixed (regulator problem) or variant with time (servo problem). After determining the optimum setpoints through an optimization procedure, the control system is designed to make the output follow the setpoint as closely as possible, as discussed in Sect. 2. Setpoint optimization problems can generally be classified into two categories: constant setpoint optimization and trajectory setpoint optimization.

Constant setpoint optimization involves finding the best constant values for the operating variables, which leads to optimum performance as measured by a given performance index. For example, the determination of the optimal steady-state temperature (T), pH, and dilution rate (D) that maximize the cellular productivity (Dx) of a steady-state continuous culture is a constant setpoint optimization problem. Here the performance index is Dx. Thus, constant setpoint optimization problems are problems in ordinary calculus, requiring the taking of partial derivatives of the performance index with respect to the operating variables (T, pH, D), setting them to zero, and solving the resulting equations for the best constant values of the operating variables. To do this it is necessary to obtain a mathematical model relating Dx to T, pH, and D. The required model is derived from the steady-state mass balance equations for a continuous bioreactor. Thus, one must be able to describe completely the continuous bioreactor by mass balance equations with all necessary kinetic rate expressions having kinetic parameters as functions of T and pH. If no model relating the performance index to the operating variables is available, or if the dependence of any one of the kinetic parameters on T and pH is missing, the above-described analytical method cannot be applied. An experimental search technique with on-line optimization must then be used to obtain the solution. This type of problem is considered in detail in Sect. 5.1.

Trajectory optimization refers to the determination of temporal or spatial functions, not constant values, which lend an optimal value to the given performance index. For example, the determination of pH or temperature profile as a function of time during fermentation to maximize the product concentration at the end of batch or fedbatch fermentations is a trajectory optimization. There is no obvious reason to expect that temperature or pH should be kept constant for batch or fedbatch bioreactors; what is best for cell growth may not be best for product formation. The performance index is a functional, a function of variables that themselves are also functions. Therefore, a trajectory optimization that requires determination of functions that optimize the performance index is a problem in the so-called variational calculus.

3.1 Constant Setpoint Optimization for Low-Level Control and Continuous Bioreactors

As stated above, constant setpoint optimization is concerned with finding the best set of constant values of the operating variables. For the above example the optimal constant temperature, pH, dilution rate, and so on are: $x = (T, \text{pH}, D)^T = (x_1, x_2, x_3, \ldots x_n)^T$, which lead to the optimum value of the given performance index, p, that in turn depends on many operating variables, $p(x) = Dx = p(T, \text{pH}, D)$. This is a problem in ordinary calculus. To obtain the solution the performance index must be expressed in terms of the operating variables, all the partial derivatives of the performance index with respect to each operating variable must be set to zero, and the result-

ing n algebraic equations for the n unknowns must be solved for optimal operating conditions:

$$\partial p(x)/\partial x_i = 0, \text{ for } i = 1, 2, 3, \ldots n \tag{8}$$

When a quantitative model relating the cellular productivity (Dx) of the continuous culture to these operating variables, T, pH, and D, is known ($Dx(T, \text{pH}, D)$) it is simple to obtain the best steady-state values:

$$\partial(Dx)/\partial T = 0, \quad \partial(Dx)/\partial(\text{pH}) = 0,$$
$$\partial(Dx)/\partial D = 0 \tag{9}$$

which must be solved simultaneously to obtain the optimum operating temperature (T), pH, and dilution rate (D).

For simplicity we consider as an example the determination of the dilution rate that best maximizes the cellular productivity, Dx, of a continuous culture. Consider an ideal continuous bioreactor that produces cell mass as its product. The steady-state cell mass and substrate balance equations are:

$$Dx - \mu x = 0 \tag{10}$$

$$D(s_F - s) - \mu x/Y = 0 \tag{11}$$

where D is the dilution rate, s and x are the substrate and cell mass concentrations, respectively, s_F is the feed substrate concentration, and Y is the yield factor. First, we must express the performance index, Dx, solely in terms of D before we take the total derivative of Dx with respect to D and set it to zero. From Eqs. (10) and (11) we have:

$$Dx = \mu x = YD(s_f - s) \tag{12}$$

in which s must be eliminated in terms of D by solving Eq. (10), $\mu(s) = D$, for s. Denoting this by:

$$s = [\mu]^{-1} D \tag{13}$$

so that for the Monod model, ($\mu = \mu_{max} s/(K + s)$, $s = [\mu]^{-1} D = DK/(\mu_{max} - D)$, and Eq. (12) is now written as:

$$Dx = YD(s_f - DK/(\mu_{max} - D)) \tag{14}$$

and $d(Dx)/dD = 0$ yields:

$$D_{opt} = \mu_{max}[1 - (K^{1/2}/(K + s_F)^{1/2}] \qquad (15)$$

The optimum dilution rate given by Eq. (15) results in maximum cell-mass productivity.

It is now desired to obtain the optimum temperature that results in maximum productivity. This is obtained by setting to zero the partial derivative of Dx with respect to T, $\partial(Dx)/\partial T = 0$:

$$K\partial\mu_{max}/\partial T = (\mu_{max} - D)\partial K/\partial T \qquad (16)$$

which must be solved together with Eq. (15) to obtain the optimum temperature and dilution rate. Eq. (16) implies that the dependence on temperature of the kinetic parameters, K and μ_{max}, must be known.

Various numerical and search techniques are available for unconstrained and constrained multivariable optimization (EDGAR and HIMMELBLAU, 1988), and the reader is encouraged to consult the original references therein. These techniques can be used with or without an analytical model. When no analytical model is available, the performance index must be evaluated experimentally by perturbing the system with the manipulated variables and observing the response of the system.

In practice, models relating the performance index to all operating variables are seldom available and may be difficult to obtain. In this scenario one has to resort to an experimental search technique or a combination of modelling and experimental techniques to maximize the performance index. This is illustrated in Sect. 5.1 in which adaptive optimization of continuous bioreactors is treated.

3.2 Profile Optimization

Profile optimization refers to situations in which the manipulated variables are functions of time or location (temporal or spatial functions) such as feed flow rate, temperature, or pH profiles as functions of time for a batch or fedbatch bioreactor. We seek the best profiles of various kinds for batch and fedbatch fermentation. We provide in this section the essence of the Maximum Principle of PON-

TRYAGIN et al. (1962) and singular control theory and seek solutions by their use.

Briefly, the essence of the Maximum Principle is as follows. Consider an ideal bioreactor that can be described by four unsteady-state mass balance equations for cell mass, substrate(s), product, and overall mass:

$$\frac{d}{dt}(xv) = Fx_F - F_0x + \mu xv \qquad x(0) = x_0 \quad (17)$$

$$\frac{d}{dt}(sv) = Fs_F - F_0s - \sigma xv \qquad s(0) = s_0 \quad (18)$$

$$\frac{d}{dt}(pv) = Fp_F - F_0p + \pi xv - kpv \quad p(0) = p_0 \quad (19)$$

$$\frac{d}{dt}(v) = F - F_0 \qquad v(0) = v_0 \quad (20)$$

where μ, σ, and π are the specific rates for growth, substrate consumption, and product formation, respectively, and x, s, and p denote the concentrations of cell mass, substrate, and product, respectively. The fermentor volume, the feed flow rate, the withdrawal rate, and the feed substrate and cell concentrations are denoted by v, F, F_0, s_F, and x_F, respectively. The initial conditions are assumed to be specified. The specific rates, μ, σ, and π, are assumed to be arbitrary functions of substrate, cell, and product concentrations, and also depend on temperature and pH. The product is assumed to decay in accordance with a first-order rate constant, k. Eqs. (17) through (20) are general mass-balance equations and can describe batch, fedbatch, or continuous bioreactors. The above mass-balance equations describe an unsteady-state continuous bioreactor. The same equations with $F_0 = 0$ describe a fedbatch bioreactor for which harvesting is undertaken only after the completion of fermentation. A batch bioreactor is described by the above equations with $F = F_0 = 0$. The objective is to maximize a performance index that represents a measure of profit associated with the final outcome of fermentation, $P(x_f, p_f, s_f)$, where the subscript f denotes the final time. In other words, the final outcome may depend on the final cell, product, and substrate concentrations. This is to be achieved by varying optimally the temperature, T, pH, and feed (F)

and withdrawal (F_0) rates as functions of time. The fermentation time, t_f, may be assumed to be given (fixed) or open to determination (free). Thus the objective is to maximize the performance index P by choosing optimally the temperature, pH, and feed and withdrawal rates as functions of time:

$$\underset{T(t),\,\mathrm{pH}(t),\,F(t),\,F_0(t)}{\mathrm{Max}} = P[x_f, p_f, s_f] \qquad (21)$$

Constraints are imposed on the final bioreactor volume, and the substrate feed rate is constrained as indicated below:

$$v(t_f) = v_{\max} \qquad (22)$$

$$0 \le F(t) \le F_{\max} \qquad (23)$$

Eqs. (17) through (20) can be written in the compact vector form

$$\frac{d}{dt}\begin{bmatrix} xv \\ sv \\ pv \\ v \end{bmatrix} = \begin{bmatrix} \mu xv \\ -\sigma xv \\ \pi xv - kpv \\ 0 \end{bmatrix} + \begin{bmatrix} x_F \\ s_F \\ p_F \\ 1 \end{bmatrix} F - \begin{bmatrix} x \\ s \\ p \\ 1 \end{bmatrix} F_0 \qquad (24)$$

or

$$\frac{d}{dt}(y) = g(y, T, \mathrm{pH}) + aF - (y/y_4)F_0 \qquad (25)$$

and

$$y(0) = y_0 \qquad (26)$$

where $y = (xv, sv, pv, v)^T$, $g(y, T, \mathrm{pH}) = (\mu xv, -\sigma xv, \pi xv - kpv, 0)^T$, $a = (x_F, s_F, p_F, 1)^T$, and $y_0 = (xv_0, sv_0, pv_0, v_0)^T$. Temperature and pH affect the specific rates in nonlinear fashion through the kinetic parameter values. In other words, g is a nonlinear vector function of T and pH. The manipulated variables, T, pH, F, and F_0 affect the outcome of fermentation, P.

The Maximum Principle states (PONTRYAGIN et al., 1962) that the maximization of P is equivalent to maximizing the Hamiltonian given below:

$$H = \lambda^T g(y, T, \mathrm{pH}) + \lambda^T aF - \lambda^T \left(\frac{y}{y_4}\right) F_0 \qquad (27)$$

where the adjoint vector λ must satisfy the following ordinary differential equations:

$$\dot{\lambda} = -\frac{\partial H}{\partial y} = -\left(\frac{\partial g}{\partial y}\right)^T \lambda - \left(\frac{\partial}{\partial y}\left(\frac{y}{y_4}\right)\right)^T \lambda \qquad (28)$$

with the final conditions depending on the functional form of P and given by:

$$\lambda(t_f) = \left(\frac{\partial P}{\partial y(t_f)}\right)^T \qquad (29)$$

When the final fermentation time, t_f, is fixed, the Hamiltonian is a constant, H^*, while the Hamiltonian is zero when the final time is free. The problem is to maximize the Hamiltonian Eq. (27) by properly determining the manipulated variables (T, pH, F, F_0) as functions of time. Thus, the necessary condition for maximization is:

$$\underset{T(t),\,\mathrm{pH}(t),\,F(t),\,F_0(t)}{\mathrm{Max}}$$
$$\cdot H = \lambda^T g(y, T, \mathrm{pH}) + \lambda^T aF - \lambda^T \left(\frac{y}{y_4}\right) F_0 \qquad (30)$$

Optimization of Temperature and pH

We shall consider the optimization of bioreactors by manipulating the temperature and pH as functions of time. Since T and pH appear nonlinearly in the Hamiltonian, the latter is maximized by forcing to zero the partial derivatives with respect to T and pH when the constraints on T and pH are inactive:

$$\frac{\partial H}{\partial T} = \frac{\partial(\lambda^T g)}{\partial T} = 0 \text{ and } \frac{\partial H}{\partial \mathrm{pH}} = \frac{\partial(\lambda^T g)}{\partial \mathrm{pH}} = 0 \qquad (31)$$

or, they could be on the boundary when the constraints on T and pH are active:

$$T = T_{\min} \text{ or } T_{\max} \text{ and } \mathrm{pH} = \mathrm{pH}_{\min} \text{ or } \mathrm{pH}_{\max} \qquad (32)$$

Therefore, to maximize P by manipulating T and pH as functions of time, it is necessary to solve the state equation with the given initial conditions, Eqs. (25) and (26), and the adjoint equation with the final conditions given by Eqs. (28) and (29) using decisions made on the basis of Eq. (31) or (32). This type of problem is called a two-point split boundary-value problem in the sense that for the state vector, y, the initial conditions are known, while for the adjoint vector, λ, the final conditions are known. There are two numerical techniques that can be used to solve this problem: a boundary condition iteration and a control vector iteration.

Boundary condition iteration refers to a technique in which one set of missing boundary conditions is estimated and iterated until the calculated boundary conditions agree with the specified boundary conditions. For example, the final conditions on the state vector, $y^i(t_f)$, are first estimated, and then the state and adjoint vector differential equations, Eq. (25) and (28), are integrated backward starting with the estimated final conditions on the state vector and the specified final conditions on the adjoint vector, Eq. (29), with the control vector (T, pH) selected by the necessary conditions, Eq. (31) or (32). This backward integration is continued until the initial time $(t=0)$, and the resulting calculated initial conditions on the state vector, $y^i(0)$, are compared with the given initial conditions, y_0, Eq. (26). If the calculated initial conditions do not agree with those given, a new set of final conditions on the state vector, $y^{i+1}(t_f)$, is proposed and the procedure is repeated. An iterative procedure is applied until the calculated initial conditions on the state vector agree with the given initial conditions. Then the resulting control vector (T, pH) profiles are the optimal profiles that maximize the Hamiltonian function, Eq. (30), and the performance index, Eq. (21). The problem of how to improve on the final conditions after an unsuccessful estimation is now considered. This is based on the variational equations for δy. It can be shown (KOPPEL, 1968) that:

$$y^{i+1}(t_f) = y^i(t_f) + \Phi(t_f; 0)[y_0 - y^i(0)] \qquad (33)$$

where $\Phi(t_f; 0)$ is the so-called transition matrix starting at $t=0$ and ending at $t=t_f$, which must satisfy the following differential equations:

$$d\Phi(t_f; t)/dt = -\Phi(t_f; t)\left(\frac{\partial g}{\partial y}\right) \quad \Phi(t_f; t_f) = I \quad (34)$$

To solve this equation for $\Phi(t_f; 0)$ we must solve the n^2 equations in Eq. (34), starting at $t=t_f$, with the final condition $\Phi(t_f; t_f)=I$, and integrating backward in time to $t=0$.

It is now largely accepted that the boundary condition iteration procedure outlined above is generally inferior to the control vector iteration procedure described below. There are exceptions, however. There are many reasons for this statement, the main one being that the backward integration of generally stable state equations results in an unstable system, so that a small error in the estimation of the final conditions on the state vector can lead to a very large discrepancy between the calculated and given initial state vector values. This makes the conversion very slow.

The control vector iteration starts with an estimated control vector, for, e.g., pH and temperature as functions of time, $m(t)=(\text{pH}, T)$. Using the estimated control vector, $m^i(t)$, and the known initial conditions, Eq. (26), the state equations, Eq. (25), are integrated forward in time from $t=0$ to $t=t_f$, while the adjoint equations, Eq. (28), are integrated backward from $t=t_f$ to $t=0$ using the specified final conditions, Eq. (29). Then the Hamiltonian function, H^i, Eq. (27), is evaluated, and a new improved control vector is calculated using the necessary conditions, Eq. (31) or (32). The process is repeated with the improved control vector until there is negligible change in the Hamiltonian and therefore also negligible change in the control vector or the performance index. The question of how to obtain an improved control vector for the iteration is provided by the following equation:

$$m^{i+1}(t) = m^i(t) + W(t)\frac{\partial H^i}{\partial m} \qquad (35)$$

where $W(t)$ is a positive definite weighting matrix. If the weighting matrix, $W(t)$, is chosen to be as follows, the method is known as the steepest ascent:

$$W(t) = \frac{G^{-1}(t)\delta s}{\left[\int_0^{t_f} \left(\frac{\partial H}{\partial m}\right)^T G^{-1} \left(\frac{\partial H}{\partial m}\right) dt\right]^{1/2}} \qquad (36)$$

where $G(t)$ is a symmetric positive definite matrix defining the distance that is the finite distance by which m is to be moved:

$$(\delta s)^2 = \int_0^{t_f} (\delta m)^T G(\tau) \delta m \, d\tau \qquad (37)$$

This steepest ascent method has been found effective for a large class of problems encountered in profile optimization.

Optimization of Feed Flow Rates, $F(t)$, for Fedbatch Bioreactors

We now consider maximizing the performance index by manipulating the flow rate, $F(t)$, for fedbatch bioreactors ($F_0(t) = 0$). The feed flow rate $F(t)$ appears linear so that maximization of the Hamiltonian depends on the sign of the coefficient of F, $\lambda^T a$. That is, if $\lambda^T a$ is positive we choose the maximum flow rate and if it is negative we take the smallest F ($F = 0$, or a batch period). But, if $\lambda^T a$ is identically zero over a finite period of time, the Maximum Principle fails to yield a solution. This time period is called the singular interval, and the feed rate is called the singular control (intermediate values). Therefore, the optimal feed rate profile consists of periods of maximum flow rate (F_{max}), minimum flow rate ($F = 0$, or a batch period), and intermediate flow rate, F_s:

$$F(t) = \begin{cases} F_{max} & \lambda^T a > 0 \\ F_s & \lambda^T a = 0 \quad t_j \le t \le t_{j+1} \\ 0 & \lambda^T a < 0 \end{cases} \qquad (38)$$

The exact sequence and the times at which the flow rate shifts from one period to another are yet to be determined. Since $\lambda^T a$ is zero over the finite time interval, its time derivatives (the first, second, and so on) also must vanish. In general, for fedbatch bioreactors the second-order derivative expression results in a term that contains $F(t)$ explicitly, so that $\lambda^T a = 0$ and the first two derivatives are sufficient to allow the determination of the feed flow rate during the singular interval in terms of the state and adjoint vectors, $y(t)$ and $\lambda(t)$. Since the details of problem formulation, necessary conditions, and numerical examples are given in Sect. 3.2.2 we shall not go into them here.

Optimization of Feed Flow Rates, $F(t)$, for Continuous Bioreactors

For steady-state continuous bioreactors the feed flow rate and the withdrawal rate are equal, $F(t) = F_0(t)$. The performance index must here be defined. If a product other than cells is involved, the volumetric productivity may be chosen as the performance index, Dp. In addition to the flow rate, the steady state temperature and pH also may be optimized. The details of this optimization scheme are given in Sect. 3.2.2 and are therefore not repeated here.

3.2.1 Batch Bioreactors

Assuming that the medium composition has already been optimized, the variables that can be manipulated to optimize the performance of batch bioreactors include temperature, pH, initial inoculum size, initial substrate concentration, and fermentation time. Of these, only temperature and pH are functions of time, and the rest are constant. The basics involved in optimizing temperature and pH profiles have been presented above. Here we give an example of temperature profile optimization reported by CONSTANTINIDES et al. (1970a, b) for batch penicillin fermentation in the context of the Maximum Principle given above. In their work optimal temperature profiles were determined for a batch penicillin fermentation model.

Optimum Temperature Profile for a Batch Penicillin Fermentation Model

The following model is based on the mass balance equations of the cell and penicillin while ignoring the substrate concentration effect altogether:

$$\dot{x} = k_1 [1 - (x/k_2)] x \qquad (39)$$

$$\dot{p} = k_3 x - k_4 p \qquad (40)$$

where x and p represent cell mass concentration and penicillin concentration, respectively, and k_1, k_2, k_3, and k_4 are empirical constants whose temperature dependencies are given below:

$$k_1 = c_1 \left[\frac{1 - c_2(T - c_3)^2}{1 - c_2(25 - c_3)^2} \right] = k_2 \frac{c_1}{c_4} \qquad (41)$$

$$k_3 = c_5 \left[\frac{1 - c_2(T - c_6)^2}{1 - c_2(25 - c_6)^2} \right] \qquad (42)$$

$$k_4 = c_7 e^{\frac{-c_8}{R} \left(\frac{1}{T + 273.1} - \frac{1}{298.1} \right)} \qquad (43)$$

where T is temperature in °C and the numerical values for constants c_1 through c_8 are reported elsewhere (CONSTANTINIDES et al., 1970b). It should be stated here that most penicillin fermentations are carried out in fed-batch culture, in which the substrate feed rate is varied with time. Assuming that for a particular strains the model proposed is valid, we proceed to look at the details involved.

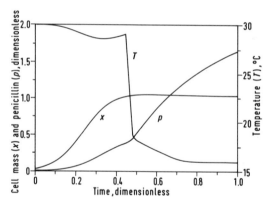

Fig. 8. Optimal profiles for cell mass, penicillin, and temperature (CONSTANTINIDES et al., 1970b).

The optimal control problem is to determine the temperature profile that maximizes a performance index, the penicillin concentration at the final time $p(t_f)$.

$$\underset{T}{\text{Max}} [P] = \underset{T}{\text{Max}} [p(t_f)] \qquad (44)$$

The Hamiltonian given by Eq. (27) is:

$$H = \lambda_1 [k_1 x - (k_1/k_2)x^2] + \lambda_2 [k_3 x - k_4 p] \qquad (45)$$

The necessary condition for maximum given by Eq. (30) is:

$$\frac{\partial H}{\partial T} = \lambda_1 x \frac{\partial k_1}{\partial T} - \lambda_1 x^2 \frac{\partial}{\partial T} \left(\frac{k_1}{k_2} \right) +$$
$$+ \lambda_2 x \frac{\partial k_3}{\partial T} - \lambda_2 p \frac{\partial k_4}{\partial T} \qquad (46)$$

The adjoint equations given by Eq. (28) and the final conditions given by Eq. (29) are:

$$\dot{\lambda}_1 = -\lambda_1 k_1 + 2\lambda_1 (k_1/k_1)x - \lambda_3 k_3 \quad \lambda_1(t_f) = 0 \quad (47)$$

$$\dot{\lambda}_2 = \lambda_2 k_4 \qquad \lambda_2(t_f) = 1 \quad (48)$$

If decreasing cell mass is not allowed, a constraint is imposed on cell mass concentration:

$$x \leq k_2 \qquad (49)$$

The value of k_2 is equivalent to the cell concentration at infinite time, i.e., the maximum growth for constant temperature. This constraint implies that the net rate of cell formation becomes zero when the temperature drops below a certain level. When the constraint is reached, Eq. (39) becomes:

$$\dot{x} = 0 \quad \text{when } x \geq k_2 \qquad (50)$$

The Hamiltonian then changes to:

$$H = \lambda_2 k_3 x - \lambda_2 k_4 p \qquad (51)$$

and the adjoint equation is modified to:

$$\dot{\lambda}_1 = -\lambda_2 k_3 \qquad (52)$$

As shown above, the application of continuous Maximum Principle results in a split boundary value problem. The authors tried the two different methods discussed above, the boundary value iteration and the control variable iteration. The control variable iteration was found to be the most rapid and effective.

Results are shown in Fig. 8. The optimum temperature profile began at 30 °C, which favored growth of the organism, decreased slowly, and then increased slightly. The optimal temperature remained above 28.6 °C through the first part of the process, resulting in a high cell concentration. When the constraint was reached, i.e., when the rate of cell formation reached zero, the optimal temperature shifted rapidly to a lower level, approximately 18 °C, maximizing the net rate of penicillin formation (the difference between the rate of formation and the rate of degradation). After that, the temperature decreased gradually, remaining below 18 °C. The final penicillin potency was 76.6% higher than that obtained with the best constant temperature of 25 °C.

3.2.2 Profile Optimization for Fedbatch Bioreactors

Early in the 20th century it was found that yeast production was maximized by adding wort at intervals of time rather than all at once (WHITAKER, 1980). Since then, many industrially important fermentation processes have been carried out in a semibatch manner. Alcohols, amino acids, antibiotics, enzymes, microbial cells, organic acids, vitamins, and various recombinant cell products are among the products for which semibatch operations have been used or tested (MODAK et al., 1986). In a typical semibatch mode of operation, the so-called *fedbatch* operation, the nutrients necessary for cell growth, the precursors for product formation, or the inducers are fed intermittently, continuously, or in a lump during an otherwise batch operation, and the fermentation broth is harvested either fully or partially at the end. The whole process may be repeated either with a fresh inoculum when the harvest is complete, or with the remaining cells acting as the inoculum for the next cycle when the harvesting is partially done.

Fedbatch operation has been found particularly effective for processes in which effects such as substrate inhibition, catabolite repression, product inhibition, glucose effects, and auxotrophic mutation are important (MODAK et al., 1986). These phenomena lead to unimodal reaction rate expressions that exhibit a maximum with respect to a single reactant concentration or in terms of two or more reactant concentrations. A living cell possesses a complex internal control system involving opposing phenomena such as activation and inhibition and induction and repression, which lead to unimodal rate expressions. Some chemical reactions also lead to unimodal rate expressions; these are autocatalytic, adiabatic exothermal, and Langmuir–Hinshelwood type catalytic reactions. Such reactions have not been subjected to thorough optimization in terms of reactor operations – batch, continuous, or semibatch operations – until recently (WAGHMARE and LIM, 1981).

Isothermal reactor operations for simple reactions have been optimized (WAGHMARE and LIM, 1981) by optimal control theory. It has been shown that whenever the rate expression goes through a maximum, i.e., the rate expression is a non-monotonic (unimodal) function of the reactant, a semibatch reactor (a variable-volume batch reactor with a programmed feed, i.e., a so-called fedbatch culture) is likely to outperform either a continuous reactor or a batch reactor. In other words, if the specific rate is a non-monotonic function of the substrate concentration, a fedbatch bioreactor may lead to a better fermentation result than that achievable with a batch bioreactor or a continuous bioreactor. Thus, one should fully explore the possibility of using fedbatch cultures and apply optimization theory as developed in this chapter to determine the best feed-rate profile as a function of time. A logical deduction from this is to anticipate that whenever there are two opposing effects such as activation and inhibition or induction and repression, a semibatch operation (or fedbatch) is a prime candidate to be considered for optimum yield or productivity. This anticipation had been proven correct. It so happens that the general category of autocatalytic reactions, of which fermentation involving free cells is a classic example, is also appropriate for semibatch operations. The problem is then the determination of the optimum feed rate of substrate as a function of time, which optimizes the given performance index such as productivity, yield, or profit.

The nutrient limiting the growth of cells in a fedbatch bioreactor provides an excellent means of controlling the growth rate and the metabolism of the cell. Thus, fedbatch bioreactors may be operated in a variety of ways by regulating the feed rate in a predetermined manner (feedforward control) or using a feedback control. The most commonly used are constantly fed, exponentially fed, extended and repeated fed-batch cultures. In extended fedbatch culture, the feed rate is regulated to maintain the substrate concentration constant until the bioreactor is full. These modes of operations are limiting cases of the complex optimal feed-rate profiles. In a repeated fed-batch culture a part of the broth remaining after a partial removal at the end of a cycle is used as an inoculum for the next cycle.

For recombinant cells that normally contain regulated promoters, fedbatch operation provides not only the means to regulate the promoters, but also the environment necessary to maximize the formation of product while minimizing the production of intermediates detrimental to cell growth and product formation, such as acetic acid in *Escherichia coli* and ethanol in yeast. The problem of such intermediates becomes more critical as one strives to maximize productivity by achieving high cell concentrations, as in industrial fermentation. Just as recombinant technology is used to optimize product formation at the molecular level, bioreactor optimization is aimed at product optimization at the bioreactor environmental level, the culture environment. In this section we consider a class of simple fedbatch cultures (MODAK et al., 1986). We will first consider the theoretical development and then describe the general characteristics of the optimal feed-rate profiles that maximize a profit function. This will be followed by examples of optimal feed-rate profiles for penicillin, yeast, and α-amylase fermentations.

Formulation of a Fedbatch Optimization Problem

An ideal case is considered in which cell growth and product formation are limited by a single substrate fed continuously in some fashion into the bioreactor. The mass balance equations for the cell substrate, product, and overall mass for an isothermal and constant density bioreactor are:

$$\frac{d}{dt}(xv) = \mu x v \qquad x(0) = x_0 \qquad (53)$$

$$\frac{d}{dt}(sv) = -\sigma x v + F s_F \qquad s(0) = s_0 \qquad (54)$$

$$\frac{d}{dt}(pv) = \pi x v - k p v \qquad p(0) = p_0 \qquad (55)$$

$$\frac{d}{dt}(v) = F \qquad\qquad v(0) = v_0 \qquad (56)$$

where the notations adopted in Sect. 3.2 have been used. The objective is to determine the optimal feed profile as a function of time, $F^*(t)$, which maximizes a performance index reflecting the final outcome of fermentation, $P[f] = g(x_f, p_f, s_f, t_f)$; the subscript f denotes the final conditions, and the fermentation time, t_f, is assumed to be given (fixed) or to be open to determination (free). Constraints are imposed on the final bioreactor volume, and the substrate feed rate is constrained as indicated below:

$$v(t_f) = v_{max} \qquad (57)$$

$$0 \leq F(t) \leq F_{max} \qquad (58)$$

Inspection of the above mass balance equations suggests that it would be convenient to work with total amounts instead of concentrations. Thus, we introduce state variables, $x_1 = xv$, $x_2 = sv$, $x_3 = pv$, $x_4 = v$, and $x_5 = t$, and rewrite Eqs. (53) through (56) in compact form:

$$\frac{d}{dt}\begin{bmatrix} x_1 \\ x_2 \\ x_3 \\ x_4 \\ x_5 \end{bmatrix} = \begin{bmatrix} \mu x_1 \\ -\sigma x_1 \\ \pi x_1 - k x_3 \\ 0 \\ 1 \end{bmatrix} + \begin{bmatrix} 0 \\ s_F \\ 0 \\ 1 \\ 0 \end{bmatrix} F \qquad (59)$$

or

$$\frac{d}{dt} x = a(x) + b F \qquad x(0) = x_0 \qquad (60)$$

where $x = (x_1, x_2, x_3, x_4, x_5)^T$, $a = (\mu x_1, -\sigma x_1, \pi x_1 - k x_3, 0, 1)^T$, $b = (0, s_f, 0, 1, 0)^T$, and $x_0 = (x_{10}, x_{20}, x_{30}, x_{40}, 0)^T$. The performance index is given by:

$$P[F] = g[x(t_f)] \qquad (61)$$

The Maximum Principle states (PONTRYAGIN et al., 1962) that maximizing $P[F]$ is equivalent to maximizing the Hamiltonian given below:

$$H = \lambda^T [a(x) + bF] = \eta(x, \lambda) + \theta(\lambda) F \qquad (62)$$

where the adjoint vector, λ, must satisfy the following ordinary differential equations:

$$\dot{\lambda} = -\frac{\partial H}{\partial x} = -\left(\frac{\partial a}{\partial x}\right)^T \lambda = -a_x^T \lambda \qquad (63)$$

with the final condition depending on the functional form of P given by:

$$\lambda(t_f) = \left(\frac{\partial P}{\partial x(t_f)}\right)^T \qquad (64)$$

When the final fermentation time is fixed, the Hamiltonian is a constant, H^*, while it is zero when the final time is free. Since the flow rate, $F(t)$, appears linearly, maximization of the Hamiltonian depends on the sign of the coefficient of F, $\theta = \lambda^T b$; i.e., if θ is positive we choose the maximum flow rate, and if θ is negative we take the smallest F ($F = 0$, or a batch period). But if θ is identically zero over a finite period the Maximum Principle fails to yield a solution:

$$F(t) = \begin{cases} F_{\max} & \theta(t) > 0 \\ F_s & \theta(t) = 0 \quad t_j \le t \le t_{j+1} \\ 0 & \theta(t) < 0 \end{cases} \qquad (65)$$

where F_s is an intermediate flow rate yet to be determined in the finite interval $t_j \le t \le t_{j+1}$ over which θ is identically zero. All that is known at this point is that the optimal profile consists of periods of maximum flow rate ($F_{\max}$), of minimum flow rate ($F = 0$, or a batch period), and of intermediate flow rate (F_s). The exact sequence and timing of flow-rate shifts from one period to another are yet to be determined. The interval over which θ is identically zero is known as the singular interval and the control as the singular control, F_s. Since $\theta(t)$ is zero over the finite interval, its time derivatives (the first, second, and so on) also must vanish:

$$\theta(t) = \lambda^T b = 0 \quad t_j \le t \le t_{j+1} \qquad (66)$$

$$\frac{d\theta(t)}{dt} = \dot{\lambda}^T b = -\lambda^T a_x b = \lambda^T c = 0 \quad t_j \le t \le t_{j+1} \qquad (67)$$

and

$$\frac{d^2\theta(t)}{dt^2} = \dot{\lambda}^T c + \lambda^T \dot{c} = \lambda^T (c_x a - a_x c) + \\ + \lambda^T c_x bF = 0 \quad t_j \le t \le t_{j+1} \qquad (68)$$

Since the feed rate appears explicitly in Eq. (68), the required singular flow rate is obtained from Eq. (68):

$$F_s = \frac{\lambda^T (a_x c - c_x a)}{\lambda^T c_x b} \qquad (69)$$

which is a non-linear feedback control law for the feed rate in terms of the state x and the adjoint variables λ. Therefore, it is necessary to solve both the state equation, Eq. (60), with the known initial conditions and the adjoint equation, Eq. (63), with the known final conditions, Eq. (64). This type of problem is known as a two-point boundary value problem. For free final time problems the computational effort may be reduced by noting that during the singular interval $H = 0$:

$$H = \lambda^T a = 0$$
$$\theta(t) = \lambda^T b = 0 \qquad (70)$$
$$\frac{d\theta(t)}{dt} = \lambda^T c = 0 \quad t_j \le t \le t_{j+1}$$

Since Eq. (70) represents three linear equations in five adjoint variables, only two adjoint variables need be determined to completely describe the remaining three. Although the information obtained is not sufficient to deduce the optimal time profile, use of Eq. (70) considerably reduces the computational burden. Analysis involving asymptotic behavior and limiting cases directly amenable to analytical solutions allow deductions to general situations. Here we will briefly present the general characteristics of optimum feed-rate profiles and computational schemes to obtain them. Details are available elsewhere (MODAK et al., 1986; LIM et al., 1986).

Characteristics of Optimum Feed-Rate
Profiles

As stated in the introduction, the advantage
of fedbatch culture over other cultures may be
realized whenever the specific rates are non-
monotonic. It can also be shown that singular
feed-rate profiles are feasible whenever the
specific rates, μ and π, exhibit non-monotonic
behavior. It is therefore convenient to classify
fermentation processes into three types; (1)
monotonic μ and non-monotonic π, (2) non-
monotonic μ and monotonic π, and (3) non-
monotonic μ and π, and provide the general
characteristics of the feed-rate profiles resulting from the above analysis and computational
schemes.

*Type I. Monotonically Increasing Specific
Growth Rate, μ, and Non-monotonic Specific
Product Formation Rate, π, Exhibiting a
Maximum*

The specific growth rate, μ, increases with
substrate concentrations, while the specific
product formation rate, π, first increases and
then decreases with substrate concentration.
Many industrial fermentation processes – production of antibiotics such as penicillins and
cephalosporins and amino acids such as lysine
and phenylalanine – belong to this type. In
fact, this is the most common type for which
fedbatch operations have been used. Typical
profiles deduced from the above analysis are
given in Fig. 9, where four different initial
conditions are depicted. Case (a) represents situations in which the amount of inoculum
($xv(0)$) and the initial substrate concentration
(s) are small; the optimal feed-flow rate consists of a period of maximum flow rate (F_{max})
followed by a period of minimum flow rate
($F=0$, a batch period), a period of singular
flow rate (F_s) until the fermentor is full, and
finally a batch period until the specified fermentation time is reached, or until it is coun-
terproductive to carry on fermentation any
further, as in free fermentation time. We may
call this sequence "bang-bang-singular-batch".
The singular flow rate normally increases with
time, exhibiting an approximately exponential
form. But, in certain cases it may decrease

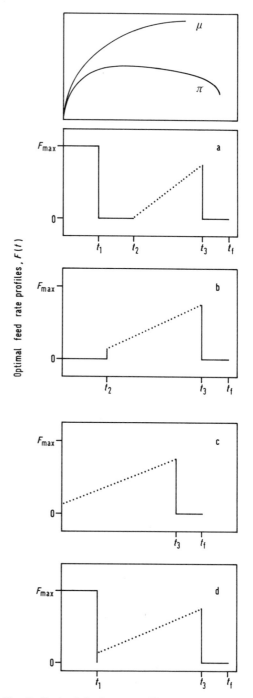

Fig. 9. Optimal feed rate profiles, $F(t)$, for type I
fermentation processes: (a) Low cell mass and low
glucose, (b) high glucose, (c) appropriate cell mass
and glucose, (d) high cell mass and low glucose.

with time toward the end of fermentation, especially when there is another carbon source formed during fermentation that can be utilized by the microorganism. It can be deduced from further analysis that the initial period of maximum flow rate followed by a batch period, the so-called "bang-bang period", is used to place the bioreactor into the singular period in minimum time. Once the bioreactor is placed into the singular period, singular control is used to maximize product formation until the bioreactor is full. In other words, the optimal feed-rate profile forces the bioreactor to reach the singular arc in a minimum time, and the singular feed-rate profile keeps the bioreactor on the singular arc as long as possible (until the fermentor is full). From a practical point of view this can be interpreted by saying that cell growth initially is maximized by the bang-bang type feed-rate profile, and the total rate of production of product $(\pi x v)$ is then maximized by the singular flow-rate profile.

When the initial substrate concentration (s) is high, the first period of maximum flow rate disappears (case (b) in Fig. 9), so that the sequence is bang (batch)-singular-batch. In other words, if the initial substrate concentration is high, one need not supply additional amounts of substrate, and all that is needed to reach the singular arc is to grow more cells, which is accomplished by utilizing a batch period. Once the proper levels of cells and substrate concentration are achieved, the singular feed-rate profile takes over, and the total product formation rate is again maximized.

When the amount of inoculum and the substrate concentration are appropriately chosen (*a priori* unknown; these must be determined through optimization), the optimum profile consists entirely of a period of singular flow rate (case (c) in Fig. 9), without a period of a maximum feed rate or a batch period. Thus, the sequence is singular-batch. Case (c) represents the ideal situation in which the initial conditions are proper to place the system on the singular arc from the start, yielding the best results. It usually requires, however, considerable inoculum, and therefore may be impractical or demand a series of inoculum transfer operations. In fact, in many industrial fermentations a series of inoculum transfer oper-

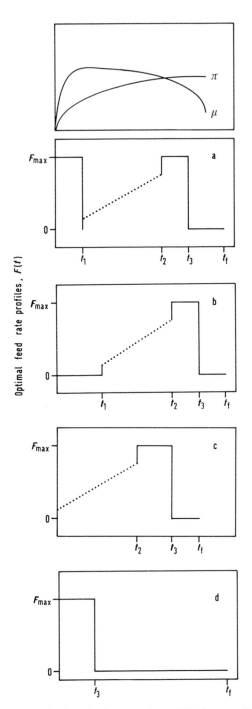

Fig. 10. Optimal feed rate profiles, $F(t)$, for type II fermentation processes: (a) Low cell mass and low glucose, (b) high glucose, (c) appropriate cell mass and glucose, (d) high cell mass and low glucose.

ations is used to provide inoculum for large bioreactors.

Finally, there is the purely theoretical situation in which the amount of inoculum is very large and the initial substrate concentration is low (case (d) in Fig. 9). For this case the optimal profile can miss a batch period after the period of maximum flow, resulting in a period of maximum flow rate followed by a period of singular control and then a batch period, bang (maximum)-singular-batch.

Type II. Specific Growth Rate Exhibiting a Maximum and Monotonically Increasing Specific Product Formation

A less common situation than type I is that in which the specific growth rate exhibits a maximum while the specific product formation rate increases with substrate concentration. Reports suggesting this may be the case include glutamic acid fermentation on ethanol and vitamin B_{12} fermentation. Single-cell protein production involving microbial cell mass is a limited version of this type, since the product itself is cell mass. The initial conditions ((a) through (d) in Fig. 10) dictate the optimal profile sequences, as we have seen above. When the amount of inoculum and the initial substrate concentrations are small, a situation denoted by case (a) in Fig. 10, the optimal profile consists of a period of maximum flow followed by a period of singular flow, a period of maximum flow, and a batch period. A simple physical explanation can be provided here. Since the specific product formation rate is a monotonically increasing function of substrate concentration, whereas specific growth first increases and then decreases with substrate concentration, going through a maximum, all that is needed to maximize the total product formation rate, $\pi x v$, is to force the substrate concentration to reach the value that maximizes μ and keeps it there as long as possible. This is accomplished by applying the feed at the maximum rate and then switching the singular flow rate to maintain the substrate concentration at a value corresponding to the maximum specific growth rate. This physical interpretation is correct if the yield coefficients are constant. When the yield coefficients are not constant,

the flow rate sequence just described does not maintain the substrate concentration constant at the value corresponding to the maximum specific growth rate, but results instead in variable substrate concentrations that maximize the total product formation rate, $\pi x v$. As with type I when the initial substrate concentration is high (case (b)), the initial period of maximum flow rate disappears, and the optimal profile consists of a batch period followed by a period of maximum flow rate and a batch period, bang (maximum)-singular-batch. In an ideal situation in which the amount of inoculum and the initial substrate concentration are chosen just right (case (c)), the optimum feed profile begins with a period of singular flow rate followed by a batch period.

Type III. Both Specific Growth and Product Formation Rates Exhibiting Maxima

Since μ and π both show maxima, the feed-rate profile must take advantage of this fact. This is the least common type of fermentation. An example is ethanol fermentation from fructose. Typical profiles are shown in Fig. 11. When the initial amount of inoculum and the substrate concentrations are low (case (a)), the feed profile resembles that of case (d), type I-a period of maximum flow rate followed by a period of singular flow rate and a batch period, bang (maximum)-singular-batch. A physical interpretation can be envisioned by considering a limiting case in which both μ and π show their maxima at the same substrate concentration and the yield coefficients are also constant. Here one should apply the maximum flow rate so that the substrate concentration in the bioreactor reaches in a minimum time the value that maximizes the specific rates; one should then apply a singular feed rate to maintain the substrate concentration at this level until the reactor is full. After that a batch operation should be continued until it is no longer economical, i.e., the maintenance cost outweighs the profit realized by converting the remaining substrate to product. Apparently, when the peaks in the specific rates do not coincide, and when the yield coefficients are not constant but vary with substrate concentration, the singular feed rate maximizes total

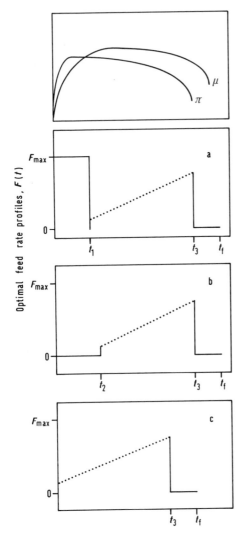

Fig. 11. Optimal feed rate profiles, $F(t)$, for type III fermentation processes: (a) Low cell mass and low glucose, (b) high glucose, (c) appropriate cell mass and glucose.

production formation, $\pi x v$, by varying the substrate concentration in the bioreactor. When the initial substrate concentration is high (case (b)), the optimum flow-rate profile consists of a batch period followed by a period of singular flow rate and a batch period, bang (minimum)-singular-batch. Finally, when the amount of inoculum and the substrate concentrations are just right, the optimum profile

may begin with a singular period followed by a batch period when the bioreactor is full.

A Computational Scheme to Determine the Optimal Feed-Rate Profile

Type I Fermentation Processes

The optimal feed-rate profile conjectured for type I fermentation processes with low initial substrate and cell mass concentrations is shown in Fig. 9(a). It consists of a period of maximum flow rate, a period of minimum flow rate (a batch period), a period of singular flow rate, and a period of minimum flow rate (batch period). The parameters to be determined are three switching times and the final times, t_1, t_2, t_3, and t_f. Switching time t_3 is the time at which the bioreactor is full and therefore is known during numerical computation. The final time is either specified *a priori* or determined by satisfying a specified condition. Only two switching times, t_1 and t_2, are unknown. The singular flow rate, F_s, is given by Eq. (69), and the computational aspect of F_s through Eq. (70) has been discussed above. The computational algorithm given below is based upon a case in which productivity (p_f/t_f) is to be maximized with a free final time.

1. Choose t_1 and t_2, $t_1 < t_2$.
2. Integrate forward the state equation with the given initial conditions, Eq. (60), using $F = F_{max}$ until $t = t_1$.
3. Continue integrating Eq. (60) from t_1 to t_2 using $F = F_{min} = 0$.
4. Estimate the unknown variables at t_2, $\lambda_3(t_2)$, and $\lambda_5(t_2)$.
5. Continue integrating until the bioreactor is full, $t = t_3$, the two adjoint equations, $\dot{\lambda}_3$ and $\dot{\lambda}_5$, Eq. (63), using the singular flow rate, F_s, given by Eq. (69), with the integrated state variables, the two integrated adjoint variables, and the remaining three calculated adjoint variables from Eq. (70).
6. At $t = t_3$ integrate the complete sets of state and adjoint equations using $F = F_{min} = 0$ until the specified final time, or until the specified final condition is met, $t = t_f$.

7. Compare the calculated adjoint variables at the final time, $\lambda_3(t_f)$ and $\lambda_5(t_f)$, with the specified values, Eq. (64), $\lambda_3(t_f) = 1/t_f$, and $\lambda_5(t_f) = -p_f/t_f^2$.

8. If the calculated values do not agree with the specified values, improve the estimated values of $\lambda_3(t_2)$ and $\lambda_5(t_2)$ by a Newton–Raphson method and go to step 5.

9. If there is agreement between the calculated and the specified values of $\lambda_3(t_2)$ and $\lambda_5(t_2)$, store the values of the switching times, t_1 and t_2, and the corresponding value of the performance index, $g[x(t_f)]$.

10. Change or increment t_2 by Δt_2, and if the entire range of t_2 has not been covered, go to step 2. If covered, go to step 11.

11. Change or increment t_1 by Δt_1, and if the entire range of t_1 has not been covered, go to step 2. If covered, go to step 12.

12. After sets of switching times have been tried, choose the one that yields the maximum value, or improve switching times by further narrowing the range.

The above approach to searching over entire ranges of t_1 and t_2 is based on the assumption that the ranges of t_1 and t_2 are known. For a given maximum flow rate, t_1 has an upper limit, and it is the time required to fill the bioreactor with the maximum flow rate, i.e., $t_{1\,\text{max}} \leq v_{\text{max}}/F_{\text{max}}$. There is also an upper limit of t_2, which is the time at which the substrate concentration becomes very small as the singular feed rate must initiate before the cells starve or near the substrate concentration corresponding to the optimum product formation rate. In actual computations the range of t_1 and t_2 can be narrowed further through analytical results as previously discussed (MODAK et al., 1986).

For other special initial conditions the optimal feed-rate profiles degenerate, as shown in Fig. 9(b)–(d), and thus the above algorithms need to be modified. Actually, these situations require simplification of the above general algorithm. Fig. 9(b) represents a situation in which the initial substrate concentration is high. The optimal control sequence now lacks

a period of maximum feed rate, $t_1 = 0$. Therefore, we can set $t_1 = 0$ in step 1 and skip steps 2 and 11. For the situation shown in Fig. 9(c), in which initial conditions are such that the process is on the singular arc, we can set $t_1 = t_2 = 0$ and skip 2, 3, 10, and 11. For Fig. 9(d) we can set $t_2 = t_1$ and skip steps 3 and 10.

Type II Fermentation Processes

We will develop a computational algorithm for the general profile given in Fig. 10(a). Algorithms for other profiles are then obtained by modifying that for Fig. 10(a). As in type I fermentation we will use as the performance index the maximization of product productivity with a free final time.

1. Choose t_1 and t_2, $t_1 < t_2$.

2. Integrate forward the state equation with the given initial conditions, Eq. (60), using $F = F_{\text{max}}$ until $t = t_1$.

3. Estimate the unknown variables at t_1, $\lambda_3(t_1)$, and $\lambda_5(t_1) = \lambda_5(t)$.

4. Integrate, λ_3 equation, Eq. (63), and the state equations, Eq. (60), forward from t_1 to t_2 using the singular flow rate, F_s, given by Eq. (69), as calculated by the integrated λ_3, the assumed $\lambda_5(t)$, the integrated state variables, and the remaining three calculated adjoint variables calculated from Eq. (70).

5. Using $F = F_{\text{max}}$, integrate forward the entire sets of the state and adjoint equations, Eqs. (60) and (63), from t_2 to t_3, the time at which the bioreactor is full.

6. Continue integrating forward using $F = F_{\text{min}} = 0$ from t_3 to t_f, at which either the specified final time or condition is met.

7. Compare the calculated adjoint variables, $\lambda_3(t_f)$ and $\lambda_5(t_f)$, with the specified values, $\lambda_3(t_f) = 1/t_f$, and $\lambda_5(t_f) = -p_f/t_f^2$.

8. If there is no agreement between the calculated values and the specified values, improve the estimated values of $\lambda_3(t_1)$ and $\lambda_5(t_1)$ by a Newton–Raphson method and go back to step 4.

9. If there is agreement, store switching

times t_1 and t_2 and the corresponding value of the performance index, $g[x(t_f)]$.

10. Change or increment t_2 by Δt_2, and if the entire range of t_2 has not been covered, go to step 2. If covered, go to step 11.
11. Change or increment t_1 by Δt_1, and if the entire range of t_1 has not been covered, go to step 2. If covered, go to step 12.
12. After sets of switching times have been tried over the entire feasible range, choose the one that yielded the maximum value of performance index. If necessary, further narrow the grid points.

When the optimal feed-rate profile is given by Fig. 10(b), it is only necessary to modify step 2 using $F = F_{min} = 0$ instead of $F = F_{max}$. For the optimal feed profile given by Fig. 10(c), we set $t_1 = 0$ in step 1 and skip steps 2 and 11. The profile given in Fig. 10(d) is a constant fedbatch process, and switching time $t_3 = [v_{max} - v(0)]/F_{max}$ is predetermined.

Type III Fermentation Processes

Again we give an algorithm for the profile depicted in Fig. 11(a) and modifications necessary for the profiles in Fig. 11(b) and (c). The performance index is a maximization of product productivity with a free final time. The algorithm is the same as that for the profile given in Fig. 9(d) and is as follows:

1. Choose t_1.
2. Integrate forward the state equation with the given initial conditions, Eq. (60), using $F = F_{max}$ until $t = t_1$.
3. Estimate the unknown variables at t_1, $\lambda_3(t_1)$, and $\lambda_5(t_1) = \lambda_5(t)$.
4. Integrate forward the λ_3 equations, Eq. (63), and the state equations, Eq. (60), from t_1 until the fermentor is full, $t = t_3$, using $F = F_s$ given by Eq. (69), as calculated from Eq. (70).
5. Continue to integrate the state equations, Eq. (60), and the entire sets of adjoint equations, Eq. (63), from $t = t_3$

until the final time or condition is met, $t = t_f$.

6. Compare the calculated adjoint variables, $\lambda_3(t_f)$ and $\lambda_5(t_f)$, with the specified values, $\lambda_3(t_f) = 1/t_f$ and $\lambda_5(t_f) = -p_f/t_f^2$.
7. If there is no agreement, improve the estimated values of $\lambda_3(t_1)$ and $\lambda_5(t_1)$ by a Newton–Raphson method and a go back to step 4.
8. If there is agreement, store the switching time t_1 and the corresponding value of the performance index, $g[x(t_f)]$.
9. Increment t_1 by Δt_1, and if the entire range of t_1 has not been covered, go to step 2. If covered, go to step 10.
10. After various values of t_1 have been tried, choose the one that gives the maximum value for the performance index. If necessary, further narrow the interval, say by a golden search.

For the profile given in Fig. 11(b), use $F = F_{min} = 0$ and $t = t_2$ in step 2. For the profile in Fig. 11(c), steps 1 and 2 are skipped, t_1 is set to zero in step 3, and steps 9 and 10 are also skipped.

Examples

Examples of penicillin, baker's yeast, and α-amylase fermentations are given in some de-

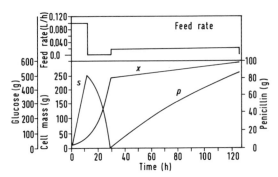

Fig. 12. Optimal glucose feed rate and corresponding cell, glucose, and penicillin profiles; low initial glucose concentration.
$x_0 = 10.5$ g, $s_0 = 10^{-6}$ g, $p_0 = 0$ g, $v_0 = 7$ L, $v_f = 10$ L, $F_{max} = 10$ mL/h, $s_F = 500$ g/L (Lim et al., 1986).

tail. The computational techniques are summarized above (the details are given elsewhere (LIM et al., 1986), and these are used to generate the numerical results below.

Penicillin Fermentation by *Penicillium chrysogenum*

The penicillin fermentation model of BAJPAI and REUSS (1981) given below has three component mass balance equations for substrate (glucose), cell mass, and penicillin, and one overall mass balance equation, Eqs. (53) through (56), with the following specific rates:

$$\mu = 0.11 \, s/(0.006x + s)$$
$$\pi = 0.004 \, s/(0.0001 + s + 10 \, s^2)$$
$$\sigma = \mu/0.47 + \pi/1.2 + 0.029$$
$$k = 0.01 \text{ h}^{-1}$$
(71)

The authors state that the model with the above rates adequately describes the experimental data available in the literature (PIRT and RIGHELATO, 1967; HOSLER and JOHNSON, 1953; MOU, 1983). The performance index to be maximized is the amount of penicillin at an unspecified final time. This is a free time problem, so the optimization procedure must be used to determine the optimum time profile for feed rate that maximizes the total mount of penicillin produced at various final times and the one must be chosen that gives the largest amount of penicillin.

This is a typical example of type I fermentation due to the non-monotonic specific penicillin production rate, π. The optimal glucose feed-rate profiles for glucose, cell mass, and penicillin are given in Fig. 12. The optimal glucose feed profile consists of a period (11.2 h) of maximum flow rate, a batch period (17.6 h), a singular flow rate period (95 h), and a negligibly small batch period. Note that the profiles clearly show the experimentally observed phenomena: a tropophase in which almost no penicillin is produced, followed by an ideophase in which penicillin is produced and very little cell growth takes place.

Production of Baker's Yeast *Saccharomyces cerevisiae*

The model developed by MODAK (1988) for baker's yeast is given below, consisting of mass balance equations for glucose, ethanol, and cell mass, and the overall mass balance equations, along the line of Eqs. (53) through (56):

$$\frac{\mathrm{d}(xv)}{\mathrm{d}t} = \mu(xv) \tag{72}$$

$$\frac{\mathrm{d}(Gv)}{\mathrm{d}t} = Fs_F - \sigma(xv) \tag{73}$$

$$\frac{\mathrm{d}(Ev)}{\mathrm{d}t} = (\pi - \eta)(xv) \tag{74}$$

$$\frac{\mathrm{d}(v)}{\mathrm{d}t} = F \tag{75}$$

where G and E stand for glucose and ethanol concentrations, respectively; various rates are given below:

$$\mu = \mu_G + \mu_E = \frac{k_1 G + k_2 G^2}{k_3 + k_4 G + G^2} + \\ + \frac{k_5 E}{(k_2 + k_7 \sigma + E)(1 + k_7 \sigma)} \tag{76}$$

$$\sigma = \mu_G/k_{11} R + k_{12}(1 - R) \tag{77}$$

$$R(G) = \frac{1 + k_9 G^2}{k_{10} + k_9 G^2} \tag{78}$$

$$\pi = k_{13} \sigma R \tag{79}$$

$$\eta = \mu_E/k_{14} \tag{80}$$

The numerical values of these parameters are given in Tab. 5.

Tab. 5. Kinetic Model Parameters (in g/g) (MODAK, 1988)

$k_1 = 0.079$	$k_6 = 0.005$	$k_{11} = 0.1776$
$k_2 = 0.45$	$k_7 = 0.59$	$k_{12} = 0.56$
$k_3 = 0.0011$	$k_8 = 1.43$	$k_{13} = 0.459$
$k_4 = 0.351$	$k_9 = 20712.9$	$k_{14} = 0.625$
$k_5 = 0.15$	$k_{10} = 100$	

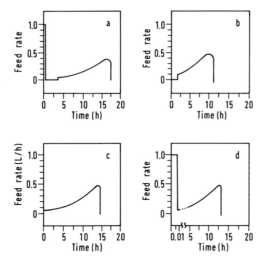

Fig. 13. Optimal feed rate profiles for baker's yeast fermentation: (a) Low cell mass and low glucose, (b) high glucose, (c) appropriate cell mass and glucose, (d) high cell mass and low glucose (MODAK, 1988).

The performance index chosen here is the profit, as represented by the difference between cell mass price and operating costs. The operating cost is assumed to be proportional to fermentation time, t_f. The glucose feed rate is optimized to maximize profit. The results are shown in Fig. 13 for four different initial conditions. Unlike the case of penicillin fermentation, for which the singular flow rate is monotonically increasing, the singular flow rate that maximizes the measure of profit for yeast fermentation is non-monotonic; i.e., it goes through a peak. The reason is that with baker's yeast the singular flow rate decreases toward the end of fermentation because the optimal policy makes use for cell mass of the residual ethanol that is generated from glucose, thus requiring less glucose feed.

Optimal Control of Carbon and Nitrogen Sources for the Production of α-Amylase

Many industrially important fermentation processes involve microbial cells that require more than one substrate for their growth. It has long been realized that the production of antibiotics and enzymes requires precise con-

trol of the nitrogen source in addition to the carbon source. Manipulation of more than one control variable for optimizing the performance of bioreactors is now becoming more common. For example, the use of recombinant plasmid cultures for production of valuable products has led to increasing use of a variety of inducers and promoters. In addition, recent reports (AIBA and KOIZUMI, 1984; KOIZUMI et al., 1985) show the use of temperature-sensitive plasmids to construct recombinant cell systems. A shift in temperature acts as an inducer for the cells to produce a desired metabolite. These advances in microbial systems pose challenging problems. They require manipulating more than one control variable, namely the bioreactor temperature and the substrate feed rate, or the feed rates of two growth-limiting nutrients. We have chosen a practical example, the optimal control of carbon and nitrogen supplies for α-amylase production.

In this example, we consider the computation of optimal feed rates of carbon and nitrogen sources for the production of α-amylase. PAZLAROVA et al. (1984) reported a kinetic model of α-amylase production by *Bacillus subtilis* using starch (s_1) and caseinate (s_2) as carbon and nitrogen sources, respectively. The kinetic model is described by the following set of differential mass balance equations:

$$\frac{d(xv)}{dt} = \mu(s_1, s_2)xv \tag{81}$$

$$\frac{d(s_1 v)}{dt} = F_1 s_{1F} - \sigma_1(s_1, s_2)xv \tag{82}$$

$$\frac{d(s_2 v)}{dt} = F_2 s_{2F} - \sigma_2(s_1, s_2)xv \tag{83}$$

$$\frac{d(pv)}{dt} = \pi(s_1, s_2)xv - kpv \tag{84}$$

$$\frac{dv}{dt} = F_1 + F_2 \tag{85}$$

where x, s_1, s_2, and p represent the concentrations of cells, starch, caseinate, and α-amylase, respectively. v is the bioreactor volume, F_1 and F_2 the feed rates of starch and caseinate, s_{1F} and s_{2F} the feed concentrations of starch and caseinate, and k the α-amylase hydrolysis rate constant. The specific rates μ, σ_1, σ_2, and π are functions of starch and caseinate concentrations:

$$\mu = \frac{0.086\, s_1 s_2}{(2.0 + s_1 + s_1^2/33.0)} \qquad \sigma_1 = \frac{\mu}{0.68}$$

$$\pi = 117.7\, \exp^{-0.311 s_2}\mu \quad \sigma_2 = \frac{\mu}{1.05} \tag{86}$$

and $k = 0.18$. The operating conditions used in the optimization study are

$$x(0) = 0.1 \text{ g/L, } s_2(0) = 0 \text{ g/L}$$
$$P(0) = 0 \text{ g/L, } v(0) = 1 \text{ L}$$
$$v_{max}(0) = 5 \text{ L, } F_{max} = 1 \text{ L/h} \tag{87}$$
$$s_{2F} = 10 \text{ g/L, } e = 5.0$$

The objective is to maximize the profit realized for α-amylase production with a penalty for operating costs that is assumed to be proportional to the fermentation operating period, t_f. The objective function, P, to be maximized is represented by Eq. (88):

$$\underset{F_1, F_2}{\text{Max}} = \{P = Pv(t_f) - et_f\} \tag{88}$$

where e is the operating cost relative to the selling price of the product and the final time, t_f, (fermentation operating period) is free.

The optimization problem posed by Eqs. (81) through (85) and (88) is a five-dimensional singular control problem with two control variables appearing linearly. We propose an alternate approach by modifying the original control variables. The original problem involves the feed rates of carbon and nitrogen sources as the control variables. In the new formulation we choose the concentration of carbon source (starch) in the bioreactor and the feed rate of caseinate as new control variables. Note that the former appears non-linearly while the latter appears linearly in Eqs. (81) through (85). In making such a control variable transformation, it is implicitly assumed that the carbon source is available as a highly concentrated solution so that its addition does not significantly change the volume of the bioreactor compared to the volume changes due to the addition of nitrogen source

($F_1 \ll F_2$). Therefore, the feed rate of carbon source, F_1, is neglected in the overall mass balance equation of Eq. (88).

We define as state variables $x_1 = xv$, $x_2 = s_2 v$, $x_3 = pv$, $x_4 = v$, and $x_5 = t$ and as control variables $m_1 = F_2$ and $m_2 = s_1$. Eqs. (81) through (85) can be expressed in terms of newly defined state and control variables as:

$$\frac{dx}{dt} = a(x, m_2) + b m_1 \tag{89}$$

and

$$a = [\mu x_1 - \sigma_2 x_1\ \pi x_1 - kx_3\ 0\ 1]^T \tag{90}$$
$$b = [0\ s_{2F}\ 0\ 1\ 0]^T \tag{91}$$

The objective function, Eq. (88), can be written as:

$$\underset{m_1, m_2}{\text{Max}} = \{P = x_3(t_f) - ex_5(t_f)\} \tag{92}$$

The necessary conditions of optimality can be developed by defining the Hamiltonian function, H:

$$H = \lambda_1 \mu x_1 - \lambda_2 \sigma_2 x_1 + \lambda_3 (\pi x_1 - kx_3) + \lambda_5 + \\ + (\lambda_2 s_{2F} + \lambda_4) F_2 = \psi(x, s_1) + \Phi_1 F_2 \tag{93}$$

Since the feed rate of caseinate F_2 appears linearly as in Eq. (93), its profile is accordingly dictated by Eq. (65), bang-bang-singular-bang, whereas the starch concentration appears non-linearly, so the optimality condition is:

$$\frac{\partial H}{\partial s_1} = \left(\lambda_1 \frac{\partial \mu}{\partial s_1} - \lambda_2 \frac{\partial \sigma_2}{\partial s_1} + \lambda_3 \frac{\partial \pi}{\partial s_1}\right) k_1 = 0 \tag{94}$$

or $\quad \dfrac{\partial \mu}{\partial s_1} = 0$

Condition (94) implies that optimal control policy is to maintain the starch concentration constant at the level that maximizes the specific growth rate of the cells. Since the specific α-amylase production rate is proportional to specific growth rate, μ, the optimal control policy that maximizes μ also maximizes π. A starch concentration of 8.12 g/L maximizes the specific growth rate, μ, and the optimal control policy is to maintain it constant at 8.12

g/L. By this fact the original optimization problem with two control variables is reduced to an optimization problem with only one (caseinate feed rate). This is a standard four-dimensional singular control problem with a single control variable. Interested readers are referred to our previous publications (MODAK et al., 1986; LIM et al., 1986; MODAK and LIM, 1989) for details of the computational algorithm.

Fig. 14 shows the optimal caseinate feeding policy for the α-amylase production process. The optimal feed rate has a period of maximum feed rate (0–2.1 h), a period of minimum (batch) feed rate (2.1–16.5 h), a period of singular feed rate (16.5–22.51 h), and a period of

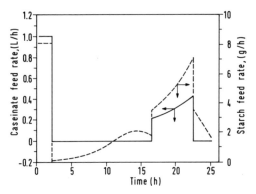

Fig. 14. Optimal caseinate and starch feed rate profiles for α-amylase production (MODAK and LIM, 1989).

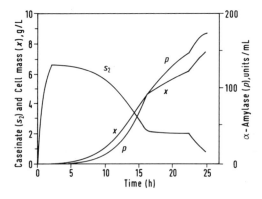

Fig. 15. Cell mass, caseinate, and α-amylase profiles with optimal starch and caseinate feeding strategy (MODAK and LIM, 1989).

minimum feed rate (22.51–24.89 h). For α-amylase production the specific growth rate of the cells increases monotonically with increases in caseinate concentration, whereas product yield is inhibited at high caseinate concentrations. Therefore, rapid cell growth can be achieved by supplying caseinate as rapidly as possible (maximum feed rate) and then shutting off the feeding (minimum feed rate) to allow the cells to grow on caseinate. This is followed by a period of singular feed rate to maintain the caseinate concentration at the level that does not inhibit product yield. The last period of minimum feed rate is a result of the constraint on the volume of the bioreactor. The concentration profiles resulting from the optimal feed rate are shown in Fig. 15. As expected, the high concentration of caseinate in the initial period (0–16.5 h) allows cells to grow rapidly, whereas during the singular interval (16.5–22.51 h), caseinate concentration is lower in order to achieve a higher product yield. The rate of addition of starch required to maintain a constant level (8.12 g/L) of starch can be calculated by rearranging the starch mass balance equation from the set of mass balance equations:

$$Q_1 = F_1^* s_{1F} = (\sigma_1^* (xv)^* + F_2^* s_1^*) \qquad (95)$$

where Q_1 is the starch addition rate and the superscript * denotes quantities evaluated with the optimal caseinate addition rate. Fig. 14 shows the optimal starch addition rate evaluated with Eq. (95).

4 Estimation Techniques

For better control and optimization of fermentation processes it is essential to measure on-line many key physiological parameters. Such measurements are difficult, if not impossible, for many of these parameters. Therefore, any control or optimization based on key physiological parameters cannot be implemented unless values can be measured on-line to provide the necessary information required by the controller or optimizer. Although efforts to develop such sensors are underway, the availability and reliability of instruments are very limited. There is thus an urgent need to provide the necessary information by estimating these parameters from others that are simpler and easier to measure. In this section we present parameter and state estimation algorithms that can be used to estimate a key parameter or a physiological state of a fermentor.

It is true for any processing industry that balance equations can provide much needed information. Material balance around a fermentor is extremely valuable in this respect. Energy balance equations have not played as important a role as material balance equations because fermentation processes take place in aqueous solutions, and because the heat of fermentation is usually rather small. Balances of elements such as C, H, N, O, and others can be used with off-gas data to estimate some quantities not directly measured, such as biomass or product concentrations. Since the state of instrumentation is still unsatisfactory, many measurements contain a high level of noise. Thus, raw measurement signals should be passed through suitable filters before they are made available to control bioreactors.

4.1 Indirect Measurements and Correlations

Certain data collected from bioreactors can be used directly, while others have very little physical significance by themselves and must therefore be combined with other measurements to provide physically significant information. For example, low level measurements such as temperature, pH, and dissolved oxygen can be directly fed back to actuate control such devices as the heater/cooler, pumps for acid/base addition, and controllers for the gas flow rate or the impeller rotation speed. On the other hand, gas flow rate data are somewhat uninformative unless they are combined with other measurements to form so-called "gateway" sensors (HUMPYHREY, 1971). Indirect measurements and gateway sensors provide information on cellular metabolism and fermentation conditions. The rates of con-

sumption or production of many species can be calculated by taking a simple species balance around the fermentor. Among the quantities most easily acquired on-line and most frequently calculated are the oxygen uptake rate (*OUR*), the carbon dioxide evolution rate (*CER*), and the respiratory quotient (*RQ*). These provide very useful information and are very good indicators of cellular respiratory activities. Other variables obtained from indirect measurements include the overall oxygen mass-transfer rate, metabolic heat-evolution rates, the specific growth rate, cell yields, substrate utilization rates, and secondary metabolite production rates. A list of the calculated variables (ARMIGER and HUMPHREY, 1979) and a set of straightforward step-by-step data analysis schemes are available (NYIRI, 1971, 1972). We discuss below a few schemes that are more practical.

The specific growth rate was measured indirectly in a turbidostat system by controlling the cell concentration between the upper and low limits via a continuous optical density measurement (VERES et al., 1981). The specific product formation rate can be determined by applying a similar technique to a pH control scheme and by carefully monitoring the cumulative amount of acid/base added to achieve neutralization. Acetic acid production by *Escherichia coli* (SAN and STEPHANOPOULOS, 1984) and gluconic acid production (VERES et al., 1981) have been estimated in this way.

Another important indirect measurement is the volumetric oxygen transfer coefficient, $k_L a$, as described by the following equation:

$$q_{O_2}(t) = (k_L a)(t)[C_{O_2}^*(t) - C_{O_2}(t)] \qquad (96)$$

where $q_{O_2}(t)$ is the volumetric oxygen transfer rate, $C_{O_2}^*(t)$ is the liquid-phase oxygen concentration in equilibrium with the gas phase, and $C_{O_2}(t)$ is the liquid-phase oxygen concentration. It is important to note that $k_L a$ is a function of time due to changes in fermentor conditions such as agitation, air dispersion, and rheological properties; it therefore needs to be estimated and constantly updated by combining the measurements of gas flow rate, gasphase oxygen concentration, and dissolved oxygen concentration.

Two general methods are used for estimation of $k_L a$: static and dynamic methods. Both methods assume that the bioreactor is well mixed, that $k_L a$ is a function of time but not space, and that the dynamics of oxygen are much faster than the dynamics of biomass, substrate, and products. The last assumption enables one to employ a quasi-steady-state approximation for the liquid phase. In the static method, $q_{O_2}(t)$ is calculated from gas flow rate and oxygen concentrations in the inlet and exit streams. By inferring $C_{O_2}^*(t)$ from Henry's law, and by directly measuring $C_{O_2}(t)$, one can calculate $k_L a$ continuously (SPRIET et al., 1982; SIEGELL and GADEN, 1962). In the dynamic method, the transient response of the dissolved oxygen is monitored continuously by interrupting the air supply for a short period (OHASHI et al., 1979; HILL and ROBINSON, 1974; DUNN and DANG, 1976).

4.2 Estimation through Macroscopic Balances

The principles and applications of macroscopic material and energy balances to fermentation fields have been summarized by ROELS (1980, 1981). These material balancing methods depend heavily on the measurement of gas exchange conditions in a bioreactor, and the availability of continuous gas analyzers for oxygen and carbon dioxide has therefore been essential for successful applications. Various on-line calculations based on elemental balances have been published (NYIRI et al., 1975; WANG et al., 1977, 1979a, b; ZABRISKIE and HUMPHREY, 1978; SWARTZ and COONEY, 1979; HIRAMA and HUMPHREY, 1980; CONSTANTINIDES and SHAO, 1981; STEPHANOPOULOS and SAN, 1982; COONEY and MOU, 1982; BRAVARD et al., 1979; MOU and COONEY, 1983a, b; SWARTZ et al., 1976; ERICKSON, 1979).

There are two variations to the material balance approach. The first method, adopted by COONEY et al. (1977) and WANG et al. (1977), is based on the concept of conservation of mass and overall chemical reaction stoichiometry. To illustrate this approach we consider perhaps the simplest fermentation, one in which no metabolite is produced, and therefore the product is cell mass. The basic feature of the method is to represent the biological conservation of substrate to cell mass by an overall chemical reaction as follows:

$$a\,C_\alpha H_\beta O_\gamma + b\,O_2 + c\,NH_3 \rightarrow$$
substrate
$$\rightarrow C_\delta H_\varepsilon O_\zeta N_\eta + d\,CO_2 + e\,H_2O \qquad (97)$$
cell mass

The stoichiometric coefficient for biomass has been normalized to 1. All the chemical formulas are assumed known (α through η are known) and constant, although the chemical composition of cells may be affected by growth rates and by the nature and composition of the medium (HERBERT, 1976; DEKKERS et al., 1981). Usually, it is only the ratio, not the absolute value, of the composition of the cell biomass that can be determined; therefore, δ can be set equal to 1 without loss of generality. Note that a, b, c, d, and e are the five unknown stoichiometric coefficients. To determine these five unknowns requires five equations, four of which are obtained from the elemental balance equations for carbon, hydrogen, oxygen, and nitrogen:

carbon: $\alpha a = \delta + d$ \qquad\qquad (98)

hydrogen: $\beta a + 3c = \varepsilon + 2e$ \qquad (99)

oxygen: $\gamma a + 2b = \zeta + 2d + e$ \quad (100)

nitrogen: $c = \eta$ \qquad\qquad\qquad (101)

By monitoring the offgas for oxygen and carbon dioxide one can calculate the carbon dioxide evolution rate (*CER*) and oxygen uptake rate (*OUR*) to yield the fifth equation needed, i.e.:

$$CER/OUR = d/b \qquad (102)$$

Thus, Eqs. (98) through (102) represent five equations with five unknowns. In other words, by monitoring the gas exchange rates (*CER* and *OUR*), the stoichiometric coefficients can be estimated on-line. The stoichiometric relationships are then used to calculate various rates such as substrate consumption rate (*SCR*), cell growth rate (*CGR*), *OUR*, and

CER, or the concentrations of various species over a short period during which the stoichiometric coefficients are assumed to be constant:

$$\Delta(\text{substrate})/-a = \Delta(\text{cell mass}), \text{ and}$$
$$SCR/a = OUR/b = CGR = CER/d \tag{103}$$

where $SCR = \sigma x v$ and $CGR = \mu x v$. Thus, the stoichiometric approach can be used to estimate rates that may be difficult to measure on-line, such as cell growth rate and substrate consumption rate, from the offgas exchange rates and species concentrations that may be difficult to measure on-line such as substrate and cell concentrations. For example, for a simple batch fermentation that can be described by cell and substrate balance equations:

$$d(xv)/dt = \mu x v = CER/d, \text{ or}$$
$$xv(t) = xv(0) + \int_0^t (CER/d)dt \tag{104}$$

$$d(sv)/dt = -\sigma x v = -CER(a/d), \text{ or}$$
$$sv(t) = sv(0) + \int_0^t (CER(a/d))dt \tag{105}$$

This approach is still valid in principle for fermentations involving one or more products and one or more substrates. In practice, however, additional measurements are necessary. For example, consider the case involving one product, given by the following stoichiometric equation:

$$a C_\alpha H_\beta O_\gamma + b O_2 + c NH_3 \rightarrow C_\delta H_\varepsilon O_\zeta N_\eta$$
substrate   cell mass
$$+ d C_\theta H_\iota O_\kappa N_\lambda + e CO_2 + f H_2O \tag{106}$$
   product

Note that *a, b, c, d, e,* and *f* are the six unknown stoichiometric coefficients. Elemental balances give four equations:

carbon: $\alpha a = \delta + \theta d + e$ (107)

hydrogen: $\beta a + 3c = \varepsilon + \iota d + 2f$ (108)

oxygen: $\gamma a + 2b = \zeta + \kappa d + 2e + f$ (109)

nitrogen: $c = \eta + \lambda d$ (110)

Two more equations are needed to solve for the six unknown stoichiometric coefficients *a* through *f*. As above, the offgas exchange information provides another equation:

$$CER/OUR = f/b \tag{111}$$

An additional relationship is still needed to complete the solution. Among the possible measurements are the concentration of nitrogen, substrate, or product, and the heat of fermentation. Although some of these species concentrations may be measurable in certain cases, they cannot be used directly to solve for the stoichiometric coefficients in Eq. (106). It is actually their time rates of change that must be monitored with a certain degree of accuracy. A simple difference between two consecutive measurements cannot be used to calculate the time rate of change because of the presence of noise inherent in any real measurement.

This first method was applied (CONSTANTINIDES and SHAO, 1981) to a batch glutamic acid fermentation by *Brevibacterium flavum*. Besides *OUR* and *CER*, the substrate (glucose) concentration was chosen as the additional measurement. The same method was also used (SWARTZ and COONEY, 1979) to monitor the growth of *Hansenula polymorpha* on methanol in a continuous fermentor.

The second method was advanced (ZABRISKIE et al., 1976, 1977) to estimate biomass concentration and growth rate through on-line material balances. This approach is based on the material balance of only one chemical component and a mathematical kinetic model that relates biomass growth to the chemical component. Therefore, the accuracy depends heavily on the validity of the mathematical model. Oxygen is an example of a chemical component that is well suited for this purpose; therefore, *OUR* is used to illustrate this method. The biomass concentration, *x*, is estimated in real time by integrating the following limiting substrate yield and maintenance model of PIRT (1965):

$$OUR = \frac{1}{Y_{O_2}} \frac{dx}{dt} + m_{O_2} x \tag{112}$$

The specific growth rate can be estimated by rewriting Eq. (24) as:

$$\mu = \frac{1}{x}\frac{dx}{dt} = \frac{Y_{O_2}}{b} OUR - m_{O_2} Y_{O_2} \qquad (113)$$

The disadvantage of this method is that instead of assuming a known constant biomass composition as in the first method, constant Y_{O_2} and m_{O_2} are assumed and determined from previous experiments run under similar conditions. Therefore, the accuracies depend heavily on the reproducibility of experimental conditions so that the same Y_{O_2} and m_{O_2} can be obtained. This second method was applied to batch cultures of *Thermoactinomyces* sp., *Streptomyces* sp., and *Saccharomyces cerevisiae* (ZABRISKIE and HUMPHREY, 1978). On-line correlations were successfully accomplished for simple organisms such as *Thermoactinomyces* and *Streptomyces*. However, in the case of more complex organisms such as *S. cerevisiae*, the values of Y_{O_2} and m_{O_2} varied, and a correction factor was needed to obtain good agreement between the estimated and actual biomass data. The correction factor, expressed as a function of *OUR* and *CER* (or *RQ*), was derived from a reference sequence of yeast metabolic energy-producing pathways, the Embden-Meyerhof-Parnas (EMP) pathways and the tricarboxylic acid (TCA) cycle.

4.3 Modern Estimation Techniques

Both balancing methods described above suffer from the inaccuracies of available instruments. The errors in the primary measurement are often large, and these errors can have profound effects on the accuracy of the estimates. Propagation of measurement errors can compound the deviation of the on-line biomass estimates from off-line assay values as fermentation progresses (CONSTANTINIDES and SHAO, 1981). Indeed, it was sometimes necessary to reinitialize the biomass concentration in the midst of fermentation as the deviation became unacceptably large (WANG et al., 1977) due to noisy oxygen measurements. Also, these balancing methods must be sup-

plied with initial conditions that are often at best rough guesses. Thus, a good noise filtration algorithm should be employed to improve the reliability of the estimated values before they are used for control purposes.

4.3.1 Deterministic Techniques

The *moving average method* (NYIRI et al., 1975), due to its simplicity, is frequently used at the instrument level to reduce noise. Oxygen and carbon dioxide concentrations in the exit gas, for example, are routinely scanned at a higher rate and then averaged for a certain number of times before being used. But this type of simple averaging is not adequate for estimating rates, such as specific growth rate, substrate uptake rate, and product formation rate.

Often the first-order *exponential filter* or its discrete version is used to remove high-frequency noise. The equation for single-exponential smoothing is:

$$\bar{c}_n = \alpha c_n + (1 - \alpha)\bar{c}_{n-1} \qquad (114)$$

where c_n is the current sampled value of the variable, $\bar{c}_n$ is the smoothed value, and $\bar{c}_{n-1}$ is the smoothed value from the previous sampling. The smoothing constant, α, takes a value from 0 to 1 (KOPPEL, 1968). A high value of α places more weight on the current measured value, with a value of 1 corresponding to no smoothing. The equation for double-exponential smoothing is:

$$\bar{\bar{c}}_n = \alpha \bar{c}_n + (1 - \alpha)\bar{\bar{c}}_{n-1} \qquad (115)$$

where $\bar{c}_n$ is obtained from Eq. (89), $\bar{\bar{c}}_n$ is the double-smoothed value of the sampled variable, and $\bar{\bar{c}}_{n-1}$ is the double-smoothed value from the previous sampling. A smoothed time derivative, $d\hat{c}_n/dt$, may be estimated as:

$$\frac{d\hat{c}_n}{dt} = \frac{1}{\Delta T}\left(\frac{\alpha}{1-\alpha}\right)(\bar{c}_n - \bar{\bar{c}}_n) \qquad (116)$$

where ΔT is the sampling period.

For example, the specific growth rate in a fed-batch fermentation can be expressed as:

$$\mu = \frac{1}{x}\frac{\mathrm{d}x}{\mathrm{d}t} + \frac{f}{v} \qquad (117)$$

where μ, x, v, and f are the specific growth rate, the dry cell weight concentration, the culture volume, and the feed flow rate, respectively. The specific growth rate can be smoothed by applying single exponential smoothing of f/v and x and double-exponential smoothing or time-derivative smoothing to $\mathrm{d}x/\mathrm{d}t$.

Without any available model, an empirical equation with constant parameters may be proposed and a *recursive least-square estimation technique* applied to determine the parameters. Thus, it is well suited for the smoothing of a series of raw measurements. This technique was used to smooth noisy biomass optical density measurements (JEFFERIS et al., 1979). The biomass concentration was expressed as a second-order polynomial function of time. The coefficients in the polynomial were estimated using exponentially-weighted discrete measurements. The growth rate was then simply calculated from the derivative of the polynomial. A similar algorithm was applied to other raw measurements such as the weight of nutrient reservoir and the flow rate (JEFFERIS, 1979). A more rapid response (JEFFERIS et al., 1979) can be obtained by incorporating a kinetic model into the least-square scheme, or the least-square scheme may be combined with a Kalman filter to reach better estimates for certain variables when empirical models can be applied to raw measurements (JEFFERIS, 1979). In another algorithm (REUSS et al., 1977), non-linear model equations were quasi-linearized, and the unknown parameters were assigned dynamic equations that were set to zero. Thus, the parameters were treated as constants. The oxygen uptake rate was used to predict biomass concentrations on-line in a batch experiment (REUSS et al., 1977).

4.3.2 Stochastic Estimation Techniques

Kalman filters (KALMAN, 1960) are very useful and powerful when a process is linear and a model is available. Kalman filters allow

estimation not only of the unknown parameters in the model but also of state variables that may be impossible to measure on-line. This technique requires a model for the system, however, together with knowledge of certain stochastic properties of measurement and disturbance noises. For non-linear systems this technique requires linearization of the model and is known as an *extended Kalman filter*. Since its inception in the early 1960s, the Kalman filter has been applied to chemical reactors (SEINFELD et al., 1969; SEINFELD, 1970). The extended Kalman filter has been applied to non-linear bioreactor state estimation and kinetic model parameter identification (SVRCEK et al., 1974).

Discrete Linear Kalman Filter

The essence of the Kalman filter is briefly summarized below. We shall begin with the discrete form of the algorithm. The system model is described by the linear vector difference equation, and the measurement model is given by a linear algebraic relation:

system model:
$$x(k+1) = \Phi(k+1, k)\,x(k) + \Gamma(k)\,w(k) \qquad (118)$$

measurement model:
$$z(k) = H(k)\,x(k) + v(k) \qquad (119)$$

where $x(k)$ is the state vector, $w(k)$ is the random disturbance vector, and $v(k)$ is the random error vector in the measurement $z(k)$. The standard assumptions and definitions are given in Tab. 6. These assumptions state that $w(k)$ and $v(k)$ are zero mean white noises with covariance matrices $Q(k)$ and $R(k)$, respectively. The noises, $w(k)$ and $v(k)$, are uncorrelated with each other, and they are also uncorrelated with the initial condition random vector $x(0)$. The mean and covariance of $x(0)$ are given the designations $m(0)$ and $P_x(0)$, respectively. $P_x(0)$ is different from a covariance matrix of estimation errors that is part of the filter formulation. If $m(0)$ is used as an initial state estimate, the filter covariance $P(0)$ is equal to the initial state covariance, $P_x(0)$.

For the corrupted model equation and corrupted measurement data in Eqs. (118) and

(119), the standard Kalman filter problem is based on a set of sequential observations $Z(k) = \{z(1), z(2), \ldots, z(k)\}$. The objective is to find linear, unbiased, minimum-variance estimates of the true state $x(j)$, denoted by $\hat{x}(j|k)$. Depending on the relative values of j and k, the estimation is referred to as prediction $(j > k)$, filtering $(j = k)$, or smoothing $(j < k)$. $\hat{x}(k|k)$ will be denoted by $\hat{x}(k)$.

Tab. 6. Standard Assumptions and Definitions of the Discrete Kalman Filter

---

$E[w(k)] = 0$
$\text{cov}[w(k), w(j)] = Q(k)\delta_{jk}$ [a]

$E[v(k)] = 0$
$\text{cov}[v(k), v(j)] = R(k)\delta_{jk}$

$\text{cov}[x(0), w(k)] = 0$, for all k
$\text{cov}[x(0), v(j)] = 0$, for all j
$\text{cov}[w(k), v(j)] = 0$, for all j, k

$E[x(0)] = m(0)$
$\text{cov}[x(0), x(0)] = P_x(0)$

---

[a] δ_{jk} is the Kronecker delta function

The Kalman estimation is described by the following set of recursive vector filtering difference equations:
Predictor:
prediction of state

$$\hat{x}(k|k-1) = \Phi(k)\hat{x}(k-1|k-1) = \Phi(k)\hat{x}(k-1) \tag{120}$$

prediction of error covariance matrix

$$P(k|k-1) = \Phi(k)P(k-1|k-1)\Phi^T(k) + Q(k-1) \tag{121}$$

Corrector:
state estimate

$$\hat{x}(k|k) = \hat{x}(k|k-1) + K(k) \cdot [z(k) - H(k)\hat{x}(k|k-1)] \tag{122}$$

error covariance matrix

$$P(k|k) = [I - K(k)H(k)]P(k|k-1) \tag{123}$$

Kalman gain matrix

$$K(k) = P(k|k-1)H^T(k) \cdot [H(k)P(k|k-1)H^T(k) + R(k)]^{-1} \tag{124}$$

initial conditions

$$\hat{x}(0) = m(0) \tag{125}$$

$$P(0) = P_x(0) \tag{126}$$

where $P(k)$ is the symmetric error covariance matrix of the estimation, defined by:

$$P(k) = [x(k) - \hat{x}(k)][x(k) - \hat{x}(k)]^T = \tilde{x}(k)\tilde{x}^T(k) \tag{127}$$

A schematic diagram depicting implementation of the above equations to obtain the best estimate is given in Fig. 16. The error covariance matrix $P(k)$ is obtained by integrating

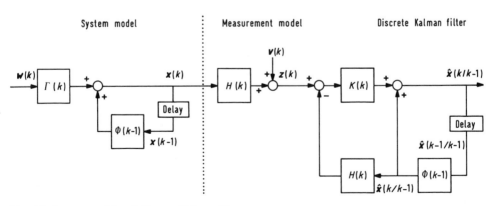

Fig. 16. System model and discrete Kalman filter.

Eqs. (121) and (123), and the Kalman gain matrix $K(t)$ is calculated from Eq. (124). The estimated noise, the difference between the corrupted measurement $z(k)$ and the estimated noise-free measurement $H\hat{x}(k)$ is multiplied by the Kalman gain and added to the noise-free predicted value, $\hat{x}(k|k-1)$, which is calculated from Eq. (120) using the best estimate of the previous step.

Extended Kalman Filter

We consider a more general, non-linear dynamic model and a discrete measurement model:

system model:

$$\dot{x} = f(x(t), t) + G(x(t), t) w(t) \qquad (128)$$

measurement model:

$$z(t_i) = h(x(t_i), t_i) + v(t_i) \qquad (129)$$

where f is a vector of a non-linear function of $x(t)$ and t, and h is a vector of non-linear functions of $x(t_i)$ and t_i. The standard assumptions and definitions are presented in Tab. 7.

Tab. 7. Standard Assumptions and Definitions of the Extended Kalman Filter

---

$E[w(t)] = 0$
$\text{cov}[w(t), w(t)] = Q(t)\delta(t-\tau)$[a]

$E[v(i)] = 0$
$\text{cov}[v(i), v(j)] = R(i)\delta_{ij}$

$\text{cov}[x(t_0), w(t)] = 0$, for all $t \geq t_0$
$\text{cov}[x(t_0), v(t_i)] = 0$, for all j
$\text{cov}[w(t), v(t_i)] = 0$, for all $t \geq t_0$ and $t_i \geq t_0$

$E[x(t_0)] = m(0)$
$\text{cov}[x(t_0), x(t_0)] = P_x(0)$

---

[a] $\delta(\cdot)$ is the Dirac delta function

For the model given by Eqs. (128) and (129), the extended Kalman filter problem can be stated as follows: given a measurement sequence $Z(k) = \{z(t_0), z(t_1), \ldots, z(t_k)\}$, find an estimator to provide estimates of $x(t)$, denoted by $x(t|t_k)$, patterned after a linear case

Kalman filter and giving small errors, $\tilde{x}(t|t_k) = x(t) - \hat{x}(t|t_k)$. One would expect that the estimator would approach the linear case Kalman filter for vanishingly small non-linearities.

The extended Kalman filter estimation is described by the following set of vector filtering equations:
Predictor:
A prediction of estimate is produced by integration of the differential equation

$$\dot{\hat{x}}(t|t_k) = f(x(t|t_k), t) \qquad (130)$$

prediction of error covariance matrix

$$P(t_k|t_{k-1}) = \Phi(t_k, t_{k-1}) P(t_{k-1}|t_{k-1}) \cdot \\ \cdot \Phi^{\mathrm{T}}(t_k, t_{k-1}) + Q(t_{k-1}) \qquad (131)$$

Corrector:
state estimate

$$\hat{x}(t_k|t_k) = \hat{x}(t_k|t_{k-1}) + K(k) \cdot \\ \cdot [z(t_k) - h(\hat{x}(t_k|t_{k-1}, t_k)] \qquad (132)$$

error covariance matrix

$$P(t_k|t_k) = [I - K(t_k) H(t_k)] P(t_k|t_{k-1}) \qquad (133)$$

Kalman gain matrix

$$K(t_k) = P(t_k|t_{k-1}) H^{\mathrm{T}}(t_k) \cdot \\ \cdot [H(t_k) P(t_k|t_{k-1}) H^{\mathrm{T}}(t_k) + R(t_k)]^{-1} \qquad (134)$$

$$H(t_k) = \left[\frac{\partial h}{\partial x} \right]_{x = \hat{x}(t_k|t_{k-1})} \qquad (135)$$

$$\frac{\partial \Phi}{\partial t} = F(t) \, \Phi(t, t_k), \quad F(t) = \left[\frac{\partial f}{\partial x} \right]_{x = \hat{x}(t_k|t_{k-1})} \qquad (136)$$

initial conditions

$$\hat{x}(t_0|t_0) = m_0 \qquad (137)$$

$$P(t_0|t_0) = P_{x_0} \qquad (138)$$

Implementation of the extended Kalman filter is essentially identical to that of the Kalman filter. The only difference is that in the extended Kalman filter algorithms F and H are the linearized matrices given by Eqs. (135) and (136).

In these Kalman filtering schemes, not only state variables but also unknown constant parameters such as cell yields can be estimated by treating them as additional "state variables" whose time derivatives are zero. For example, consider a batch bioreactor characterized by two state variables: biomass and substrate. If the growth kinetics follow the Monod type and the yield coefficient is constant, the following *system models* are generally used for growth and substrate consumption:

the saturation constant, the yield coefficient, and the death rate coefficient, respectively. Besides the system model, the *measurement model* must be specified. If the dry weight of the biomass can be measured directly while the substrate concentration measurement is biased, one can propose the following meas-

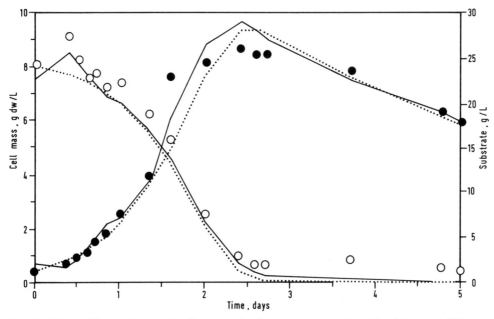

Fig. 17. Kalman filter estimates of cell mass and substrate concentration. Model ($\cdots\cdots$) Kalman filter (——) (● cell mass, ○ substrate)
Experimental data: NIHTILA and VIRKKUNEN (1977).

the growth kinetics follow the Monod type and the yield coefficient is constant, the following *system models* are generally used for growth and substrate consumption:

$$\frac{dx}{dt} = \frac{\mu_{max}s}{K_s+s}x - k_d x \qquad (139)$$

and

$$\frac{ds}{dt} = -\frac{1}{Y}\frac{\mu_{max}s}{K_s+s}x \qquad (140)$$

where x and s are the concentrations of biomass and substrate, respectively; μ_{max}, K_s, Y, and k_d are the maximum specific growth rate,

urement model (NIHTILA and VIRKKUNEN, 1977):

$$x_m = x + v_1 \qquad (141)$$

$$s_m = s + s_b + v_2 \qquad (142)$$

where x_m and s_m are the measured values of dry weight and substrate concentration, s_b is a constant bias, and v_1 and v_2 denote measurement errors. The unknown parameters μ_{max}, K_s, k_d, Y, and s_b are contained in the system and measurement equations, Eqs. (139)–(142). These parameters can be estimated with x and s by setting up an augmented dynamic equation:

$$\dot{x} = \begin{bmatrix} \dfrac{x_3 x_2}{x_4 + x_2} x_1 - x_5 x_1 \\[2mm] -\dfrac{1}{x_6} \dfrac{x_3 x_2}{x_4 + x_2} x_1 \\[1mm] 0 \\ 0 \\ 0 \\ 0 \\ 0 \end{bmatrix} + w \qquad (143)$$

$$z = \begin{bmatrix} x_1 \\ x_2 + x_7 \end{bmatrix} + v \qquad (144)$$

where $x = [xs\mu_{max}K_s k_d Ys_b]^T$ and $z = [x_m s_m]^T$. Now we have system and measurement models. Using the extended Kalman filter equations, Eqs. (130)–(138), we can simultaneously estimate cell mass and substrate concentrations and model parameters. Experimental data for a batch culture of *Trichoderma viride* grown

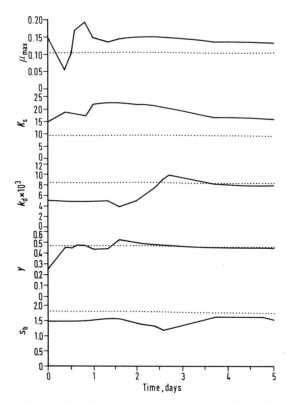

Fig. 18. Model parameter estimates by a Kalman filter. True values ($\cdots\cdots$) estimates (——)

on glucose as the main carbon source (NIHTILA and VIRKKUNEN (1977)) are used as the measured values of dry cell mass and substrate concentrations (circles and filled circles in Fig. 17). To see the filtering capability the initial values of states and parameters are set ±20–50% of the true values. The estimation results are shown in Figs. 17 and 18. As is apparent from Fig. 17, the state variables are fairly well tracked by this technique. Fig. 18 shows all the model parameters converging on the true values toward the end of the run.

As stated earlier, this general scheme can be applied in many situations to estimate continuously not only the states but also the associated kinetic parameters on-line, since they are observable. This scheme is especially useful for estimating highly noisy derivative variables such as the specific growth rate. A general drawback of these filters is that judicious choices of the noise covariance matrices, Q and R, and the initial error covariance matrix, P_0, are required to achieve proper filtering. This may sometimes be a difficult task.

5 Adaptive Optimization and Control

Whereas in adaptive control the operating points (setpoints) are assumed to remain constant, the objective of adaptive optimization is to determine optimal operating conditions for bioreactors that may be unknown or may change with time. For example, for a continuous culture one needs to determine on-line such optimal operating conditions as dilution rate, temperature, and pH. This is done by perturbing the culture, observing the output (perhaps the productivity), and continually improving the output. For batch and fedbatch cultures there is no reason to hold any operating conditions constant, since optimal conditions for growth may not necessarily be optimal for product formation. Consequently it is desirable to determine on-line the changing characteristics of cultures and reoptimize bioreactor operating conditions. Due to the complexity of microbial cultures, it is difficult to understand and quantify the complex intricate networks of biochemical reactions present in a living cell with sophisticated regulatory systems governing the rate of growth and synthesis of various products. Therefore, it is practically impossible to develop a model that can adequately describe the process over a wide range of operating conditions and yet be simple enough to be used for optimization calculations. In addition, due to cellular adaptation, selection, enrichment, as well as variations in the crude feed, optimum conditions cannot be determined off-line. Thus, it is important to track closely changes in the culture and operating conditions and reflect them in the optimization. Therefore, an adaptive feature is required in the optimization of bioreactors. A typical adaptive optimization scheme is shown in Fig. 19. The process optimizer consists of a process identifier that identifies changes in the culture and operating conditions through a model with adjustable parameters and an optimization scheme for the setpoint, which uses the model with updated parameter values. We

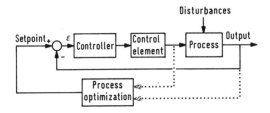

Fig. 19. Adaptive optimization.

shall discuss in some detail the adaptive optimization of a continuous culture to maximize the productivity of a baker's yeast culture by manipulating the dilution rate and temperature.

5.1 Adaptive Optimization of Continuous Bioreactors

In this section adaptive optimization strategy is implemented to maximize the productivity of a continuous culture. Specifically, steady-state cellular productivity of a contin-

uous culture of *Saccharomyces cerevisiae,* more commonly known as baker's yeast, is maximized by manipulating both temperature and dilution rate. This system is of industrial importance due to the use of yeast in the baking and brewing industries and in the manufacture of various yeast products using non-recombinant and recombinant yeast strains. A continuous bioreactor can be optimized using steady-state optimization techniques based on steady-state data. However, these techniques are usually extremely time-consuming since the time required for the continuous cultures to reach a new steady state after step changes in the manipulated variables may be over 20 hours, resulting in optimization times of the order of days and weeks. Therefore, it is essential to develop a speedy optimization algorithm based on short transient data rather than to wait for steady-state data.

An adaptive optimization algorithm based on dynamic model identification has been developed for continuous cultures (ROLF and LIM, 1985; CHANG et al., 1988; SEMONES and LIM, 1989; CHANG and LIM, 1989). This method requires only minimal *a priori* information, such as model structure, parameter values, and feed conditions. A simple empirical dynamic input–output model is used, and its parameters are recursively estimated from transient data to track the process on-line. A recursive least-square method was used for parameter estimation. In determining the control action to improve the performance of the continuous culture, the steady-state portion of the dynamic model was used instead of waiting for actual steady-state information.

5.1.1 Problem Formulation

The performance index to be maximized is the productivity of a continuous culture of baker's yeast, i.e., the product of the dilution rate, D, and the cell concentration x:

$$P = Dx \tag{145}$$

The manipulated variables for the optimization are temperature, T, and dilution rate, D:

$$u^T = [T, D] \tag{146}$$

The output variable is the cell concentration:

$$y = x \qquad (147)$$

The gradient optimization algorithm is used:

$$u(k+1) = u(k) + a\,\partial P/\partial u \qquad (148)$$

which for this problem reduces to:

$$D(k+1) = D(k) + a_D[x + D\,\partial x/\partial D] \qquad (149)$$

$$T(k+1) = T(k) + a_T D\,\partial x/\partial T \qquad (150)$$

where a_D and a_T are the gains (step sizes) for the dilution rate and temperature optimization, respectively.

The optimization scheme can be easily implemented if one can determine the steady-state gains, $\partial x/\partial D$ and $\partial x/\partial T$. These can be calculated on-line using an input–output dynamic model of the process and updated parameter values.

5.1.2 Process Model

Proper choice of a process model is critical in applying adaptive process identification algorithms (ROLF and LIM, 1985; SEMONES and LIM, 1989). Based on the accuracy of the steady-state gain estimates, matrix invertibility, and sensitivity to data-sampling time, the model selected is a second-order linear model given by:

$$
\begin{aligned}
x(k) =\ & -a_1 x(k-1) - a_2 x(k-2) + b_{10} D(k) + \\
& + b_{11} D(k-1) + b_{12} D(k-2) + b_{20} T(k) + \\
& + b_{21} T(k-1) + b_{22} T(k-2) + c
\end{aligned} \qquad (151)
$$

where k is the discrete sampling-time index and the constants a_i, b_{ij}, and c are the model parameters to be determined on-line using transient data. Eq. (151) may be also written as:

$$y(k) = q^T x(k) \qquad (152)$$

where

$$q^T = [a_1, a_2, b_{10}, b_{11}, b_{12}, b_{20}, b_{21}, b_{22}, c] \qquad (153)$$

and

$$
x^T(k) = [-y(k-1),\ -y(k-2),\ D(k),\ D(k-1),\\
D(k-2),\ T(k),\ T(k-1),\ T(k-2),\ 1] \qquad (154)
$$

Steady-state gains are then obtained from the steady-state solution of Eq. (151):

$$
\begin{aligned}
y(k) =\ & [(b_{10} + b_{11} + b_{12})\,D(k) + \\
& + (b_{20} + b_{21} + b_{22})\,T(k)]/(1 + a_1 + a_2)
\end{aligned} \qquad (155)
$$

by evaluating the partial derivatives:

$$
\begin{aligned}
\partial x/\partial D &= (b_{10} + b_{11} + b_{12})/(1 + a_1 + a_2) \\
\partial x/\partial T &= (b_{20} + b_{21} + b_{22})/(1 + a_1 + a_2)
\end{aligned} \qquad (156)
$$

The remaining problem is to determine the nine constants in Eq. (151). Knowing these, the steady-state gains can be calculated from Eq. (156). Eqs. (149) and (150) are then used to calculate appropriate changes in the dilution rate and temperature to optimize continuous culture.

5.1.3 Parameter Estimation

The initial estimates of the nine constants in Eq. (151) are established by using a standard least-square technique. During initialization the culture is excited by introducing test signals in the manipulated variables to generate an initial data set that will yield good first estimates of the parameters. The test signal chosen is a weighted m-sequence (KASHIWAGI, 1974) that is a low frequency signal with an impulse-like autocorrelation function. Staggered m-sequence signals are used for temperature and dilution rate as shown in Fig. 20. A one-shot least-square algorithm is then applied to the data set to obtain the initial parameter estimates:

$$q = (X^T X)^{-1} X^T y \qquad (157)$$

where

$$X^T = [x^T(i),\ x^T(i+1),\ \ldots,\ x^T(i+j)] \qquad (158)$$

and

$$y^T = [y(i),\ y(i+1),\ \ldots,\ y(i+j)] \qquad (159)$$

where j is the length of the data set. After

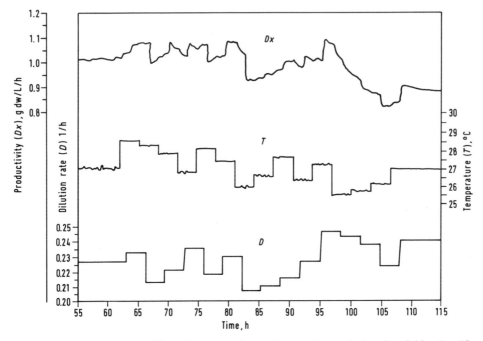

Fig. 20. The *m*-sequence profile with staggered step changes in manipulated variables ($n = 15$, $\Delta t = 96$ min, SEMONES and LIM, 1989).

making the first step changes in the manipulated variables, the parameters are updated to improve the accuracy of the linear model. This is done by using a recursive least-square algorithm (HSIA, 1977):

$$q(k+1) = q(k) + \gamma(k+1)\, P(k)\, x(k+1) \cdot \\ \cdot [y(k+1) - x^{\mathrm{T}}(k+1)\, q(k)] \qquad (160)$$

$$\gamma(k+1) = 1/[1 + x^{\mathrm{T}}(k+1)\, P(k)\, x(k+1)] \qquad (161)$$

$$P(k+1) = [P(k) - \gamma(k+1)\, P(k)\, X^*\, X^{*\,\mathrm{T}} \cdot \\ \cdot (k+1)\, P(k)]/\lambda \qquad (162)$$

$$X^*(k) = [x(k), 0, \ldots, 0] \qquad (163)$$

$$P(0) = (X^{\mathrm{T}} X)^{-1} \qquad (164)$$

where λ is the forgetting factor, $0 < \lambda \leq 1$ and P is the covariance matrix of the input–output data. This algorithm is sensitive to the choice of the forgetting factor. The forgetting factor can be set at a constant value, adaptively tuned (HSIA, 1977), or adjusted by other criteria.

Our experience has been best with the so-called bilevel forgetting factor method in which the forgetting factor is placed either at a maximum level, $\lambda_{\max}$, when the error between the measured and predicted output values is smaller than a preset value, $e^2(k) = [y(k) - y(k)]^2$ < vartol, or at a minimum value, $\lambda_{\min}$, when the error is greater than the preset value (CHANG et al., 1988; CHANG and LIM, 1989). The details and the exact experimental conditions are available elsewhere and are therefore omitted here, where we instead give a brief description of its implementation.

5.1.4 Implementation

The procedure consists of a sequence of initialization, optimization, and supervision/reoptimization. In the initialization stage, staggered *m*-sequences in dilution rate and temperature are introduced to generate an initial set of data, from which the initial estimates of the input–output model parameters and the steady-state gains are made. These initial esti-mates are used to initiate recursive parameter estimation by sampling the data and introduc-ing step changes in the manipulated variables at several sampling times. This process of pa-rameter updating and making step changes in the manipulated variables continues until the algorithm is judged to have converged by the smallness of the gradients in the performance index.

At convergence the step changes are halted, but parameter updating is continued (supervi-sion). In the event a process variation or dis-turbance occurs, the gradients will increase and trigger a restart of the optimization (reop-timization).

5.1.5 Results and Discussion

Fig. 21 shows the initialization and optimi-zation stages, and Fig. 22 shows the reoptimi-zation phase. The initialization was started at 43.7 hours into a steady-state operation at

$D_0 = 0.27 \text{ h}^{-1}$ and $T_0 = 28.0 \text{ °C}$. The initializa-tion phase took 45 hours, which was then fol-lowed by an optimization phase of about 100 hours. At 210.5 hours a very large step change was introduced in the pH (from 5.5 to 7.0) to test the adaptability to extreme external distur-bance and the reoptimization capability of the algorithm (Fig. 22). The gains in the steepest ascent were increased for faster optimization results. This large change in pH caused the growth rate to decrease substantially; as a re-sult, the dilution rate decreased rapidly at the beginning to save the culture from washout. After about 40 hours the culture converged to a new steady state: $D = 0.25 \text{ h}^{-1}$, $T = 28.6 \text{ °C}$, and $Dx = 0.79 \text{ g/L/h}$. The algorithm can ap-parently detect changes, and it quickly finds the corresponding optimum conditions and maintains the operation at this optimum. To see if more effective perturbation of the cul-ture in the initiation stage would improve the initial estimate and therefore shorten the op-timization time, another run was made using larger step changes in temperature. In addi-

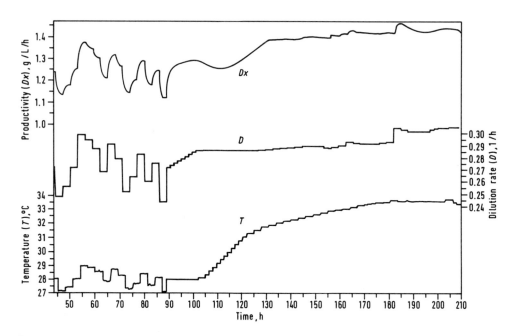

Fig. 21. Multivariable on-line optimization (initialization and optimization periods) (CHANG and LIM, 1989).
$AD = 0.026 \text{ h}^{-1}$, $AT = 1.0 \text{ °C}$, $m(i) = -1, -1, -1, -1, -1$ for D and T, $\lambda_{min} = 0.94$, $\lambda_{max} = 0.995$, $\alpha_D = 0.0005$, $\alpha_T = 4.0$.

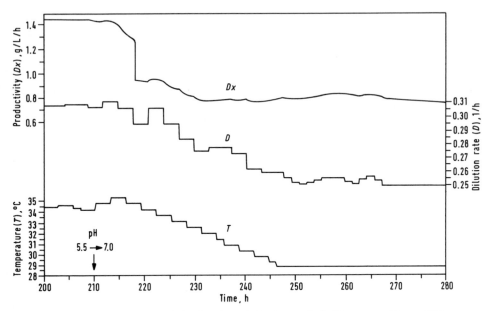

Fig. 22. Multivariable on-line optimization (reoptimization period) (CHANG and LIM, 1989). $\alpha_D = 0.001$, $\alpha_T = 7.0$

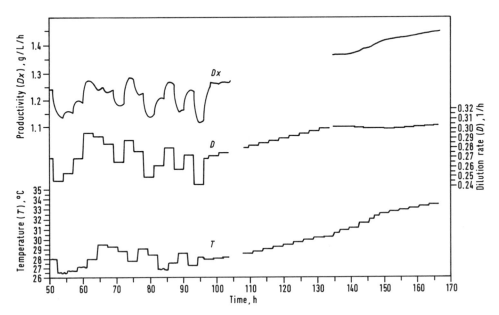

Fig. 23. Multivariable on-line optimization (CHANG and LIM, 1989).
$AD = 0.026 \, \text{h}^{-1}$, $AT = 1.5 \, °\text{C}$, $m(i) = -1, -1, -1, -1, -1$ for D and $m(i) = 1, -1, -1, -1, -1$ for T, $\lambda_{\min} = 0.95$, $\lambda_{\max} = 0.995$, α_D changed at 137.7 hours from 0.001 to 0.00025, $\alpha_T = 4.0$.

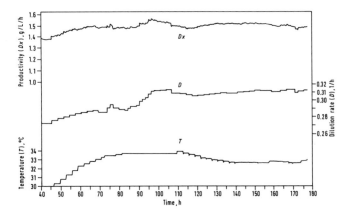

Fig. 24. Multivariable on-line optimization (CHANG and LIM, 1989). No initialization stage, $\lambda_{min} = 0.95$, $\lambda_{max} = 0.995$, α_D changed at 97.5 hours from 0.001 to 0.00025, α_T changed at 92.0 hours from 10 to 2.5.

tion, unlike the first run in which the same *m*-sequence but staggered signals were used, different *m*-sequences were employed. The results are shown in Fig. 23. The initial estimates were more accurate, but the overall results were difficult to judge due to the many tuning constants in the algorithm. Since considerable time is spent in making the initial estimate, another run was made in which the initialization stage was eliminated by using the initial estimates from a different run. The results (Fig. 24) show no ill effect, and this implies that the algorithm can accommodate considerable error in the initial estimate.

5.2 Adaptive Optimization of Batch and Fedbatch Bioreactors

Adaptive optimization of batch and fedbatch bioreactors is theoretically feasible, especially for those situations in which the period of bioreactor operation is long, for example, a penicillin fermentation that may last 5 days or more. In these situations the bioreactor operation period is broken into a number of subintervals, the number depending on the process dynamics. The bioreactor operation during the first subinterval is used to collect information necessary for parameter updating and subsequent optimization calculations. If the computational burden is great, the bioreactor operation may continue over a short interval using operational conditions prescribed perhaps by the previous run, during the pa-

rameter updating and optimization calculations. The calculated optimization results are then implemented for the next interval. During this interval the necessary information is collected again for parameter updating and optimization calculations. The results are implemented in the subsequent subinterval. This procedure may be repeated until the end of the bioreactor operation period. Therefore, as long as one can monitor on-line the key state variables that make up the mass and energy balance equations, one can update the bioreactor model parameters through an identification scheme and then reoptimize the operating conditions. If one or more of the key state variables are measurable on-line, other readily accessible parameters should be measured, which are then used in an estimation scheme to provide the necessary information to update the bioreactor model parameters. Alternatively, the extended Kalman filter may be used to obtain simultaneously an estimate of the missing key state variables and the bioreactor model parameters, as demonstrated in Sect. 4.

On the other hand, the operating period of bioreactors may be rather short, sometimes only several hours, relative to the repeated on-line parameter updating and optimization computation required for adaptive optimization. In these situations the computational burden may be too great to subdivide the operation period into several subintervals, and on-line adaptive optimization may not be realistic. One approach based upon the bioreactor operation time and computational requirement is the so-called cycle-to-cycle optimization.

The basic principle behind *cycle-to-cycle optimization* is simple and is often practiced in industry. The idea is to make a batch or fed-batch run and generate the data required to update the bioreactor parameters. Parameter estimation and optimization calculation are done off-line, and the results are implemented for the next cycle. This cycle then generates the necessary information for off-line parameter update and optimization calculation. The calculated results are then implemented in the next cycle. This procedure is repeated until there is negligible change predicted by the optimization calculation. This approach has apparently been successfully implemented for penicillin (CHITTUR, 1989) and baker's yeast fermentations (MODAK, 1988). It was found that the procedure would converge on the optimal results in two to three iterations. However, if the structural form of the model is incorrect the result may not converge. Interested readers are advised to consult the original sources.

6 Future Prospects

In recent years computer technology has had a significant impact on fermentation. In the past, integration of computers into fermentation processes was generally justified by the capacity of computers for process monitoring, data acquisition, data storage, low-level control, and error detection. However, computer capabilities are now further exploited through the implementation of advanced control and optimization strategies. This area can provide significant improvement in product quality, yield, and productivity.

The extent of computer application depends on the development of better on-line instrumentation that can monitor cellular physiology, as well as on reliable mathematical models that can describe various cellular dynamics. When the availability of on-line sensors is limited, mathematical models and estimation techniques can play very important roles in optimizing bioreactors. Presently, the application of closed-loop computer control is considerably limited by the scarcity of reliable on-line sensors for monitoring key parameters in a fermentor. Availability of more on-line sensors will help in utilizing computers more fully for optimization and control of fermentation processes.

Empirical correlations capable of predicting effects of small changes in control variables on the response of microbial systems are very useful, and they can be effectively utilized in adaptive control. The next logical challenge is a broadly applicable adaptive optimization algorithm that is capable of self-optimization. We also look forward to seeing more effective applications of artificial intelligence and expert systems.

The intent of this chapter has been to provide necessary information concerning the types of control techniques available and their significance to fermentation processes. Specific examples of the application of some of these techniques can be found in the reviews mentioned earlier. Finally, it must be stressed that control considerations should be taken into account early in the design stage of a fermentation process. This will assure both controllability of the fermentation process and overall optimal performance of the entire system.

7 References

AIBA, S., KOIZUMI, J. (1984), *Biotechnol. Bioeng.* **26**, 1026.

ARMIGER, W. B., HUMPHREY, A. E. (1979), in *Microbial Technology* (PEPPLER, H. J., PERLMAN, D., Eds.) Vol. 2, pp. 375–401, New York: Academic Press.

ASTROM, K. J., HAGGLUND, T. (1988), Research Triangle Park, NC: Instrum. Society of America.

ASTROM, K. J., WITTENMARK, B. (1988), *Adaptive Control Systems,* Reading, MA: Addison-Wesley.

BAJPAI, R. K., REUSS, M. (1981), *Biotechnol. Bioeng.* **23**, 717.

BRAVARD, J. P., CORDONNIER, M., KERNEVEZ, J. P., LEBEAULT, J. M. (1979), *Biotechnol. Bioeng.* **21**, 1239.

BRISTOL, E. H. (1966), *IEEE Trans. Autom. Control* **AC-11**, 133.

BRISTOL, E. H. (1977), *Control Eng.* **24**(1), 24.

BULL, D. N. (1983), *Annu. Rep. Ferment. Processes* **6**, 359–375.

CHANG, Y. K., LIM, H. C. (1989), *Biotechnol. Bioeng.* **34**, 577.

CHANG, Y. K., PYUN, Y. R., LIM, H. C. (1988), *Biotechnol. Bioeng.* **31**, 944.

CHITTUR, V. (1989), *Ph. D. Thesis,* Purdue University, West Lafayette, IN.

COHEN, G. H., COON, G. A. (1953), *Trans. ASME* **75**, 827.

CONSTANTINIDES, A., SHAO, P. (1981), *Ann. N. Y. Acad. Sci.* **369**, 167.

CONSTANTINIDES, A., SPENCER, J. L., GADEN, E. L. (1970a), *Biotechnol. Bioeng.* **12**, 803.

CONSTANTINIDES, A., SPENCER, J. L., GADEN, E. L. (1970b), *Biotechnol. Bioeng.* **12**, 1081.

COONEY, C. L., MOU, D. G. (1982), in *Computer Applications in Fermentation Technology,* p. 219, London: Society of Chemical Industry.

COONEY, C. L., WANG, H. Y., WANG, D. I. C. (1977), *Biotechnol. Bioeng.* **19**, 55.

COUGHANOWR, D. R., KOPPEL, L. B. (1965), *Process Systems Analysis and Control,* New York: McGraw-Hill.

COURT, J. R. (1988), *Progress in Industrial Microbiology* (BUSHELL, M. E., Ed.), pp. 1–45. New York: Elsevier.

DEKKERS, J. G. J., DE KOK, H. E., ROELS, J. A. (1981), *Biotechnol. Bioeng.* **23**, 1023.

DORBY, D. D., JOST, J. L. (1977), *Annu. Rep. Ferment. Processes* **1**, 95–114.

DOUGLAS, J. M. (1972), *Process Dynamics and Control,* Englewood Cliffs, NJ: Prentice-Hall.

DUNN, I. J., DANG, N. D. P. (1976), in *Abstr. Papers 5th Int. Fermentation Symp.* (DELLWEG, H., Ed.), p. 59, Berlin.

EDGAR, T. F., HIMMELBLAU, D. M. (1988), *Optimization of Chemical Processes,* New York: McGraw-Hill.

ERICKSON, L. E. (1979), *Biotechnol. Bioeng. Symp.* **9**, 49.

GOFF, K. W. (1966), *ISA J.,* December, 44.

HAMPEL, W. A. (1979), *Adv. Biochem. Eng.* **13**, 1–33.

HATCH, R. T. (1982), *Annu. Rep. Ferment. Processes* **5**, 291–311.

HERBERT, D. (1976), in *Continuous Culture 6: Applications and New Fields* (DEAN, A. C. R., ELLWOOD, D. C., EVANS, C. G. T., MELLING, J., Eds.), London: Ellis Horwood.

HILL, G. A., ROBINSON, C. W. (1974), *Biotechnol. Bioeng.* **16**, 531.

HIRAMA, T., HUMPHREY, A. E. (1980), *Biotechnol. Bioeng.* **22**, 821.

HOSLER, P., JOHNSON, M. J. (1953), *I & EC* **45**, 871.

HSIA, T. C. (1977), *System Identification,* Toronto: Lexington.

HUMPHREY, A. E. (1971), *Proc. LABEX Symp. Computer Control of Fermentation Processes, London,* p. 1.

JEFFERIS, R. P. (1975), *Process Biochem.* **10**(3), 15.

JEFFERIS, R. P., III (1979), *Ann. N. Y. Acad. Sci.* **326**, 241.

JEFFERIS, R. P., III, WINTER, H., VOGELMAN, H. (1977), in *Workshop Computer Applications in Fermentation Technology, GBF Monogr. Ser.* **3**, 141.

JUTAN, A., RODRIGUEZ, E. S. (1984), *Can. J. Chem. Eng.* **62**, 802.

KALMAN, R. E. (1960), *J. Basic Eng.,* 35.

KASHIWAGI, H. (1974), *Proc. Fifteenth Joint Automatic Control Conference,* p. 285, AIChE Publication.

KOIZUMI, J., MODEN, Y., AIBA, S. (1985), *Biotechnol. Bioeng.* **27**, 721.

KOPPEL, L. B. (1968), *Introduction to Control Theory,* Englewood Cliffs, NJ: Prentice-Hall.

LIM, H. C., TAYEB, Y. J., MODAK, J. M., BONTE, P. (1986), *Biotechnol. Bioeng.* **28**, 1408.

MODAK, J. M. (1988), *Ph. D. Thesis,* Purdue University, West Lafayette, IN.

MODAK, J. M., LIM, H. C. (1989), *Chem. Eng. J.* **42**, B15.

MODAK, J. M., LIM, H. C., TAYEB, Y. L. (1986), *Biotechnol. Bioeng.* **28**, 1396.

MOORE, C. F. (1986), *Proc. Am. Control Conf.,* Seattle, p. 643.

MOU, D. G. (1983), *Ph. D. Thesis,* Massachusetts Institute of Technology, Cambridge, MA.

MOU, D. G., COONEY, C. L. (1983a), *Biotechnol. Bioeng.* **25**, 225.

MOU, D. G., COONEY, C. L. (1983b), *Biotechnol. Bioeng.* **25**, 257.

NIHTILA, M., VIRKKUNEN, J. (1977), *Biotechnol. Bioeng.* **19**, 1831.

NYIRI, L. K. (1971), *Proc. LABEX Symp. Computer Control of Fermentation Processes, London,* p. 16.

NYIRI, L. K. (1972), *Adv. Biochem. Eng.* **2**, Chap. 2.

NYIRI, L. K., TOTH, G. M., CHARLES, M. (1975), *Biotechnol. Bioeng.* **17**, 1663.

OHASHI, M., WATABE, T., ISHIKAWA, T., WATANABE, Y., MIWA, K., SHODA, M., ISHIKAWA, Y., ANDO, T., SHIBATA, T., KITSUNAI, T., KAMIYAMA, N., OIKAWA, Y. (1979), *Biotechnol. Bioeng. Symp.* **9**, 103.

PAZLAROVA, J., BAIG, M. A., VOTRUBA, J. (1984), *Appl. Microbiol. Biotechnol.* **20**, 331.

PERRY, R. H., GREEN, D. (Eds.) (1984), *Perry's Chemical Engineers' Handbook,* 6th Ed., pp. 22–26, New York: McGraw-Hill.

PIRT, S. J. (1965), *Proc. R. Soc. London Ser. B* **163**, 224.

PIRT, S. J., RIGHELATO, R. C. (1967), *Appl. Microbiol.* **15**, 1284.

PONTRYAGIN, L. S., BOLTYANSKII, Y. G., GAMKRELIDZE, R. V., MISHCHENKO, E. F. (1962), *The Mathematical Theory of Optimal Processes,* New York: Wiley-Interscience.

RADJAI, M. K., HATCH, R. T., CADMAN, T. W. (1984), *Biotech. Bioeng. Symp.* **14**, 657.

REUSS, M., JEFFERIS, R. P., III, LEHMANN, J. (1977), in *Workshop Computer Applications in Fermentation Technology, GBF Monogr. Ser.* **3**, 107.

ROELS, J. A. (1980), *Biotechnol. Bioeng.* **22**, 2457.

ROELS, J. A. (1981), *Ann. N. Y. Acad. Sci.* **369**, 113.

ROLF, M. J., LIM, H. C. (1982), *Enzym. Microb. Technol.* **4**, 370.

ROLF, M. J., LIM, H. C. (1985), *Biotechnol. Bioeng.* **27**, 1236.

SAN, K. Y., STEPHANOPOULOS, G. (1984), *Biotechnol. Bioeng.* **26**, 1209.

SEBORG, D. E., EDGAR, T. F., MELLICHAMP, D. A. (1989), *Process Dynamics and Control,* New York: John Wiley & Sons.

SEINFELD, J. H., GAVALAS, G. R., HWANG, M. (1969), *Ind. Eng. Chem. Fundam.* **8**, 257.

SEINFELD, J. H. (1970), *AIChE J.* **16**, 1016.

SEMONES, G. B., LIM, H. C. (1989), *Biotechnol. Bioeng.* **33**, 16.

SHINSKEY, F. G. (1979), *Process-Control Systems, Application, Design, Adjustment,* New York: McGraw-Hill.

SIEGELL, S. D., GADEN, E. L. (1962), *Biotechnol. Bioeng.* **4**, 345.

SMITH, C. R., MOORE, C. F., BRUNS, D. D. (1981), *Paper TA-7* in *Joint Autom. Control Conf.,* Charlottesville, VA.

SPRIET, J. A., BOTTERMAN, J., DE BUYSER, D. R., DE VISSCHER, P. L., VANDAMME, E. J. (1982), *Biotechnol. Bioeng.* **24**, 1065.

STEPHANOPOULOS, G., SAN, K.-Y. (1982), in *Chemical Engineering-Boston, ACS Symp. Ser.* **196**, 155.

SVRCEK, W. Y., ELLIOTT, R. F., ZAJIC, J. E. (1974), *Biotechnol. Bioeng.* **16**, 827.

SWARTZ, J. R., COONEY, C. L. (1979), *Biotechnol. Bioeng. Symp.* **9**, 95.

SWARTZ, J. R., WANG, H. Y., COONEY, C. L., WANG, D. I. C. (1976), in *Abstr. Papers, 5th Int. Fermentation Symp.* (DELLWEG, H., Ed.), p. 29, Berlin.

TAGUCHI, H., HUMPHREY, A. E. (1966), *J. Ferment. Technol.* **44**, 881.

VERES, A., NYESTE, L., KURUCZ, I., KIRCHKNOPF, L., SZIGETI, L., HOLLO, J. (1981), *Biotechnol. Bioeng.* **23**, 391.

WAGHMARE, R. S., LIM, H. C. (1981), *Ind. Eng. Chem. Fundam.* **20**, 361.

WANG, H. Y., COONEY, C. L., WANG, D. I. C. (1977), *Biotechnol. Bioeng.* **19**, 69.

WANG, H. Y., COONEY, C. L., WANG, D. I. C. (1979a), *Biotechnol. Bioeng. Symp.* **9**, 13.

WANG, H. Y., COONEY, C. L., WANG, D. I. C. (1979b), *Biotechnol. Bioeng.* **21**, 975.

WEBB, J. C., *DDC Hardware and Software,* General Electric Company.

WEIGAND, W. A. (1978), *Annu. Rep. Ferment. Processes* **2**, 43–72.

WHITAKER, A. (1980), *Process Biochem.* **15**, 10.

WILLIAMS, T. J., MOWLE, F. L., WEIGAND, W. A., REKLAITIS, G. V., GOODSON, R. E., LIM, H. C. (1973), *Digital Computer Applications to Process Control,* Purdue University, West Lafayette, IN.

YUWANA, M., SEBORG, D. E. (1982), *AIChE J.* **28**, 434.

ZABRISKIE, D. W. (1979), *Ann. N. Y. Acad. Sci.* **326**, 223.

ZABRISKIE, D. W., HUMPHREY, A. E. (1978), *AIChE J.* **24**, 138.

ZABRISKIE, D. W., ARMIGER, W. B., HUMPHREY, A. E. (1976), in *Abstr. Papers, 5th Int. Fermentation Symp.* (DELLWEG, H., Ed.), p. 91, Berlin.

ZABRISKIE, D. W., ARMIGER, W. B., HUMPHREY, A. E. (1977), in *Workshop Computer Applications in Fermentation Technology, GBF Monogr. Ser.* **3**, 59.

ZIEGLER, J. G., NICHOLS, N. B. (1942), *Trans. ASME* **64**, 759.

17 Automation in Biotechnology

ANDREAS LÜBBERT

Hannover, Federal Republic of Germany

1 Introduction

In most practical cases the ultimate goal of industrial activity is to earn money, i.e., to maximize profit. This is realized by optimizing the product quality, thus its value, and by minimizing costs under the actual boundary conditions set by the market, the competitors, and the environmental circumstances or pressures. Production costs are decisive when constructing industrial fermentation plants, and one way to reduce them is thought to be automation of the plants. But there are also some further considerations that stimulate interest in automation. Many processes are extremely sensitive to changes in special process variables and will not work without rigorous automatic control, which can be performed optimally with computers. There are other examples, including complex measuring techniques that are too complicated for manual operation by plant personnel, and thus demand at least open loop control.

Automation by means of modern computer systems allows for managing very large production units, where it would be extremely difficult for an individual to keep track. Such larger systems are often widely distributed in the field, while being controlled from single observation points. Modern automation systems tend to control the production process in a wider sense than formerly. In biotechnological production processes, not only the fermenters, but also the substrate preparation processes and downstream processing are included. Development proceeds in the direction of including more processes, namely logistic and even commercial ones. It is generally assumed that the profit obtainable from a production plant can be increased by integrating information about all aspects of manufacturing. Hardware and especially software technologies are being developed to implement this concept in the form of a computer network which acts as a single plant-wide entity. This is the idea behind "computer integrated manufacturing" (CIM), put forward mainly by big companies in manufacturing industries (e.g., General Motors).

All new plants in the chemical and biotechnological industry are equipped with process control systems. As development in automation and plant control proceeds, a much larger number of plants will be controlled by computer systems within a few years. Automation must be regarded as a part of the production process equipment. Since modern controllers and especially the analyses on which they are based are specific to the system in question, they must be adapted to the processes individually, in essentially the same way as other system components, e.g., the bioreactor. Not only must the reactor be chosen from a variety of possible types, but the same also applies to the strategy and implementation of process control. In this respect the choice of a process control system, in some critical systems, becomes equally important to that of the reactor. Thus, up to 15% or more of the investment costs of a new plant are put into control. Both components of bioprocess engineering, the reactor and its control system, are dependent on each other and should thus be developed simultaneously.

Automation of biotechnology can be regarded as a task similar to that of the automation of chemical production plants, but it is quite different from automation in construction industries. The difference is due to the nature of the systems to be controlled. In construction engineering one usually has deterministic systems, whereas in chemical and biochemical process industries the systems have many stochastic features. Neither the catalytic reaction nor the mass transport mechanisms are deterministic. Furthermore, the necessary measuring techniques are not in a comparable stage of development. All this has immediate consequences for measurement and control, which are key elements of process automation. Fortunately, there is one point which makes automation in biotechnology easier than automation in chemical industries: bioprocesses are usually slower by chemical engineering standards.

During the development of a new process, one usually starts with a standardized stirred tank fermenter. This type of bioreactor has proven to be most easily adaptable to different tasks and, hence, is known as a universal tool of high flexibility. If one looks for a computer control system of similar universality and simplicity in order to directly install control strategies, one cannot find an analog. Hence, in order to develop process control, a tool is necessary which is so easy to use that the biotechnologist himself can use it directly to implement his ideas on process control. It must be a main goal in the development of computer applications in fermentation technology to make a universal system of this type available that has easily mastered user interfaces similar to PC-based text processing software. Simultaneously, the aim must be to reduce the fermentation operator's work load, not to replace him. The latter will not be possible in the foreseeable future.

The term automation in biotechnology deserves some comment. Due to the very nature of the processes discussed, one usually cannot automate chemical production as fully as a car manufacturing assembly. Most computer activities are therefore directed toward supporting the operators in keeping track of their systems, providing them with all information necessary for decision-making, and acquiring the capabilities for manipulating the process if something is going wrong.

Biotechnology has its own problems specific to its tasks. For essentially the same reason that one cannot call for mathematicians to model biochemical kinetics, it is not possible to look for an information specialist to solve problems of computer-aided bioprocess control. Consequently, biotechnology must build its own measuring and control systems, but based on general developments investigated in specialized fields and with the help of measuring theory, physics, and control engineering. Hence, it is necessary for a biotechnologist to gain some general knowledge in these fields.

2 Experiences from Initial Applications

The general aims in the initial application of computers to biotechnological practice were to improve the quality of information on the actual process state, and on that basis to improve the fermentation control, to enhance the flexi-

bility of control hardware, and to reduce its costs.

One of the first goals was to exploit the most prominent advantage of computers, their computational power. The basic idea behind this was to improve actual information about the process by model-supported measurements.

Controllers were formerly based on analog electronics. With the advent of devices based on software running in a computer it seemed furthermore possible to install a nearly unlimited number of controllers simply by copying the program code. With a sufficiently large number of controllers the expensive computer was thought to pay for itself. Consequently, one tried to apply this idea by drastically enhancing the flexibility of the fermentation control system.

Initial implementations in industry and scientific research laboratories, however, revealed numerous difficulties with the new technique. Most of the difficulties seem now to be forgotten, but they are still of importance for today's applications.

2.1 Multi-Control Applications

The development of computer control began during the mid-sixties. MURANO and YAMASHITA (1967) were the first to discuss some kind of supervisory monitoring of fermentations by means of computers. Distra Products Ltd. was a pioneer in its decision to install a computer providing automatic control of an existing plant for antibiotics production and of some new large-capacity fermenters. A total of 114 control loops were implemented at 26 production fermenters and 10 seed vessels. GRAYSON (1969) reported that, in comparison with conventional controllers, the main criteria for the choice of this novel technique, were:

- greater reliability
- more precise control
- greater flexibility with computer control algorithms
- additional data logging and alarm facilities provided by computers
- the possibility of model-supported control.

The company started with a 13-bit ARCH-102 process computer from ELLIOTT AUTOMATION (1967) with a basic ferrite core memory of only 8 kWords. Calculation speed, store access times, and memory size were very small, even by today's home computer standards. Moreover, neither higher language compilers, nor an operating system was available. The machine was programmed in ARCH-102 machine code. Even an assembler was not available. Since at that time computers were known to be unreliable, one did not rely on computer control alone, but installed additional conventional instrumentation and valves so that manual control could be used to run the process during hardware or software breakdowns.

It was found that control deviations were significantly smaller than those of conventional controllers, and data logging as well as accompanying alarm facilities were reported to be advantageous. However, there also appeared some significant disadvantages. The main severe problems of this system were:

- Insufficient reliability of some components of the computer, especially the electromechanical ones. This pertained to the main I/O-interfaces – i.e., the multiplexers – which were relay-based at that time.
- The program coding took a great deal of time on account of very unsatisfactory software tools.
- Moreover, since computer memory was limited, software could not be structured appropriately, but had to be optimized to minimal code length. Software tools could not be produced. Consequently, the machine-coded programs became very complicated and their modification extremely difficult and time-consuming.

2.2 Computer-Aided Measuring and Control

At the same time certain academics started to take advantage of computers in two ways:

- Model-supported measurement of quantities, which could not be measured directly. The most interesting was biomass.
- Model-supported control on the basis of indirectly measured quantities.

Two pioneering groups that began this research were the group around A. E. HUMPHREY and his collaborators (e.g., JEFFERIS and HUMPHREY, 1972; ZABRISKIE, 1976) and the one around D. I. C. WANG (COONEY et al., 1968, 1977; WANG et al., 1977; SWARTZ and COONEY, 1978). These groups saw the most prominent advantage in computer applications to be the extension of measuring and control potential by using the computational power of the machines. They tried to obtain data on key parameters not measurable on-line. Since the early efforts to design instruments capable of measuring some of the most fundamental cultivation process variables such as biomass concentration and growth rate, attention has been focused on numerical procedures that indirectly estimate these parameters by means of process variable data which can be measured on line and which are associated with culture growth. The latter include viscosity (SWARTZ et al., 1971), heat evaluation (COONEY et al., 1968), carbon dioxide evolution (STOUTHAMER, 1971), oxygen uptake (JEFFERIS and HUMPHREY, 1972), and substrate consumption and product synthesis (COONEY et al., 1977). The primary relationships were based on stoichiometric balances of all principal chemical components that compose biomass and products (COONEY et al., 1977), or on relationships between a single chemical component and culture growth (ZABRISKIE, 1976; ZABRISKIE and HUMPHREY, 1978).

The same applies to control algorithms (WANG et al., 1979). To optimize control actions, it is necessary to know how the system will react. This can only be predicted by means of a suitable process model, which necessarily must be a special one for each individual process. An adequate evaluation of such a model can only be done by means of a computer.

2.3 Conclusions from Initial Computer Applications

Interestingly, nearly all the aims still on the agenda of workers active in automation in biotechnology were already formulated by the first investigators (HUMPHREY, 1977a, b):

- Data acquisition, storage, analysis, and reduction
- Monitoring of fermentation plants by multichannel data logging connected with a flexible automatic alarm system
- Extension of the limited on-line measuring possibilities by model-supported methods, especially for the on-line measurement of biomass
- Direct digital control with many control loops working simultaneously, coupled or running independently
- Model-supported control to enhance the performance of conventional controllers
- On-line process optimization.

From the fact that some of these goals have not yet been reached in a generally satisfactory way, it is evident that there are many practical obstacles. On the other hand, even during first attempts it proved possible to operate a large number of control loops simultaneously in a single computer system. GRAYSON (1969) stated that control accuracy had already been improved significantly compared to conventional analog computers.

Nowadays, the early assumption that digital control could be installed more cheaply than analog techniques is accepted as true, but it was not so at that time, not if one took into account the overall cost of single central process computers, including their installation, software development, and maintenance. As time progressed, the probability of failure of microelectronic components dropped exponentially by several orders of magnitude. Thus, this risk has become significantly smaller, but it still exists and must be considered when developing a new computer-based control system. As compared to today's computer prices, the early machines were extremely expensive tools; investments were thus very high. Compensation by a corresponding return was not possible for several reasons. The main disadvantage of the early systems was unreliability. The hoped-for reduction in the number of process operators could therefore not be reached. On the contrary, additional, better educated personnel were required to keep machines running and to adapt the machines to the processes. Since computer downtime meant breakdown of all controllers, it was not possible to keep product quality as uniform as was originally expected.

Today, most controllers work internally with digital electronic components, and they are once again installed in separate units, which are incorporated into a system of front-end processors. Internally, however, they work on the same basis as analog PID-controllers. This computer control often degenerated to an at best more economical substitute for the early analog controllers. Only very recently have more advanced controllers (as compared to simple PID-controllers) become available for process automation systems.

The limited control obtained in the first applications was obtained at high cost in money as well as in manpower. The cost was much too high for most companies in the fermentation industry. They also saw that the use of a single computer to control a whole plant would constitute a considerable risk of production breakdown. It was essentially this low reliability together with the high cost in manpower that led to fierce resistance in industry against a wide application of computer control in fermentation plants. Another significant reason for the slow acceptance of the rapidly developing computers in fermentation plants was that on-line sensors were missing for many essential process variables such as biomass, substrate and product concentrations etc. (CHATTAWAY and STEPHANOPOULOS, 1989). Moreover, difficulties in analytically modelling fermentation processes prevented playing the highest trump in computers for any process automation: model-supported measuring and control.

In general, the more sophisticated aims were unreachable with the computer systems available at that time. For example, model-supported measurements of biomass proved to be much more complicated than originally thought, since correct and sufficiently detailed models were not available, and if they were, the necessary data or other process parameters could often not be supplied with sufficient accuracy. Thus, smaller steps in the development of bioreactor automation had to be chosen with aims that could be reached within reasonable time intervals and that were easy to survey. To change the control strategy of a whole plant with a new technique, as was done by Dista Products, was then and remains a high-risk endeavor. From the industrial point of view, the advanced methods of measuring and control seemed to be unsuccessful.

However, first experiences were not sufficiently discouraging to stop further attempts at improving computer-based process control. Especially desirable was the possibility of running a large production system from a central monitoring system with improved control quality.

3 Actual Industrial Installations

Actual working industrial installations have been constructed by responding to the main failures suffered by the pioneers. Since the drawbacks of the first installations were severe, drastic measures were chosen to reduce cost and to enhance reliability. Essentially two paths were taken in the hardware field. The first was to distribute several parts of the systems' load to different processors, and the second was to install system redundancy wherever it could be paid for.

From these facts, the following two hardware philosophies emerged:

- The first saw the advantages of powerful process computers in their computing power and stressed the software arguments. A simple redundancy approach was preferred, which was realized through tandem computers.
- The second preferred widely distributed systems consisting of smaller computers. These have the advantages of hardware building-block systems, but suffer from

having less software flexibility. Redundancy must additionally be applied in these systems.

In the software of such systems the term *free programming* was eliminated as far as possible, since programming by users was identified as a major source of failures. Instead, the software was distributed in a closed form, containing routines necessary to handle standard tasks. For a concrete application, the software modules must be linked together and supplied with actual process parameters. This activity was called *configuration* and was carefully distinguished from programming.

3.1 Quantitative Measures of Reliability

Reliability and performance of the computers turned out to be the main obstacle to wide acceptance of computers in industry. Control engineering companies developed new systems to respond to the bad experiences suffered in industrial practice. These ideas, subsequently developed, still rule today's computer installations in fermentation plant control.

Since reliability is the decisive demand of an industrial control system, this topic deserved to be discussed first. The breakdown of an automation system is a random event. Hence, a quantitative treatment of reliability can only be made from a statistical base. As generally accepted, one takes the mean relative time, A, during which the particular system is running correctly, to quantify reliability. This is termed *availability A*:

$$A = MTBF/(MTBF + MTTR)$$

where $MTBF$ represents the initial letters of "mean time between failures", and $MTTR$ the "mean time to repair" the system. It is obvious that $MTBF$ must be maximized and $MTTR$ minimized in order to obtain optimal availability.

To reduce $MTTR$, one not only needs well-educated field service specialists who find and repair failures very quickly, but also a stock that contains all required components that may need to be replaced. It is obvious that modular construction of the hardware facilitates the replacement of defective components, therefore assuring brief repair times.

Since the reasons behind errors can be extremely difficult and time-consuming to find, much developmental effort has been focused on automatic system tests and failure analysis procedures. Self-test and analysis features are becoming essential criteria for hardware and software components. Especially in small-scale automation, e.g., in automated measuring systems, self-test procedures of high sophistication have already been implemented.

There are essentially three methods of increasing *MTBF*. First, it is desirable to construct the systems by using sufficiently tested, highly reliable components. The second general way is to distribute the whole system among many components that work as independently as possible. In this way the burden is put on many shoulders, so that the system will not completely break down at every failure in hard- or software. The third possibility is to use redundancy (e.g., MAEHLE, 1988). This means that less reliable components are installed more than once. In the case of failure of one, the work-load can be taken over by another. This also implies that the system architecture must be designed modularly in such a way that some of the modules do – or at least can – take over the same activities.

From this concept, two main hardware design routes appeared in practice:

- Redundant process computer systems, and
- widely distributed process control systems.

Most of today's computer-aided automation systems have been built on the basis of these general ideas. In the next section these developments are discussed in more detail. The discussion will reveal that the choice between the two main hardware design routes for a special application depends on the importance assigned to the software aspects.

3.2 Distributed Control

An immediate consequence, drawn from the limited success of using advanced measuring and control concepts, was to cancel all those activities and to concentrate on very simple data acquisition, monitoring, storage and elementary control activities. Only the most important process parameters were treated.

In the distributed digital automation system approach, *hardware modularity* has been placed in the foreground. The central idea is a system consisting of a set of different modules that are flexibly interconnectable to fit the special demands within a concrete production system. Here, microprocessor-based hardware units are used to set up special process stations, e.g., for closed-loop control, open-loop control, and monitoring the process in question.

If the hardware components of such modular systems are well-balanced, reliable automation systems can be constructed using only a few different modules. In this way the configuration can easily be changed or extended if the process is extended or the goals are changed. This idea has been favored by most vendors of automation systems during the last few years. In most cases, the modules of such distributed computer systems were arranged along a linear bus system interconnecting the parts. Such a modular setup reduces the maintenance cost, since, as already mentioned, defective modules can simply be exchanged as a whole and be repaired after the system is working again. Additionally, investments can be kept within reasonable limits by using only a few standard hardware components that can be fitted together flexibly. The cost of the building blocks in a construction system is obviously lower the more such standard modules are produced.

Another significant advantage of such distributed control systems is the ability to directly locate the front-end-processors at the plant sites in the area where the measuring and actuator devices are installed. This reduces considerably the cost of connections and cabling.

In most real installations, the software running on the different hardware components consists of ready-made modules which need only to be configured and parametrized. The software is composed of well-tested but fixed modules. In nearly all instances it is not possible to incorporate user-written programs into such systems (and most applicants were not interested in doing so anyway). It is argued that all changes in software would be a risk that might decrease the availability of the system. Primary interest in the reliability of the system thus eliminates the most prominent feature of a computer: the flexible programming of software.

Automated control in larger biotechnological production plants is dominated by the same distributed control systems that are used in the chemical industry. A noteworthy example of such a distributed process control system with a classical architecture was implemented recently in the t-PA (tissue plasminogen activator) production factory of Dr. Karl Thomae in Biberach, West Germany. This process control system is based on the Teleperm M of Siemens AG. It controls the main production units, substrate preparation, fermentation, and downstream processing, as well as such peripheral units as the central power supply, the supply of extremely pure water, pressurized air, etc.

The general layout of the system (SCHNABEL and AUER, 1987; ANONYMOUS, 1988) is depicted in Fig. 1. It contains 11 "AS230" units with up to 240 batch programs, each one connected to 21 operator terminals. These components are interconnected via a "Data Bus System CS 275". Essentially separate from the "AS230" systems, 6 "Simatic-S5" units (115U and 135U) have been installed as open loop programmable control systems for complex interlocking control tasks in the auxiliary equipment. The software of these systems is necessarily also modular (MATHER et al., 1988).

A typical drawback of many installed distributed control systems can be recognized. Open- and closed-loop control are features not equally well developed in most systems. Hence, components of two different systems which may not harmonize well with each other must be put together to match the requirements. Such a separated architecture is neither necessary nor advantageous. It merely results from different development paths associated

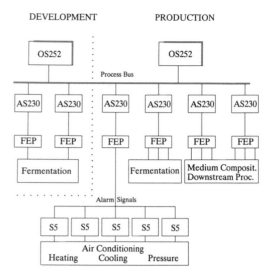

Fig. 1. General layout of the distributed control system installed in the t-PA production plant of Dr. Karl Thomae in Biberach, Germany (SCHNABEL and AUER, 1987).

with the two components. While open-loop control was developed by electrical engineers, closed-loop control units were historically developed by control engineers. The rivalry between both groups in industrial practice impeded the development of integral systems for a long time. Only very recently have these indispensible system parts begun to merge.

3.3 Redundancy Approach

Since a breakdown of an essential module in a simple distributed control system would block the whole system, redundancy is often applied to obtain the desired reliability. Redundancy means that the set of units that can be used for a particular function in question is installed more than once. The number is chosen according to the function's importance and is the reciprocal of the mean availability of the system component concerned.

Those components of primary importance to the system's availability are most often laid out redundantly. Initially, redundancy was implemented by simply doubling the important system components whose probabilities of failure were larger than acceptable under given circumstances. In distributed control systems, the central bus connecting the different modules usually has a backup, whereas for front-end-processors it may be sufficient to install one reserve module for several (e.g., four) actively working units (LITZ, 1990).

In production environments, in which a bigger process computer was thought to be necessary, one often simply doubled the central computer. There are essentially two modes in which tandem processors are operated. In the first possible mode one computer carries the control burden while the other is free for other tasks, e.g., report generation, program development, etc. If something is malfunctioning in the active unit, the activities of the other one is stopped immediately, and the second computer takes over the control.

The design of such systems (e.g., MAEHLE, 1988) is somewhat more complicated, since there must be an arbiter that decides whether the active computer is working correctly or not. For this task special watch-dog units have been implemented. If they recognize an error within the active computer, they immediately force the back-up system to drop all current activities and take over process control.

In the other mode, the second system works as a so-called "hot stand by" system in the sense that it receives the same process information as the active one, and also performs every calculation. Comparisons between results obtained in both computers can be used to detect failures. Again, the problem is to decide which system delivers the correct results. Thus, additional test calculations become necessary when differences appear.

Most often the hardware of such tandem systems consists of powerful and expensive process computers. They only pay off where the process necessitates sophisticated software that requires well-performing minicomputers. In such computer systems emphasis is on software performance. These computers basically run under real-time operating systems. The application software packages in such machines are called "process management systems" (PMS).

The most notable examples of control system installations based on double process computer systems running as well-performing process management software are the "Argus" systems developed by Ferranti. These automation systems have been installed with some frequency in biochemical production plants, and are still doing their work with sufficient reliability, e.g., in plants of Pfizer or Hoechst. The basic sophisticated software systems, such as that of Ferranti, would not have been possible if they had been developed for chemical or biochemical plant automation only. Significant parts of the systems have been taken over from developments in other projects of this well-known company. To reduce software costs in the future, it generally will be necessary to apply appropriate software components from other computer application fields.

Although these relatively powerful double processor systems have been overshadowed by distributed systems during the last few years, mainly because of the high cost of this solution, the general idea is not outdated. Since, as opposed to software, hardware costs are drastically falling, full one-to-one redundancy of process computer hardware will become economically more justifiable in the near future. Today, however, the construction is somewhat different. Machines offered by the manufacturers are fully redundant internally. Such computers are sold under the name "failure redundant systems" (DEC, 1989). They appear as single processor machines for software.

3.4 Software Analogs

Nearly the same arguments raised on behalf of hardware construction apply to software components. Modular construction leads to a transparent software which can be supervised more efficiently, permitting errors to be detected and eliminated more easily. Needless to say, all modules must be tested as carefully as possible before they can be used in automation systems. Special program verification methods are being developed under the heading "computer aided software engineering".

Redundancy concepts have always been used by serious engineers. Test calculations with different independent solving algorithms are necessary to check for calculation errors. The same must be applied to computer algorithms. However, software redundancy is rarely applied in practice.

Unfortunately, it is unrealistic to expect extensive software packages to be completely free of errors. Thus, the software itself must be prepared to detect failures which may appear during the run time. Such failures can appear randomly with unusual data combinations, wrong inputs by the operators, or hardware errors. Fault-tolerant programming is one means of coping with this problem. Fault diagnosis is its first step. A review of the different concepts of model-based fault detection and isolation has been given by FRANK (1990). ISERMANN (1988) described modern fault detection using expert system techniques. An approach to fault diagnosis by means of artificial neural networks was described by WATANABE et al. (1989). Fault-tolerant programming aims at correct execution of a specific program, even if faults occur. This only works well if the potentially detrimental consequences of the defects are overcome by employing special software techniques. One of these is filtering all operator inputs for systematic failures.

A basic means of obtaining software reliability is to make available a highly modular software system consisting of well-tested modules for data acquisition and analysis, process control, and process management. Breaking down the software into several logical pieces makes the design of a special application much easier, since the software modules can be kept smaller and, therefore, easier to write, debug, and maintain than larger programs. In most cases, only ready-to-use modules are supplied to the users. Modularization by itself is not sufficient; subroutine libraries of reliable modules have been used since the earliest beginnings of software development. Many errors can be made, however, in linking them for special applications and supplying them with appropriate parameters.

Thus, special programs were developed to aid the application engineer to configure the software needs of a special plant and to parametrize all the controllers. This drastically reduces the knowledge of programming required by the personnel, and ensures that the plant control can be configured and installed internally so that the secrecy demands of companies can be fulfilled. These developments can be regarded as the beginning of user interfaces to support at least the control engineer,

by adapting the software to the language of control instead of forcing the applicant to learn special computer details. Unfortunately, the software of process control systems has been developed by control engineers, and user interfaces have been constructed and optimized to support mainly their needs; these may be much different from those of biotechnologists.

The trend has been to claim that it is not necessary for a user to learn computer programming. Over many years software has been withheld from users within plants, because it was believed that sufficient software stability could not be guaranteed if users were able to make changes. The fact is, however, that this has only served to divert from the main weakness of these systems, i.e., they are not failure-tolerant enough.

The result of this development was that many of the arguments raised for the application of computers in control systems, i.e., to exploit the knowledge of special models for the processes to be controlled, were discarded in order to increase reliability. Special knowledge of a particular biotechnological process cannot be found in general automation systems, but must be coded separately. The same applies to special advanced model-supported measuring techniques.

3.5 Concluding Remarks

Most installations of process control systems currently working in industrial plants are based on distributed hardware systems. In nearly all systems some redundancy can be found (POLKE, 1988). The dominating design aspect has been availability, which is most often obtained at the cost of software flexibility, advanced control activities, and further sophisticated applications, such as optimization, etc. Large investments are put into industrial control systems; in new plants it may be 15% or more of the total investment. The two main process control hardware architectures, distributed systems and tandem process computers, both provide remarkable stability. In comparison to sensors, valves, and associated tubings, their availability is no longer the limiting component in process control.

The major justification for large investments in process control systems, including the required on-line field measuring devices, is reduction of overall costs by dispensing with shift-workers, reducing substrates and energy, and avoiding costly plant downtimes as far as possible (LITZ, 1989). The reduction of personnel costs is often the most directly accountable and often the biggest point. Advantages in energy and substrate consumption are also predictable. Some advantages are not easy to quantify in concrete numbers, i.e., the greater flexibility with respect to changes to other products, enhancement of product quality, and the achievement of more uniform quality, since they influence the overall result only indirectly.

One very prominent feature of the control systems described is that they rely predominantly on ready-to-use software. Configuration has replaced programming in order to reflect the often-claimed simplicity with which system activities, provided for in the software, could be made available. Many standard control actions can be implemented in this way. Inspection of systems in actual use shows that closed-loop control is more often supported, sometimes at the cost of open-loop control. Free programming possibilities, however, were restricted or disallowed. This makes the integration of advanced measuring and control applications difficult. The same applies to the integration of all other user-defined software components. Basically, most plants are controlled as they were some 40 years ago using PID algorithms adjusted according to ZIEGLER and NICHOLS (1943) or variants.

The application of the described systems is suited for productions that are already optimized and are not designed for continuous improvement. However, in biotechnology, especially in research and development departments, things are always changing. Thus, flexibility of the automation systems is at least as important as availability.

One of the main drawbacks of the systems discussed above is that they cannot easily be combined with hardware and software components of other manufacturers in process automation. A case in point might be the coupling of a general purpose computer for process simulation or optimization. The main problem is that the bus systems within most distributed control systems do not meet international standards. Moreover, the software does not support the integration of components from different vendors. This situation came about because most current standards did not exist at the time the existing control systems were developed. The development of classical automation systems dates back to the seventies and the early eighties.

Much money has been invested by many companies in developing their own special standards which are not compatible with others (e.g., DOHMEN, 1990). It is evident that many do not like to cancel their developments and change direction to meet international standards. This is a disadvantage to customers, since these systems are more or less closed forcing consumers to buy components of one vendor's system only. Moreover, often only a very limited number of basic components is available. This restricts the customer's flexibility in building an optimal system, since he cannot combine the components best suited to his special needs. In this way there arises a dependency on the initial vendor.

For new installations it is therefore recommended to look for systems based as far as possible on universal hardware and software components which meet international industrial standards. There are several such choices now on the market.

4 Actually Developed Systems

4.1 General Requirements

Automation systems currently installed in the industry are optimized for high reliability. This was mainly achieved through two design principles. The first is modular design, which led to distributed automation systems. The second, supported by this modularity, is the redundancy concept. The advantages of modularity are not restricted to the hardware, however, but can also be extended to the software domain. Based on the idea of a building block system, modular software systems contain a fixed number of modules, which can be configured and parametrized in order to adapt the software to a given automation task. Such systems can fulfill the requirements of system reliability.

In actual process automation systems, much more data about the processes are accumulated and generated than was formerly possible. In order to use these data adequately, intelligent management is required. Information processing also rapidly developed in various other parts of industrial companies, e.g., in the sales departments. There are many data of importance to more than one department in a company. Thus, it is advantageous to integrate computers into company-wide information processing systems. The structure of such comprehensive systems is everywhere similar to that proposed by NAMUR, as sketched in Fig. 2. Process automation, then, is only one part of the whole. Since the individual systems within the different levels are all distributed systems, development proceeds in the direction of creating networks within networks. The connection of the local area networks installed within different levels of a company's hierarchy requires efficient interfaces between the individual systems in order to ensure that benefit

is derived from the information gained on different levels.

The spectrum of different tasks processed in modern automation systems is still expanding. At the same time, the computational requirements for individual tasks, e.g., process control, do not become smaller. This expansion in quality and quantity of tasks to be handled brought the conventional process automation systems to their limits. The tasks within automation systems require more computing power than current computer nodes can supply. A straightforward way to extend the performance of these systems is to use more powerful nodes in the distributed systems (e.g., FEWKES, 1988) and to increase the number of "intelligent" components. This way, however, is circumscribed, since the nodes of most current automation systems are constructed by the vendors and cannot be freely exchanged by units of higher performance. Even if this would be possible on the hardware level, most vendor companies would not be able to extend their software bases, which usually are also individually written.

It can be seen from the software developments of the last ten years that even very large software development companies are not able to develop basic software systems at a rate comparable to the development of new-generation processors in microelectronics. Hardware developments thus orient themselves to standard software bases, e.g., new generation processors must be operated with established operating systems. Consequently, one must base new automation systems on hardware and software components that are involved in these development flows. The development of process automation systems will probably move in the direction of using more and more standard components, which may be developed in completely different environments and are supported by large interest groups rather than by comparatively small control engineering companies.

Performance is the primary criterion for the choice of individual nodes within a distributed process automation system. It is highly improbable that one vendor can supply all components that are optimal in performance and meet all needs in a special application. Often, programmable open-loop controllers are better

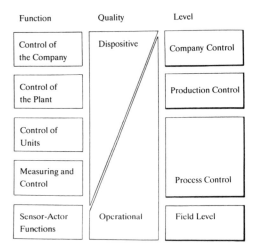

Fig. 2. General structure of a comprehensive automation system as proposed by NAMUR. Process automation is one significant part of the whole.

developed by one company, while an advanced closed-loop controller of another firm may be more suitable. In most instances it is better to acquire a database system from a third supplier. Hence, automation systems are required in which a customer is to put those components together which optimally meet his demands. Such systems are referred to as "open systems".

It is obvious that such open systems can only work if certain standards are observed. This does not necessarily mean that the different nodes must be standard components. That would restrict development too severely. It merely denotes that communication between the system's components must be standardized, a prerequisite if cooperation is to be efficient.

As compared with modern software components available to users of PCs and workstations, the active user support in currently used process automation systems is highly unsatisfactory. The ease with which automation systems can be handled by process personnel must be drastically increased. It is essential that the user be supported by the system and not hindered by complex operating procedures. Much can be learned from developments in the PC field concerning convenient user interfaces, especially graphical visualization techniques, to mention only one aspect. Summarizing, one can state that systems must be optimized to support the user.

Current development will be discussed by means of some concrete examples, in which the process management system CIF will be the main accompanying example to guide the discussion.

Accompanying example: CIF

CIF is a modern process automation management system that can be used to automate medium-scale biotechnical pilot and production units. CIF stands for "computer integrated fermentation". It is set up to suit the special requirements of biotechnology; thus the idea behind it is to enhance the performance of universal automation systems by concentrating on special problem-oriented processing concepts of biotechnology and support-

ing the biotechnologists. Integration plays a special role in this system. In other words, data from different sources are integrated by CIF, measuring and control subsystems of different vendors can be integrated, and data analysis software from different developers can be readily adapted. Finally, CIF itself can be integrated into a company-wide information processing system.

Although several biotechnological cultivations can be controlled completely independently of each other, it is also possible to interconnect different applications. CIF is a strictly modular software package running under the operating system VMS. Its basic hardware components are VAX computers, which are the recognized standards among more powerful 32-bit computers. These computers are widely distributed in chemical and biochemical industries. CIF software runs on single computers and can be distributed onto different nodes in a local area network (LAN) as well. For initial installations, one can use a single workstation, and one can expand the system later by setting up additional computers.

The different components in a system containing more than one computer are arranged along a serial bus system based on Ethernet technology (IEEE 802.3). This network system is open to computers of different vendors, since Ethernet has emerged as the most widely accepted standard for such local area networks. Many vendors supply devices with Ethernet interfaces and appropriate network software, which can be used to integrate the devices into VAX computer networks.

In an actual installation of CIF, the VAX computers must be complemented by "front-end processors" optimized to data acquisition and basic control tasks. Front-end processors developed by individual fermenter manufacturers can be used, e.g., the DCU of B. Braun Melsungen AG. Often, programmable controllers such as the Simatic S5, supplied by Siemens, are still installed at the plants, and it is important to be able to integrate them into the automation system. Many other types of front-end processors used in biotechnical plants can be adapted.

Fig. 3 depicts the general software structure of CIF. Central components are the real-time database called POOL, together with its data

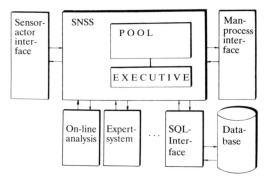

Fig. 3. Software structure of CIF. Many functional modules are grouped around the central core of the software, which consists of the real-time data base POOL, its manipulation program EXECUTIVE, and the general software interface SNSS. For details see the text.

manipulation and management program EXECUTIVE. The latter operates information-oriented, i.e., it reacts on the data which it finds in the POOL. These may be measuring data, results from calculations involving data, or inputs by the user. Both central components are shielded from all other functional modules via the universal software interface, SNSS. A properly-performing data security system within the SNSS controls all access to the central real-time database of CIF.

Additonal functional modules are arranged around this central core as depicted in Fig. 3. Two, the modules "Man/Process Interface" and "Sensor/Actor Interface" are necessary in every application. Both are designed as virtual interfaces. The reason is that they should be open to any hardware devices chosen by the applicant. All other functional modules are attached to the core system if necessary in a special application.

Since the SNSS interface is network-transparent, all functional modules can be transferred to other computers within the local area network. Thus, the functional modules can be run either on the same machine or, if they require more computing power than is available there, or if they require special hardware facilities, the modules can be transferred onto other nodes within the local area network. Hence, there are two means of expanding the computational performance of the whole system. The

first is to replace the computer nodes by more powerful ones; the second is to expand the number of nodes and distribute the software components accordingly.

4.2 Architectural Considerations of Modern Automation Systems

A common experience based on currently working automation systems is that it is advantageous to use modular distributed computer systems. This idea is being adopted in nearly all newly-developed systems. Such modular systems have the additional advantage of supplying a means of realizing truly parallel computing, thus increasing the overall computational power of the systems. Furthermore, they readily permit the incorporation of redundancy into the systems if it becomes necessary. However, any distribution of system activities to different computers requires a significant amount of information transfer and synchronization effort. This means that distributing system activities does not necessarily guarantee a net gain in every case. Hence, the division of all system activities into parts to run on different computers cannot be arbitrary. It is necessary to optimize the partition of tasks. Groups must be formed between which communication and synchronization needs are minimal. Hence, it is necessary to divide the software into separate parts that are as self-sufficient as possible. In an automation system such tasks may include supervisory control, process simulation, off-line data analysis, database operations, etc. In the accompanying example, CIF, such tasks are symbolized by separated blocks attached to the system, as depicted in Fig. 4.

Modern automation systems in biotechnology must also be able to incorporate advanced complex measuring systems, e.g., GC, HPLC, MS, FIA, etc., together with their basic control and data analysis hardware and software. From the point of view of process management, one is only interested in the final results, such as the concentration of special chemical substances. All efforts to obtain these data are of at most secondary importance relative to the management system. Such analytical in-

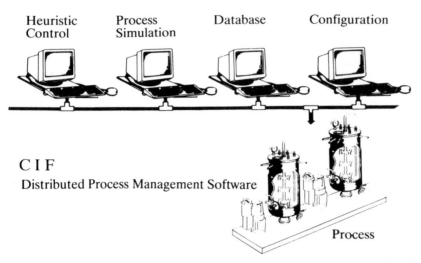

Fig. 4. Distribution concept of the process management system CIF. Encapsulated process functions are distributed to the different nodes of the local area network that constitutes the process automation system. The distribution of the different modules is not fixed to the nodes but can be reshuffled according to the special needs in an actual situation.

strument systems are often automated in themselves. It is quite clear that these small-scale automation systems must be incorporated into the larger process automation system as a whole, instead of trying to develop new automation software for these devices within the larger system. Hence, smaller systems must be capable of integration into larger ones.

The same reasoning applies to the development of software for other basic automation system components, e.g., for programmable controllers. System integration is of increasing importance, since there is a general trend of migration of microprocessors into measuring and control devices, i.e., closer to the process. It is necessary to make the process automation management systems open for the incorporation of many different internally automated components from different vendors. The choice of measuring and control components must above all be guided by their performance in relation to the task to be performed, and not primarily by other arguments. Hence, a modern process automation management system must supply a common base for integration of a wide variety of different devices. This is essentially a question of communication between the base and its components.

Information exchange between different items requires an agreement about a common base of communication. For the communication between automation system components, several standards have emerged. They are of widely differing performance, and so cannot be applied equally efficiently throughout the automation system. For data transmission between a measuring device and its host computer, a 4–20 mA serial connection according to RS232 may be sufficient in many cases, whereas such a connection is completely inappropriate if used to connect different 32-bit computers. Connections between computers are not only different in their performance, but also in cost. Numerous solutions have been discussed within the standards organizations, but no general agreement is in sight. In the meantime, customers could not suspend the development of their automation systems, so they created *de-facto* standards simply by their choices. For communication between computers in local area networks, only a very small number of network systems in practice proved to work with sufficient reliability; of these, Ethernet (IEEE 802.3) and SNA (IEEE 802.5) emerged as *de-facto* standards (CLEAVELAND, 1989), although they were once severely criticized by network experts. Such a *de-facto* standardization is not new on the computer scene. The IBM-PC is another example, which, as is well known, became an industrial standard owing to the fact that numerous customers chose it as their favorite system.

Although actual development moves strictly in the direction of open systems, the number of different components in a network must be kept minimal for economic reasons, since maintenance costs rise with the number of different system components. Thus, there are conflicting requirements, for each of which a compromise must be found.

The hierarchy in modern automation systems is of a logical nature. Concerning network hardware, democratic networks such as Ethernet are preferred. In Ethernet, all nodes are connected to a bus system and have the same chance to access it. The hierarchy is established by an authorizational levels system for persons logged in at the workstation nodes and by a priority system for different software modules. To establish a well-structured priority system within a distributed process management system, powerful operating systems are required on the nodes that support such structure.

The large centralized control rooms, characteristic of current automation systems, are becoming obsolete, since in the new network systems all information and manipulation capabilities are available everywhere within the network. Control rooms are only necessary where the operators must be protected from extreme environmental conditions, such as humidity, excessive noise, or extreme temperatures. Modern local area networks permit the process operator or the bioprocess engineer to plug into workstations wherever needed. One day they become necessary near the reactor, the next near a downstream processing unit; at other times they must be used in a lab, or in the process engineer's office. This flexibility is a primary criterion that saves considerable time for the personnel.

In the accompanying example, CIF, Ethernet was chosen as the basic network technology. It actually provides the broadest base for different subsystems available on the market.

4.2.1 Nodes in the Local Area Network

The choice of the basic computer components of an automation system must be guided by practical considerations. The application software providing the functionality of computer automation systems must be the first criterion in the choice of a computer system. This software determines what computers are required.

In smaller applications that can be solved with small software packages, PC-systems are suitable. There are many small automation systems on the market written for single PC-systems. These systems can be classified by the operating systems under which they run. The smallest systems are based on the well-known PC-operating system MS-DOS, which is a single-task operating system. More flexible systems use multitask operating systems of varying performance. Concurrent DOS is the smallest of these, OS/2 is a more extended one, and some PCs even use UNIX. Since convenience in operating systems usually can be achieved only at the expense of computing power, the latter systems require more powerful PC-versions, e.g., ones based on Intel's 80386/80486 or Motorola's 68020/68030 processors.

There is a continuous transition between UNIX-based PCs and workstations. Workstations can be viewed simply as more powerful single-user multitask systems with advanced graphic capabilities. In this area the UNIX-systems provided by different vendors are in strong competition with Apple's Macintosh and especially DEC's VAXstations, both running under their own powerful operating systems. The different workstation operating systems differ greatly in performance, so different criteria are applied in choosing among them. The system of the Macintosh has been optimized for user communication. Apple set an important landmark on this subject, much discussed and further developed by many other software developers. The VMS-operating system of DEC is the most convenient of the general operating systems and has the shortest real-time response time of the systems mentioned. UNIX, however, has the advantage of being the system with the best portability features among the multitask systems discussed above.

In CIF, the VMS operating system was chosen, since comfort and real-time performance of the operating system were rated higher than portability. Furthermore, for VAX computers a well-performing network software, DECnet, is available. Moreover, VAX computers have been particularly favored in industrial applications. The computer family concept of DEC is of great practical advantage. From workstations up to mainframe computers, many VAX computers of different performance levels are available, and all can be operated under the same operating system, VMS, so they do not require changes in the application software. Another criterion is that these computers have been proved to be compatible with successive developments for the last 15 years. Such a continuous development is of considerable economic importance, since the system can be extended or adapted stepwise to the actual development over many years.

The choice of a special operating system loses its significance as a result of the latest development in local area network software. In a network of VAX computers it is possible simultaneously to use some nodes running under VMS and others under DEC's version of UNIX, known as ULTRIX, or even MS-DOS machines (cf. Fig. 5). The general direction of development is to further reduce the operating system differences in local area networks by supplying universal software environments that support the exchange of information between different tasks running on different nodes under different operating systems. An important example is X-Windows (SCHEIFLER, 1987), which is becoming the standard base for user interfaces within local area networks.

4.2.2 User Interfacing

The fact that the system operator is indispensible in the control of biotechnical cultivation processes leads to the conclusion that he must be supported by the computer system. Most of today's digital control systems (DCS) do not have adequate operator interface capabilities. To facilitate the contact between man and machine, the so-called user surface or interface must be optimized to reduce the infor-

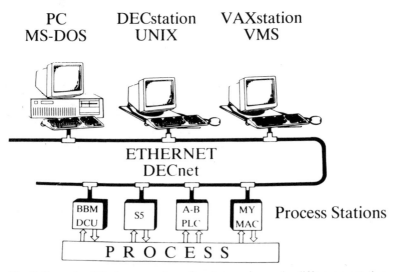

PC   DECstation   VAXstation
MS-DOS   UNIX   VMS

ETHERNET
DECnet

| BBM DCU | S5 | A-B PLC | MY MAC | Process Stations |

P R O C E S S

Fig. 5. Example of the incorporation of nodes running under different operating systems in the CIF-network in the context of an open network. The nodes belong to the nodes on the process data management level as well as to the front-end computer level.

mation transmission lost between them. The operator should not be overloaded with data. This means that he should receive only that information necessary to recognize the process state and to decide on the questions which must be resolved in the immediate context, and nothing more. It has not been generally realized that the computer must be adapted to the user. Since this fact became apparent, special tools have been developed to simplify the user interface with the computer.

The arrival of workstations allows for an integration of new technologies from computer graphics, bit-mapped video displays, etc., into components of automation systems. However, these new techniques, to be fully exploited by the developers of application software, require software tools (supplied by subroutine packages) in order to keep the development cost of high-performance graphical user interfaces within tolerable limits.

X-Windows is a software tool that supports the development and application of open user surfaces on engineering workstations. It serves as a software baseline system to exploit the many new possibilities in modern graphic-oriented user surfaces. As the name indicates, its operating environment supports multiple overlapping windows on color and mono-chrome workstations. It further supports mouse operations and thus allows for fast symbol-oriented command modes. From the point of view of process data management systems, its most interesting feature is that it provides a portable standard environment for man–process interfaces which can be used in any application software. Applications that use X-Windows can easily be run on a variety of workstations, PCs, and even on some terminals from different vendors coupled together in a computer network. It was developed primarily by SCHEIFLER and GETTYS (1986) at MIT during project 'Athena', sponsored by IBM and DEC, in which open computer networks were investigated.

Since modern user interfaces are expensive to develop even if tools such as X-Windows are used, different workstations in a computer network must not demand separate software development. Thus, X-Windows was made as hardware-independent as possible. The goal is that X-Windows applications running on one

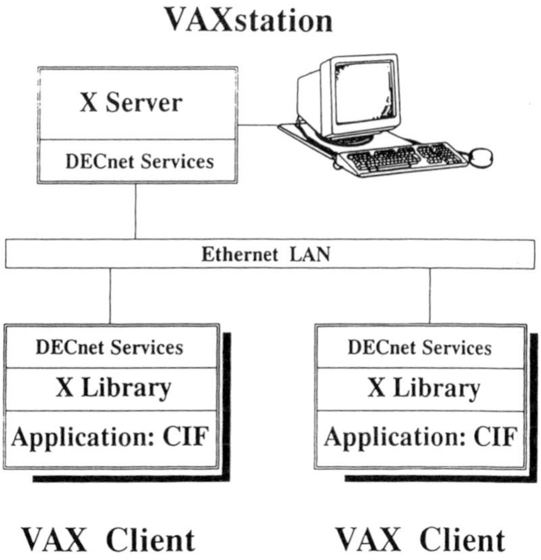

Fig. 6. Principle of the client–server architecture of an X-Windows application in an Ethernet-based local area network. The application program, e.g., CIF, is assumed to run on VAX computers in a local area network. The user interface is considered to run on the screen of a workstation within the network. The server program within the VAXstation generates the window display. The transfer of data is managed by DECnet services.

CPU in a local area network (Fig. 6), called the client, be capable of showing their output on the screen of the same or any other computers called the server. A single program working with X-Windows can display output windows simultaneously on different workstation screens. Network transparency also allows an application which needs much computational power to run on a more powerful network-connected computer, while user communication runs on a remote workstation. Standard software such as X-Windows, designed in such a client-server architecture, promises secure network transmission of data between different nodes in a local area network, whether they are PCs, workstations, minis, or even mainframes. The main feature that distinguishes X-Windows from other graphics systems is its network support.

The communication between server and library takes place by an asynchronous network-transparent interprocess communication protocol, which separates the virtual interface of the application program from the server implementation. The latter does not necessarily run on the same computer, and it alone must be specialized for the concrete display hardware. With such a separation of the (virtual) I/O of an application program from the software serving a special display hardware, the application becomes more portable. In the case of process management systems, this feature can be exploited in two directions. First, a special application task can be interchanged between the various computers within the network, while the operator stays at the same video front end. Secondly, a process can be observed from different nodes within the network using a man–process interface, which has the same appearance for the operator.

Windowing allows operators to shuttle between different application programs. Windows can be depicted on a video screen in overlapping or tiled fashion. This allows for more transparent process monitoring. But X-Windows does more than windowing. X-Windows is a tool that can be used as the basis for the development of network-wide user interfaces regardless of operating system compatibility among the nodes in the network. It therefore promotes uniformity of user interface functionality in different application programs in order to present them via similar screens, menus, and graphics.

All these convenient features and the flexibility of operation naturally come at a price. Workstations for X-Windows applications need large, high-resolution video displays for the simultaneous representation of many windows. Since the on-screen windows can, if necessary, be superimposed upon other windows, their contents must be backed up. Thus, large memories are required to hold simultaneously the overlapped parts of the display in the background in order to allow for a rapid restoration of hidden parts if the foreground windows are removed. If one uses several windows on a high resolution video display, several MBytes of working memories are required to hold the graphic displays in RAM. Hence, more than 10 MBytes of memory is not unusual within a modern workstation. It should be mentioned that such window-based systems become very slow if the server does not have enough working memory, forcing the windows to be stored on the system disk. The degree of user friendliness, depends heavily on the speed with which the process information can be displayed on the video screens (e.g., BAUMAN, 1989).

The ability to display on the user's screen only the information necessary for judging the actual process state is a matter of experience with the process in question. A context-sensitive selection of information requires at least heuristic knowledge of the process. Several groups are working on the support of man–process interfaces by expert-knowledge-based systems (BÄR and ZEITZ, 1989; NAKAMORI, 1989), as mentioned later in this chapter.

Such standardized user interfaces eliminate many differences in the appearance of operating systems to the users, and they reduce differences in application systems to functional ones. In this way different software components are virtually integrated from the user's viewpoint. Since there is no operating system available that can be used at all levels of an automation system, it is more important to have a standardized man–machine interface than to have a standard operating system. Thus, the days of lengthy discussions on compatible operating systems seem to be numbered. Only those will survive which support standard user interface systems. It now appears that X-Windows or OSF Motif will be the standard for the next few years.

The advantage of X-Windows-based user interface stations has been recognized by many companies, and they are actively working to incorporate them into their systems. One current example is the series 40 system of Bailey Controls (BAILEY CONTROLS, 1989).

The user interface of CIF has been developed directly with X-Windows and thus exploits its full functionality. It allows the user to design his own individual man–process interface without restrictions as to special terminals and PC- or workstation screens. Moreover, the management of network-transparent access to the different application programs is taken over by CIF.

As a classic example of a non-trivial X-Windows application in CIF, a typical display is shown in Fig. 7. The operator can acquire a quick overview of the system by arranging a small window for every key process signal on

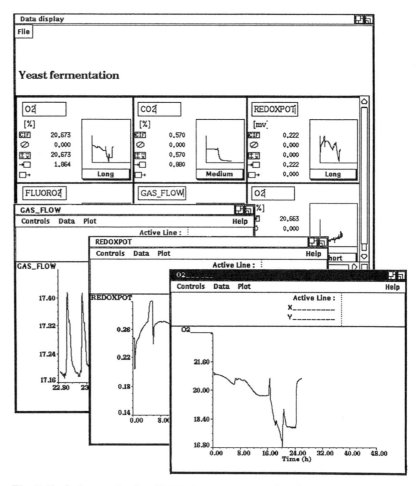

Fig. 7. Typical example of an X-Windows operator surface in CIF. The upper window contains smaller windows through which the operator can monitor key variables. These windows contain so-called push buttons, i.e., miniaturized plots of the signal records taken over short, medium, or long time segments. These are continuously updated. These small windows also contain the current values of the variables. If the push buttons are clicked with the mouse, the plots can be drawn to an arbitrary size as shown in the lower part of the figure. The plots are then fully scaled and captioned.

the video screen. These windows in turn contain smaller windows, each of which contains a history of the corresponding signal in the form of a chart recorder strip observed from a distance. These miniaturized histories are continuously updated. In this way trends can easily be recognized from the curves. The windows also contain the current value of the measuring variable. If the operator is interested in more

detail, he can, with a mouse click at the small window push-button icon, immediately zoom the signal representation to any size representable on his video screen. These larger representations are then fully scaled, and by means of cross-hairs the operator can examine any particular signal value within the display.

X-Windows supports the flexibility of CIF to run on different hardware configurations, both on a single VAX computer and on different systems within a local area network. Of course, all the software (including the X server) can run on a single VAXstation smaller applications. The second possibility is that the CIF system runs on one VAX and the server on another station. Finally, it is also possible to distribute the CIF software on different nodes in a local area network independently of the distribution of the X servers.

4.2.3 PCs as Front-End Processors to the User

It is a general trend in the development of distributed process control systems to integrate PCs as front-end processors for the user (e.g., Zoz, 1990). One reason is to make their well-developed graphical user interfaces available to the larger systems, e.g., window surfaces. High-resolution video screens like the VGA-standard in PCs are unusual in currently working process control systems (LITZ, 1990). Because of the large market, much more manpower can usually be put into the development of PC-software. In the future, many software components familiar to the growing community of PC-users will be available in process control systems, e.g., spread sheet programs such as Lotus 1-2-3 or Symphony, either as software packages on the normal nodes of the systems or via a software interface on PCs within the local area network.

For the VMS system, on which CIF is based, a fully PC-compatible Lotus 1-2-3 is available. This is compatible with DEC windows. Other software packages such as the often-used text editing system MS-Word can be run on VAX/VMS-systems with the help of an MS-DOS simulator. Using the tool PCSA, PC-files can be transformed transparently into VMS-files, which then can be transferred through the network.

4.3 Special Functionality

A high degree of functionality is required of automation systems in biotechnical applications. Model-supported measurements are ur-

gently required, since many state variables cannot be measured directly. Furthermore, many key variables can only be measured off-line; the measuring results then must be returned into the system and used for process control as promptly as possible.

4.3.1 Processing of Mathematical Relationships between Real-Time Data

The performance of a process management system is determined not by its ability to acquire, display, and store the process data appropriately, but primarily by its capability of manipulating the data in order to extract more valuable information about the process state, and to react to deviations from the predefined path of operation by calculating control actions and performing them. This principally requires a means to relate data from one, two, or several measured signals by mathematical functions or algorithms.

Such on-line calculations between process signals must be possible without a change in the basic software of the application system. The relationships necessary in a special application cannot be contained in a general automation system. They may change from application to application. Hence, they must be defined by the user and be made available to the system. There are essentially two principal means by which relationships can be implemented: The first is to attach a separate user-supplied task to the system, which is able to access the process data, make the manipulations on them, and transfer the results back into the base-line system.

The second means is a math-function processing component in the automation system, which can be supplied with the functional relationships to be processed at run time. It is advantageous, especially in pilot-scale applications, to install, change, and remove such relationships as a run-time activity of the system.

Such real-time math-function signal-processing facilities are available in several systems. Usually they work as interpreters, processing the instructions introduced by the operator into a prepared buffer. Unfortunately, in-

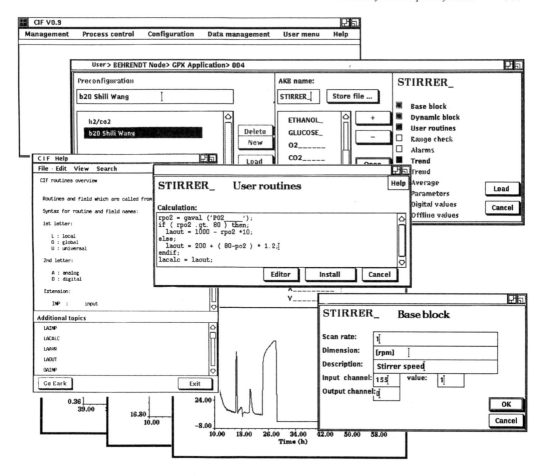

Fig. 8. User interface provided in the process control system CIF to easily program math-functions. They are directly written into the window at the center of the video screen depicted here using a higher-level language such as FORTRAN. The other windows around provide additional information for the user.

terpreters do not have good real-time properties. Thus, to decrease the execution time for the statements by the formula interpreter, the syntax of the code must be optimized. In order to obtain short real-time response times, it was most common to choose a reversed Polish notation in process control systems, recalling the programming of HP pocket calculators. Such coding may be carried out quickly, but it is not simple to use if larger code lengths are required. Thus, higher-level programming languages were preferred by the users. To increase

the interpretation speed for such source code segments, precompilers have been developed to translate higher-level language code into a compact, quickly interpretable auxiliary code. An additional way was developed for CIF, the cited example. Exploiting the convenient operating system VMS, it was possible to find a way by which modules containing the mathematical functional relationships between signals and all other data available could be written in higher programming languages, supported by a comfortable user interface as shown in Fig. 8. All languages available in VAX computers are usable. The source code segments are automatically complemented by code to produce complete subroutines, which can be compiled with standard compilers and then linked into the running automation system. Such compiled code is the fastest way to process statements, permitting the relationships to be processed much faster than any interpreter can work.

All on-line and off-line data, as well as all parameter values, can be used in CIF's function processing module. Even complicated, special higher level control algorithms can be installed. They not only can be caused to generate key auxiliary variables that more clearly depict the system's state, but also can reduce the redundancy of data and compare the system's actual state with the required one. In this way the math-function processing facility of CIF can be used as a decision-support tool.

4.3.2 Off-Line Data

Since many key variables in fermentations are measured off-line, it is necessary to feed them back into the automation system as promptly as possible to ensure their proper use. In smaller laboratories, the off-line data must be entered into the computer manually via the computer terminal; in larger ones, it is necessary to couple the laboratory information management systems (LIMS), usually operative in analytical laboratories, to the process automation management systems.

For manual off-line data input it is first of all necessary to have a suitable user interface with which the plant personnel can enter the data in a way that avoids input failures as completely as possible. A simple-to-use menu should be displayed on the video screen, and the software behind it should be failure-tolerant. This means that all inputs must be checked immediately for completeness and plausibility, so that the operator can directly recognize possible input failures and correct them. Moreover, it is desirable to inform the control system about every sampling event during the off-line analysis, to record the exact sampling time, and to remind the operator after a certain period of time to look for the result of the off-line analyses.

Of course, the process management system must be able to utilize the off-line data immediately. First of all, the data must be stored in essentially the same way as on-line data taken at larger sampling intervals and arriving at the computer with larger time delays. Consequently, they must be treated just as the latter are. Data with low sampling frequencies and long time delays often arise in biotechnology. Nearly all the more sophisticated measuring techniques, e.g., HPLC, FIA, autoanalyzer, etc., exhibit this disadvantage.

In CIF, the off-line data are stored in the same way as those measuring values that stay fairly stable during the fermentation run. Their values can be used directly by the math-function processing system of CIF, which in the evaluation of formulas takes the most current measuring value available. Where this is not possible, extrapolations can be performed on the basis of user-defined model equations.

4.3.3 Database Operation

In addition to the real-time databases in process data management systems, considerable benefit can be derived by installing conventional, large mass-storage-based relational databases to make proper use of much more data than can be stored in the working memory of the automation system.

Since process control systems generate increasing quantities of data, a problem arises in storing data in such a way that they can be recovered quickly when needed in subsequent fermentations or during an off-line analysis. It is important to be able to use standard database systems, which can guarantee a high degree of data security with respect to access and modification of data. The informational value of the process data from fermentations accumulated over some years should not be underestimated.

It is often desirable to compare data from earlier fermentations with those arising during a current run. This requires the display of historical data onto the video screens during the monitoring of the actually sampled data. Histories must thus be accessible in a simple way via an appropriate easy-to-use operator surface.

To facilitate access to the data, it is imperative to use a standard interface that is independent of the underlying database systems. This permits the specific system employed in the user's company to be easily coupled. Such an interface must, therefore, meet common standards. The "standard query language" (SQL) provides such a standard for the most frequently used relational database systems such as Oracle, RDB, d-Base, etc. (e.g., JONES, 1989). It decreases the sensitivity of application programs to details of the data storage techniques and *vice versa,* since it acts as a buffer between the two. The present version of SQL is not very convenient, but new versions are to be expected in the near future.

Since process control applications place high demands with respect to the mass of data to be stored and the required speed of data access, specific data-storage techniques must be used. In CIF the database is assumed to contain only that information necessary to find the real-time data and to supply additional information required for their analyses, while the original sampled measuring values are stored in files accessible at a much higher speed.

CIF provides a functional module for access to the data within a relational database by means of an SQL-based interface (cf. Fig. 3). Hence, this module is not restricted to a specific database system. This user interface is mainly used to quickly find all relevant data on the currently running fermentations as well as from all completed fermentations for comparison purposes. The data in the database can be accessed and extended, even beyond the end of the fermentation.

4.3.4 Control Profile Generation

One of the tasks most often required of the bioprocess engineer in pilot plants is to improve the control sequence for the next fermentation in order to optimize the process. In most cases the control profiles are determined empirically according to experience from previous cultivations. Often a model fermentation is chosen from preceding experiments as a guide for the subsequent runs. In this case it is obvious that one must try to control the adjustable parameters such as temperature, pH, and pO$_2$ along the profiles of the sample. Noisy data from older fermentations obviously cannot be used directly. Thus, the control profiles must be appropriately prepared by means of filters.

CIF, therefore, supplies a filter bank that allows the biotechnologist to simply refurbish historical fermentation data taken from the database via the SQL-interface. As indicated in Fig. 9, several filter algorithms are supplied. If none of the standard filters satisfies the user, it is possible to define the control profile manually by setting certain estimated data points into the plot using the mouse and subsequently starting a spline interpolation to finish with a smooth control profile. This can be sent to the POOL, from which it can be accessed by the controllers.

4.3.5 Implementation of Models

Advanced control is usually thought to be model-supported. Appropriate control algorithms must be specially programmed routines, since they cannot be supplied by a general process automation system. Since currently installed automation systems do not provide general programming environments, additional computers are required which must be fitted into the automation system. Their task is to supply a base for the installation of extended control programs, since such algorithms usually require a universal software base as provided by general computer operating systems such as UNIX, VMS, or comparable systems.

There are essentially two ways in which models can be attached to process management systems. The first is to supply an open software interface to a programming environment linked to the user-written application program. In this way, a fast connection between the user program and the automation system can be established. In distributed systems the interfacing routines are required to be network-transparent.

The possibility of using the math-function interpreters already discussed is very limited, since interpreters are too slow for real-time model simulations. This is because the original equations must be evaluated too often in com-

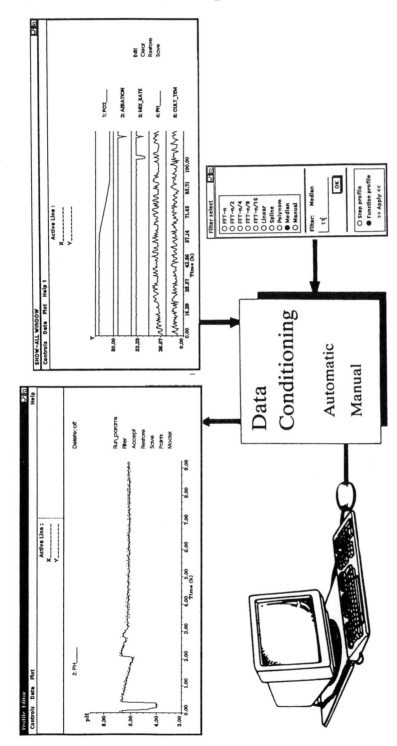

Fig. 9. Modern user interface for refurbishing signals of adjustable process variables. The resultant curves can be used as control profiles for future fermentations.

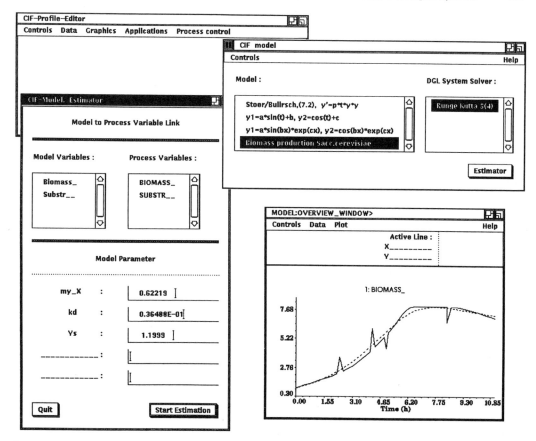

Fig. 10. Operator interface to the model bank in CIF. In the upper left window the user can click a desired model and a solver. Below the window the process data are displayed. At the right-hand side the model variables appear, which must be fitted to the measured process variables. The order of both must be matched by mouse clicks. Beneath that appear the proposed start parameters for the estimation procedure.

plex applications. Another possibility is to provide a means of incorporating model equations directly into the automation system via a prepared user interface; this is only economical if a compiled code is used in the function processing instead of the interpretation of the equation.

A notable effort to make simple models, which is discussed in the literature available to the biotechnologist in industry, is the construction of a model base running on PCs. In this development, many models which may be of interest for cultivations in a special plant can be put together in a model bank and selected for a special application by a simple interface.

These models are stored in the computer together with information on appropriate solving algorithms and a set of start values for the parameter estimation procedure. Model banks can be used on-line to obtain information on the state of the process as well as off-line for process data analysis.

In CIF, a similar model bank is available to the biotechnologist (MATHISZIK et al., 1990). It is restricted to simple models that are universal enough to be of general importance. The model base can be changed and extended by experienced users. Fig. 10 shows an example of its user interface to give an idea of the simplicity of such a tool.

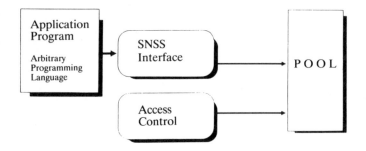

Fig. 11. Technique used to couple user-written programs to the process data management system CIF.

CIF also permits the other two methods of model evaluation. Since, as mentioned, its math-function processing module processes compiled code, it is possible to implement smaller models. It is also possible to attach freely programmed modelling routines of arbitrary complexity to the system. The access of such programs to the POOL is regulated in the same way as that of any person. The program must be logged in with the authorization privileges necessary to acquire the required access to the data (Fig. 11).

4.4 New Approaches to Implement *a priori* Knowledge

The limitations of automation systems do not concern only the software and hardware, but also the knowledge of the systems to be automated. In order to optimize an automation system, as much *a priori* knowledge as possible about the biotechnological system must be exploited to improve its control. Since most knowledge available is not representable in the form of mathematically formulated models, one must also tap other resources for *a priori* information.

4.4.1 Expert Systems

As already mentioned several times, one of the main deficiencies in bioprocess automation is the lack of analytical models for the fermentation processes. As a result, the most advantageous features of computers will not properly come into play, i.e., their capacity for in-

formation processing, e.g., by means of advanced model-supported methods in measuring and control. Apart from better sensors, nearly all the developments in measuring and control have been based on the use of *a priori* knowledge of the processes in question. Since mathematical models are scarce in production-scale systems of biotechnology, it is necessary to exploit alternative sources of knowledge.

Therefore, it will be appropriate to leave the beaten track taken in conventional process control engineering. Experienced fermenter operators can control their fermentation processes manually if they deviate from the expected paths without using deep mathematical knowledge. Until now, however, their knowledge could not be used directly in computers to aid fermenter automation, since it could not be coded in an algorithmic way. A new software technique for coping with such linguistically formulated knowledge has been developed in information science under the heading "expert systems". These provide new programming techniques that make it easier to manage problems involving large amounts of symbolic (non-numerical) data than with classical programming languages.

During the last years, much interest has been focused on expert systems. Many authors have reported sometimes too enthusiastically about possible advantages of expert systems. Unfortunately, much nonsense has been written on expert systems by people who have never worked actively in this field. In too many articles insufficient distinction has been made between mere ideas and implemented software, causing high expectations awakened in potential users, more promise than can be satisfied within the foreseeable future. Only very few concrete results have followed the papers;

on the contrary, many of the early advocates even beat a retreat, and the whole field has become somewhat discredited.

In the following section the basic ideas of expert systems in automation are discussed and illustrated by a working example. We restrict the discussion to real-time expert systems, which can be used during automation of bioreactors. Other interesting developments in the design of processes and equipment, of importance to bioprocess automation as well, cannot be enlarged upon here.

4.4.1.1 General Aspects

As mentioned frequently in this chapter, process automation requires a great deal of knowledge of the biotechnological process, especially of biochemical kinetics, bioreactors, and other branches of bioprocess engineering. This is necessary not only to predetermine the process responses to possible actuator changes during control, but also to determine the state actually encountered by the real process at a given time.

In the natural or engineering sciences, a working mathematical model of the process constitutes the best way to represent such knowledge. One is interested primarily in exact mathematical models that can be solved and used to describe the time development of the key parameters at given start and boundary conditions. Otherwise it would be too complicated to quantitatively describe the process state and its dynamic development. Unfortunately, this case is normal in biotechnology, where process states are described more qualitatively, usually in words rather than in mathematical formulas.

Previously, descriptions based on linguistic representations could not be used with computers in a straightforward manner, and people tried to translate their knowledge as far as possible into formal mathematical relationships. The most familiar method in chemical and biochemical engineering is the use of so-called correlations, which suffer from not being based on crisp dynamic models, but describe experimental experiences by artificial formulas that can easily be exploited numerically.

In many cases even this type of representation is not possible, even though enough experience may be available with respect to the particular biotechnological process in question. In many practical cases this experience can best be formulated by rules, e.g.: *"If certain conditions are met, then specific conclusions can be drawn."* Such conclusions may consist of the assumption that the system will achieve a special state, or that the system will next behave in a predefined way. This type of rule is often called "production" or "production rule".

Apart from such state descriptions, the conclusions may also refer to *actions* to be performed by the computer; should that be the case, they are applied, provided the conditions are met. Such rules are usually imitations of reactions of human operators to special process states. Thus, one often merely attempts an impersonation of an experienced operator instead of devising a linguistic analog to a mathematical model of the process.

4.4.1.2 Software Base

The essential difference between software dealing with heuristics and normal algorithmic programs is that in the first case one processes written words or, more generally, character strings rather than numbers. Conventional computer languages used so far in the natural sciences and engineering, FORTRAN, PASCAL, C, and others, are not optimized for this type of representation. Hence, special languages have been constructed to deal with strings, e.g., LISP (from "list processing"), the oldest one, or PROLOG.

Early expert systems were programmed exclusively in these languages. Today, only a few are based on such elementary languages. Most are now based on higher-level languages, which are themselves grounded on an elementary language. One example is the language OPS5, a LISP-based development of the Carnegie-Mellon University (e.g. BROWNSTON et al., 1985). Most applications of expert systems use ready-to-use expert system shells, i.e., complete user programs, into which are entered knowledge in the form of data and rules, which are then transferred via special user interfaces.

Not only are the common programming languages not optimal for list processing, but universal computers themselves are not well-suited to string processing. The by far most frequent process (and thus really time-limiting procedure in list processing) is the comparison of strings for equivalence in so-called matching routines. To enhance the processing speed of these performance-limiting routines, it was decided to relegate them to special computers constructed solely for this purpose. Such special LISP-machines have been built, e.g., by Symbolics. Today, there are also special LISP-processors on the market that can be used as coprocessors in universal computers. This might represent the more economical path for the future, since it reduces problems of interfacing with machines coupled to the processes.

All rules and facts, i.e., information on the actual state of the system being treated, are usually stored within a knowledge base. Its two parts are called the "rule memory" and the "working memory". Besides this knowledge base, there must also be an executive program. This is usually an interpreter for the rules, often called an "inference engine", which applies one rule after the other to the facts sampled in the working memory, seeking the conditions that are to be met (pattern matching).

A remarkable feature of rule-based software is that the order of rules within the rule memory does not have a strong influence on the processing sequence. This is of immediate practical effect on the maintenance of such programs as compared to classical control programs, where the order of statements is extremely critical. Such flexibility in the order of the rules allows for simple extensions of the software by simply adding additional rules or discarding others which are no longer necessary. In this way, maintenance of the software is significantly simplified. Both are advantageous in large software systems as compared to conventionally programmed automation systems. However, a cyclic search through all the rules for matching the required conditions is time-consuming. In real-time applications, which are indispensible in automation, processing time is a primary criterion. Thus, a central question becomes how to minimize the response times of real-time expert systems. Since most expert systems developed in information science were designed as consulting systems, this question was not of primary importance during the development of most existing software tools.

One practical way to obtain higher processing speeds is to reduce as far as possible the number of items in the working memory in order to reduce the number of matching operations. Another possibility, based on the practice in process management systems, is to structure the whole set of rules constituting the expert system in order to reduce the number of rules that must be considered during every cycle of the system. This can be done by context-sensitive activating or deactivating of parts of the rule volume according to different process phases.

Concerning real-time expert systems in process automation, an essential point is that heuristic knowledge is by no means the only source of knowledge that must be taken into account to control a process. Many subunits of a process can be more adequately described by mathematical models. To combine both aspects, hybrid software is necessary (Fig. 12). Thus, expert systems should invoke algorithmic and heuristic software modules. The way in which these parts communicate with each other to exchange information provides a criterion with which to judge their performance.

In expert systems built as consulting systems, the decision-making process is usually made transparent to the user by some kind of explanatory component, which depicts the line of argumentation that leads to a particular special decision. Such a feature, however, is much too time-consuming to be incorporated into a real-time expert system. The basic idea of informing the user about the system's actions, however, is of such importance to the acceptance of expert systems that one must look for alternative arrangements. One possibility is to continuously inform the user by means of a properly arranged graphical user interface onto which the process state and its possible developments are displayed. This concept is shown in the examples to be discussed.

Different routes in the development of expert systems are of interest with respect to automation in biotechnology. The first is the application to control. Under the heading "ex-

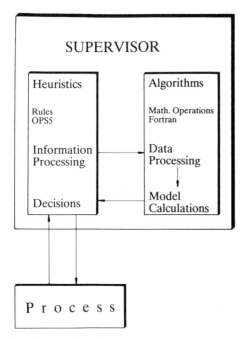

Fig. 12. Hybrid architecture of the expert system SUPERVISOR. It contains essentially two components, a heuristic part and an algorithmic one. The heuristic part is programmed in OPS5, the deterministic in FORTAN. The heuristic part controls the system. It gives instructions for computing algorithmic tasks to the algorithmic part and takes over its results, on the basis of which it makes decisions. These can lead to control actions which are transferred into the underlying process control system which is designed to execute them.

pert control" numerous investigations have been reported in the literature. The general idea, as stressed by ASTROEM et al. (1986), is that the actual engineering of control systems contains a substantial amount of heuristic logic that could be replaced by an expert system. One must think in particular of so-called "safety jackets" around simple control algorithms as a way of stabilizing them, a concept which is not above criticism (STEPHANOPOULOS, 1989). However, there is general agreement that in sufficiently complex control problems more advantageous and powerful control laws can be obtained by combining conventional control algorithms with an expert system.

Some companies have developed dedicated control modules for decentralized digital control systems, which make use of heuristic methods to automatically tune the settings of various controllers on optimal values. One proposed way (HIGHAM, 1986) is to tune PID controllers automatically, similar to the well-known methods proposed by ZIEGLER and NICHOLS (1943).

Another important route relates to process state and fault recognition. It is necessary to use the methodology of expert systems for this central task if there is no adequate mathematical model available for a state identification (ISERMANN, 1988). State identification is a prerequisite for the question of whether the system behaves in a predefined manner or not. This naturally leads to a fault recognition, which is most beneficial if it can be reached while the automation system is still within tolerable limits, and there is sufficient time to properly respond to the fault. A first step is to classify the fault; if it has been recognized as severe, diagnostic routines will be initiated. Once the cause of the fault has been identified, proper maintenance activities can be started. It is difficult to incorporate all these steps into an automation computer using normal programming languages within a reasonable time.

Similar techniques, using both artificial intelligence techniques/tools and compiled algorithmic computational modules, have been developed as batch decision support systems to enhance the operational safety of batch reactor systems by providing the operator with online problem diagnosis and real-time expert assistance (CHARPENTIER and TURK, 1986). All of the above refers to available data. Often, however, one has to cope with insufficient information, e.g., lack of data with respect to some part of the fermentation, or signals with low signal-to-noise ratios stemming from imprecisely working sensors. This leads to still another unsatisfactorily solved problem.

4.4.1.3 Examples of Process Automation Discussed in the Literature

Most real-time expert systems under development in the domain of automation in biotechnology are designed for supervisory control of the process. In most cases a conventional process control system was employed to serve as a connection with the fermentation process.

KARIM and HALME (1988) are developing a real-time expert system for monitoring and analysis, diagnosis of faults and malfunctions, and on-line optimization of fermentation processes. They have applied their results to a batch fermentation in which the enzyme α-amylase was produced with *Bacillus subtilis* in a Braun Biostat fermenter.

The expert system was written in the special language OPS83, which supports the construction of rule-based expert systems on a PDP11-computer running under the operating system RSX11M. Connections to the fermenter were made via RINTEKNO's process data management system MFCS. A closed-loop control with the expert system within the loop has not yet been accomplished.

COONEY et al. (1988) developed an expert system which they called IFCONS (intelligent fermentation control system). It is designed to act as a reliable and robust supervisor to process controllers within a biotechnological process in order to surmount the limitations of conventional control. Especially stressed are the dependency of the control algorithms on the quality of information from sensors and the inflexibility of most mathematical models to alterations in metabolic pathways. They applied heuristics in the form of rules. The basic control strategy was taken from early papers of WANG et al. (1977, 1979) and material balances from COONEY et al. (1977).

COONEY et al. (1988) took the control of a baker's yeast cultivation as a concrete example for which they developed their software. Their conventional laboratory bioreactor was controlled by means of a direct digital controller running on a PDP11/23 process computer attached to a special LISP-machine (Symbolics 3640) used for the expert system. IFCONS was written in the shell ART (automated reasoning tool), developed by Inference Corporation, which also supplied elegant user interfaces. Access to the data was achieved via the PDP11. A feedback to the process occurred via the human operator of the system. As an example of an expert system operating in a

closed loop with biotechnological processes, SUPERVISOR (LÜBBERT and HITZMANN, 1987; HITZMANN, 1988; LÜBBERT et al., 1989), a system for supervisory control of a cultivation of genetically modified *Escherichia coli* bacteria to produce fusion proteins, is discussed in more detail. SUPERVISOR was built to transfer operator activities stepwise into the computer controlling the cultivation process.

Since it does not make sense to again program into an expert system the data acquisition and basic control routines, which are already well-developed in conventional process control systems, there is a need to use an appropriate underlying automation system. The only question of interest, for all practical purposes, is how to obtain rapid access to the process measuring data. In SUPERVISOR, the basic computer control of the cultivation process was accomplished separately by the process management system CASFA (FRÜH et al., 1986).

SUPERVISOR runs on a VAX computer operated with the VMS operating system, and it is written in the knowledge-representation language OPS5 (DEC, 1988). CASFA runs on a PDP11 process computer under RSX11M+, and is coded primarily in FORTRAN77. The two computers are interconnected by a fast local area network using Ethernet technology. DECnet, the network software working on both computers, was utilized to connect the two software systems. It supplies a fast inter-task communication facility.

OPS5 proved well-suited to realizing the interplay of heuristic and conventional algorithmic software modules. In the OPS5-versions for VAX computers used in SUPERVISOR, this was achieved by the usual subroutine techniques. Subroutines written in any language for which a compiler is available for the VAX can be linked to OPS5 programs and called by simple subroutine calls within the conclusion part of any rule. The parameter transfer is accomplished accordingly.

A typical rule written in this special language, OPS5, is shown in Fig. 13. This example not only shows the usual method of programming rules in SUPERVISOR, but it also demonstrates one of the general design aspects of this expert system: by relatively simple ob-

Example for Coding a Rule

IF

Off-gas CO_2-measuring values become smaller,

THEN

Suspect a limitation

in S U P E R V I S O R

(p rule_CO2_warning limitation
  (evaluation
     ^meas_probe   CO2offgas
     ^determination slope
     ^value   < 0.0)
-->
  (MAKE limitation suspected
     CO2offgas))

Fig. 13. Typical rule written in OPS5. The example shows the usual way of programming rules in SUPERVISOR and also demonstrates one of the general design aspects of this expert system, as explained in the text.

servations in the database, suspicions are raised about changes in the actual process state. From these suspicions a hypothesis is formulated, which afterwards is tested by activating several specific reasoning lines. Their aims are to improve or to reject the hypothesis as quickly and as early as possible. Numerous interconnected questions must be answered in such a case.

A further design component of SUPERVISOR is the special user interface for providing information about the system's state and behavior. This interface makes use of modern graphic video terminals, which permit the display of different types of information on the screen. Central components of the interface are plots of the different process variables, which can be put on the screen together with results of simple models fitted to the experimental data. These fits primarily constitute the base of a prediction or estimation of future process behavior. This must be known, especially in processes with higher time constants, in order to become aware of unwanted process states prior to the time at which they become critical. All extrapolations are analyzed and formulated as warnings, which are written in short texts into the graphic representation if they indicate severe deviations from the normal process state. It is obvious that this method of analysis utilizes many algorithmic features of the software.

Such user interfaces will be given much more attention in the future of process management generally. They must be optimized for the needs and requirements of the human operator (which will still be necessary in the future) and not for the characteristics of a given computer. Such a sophisticated user interface must also be based on heuristic knowledge, on the ways in which a human operator can survey a given process state as quickly as possible. Similar developments are also apparent in related disciplines. BÄR and ZEITZ (1989) demonstrated one concept of a knowledge-based user interface, originally created for a dynamic simulator for chemical processes and plants, which meets many of the requirements of process automation as well.

4.4.1.4 Knowledge Acquisition

One of the essential limitations in building expert systems is the knowledge acquisition process. Several aspects of this deserve special mention. First, it has been observed that most process operators encounter severe difficulties in formulating linguistically the way in which they proceed in a given situation. The second problem relates to transmission losses during the transfer of information from the biotechnological expert to the person who is experienced enough to incorporate such knowledge into the expert system. Finally, it should also be mentioned that there is a natural hesitation by experts to disclose their special knowledge to others, especially if this is to be installed into a machine. In most cases these persons are afraid to make themselves dispensable.

The most realistic way to cope with the problem is to hire biotechnologists with practical knowledge in both fermentation and software. Before they can start building an expert system, however, they must actively work with the cultivation process over a long period in order to learn the difficulties of the special process. Another measure would be to develop user interfaces to expert systems that would allow biotechnologists to implement their rules directly into a given system, i.e., to develop special shells optimized for biotechnologists.

4.4.1.5 Completeness Problem

A point often criticized in expert system developments is that one does not have a real chance to sample all necessary knowledge within a given domain, e.g., the automation of bioreactors. Omniscience is not achievable. But there are two ways to cope with the problem.

The first is to reduce drastically the goals. In the accompanying example of SUPERVISOR, this has been achieved simply by restricting the support given to the operator regarding process supervision. This was accomplished by dividing the task into small slices of operator activities, e.g., the detection of very special process states such as the detection of diauxic growth of the organism. Thus, step by step, more and more activities can be added to the system. Consequently, the system is constantly in a working state, and it can be used advantageously by the operator.

The second possibility for coping with the problem is to use methods developed to handle incomplete, inexact, or even contradictory information. In such cases, rule-based systems may encounter conflicts, which must be resolved as well as possible by a robust method. Such a technique is the fuzzy theory or fuzzy reasoning. Fuzzy reasoning is the process by which a possibly imprecise conclusion is deduced from a collection of imprecise premises. The first model of fuzzy reasoning was developed by ZADEH beginning in 1965 (ZADEH, 1965, 1984). This technique has been attempted several times in biotechnology, especially in measurement (POSTLETHWAITE, 1989), process control (TONG et al., 1980; NAKAMURA et al., 1985; CZOGALA and RAWLIK, 1989), and process simulation (e.g., TURUNEN et al., 1985). Also important for biotechnolog-

ists are the papers from chemical engineering (e.g., YAMASHITA et al., 1988a, b).

4.4.1.6 Further Applications of Knowledge-Based Systems in Automation

Since the human operator will remain the decisive element in fermentation control, it is necessary to support him with as much information on the process state as possible, but not more than that. Thus, compared with the data displayed on today's video screens, the information must be reduced to what is necessary to decide how to proceed in the actual situation. This information is equivalent to a well-defined process state representation, which is only possible with a model of the system. By the same arguments stated above for model-supported control, such information can sometimes be given only in heuristic terms, often only in an incomplete and uncertain way. Thus, it is obvious that heuristic knowledge-based methods are of advantage in supporting the man–process interface if a comprehensive mathematical model is not available.

Another field in which knowledge-based systems become important is that of "measuring techniques". Here, sensor fusion is gaining increasing importance in combining information obtained through various types of sensors, which can be realized most appropriately using unified knowledge-based methods through multilevel representation combining quantitative and symbolic attributes (PAU, 1989). The primary goal is to extract better features and to achieve sensor diversity. This is an approach similar to model-supported measurement on a heuristic base, which is necessary where neither a sensor for certain key process variables nor mathematical models which relate them to easily measurable quantities are available.

4.4.1.7 Conclusion

The idea of using expert systems in automation biotechnology is very attractive to many people. Expert systems are under development at many places. However, their broad application will take considerable time.

Many people in industry would be satisfied if only the operator behavior could be assumed by a computer. The goal is not only to spare manpower, but also to ensure well-defined reactions to specific process faults. In practice it has been recognized that crews in different shifts react markedly differently to identical deviations of the process from a predefined path. This leads to unwanted variations in product quality. Such fluctuations could be reduced or eliminated if a unified behavior could be implemented and incorporated into a knowledge-based computer system. Another point raised by industrial people is that such an implementation would conserve at least some knowledge of older, highly experienced workers against loss after their retirement.

As the term "expert system" indicates, the system tries to reason heuristically, like people in situations in which no exact methods are available, or when the evaluation of an adequate solution is not possible during the time period within which a decision must be reached. Biotechnology is, therefore, an ideal application field for expert systems, since it represents a very complex field in which only a few aspects can be covered by exact mathematical models.

With expert systems in automation of bioreactors, the way ahead will be analogous to the progress already apparent in the underlying digital process control systems. Distributed systems will arise, and priority structures will be incorporated into the expert systems. Finally both universal control systems and real-time expert systems will merge.

Many critical comments on expert systems from people living in the world of exact models have missed the mark. It is well accepted that exact modelling is better in nearly any case; in most cases, however, they are not immediately practicable. Thus, to solve problems now, one must take "the bird in the hand rather than two in the bush".

4.4.2 Fuzzy Methods

In the fuzzy approach, one attempts to model the strategies of a process operator in hand-

ling complex processes for which the existing models are insufficient (TONG, 1984). It is based on fuzzy sets and fuzzy reasoning.

The general idea of fuzzy reasoning has been developed by ZADEH (1973). It can be understood as an extension of set theory, in which the essential idea is that membership of an element in a set is not a strict binary yes-or-no relationship as in conventional set theory, but is defined by a membership function that takes continuous values within the interval [0, 1]. These membership functions express to what degree the element is assumed to be a member of the fuzzy set. Since membership functions are more or less subjectively defined by experience, this is another way of incorporating heuristic knowledge into an automation system.

The development of fuzzy rule-based controllers of interest to automation systems dates back to ASSILIAN and MAMDANI (1974), who tried to establish a non-mathematical technique to control a pilot-scale steam engine.

The difference between fuzzy rule-based systems and the normal expert system approach is that the conditions are formulated by fuzzy expression or relationships, such as:

if the deviation of the measuring variable from the setpoint is of positive medium size:

then change the valve setting by a small amount in the positive direction.

Here the fuzzy variables "measuring variable" and "valve setting" take fuzzy values, e.g., "positive small", which represent their linguistic meaning. The relationships between the variables are calculated by means of fuzzy reasoning. In a concrete application the relations are then made discrete and put into a large matrix from which they are retrieved during the operation of the controller. This table look-up procedure guarantees fast processing of the fuzzy controllers.

There have been several attempts to implement such fuzzy controllers in biotechnology. TONG et al. (1980) used a fuzzy controller in an activated sludge wastewater treatment process. NAKAMURA et al. (1985) applied fuzzy control to a glutamic acid fermentation. TURUNEN et al. (1985) applied fuzzy techniques to modelling the inversion of sucrose. STANIS-KIS and KILDISAS (1989) used fuzzy control during a batch growth of *Escherichia coli* and showed that the performance of such controllers is quite good in the region of the operating point. KONSTANTINOV and YOSHIDA (1989) viewed fermentation as a process with variable structure and used fuzzy methods to recognize the state assumed by the process and then employ the most relevant control strategy from a pool of several alternatives. CZOGALA and RAWLIK (1989) simulated and controlled biological processes by fuzzy techniques, with the example of a continuous cultivation of bacteria. They demonstrated in a comparison between different controllers that their fuzzy controller was superior to classical PID controllers.

4.5 Measures to Support Maintenance of Complex Automation Systems

Since the speed of hardware development has been much faster than that of software, the software gap will probably become much more pronounced in time. One must develop new ways to speed up software development. Computer companies have developed tool kits to aid process engineers in writing application software modules, thereby shortening the time between problem recognition and solving. Tools are available which support application programming and aid in the management of program development.

4.5.1 Programming Techniques

Modularity of the components of an automation system has proved to be advantageous in hardware building block systems, and it is also of considerable importance in the software domain. One technique that supports the development of software building block systems is "object-oriented programming" (BOOCH, 1986; MEYER, 1988), which is used with increasing frequency in the development of larger and more complex software.

Object-oriented programming is one way favored by many software developers to improve the extendibility of software systems by structuring programs in a new way. Simultaneously, the reusability of software modules has been improved significantly. The original idea dates back to the 1960s, when the concept appeared in connection with the development of new programming languages (DAHL and NYGAARD, 1966), e.g., SMALLTALK.

An object is referred to as a fundamental software building block used to model some entity in an application. Objects are described by their function. They consist of an exclusive, private software module with a public interface. The module contains local data structures and local procedures to operate on the data. The main idea is to make the modules as autonomous and closed as possible. The software technical term for this is encapsulation. Nevertheless, communication with other objects is made as general as possible by means of universally usable input and output interfaces in order to ensure broad reusability. This encapsulation acts further as a protective shield for the internal data against disallowed access. From the functional point of view object-oriented programming copes with the problem of increasing complexity of software systems in essentially the same way as in microelectronics. There, advanced integrated circuits (ICs) are constructed to replace complete boards composed of several elementary components.

Hence, the software objects must be designed in such a way that elementary objects can easily be combined with more complex ones, which, again, appear to their users as separate and independent objects described by new, more comprehensive functionalities. In this way, autonomous software modules on different levels of abstraction can be generated which can be used as independent logical structures of high functionality. Since the objects are designed as independent software modules, they can be developed and tested separately. This increased transparency decreases the probability of hidden software errors. A reduction of the programming load by using such objects, however, can only be expected if information transfer and the necessary synchronization between objects is minimized.

This requires the individual modules to be self-contained and matchable. Furthermore, universal usabiltiy requires that objects be equipped with standard communication interfaces by which they can easily be connected with others.

One difference between object-oriented and conventional procedural programming arises in the handling of data. In an object-oriented environment an object's internal data structures and current values are accessible only to the methods within the object. These methods are activated through messages passed from other objects which specify the methods to be activated and the data or parameters required. Whole programs are then constituted by a hierarchy of objects and their interrelationships (CHEN, 1977).

Traditional programs rely on function-oriented approaches that organize the process through a hierarchy of functions or data-oriented approaches, which in turn emphasize a hierarchy of data structures. Object-oriented methods, on the other hand, impose the natural modularization of the process in question through an emphasis on objects with a one-to-one correspondence with the actual components in the process. This provides for easy mapping between the software and the process.

This software technique, thus, points in the same direction as the modularization of hardware components within distributed control systems, which has demonstrated high reliability with respect to errors because of its high transparency. Object-oriented programming thus exploits the advantages of distributed digital control systems.

By definition, a software object is a set of data together with the software components required to manage the set and provide access to it. This is essentially the way in which the central core of CIF works, including the central database POOL, its managing program EXECUTIVE, and the standard software interface SNSS (see Fig. 3).

This CIF-core is the fundamental software building block employed to perform the basic automation activity at the biotechnological plant. Its data and the basic data manipulation routines are completely separated from all other modules and, thus constitute an object. The SNSS acts as the only public interface for the core, which allows for communication with other objects around the CIF-core. This encapsulation technique acts as a protective shield for the internal data against unallowed access, since the SNSS-routines check every request for POOL data in terms of access authorization. The SNSS interface is network-transparent; this object can be used by other object-oriented programs running on any computer within the local area network.

This core object holds all data and provides the basic functionality of the automation system. It is used extensively by all other objects. It decouples programs for more sophisticated process management tasks, e.g., advanced model-supported supervisory process control algorithms, as well as knowledge-based process control software modules from process visualization and basic process control.

4.5.2 CASE Tools

A widely heard criticism regarding software development is that it is not done systematically enough in terms of engineering and organization. As compared to hardware development, where all stops are pulled for planning and production control, the way in which programs are developed has been compared to artistic work. It has been argued that this significantly retards development.

Consequently, to tighten development and permit the planning of larger software projects, engineering methods have been proposed and tested in many different software projects. It is obvious that one can use the computer itself to aid software development by assigning to it all tasks in software development that can be formalized. Software tools designed to support systematic program development have been developed by all of the larger computer manufacturers and software houses under the name CASE tools (computer aided software engineering).

Software engineering can be divided into three main parts. First there must be a *model* of how to proceed during the development of a software project. It must describe the sequence of steps to be performed. The second part contains the description of the *methods* to be applied within the individual steps. Finally one needs the *software tools* necessary to perform individual tasks. Software tools are now evolving into complete tools for problem solving instead of tools that can be used only to construct a solving mechanism.

CIF has been developed using CASE toolbox VAXset supplied by DEC to support program development in all its phases, starting with system design, moving to advanced language-sensitive editors, and culminating in software version management tools.

5 Conclusions and Recommendations for Future Developments

Distributed digital automation systems dominate the automation of chemical and biochemical production systems. They are composed of microprocessor-based computers linked together in local area networks. Systems will be expanded to networks within larger company-wide networks. Even in a chemical or biotechnological production unit, several levels of networks will soon appear, e.g., networks of field instruments and networks of laboratory equipment in a network for process management.

Their main advantage is that of a building block system. Powerful systems clearly organized and adapted to the special needs within an application can be flexibly built (Fig. 14). Maintenance is supported by working with a limited set of small standard components. The computing power of such distributed systems is naturally increased by true parallel computing of the individual nodes.

However, the structuring of distributed automation systems cannot be arbitrarily. The structure must resemble the logical structure of the process to be automated. Individual components must be chosen to be as internally self-contained as possible to avoid unnessessary communication and synchronization efforts that would overcompensate for the advantages of distributed systems.

To utilize the concept of distributed processing of system activities comprehensively, it is necessary that generally-accepted open standards be obeyed. These are, first of all, necessary to assemble a distributed system of the components from different vendors that are regarded as optimal for an applicant's needs. This increases the flexibility of adapting a real system to changing conditions from the viewpoint of the goals of production system as well as those of the available system components.

The realization of this concept is not so much limited by the hardware components as by the availability of appropriate software. The software is responsible for the functionality of an automation system and is thus more important. Therefore, it is necessary to choose software that meets the requirements and then the hardware on which the software can be run.

This argument becomes even more important if one realizes that the software development time constant is much longer than the innovative periods in hardware development. Today, software cost is roughly twice that of the computer on which it runs. This means that a chosen software package often must survive several hardware generations. Thus, the customer is advised to look for hardware manufacturers with a strong family concept of computers, offering confidence that the chosen software will also run for years to come on machines adapted to a more transient state of the art.

The requirements for automation software are steadily and rapidly growing. More and more functions are becoming available to process management systems, and additional economical and ecological boundary conditions must be met and integrated into process automation. These all require software extensions. Hence, the systems become larger and more difficult to control. Measures are required to keep larger software systems transparent. One promising approach is software structuring by means of object-orientated programming. This point of view is important not only for software development, but also to the user, since it significantly influences the software's maintenance and extendibility properties.

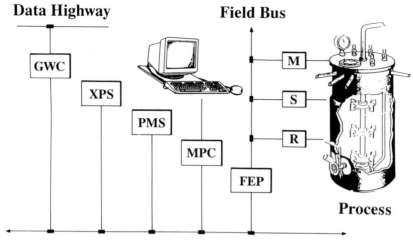

Fig. 14. Architecture of an ideal process automation system hardware. Several different components are arranged in a local area network utilizing Ethernet technology. FEP stands for front-end processors, which serve as gates to field busses to which the measuring (M), open-loop control (S), and closed-loop control (R) devices are attached. MPC represents the man–process interface workstations, PMS the process-management computers, XPS the expert system hardware, and GWC the gateway processors that provide connections to the data highways of the higher-level company-wide information processing systems.

It has become increasingly evident in recent years that the biotechnologist at the plant site is not replaceable by automation systems in the foreseeable future. He remains the decisive component in the bioprocess. Software development must therefore be focused on supporting him in his decisions as much as possible. This requires putting more of the software and hardware money into user interfaces. It is necessary to provide the biotechnologist with tools for incorporating his ideas on process control directly into the automation software.

All types of *a priori* information must be used to improve measurement and control: knowledge represented by mathematical models as well as heuristic knowledge that can only be formulated linguistically in the form of rules or other heuristic means. Conventional process control will merge with the knowledge-based automation techniques that we now call "expert systems". Both parts will be used in a complementary way, with a clear preference for mathematically exact models, wherever they exist and are solvable within the time limits set by the requirements of the real-time environment. Knowledge-based systems, for example, can build a platform for dealing with uncertain and incomplete information, one of the properties of human beings that in long-term thinking deserves to be incorporated into automation systems.

One point usually not emphasized in discussions of automation is the requirement that sufficient information on the actual state of the process be available. This, above all, is a matter of reliable sensors for the key process variables. A remaining general problem in biotechnology is the scarcity of powerful sensors for the basic quantities.

6 References

ANONYMOUS (1988), Teleperm M automatisiert biotechnische Produktionsanlage, *SENSOR Report*, 3, June/88, 26–29; *Chem. Anlagen + Verfahren* **21**, 18–26.

ASSILIAN, S., MAMDANI, E. H. (1974), An experiment in linguistic synthesis with a fuzzy logic controller, *Int. J. Man–Mach. Stud.* **7**, 1–13.

ASTROEM, K. J., ANTON, J. J., ARZEN, K. E. (1986), Expert control, *Automatica* **22**, 277–286.

BAILEY CONTROLS (1989), *Operator Interface Station Series* **40**, Bailey Controls, 29801 Euclid Av., Wickliffe, OH 44092, USA.

BÄR, M., ZEITZ, M. (1989), A knowledge-based flowsheet-oriented user interface for a dynamic process simulator, in: *Computer Applications in the Chemical Industry, DECHEMA Monogr.* **116**, 173–179.

BAUMAN, D. E. (1989), Systems design using the single-loop digital control concept, *InTech* **36**, 24–27.

BOOCH, G. (1986), Object oriented development, *IEEE Trans. Softw. Eng.* **SE-12**, 211–221.

BROWNSTON, L., FARELL, R., KANT, E., MARTIN, N (1985), *Programming Expert Systems in OPS5: An Introduction to Rule-based Programming*, Reading, Mass.: Addison-Wesley Publishing Company, Inc.

CHARPENTIER, L. R., TURK, M. A. (1986), Batch decision support system enhances safety of reactor operation, *Adv. Instrum.* **41**, 71–75.

CHATTAWAY, T., STEPHANOPOULOS, G. (1989), Adaptive estimation of bioreactors: Monitoring plasmid stability, *Chem. Eng. Sci.* **44**, 41–48.

CHEN, P. (1977), The entity-relationship approach to logical data base design, *Q.E.D. Information Science*, 1977.

CLEAVELAND, P. (1989), Industrial LANs: Where's the action?, *I&CS* **11**/1989, 27–29.

COONEY, C. L., WANG, D. I. C., MATELES, R. I. (1968), Measurement of heat evolution and correlation with oxygen consumption during microbial growth, *Biotech. Bioeng.* **11**, 269–281.

COONEY, C. L., WANG, H. Y., WANG, D. I. C. (1977), Computer aided material balancing for predicting biological parameters, *Biotech. Bioeng.* **19**, 55–66.

COONEY, C. L., O'CONNOR, G. M., SANCHEZ-RIVERA, F. (1988), An expert system for intelligent supervisory control of fermentation processes, *Proc. 8th Int. Biotechnology Symp., Paris, 1988* (DURAND, G., BOBICHON, L., FLORENT, J., Eds).

CZOGALA, E., RAWLIK, T. (1989), Modelling of a fuzzy controller with application to the control of

biological processes, *Fuzzy Sets and Systems* **31**, 13–22.

DAHL, O. J., NYGAARD, K. (1966), Simula, an algol-based simulation language, *Commun. ACM* **9**, 671–687,

DEC (1988), *Programmiersprache VAX OPS5*, Produktschrift, Digital Equipment GmbH, München, Germany.

DEC (1989), Digital Equipment Corp., Maynard, Mass., USA.

DOHMEN, W. (1990), Ein Konzept für moderne Leitsysteme, *Elektrotech. Z.* **111**, 76–79.

ELLIOT-AUTOMATION LTD. (1967), Automation in Action – Batch Fermentation, Booklet published by Elliot-Automation Ltd.

FEWKES, R. C. J. (1988), Management of process control data for bio-process analysis, in: *Proc. 4th Int. IFAC Congr. on Computer Applications in Fermentation Technology, Modelling and Control of Biotechnical Processes*, pp. 381–391 (FISH, N. M., FOX., R. I., Eds.), London: Elsevier Applied Science.

FRANK, P. M. (1990), Fault diagnosis in dynamic systems using analytical and knowledge-based redundancy – A survey and some new results, *Automatica* **26**, 459–474.

FRÜH, K., HIDDESSEN, R., NIEHOFF, J., LÜBBERT, A., SCHÜGERL, K. (1986), On-line Messung und Regelung bei der Penicillin V Produktion, *BTF-Biotech Forum* **4**, 204–210.

GRAYSON, P. (1969), Computer control of batch fermentations, *Process Biochem.* **3**, 43, 44, 61.

HIGHAM, E. H. (1986), A different approach for self-tuning in process controllers – the case for introducing an expert system, *Meas. Control* **19**, 253–257.

HITZMANN, B. (1988), SUPERVISOR, ein Realzeit-Expertensystem für die Prozeßführung einer *E. coli*-Kultivierung, *Dissertation*, Universität Hannover, Germany.

HUMPHREY, A. E. (1977a), Computer assisted fermentation developments, *Dev. Ind. Microbiol.* **18**, 58–70.

HUMPHREY, A. E. (1977b), The use of computers in fermentation systems, *Process Biochem.* **12**(3), 19–25.

ISERMANN, R. (1988), Wissensbasierte Fehlerdiagnose technischer Prozesse, *Automatisierungstechnik* **36**, 421–426.

JEFFERIS, R., HUMPHREY, A. E. (1972), Indirect measurement of cell biomass by oxygen material balancing, *Paper at Annual Meeting American Chemical Society*, Chicago.

JONES, R. (1989), SQL – problems with an emerging standard, *Inf. Software Technol.* **31**, 2–6.

KARIM, M. N., HALME, A. (1988), Reconciliation of measurement data in fermentation using on-line expert system, in: *Proc. 4th Int. IFAC Congr. on Computer Applications in Fermentation Technology, Modelling and Control of Biotechnical Processes,* pp. 37–46 (FISH, N. M., FOX, R. I., Eds.), London: Elsevier Applied Science.

KONSTANTINOV, K. B., YOSHIDA, T. (1989), An expert approach for control of fermentation processes as variable structure plants, *IFAC Workshop on Expert Systems in Biotechnology, Helsinki.*

LITZ, L. (1989), Der Unternehmer ist gefordert, *Chem. Ind.* **9**/89, 35–38.

LITZ, L. (1990), Prozeßleitsysteme – Stand der Technik und Trends, in: *Der Interkama Report 90,* pp. 20–35 (URBACH, W. E., Ed.), München: Oldenbourg Verlag.

LÜBBERT, A., HITZMANN, B. (1987), Are there any prospects for expert systems in chemical engineering? *Hung. J. Ind. Chem.* **15**, 39–45.

LÜBBERT, A., HITZMANN, B., KRACKE-HELM, A. SCHÜGERL, K. (1989), On experiences with expert systems in the control of bioreactors, in: *Computer Applications in Fermentation Technology, Modelling and Control of Biotechnical Processes,* pp. 297–302 (FISH, N. M., FOX, R. I., Eds.), London: Elsevier Applied Science.

MAEHLE, E. (1988), Architecture of fault-tolerant systems, *Informationstechnik* **30**, 169–179.

MATHER, M., KOHN, D., SCHNABEL, G. (1988), Module für Sicherheit und Automation, *Chem. Ind.* **6**/88, 89–97.

MATHISZIK, B., GOLDSCHMIDT, B., DORS, M., LÜBBERT, A. (1990), Die Modellbank Biotechnologie als Komponente von CIF, in preparation.

MEYER, B. (1988), *Object-Oriented Software,* New York: Prentice Hall.

MURAO, C. J., YAMASHITA, S. (1967), Fermentation processes, *J. Soc. Instrum. Control Eng.* (Tokyo) **6**, 735–740.

NAKAMORI, Y. (1989), Development and application of an interactive modelling support system, *Automatica* **25**, 185–206.

NAKAMURA, T., KURATANI, T., MORITA, Y. (1985), Fuzzy control application to glutamic acid fermentation, in: *Proc. 1st IFAC Symposium on Modelling and Control of Biotechnological Processes,* p. 211 (JOHNSON, A., Ed.), Oxford: Pergamon Press.

PAU, L. (1989), Knowledge representation approaches in sensor fusion, *Automatica* **25**, 207–214.

POLKE, M. (1988), Entwicklungsstand und Entwicklungstrend der Prozeßleittechnik, *atp* **30**, 205–206.

POSTLETHWAITE, B. E. (1989), A fuzzy state estimator for fed-batch fermentation. *Chem. Eng. Res. Dev.* **67**, 267–272.

SCHEIFLER, R. W. (1987), X Window System Protocol, X Version 11, Release 2, *MIT-Report.*

SCHEIFLER, W. R., GETTYS, J. (1986) The X Window system, *ACM Trans. Graph.* **5**, 79–109.

SCHNABEL, G., AUER, W. (1987), Teleperm M zur Steuerung einer biotechnischen Großanlage, *Siemens Zeitschrift Energie und Automation,* Dez. 87.

STANISKIS, J., KILDISAS, V. (1989), Fuzzy control of *E. coli* batch growth, *IFAC Workshop on Expert Systems in Biotechnology, Helsinki.*

STEPHANOPOULOS, G. (1989), Artificial intelligence in process engineering: Current state and future trends, *DECHEMA Monogr.* **116**, 3–25.

STOUTHAMER, A. H. (1971), Determination and significance of molar growth yields, in: *Methods of Microbiology* (NORRIS, J. R. RIBBONS, D. W., Eds.), Vol. 1, pp. 629–663, New York: Academic Press.

SWARTZ, J. R., COONEY, C. L. (1978), Instrumentation in computer-aided fermentation, *Process Biochem.,* Febr., 3–24.

SWARTZ, J. R., WANG, H., COONEY, C. L. (1971), Utilization of thermal and viscosimetric methods for biomass estimation, *Paper at Joint US/USSR Conference on Data Acquisition and Processing for Laboratory and Industrial Measurements in Fermentation, University of Pennsylvania,* Philadelphia.

TONG, R. M. (1984), A retrospective view of fuzzy control systems, *Fuzzy Sets Syst.* **14**, 199–210.

TONG, R. M., BECK, M. B., LATTEN, A. (1980), Fuzzy control of the activated sludge wastewater treatment process, *Automatica* **16**, 695–701.

TURUNEN, I., NYBERG, T., JÄRVELÄINEN, J., LINKO, Y. Y., LINKO, P., DOHNAL, M (1985), Fuzzy modelling in biotechnology: Sucrose inversion, *Chem. Eng. J.* **30**, B51–B60.

WANG, H. Y., COONEY, C. L., WANG, D. I. C. (1977), Computer aided baker's yeast fermentation, *Biotechnol. Bioeng.* **19**, 67–86.

WANG, H. Y., COONEY, C. L., WANG, D. I. C. (1979), Computer controlled baker's yeast production, *Biotechnol. Bioeng.* **21**, 975–995.

WATANABE, K., MATSUURA, I., ABE, M., KUBOTA, M., HIMMELBLAU, D. M. (1989), Incipient fault diagnostics of chemical processes via artificial neural networks, *AIChE J.* **35**, 1803–1812.

YAMASHITA, Y., MATSUMOTO, S., SUZUKI, M. (1988a), On-line estimation of flow regimes on perforated plates based on fuzzy reasoning, *Int. J. Eng. Fluid Mech.* **1**, 189–204.

YAMASHITA, Y., MATSUMOTO, S., SUZUKI, M. (1988b), Start-up of a catalytic reactor by a fuzzy controller, *J. Chem. Eng. Jpn.* **21**, 277–282.

ZABRISKIE, D. (1976), Real-time estimation of aerobic batch fermentation biomass concentration by component balancing and culture fluorescence, *PhD Thesis,* University of Pennsylvania, Philadelphia.

ZABRISKIE, D. W., HUMPHREY, A. E. (1978), Real-time estimation of aerobic batch fermentation – biomass concentration by component balancing, *AIChE J.* **24**, 138–146.

ZADEH, L. A.(1965), Fuzzy sets, *Inf. Control* **8**, 338.

ZADEH, L. A. (1973), Outline of a new approach to the analysis of complex systems and decision processes, *IEEE Trans.* **SMC-3**, 28–44.

ZADEH, L. A. (1984), Making computers think like people, *IEEE-Spectrum,* Aug., 26–32.

ZIEGLER, J. G., NICHOLS, N. B. (1943), Optimum settings for automatic controllers, *Trans. ASME* **65**, 433–444.

ZOZ, B. (1990), Einsatz von Industrie-PCs als lokale Kopfstation flexibler verfahrenstechnischer Anlagen, *Verfahrenstechnik* **24**, 52–58.

18 Modelling, Design, and Control of Downstream Processing

SUTEAKI SHIOYA
KEN-ICHI SUGA

Osaka, Japan

List of Symbols

A	cm^2	cross-sectional area of column
b	—	fouling constant
C	g cm^{-3}	concentration of particle in the suspension
C_b	mg cm^{-3}	concentration in bottom phase
C_B	g cm^{-3}	bulk concentration
C_G	g cm^{-3}	maximum concentration
C_m	mg cm^{-3}	concentration in mobile phase
C_m^*	mg cm^{-3}	equilibrium concentration in mobile phase
C_s	mg cm^{-3}	concentration in stationary phase
C_t	mg cm^{-3}	concentration in top phase
d	cm	fluid channel height above the membrane
d_p	cm	particle diameter
dS	g	small amount of sample
D	cm^2 s^{-1}	diffusivity
D_r	cm^2 s^{-1}	axial dispersion coefficient
D_s	cm^2 s^{-1}	diffusion coefficient into the particle
$F(r)$	—	fraction of population in rth cavity
F_s	mg cm^{-3} s^{-1}	transfer rate of solute
g	cm s^{-2}	gravitational constant
G	—	distribution coefficient
H	—	ratio of effective space $= (1-\varepsilon)/\varepsilon$
$HETP$	cm	height equivalent to a theoretical plate
J	cm^3 cm^{-2} s^{-1}	flux
J_t	cm^3 cm^{-2} s^{-1}	flux at a time t
J_1	cm^3 cm^{-2} s^{-1}	flux at $t=1$ min
k_J	cm s^{-1}	mass transfer coefficient
K	—	partition coefficient
$K_{f,a}$	s^{-1}	overall mass transfer coefficient
L	cm	length of cylindrical centrifuge
L_m	cm	length of the channel of cross-flow filtration
m	—	number of space between disks in the stack
n	—	number of fractions
N	—	number of ideal stages
P	—	fraction of materials in top phase
p_i	mg cm^{-3}	concentration of the product in ith purification stage
P_i	mg	amount of the product in ith purification stage
$\bar{P}_i$	Pa	gauge inlet pressure
$\bar{P}_o$	Pa	gauge outlet pressure
ΔP_{TM}	Pa	average transmembrane pressure drop
$\bar{q}$	—	amount of absorption per unit weight
Q	cm^3 s^{-1}	throughput of the continuous centrifuge
$\bar{Q}$	s^{-1}	absorption rate per unit weight of packed bed
r	cm	radial distance from the center of the centrifuge
r_1	cm	radius of liquid surface in the bowl of the centrifuge
r_2	cm	radius of the bowl of the centrifuge
r_m	—	peak position of objective material
R	—	ratio of the solute in the mobile phase to the total amount of the solute
RCF	—	relative centrifugal force
Re	—	Reynolds number ($d\rho_s v/\mu_s$)
R_g	cm^{-1}	resistance of the retained components

R_m	cm^{-1}	restistance of the membrane
R_s	—	resolution factor
s	—	Laplace operator
$\bar{s}$	s	sedimentation coefficient
$s_{20,w}$	s	sedimentation coefficient of particle in water at 20°C
Sc	—	Schmidt number ($\mu_s/D\rho_s$)
Sh	—	Sherwood number ($k_j d/D$)
t	s	time
t_B	s	starting time of breakthrough
t_E	s	ending time of breakthrough
t_R	s	holdup time
u	cm s^{-1}	moving velocity of the peak
v	cm s^{-1}	fluid velocity
v_c	cm s^{-1}	terminal velocity in a centrifugal field
v_g	cm s^{-1}	terminal velocity in a gravitational field
V_0	cm^3	void volume of the column
V_b	cm^3	liquid volume of top phase
V_e	cm^3	elution volume
V_t	cm^3	liquid volume of top phase or total volume of the column
$W_{1/2}$	—	half width of elution curve
W_i	—	width of the ith peak ($i=1, 2$)
X_B	—	dimensionless concentration at the start of breakthrough
X_E	—	dimensionless concentration at the end of breakthrough
Y_i	—	product yield
$Y_{s,i}$	—	recovery yield at each step
Y_T	—	overall product yield
z	cm	axial distance from the entrance
Z	cm	column length
Z_p	cm	peak position from the entrance

Greek symbols

α_i	—	purification index
β	—	parameter of absorption equilibrium
ε	—	void ratio
θ	—	dimensionless time
$\theta_{R,i}$	s	retention time of the peak i ($i=1, 2$)
μ_n	—	nth moment of the elution curve ($n=1, 2, \ldots, n$)
μ_s	g cm^{-1} s^{-1}	viscosity of the fluid
$\mu_{T,s}$	g cm^{-1} s^{-1}	viscosity of the fluid at T°C
$\mu_{20,w}$	g cm^{-1} s^{-1}	viscosity of the water at 20°C
$\bar{v}$	g^{-1} cm^3	partial specific volume
ρ_b	g cm^{-3}	density of packed bed
ρ_p	g cm^{-3}	density of the particle
ρ_s	g cm^{-3}	density of the fluid
$\rho_{T,s}$	g cm^{-3}	density of the fluid at T°C
$\rho_{20,w}$	g cm^{-3}	density of the water at 20°C
σ	—	standard deviation of the elution curve with respect to the peak r_m
τ	s	retention time
ϕ	rad	the conical half angle
ω	rad s^{-1}	rate of rotation

1 Role of Downstream Processing in Biotechnology

The growth of microorganisms and the tissue culture of plant or animal cells are at the center of all bioproduction processes. These fermentations are preceded by upstream processing which includes the preparation of microorganisms and cells (including the genetic manipulation and screening of microorganisms), preparation of raw material, and the scaling up of cell numbers and volume. The downstream process includes the separation and purification steps, such as cell separation, extraction, condensation, chromatography, crystallization and drying. Just as upstream processing is important in increasing productivity, downstream processing is important in reducing production costs. In some cases success or failure of new process development depends completely on the economical feasibility of the downstream process. Often the cost of separation and purification of finished product from crude product is the dominating cost of the whole production process. From an economical viewpoint the separation of bioproduct is one of the key steps in process optimization.

The characteristic features of the separation and purification bioprocesses are:

(1) The bioproduct is frequently unstable and sensitive to the environment; mild conditions, such as neutral pH and low temperature, are required for purification.
(2) The concentration of bioproduct is extremely low in the bulk solution.
(3) Materials quite similar to the product are frequently present in the bulk solution.
(4) The required level of product purity depends on the process or field of product application itself. For example, in a pharmaceutical product, toxic materials or proteins of different origin should be completely removed, whereas materials non-toxic to humans need not be removed.

The principles of separation and purification and the properties of bioproducts used for separation are listed in Tab. 1. Various principles and techniques have been applied to downstream processing. Several will be explained in later sections. Also, several review papers dealing with bioproduct separation have been published (e.g., JOHNSON and HEDMAN, 1982; KULA, 1985; ATKINSON et al., 1987; BRÜMMER and GUNZER, 1987).

In order to increase recovery yield, not only must partition coefficients be increased at each step, but it is also important to reduce the number of purification steps. Overall product yield Y_T is represented by

$$Y_T = \prod_{i=n} Y_{s,i} \tag{1}$$

where $Y_{s,i}$ is the recovery yield at each step. Approximately speaking, the fact that $Y_{s,i}$ is less than 1.0 will result in an increase of Y_T with the decrease of n. From this viewpoint, the bioseparation process should be designed to minimize the number of steps and be operated to maximize $Y_{s,i}$. This chapter describes modelling and design of certain downstream processes, viewed in terms of optimal design of the total system.

2 Principle of Unit Operation in Downstream Processing

2.1 Pretreatment

2.1.1 Cell Separation

Within the rapidly growing biotechnological industry, downstream processing has emerged as a major limitation of successful commercial development. The primary process of biological product recovery is the separation of cells from the culture medium. It is important in choosing a separation method whether the desired product is extracellular, intracellular, or the biomass itself. Batch or continuous centri-

Tab. 1. Principles for Bioproduct Purification and Separation

Single Process	Thermodynamic equilibrium	Production of new-phase from single-phase	G–L G–S	Evaporation Condensation Sublimation Crystallization
			L–S(L)	Sedimentation Coagulation Equipotential coagulation Sieve sedimentation Salting out
		Contact and equilibrium between more than two phases	G–L L–L	Various types of distillation Liquid–liquid extraction Partition Countercurrent partition Partition chromatography
			L–S	Absorption Ion-exchange Solid–liquid extraction Extraction by supercritical fluid Adsorption Elution
	Transfer process	Membrane	L–L	Membrane separation Liquid-membrane separation
			L–S	Filtration Expression Pressing
		Gravity Centrifugation	L–S	Centrifugation Density gradient centrifugation Precipitation
		Electric power	L–S (charged particle)	Electrophoresis
Complex Process	Concentration gradient		Interruption	Dialysis Gel-filtration
	Pressure gradient		Interruption	Ultrafiltration Reverse Osmosis
	Electric power		Interruption	Ion exchange chromatography Gel electrophoresis Electrophoresis
	Hydrophobic interaction Gravity Electric power Interruption		Interruption Electric power Bio-affinity Bio-affinity	Hydrophobic chromatography Two-dimensional electrophoresis Cell sorter Affinity chromatography Affinity ellution
	Interruption	Electric power	Concentration difference (pH)	Ion-exchange chromatography gradient elution Chromato-focusing Disk-gel electrophoresis Gel-isoelectric metering Gel-isoelectric equilibrium Isotachophoresis

fugation and filtration with filter presses or rotary vacuum filters are the processes ordinarily used. Recently, the application of cross-flow membrane filtration for the harvesting of cells has been investigated. In this filtration, the suspension is recirculated across the filter surface at high velocity to remove the retained particles. This type of filter, therefore, is different from the conventional type, in which the filtration rate decreases with the build-up of retained cells on the filter surface. In this system the filtration rate depends on (1) the transmembrane pressure, (2) the bulk cell concentration, and (3) the mass transfer coefficient. PATEL et al. (1987) reported the feasibility of harvesting yeast cells using synthetic membranes. They concluded that (1) filtration rates decreased exponentially with time, (2) fouling rates became lower at lower cell concentration, lower transmembrane pressures, and higher velocity, and (3) media components played a relatively greater role in the fouling phenomenon at low cell concentrations. Meanwhile, porous cylindrical sintered stainless steel tubes of 2 µm and 3 µm nominal pore size have been applied to the cross-flow separation of yeast cells (KAVANAGH and BROWN, 1987). In this case average filtration rates as high as $1.25 \text{ m}^3 \text{ m}^{-2} \text{ h}^{-1}$ were recorded for low cell concentrations (3 g L^{-1}). As cell concentration was increased from 3 g L^{-1} to 25 g L^{-1} there was a steady but gentle decline in average filtration rate. The flux of the tubular microfilter was found to be more than $100 \text{ L m}^{-2} \text{ h}^{-1}$ at yeast cell concentrations of 100 g L^{-1}, while the hollow-fibers had a flux of $40 \text{ L m}^{-2} \text{ h}^{-1}$ at 250 g L^{-1} (PATEL et al., 1987).

The coarse sintered stainless steel cylindrical filter element (nominal pore size of 75 µm) as a sterilizable cross-flow filtration unit was used for the separation of filtrate from a fungal *Trichoderma reesei* suspension in a sterile manner (BROWN and SALAM, 1984). This paper also reported an average filtration rate of $1.0 \text{ m}^3 \text{ m}^{-2} \text{ h}^{-1}$ and the fact that cell concentration increased from 16.4 to 47 kg m^{-3}. Further, in the harvesting of *Aspergillus niger* from a fermentation broth, the flux of the filtrate was $80 \text{ L m}^{-2} \text{ h}^{-1}$ at a mycelial dry mass concentration of 57 g L^{-1}, a velocity of 4 m s^{-1}, and an initial transmembrane pressure of 10 kPa using a tubular cross-flow microfil-

Tab. 2. Comparison of Hollow-Fiber Filtration and Centrifugation for Bacteria Harvesting (Annual)

	Hollow-Fiber Filtration	Centrifugation
Depreciation	$ 4500	$ 15000
Maintenance	1800	6000
Labor	7500	7500
Power	1500	6200
Water, chemicals	1400	1200
Membranes	10500	—
Total	$ 27200	$ 35900
Capital cost	$ 45000	$ 150000

Assumptions:
System output 5000 L/h (water removal rate)
Operation 250 d/a, 16 h/d; plus 2 h/d clean up
Depreciation 10 years S/L
Maintenance 4% of capital cost annually
Labor at $ 15.00/h
Power at $ 0.05/kWh
Membrane life of 1 year, $ 157/m^2
Membrane av. flux rate 75 L/m^2 h
Broth *Escherichia coli* simple medium, 20 g/L feed 180 g/L product

Tab. 3. Comparison of Hollow-Fiber Filtration and Centrifugation for Yeast Harvesting (Annual)

	Hollow-Fiber Filtration	Centrifugation
Depreciation	$ 7500	$ 5000
Maintenance	3000	2000
Labor	7500	7500
Power	2400	2900
Water, chemicals	2300	1500
Membranes	17400	—
Total	$ 40100	$ 18900
Capital cost	$ 75000	$ 50000

Assumptions:
System output 10000 L/h (water removal rate)
Operation 250 d/a, 16 h/d; plus 2 h/d clean up
Depreciation 10 years S/L
Maintenance 4% of capital cost annually
Labor at $ 15.00/h
Power at $ 0.05/kWh
Membrane life of 1 year, $ 157/m^2
Membrane av. flux rate 90 L/m^2 h
Broth *Saccharomyces cerevisiae*, simple medium, 40 g/L feed 180 g/L product

ter (pore size of 0.2 μm) (SIMS and CHERYAN, 1986). From the economic viewpoint a comparison of hollow-fiber filtration (0.1 μm microporous-type) and centrifugation (a continuous disc-type centrifuge) was made for both bacteria and yeast harvesting (TUTUNJIAN, 1984) as shown in Tabs. 2 and 3. It is evident that the hollow-fiber system is preferable to centrifugation when dealing with smaller cells, in which case centrifuge outputs drop significantly.

2.1.2 Cell Breakage

When the product is accumulated inside cells, the first step of separation is the breakdown of the cells, and the desired product is usually extracted into an aqueous or other phase. Methods of cell breakage are: destruction by osmotic pressure, repetitive freezing and thawing, utilization of enzymes, self-digestion or autolysis, and mechanical disruption by Dyno-Mill, sonication, French press, and other methods. Also, surface-active agents such as polyoxyethylen(n) octylphenyl ether ($n = 9 \sim 40$) are frequently effective in breaking down cells and viruses.

2.1.3 Precipitation Partitioning

In order to obtain concentrated samples and also to reduce sample volume, precipitation by denaturation of protein is utilized for separation from bulk solutions. The most popular method is salt dialysis, which depends on salt concentration and surface properties of target proteins. Isoelectric precipitation is also used, because precipitation occurs readily under the condition of low ionic strength at isoelectric pH. Precipitation by organic solvents is also used for protein partitioning.

2.2 Separation of Bioproduct

2.2.1 Centrifugation and Membrane Filtration

After cell harvesting and disruption, the recovery of intracellular products from cell debris is carried out by centrifugation and/or filtration. Cell debris is made up of particles with a size of the order of 0.1 μm, while protein molecules are two orders of magnitude smaller. Cell debris removal is the most difficult and expensive solid–liquid separation. Various techniques are available, and the choice depends on cell properties and on scale of operation. This has most commonly been performed by centrifugation. Semi-continuous type centrifuges are generally used in large-scale extraction. The insoluble cell debris is pumped continuously into the tubular-bowl centrifuge, where it is quickly accelerated to bowl speed. Solids are deposited, while the clarified liquid is discharged continuously over a dam at the top of the centrifuge bowl. Since there is no automatic solids removal system, only small concentrations of solids can be handled. Disc centrifuges have provided an excellent means of extract clarification. These types of centrifuges, however, may rotate at slow speeds, developing centrifugal forces only up to $8000 \times g$. They have a capacity of up to 20 kg of sediment. For large-scale separation, continuous flow centrifuges are expensive in capital and running costs. When the density difference between the particles and medium is small, centrifugal throughputs may need to be reduced in order to maintain high separation efficiency. Recently, tangential flow filtration has been investigated as an alternative to centrifugation for the separation of soluble intracellular products from cell debris. The performance of tangential flow filtration in the removal of cell debris is affected by a large number of factors, such as the organisms utilized, the molecular weight of the desired enzyme, the method of cell breakdown, the physicochemical properties of the membranes and their pore sizes, pressure, flow hydrodynamics, temperature and cleaning methods. These factors may be investigated by measuring the effect of flux rate, process time, and turbidity of the filtrate

on the yield of enzyme as the desired bioproduct. The molecular weight of the enzyme
could be an important parameter during microfiltration. QUIRK and WOODROW (1984) reported that enzyme yields were in inverse order
of their molecular weights using asparaginase
(M.W. 132000), carboxypeptidase (M.W.
83500), and arylamidase (M.W. 52000). Similar results were obtained (DATAR, 1985) for β-
galactosidase (M.W. 540000), carboxypeptidase Y (M.W. 65000), and human growth hormone (M.W. 21000). The Domnick-Hunter
membranes with an asymmetric structure (pore
size of 0.45 μm on the tight side and approximately 1.5–2.0 μm on the open side) gave the
same recovery of arylamidase as millipore
poly(vinylidene fluoride) membranes (0.5 μm).
Increasing the pore size of an Amicon membrane to 0.6 μm increased carboxypeptidase
yield (from 52 to 78%), but resulted in decreased quality (QUIRK and WOODROW,
1984). Yields may be increased by using membranes with larger pore sizes. However, the
passage of particles of larger sizes through the
membrane blocks the column for its use in column chromatography in subsequent downstream unit operations. Cross-flow velocity
and pressure have a significant influence on
flux rates. Flux was proportional to the feed
velocity raised to the power of 0.5 with aryl
acyl amidohydrolase lysates obtained by lysozyme treatment (LE and ATKINSON, 1985).
The effect of pressure on flux of the cell lysate
for various feed velocities was investigated by
LE and ATKINSON (1985). Flux increased with
increasing pressure up to a limiting level. Further, the transmission activity, i.e., the ratio of
enzyme activity in the filtrate to that in the
feed, was proportional to the feed velocity
raised to the power of 0.18, and it increased
rapidly with increasing pressure. Physical and
chemical (or enzymatic) methods of cell breakage had little effect on enzyme yield in the separation of arylamidase from *Pseudomonas fluorescens* (LE and ATKINSON, 1985; QUIRK and
WOODROW, 1984) and on asparaginase from
Erwinia carotovora (QUIRK and WOODROW,
1984). However, process time for filtration significantly varied with changes in the methods
of cell breakage. As process time decreased
with the different treatments, such as Dyno-
Mill breakage (480 min), lysozyme breakage

(300 min), and mercaptoethanol following lysozyme breakage (180 min) for *P. fluorescens*
homogenates, the specific activity of filtrates
decreased from 0.97 to 0.50 to 0.43, respectively (QUIRK and WOODROW, 1984). This
suggests that the characteristics of the cell debris may affect the separation performance
with respect to change of flux rate and specific
activity. It also indicates the formation of a
secondary filtration layer by accumulation of
either insoluble matter or soluble protein.

LE and ATKINSON (1985) reported the effect
of ionic strength on separation efficiency using
a lysate of *P. fluorescens* cells. The effect of
ionic strength on enzyme transmission was
said to be independent of the method of cell
treatment. However, the enzyme transmission
level increased with an increase in buffer
strength.

Tab. 4 provides the data for the separation
of formate dehydrogenase, using different
methods, from the cell debris of *Candida boidinii* (KRONER et al., 1984). The data show
that energy consumption was low for cross-
flow filtration in comparison with tubular-type
centrifugation. The clarification of the filtrate
was 100% for cross-flow filtration. As mentioned previously, this may be important for
further purification procedures, such as chromatography. The separation efficiency for soluble enzymes from cell debris by a cross-flow
filter system is presently not satisfactory in
terms of enzyme yield.

In order to improve flux rate, hollow-fiber
ultrafiltration membranes were examined for
the processing of a precipitate suspension of
isoelectric soya protein (DEVEREUX and
HOARE, 1986). These authors determined that
protein precipitation was a method that might
be used to improve permeate flux when polarization is a major limitation to membrane separation.

2.2.2 Extraction

Differences in the partition of solutes in a
liquid–liquid mixture can be utilized for extraction and separation of product. The two
phases may be water–organic solvent, water–
water, or liquid–gas under supercritical conditions.

Tab. 4. Comparison of Methods for Separation of Cell Debris from Enzymes

Method	Concentration $C_0 \rightarrow C_f$ (Vol. %)	Purity of Liquid	Enzyme Yield	Performance Factor $(L\ h^{-1}\ m^{-2})$	Energy Demand $(Wh\ L^{-1})$
Pressure filter with 5% celite	20–80	99.3	74	1.2	526
High speed tubular centrifuge	20–100	99.5	89	—	150
Cross-flow filter hollow-fiber, PC/2 10^6 Daltons	4–52	100	50	8.6	33
Cross-flow filter flat membrane cassette HVLP – 0.45 μm	12 (diafiltration)	100	58 (after 4 cycles)	12.5	77
Cross-flow filter stirred cell, HVLP – 0.45 μm	8–40	100	62	14.7	370

Enzyme: Formate dehydrogenase
Cell debris: *Candida boidinii*

Aqueous two-phase separation

The polyethylene glycol (PEG)/Dextran system has mainly been used for aqueous two-phase separation (ALBERTSON, 1960; WALTER et al., 1985). The required time for two-phase separation is not short. However, denaturation of protein or damage of product is relatively less than with organic solvent systems. Recently, many tests have been performed on aqueous two-phase separation with affinity materials.

Extraction by supercritical liquid

Superciritical fluid, which has properties of both a liquid and a gas, is used for separation in such a way that the required material is extracted into liquid while other components are vaporized. Caffeine is frequently extracted from instant coffee granules with supercritical carbon dioxide fluid because aroma components of the sample are not lost.

2.2.3 Chromatography

The separation of the components of a mixture by distribution between two immiscible liquids, either by bulk extraction or by liquid–liquid partition chromatography, is a familiar and long-established technique in chemistry. Chromatography is defined as the process that takes advantage of the difference in partition of solutes between stationary and mobile phases in the separation of target materials. Various combinations of stationary and mobile phases lead to such methods as gas, liquid, supercritical, and aqueous two-phase chromatography.

For bioproduct separation, particles of the stationary phase are packed into a column through which the liquid mobile phase is then poured. This is an example of liquid chromatography, a widely used method. Liquid chromatography is based on principles arising from interactions between the stationary and mobile phases. Many types of chromatography are listed in Tab. 1. The one most suitable for the separation of the target product and impurities should be selected from those available. Liquid chromatography involves simple equipment and simple operations. Process conditions are typically mild, permitting the retention of much of the activity of a protein. Liquid chromatography (LC) separation can be used for a wide range of procedures, from biochemical and chemical analysis to large-scale preparation.

3 Modelling of Unit Operations

3.1 Centrifugal Separation

Centrifugation is a useful method for the separation of cells, subcellular organelles, or large molecules. The basic parameters governing the separation of particles in a centrifugal field are mass, density, and frictional coefficients of the particles present in the suspension. Considering a spherical particle falling in a gravitational field such that other particles present do not hinder its fall, its velocity will reach a constant value determined by a balance between the net force acting on the particle and the frictional resistance. From Stokes' law the terminal velocity v_g in a gravitational field is given by

$$v_g = \frac{d_p^2 (\rho_p - \rho_s)}{18\,\mu_s}\, g \tag{2}$$

where ρ_p is the density of the particle,
ρ_s the density of the fluid,
d_p the diameter of the particle,
μ_s the viscosity of the fluid, and
g the gravitational constant
(980 cm s^{-2}).

The terminal falling velocity of spherical particles at radius r in a centrifugal field rotating at the rate ω is given by

$$v_c = \frac{d_p^2 (\rho_p - \rho_s)}{18\,\mu_s}\, r\omega^2 \tag{3}$$

or

$$v_c = \frac{dr}{dt} = \bar{s} r\omega^2 \tag{4}$$

In Eq. (4) $\bar{s}$ is the sedimentation velocity per unit of centrifugal force and is called the sedimentation coefficient. It is usually expressed in Svedbergs (S), equivalent to 10^{-13} s. Thus, a particle whose sedimentation coefficient is 10^{-12} s has a sedimentation coefficient of 10 S. If the density and viscosity of the solution and the temperature of the experiment are known, a sedimentation coefficient, $s_{20,w}$, which a particle has in water at 20 °C, can be calculated from the following equation

$$s_{20,w} = \bar{s} \left(\frac{1 - \bar{v}\rho_{20,w}}{1 - \bar{v}\rho_{T,s}} \right) \frac{\mu_{T,s}}{\mu_{20,w}} \tag{5}$$

where $\bar{v}$ is partial specific volume. The value of $s_{20,w}$ decreases with increasing concentration, C, of particles according to the following equation

$$s_{20,w} = \frac{s_{20,w}^0}{1 + K_s C} \tag{6}$$

where $s_{20,w}^0$ and K_s are constants. Therefore, if $s_{20,w}^0$ and $\bar{v}$ are known, the terminal velocity v_c can be estimated by Eq. (5) using the characteristics ($\rho_{T,s}$, $\mu_{T,s}$) of the liquid.

The relative centrifugal force (RCF), defined by the ratio of v_c to v_g from Eqs. (3) and (4), is commonly used to express the centrifugal field as follows

$$RCF = \frac{v_c}{v_g} = \frac{r\omega^2}{g} \tag{7}$$

$$RCF = \frac{4\pi^2 (\text{rpm})^2 r}{980 \cdot 3600} \tag{8}$$

$$RCF = 1.119 \cdot 10^{-5} (\text{rpm})^2 r \tag{9}$$

Semi-continuous or continuous type centrifuges are widely used for large-scale centrifugation. In semi-continuous centrifugation, feed that contains cells or cell debris flows into the centrifuge bowl and solids are deposited, while the clarified solution discharges continuously. The throughput, Q, of a semi-continuous tubular type centrifuge can be described as

$$Q = v_g \frac{\pi L (r_2^2 - r_1^2) \omega^2}{g \ln (r_2/r_1)} = v_g \cdot \Sigma \tag{10}$$

where L is the length of the cylindrical centrifuge, and r_2 and r_1 are the radii of bowl and liquid surface, respectively.

$$\Sigma = \frac{\pi L (r_2^2 - r_1^2) \omega^2}{g \ln (r_2/r_1)} \tag{11}$$

Σ is a parameter of the centrifuge, which is equivalent in area to a gravity settling tank that is theoretically capable of doing the same amount of work.

For the disk-bowl centrifuge

$$\Sigma = \frac{2\pi m (r_2^3 - r_1^3) \omega^2}{3g \tan \phi} \tag{12}$$

where m is the number of spaces between disks in the stack and

ϕ the conical half angle

The Σ factor can then be used as a means of comparing centrifuges.

3.2 Partition in Aqueous Two-Phase Systems

Countercurrent distribution (CCD) is an established method for carrying out repeated partition steps (TREFFRY and SHARPE, 1985). It was originally designed for separations involving aqueous–organic or organic–aqueous two-phase systems (CRAIG, 1960). Multiple extraction steps require, after each partition, that the top and bottom phases are physically separated, fresh top phase is added to the bottom phase, and fresh bottom phase is added to the separated top phase. In this way material that has partitioned into either of the phases is repartitioned. Each chamber is loaded with the volume of top and bottom phases in CCD.

Plotting the quantity of material in each chamber against the number of the chamber leads to a symmetrical distribution curve. Its peak in the chamber corresponds to half the number of partition steps carried out when the partition coefficient is 1.0.

For both soluble and particulate material, the most useful parameter for describing and predicting CCD curves is P, the fraction of the total amount of material in a chamber appearing in the top phase. As will be seen subsequently, this is the partition parameter that enters directly into the equations describing the distribution of a uniform population of molecules or particles. The other commonly used parameters are the partition coefficient,

K, and the distribution coefficient, G. For molecular distributions the partition coefficient is defined as the concentration ratio of the top phase to the bottom phase. The distribution coefficient is the ratio of the total amount of material in the top phase to that in the bottom phase. The three parameters are related as

$$K = C_t / C_b \tag{13}$$

$$G = \frac{C_t V_t}{C_b V_b} \tag{14}$$

$$P = \frac{G}{G+1} \tag{15}$$

Particles sometimes partition between the interface and the top phase. Therefore, a small volume of top phase is left with the interface in the bottom of the cavity to ensure that no adsorbed particles are carried over when the top phase is transferred. If necessary, the volume left should be taken into account for mass balance.

Once the appropriate value of P for the population of interest has been calculated from Eq. (15), the CCD curve for all the components in a sample having that P value can be predicted.

If n distributions are carried out (i.e., $n-1$ transfers), then the fraction of the total population appearing in the rth cavity, $F(r)$, will be given simply by the binomial distribution (TREFFRY and SHARPE, 1985):

$$F(r) = \frac{n!}{r!\,(n-r)!} P^r (1-P)^{n-r} \tag{16}$$

By recognizing that at the peak of the distribution $F(r)$ will be approximately equal for two adjacent cavities, i.e., $F(r_m) = F(r_m + 1)$, the location of the peak, r_m, is easily found to be

$$r_m = nP \tag{17}$$

or using Eq. (15)

$$r_m = \frac{nG}{G+1} \tag{18}$$

A useful expression for the half-width of the curve at one-half the peak height, $W_{1/2}$, can also be obtained (TREFFRY and SHARPE, 1985)

$$W_{1/2} = 1.18\,[r_m(1-r_m/n)]^{1/2} \qquad (19)$$

Therefore, if the width at the half-height of an experimental CCD peak is significantly different from $W_{1/2}$ as calculated from Eq. (19), the assumption that the population is homogeneous must be questioned. Note that the peak width divided by the peak location, Eq. (20), decreases as the number of transfers is increased according to

$$\frac{\sigma}{r_m} = \left(\frac{1-P}{np}\right)^{1/2} \qquad (20)$$

where the curve is approximated by the normal distribution with $\sigma = nP(1-P)^{1/2}$. Hence, the resolution between two peaks will increase as $n^{1/2}$.

3.3 Chromatographic Separation

Since the driving force of the separation depends on differences in the partitioning of materials between mobile and stationary phases, various types of liquid chromatography can be categorized according to the principal mechanisms of partitioning, as shown in Tab. 5. Many target bioproducts to be separated have amphipathic properties, so the ampholyte, size, shape, electrostatic, and polar properties and structural specificity of molecules have all been utilized for separation.

Elution procedures can be divided into three kinds of operations, depending on the nature of the interaction between the stationary phase and the solution used for elution. First, a solution with the same composition as the sample solution is used for elution in the case of gel or hydrophobic chromatography (isocratic elution). Second, when partition is complete or clearly defined, the composition of the elution solution is subjected to stepwise change (stepwise elution), in order to cause a gradual change in the strength of partition. Third, when the partitioning of the target component is especially complete, as in affinity chromatography, adsorption is continued until a breakthrough occurs, after which elution is accomplished with the aid of a solution that is completely different from the adsorbent. This is called a selective adsorbent.

3.3.1 Separation Characteristics by Liquid Chromatography

To predict separation potential using LC, it is necessary to know the elution volume or time required for attaining the peak of elution of each component as well as the width of the elution curve. Elution time depends on the rate of motion of the peak as established by the partitioning properties of the solute with respect to the mobile and stationary phases. Further, the width of the elution curve is governed by the diffusivity of the solute into both phases, backmixing of the liquid, and uniformity of flow. These properties can be represented by the ideal plate model or mass balance, as shown later. The former uses the ideal plate number N to explain the observed separation characteristics. The latter utilizes differential equations describing the mass transfer rate and axial diffusion of the solute. The concepts of adsorption and mass balance are also applicable to affinity chromatography, but in a slightly different way.

3.3.2 Equilibrium Model

V_t and V_0 denote the total volume of the column and volume of the void, respectively, and the mobile phase liquid flows at a space velocity u. The rate of motion of a solute peak is proportional to the ratio of the amount of solute in the mobile phase to that in the stationary phase. The velocity of the peak then equals the product of u and the ratio. The equilibrium governing the concentration of the solute in the mobile phase C_m and that in the stationary phase C_s can be described as

$$C_s = K\,C_m \qquad (21)$$

where K is the partition coefficient.

Now the ratio, R, of the solute in the mobile phase to the total amount of the solute can be written as

Tab. 5. Liquid Chromatography for Bioproduct Separation

LC	Force	Elution	Characteristics	Desired Process or Bioproduct
Gel filtration chromatography (GFC)	Size of molecules Shape	Isocratic	$K = 0 \sim 1$ Column length: medium High recovery ratio	Desalination Exchange of buffer solution Differentiating of proteins
Ion exchange chromatography (IEC)	Electrostatic power	Stepwise gradient	Applicable to a wide area Amount of treatment: large Adjustable of partition ratio by elution Condensation	From small molecules to large molecules
Hydrophobic chromatography (HIC)	Hydrophobic power	Stepwise gradient	Adsorption by high ionic strength Many elution methods can be used	Protein with conformation Cells
Chromato-focusing	Difference of isoelectricity	Gradient	High partition ratio High resolution ratio	Isozyme
Affinity chromatography	Biological affinity	Gradient	High selectivity Amount of treatment: large Need to check the elution condition	Low concentration bioactive material

$$R = \frac{C_m \varepsilon}{C_m \varepsilon + C_s(1-\varepsilon)} = \frac{1}{1+HK} \qquad (22)$$

where ε is the void ratio ($= V_0/V_t$) and $H = (1-\varepsilon)/\varepsilon$. Finally, the velocity of the solute peak can be expressed as

$$\frac{dZ_p}{dt} = \frac{u}{1+HK} \qquad (23)$$

Provided the partition coefficient, K, is constant and independent of the solute concentration, then the time required for the solute peak to emerge from the column, t_R (known as holdup time), is obtained from Eq. (23) as

$$t_R = \frac{Z(1+HK)}{u} \qquad (24)$$

where Z is the column length. The elution volume V_e is given as

$$V_e = V_0 + \frac{V_0 K(1-\varepsilon)}{\varepsilon} = V_0 + K(V_t - V_0) \qquad (25)$$

As partition of the solute into the stationary phase increases (i.e., K is large), the movement of the peak is slower and the solute emerges more slowly from the column.

3.3.3 Ideal Stage Model

The equilibrium model takes account only of the moving velocity of the solute peak. In a discussion of the separability or resolution of peaks of two components, the spread or variance of the peaks should also be considered. The variance is represented by the *HETP* (height equivalent to a theoretical plate) (VAN DEEMTER et al., 1956). In the ideal stage model, a column with the height Z is divided into N stages of equal volume, composed of mobile and stationary phases as shown in Fig. 1.

$$HETP = Z/N \qquad (26)$$

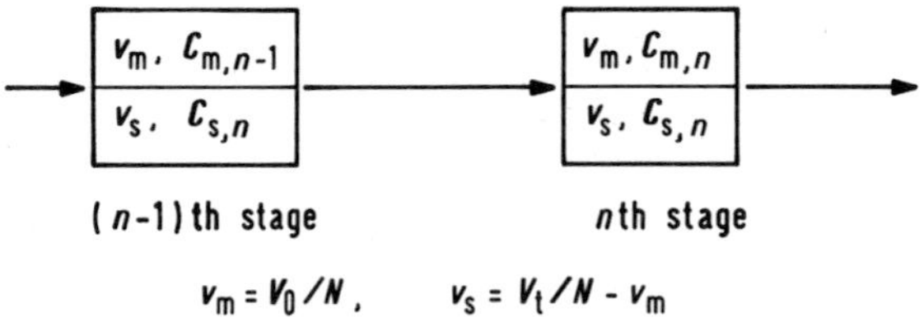

$$v_m = V_0/N, \qquad v_s = V_t/N - v_m$$

Fig. 1. Ideal strage model ($V_m = V_0/N$, $V_s = V_t/N - V_m$).

When a small amount of the sample, dS, flows continuously and at a constant rate into the column, the mass balance equation for the solute is shown to be

$$(C_{m,n-1} - C_{m,n}) dS =$$
$$= \frac{V_0}{N} dC_{m,n} + \frac{V_t - V_0}{N} dC_{s,n} \qquad (27)$$

Noting that the retention time of the liquid can be expressed as $\tau = Z/u$, Eq. (27) can be rewritten in the form

$$\left(\frac{\tau}{N} + \frac{\tau}{N} \frac{(V_t - V_0)}{V_0} K \right) \frac{dC_{m,n}}{dt} = C_{m,n-1} - C_{m,n} \qquad (28)$$

That is,

$$\frac{\tau}{N}(1+HK)\frac{dC_{m,n}}{dt} = C_{m,n-1} - C_{m,n} \qquad (29)$$

because

$$A u dt = dS \qquad C_{s,n} = KC_{m,n} \quad \text{and}$$
$$dS/V_0 = \frac{Au}{V_0} dt = \frac{u}{z} dt = \frac{dt}{\tau}$$

When a sample with a solute concentration of C_0 is applied from 0 to t_0, after which the solution no longer contains solute, the initial boundary conditions (BC) become

$$
\begin{aligned}
t &= 0 & C_{m,n} &= C_{s,n} = 0 \\
0 &< t \le t_0 & C_{m,0} &= C_0 \\
t &> t_0 & C_{m,0} &= 0
\end{aligned}
\qquad (30)
$$

Using these BCs, Eq. (29) can be solved numerically for the general case. In the particular case where the sample is very small and is added all at once, and where the number of stages N is large, the elution curve from the column is approximated by a Gaussian distribution;

$$C_{m,n} = \frac{C_0 \theta_0}{\sqrt{2\pi (1+HK)^2/N}} \cdot$$

$$\cdot \exp\left[-\frac{\{\theta - (1+HK)\}^2}{2\frac{(1+HK)^2}{N}} \right] \qquad (31)$$

where $\theta = t/\tau$ and $\theta_0 = t_0/\tau$. The holdup time θ_R is given as $\theta_R = 1 + HK$, and the variance of the curve is $(1+HK)^2/N$.

If the maximum value of the curve is defined as C_{max}, the distance between the points at which the height of the curve is $C_{max}e^{-1/2}$ is 2σ. The width associated with the intersections of the elution curve gradient with the x-axis becomes 4σ, i.e.:

$$W = 4\sigma = \frac{4(1+HK)}{\sqrt{N}} \qquad (32)$$

$$N = 16(\theta_R/W)^2 \qquad (33)$$

As the number of ideal stages increases, the peak width becomes narrower and separation characteristics can be described as good. Another expression derived by GIDDINGS (1965) is useful to be considered in scaling up a chromatographic separation.

3.3.4 Diffusion Model

The diffusion model takes into account mass transfer between mobile and stationary phases as well as axial mixing along the column, and these can also be used for the analysis of liquid chromatography (KUBIN, 1975). Consider the case in which the column is filled with uniform particles of diameter d_p and the sample is applied from time θ to t_0, following which elution begins. The mass balance equation along the column length becomes

$$\frac{\partial C_m}{\partial t} = D_r \frac{\partial^2 C_m}{\partial z^2} - u \frac{\partial C_m}{\partial z} - HF_s \tag{34}$$

where C_m is the solute concentration in the mobile phase,
D_r the axial diffusion coefficient, and
F_s the transfer rate of solute per unit volume of the stationary phase.

The diffusion of solute C_s into a particle can be written as

$$\frac{\partial C_s}{\partial t} = D_s \left(\frac{\partial^2 C_s}{\partial r^2} + \frac{2}{r} \frac{\partial C_s}{\partial r} \right) \tag{35}$$

where D_s is the diffusion coefficient into the particle; that is,

$$HF_s = H \frac{6D_s}{d_p} \frac{\partial C_s}{\partial r} \bigg|_{r=d_p/2} \tag{36}$$

If the liquid phase mass-transfer resistance can be neglected, then the initial and boundary conditions become

$$
\begin{array}{llll}
t = 0 & z > 0 & C_m = C_s = 0 \\
t > 0 & r = d_p/2 & C_s = K C_m \\
& r = 0 & \partial C_s/\partial r = 0 & (37) \\
0 < t \le t_0 & z = 0 & C_m = C_0 \\
t > t_0 & z = 0 & C_m = 0
\end{array}
$$

By utilizing these basic partial differential equations, the correct elution curve can be determined, although doing so entails many time-consuming calculations. However, the average elution time and the *HETP* can be de-

rived more simply using the moment method. The moment is defined as

$$\mu_1' = \int_0^\infty t C_m(z, t) \, dt \bigg/ \int_0^\infty C_m(z, t) \, dt \tag{38}$$

μ_1' corresponds to the average retention time, and the variance around μ_1' is given by

$$\mu_2 = \int_0^\infty (t - \mu_1')^2 C_m(z, t) \, dt \bigg/ \int_0^\infty C_m(z, t) \, dt \tag{39}$$

Here we can use the following relations in the Laplace transformation

$$\mu_n' = (-1)^n \lim_{s \to 0} (d/ds)^n \bar{C}(z, s) \big/ \lim_{s \to 0} \bar{C}(z, s) \tag{40}$$

where s is the Laplace operator defined as

$$\bar{C}(z, s) = \int_0^\infty C(z, t) e^{-st} \, dt \tag{41}$$

Using Eq. (40), the average retention time from the basic equation (34) can be derived as

$$\mu_1' = \tau(1 + HK) + \frac{t_0}{2} \tag{42}$$

$$\mu_2 = 2\tau \left\{ \frac{D_r(1 + HK)^2}{u^2} + \frac{d_p^2 HK}{60 D_s} \right\} + \frac{t_0^3}{12} \tag{43}$$

Then

$$
\begin{aligned}
HETP &= Z\mu_2/(\mu_1')^2 \\
&= \frac{2D_r}{u} + \frac{d_p^2 u}{30 D_s} \frac{HK}{(1 + HK)^2}
\end{aligned} \tag{44}
$$

From this equation it can be seen that there is an optimum u that minimizes *HETP*.

3.3.5 Mass Transfer in Affinity Adsorption

In this case the third term of Eq. (34) can be rewritten as

$$HF_s = \frac{\rho_b}{\varepsilon} \bar{Q} \tag{45}$$

by neglecting the diffusion inside the particle, where ρ_b is the density of the packed bed and $\bar{Q}$ is the absorption rate per unit weight of the packed bed. The following equation for the absorption rate can be assumed to apply except during the initial phase of absorption:

$$\rho_b \bar{Q} = K_{fa}(C_m - C_m^*) = \rho_b \frac{\partial \bar{q}}{\partial t} \qquad (46)$$

where K_{fa} is the overall mass transfer coefficient,

$\quad \bar{q}$ the average amount of absorption per unit weight, and

$\quad C_m^*$ the mobile phase equilibrium concentration corresponding to $\bar{q}$.

Thus, the absorption rate is proportional to the displacement from equilibrium. This relationship is known as the absorption equilibrium and is expressed as

$$\rho_b \bar{q} = K C_m^\beta \qquad (47)$$

as in the Freundlich adsorption equation. In the case of elution, on the other hand, Henry's equation is used with $\beta = 1$, which is similar to Eq. (21). In general, K_{fa} changes during the course of absorption. However, if β is small, K_{fa} can be calculated using the shape of the breakthrough curve:

$$K_{fa} = \frac{\rho_b \bar{q}_0}{C_0(t_E - t_B)} \cdot$$
$$\cdot \left\{ \ln(X_E/X_B) + \frac{\beta}{1-\beta} \ln \frac{1 - X_B^{(1-\beta/\beta)}}{1 - X_E^{(1-\beta/\beta)}} \right\} \qquad (48)$$

Equations (34) and (46), with K_{fa} obtained from Eq. (48), provide breakthrough curves and elution curves by numerical calculations, where t_B and X_B represent time and dimensionless concentration at the start of breakthrough, and t_E and X_E are time and dimensionless concentration at the end of breakthrough, respectively. Moreover, when the affinity constant between the stationary phase and solute is very high, as in antigen-antibody system ($\beta \ll 0.1$), equilibrium can be approximated by an irreversible equilibrium and a breakthrough curve can be easily calculated.

3.3.6 Factors Affecting the Separation

Usually the extent of separation can be measured by the resolution factor, R_s, which represents the extent of separation between a given peak and the closest neighboring peak. The resolution factor is defined as

$$R_s = (\theta_{R,2} - \theta_{R,1}) / \{(W_1 + W_2)/2\} \qquad (49)$$

where the subscripts 1 and 2 refer to peaks 1 and 2. High performance separation means that R_s is large. To achieve this end the difference of retention times $(\theta_{R,2} - \theta_{R,1})$ should be large and the width of the peak (the denominator of Eq. (49)) should be as small as possible. Substituting Eqs. (24), (26), and (31) into Eq. (49), and if $W_1 = W_2$, then

$$R_s = H(K_2 - K_1)\sqrt{N}/4(1 + HK_1) \qquad (50)$$

or

$$R_s = H(K_2 - K_1)\sqrt{N}/4(1 + HK_1)\sqrt{HETP_1} \qquad (51)$$

Once again, the smallest possible $HETP$ and the largest possible difference in K values are required to obtain a large R_s.

3.4 Membrane Separation

Membrane separations commonly used in the bioindustry include reverse osmosis (RO), ultrafiltration (UF), microfiltration (MF), dialysis (DS), electrodialysis (ED), and pervaporation (PV). Separations by RO, UF, and MF are accomplished by forcing a fluid through a porous membrane. Solid particles and/or large protein molecules build up as a layer on the surface of the membrane. Thus, to attain a reasonable throughput, the pressure drop (the so-called transmembrane pressure) should be increased or the resistance to permeation decreased. Cross-flow filtration aims at the prevention of cake formation in order to decrease the resistance due to retained feed components on the membrane surface. The gel-polarization or concentration-polarization model has been commonly applied to the calculation of the performance for cross-flow filtration. The average transmembrane pressure drop, ΔP_{TM}, is defined as

$$\Delta P_{TM} = \frac{\bar{P}_i + \bar{P}_o}{2} \qquad (52)$$

where $\bar{P}_i$ is the inlet gauge pressure and $\bar{P}_o$ is the outlet gauge pressure of cross-flow filtration.

The permeate flux, J, can be expressed as

$$J = \frac{\Delta P_{TM}}{\mu_s (R_m + R_g)} \qquad (53)$$

where R_m is the resistance of the membrane,
R_g the resistance of the retained components, and
μ_s the viscosity of the fluid.

After polarization takes place, the flux is given by

$$J = k_J \ln \frac{C_G}{C_B} \qquad (54)$$

where C_B is the bulk concentration,
C_G the maximum concentration, and
k_J the mass transfer coefficient.

k_J is given by the Sherwood equation:

$$Sh = \frac{k_J d}{D} = 1.62 \left(Re \, Sc \, \frac{d}{L_m} \right)^{1/3} \text{ in laminar flow} \qquad (55)$$

$$= 0.04 \, Re^{3/4} Sc^{1/3} \text{ in turbulent flow} \qquad (56)$$

where Sh is the Sherwood number,
Re the Reynolds number $(d\rho_s v/\mu_s)$,
Sc the Schmidt number $(\mu_s/D\rho_s)$,
D the diffusivity,
d the fluid channel height above the membrane,
μ_s the viscosity of the fluid, and
v the fluid velocity.

Dependence of k_J on fluid velocity can be expressed by Eqs. (55) or (56) for protein particles or smaller particles. However, for solid particles such as cells, the dependence on fluid velocity is greater than the value calculated from Eqs. (55) and (56).

The phenomenon of decline of flux rate with time is unavoidable in cross-flow filtration. PATEL et al. (1987) reported that the fouling process could be assumed to follow a first order mechanism:

$$J_t = J_1 t^{-b} \qquad (57)$$

where J_t is the flux at time t, J_1 is the flux at $t = 1$ min, and b is the fouling constant. Their conclusions concerning the fouling process may be summarized as follows:

a) Higher flow rates reduce the rate of flux decline.
b) Higher pressures increase the rate of fouling.
c) Higher cell concentration leads to a greater fouling rate.

4 Design and Control of Separation Systems

4.1 Synthesis of Separation Processes

When the raw materials to be separated and the desired quantities of final products are prescribed, a rational separation system is usually designed on the basis of the following three steps.

(1) Process synthesis: the structure of the process and the configuration of the separation scheme are established in such a way as to satisfy the imposed limits with regard to production and operation and to meet the requirements for the most economical and effective production.
(2) Process flow-sheeting: in this step the input–output relationships are determined or calculated in order to satisfy the energy and mass balance with respect to a given process configuration.
(3) Optimization: if there is some lati-

tude in the second step, this flexibility should be exploited for process optimization or process safety; i.e., process optimization can be affected by changing various design variables and/or operating or manipulating variables.

Of course no step can be taken in isolation. It is readily apparent that if one wants to choose the best configuration, then optimal conditions are required based on correct input–output relationships. Viewed in this way, steps 2 and 3 are already included in step 1.

4.1.1 Process Synthesis

Extensive experience and intuitive comprehension of current techniques are required to properly determine process configuration. In other words, no systematic way to solve such problems has yet been developed. One possible approach is to choose one of the candidates amenable to solution and treat it as a combinatorial problem.

In any case, process configuration or alternative structures of purification systems should be taken into account if systems are to be investigated that are more effective and economical than current ones. For example, penicillin G (PenG) has been industrially purified by the system shown in Fig. 2. PenG is extracted from the broth into butyl acetate (BA) and then re-extracted from butyl acetate into a water phase by chemical reaction with K_2CO_3.

The condensed PenG-K solute is again extracted into butanol (BuOH), with the PenG becoming purified by crystallization because the solubility of penicillin in BuOH is low. In an alternative configuration for the separation process, PenG can be separated from other amino acids by LC using absorption column (preparative) chromatography rather than through extraction by BA and concentration of PenG into PenG-K.

Another example, the sequence of protein purification, is generally reported as involving the series of steps: homogenization, cell precipitation, ion-exchange LC affinity chromatography, and finally gel filtration (BONNERJEA et al., 1986). Corresponding data were obtained from 100 papers on protein purification published during 1984 in eight journals. Ion-exchange chromatography was the most common reported method. Although there is no strict sequence for application of the methods, a distinct trend is obvious. Homogenization is generally followed by precipitation, then ion-exchange chromatography, affinity separation, and finally gel filtration. Sometimes the sequence of purification techniques is not made explicit. However, it is a logical one.

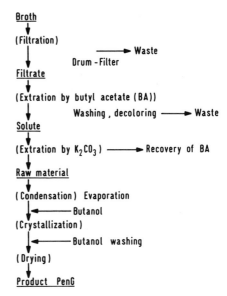

Fig. 2. Flow-sheet of a separation and purification system of penicillin G.

Precipitation is appropriate for dealing with large quantities of material, and it is less affected by interfering non-protein material than adsorption and chromatographic procedures. Affinity methods can be applied at an earlier stage, but the materials are expensive; it makes sense to initially use less costly ion-exchange media to reduce protein loads and to remove remaining fouling substances. Gel filtration has the least capacity for loaded protein, but it serves an important function in removing self-aggregates of otherwise purified proteins, and it also allows for changes of buffer. In any case, experience or expertise has an important decision-making role at this stage. In many cases it cannot be strictly determined whether or not the structure selected was really optimal, because not all possible structures can be tested in practical calculation and design.

4.1.2 Process Flow-Sheeting and Optimization

In the stage of input–output calculation, where material and energy balances can be taken into account (flow-sheeting), it is preferable to utilize commercial program packages. Several packages are already available or under development (e.g., BPS by ASPEN Tech).

As already mentioned, the optimization stage cannot be done separately, and it requires cooperative effort. Optimization can be carried out based on flow-sheeting calculations.

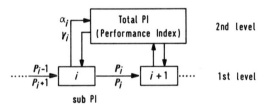

Fig. 3. Multilevel technique for optimization.

Many numerical iteration methods have been developed for the optimization technique. In this separation system, however, the multilevel technique shown in Fig. 3 is preferable because the overall calculation to achieve optimization is readily understood. In this configuration, two indices have important roles, the purification index (or partition coefficient) α_i, defined as

$$\alpha_i \triangleq p_{i-1}/p_i \tag{58}$$

and the product yield Y_i, defined as

$$Y_i = P_{i-1}/P_i \tag{59}$$

The investment cost per unit can probably be estimated from α_i and Y_i at the 2nd level, At

the 1st level other conditions can help to minimize the extractor cost when α_i and Y_i are fixed, as in the case of pH for the extraction of PenG into butyl acetate during penicillin purification (Fig. 2). Note that at low pH PenG will be inactivated more in the water phase than in the organic phase. However, the partition coefficient, K, of PenG between the butyl acetate and aqueous phases depends heavily on the pH, and a high pH is preferable for increasing K. Use of the appropriate pH will thus maximize the overall recovery of PenG in a continuous extractor. The solution to the problem is therefore obtained at the first stage.

4.2 Control of Separation Processes

A control system is necessary for each unit used in the separation during industrial operation. There are many items that should be considered in terms of the control of the separation process. We will not touch upon this problem here, however, because space is limited. Only the following point is stressed: if one wants to control a system, the output

Tab. 6. Principles of On-Line Sensors in Separation Processes

Absorbance (in visible range)
UV Absorbance
Refractive index
Fluorescence
NMR spectrum
Mass spectrum
Others

should be measured in-line or on-line. Accepting the viewpoint that output should be measured on-line, the separation process remains difficult to control because the output frequently cannot be measured. Tab. 6 shows the principle as it is currently utilized. In the near future more items will be subject to measurement with newly developed sensors.

4.3 Necessity of Integration of Upstream and Downstream Processing

The design of a separation system inevitably involves upstream processing. For the sophisticated design of a reactor or fermentor, both downstream and upstream processing should be taken into account, as shown in the following examples.

Several methods have been developed for attaining high cell density with mammalian cell cultures. High cell concentration results in high concentrations of waste as well as products excreted from the cells. The reported methods claim to effectively separate these components. For example, effective separation is achieved by the hollow-fiber system, which retains cells at high concentration on one side of the membrane tube. On the other side essential nutrients are added, and undesirable wastes are removed through the membrane. The concentration of bioproduct produced by the cells becomes high because it is retained inside the reactor. Therefore, this system plays the dual roles of bioreactor and separator of product from the medium containing nutrients and wastes. The hollow-fiber system thus stands as an impressive example of an integrated system of reactor and downstream process.

Another example utilizes gene fusion to a gene encoding staphylococcal protein A, which can serve as an affinity "tail" through its strong affinity to IgG, thereby facilitating protein purification (MOKS et al., 1987). The separation process includes affinity chromatography on IgG Sepharose Fast Flow and also site-specific chemical cleavage of the fusion protein. However, the key point is that fusion, or so-called chimera protein, is produced by genetic manipulation, and the properties of the chimera protein are used for effective separation and purification.

5 Concluding Remarks

Today, downstream processing, the separation and purification of the bioproduct, is very important in the development of biotechnological processes. Many novel separation processes based on various principles have been developed, and others have yet to be investigated. The essential point is to choose the most suitable configuration of separate sequences for a given bioproduction process, keeping also in mind the importance of optimal design and operation of the separation process. Integration of upstream and downstream processing will result in the most desirable mode of bioproduction.

6 References

ALBERTSON, P. A. (1960), *Partition of Cell Particles and Macromolecules.* Stockholm: Almgrist & Wiksell. New York: Wiley.

ATKINSON, T., SCAWEN, M. D., HAMMOND, P. M. (1987), Large-scale industrial techniqus of enzyme recovery, in: *Biotechnology* (REHM, H. J., REED, G., Eds.), 1st Ed., Vol. 7a, pp. 279-323. Weinheim-New York-Basel-Cambridge: VCH.

BONNERJEA, J., OH, S., HOARE, M., DUNNILL, P. (1986), Protein purification: right step at the right time, *Bio/Technology* **4**, 954-958.

BROWN, D. E., SALAM, F. R. A. (1984), A new filter for cell separation, *Biotechnol. Lett.* **6**, 401-406.

BRÜMMER, W., GUNZER, G. (1987), Laboratory techniques for enzyme recovery, in: *Biotechnology* (REHM, H. J., REED, G., Eds.), 1st Ed., Vol. 7a, pp. 213-278. Weinheim-New York-Basel-Cambridge: VCH.

CRAIG, L. C. (1960), Partition, in: *A Laboratory Manual of Analytical Methods of Protein Chemistry* (ALEXANDER, P., BLOCK, R. J., Eds.), Vol. 1, pp. 122-160. Oxford: Pergamon.

DATAR, R. (1985), Studies on the separation of intracellular soluble enzymes from bacterial cell debris by tangential flow membrane filtration, *Biotechnol. Lett.* **7**, 471-476.

DEVEREUX, N., HOARE, M. (1986), Membrane separation of protein precipitates: studies with cross flow in hollow fibers, *Biotechnol. Bioeng.* **28**, 422-431.

VAN DEEMTER, J. J., ZUIDERWERG, F. J., KLINKENBERG, A. (1956), *Chem. Eng. Sci.* **5**, 271.

GIDDINGS, J. C. (1965), *Dynamics of Chromatography*, Part 1: *Principles and Theory*. New York: Marcel Dekker.

JANSON, J. C., HEDMAN, P. (1982), Large-scale chromatography of protein, *Adv. Biochem. Eng.* **25**, 43–99.

KAVANAGH, P. R., BROWN, D. E. (1987), Cross-flow separation of yeast cell suspensions using a sintered stainless steel filter tube, *J. Chem. Tech. Biotechnol.* **38**, 187–200.

KRONER, K. H., SCHÜTTE, H., HUSTEDT, H., KULA, M. R. (1984), Cross-flow filtration in the downstream processing of enzymes, *Process Biochem.*, April, 67–74.

KUBIN, M. (1975), *J. Chromatogr.* **108**, 1.

KULA, M. R. (1985), Recovery operations, in: *Biotechnology* (REHM, H. J., REED, G., Eds.), 1st Ed., Vol. 2, pp. 725–760. Weinheim–Deerfield Beach/Florida–Basel: VCH.

LE, M. S., ATKINSON, T. (1985), Crossflow microfiltration for recovery of intracellular products, *Process Biochem.*, February, 26–31.

MOKS, T., ABRAHMSEN, L., OSTERLOF, B., JOSEPHSON, S. NILSSON, B. (1987), Large-scale affinity purification of insulin-like growth factor I from culture medium of *E. coli*, *Bio/Technology* **5**, 379–382.

PATEL, P. N., MEHAIA, M. A., CHERYAN, M. (1987), Cross-flow membrane filtration of yeast suspension, *J. Biotechnol.* **5**, 1–16.

QUIRK, A. V., WOODROW, J. R. (1984), Investigation of the parameters affecting the separation of bacterial enzymes from cell debris by tangential flow filtration, *Enzyme Microb. Technol.* **6**, 201–206.

SIMS, K. A., CHERYAN, M. (1986), Cross-flow microfiltration of *Aspergillus niger* fermentation broth, *Biotechnol. Bioeng. Symp.* **17**, 495–505.

TREFFRY, T. E., SHARPE, P. T. (1985), Thin-layer countercurrent distribution and apparatus, in: *Partitioning in Aqueous Two-phase Systems* (WALTER, H., BROOKS, D. E., FISHER, D., Eds.), pp 131–159. London: Academic Press.

TUTUNJIAN, R. S. (1984), Cell separation with hollow fiber membranes, *Dev. Ind. Microbiol.* **25**, 415–435.

WALTER, H., BROOKS, D. E., FISHER, D. (1985), *Partitioning in Aqueous Two-phase Systems.* London: Academic Press.

19 Expert Systems for Biotechnology

AARNE HALME

Helsinki, Finland

NAZMUL KARIM

Fort Collins, Colorado 80523, U.S.A.

1 Introduction

Recent developments in process control emphasize the utilization of human expert knowledge. Artificial intelligence and knowledge-based systems are tools that make possible the organized use of schematic information and logical descriptions often employed by people dealing with problems. The phenomena most naturally treated in this way are those that are difficult to model using mathematical formulas and numerical information. Biotechnical processes typically include many features that are too complex to be modelled mathematically, or else the mathematical models available are too simple to describe the processes with fidelity.

Expert systems are of potential interest in many biotechnological applications, although they have been little used thus far. These applications may be classified according to their relation to process development or, alternatively, to process operation. In process development, expert systems can be used to plan experiments, interpret experimental results, and to support the design of new organisms or processes. In process operations, applications such as process fault diagnosis and planning of control actions have the greatest potential. In all cases the expert system is a tool that helps people other than the experts who constructed it to solve problems. Usually such systems are interactive, meaning that the user "discusses" the problem with the system to arrive at a conclusion. Automatic operations, especially those that control processes, are extremely rare. This is because knowledge-based systems are difficult to make 100% reliable, in as much as the knowledge base itself is limited in applicability to the situations considered in its construction, and because it is difficult to debug knowledge bases to be completely error-free. Human interpretation and/or validity checking is usually necessary before the results can be applied.

2 AI and Expert Systems

2.1 Basic Concepts

Artificial Intelligence (AI) is not easily defined with precision. All techniques that can be used to make computers solve problems can be classified as AI in the broad sense. The capacity for symbolic reasoning is another characteristic property. The programming of such a property is commonly done by defining the desired result rather than by giving detailed instructions on its computational steps. Special programming languages have been developed for AI such as LISP and PROLOG that permit the programmer to tell the computer how to reason. These languages are very different from conventional procedural languages such as FORTRAN or PASCAL. It is predicted that half of the computers sold in the middle 1990s will contain artificial intelligence rather than arithmetical components.

Expert systems form a special class of AI. They handle problems by using reasoning based on a knowledge base generated by a human expert (or experts) possessing a depth of knowledge concerning the problem at hand. The knowledge base represents known facts, their logical relations, and other information about the target process that is necessary to handle the problem. In addition to a knowledge base, a database is also generally needed to store the process and environmental data. Reasoning is done with the aid of a so-called inference engine that resolves the logic built into the knowledge base. In principle, reasoning can be executed in two ways, by using either backward or forward chaining of logical connections. In backward chaining the system tries to answer to a logic statement concerning the process or its behavior if it is true or not and "why" it is so. In forward chaining all logical consequences of a certain assertion are determined; fundamentally the system answer to the question "what?" consists of the logical consequences from these premises. In practice both types of reasoning capabilities are often needed in the same expert system. Knowledge can be represented formally in many different ways. The most common way is to represent it by using *if ... then ... else*-type rules. Other methods that are sometimes used are representations on the basis of frames or schematic nets (HAYES-ROTH et al., 1983). A special class of expert system is the system that operates with real-time data. Most of the applications related to process control systems have this feature.

2.2 How to Build an Expert System

Since an expert system is essentially a computer program, the basic elements needed to construct it are a computer and proper software tools. Although special computer hardware exists for AI applications, it is current practice to use standard hardware such as engineering workstations and personal computers. Special "development shells" are available on the market for software development.

The part of an expert system most laborious to construct is the knowledge base. It usually requires careful analysis of the problem by an expert and a knowledge engineer. The expert (or experts) is a person who knows the problem thoroughly and can solve it by applying his specialized knowledge. The knowledge is in the expert's head, often as a complex mixture of schematic descriptions, rules, and important values. The knowledge engineer is a person responsible for gathering this knowledge and organizing it into a rational form for the computer system. The first step is usually to interview the expert. This is done iteratively, proceeding in several steps, each consisting of a refinement of the knowledge obtained in the previous step. If the knowledge is represented in the form of rules, it means that the rule set must be modified several times in order to agree with the expert's thinking. The second step is to run the inference engine on the prototype knowledge base together with a proper database, and ascertain whether the conclusions drawn by the system are in accord with the expert's opinions. If the knowledge base is constructed of rules, the conclusions may be reached by chaining the rules backward or forward. Backward chaining typically explains why a statement is true or false. Forward chaining, on the other hand, determines possi-

ble consequences of a statement supposed to be true or false.

The last step in development is to validate the expert system. This is done basically by determining how well the system knows its field of expertise and how logically it can "think". In practice, the only way validation can be achieved is to test the system thoroughly with the expert. Although the knowledge base was constructed with the expert's assistance, it does not necessarily imply that the inferred results must be rational.

A special feature of expert systems used in connection with process control systems is that they operate with real-time data such as sensor readings. The proper connection of data of this type with the knowledge base requires special tools, such as event-handling mechanisms and temporal logic to ensure that time-dependencies of data and events are taken correctly into account. Many expert systems applied to bioreactor control are of this type, and thus require these special features when implemented.

2.3 What Kinds of Problems Should be Solved with Expert Systems?

It is important to realize that an expert system is not always the best solution to a problem. If valid mathematical or biochemical models representing the main phenomena of the problem are available, it is wiser to use them than to build a knowledge-based system that explains the same phenomena on the logical level. In fact, such models represent high-level and concentrated knowledge that is difficult to replace with rule sets or similar knowledge commonly used in expert systems. Therefore, the problem is best treated by employing conventional methods such as mathematical calculation, simulation, and parameter identification. If the models or their usages are complicated, however, an expert system may be helpful in guiding a non-specialist in their use.

3 Expert Systems for Bioreactors

Expert systems for bioreactor control may be designed for several different purposes. Examples of specific tasks that an expert system could help to perform are:

- Checking overall process conditions
- detecting and localizing non-trivial faults in instruments and process equipment
- clearing alarm blocks in startup/shutdown operations
- finding the most "appropriate" way to run the reactor under certain conditions
- detecting abnormalities in a fermentation, e.g., unexpected growth, contamination, etc.
- predicting the most suitable harvest time
- guiding the operator in taking samples when process conditions vary.

An experienced operator can do all of the above tasks well, but a less experienced one may have problems. Also, the tasks are difficult to perform by mathematical methods because the information available is unorganized and partly schematic. The problems themselves are not unique to bioreactors, and can be easily generalized for similar problems in chemical process control. Few prototype systems have been constructed thus far. Examples reported in the literature include AARTS et al. (1989), CHEN-QUI et al. (1989), KARIM and HALME (1989), and LÜBBERT et al. (1989). Research in the field is presently very active, and new applications are reported in conferences every year. The first overview of the possibilities in the area of process development was presented by STEPHANOPOULOS and STEPHANOPOULOS (1986). A similar but more generally oriented overview may be found in LÜBBERT and HITZMANN (1987).

4 A Practical Example

Until now the reader may have found the expert system to be an abstract concept. This is of course true, and the best way to make it more concrete is to consider an example. The following example is of a prototype real-time expert system developed in the Laboratory of Automation Technology, Helsinki University of Technology, for monitoring and analysis, diagnosis of faults and malfunctions, and on-line optimization of batch and fed-batch fermentation processes (KARIM and HALME, 1989). The main purpose of the expert system is to assist the operators in running and improving fermentation conditions with minimum supervision. The system is interfaced with a computer-based data acquisition and control system which gives the basic on-line information about the process. In addition, the operator can display off-line data, which may include physical attributes of the process (color and general appearance of the fermentation broth) and microscopic states of the microorganisms. Computational information such as estimates of variables can also be included. The expert system has a knowledge base that incorporates typical fermentation data (characteristic curves) of commonly run fermentation, culture data, operational data, calibration sensors, and rule sets that make inferences. The main functions of the system are related to its fault detection capabilities in making intelligent decisions about the abnormalities of fermentation experiments. The underlying goal is to identify automatically the so-called functional states of the bioreactor (HALME, 1989). The functional states indicate, roughly, the overall biochemical state of the fermentation. Having this information available on-line makes it possible to further analyze the measurement data by using rule-based logic. The overall configuration of the expert system is shown in Fig. 1. The system has been tested on the batch fermentation of genetically engineered *Bacillus subtilis* strains producing α-amylase. The prototype work was completed in 1988.

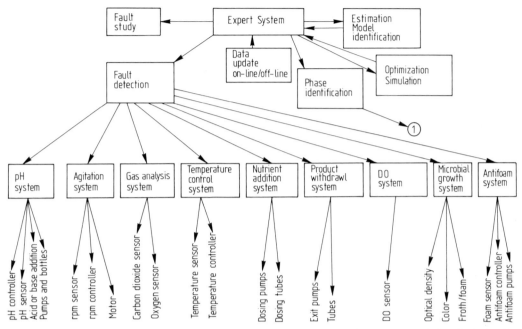

Fig. 1. Symptoms and rules, inference engine, connectivities, etc. The rules are generated to interconnect subsystems for expert analysis. ① Rules for phase determination.

4.1 Computer System and the Programming Environment

In this example the expert system was designed to run in a separate PC/386 connected to the computerized fermentation unit (BIO-STAT with Rinteknos MFCS) with a serial computer link (RS-232C). The fermentation computer system provides on-line measurement data as well as the laboratory analytical data available after sample analysis. The expert system PC picks up the data it needs for reasoning and informs the operator of its conclusions via an interactive user interface. All process control operations are performed via the control computer interface.

Program development of the expert system was done under OPS 83, which is a rule-based on-line programming tool to develop expert systems. The following features and requirements were specified for the system:

1. All the decisions are to be based on real-time data, although off-line sample analysis may enter the decision process if the expert system is unable to make a firm decision based on real-time data. Data-producing state and parameter estimation algorithms may be used to assist the expert system to come to the "correct" conclusion.

2. Data from sensors must be interpreted and reasoned about to identify types of events and "objects" known to the knowledge base, e.g., different data and rules are used during different phases of fermentation. The data are time-dependent; i.e., the information included may be a function of the time at which the data are used in reasoning. If the data are ambiguous or inaccurate, the reasoning system must make "reasonable" and "safe" decisions.

3. Characteristic curves are to be included in the knowledge base, with variances representing the most "deviant" behaviors observed in previous experiments.

4. The knowledge base is dynamic rather than static, as is the case with most expert systems. This means that as more experience is acquired about a fermentation, the knowledge base and the rule sets are modified to account

for the peculiarities associated with different fermentation conditions. Therefore, situations analyzed in the past can provide the line of reasoning for the analysis of the current state of the fermentation without a need to repeat the reasoning.

5. The expert system uses a sliding window concept in which the system moves the domain of observation and inference from one phase of fermentation to another. This permits the use of comparatively modest hardware, such as a PC, in the project.

4.2 Example of Rule-Based Programming

As an example of programming procedure in rule-based OPS 83, we will consider fault diagnosis in the pH measurement and regulation system. From the process point of view the pH control system is very important. Fig. 2 shows in detail the pH-control loop in the fermenter used. The pH-control system has an accuracy of ±0.1 units. All rules that check the overall

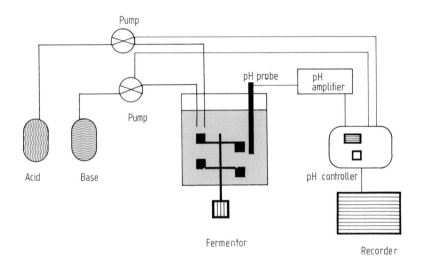

Fig. 2. Measurement and regulation of pH systems.

6. Once the identification of a phase is completed (this may be based on a small subset of the on-line measurements), the total set of measurements is analyzed for faults in the process and for contamination. If contamination is suspected, the operator is guided in the taking of a sample and having microscopic analysis performed on the sample in order to determine the nature of contamination. In this sense the decision made by the expert system is not necessarily a binary one.

7. If and when the measurement signals are noisy, a moving average filter is used before the measurements are processed. The expert system decides which variable needs filtering and it does it in real time.

condition of the measurement and control functions in the fermentation system are defined as "Level 1 Rules". The expert system continuously checks these rules, and if a fault is suspected in any element of the measurement and control loops, it will activate the "Level 2 Rules", which then analyze the specific fault and make recommendations to the operator concerning possible courses of action to alleviate the situation. The basic strategy for finding faults in the pH controller is shown in Fig. 3. Here it is assumed that any inaccuracy of ±0.3 is due to a malfunction of the controller. Fig. 4 shows some of the OPS 83 codes for this strategy. Similar rules can be formulated for the pH sensor and acid/base pumps, as well as for the other sensors and controllers. The windowing technique is used in operator

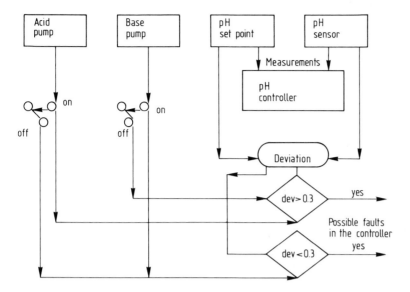

Fig. 3. Logic for finding faults in the controller.

```
  Rule            Fault in pH controller

  This rule checks to see if pH-controller is faulty.
```

```
{
  &goal (goal  type = find_fault;  level + 1;);

  -->

local
        &dev  : real,
        &base_on  : logical,
        &acid_on  : logical;

        &dev = &pH.value[1] - &pH.stand[1];
        if (&base.value[1] - &base.value[2] > 0.0)
          &base_on = 1b
        else &base_on = 0b
          if (&acid.value[1] - &acid.value[2] > 0.0)
            &acid_on = 1b
        else &acid_on = 0b;

        if (( &dev > 0.3) ∧ (&base_on ∧ not (&acid_on)))
            make (symptom name = controller; circuit = pH);
        if (( &dev < -0.3) ∧ (&acid_on ∧ not (&base_on)))
            make (symptom name = controller; circuit = pH);

};
```

Fig. 4. OPS 83 rules for finding faults in the pH controller.

interfacing. Different windows are created to show multiple menus simultaneously. Special care is taken to give meaningful messages to the operators. Color-coded information is provided to distinguish the quality of information for the operator, e.g., when a fault is detected a message in red is shown in one of the windows. When the operator acknowledges the alarm, the message stays on the screen, changing to yellow; when corrective action is taken the message is shown in green. The time at which the operator took action is also displayed for record keeping purposes.

4.3 Materials and Methods

The prototype expert system was tested in the laboratory environment with several parallel fermentations of *Bacillus subtilis* producing α-amylase. Materials and methods used in these experiments are summarized below.

Microorganism: *Bacillus subtilis* BRB 423 (glucose repression-negative) carrying the *B. amyloliquefaciens* α-amylase gene as a single-copy chromosomal integrate.

Media and growth conditions: five Braun Biostat M fermenters with working volumes of 1.0 L were available for this study. Fermentation media consisted of 1.0% glucose, 1.0% yeast extract, 2.0% tryptone, 85 mmol NaCl, 17 mmol K_2HPO_4, and 5 mg/mL of antibiotics. Fermentations were carried out at 37°C. The pH was maintained at 7.2, the aeration rate at 1.2 vvm, and the stirrer speed at 800 rpm.

Instrumentation: Each fermenter was equipped with a dissolved oxygen probe (Ingold), pH probe (Ingold), temperature element, and foam detection probe. The control of pH, temperature, stirrer speed, and antifoam addition was accomplished by set-point control with a Braun Biostat system. A gas analyzer (Siemens) measured carbon dioxide and oxygen from the exhaust gas.

Process control and data acquisition were performed by the MFCS software package (Rintekno), which was especially designed for fermentation processes. The MFCS package runs on the DEC (Digital Equipment Corp.) Micro PDP/11 computer using the Micro RSX operating system.

Analytical methods: α-amylase activity was determined by the Phadebas (Pharmacia) amylase test. Turbidity was measured with a Klett–Summerson photoelectric colorimeter. Dry-cell weight was determined by freeze-drying centrifuged cells to constant weight. Glucose concentration was measured with a Beckman glucose analyzer.

4.4 Examples of Experimental Tests

Test 1. The following is an analysis of runs 870115 and 870116, conducted simultaneously under supposedly identical operating conditions. Due to a hidden equipment fault, however, the operating conditions were slightly different. Different fermentation behaviors probably resulted, although the indicators of main variables of fermentation conditions remained the same. It was therefore difficult for the usual operator to invoke corrective control operations during the runs.

However, a skilled operator might have found the following symptom that was detected by the expert system. Figs. 5 and 6 show acid and base additions and CO_2 and dry weight measurements for the two fermentations. In run 870115 much more base was added than in run 870116. Since the acid and base additions represented cumulative plots of the duration that pumps were open (calibrated to mL of acid/base addition), it is possible that the pumps were pumping air in run 870115. Fig. 7 shows the results of fault analysis of run 870115. The expert system predicts a possible fault in the base pump at 3.33 hours into the fermentation. After detecting this possible fault, the operator should use the fault study menu to determine the reasoning employed by the expert system in its decision. He may accept the decision, but he may also ignore it by acknowledging that he has seen the message and elected to take no action.

In this instance the expert system predicted that the fault was probably due to pumping air through the system, and, in fact, it was confirmed that the pump in real time was intermit-

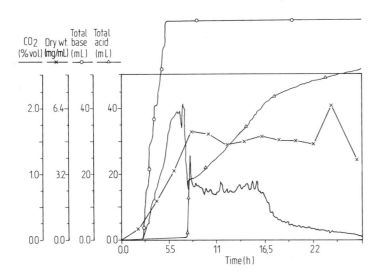

Fig. 5. Batch fermentation of *Bacillus subtilis,* run 870115. Base pump intermittently pumping air.

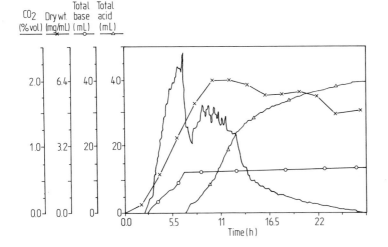

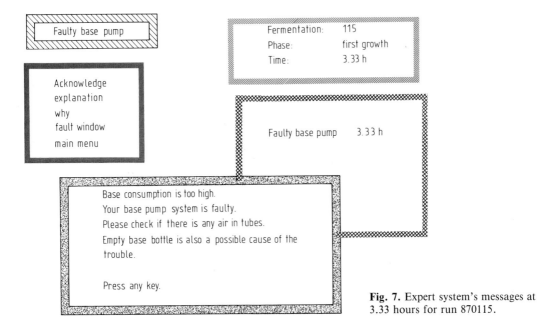

Fig. 6. Batch fermentation of *Bacillus subtilis,* run 870116. All instruments apparently working correctly.

Fig. 7. Expert system's messages at 3.33 hours for run 870115.

tently pumping air rather than base. It could be concluded, although not conclusively, that the difference between the two runs was mainly in the faulty base and acid pumps in run 870115. The result was lower productivity of α-amylase in that run. The use of an on-line expert system in an actual production process could have resulted in a faster response from the operator at about 3.33 hours into the fer-

mentation, thus preventing the operation from proceeding with faulty equipment.

Test 2. In another test, run 880034, deliberate faults were introduced at 3.5 hours in the pH controller and at 6.35 hours in the CO_2 sensor. The pH fault was introduced by adding a bias in the acid pump, and the CO_2 fault originated with the addition of a pure carbon dioxide pulse into the carrier stream. Figs. 8

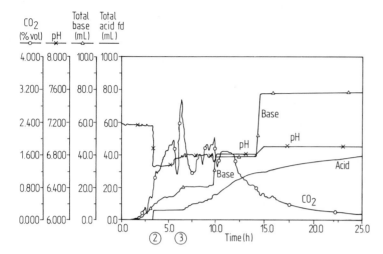

Fig. 8. Expert system diagnosis, run 880034.

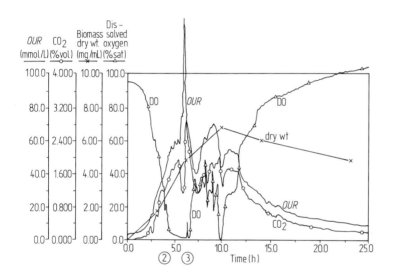

Fig. 9. Expert system diagnosis. At point ② fault in acid pump was detected, at point ③ fault in DO system was the inference, run 880034.

1	Beginning of first growth phase	1.3 h	
2	Faulty acid pump; check bottles	3.57 h	
3	Faulty dissolved oxygen probe or rpm control malfunction (!)	6.4 h	
4	Stationary phase 1	6.7 h	
5	Beginning of second growth phase	7.55 h	

Fig. 10. Expert system diagnosis, run 880034.

and 9 show the plots from the MFCS system. Fig. 10 indicates the analysis by the expert system. The faults were detected about two minutes after they were introduced. The introduction of CO_2 into the carrier stream perturbed the *OUR* measurements and upset the *OUR* reading. This, however, created problems for the expert system, as it did not know that the faults were simulated. The expert system did predict the faults in the pH control system correctly, but it failed to detect the fault in the CO_2 system. Instead, using heuristic "two out of three" rules, it found that both *OUR* and CO_2 sensors had behaved in an identical manner (sharp increase), indicating rapid growth. The DO probe did not show this kind of activity; it therefore reasoned that the gas sensors were performing properly and that the r.p.m. control system was probably faulty. Of course this analysis was wrong. However, since the faults were not real ones, and merely scanning the figures could not result in a determination of the cause of the abnormalities, the decision of the expert system was logical. This would also have been the conclusion of a human expert.

4.5 Some Practical Aspects

The knowledge base in the example above consisted of about 300 rules, an average number. However, as a prototype system, the knowledge base did not cover all practical cases, but rather demonstrated the functionality of the system. Even with a knowledge base of this size, inferencing with a PC takes time. It helps considerably to divide the rules into various levels and to limit the rule set by "zooming" the inferencing in the lower level. The number of rules for a practical situation could easily by 500–700, depending on the complexity of the fermentation facility. A single PC may not be adequate in such a case.

Experimental tests have shown that the reasoning of the system is quite reliable provided the underlying rules are robust in the sense that they explain logical rather than numerical behavior. It is not always easy to devise rules in which this condition is valid. When interpreting the thinking of an expert and formulating the rules, this precept is good to keep in mind, because the same phenomena can be described in several different ways.

The problem of updating the knowledge base was not considered during the testing of the prototype system. In practical cases the updating problem should always be taken into account. Even if the general philosophy of knowledge base construction is supposed to be independent of changes in the organism, media, or the process equipment, experience shows that updating the database is usually insufficient, and that rules must also be changed. This is a feature of expert systems that can cause severe problems in practice. Changes in the knowledge base should, in principle, be made by the same expert who contributed to its creation. In practice, however, this problem may be overcome in well-designed software by extracting those rules and data values that appear to require it, and describing the necessary changes as clearly as possible.

5 References

AARTS, R. J., SUVIRANTA, A., RAUMAN-AALTO, P., LINKO, P. (1989), An expert system in enzyme production control, in: *Proc. Int. Conf. on Biotechnology and Food,* February, Stuttgart, FRG, pp. 20–24.

CHEN-QI, SHU-QING, W., JI-CHENG, W. (1989), Application of expert system to the operation and control of industrial antibiotic fermentation process, in: *Proc. 4th Int. Congr. on Computer Applications in Fermentation Technology* (FISH, N. M., FOX, R. I., THORNHILL, N. F., Eds.), London: Elsevier.

HALME, A. (1989), Expert system approach to recognize the state of fermentation and to diagnose faults in bioreactors, in: *Proc. 4th Int. Congr. on Computer Applications in Fermentation Technology* (FISH, N. M., FOX, R. I., THORNHILL, N. F., Eds.), London: Elsevier.

HAYES-ROTH, F., WATERMAN, D. A., LENAT, D. B. (1983), *Building Expert Systems,* Reading, Mass.: Addison-Wesley

KARIM, M. N., HALME, A. (1989), Reconciliation of measurement data in fermentation using on-line expert system, in: *Proc. 4th Int. Congr. on Computer Applications in Fermentation Technology* (FISH, N. M., FOX, R. I., THORNHILL, N. F., Eds.), London: Elsevier.

LÜBBERT, A., HITZMANN, B. (1987), Are there any prospects for expert systems in chemical engineering?, *Hung. J. Ind. Chem.* **15**, 39–45.

LÜBBERT, A., HITZMANN, B., KRACKE-HELM, H.-A., SCHÜGERL, K. (1989), On experiences with expert systems in the control of bioreactors, in: *Proc. 4th Int. Congr. on Computer Applications in Fermentation Technology* (FISH, N. M., FOX, R. I., THORNHILL, N. F., Eds.), London: Elsevier.

STEPHANOPOULOS, G., STEPHANOPOULOS, G. N. (1986), Artificial intelligence in the development and design of biochemical processes, *Trends Biotechnol.*, 241–249.

Index